Microprocessor Architecture, Programming, and Applications

with the 8085

FOURTH EDITION

Ramesh S. Gaonkar

STATE UNIVERSITY OF NEW YORK,
O.C.C. CAMPUS AT SYRACUSE

Prentice Hall
Upper Saddle River, New Jersey   *Columbus, Ohio*

7/01   # 40043629

Library of Congress Cataloging-in-Publication Data

Gaonkar, Ramesh S.
    Microprocessor architecture, programming, and applications with
the 8085 / Ramesh S. Gaonkar. — 4th ed.
       p. cm.
    First-2nd eds. published under title: Microprocessor architecture,
programming, and applications with the 8085/8080A.
    Includes index.
    ISBN 0–13–901257–5 (alk. paper).
    1. Microprocessors. 2. Computer architecture. I. Gaonkar,
Ramesh S. Microprocessor architecture, programming, and applications
with the 8085/8080A, II. Title.
TK7895.M5G36 1999   98–45804
004.165—dc21   CIP

Cover photo: © H. Armstrong Roberts
Editor: Charles E. Stewart, Jr.
Production Editor: Alexandrina Benedicto Wolf
Design Coordinator: Karrie M. Converse
Cover Designer: Jason Moore
Production Manager: Deidra M. Schwartz
Marketing Manager: Ben Leonard

This book was set in Times Roman by The Clarinda Company and was printed and bound by R. R.
Donnelley & Sons Company. The cover was printed by Phoenix Color Corp.

 ©1999, 1996 by Prentice-Hall, Inc.
Simon & Schuster/A Viacom Company
Upper Saddle River, New Jersey 07458

Earlier editions, entitled *Microprocessor Architecture, Programming, and Applications with the
8085/8080A,* ©1989 by Macmillan Publishing Company, and © 1984 by Merrill Publishing
Company.

Printed in the United States of America

10 9 8 7 6 5 4 3

ISBN: 0-13-901257-5

Prentice-Hall International (UK) Limited, *London*
Prentice-Hall of Australia Pty. Limited, *Sydney*
Prentice-Hall Canada, Inc., *Toronto*
Prentice-Hall Hispanoamericana, S. A., *Mexico*
Prentice-Hall of India Private Limited, *New Delhi*
Prentice-Hall of Japan, Inc., *Tokyo*
Simon & Schuster Asia Pte. Ltd., *Singapore*
Editora Prentice-Hall do Brasil, Ltda., *Rio de Janeiro*

Preface

By the time this fourth edition is published, this microprocessor textbook will have been in the field for fifteen years. It is gratifying to see such acceptance of the integrated approach to teaching microprocessor concepts. The text is intended for introductory microprocessor courses at the undergraduate level in technology and engineering. It is a comprehensive treatment of the microprocessor, covering both hardware and software based on the 8085 microprocessor family. The text assumes a course in digital logic as a prerequisite; however, it does not assume any background in programming. At the outset, though, we need to answer the following three critical questions.

1. *In the late 1990s, is an 8-bit microprocessor an appropriate device through which to teach microprocessor concepts when 32-bit microprocessors are readily available?* If we consider the worldwide sales volume of microprocessor chips, the answer is a resounding *yes*: 8-bit microprocessors (including single-chip microcontrollers) account for more than 85 percent of the total. Consider an analogy from the auto industry. For transportation, we need trucks, sports cars, family cars, and compact cars. Each serves a different purpose. The 8-bit microprocessor has already established its market in the areas of industrial control, such as machine control, process control, instrumentation, and consumer appliances; these systems that include a microprocessor are known as embedded systems or microprocessor-based products. The recent 32-bit microprocessors are used primarily in microcomputers and workstations; they are so powerful that their applications are better suited in such areas as high-speed data processing, CAD/CAM, multitasking, and multiuser systems. The 32-bit microprocessors are less likely to replace 8-bit microprocessors in industrial control applications. In many applications, even 8-bit microprocessors are utilized at less than 50 percent of their capacity.

We are interested in teaching the basic concepts underlying a programmable device, such as buses, machine cycles, various processes of data flow (parallel, serial, interrupts, and DMA), internal register architecture, programming, and interfacing. A general-purpose 8-bit microprocessor is an ideal device to teach these concepts, especially in a rapidly changing technological environment. When students master the basic concepts, they will be able to apply those concepts in such an environment, whether it is based on a microcontroller, an 8-bit processor with a different set of instructions, or a 32-bit processor.

2. *Why shouldn't we focus on the Intel 16-bit processor when PCs (personal computers) are commonly available in college laboratories?* To teach basic concepts, we need a simple processor with an adequate instruction set. The Intel 16-bit processors are too complex at the introductory level because of the concepts involved in memory segmentation and a large instruction set, which is suited for high-level languages. These 16-bit processors were used primarily in PCs, and they are being replaced by 32-bit processors. Technologically, they are obsolete as general-purpose processors; they may be revived as 16-bit microcontrollers.

3. *Why teach the 8085 microprocessor?* Any commonly available 8-bit microprocessor will meet the teaching criteria, and the 8085 is one of the most widely used microprocessors in college laboratories. It has simple architecture and an adequate instruction set, which enable instructors to teach necessary programming concepts. It is inconsequential which microprocessor is selected as the focus; the concepts are easily transferable from one device to another. Having learned basic concepts with the 8085 microprocessor, students can adapt to the microcontroller (such as the Intel 8051 or Motorola 68HC11) environment or to the PC environment. Furthermore, peripheral devices (such as the 8255A, 8254, and 8259) are used commonly in the PC environment. One can argue for a microcontroller as a basis for an introductory course. However, the experiences of many institutions suggest that a microcontroller is an appropriate device for a higher-level course; at an introductory level, the pedagogy becomes quite cumbersome. Furthermore, general-purpose 8-bit processors are being used in small systems. As an example, some Texas Instruments graphic calculators use the 8-bit Z80 processor.

PEDAGOGICAL APPROACH AND TEXT ORGANIZATION

The microprocessor is a general-purpose programmable logic device. A thorough understanding of the microprocessor demands concepts and skills from two different disciplines: hardware concepts from electronics and programming skills from computer science. Hardware is the physical structure of the microprocessor, and programming makes it come alive; one without the other is meaningless. Therefore, this text presents an integrated approach to hardware and software in the context of the 8085 microprocessor. Part I focuses on microprocessor architecture and interfacing; Part II introduces programming; and Part III integrates hardware and software concepts from the earlier sections in interfacing and designing microprocessor-based products. Each topic is covered in depth from basic concepts

to industrial applications and is illustrated by numerous examples with complete schematics. The material is supported with assignments of practical applications.

Part I has four chapters dealing with the hardware aspects of the microcomputer as a system, presented with the spiral approach similar to the view from an airplane that is getting ready to land. As the plane circles around, what one observes is a view without any details. As the plane starts descending, one begins to see the same view but with more details. This approach is preferable because students need to use a microcomputer as a system in their laboratory work in the early stages of a course, without having an understanding of all aspects of the system. Chapter 1 presents an overview of microprocessor-based systems. It presents the 8-bit microprocessor as a programmable device and an embedded controller, rather than a computing device or CPU used in computers. Chapters 2, 3, and 4 examine microprocessor architecture, memory, and I/O, with increasing depth in each chapter, from registers to instruction timing and interfacing.

Part II has seven chapters dealing with 8085 instructions, programming techniques, program development, and software development systems. The contents are presented in a step-by-step format. A few instructions that can perform a simple task are selected. Each instruction is described fully with illustrations of its operations and its effects on the selected flags. Then, these instructions are used in writing programs, accompanied by programming techniques and troubleshooting hints. Each illustrative program begins with a problem statement, provides the analysis of the problem, illustrates the program, and explains the programming steps. Chapters conclude with reviews of all the instructions discussed. The contents of Part II are presented in such a way that, in a course with heavy emphasis on hardware, students can teach themselves assembly language programming if necessary.

Part III synthesizes the hardware concepts of Part I and the software techniques of Part II. It deals with interfacing of I/Os, with numerous industrial and practical examples. Each illustration analyzes the hardware, includes software, and describes how hardware and software work together to accomplish given objectives. Chapters 12 through 16 include various types of data transfer between the microprocessor and its peripherals such as interrupts, interfacing of data converters, I/O with handshake signals using programmable devices, and serial I/O. Chapter 14 discusses special-purpose programmable devices used primarily with the 8085 systems (such as the 8155), while Chapter 15 discusses general-purpose programmable devices (such as the 8255A, 8254, 8259, and 8237). Chapter 17 deals primarily with the project design of a single-board microcomputer that brings together all the concepts discussed in the text. Chapter 18 focuses on how to extend 8-bit microprocessor concepts to higher level processes and microcontrollers. It also discusses trends in microprocessor technology ranging from recent microcontrollers to the latest general-purpose 32-bit microprocessors.

NEW AND IMPROVED FEATURES IN THE FOURTH EDITION

The fourth edition preserves the focus as described and includes the following changes and additions, suggested by reviewers and by faculty who have used the book in their classrooms:

1. Chapter 1 is revised to include the most recent technological changes.
2. Part II (Chapters 5 through 11) has few changes in the content, except in Chapter 11, which is revised to include technological changes.
3. In Chapter 15, the DMA controller 8257 is replaced by the 8237, which is commonly used in newer systems. The 8237 is thoroughly discussed with illustrations.
4. Chapter 17 includes an additional interfacing application: how to interface an LCD module, which has become a popular display device in industrial systems.

5. Chapter 18 is updated to include the latest technological changes in 32-bit microprocessors.
6. In Appendix B, the Intel SDK-85 system is replaced by the EMAC Primer, a stand-alone single-board microcomputer system with a Hex keyboard and LED displays. Its enhanced version can be used with a PC. Students can write assembly language programs using an editor on a PC, assemble the programs, download the binary code from a PC to the Primer trainer, and execute the programs. EMAC Primer is a complete development system, ideally suited to an educational environment.
7. In Appendix D, complete data sheets for the 8237, 8259, and an LCD panel are included. These data sheets will enable students to perform many experiments outside the scope of this introductory text.

A WORD WITH FACULTY

This text is based on my teaching experience, my course development efforts, and my association with industry engineers and programmers. It is an attempt to share my classroom experiences and my observations in industrial practices. Some of my assumptions and observations of 15 years ago appear still valid today:

1. Software (instructions) is an integral part of the microprocessor and demands emphasis equal to that of the hardware.
2. In industry, for development of microprocessor-based projects, 70 percent of the effort is devoted to software and 30 percent to hardware.
3. Technology and engineering students tend to be hardware oriented and have considerable difficulty in programming.
4. Students have difficulty in understanding mnemonics and realizing the critical importance of flags.

In the last fifteen years, numerous faculty members shared their classroom experiences, concerns, and student difficulties with me through letters and e-mail messages. I have made every effort to incorporate those concerns and suggestions in the fourth edition. This revised edition can be used flexibly to meet the objectives of various courses at the undergraduate level. If used for a one-semester course with 50 percent hardware and 50 percent software emphasis, the following chapters are recommended: Chapters 1 through 4 for hardware lectures and Chapters 5 through 9 and selected sections of Chapter 10 for software laboratory sessions. For interfacing, the initial sections of Chapters 12 and 16 (introduction to interrupts and serial I/O) are recommended. If the course is heavily oriented toward hardware, Chapters 1 through 4 and Chapters 12 through 17 are recommended, and necessary programs can be selected from Chapters 5 through 9. If the course is heavily oriented toward software, Chapters 1 through 11 and selected portions of Chapters 12 and 16 can be used. For a two-semester course, it is best to use the entire text.

A WORD WITH STUDENTS

Microprocessor technology is an exciting, challenging, and growing field; it will pervade industry for decades to come. To meet the challenges of this growing technology, you will have to be conversant with the programmable aspect of the microprocessor. Programming is a process of problem solving and communication in a strange language of mnemonics. Most often, hardware-oriented students find this communication process very difficult. One of the questions frequently asked by students is: How do I get started in a given programming assignment? One approach to learning programming is to examine various types of programs and imitate them. You can begin by studying the illustrative program relevant to an assignment, its flowchart, its analysis, program description, and particularly the comments. Read the instructions from Appendix F as necessary and pay attention to the flags. This text is written in such a way that simple programming aspects of the microprocessor can be self-taught. Once you master the elementary programming techniques, then interfacing and design become exciting and fun.

ACKNOWLEDGMENTS

My sincere thanks to my family members: my wife, Shaila, for her unwavering support and my daughters, Nelima and Vanita, for their assistance in completing various tasks. Several persons have made valuable contributions to this text. I would like to extend my sincere appreciation to my colleagues Charles Abate, James Delaney, and John Merrill, who offered many suggestions throughout the project, and Chris Conty, who initiated the project. Similarly, I appreciate the efforts and numerous suggestions of my reviewers: John Morgan from DeVry Institute, Peter Holsberg from Mercer Community College, David Hata from Portland Community College, and David Delkar from Kansas State University. If this text reads well, the credit goes to Gnomi Gouldin and to my colleague from the English Department, Dr. Kathy Forrest, who devoted painstaking hours to editing the rough draft of the first edition. For this fourth edition, I would like to express my sincere appreciation to the following reviewers of this edition who provided me with valuable comments and suggestions: Cheryl Schmidt, Florida Community College at Jacksonville; Michael Pelletier, Northern Essex Community College; and Ted Nguyen, San Jose City College. I also thank Alex Wolf, my production editor at Prentice Hall, and copy editor Sheryl Rose for their contributions to the text. I would appreciate any communication about the text from the reader. Please feel free to write or send an e-mail message.

Ramesh Gaonkar
State University of New York
O.C.C. Campus at Syracuse
Syracuse, New York 13215
E-mail: gaonkarr@sunyocc.edu

Contents

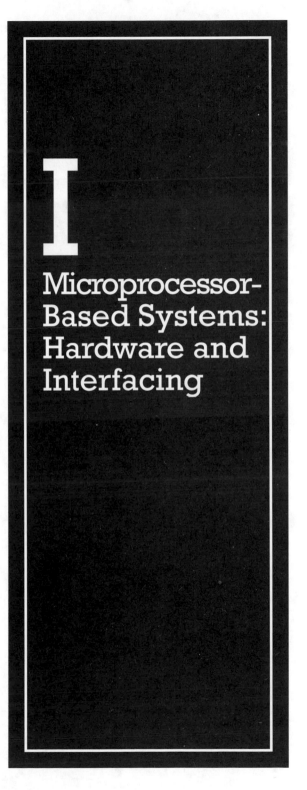

I
Microprocessor-Based Systems: Hardware and Interfacing

Part I of this book is concerned primarily with the microprocessor architecture in the context of microprocessor-based products. The microprocessor-based systems are discussed in terms of the three components—microprocessor, memory, and I/O (input/output)—and their communication process. The role of the programming languages, from machine language to higher-level languages, is presented in the context of the system.

The material is presented in a format similar to the view from an airplane that is getting ready to land. As the plane circles around, a passenger observes a view without any details. As the plane starts descending, the passenger begins to see the same view but with more details. In the same way, Chapter 1 presents the microprocessor from two points of view: the microprocessor as a programmable embedded device in a product and as an element of a computer system, and how it communicates with memory and I/O. The chapter also discusses the role of assembly language in microprocessor-based products and presents an overview of various types of computers—from large computers to microcomputers and their applications.

Chapter 2 provides a closer look at a microcomputer system in relation to the 8085 microprocessor. Chapter 3 examines the details of the 8085 microprocessor and memory interfacing. Chapter 4 discusses the interfacing of input/output (I/O) devices.

PREREQUISITES

The reader is expected to know the following concepts:

☐ Number systems (binary, octal, and hexadecimal) and their conversions.
☐ Boolean algebra, logic gates, flip-flops, and registers.
☐ Concepts in combinational and sequential logic.

Microprocessors, Microcomputers, and Assembly Language

The microprocessor plays a significant role in the everyday functioning of industrialized societies. The microprocessor can be viewed as a programmable logic device that can be used to control processes or to turn on/off devices. On the other hand, the microprocessor can be viewed as a data processing unit or a computing unit of a computer. The **microprocessor** is a programmable integrated device that has computing and decision-making capability similar to that of the central processing unit (CPU) of a computer. Nowadays, the microprocessor is being used in a wide range of products called microprocessor-based products or systems. The microprocessor can be embedded in a larger system, can be a stand alone unit controlling processes, or it can function as the CPU of a computer called a **microcomputer.** This chapter introduces the basic structure of a microprocessor-based product and shows how the same structure is applicable to microcomputers and other large (mini- and mainframe) computers. Later in the chapter, microprocessor applications are presented in the context of the entire spectrum of various computer applications.

The microprocessor communicates and operates in the binary numbers 0 and 1, called **bits.** Each microprocessor has a fixed set of instructions in the form of binary patterns called a **machine language.**

However, it is difficult for humans to communicate in the language of 0s and 1s. Therefore, the binary instructions are given abbreviated names, called **mnenomics,** which form the **assembly language** for a given microprocessor. This chapter explains both the machine language and the assembly language of the microprocessor, known as the 8085. The advantages of assembly language are compared with high-level languages (such as BASIC, FORTRAN, C, and C++).

OBJECTIVES

☐ Draw a block diagram of a microprocessor-based system and explain the functions of each component: microprocessor, memory, and I/O, and their lines of communication (the bus).

☐ Explain the terms *SSI, MSI,* and *LSI.*

☐ Define the terms *bit, byte, word, instruction, software,* and *hardware.*

☐ Explain the difference between the machine language and the assembly language of a computer.

☐ Explain the terms *low-level* and *high-level languages.*

☐ Explain the advantages of an assembly language over high-level languages.

☐ Define the term *ASCII* code and explain the relationship between the binary code and alphanumeric characters.

☐ Define the term *operating system.*

☐ List components and peripherals of a typical personal computer (PC).

1.1   MICROPROCESSORS

A microprocessor is a multipurpose, programmable, clock-driven, register-based electronic device that reads binary instructions from a storage device called *memory,* accepts binary data as input and processes data according to those instructions, and provides results as output. A typical programmable machine can be represented with three components: microprocessor, memory, and I/O as shown in Figure 1.1. These three components work together or interact with each other to perform a given task; thus, they comprise a system. The physical components of this system are called **hardware.** A set of instructions written for the microprocessor to perform a task is called a **program,** and a group of programs is called **software.** The machine (system) represented in Figure 1.1 can be programmed to turn traffic lights on and off, compute mathematical functions, or keep track of a guidance system. This system may be simple or sophisticated, depending on its applications, and it is recognized by various names depending upon the purpose for which it is designed. The microprocessor applications are classified primarily in two categories: reprogrammable systems and embedded systems. In reprogrammable systems, such as microcomputers, the microprocessor is used for computing and data processing. These systems include general-purpose microprocessors capable of handling large data, mass storage devices (such as disks and CD-ROMs), and peripherals such as printers; a personal computer (PC) is a typical illustration. In embedded systems, the microprocessor is a part of a final product and is not available for reprogramming to the end user. A copy-

FIGURE 1.1
A Programmable Machine

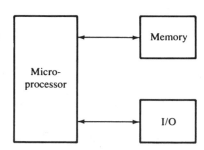

ing machine is a typical example of an embedded system. The microprocessors used in these systems are generally categorized as: (1) **microcontrollers** that include all the components shown in Figure 1.1 on one chip, and (2) general-purpose microprocessors with discrete components shown in Figure 1.1. Embedded systems can also be viewed as products that use microprocessors to perform their operations; they are known as microprocessor-based products. Examples include a wide range of products such as washing machines, dishwashers, automobile dashboard controls, traffic light controllers, and automatic testing instruments.

BINARY DIGITS

The microprocessor operates in binary digits, 0 and 1, also known as bits. **Bit** is an abbreviation for the term *binary digit*. These digits are represented in terms of electrical voltages in the machine: Generally, 0 represents one voltage level, and 1 represents another. The digits 0 and 1 are also synonymous with low and high, respectively.

Each microprocessor recognizes and processes a group of bits called the *word,* and microprocessors are classified according to their word length. For example, a processor with an 8-bit word is known as an 8-bit microprocessor, and a processor with a 32-bit word is known as a 32-bit microprocessor.

A MICROPROCESSOR AS A PROGRAMMABLE DEVICE

The fact that the microprocessor is programmable means it can be instructed to perform given tasks within its capability. A piano is a programmable machine; it is capable of generating various kinds of tones based on the number of keys it has. A musician selects keys depending upon the musical score printed on a sheet. Similarly, today's microprocessor is designed to understand and execute many binary instructions. It is a multipurpose machine: It can be used to perform various sophisticated computing functions, as well as simple tasks such as turning devices on or off. A programmer can select appropriate instructions and ask the microprocessor to perform various tasks on a given set of data.

The person who designs a piano determines the frequency (tone) for a given key and the scope of the piano music. Similarly, the engineers designing a microprocessor determine a set of tasks the microprocessor should perform and design the necessary logic circuits, and provide the user with a list of the instructions the processor will understand. For example, an instruction for adding two numbers may look like a group of eight binary digits, such as 1000 0000. These instructions are simply a pattern of 0s and 1s. The user (programmer) selects instructions from the list and determines the sequence of execution for a given task. These instructions are entered or stored in storage, called *memory,* which can be read by the microprocessor.

MEMORY

Memory is like the pages of a notebook with space for a fixed number of binary numbers on each line. However, these pages are generally made of semiconductor material. Typically, each line is an 8-bit register that can store eight binary bits, and several of these registers are arranged in a sequence called memory. These registers are always grouped together in powers of two. For example, a group of $1024(2^{10})$ 8-bit registers on a semiconductor chip is known as 1K byte of memory; 1K is the closest approximation in thousands.* The user writes the necessary instructions and data in memory through an input device (described below), and asks the microprocessor to perform the given task and find an answer. The answer is generally displayed at an output device (described below) or stored in memory.

INPUT/OUTPUT

The user can enter instructions and data into memory through devices such as a keyboard or simple switches. These devices are called **input devices.** The microprocessor reads the instructions from the memory and processes the data according to those instructions. The result can be displayed by a device such as seven-segment LEDs (Light Emitting Diodes) or printed by a printer. These devices are called **output devices.**

MICROPROCESSOR AS A CPU (MPU)

We can also view the microprocessor as a primary component of a computer. Traditionally, the computer is represented in block diagram as shown in Figure 1.2(a). The block diagram shows that the computer has four components: Memory, Input, Output, and the central processing unit (CPU), which consists of the Arithmetic/Logic Unit (ALU) and Control Unit. The CPU contains various registers to store data, the ALU to perform arithmetic and logical operations, instruction decoders, counters, and control lines. The CPU reads instructions from the memory and performs the tasks specified. It communicates with input/output devices either to accept or to send data. These devices are also known as peripherals. The CPU is the primary and central player in communicating with devices such as memory, input, and output. However, the timing of the communication process is controlled by the group of circuits called the **control unit.**

In the late 1960s, the CPU was designed with discrete components on various boards. With the advent of the integrated circuit technology, it became possible to build the CPU on a single chip; this came to be known as a microprocessor, and the traditional block diagram shown in Figure 1.2(a) can be replaced by the block diagram shown in Figure 1.2(b). It is also known as an MPU (microprocessor unit).

1.11 Advances in Semiconductor Technology

In the last forty years, semiconductor technology has undergone unprecedented changes. After the invention of the transistor, integrated circuits (ICs) appeared on the scene at the end of the 1950s; an entire circuit consisting of several transistors, diodes, and resistors could be designed on a single chip. In the early 1960s, logic gates known as the 7400 series were commonly available as ICs, and the technology of integrating the circuits of a logic gate on a single chip became known as small-scale integration (SSI). As semiconductor technology advanced, more than 100 gates were fabricated on one chip; this was called medium-scale integration (MSI). A typical example of MSI is a decade counter (7490). Within a few years, it was possible to fabricate more than 1000 gates on a single chip; this came to be known as large-scale integration (LSI). Now we are in the era of

---

*In computer terminology, 1K is equal to 1024. In scientific terminology, 1k is equal to 1000.

FIGURE 1.2

Traditional Block Diagram of a Computer (a) and Block Diagram of a Computer with the Microprocessor as CPU (MPU) (b)

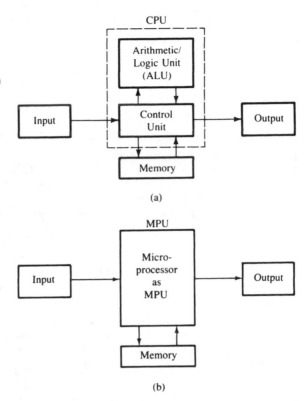

very-large-scale integration (VLSI) and super-large-scale integration (SLSI). The lines of demarcation between these different scales of integration are rather ill defined and arbitrary.

As the technology moved from SSI to LSI, more and more logic circuits were built on one chip, and they could be programmed to do different functions through hard-wired connections. For example, a counter chip can be programmed to count in Hex or decimal by providing logic 0 or 1 through appropriate pin connections. The next step was the idea of providing 0s and 1s through a register. The necessary signal patterns of 0s and 1s were stored in registers and given to the programmable chip at appropriate times; the group of registers used for storage was called memory. Because of the LSI technology, it became possible to build many computing functions and their related timing on a single chip.

The Intel 4004 was the first 4-bit programmable device that was primarily used in calculators. It was designed by Intel Corporation and became known as the 4-bit microprocessor. It was quickly replaced by the 8-bit microprocessor (the Intel 8008), which was in turn superseded by the Intel 8080. In the mid-1970s, the Intel 8080 was widely used in control applications, and small computers also were designed using the 8080 as the CPU; these computers became known as microcomputers. Within a few years after the emergence of the 8080, the Motorola 6800, the Zilog Z80, and the Intel 8085 microprocessors were developed as improvements over the 8080. The 6800 was designed with a different architecture and the instruction set from the 8080. On the other hand, the 8085 and the Z80 were designed as **upward software compatible** with the 8080; that is, they included all the instructions of the 8080 plus additional instructions. As the microprocessors began to acquire more and more computing functions, they were viewed more as CPUs rather than as programmable logic devices. Most microcomputers are now built with 32- and 64-bit microprocessors. Each microprocessor has begun to carve a niche for its own applications. The 8-bit microprocessors are being used as programmable logic devices in control applications, and more powerful microprocessors are being used for mathematical computing (number crunching), data processing, and computer graphics applications. Our focus here is on using 8-bit microprocessors as programmable devices.

1.12   Organization of a Microprocessor-Based System

Figure 1.3 shows a simplified but formal structure of a microprocessor-based system or a product. Since a microcomputer is one among many microprocessor-based systems, it will have the same structure as shown in Figure 1.3. It includes three components: *microprocessor, I/O (input/output),* and *memory* (read/write memory and read-only memory). These components are organized around a common communication path called a **bus.** The entire group of components is also referred to as a system or a microcomputer system, and the components themselves are referred to as sub-systems. At the outset, it is necessary to differentiate between the terms *microprocessor* and *microcomputer* because of the common misuse of these terms in popular literature. The microprocessor is one component of the microcomputer. On the other hand, the microcomputer is a complete computer similar to any other computer, except that CPU functions of the microcomputer are performed by the microprocessor. Similarly, the term   **peripheral** is used for input/output devices. The various components of a microprocessor-based product or a microcomputer are shown in Figure 1.3 and their functions are described in this section.

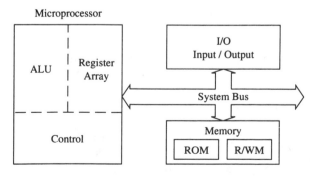

FIGURE 1.3
Microprocessor-Based System with Bus Architecture

MICROPROCESSOR

The microprocessor is a clock-driven semiconductor device consisting of electronic logic circuits manufactured by using either a large-scale integration (LSI) or very-large-scale integration (VLSI) technique. The microprocessor is capable of performing various computing functions and making decisions to change the sequence of program execution. In large computers, a CPU implemented on one or more circuit boards performs these computing functions. The microprocessor is in many ways similar to the CPU, but includes all the logic circuitry, including the control unit, on one chip. The microprocessor can be divided into three segments for the sake of clarity, as shown in Figure 1.3: arithmetic/logic unit (ALU), register array, and control unit.

Arithmetic/Logic Unit This is the area of the microprocessor where various computing functions are performed on data. The ALU unit performs such arithmetic operations as addition and subtraction, and such logic operations as AND, OR, and exclusive OR.

Register Array This area of the microprocessor consists of various registers identified by letters such as B, C, D, E, H, and L. These registers are primarily used to store data temporarily during the execution of a program and are accessible to the user through instructions.

Control Unit The control unit provides the necessary timing and control signals to all the operations in the microcomputer. It controls the flow of data between the microprocessor and memory and peripherals.

Now the question is: What is the relationship among the programmer's instruction (binary pattern of 0s and 1s), the ALU, and the control unit? This can be explained with the example of a full adder circuit. A full adder circuit can be designed with registers, logic gates, and a clock. The clock initiates the adding operation. Similarly, the bit pattern of an instruction initiates a sequence of clock signals, activates the appropriate logic circuits in the ALU, and performs the task. This is called microprogramming, which is done in the design stage of the microprocessor. The bit patterns required to initiate these microprogram operations are given to the programmer in the form of the instruction set of the microprocessor. The programmer selects appropriate bit patterns from the set for a given task and enters them sequentially in memory through an input device. When the CPU reads these bit patterns one at a time, it initiates appropriate microprograms through the control unit, and performs the task specified in the instructions.

At present, various microprocessors are available from different manufacturers. Examples of widely used 8-bit microprocessors include the Intel 8085, Zilog Z80, and Motorola 68008. Earlier microcomputers were designed around the 8-bit microprocessors; now these processors are generally used in embedded systems. The recent versions of IBM personal computers are designed around the Intel 32- or 64-bit microprocessors. Single-board microcomputers such as the SDK-85 (Intel), The Primer (EMAC Inc.), and the Micro-Professor (Multitech) are commonly used in college laboratories; the SDK-85 and The Primer (described in Appendix B) are based on the 8085 microprocessor, and the Micro-Professor is based on the Z80 microprocessor.

MEMORY

Memory stores such binary information as instructions and data, and provides that information to the microprocessor whenever necessary. To execute programs, the microprocessor reads instructions and data from memory and performs the computing operations in its ALU section. Results are either transferred to the output section for display or stored in memory for later use. The memory block shown in Figure 1.3 has two sections: **Read-Only memory** (ROM) and **Read/Write memory** (R/WM), popularly known as **Random-Access memory** (RAM).

The ROM is used to store programs that do not need alterations. The monitor program of a single-board microcomputer is generally stored in the ROM. This program interprets the information entered through a keyboard and provides equivalent binary digits to the microprocessor. Programs stored in the ROM can only be read; they cannot be altered.

The Read/Write memory (R/WM) is also known as *user memory*. It is used to store user programs and data. In single-board microcomputers, the monitor program monitors the Hex keys and stores those instructions and data in the R/W memory. The information stored in this memory can be easily read and altered.

I/O (INPUT/OUTPUT)

The third component of a microprocessor-based system is I/O (input/output); it communicates with the outside world. I/O includes two types of devices: input and output; these I/O devices are also known as *peripherals.*

The input devices such as a keyboard, switches, and an analog-to-digital (A/D) converter transfer binary information (data and instructions) from the outside world to the microprocessor. Typically, a microcomputer used in college laboratories includes either a hexadecimal keyboard or an ASCII keyboard as an input device. The hexadecimal (Hex) keyboard has 16 data keys (0 to 9 and A to F) and some additional function keys to perform such operations as storing data and executing programs. The ASCII (the term is explained in Section 1.2) keyboard is similar to a typewriter keyboard, and it is used to enter programs in an English-like language. Although the ASCII keyboard is found in most microcomputers (PCs), single-board microcomputers generally have Hex keyboards, and microprocessor-based products such as a microwave oven have decimal keyboards.

The output devices transfer data from the microprocessor to the outside world. They include devices such as light emitting diodes (LEDs), a cathode-ray tube (CRT) or video screen, a printer, X-Y plotter, a magnetic tape, and digital-to-analog (D/A) converter. Typically, single-board microcomputers and microprocessor-based products (such as a dishwasher or a microwave oven) include LEDs, seven-segment LEDs, and alphanumeric LED displays as output devices. Microcomputers (PCs) are generally equipped with output devices such as a video screen (also called a monitor) and a printer.

SYSTEM BUS

The system bus is a communication path between the microprocessor and peripherals; it is nothing but a group of wires to carry bits. In fact, there are several buses in the system that will be discussed in the next chapter. All peripherals (and memory) share the same bus; however, the microprocessor communicates with only one peripheral at a time. The timing is provided by the control unit of the microprocessor.

1.13   How Does the Microprocessor Work?

Assume that a program and data are already entered in the R/W memory. (How to write and execute a program will be explained later.) The program includes binary instructions to add given data and to display the answer at the seven-segment LEDs. When the microprocessor is given a command to execute the program, it reads and executes one instruction at a time and finally sends the result to the seven-segment LEDs for display.

This process of program execution can best be described by comparing it to the process of assembling a radio kit. The instructions for assembling the radio are printed in a sequence on a sheet of paper. One reads the first instruction, then picks up the necessary components of the radio and performs the task. The sequence of the process is *read, interpret,* and *perform.* The microprocessor works the same way. The instructions are stored sequentially in the memory. The microprocessor fetches the first instruction from its memory sheet, decodes it, and executes that instruction. The sequence of *fetch, decode,* and *execute* is continued until the microprocessor comes across an instruction to *stop.* During the entire process, the microprocessor uses the system bus to fetch the binary instructions and data from the memory. It uses registers from the register section to store data temporarily, and it performs the computing function in the ALU section. Finally, it sends out the result in binary, using the same bus lines, to the seven-segment LEDs.

1.14   Summary of Important Concepts

The functions of various components of a microprocessor-based system can be summarized as follows:

1. The microprocessor
   - □ reads instructions from memory.
   - □ communicates with all peripherals (memory and I/Os) using the system bus.
   - □ controls the timing of information flow.
   - □ performs the computing tasks specified in a program.
2. The memory
   - □ stores binary information, called instructions and data.
   - □ provides the instructions and data to the microprocessor on request.
   - □ stores results and data for the microprocessor.
3. The input device
   - □ enters data and instructions under the control of a program such as a monitor program.
4. The output device
   - □ accepts data from the microprocessor as specified in a program.
5. The bus
   - □ carries bits between the microprocessor and memory and I/Os.

1.2 MICROPROCESSOR INSTRUCTION SET AND COMPUTER LANGUAGES

Microprocessors recognize and operate in binary numbers. However, each microprocessor has its own binary words, meanings, and language. The words are formed by combining a number of bits for a given machine. The **word** (or word length) is defined as the number of bits the microprocessor recognizes and processes at a time. The word length ranges from four bits for small, microprocessor-based systems to 64 bits for high-speed large computers. Another term commonly used to express word length is byte. A **byte** is defined as a group of eight bits. For example, a 16-bit microprocessor has a word length equal to two bytes. The term **nibble,** which stands for a group of four bits, is found also in popular computer magazines and books. A byte has two nibbles.

Each machine has its own set of instructions based on the design of its CPU or of its microprocessor. To communicate with the computer, one must give instructions in binary language **(machine language).** Because it is difficult for most people to write programs in sets of 0s and 1s, computer manufacturers have devised English-like words to represent the binary instructions of a machine. Programmers can write programs, called **assembly language** programs, using these words. Because an assembly language is specific to a given machine, programs written in assembly language are not transferable from one machine to another. To circumvent this limitation, such general-purpose languages as BASIC and FORTRAN have been devised; a program written in these languages can be machine-independent. These languages are called **high-level languages.** This section deals with various aspects of these three types of languages: machine, assembly, and high-level. The machine and assembly languages are discussed in the context of the 8085 microprocessor.

1.21 Machine Language

The number of bits in a word for a given machine is fixed, and words are formed through various combinations of these bits. For example, a machine with a word length of eight bits can have 256 (2^8) combinations of eight bits—thus a language of 256 words. However, not all of these words need to be used in the machine. The microprocessor design engineer selects combinations of bit patterns and gives a specific meaning to each combination by using electronic logic circuits; this is called an **instruction.** Instructions are made up of one word or several words. The set of instructions designed into the machine makes up its machine language—a binary language, composed of 0s and 1s—that is specific to each computer. In this book, we are concerned with the language of a widely used 8-bit microprocessor, the 8085, manufactured by Intel Corporation. The primary focus here is on the microprocessor because the microprocessor determines the machine language and the operations of a microprocessor-based system.

1.22 8085 Machine Language

The 8085 is a microprocessor with 8-bit word length: its **instruction set** (or language) is designed by using various combinations of these eight bits. The 8085 is an improved version of the earlier processor 8080A.

An *instruction* is a binary pattern entered through an input device in memory to command the microprocessor to perform that specific function.

For example:

0011 1100   is an instruction that increments the number in the register called the **accumulator** by one.

1000 0000   is an instruction that adds the number in the register called B to the number in the accumulator, and keeps the sum in the accumulator.

The 8085 microprocessor has 246 such bit patterns, amounting to 74 different instructions for performing various operations. These 74 different instructions are called its instruction set. This binary language with a predetermined instruction set is called the 8085 machine language.

Because it is tedious and error-inducive for people to recognize and write instructions in binary language, these instructions are, for convenience, written in hexadecimal code and entered in a single-board microcomputer by using Hex keys. For example, the binary instruction 0011 1100 (mentioned previously) is equivalent to 3C in hexadecimal. This instruction can be entered in a single-board microcomputer system with a Hex keyboard by pressing two keys: 3 and C. The monitor program of the system translates these keys into their equivalent binary pattern.

1.23   8085 Assembly Language

Even though the instructions can be written in hexadecimal code, it is still difficult to understand a program written in hexadecimal numbers. Therefore, each manufacturer of a microprocessor has devised a symbolic code for each instruction, called a **mnemonic.** (The word *mnemonic* is based on the Greek word meaning *mindful;* that is, a memory aid.) The mnemonic for a particular instruction consists of letters that suggest the operation to be performed by that instruction.

For example, the binary code 0011 1100 ($3C_{16}$ or 3CH* in hexadecimal) of the 8085 microprocessor is represented by the mnemonic INR A:

INR A   INR stands for increment, and A represents the accumulator. This symbol suggests the operation of incrementing the accumulator contents by one.

Similarly, the binary code 1000 0000 (80_{16} or 80H) is represented as

ADD B   ADD stands for addition, and B represents the contents in register B. This symbol suggests the addition of the contents in register B and the contents in the accumulator.

Although these symbols do not specify the complete operations, they suggest its significant part. The complete description of each instruction must be supplied by the manufacturer. The complete set of 8085 mnemonics is called the 8085 assembly language, and a program written in these mnemonics is called an assembly language pro-

---

*Hexadecimal numbers are shown either with the subscript 16, or as a number followed by the letter H.

gram. (Again, the assembly language, or mnemonics, is specific to each microprocessor. For example, the Motorola 6800 microprocessor has an entirely different set of binary codes and mnemonics than the 8085. Therefore, the assembly language of the 6800 is far different from that of the 8085.) An assembly language program written for one microprocessor is not transferable to a computer with another microprocessor unless the two microprocessors are compatible in their machine codes.

  Machine language and assembly language are microprocessor-specific and are both considered **low-level languages.** The machine language is in binary, and the assembly language is in English-like words; however, the microprocessor understands only the binary. How, then, are the assembly language mnemonics written and translated into machine language or binary code? The mnemonics can be written by hand on paper (or in a notebook) and translated manually in hexadecimal code, called **hand assembly,** as explained in Section 1.25. Similarly, the mnemonics can be written electronically on a computer using a program called an Editor in the ASCII code (explained in the next section) and translated into binary code by using the program called an **assembler.**

1.24 ASCII Code

A computer is a binary machine; to communicate with the computer in alphabetic letters and decimal numbers, translation codes are necessary. The commonly used code is known as **ASCII**—American Standard Code for Information Interchange. It is a 7-bit code with 128 (2^7) combinations, and each combination from 00H to 7FH is assigned to either a letter, a decimal number, a symbol, or a machine command (see Appendix E). For example, hexadecimal 30H to 39H represent 0 to 9 decimal digits, 41H to 5AH represent capital letters A through Z, 20H to 2FH represent various symbols, and initial codes 00H to 1FH represent such machine commands as carriage return and line feed. In microcomputer systems, keyboards (called ASCII keyboards), video screens, and printers are typical examples of devices that use ASCII codes. When the key "9" is pressed on an ASCII keyboard, the computer receives 39H in binary, called an ASCII character, and the system program translates ASCII characters into appropriate binary numbers.

  However, recent computers use many more characters than the original 128 combinations; this is called Extended ASCII. It is an 8-bit code that provides 256 (2^8) combinations; the additional 128 combinations are assigned to various graphics characters.

1.25 Writing and Executing an Assembly Language Program

As we explained earlier, a program is a set of logically related instructions written in a specific sequence to accomplish a task. To manually write and execute an assembly language program on a single-board computer, with a Hex keyboard for input and LEDs for output, the following steps are necessary:

1. Write the instructions in mnemonics obtained from the instruction set supplied by the manufacturer.
2. Find the hexadecimal machine code for each instruction by searching through the set of instructions.

3. Enter (load) the program in the user memory in a sequential order by using the Hex keyboard as the input device.

4. Execute the program by pressing the Execute key. The answer will be displayed by the LEDs.

This procedure is called either **manual** or **hand assembly.**

When the user program is entered by the keys, each entry is interpreted and converted into its binary equivalent by the monitor program, and the machine code is stored as eight bits in each memory location in a sequence. When the Execute command is given, the microprocessor fetches each instruction, decodes it, and executes it in a sequence until the end of the program.

The manual assembly procedure is commonly used in single-board microcomputers and is suited for small programs; however, looking up the machine codes and entering the program is tedious and subject to errors. The other process involves the use of a computer with an ASCII keyboard and an assembler.

The **assembler** is a program that translates the mnemonics entered by the ASCII keyboard into the corresponding binary machine codes of the microprocessor. Each microprocessor has its own assembler because the mnemonics and machine codes are specific to the microprocessor being used, and each assembler has rules that must be followed by the programmer. Personal Computers (PCs—see Section 1.3) are commonly available on college campuses. These computers are based on 16- or 32-bit microprocessors with different mnemonics than the 8085 microprocessor. However, the programs known as **cross-assemblers** can be used to translate the 8085 mnemonics into appropriate machine codes. (Assemblers and cross-assemblers are discussed in Chapter 11.)

1.26 High-Level Languages

Programming languages that are intended to be machine-independent are called **high-level languages.** These include such languages as FORTRAN, BASIC, PASCAL, C, and C++, all of which have certain sets of rules and draw on symbols and conventions from English. Instructions written in these languages are known as **statements** rather than mnemonics. A program written in BASIC for a microcomputer with the 8085 microprocessor can generally be run on another microcomputer with a different microprocessor.

Now the question is: How are words in English converted into the binary languages of different microprocessors? The answer is: Through another program called either a **compiler** or an **interpreter.** These programs accept English-like statements as their input, called the *source code.* The compiler or interpreter then translates the source code into the machine language compatible with the microprocessor being used in the system. This translation in the machine language is called the *object code* (Figure 1.4). Each microprocessor needs its own compiler or an interpreter for each high-level language. The primary difference between a compiler and an interpreter lies in the process of generating machine code. The compiler reads the entire program first and translates it into the object code that is executed by the microprocessor. On the other hand, the interpreter reads one instruction at a time, produces its object code (a sequence of machine actions), and executes the instruction before reading the next instruction. M-Basic is a common example

FIGURE 1.4

Block Diagram: Translation of High-Level Language Program into Machine Code

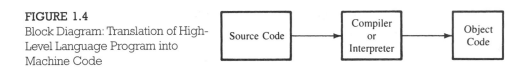

of an interpreter for BASIC language. Compilers are generally used in such languages as FORTRAN, PASCAL, C, and C++.

Compilers and interpreters require large memory space because an instruction in English requires several machine codes to translate it into binary. On the other hand, there is one-to-one correspondence between the assembly language mnemonics and the machine code. Thus, assembly language programs are compact and require less memory space. They are more efficient than the high-level language programs. The primary advantage of high-level languages is in troubleshooting (**debugging**) programs. It is much easier to find errors in a program written in a high-level language than to find them in a program written in an assembly language.

In certain applications such as traffic control and appliance control, where programs are small and compact, assembly language is suitable. Similarly, in such real-time applications as converting a high-frequency waveform into digital data, program efficiency is critical. In real-time applications, events and time should closely match each other without significant delay; therefore, assembly language is highly desirable in these applications. On the other hand, for applications in which programs are large and memory is not a limitation, high-level languages may be desirable. Typical examples of applications programs are word processors, video games, tax-return preparation, billing, accounting, and money management. These programs are generally written by professionals such as programmers in high-level languages. The advantage of time saved in debugging a large program may outweigh the disadvantages of memory requirements and inefficiency. Now we need to examine the relationship and the interaction between the hardware (microprocessor, memory, and I/O) and software (languages and application programs).

1.27   Operating Systems

The interaction between the hardware and the software is managed by a set of programs called an **operating system** of a computer; it oversees all the operations of the computer. The computer transfers information constantly between memory and various peripherals such as printer, keyboard, and video monitor. It also stores programs on disk. The operating system is responsible primarily for storing information on the disk and for the communication between the computer and its peripherals. The functional relationship between the operating system and the hardware of the computer is shown in Figure 1.5(a).

Figure 1.5(b) shows the relationship and the hierarchy among the hardware, the operating system, high-level languages, and application programs. The operating system is closest to the hardware and application programs are farthest from the hardware. When the computer is turned on, the operating system is in charge of the system and stays in the background and provides channels of communications to application programs. Each computer has its own operating system. In the 1970s, CP/M (Control Monitor Program) was a widely used operating system; it was designed for 8-bit processors such as Z80 and 8085/8080A. In 16-bit microcomputers, such as personal computers, MS-DOS (Microsoft Disk Operating System) has become an industry standard. Similarly, Apple and Macintosh computers have their own operating systems. In recent 32- and 64-bit micro-computers, operating systems such as UNIX, OS/2 (Operating System 2), Windows 95/(98), and Windows NT are commonly used.

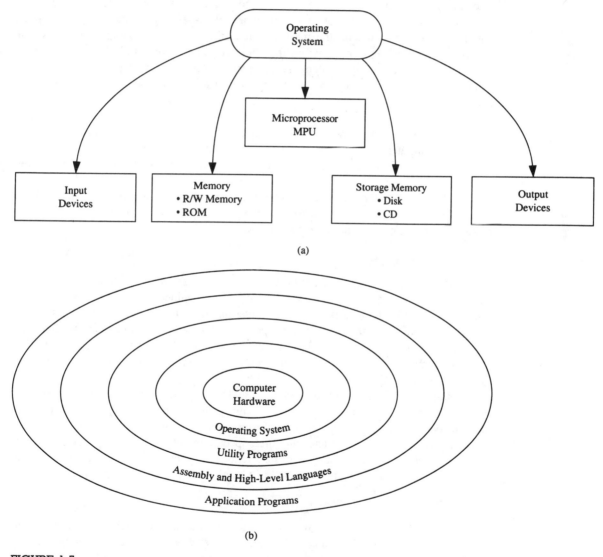

(a)

(b)

FIGURE 1.5
(a) Operating System and Its Functional Relationship with Various Hardware Components
(b) Hierarchical Relationship between Computer Hardware and Software

1.3 FROM LARGE COMPUTERS TO SINGLE-CHIP MICROCONTROLLERS

In the last forty years, advances in semiconductor technology have had an unprecedented impact on computers. In the 1960s, computers were accessible only to big corporations, universities, and government agencies. Now, "computer" has become a common word. The range of computers now available extends from such sophisticated, multimillion-dollar machines as the Cray computers to the less-than-$1000 personal computer. All the computers now available on the market include the same basic components shown in Figure 1.3. Nevertheless, it is obvious that these computers are not all the same.

Different types of computers are designed to serve different purposes. Some are suitable for scientific calculations, while others are used simply for turning appliances on and off. Thus, it is necessary to have an overview of the entire spectrum of computer applications as a context for understanding the topics and applications discussed in this text. Until the 1970s, computers were broadly classified in three categories: mainframe, mini-, and microcomputers. Since then, technology has changed considerably, and the distinctions between these categories have been blurred. Initially, the microcomputer was recognized as a computer with a microprocessor as its CPU. Now practically all computers have various types of microprocessors performing different functions within the large CPU. For the sake of convenience, computers are classified here as large computers, medium-size computers, and microcomputers.

1.31 Large Computers

These are large, general-purpose, multi-user, multitasking computers designed to perform such data processing tasks as complex scientific and engineering calculations, and handling of records for large corporations or government agencies. These computers can be classified broadly into mainframes and supercomputers, and mainframes are further classified according to their sizes. The prices range from $100,000 to millions of dollars. Typical examples of these computers include the IBM System/390 series and Cray computers (Cray-2, Y-MP).

Mainframes are high-speed computers, and their word length generally ranges from 32 to 64 bits. They are capable of addressing megabytes of memory and handling all types of peripherals and a large number of users. Supercomputers such as the Cray-2 and Y-MP are 64-bit high-performance and high-speed computers. They are the fastest computers, capable of executing billions of instructions per second, and are used primarily in research dealing with problems in areas such as global climate and high-energy physics.

1.32 Medium-Size Computers

In the 1960s, these computers were designed to meet the instructional needs of small colleges, the manufacturing problems of small factories, and the data processing tasks of medium-size businesses, such as payroll and accounting. These were called minicomputers. These machines were slower and smaller in memory capacity than mainframes. The price range used to be anywhere from $25,000 to $100,000. Typical examples include such computers as the Digital Equipment PDP 11/45 and the Data General Nova.

However, current low-end mainframes and high-end microcomputers (described in the next section) overlap considerably in price, performance, and applications with traditional minicomputers. Therefore, the term **minicomputer** is becoming almost obsolete.

1.33 Microcomputers

The 4-bit and 8-bit microprocessors became available in the mid-1970s, and initial applications were primarily in the areas of machine control and instrumentation. As the price of the microprocessors and memory began to decline, the applications mushroomed in almost all areas, such as video games, word processing, and small-business applications. Early microcomputers were designed around 8-bit microprocessors. Since then, 16-, 32- and 64-bit microprocessors, such as the Intel 8086/88, 80386/486, Pentium, Pro-Pentium, and Motorola 68000, and the Power PC series have been introduced, and recent microcomputers are being designed around these microprocessors. Present-day microcomputers can be classified in four groups: personal (or business) computers (PC), workstations, single-board, and single-chip microcomputers.

PERSONAL COMPUTERS (PC)

These microcomputers are single-user systems and are used for a variety of purposes, such as payroll, business accounts, word processing, legal and medical record keeping, personal finance, and instruction. They are also known as personal computers (PC) or desktop computers. Typically, the price ranges from $1000 to $5000 for a single-user system. Examples include such microcomputers as the IBM Personal Computer (Aptiva series), the Hewlett-Packard Pavilion series, and the Apple Macintosh series. Figure 1.6 shows an example.

At the low end of the microcomputer spectrum, a typical configuration includes a 32-bit (or 64-bit) microprocessor, 32 to 64 MB (megabytes) of system memory, a video screen (monitor), a 3½″ high-density floppy disk, and a hard disk with storage capacity of more than

FIGURE 1.6

Microcomputer with Disk Storage:
IBM PC

SOURCE: Photograph courtesy of International
Business Machines Corporation.

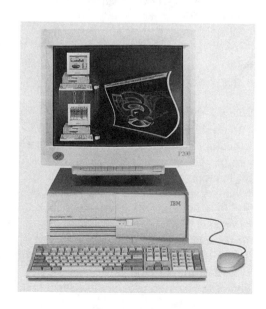

4 gigabytes. The **floppy disk** is a magnetic medium similar to a cassette tape except that it is round in shape, like a record. Information recorded on these disks can be accessed randomly using disk drives. Conversely, information stored on a cassette tape is accessed serially. In order to read information at the end of the tape, the user must run the entire tape through the machine. The **hard disk** is similar to the floppy disk except that the magnetic material is coated on a rigid aluminum base that is enclosed in a sealed container and permanently installed in a microcomputer. The hard disk and the floppy disks are used to store programs semipermanently, i.e., the binary information does not disappear when the power is turned off. However, the microprocessor does not have direct access to this information; it must copy this information (programs) into system memory to modify or execute these programs. The hard disk has a large storage capacity; therefore, large and frequently used programs such as compilers, interpreters, system programs, and application programs are stored on this disk. The floppy disk is generally used for user programs and to make backup copies.

The microcomputers are further classified according to their size, weight, and portability. They are called laptop and notebook. The laptop computer is a portable microcomputer that has a flat screen, a hard disk, and a $3\frac{1}{2}''$ floppy disk, and usually weighs around ten pounds. These computers can be battery operated or use AC power and are carried easily from place to place. These are called laptop (instead of desktop) because the size is small enough to place them in one's lap (if necessary). The notebook computer is a portable microcomputer of a notebook size ($8\frac{1}{2}'' \times 11'' \times 2''$) and weighs around five pounds. A microcomputer smaller than the notebook computers, called a subnotebook, is also available.

WORKSTATIONS

These are high-performance cousins of the personal computers. They are used in engineering and scientific applications such as computer-aided design (CAD), computer-aided engineering (CAE), and computer-aided manufacturing (CAM). They generally include system memory larger than 200 MB, storage (hard disk) memory in gigabytes, and a high-resolution screen.

The workstations are designed around RISC (reduced instruction set computing) processors (described in Chapter 18). The RISC processors tend to be faster and more efficient than the processors used in personal computers. Some of the workstations have better performance than that of the low-end minicomputers.

SINGLE-BOARD MICROCOMPUTERS

These microcomputers are primarily used in college laboratories and industries for instructional purposes or to evaluate the performance of a given microprocessor. They can also be part of some larger systems. Typically, these microcomputers include an 8- or 16-bit microprocessor, from 256 bytes to 8K bytes of user memory, a Hex keyboard, and seven-segment LEDs as display. The interaction between the microprocessor, memory, and I/Os in these small systems is looked after by a program called a system **monitor program,** which is generally small in size, stored in less than 2K bytes of ROM. When a single-board microcomputer is turned on, the monitor program is in charge of the system; it monitors the keyboard inputs, interprets those keys, stores programs in memory, sends system displays to the LEDs, and enables the execution of the user programs. The function of the monitor program in a small system is similar to that of the operating system in

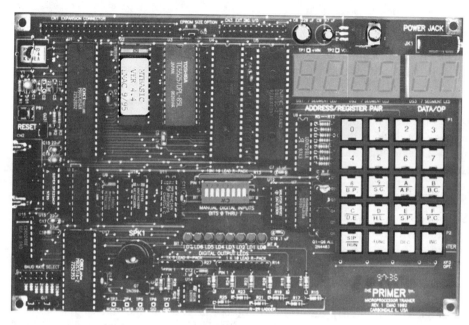

FIGURE 1.7

Single-Board Microcomputer: The Primer

SOURCE: Courtesy of EMAC Inc.

a large system. The prices of these single-board computers range from $100 to $800, the average price being around $300.

Examples of these computers include such systems as the Intel SDK 85, SDK 86, and the EMAC Primer (Figure 1.7). These are generally used to write and execute assembly language programs and to perform interfacing experiments.

SINGLE-CHIP MICROCOMPUTERS (MICROCONTROLLERS)

These microcomputers are designed on a single chip, which typically includes a microprocessor, 64 bytes of R/W memory, from 1K to 2K bytes of ROM, and several signal lines to connect I/Os. These are complete microcomputers on a chip; they are also known as **microcontrollers.** They are used primarily for such functions as controlling appliances and traffic lights. Typical examples of these microcomputers include such chips as the Zilog Z8, Intel MCS 51 and 96 series, and Motorola 68HC11.

SUMMARY

The various concepts and terms discussed in this chapter are summarized here:

Computer Structure

☐ **Digital computer**—a programmable machine that processes binary data. It is traditionally represented by five components: CPU, ALU plus control unit, memory, input, and output.

☐ **CPU**—the central processing unit. The group of circuits that processes data and provides control signals and timing. It includes the arithmetic/logic unit, registers, instruction decoder, and the control unit.

☐ **ALU**—the group of circuits that performs arithmetic and logic operations. The ALU is a part of the CPU.

☐ **Control unit**—the group of circuits that provides timing and signals to all operations in the computer and controls data flow.

☐ **Memory**—a medium that stores binary information (instructions and data).

☐ **Input**—a device that transfers information from the outside world to the computer.

☐ **Output**—a device that transfers information from the computer to the outside world.

Scale of Integration

☐ **SSI** (small-scale integration)—the process of designing a few circuits on a single chip. The term refers to the technology used to fabricate discrete logic gates on a chip.

☐ **MSI** (medium-scale integration)—the process of designing more than a hundred gates on a single chip.

☐ **LSI** (large-scale integration)—the process of designing more than a thousand gates on a single chip. Similarly, the terms *VLSI* (very-large-scale integration) and *SLSI* (super-large-scale integration) are used to indicate the scale of integration.

Microcomputers

☐ **Microprocessor (MPU)**—a semiconductor device (integrated circuit) manufactured by using the LSI technique. It includes the ALU, register arrays, and control circuits on a single chip. The term *MPU* is also synonymous with the microprocessor (see Section 2.2 for additional details).

☐ **Microprocessor-based product**—a machine or product that uses a microprocessor to run or execute its operations. It is represented by three components: microprocessor, memory, and I/O (input/output).

☐ **Microcontroller**—a device that includes microprocessor, memory, and I/O signal lines on a single chip, fabricated using VLSI technology.

☐ **Microcomputer**—a computer that is designed using a microprocessor as its CPU. It includes microprocessor, memory, and I/O (input/output).

☐ **Bus**—a group of lines used to transfer bits between the microprocessor and other components of the computer system.

☐ **ROM** (Read-Only memory)—a memory that stores binary information permanently. The infcrmation can be read from this memory but cannot be altered.

☐ **R/WM** (Read/Write memory)—a memory that stores binary information during the operation of the computer. This memory is used as a writing pad to write user programs and data. The information stored in this memory can be read and altered easily.

Computer Languages

☐ **Bit**—a binary digit, 0 or 1.

☐ **Byte**—a group of eight bits.

- ☐ **Nibble**—a group of four bits.
- ☐ **Word**—a group of bits the computer recognizes and processes at a time.
- ☐ **Instruction**—a command in binary that is recognized and executed by the computer to accomplish a task. Some instructions are designed with one word, and some require multiple words.
- ☐ **Mnemonic**—a combination of letters to suggest the operation of an instruction.
- ☐ **Program**—a set of instructions written in a specific sequence for the computer to accomplish a given task.
- ☐ **Machine language**—the binary medium of communication with a computer through a designed set of instructions specific to each computer.
- ☐ **Assembly language**—a medium of communication with a computer in which programs are written in mnemonics. An assembly language is specific to a given computer.
- ☐ **Low-level language**—a medium of communication that is machine-dependent or specific to a given computer. The machine and the assembly languages of a computer are considered low-level languages. Programs written in these languages are not transferable to different types of machines.
- ☐ **High-level language**—a medium of communication that is independent of a given computer. Programs are written in English-like words, and they can be executed on a machine using a translator (a compiler or an interpreter).
- ☐ **Source code**—a program written either in mnemonics of an assembly language or in English-like statements of a high-level language (before it is assembled or compiled).
- ☐ **Compiler**—a program that translates English-like words of a high-level language into the machine language of a computer. A compiler reads a given program, called a source code, in its entirety and then translates the program into the machine language, which is called an object code.
- ☐ **Interpreter**—a program that translates the English-like statements of a high-level language into the machine language of a computer. An interpreter translates one statement at a time from a source code to an object code.
- ☐ **Assembler**—a computer program that translates an assembly language program from mnemonics to the binary machine code of a computer.
- ☐ **Manual assembly**—a procedure of looking up the machine codes manually from the instruction set of a computer and entering those into the computer through a keyboard.
- ☐ **ASCII**—American Standard Code for Information Interchange. This is a 7-bit alphanumeric code with 128 combinations. Each combination is assigned to either a letter, decimal digit, a symbol, or a machine command.
- ☐ **Extended ASCII**—an 8-bit code with 256 combinations. The ASCII code is extended from seven bits to eight bits to include additional graphic symbols.
- ☐ **Operating system**—a set of programs that manages interaction between hardware and software. It is responsible primarily for storing information on disks and for communication between microprocessor, memory, and peripherals.
- ☐ **Monitor program**—a program that interprets the input from a keyboard and converts the input into its binary equivalent.

LOOKING AHEAD

This chapter has given a brief introduction to computer organization and computer languages, with emphasis on the 8085 microprocessor and its assembly language. The last section has provided an overview of the entire spectrum of computers, including their salient features and applications. The primary focus of this book is on the architectural details of the 8085 microprocessor and its industrial applications. Heavy emphasis also is put on assembly language programming in the context of these applications. In the microcomputer field, little separation is made between hardware and software, especially in applications where assembly language is necessary. In designing a microprocessor-based product, hardware and software tasks are carried out concurrently because a decision in one area affects the planning of the other area. Some functions can be performed either through hardware or software, and a designer needs to consider both approaches. This book focuses on the tradeoffs between the two approaches as a design philosophy.

Chapter 2 expands on the architectural concepts of microcomputers introduced in this chapter; it deals in detail with each component of the block diagram shown in Figure 1.3. Chapter 3 focuses on the architectural details of the 8085 microprocessor and memory interfacing, and Chapter 4 discusses I/O interfacing.

QUESTIONS AND PROBLEMS

1. List the components of a computer.
2. Explain the functions of each component of a computer.
3. What is a microprocessor? What is the difference between a microprocessor and a CPU?
4. Explain the difference between a microprocessor and a microcomputer.
5. Explain these terms: *SSI, MSI,* and *LSI.*
6. Define *bit, byte, word,* and *instruction.*
7. How many bytes make a word of 32 bits?
8. Explain the difference between the machine language and the assembly language of the 8085 microprocessor.
9. What is an assembler?
10. What are low- and high-level languages?
11. Explain the difference between a compiler and an interpreter.
12. What are the advantages of an assembly language in comparison with high-level languages?
13. What is an ASCII code?
14. Identify the difference between the ASCII and the extended ASCII codes.
15. Find the ASCII codes for the letters "A," "Z," and "m" from the ASCII table in Appendix E.
16. What is an operating system?

Microprocessor Architecture and Microcomputer Systems

A microcomputer system consists of three components—the microprocessor, memory, and I/O (in-
put/output)—as discussed in the previous chapter.
The microprocessor manipulates data, controls the
timing of various operations, and communicates
with such peripherals (devices) as memory and
I/O. The internal logic design of the microprocessor, called its **architecture,** determines how and
when various operations are performed by the microprocessor. The system bus provides paths for
the flow of binary information (data and instructions).

    This chapter expands on the bus concept dis-
cussed in the previous chapter and shows how binary information flows externally among the components of the system. The chapter deals with the
internal architecture and various operations of the
microprocessor in the context of the 8085. It also
expands on topics such as memory and I/O, and reviews interfacing devices, such as buffers, decoders,
and latches.

OBJECTIVES

☐ List the four operations commonly performed by
the microprocessor or the microprocessing unit
(MPU).

☐ Define the address bus, the data bus, and the control bus, and explain their functions in reference
to the 8085 microprocessor.

☐ List the registers in the 8085 microprocessor, and
explain their functions.

☐ Explain the functions Reset, Interrupt, Wait, and
Hold.

☐ Explain memory organization and memory map, and explain how memory addresses are assigned to a memory chip.

☐ List the types of memory and their functions.

☐ Explain the difference between the peripheral I/O (also known as I/O-mapped I/O) and the memory-mapped I/O.

☐ Describe the steps in executing an instruction in a bus-oriented system.

☐ Define tri-state logic and explain the functions of such MSI devices as buffers, decoders, encoders, and latches.

2.1 MICROPROCESSOR ARCHITECTURE AND ITS OPERATIONS

The microprocessor is a programmable digital device, designed with registers, flip-flops, and timing elements. The microprocessor has a set of instructions, designed internally, to manipulate data and communicate with peripherals. This process of data manipulation and communication is determined by the logic design of the microprocessor, called the **architecture.**

The microprocessor can be programmed to perform functions on given data by selecting necessary instructions from its set. These instructions are given to the microprocessor by writing them into its memory. Writing (or entering) instructions and data is done through an input device such as a keyboard. The microprocessor reads or transfers one instruction at a time, matches it with its instruction set, and performs the data manipulation indicated by the instruction. The result can be stored in memory or sent to such output devices as LEDs or a CRT terminal. In addition, the microprocessor can respond to external signals. It can be interrupted, reset, or asked to wait to synchronize with slower peripherals. All the various functions performed by the microprocessor can be classified in three general categories:

☐ Microprocessor-initiated operations
☐ Internal operations
☐ Peripheral (or externally initiated) operations

To perform these functions, the microprocessor requires a group of logic circuits and a set of signals called control signals. However, early processors did not have the necessary circuitry on one chip; the complete units were made up of more than one chip. Therefore, the term *microprocessing unit* (MPU) is defined here as a group of devices that can perform these functions with the necessary set of control signals. This term is similar to the term *central processing unit* (CPU). However, later microprocessors include most of the necessary circuitry to perform these operations on a single chip. Therefore, the terms *MPU* and *microprocessor* often are used synonymously.

The microprocessor functions listed above are explained here in relation to the 8085 MPU but without the details of the MPUs. However, the general concepts discussed here are applicable to any microprocessor. The devices necessary to make up the 8085 MPUs will be discussed in the next chapter.

2.11 Microprocessor-Initiated Operations and 8085 Bus Organization

The MPU performs primarily four operations:*

1. Memory Read: Reads data (or instructions) from memory.
2. Memory Write: Writes data (or instructions) into memory.
3. I/O Read: Accepts data from input devices.
4. I/O Write: Sends data to output devices.

All these operations are part of the communication process between the MPU and peripheral devices (including memory). To communicate with a peripheral (or a memory location), the MPU needs to perform the following steps:

Step 1: Identify the peripheral or the memory location (with its address).
Step 2: Transfer binary information (data and instructions).
Step 3: Provide timing or synchronization signals.

The 8085 MPU performs these functions using three sets of communication lines called buses: the address bus, the data bus, and the control bus (Figure 2.1). In Chapter 1, these buses are shown as one group, called the system bus.

ADDRESS BUS

The address bus is a group of 16 lines generally identified as A_0 to A_{15}. The address bus is **unidirectional:** bits flow in one direction—from the MPU to peripheral devices. The MPU uses the address bus to perform the first function: identifying a peripheral or a memory location (Step 1).

---

*Other operations are omitted here for clarity and discussed in the next chapter.

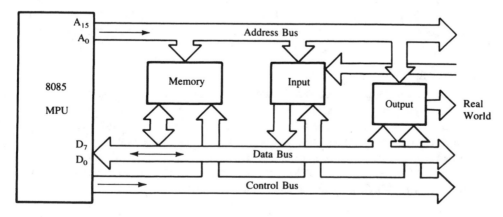

FIGURE 2.1
The 8085 Bus Structure

In a computer system, each peripheral or memory location is identified by a binary number, called an **address,** and the address bus is used to carry a 16-bit address. This is similar to the postal address of a house. A house can be identified by various number schemes. For example, the forty-fifth house in a lane can be identified by the two-digit number 45 or by the four-digit number 0045. The two-digit numbering scheme can identify only a hundred houses, from 00 to 99. On the other hand, the four-digit scheme can identify ten thousand houses, from 0000 to 9999. Similarly, the number of address lines of the MPU determines its capacity to identify different memory locations (or peripherals). The 8085 MPU with its 16 address lines is capable of addressing $2^{16} = 65,536$ (generally known as 64K) memory locations. As explained in Chapter 1, 1K memory is determined by rounding off 1024 to the nearest thousand; similarly, 65,536 is rounded off to 64,000 as a multiple of 1K.

Most 8-bit microprocessors have 16 address lines. This may explain why microcomputer systems based on 8-bit microprocessors have 64K memory. However, not every microcomputer system has 64K memory. In fact, most single-board microcomputers have less than 4K of memory, even if the MPU is capable of addressing 64K memory. The number of address lines is arbitrary; it is determined by the designer of a microprocessor based on such considerations as availability of pins and intended applications of the processor. For example, the Intel 8088 processor has 20 and the Pentium processor has 32 address lines.

DATA BUS

The data bus is a group of eight lines used for data flow (Figure 2.1).* These lines are **bidirectional**—data flow in both directions between the MPU and memory and peripheral devices. The MPU uses the data bus to perform the second function: transferring binary information (Step 2).

The eight data lines enable the MPU to manipulate 8-bit data ranging from 00 to FF ($2^8 = 256$ numbers). The largest number that can appear on the data bus is 11111111 (255_{10}). The 8085 is known as an 8-bit microprocessor. Microprocessors such as the Intel 8086, Zilog Z8000, and Motorola 68000 have 16 data lines; thus they are known as 16-bit microprocessors. The Intel 80386/486 have 32 data lines; thus they are classified as 32-bit microprocessors.

CONTROL BUS

The control bus is comprised of various single lines that carry synchronization signals. The MPU uses such lines to perform the third function: providing timing signals (Step 3).

The term **bus,** in relation to the control signals, is somewhat confusing. These are not groups of lines like address or data buses, but individual lines that provide a pulse to indicate an MPU operation. The MPU generates specific control signals for every operation (such as Memory Read or I/O Write) it performs. These signals are used to identify a device type with which the MPU intends to communicate.

---

*The term *data* refers to any binary information that may include an instruction, an address, or a number.

To communicate with a memory—for example, to read an instruction from a memory location—the MPU places the 16-bit address on the address bus (Figure 2.2). The address on the bus is decoded by an external logic circuit, which will be explained later, and the memory location is identified. The MPU sends a pulse called Memory Read as the control signal. The pulse activates the memory chip, and the contents of the memory location (8-bit data) are placed on the data bus and brought inside the microprocessor.

What happens to the data byte brought into the MPU depends on the internal architecture of the microprocessor, which we will describe in the next section.

2.12 Internal Data Operations and the 8085 Registers

The internal architecture of the 8085 microprocessor determines how and what operations can be performed with the data. These operations are:

1. Store 8-bit data.
2. Perform arithmetic and logical operations.
3. Test for conditions.
4. Sequence the execution of instructions.
5. Store data temporarily during execution in the defined R/W memory locations called the stack.

To perform these operations, the microprocessor requires registers, an arithmetic/logic unit (ALU) and control logic, and internal buses (paths for information flow). Figure 2.3 is a simplified representation of the 8085 internal architecture; it shows only those registers that are programmable, meaning those registers that can be used for data manip-

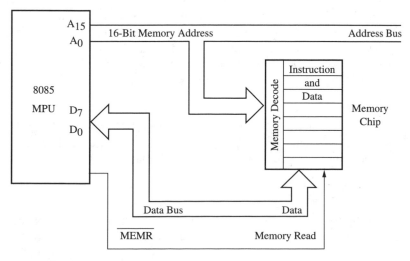

FIGURE 2.2
Memory Read Operation

FIGURE 2.3

The 8085 Programmable Registers

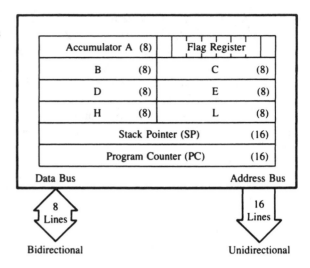

ulation by writing instructions. These registers are described in reference to the five operations previously listed.

REGISTERS

The 8085 has six general-purpose registers to perform the first operation listed above; that is, to store 8-bit data during program execution. These registers are identified as B, C, D, E, H, and L, as shown in Figure 2.3. They can be combined as register pairs—BC, DE, and HL—to perform some 16-bit operations.

These registers are **programmable,** meaning that a programmer can use them to load or copy data from the registers by using instructions. For example, the instruction MOV B,C copies the data from register C to register B. Conceptually, the registers can be viewed as memory locations, except they are built inside the microprocessor and identified by specific letters for user convenience. Some microprocessors do not have these types of registers; instead, they use memory space as their registers.

ACCUMULATOR

The accumulator is an 8-bit register that is part of the arithmetic/logic unit (ALU). This register is used to store 8-bit data and to perform arithmetic and logical operations. The result of an operation is stored in the accumulator. The accumulator is also identified as register A.

FLAGS

The ALU includes five flip-flops that are set or reset according to the result of an operation. The microprocessor uses them to perform the third operation; namely, testing for data conditions.

For example, after an addition of two numbers, if the sum in the accumulator is larger than eight bits, the flip-flop that is used to indicate a carry, called the **Carry** (CY)

flag, is set to one. When an arithmetic operation results in zero, the flip-flop called the **Zero** (Z) **flag** is set to one. The 8085 has five flags to indicate five different types of results or data conditions. They are Zero (Z), Carry (CY), Sign (S), Parity (P), and Auxiliary Carry (AC) flags. The most commonly used flags are Sign, Zero, and Carry; the others will be explained as necessary.

Figure 2.3 shows an 8-bit register, called the **flag register,** adjacent to the accumulator. It is not used as an 8-bit register; five bit positions, out of eight, are used to store the outputs of the five flip-flops. The flags are stored in the 8-bit register so that the programmer can examine these flags (data conditions) by accessing the register through an instruction. In the instruction set, the term *PSW* (Program Status Word) refers to the accumulator and the flag register. This term will be discussed again in Chapter 9, "Stack and Subroutines."

These flags have critical importance in the decision-making process of the microprocessor. The conditions (set or reset) of the flags are tested through software instructions. For example, the instruction JC (Jump On Carry) is implemented to change the sequence of a program when the CY flag is set. The importance of the flags cannot be emphasized enough; they will be discussed again in applications of conditional jump instructions.

PROGRAM COUNTER (PC)

This 16-bit register deals with the fourth operation, sequencing the execution of instructions. This register is a **memory pointer.** Memory locations have 16-bit addresses, and that is why this is a 16-bit register (see Section 2.22 for memory address).

The microprocessor uses this register to sequence the execution of instructions. The function of the program counter is to point to the memory address from which the next byte is to be fetched. When a byte (machine code) is being fetched, the program counter is incremented by one to point to the next memory location.

STACK POINTER (SP)

The stack pointer is also a 16-bit register used as a memory pointer; initially, it will be called the stack pointer register to emphasize that it is a register. It points to a memory location in R/W memory, called the **stack.** The beginning of the stack is defined by loading a 16-bit address in the stack pointer (register).*

2.13 Peripheral or Externally Initiated Operations

External devices (or signals) can initiate the following operations, for which individual pins on the microprocessor chip are assigned: Reset, Interrupt, Ready, Hold.

---

*The concept of the stack memory is difficult to explain at this time; it is not necessary for the reader to understand the stack memory until subroutines are discussed. It is included here only to provide continuity to the description of programmable registers and microprocessor operations. This concept will be explained more fully in Chapter 9.

□ Reset: When the reset pin is activated by an external key (also called a reset key), all internal operations are suspended and the program counter is cleared (it holds 0000H). Now the program execution can again begin at the zero memory address.

□ Interrupt: The microprocessor can be interrupted from the normal execution of instructions and asked to execute some other instructions called a **service routine** (for example, emergency procedures). The microprocessor resumes its operation after completing the service routine (see Chapter 12).

□ Ready: The 8085 has a pin called READY. If the signal at this READY pin is low, the microprocessor enters into a Wait state. This signal is used primarily to synchronize slower peripherals with the microprocessor.

□ Hold: When the HOLD pin is activated by an external signal, the microprocessor relinquishes control of buses and allows the external peripheral to use them. For example, the HOLD signal is used in Direct Memory Access (DMA) data transfer (see Chapter 15).

These operations are listed here to provide an overview of the capabilities of the 8085. They will be discussed in Part III.

2.2 MEMORY

Memory is an essential component of a microcomputer system; it stores binary instructions and data for the microprocessor. There are various types of memory, which can be classified in two groups: prime (or main) memory and storage memory. In the last chapter, we discussed briefly two examples of prime memory: Read/Write memory (R/WM) and Read-Only memory (ROM). Magnetic tapes or disks can be cited as examples of storage memory. First, we will focus on prime memory and then, briefly discuss storage memory when we examine various types of memory.

The R/W memory is made of registers, and each register has a group of flip-flops or field-effect transistors that store bits of information; these flip-flops are called memory cells. The number of bits stored in a register is called a **memory** word; memory devices (chips) are available in various word sizes. The user can use this memory to hold programs and store data. On the other hand, the ROM stores information permanently in the form of diodes; the group of diodes can be viewed as a register. In a memory chip, all registers are arranged in a sequence and identified by binary numbers called memory addresses. To communicate with memory, the MPU should be able to

□ select the chip,
□ identify the register, and
□ read from or write into the register.

The MPU uses its address bus to send the address of a memory register and uses the data bus and control lines to read from (as shown in Figure 2.2) or write into that register. In the following sections, we will examine the basic concepts related to memory: its structure, its addressing, and its requirements to communicate with the MPU and build a model for R/W memory. However, except for slight differences in Read/Write control signals, the discussion is equally applicable to ROM.

2.21   Flip-Flop or Latch as a Storage Element

What is memory? It is a circuit that can store bits—high or low, generally voltage levels or capacitive charges representing 1 and 0. A flip-flop or a latch* is a basic element of memory. To write or store a bit in the latch, we need an input data bit (D_{IN}) and an enable signal (EN), as shown in Figure 2.4(a). In this latch, the stored bit is always available on the output line D_{OUT}. To avoid unintentional change in the input and control the availability of the output, we can use two tri-state* buffers on the latch, as shown in Figure 2.4(b). Now we can write into the latch by enabling the input buffer and read from it by enabling the output buffer. Figure 2.4(b) shows the Write signal as $\overline{WR}$ and the Read signal as $\overline{RD}$; these are active low signals indicated by the bar. This latch, which can store one binary bit, is called a memory cell. Figure 2.5(a) shows four such cells or latches grouped together; this is a register, which has four input lines and four output lines and can store four bits; thus the size of the memory word is four bits. The size of this register is specified either as 4-bit or 1×4-bit, which indicates one register with four cells or four I/O lines. Figures 2.5(b) and (c) show simplified block diagrams of the 4-bit register.

   In Figure 2.6, four registers with eight cells (or an 8-bit memory word) are arranged in a sequence. To write into or read from any one of the registers, a specific register should be identified or enabled. This is a simple decoding function; a 2-to-4 decoder can perform that function. However, two more input lines A_1 and A_0, called address lines, are required to the decoder. These two input lines can have four different bit combinations (00, 01, 10, 11), and each combination can identify or enable one of the registers named as Register 0 through Register 3. Thus the Enable signal of the flip-flops in Figure 2.5 is replaced by two address lines in Figure 2.6. Figure 2.6(a) has 8-bit registers and Figure

---

*If you are not familiar with these devices, review Section 2.5; flip-flops (latches), tri-state buffers, and decoders are discussed briefly in Section 2.5.

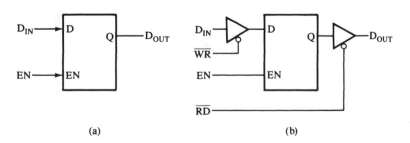

(a)   (b)

FIGURE 2.4
Latches as Storage Element: Basic Latch (a) and Latch with Two Tri-State Buffers (b)

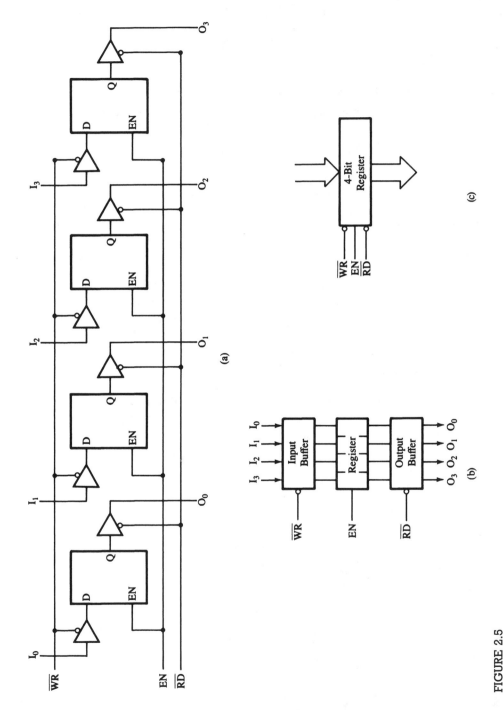

FIGURE 2.5
Four Latches as a 4-Bit Register (a) and Block Diagrams of a 4-Bit Register (b and c)

34

2.6(b) has two chips with 4-bit registers. This is an illustration of how smaller word size chips can be connected to make up an 8-bit word memory size. Now we can expand the number of registers. If we have eight registers on one chip, we need three address lines, and if we have 16 registers, we need four address lines.

An interesting problem is how to deal with more than one chip; for example, two chips with four registers each. We have a total of eight registers; therefore, we need three address lines, but one line should be used to select between the two chips. Figure 2.7(b) shows two memory chips, with an additional signal called Chip Select ($\overline{CS}$), and A_2 (with an inverter) is used to select between the chips. When A_2 is 0 (low), chip M_1 is selected, and when A_2 is 1 (high), chip M_2 is selected. The addresses on A_1 and A_0 will determine the registers to be selected; thus, by combining the logic on A_2, A_1, and A_0, the memory addresses range from 000 to 111. The concept of the Chip Select signal gives us more flexibility in designing chips and allows us to expand memory size by using multiple chips.

Now let us examine the problem from a different perspective. Assume that we have available four address lines and two memory chips with four registers each as before. Four address lines are capable of identifying 16 (2^4) registers; however, we need only three address lines to identify eight registers. What should we do with the fourth line? One of the solutions is shown in Figure 2.8. Memory chip M_1 is selected when A_3 and A_2 are both 0; therefore, registers in this chip are identified with the addresses ranging from 0000 to 0011 (0 to 3). Similarly, the addresses of memory chip M_2 range from 1000 to

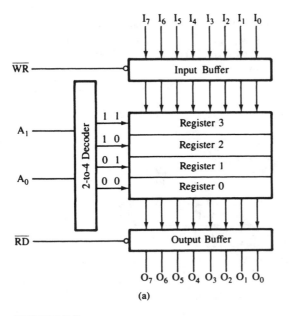

(a)

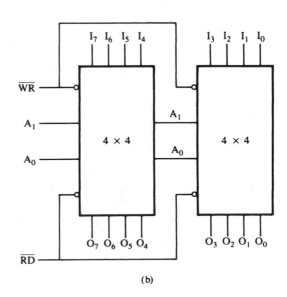

(b)

FIGURE 2.6
4 × 8-Bit Register

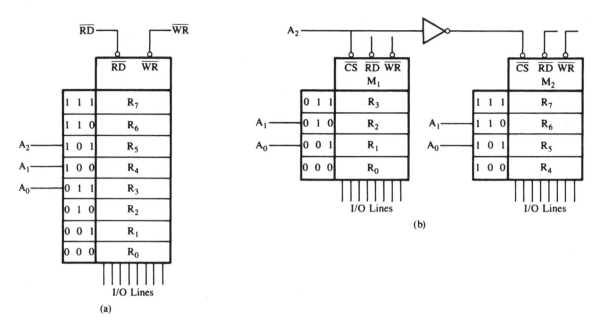

FIGURE 2.7
Two Memory Chips with Four Registers Each and Chip Select

1011 (8 to B); this chip is selected only when A_3 is 1 and A_2 is 0. In this example, we need three lines to identify eight registers: two for registers and one for Chip Select. However, we used the fourth line for Chip Select also. This is called complete or absolute decoding. Another option is to leave the fourth line as don't care; we will further explore this concept later.

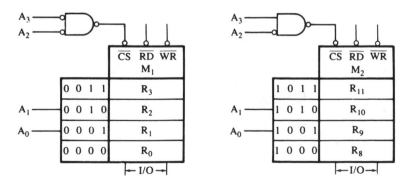

FIGURE 2.8
Addressing Eight Registers with Four Address Lines

After reviewing the above explanation, we can summarize the requirements of a memory chip, then build a model and match the requirements with the microprocessor bus concepts:

1. A memory chip requires address lines to identify a memory register. The number of address lines required is determined by the number of registers in a chip (2^n = Number of registers where n is the number of address lines). The 8085 microprocessor has 16 address lines. Of these 16 lines, the address lines necessary for the memory chip must be connected to the memory chip.

2. A memory chip requires a Chip Select ($\overline{CS}$) signal to enable the chip. The remaining address lines (from Step 1) of the microprocessor can be connected to the $\overline{CS}$ signal through an interfacing logic.

3. The address lines connected to $\overline{CS}$ select the chip, and the address lines connected to the address lines of the memory chip select the register. Thus the memory address of a register is determined by the logic levels (0/1) of all the address lines (including the address lines used for $\overline{CS}$).

4. The control signal Read ($\overline{RD}$) enables the output buffer, and data from the selected register are made available on the output lines. Similarly, the control signal Write ($\overline{WR}$) enables the input buffer, and data on the input lines are written into memory cells. The microprocessor can use its Memory Read and Memory Write control signals to enable the buffers and the data bus to transport the contents of the selected register between the microprocessor and memory.

A model of a typical memory chip representing the above requirements is shown in Figure 2.9. Figure 2.9(a) represents the R/W memory and Figure 2.9(b) represents the Read-Only memory; the only difference between the two as far as addressing is concerned is that ROM does not need the $\overline{WR}$ signal. Internally, the memory cells are arranged in a matrix format—in rows and columns; as the size increases, the internal de-

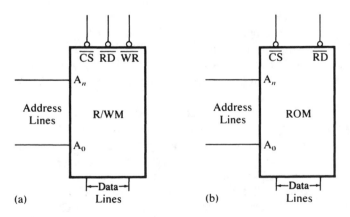

FIGURE 2.9
R/W Memory Model (a) and ROM Model (b)

coding scheme we discussed becomes impractical. For example, a memory chip with 1024 registers would require a 10-to-1024 decoder. If the cells are arranged in six rows and four columns, the internal decoding circuitry can be designed with two decoders, one for selecting a row and the other for selecting a column. However, we will not be concerned about the internal row and column arrangement because it does not affect our external interfacing logic, which is explained in the next section.

2.22   Memory Map and Addresses

Typically, in an 8-bit microprocessor system, 16 address lines are available for memory. This means it is a numbering system of 16 binary bits and is capable of identifying 2^{16} (65,536) memory registers, each register with a 16-bit address. The entire memory addresses can range from 0000 to FFFF in Hex. A memory map is a pictorial representation in which memory devices are located in the entire range of addresses. Memory addresses provide the locations of various memory devices in the system, and the interfacing logic defines the range of memory addresses for each memory device. The concept of memory map and memory addresses can be illustrated with an analogy of identical houses built in sequence and their postal addresses, or numbers.

Let us assume that houses are given three-digit decimal numbers, which will enable us to number one thousand houses from 000 to 999. Because it is cumbersome to direct someone to houses with large numbers, the numbering scheme can be devised with the concept of a row or block. Each block will have a hundred houses to be numbered with the last two digits from 00 to 99. Similarly, the blocks are also identified by the first decimal digit. For example, a house with the number 247 is house number 47 in block 2. With this scheme, all the houses in block 0 will be identified from 000 to 099, in block 2 from 200 to 299, and in block 9 from 900 to 999. This numbering scheme with three decimal digits is capable of giving addresses to one thousand houses from 000 to 999 (10 blocks of 100 houses each). Let us also assume that all houses are identical and have eight rooms.

The example of numbering the houses is directly applicable to assigning addresses to memory registers. In the binary number system, 16 binary digits can have 65,536 (2^{16}) different combinations. In the hexadecimal number system, 16 binary bits are equivalent to four Hex digits that can be used to assign addresses to 65,536 (0000H to FFFFH) memory registers in various memory chips. In our analogy, a memory chip is similar to a block in a housing development and a register can be viewed as a house with eight identical rooms.

Let us assume that we have a memory chip with 256 registers. Therefore, we need only 256 numbers (out of 65,536) that require eight address lines ($2^8 = 256$). Now the question is what we should do with the remaining eight address lines of the microprocessor. We can find a clue in our housing analogy. Let us assume that we have only 100 houses in block five. They will be numbered as 500 to 599; the first digit 5 remains constant and the next two digits vary from 00 to 99. Similarly, we can use the remaining eight address lines to assign fixed logic to generate a constant (fixed) number. This can be accomplished by using the remaining eight lines for the Chip Select through appropriate logic gates, as shown in Example 2.1.

As mentioned previously, in computer systems, we define 1024 as 1K; therefore a 1K-byte memory chip has 1024 registers with 8 bits each. Similarly, a group of 256 registers is defined as one *page* and each register is viewed as a *line* to write on. This is analogous to a notebook containing various pages, with each page having a certain number of lines. With this analogy, we can view 1K-byte memory as a chip with four pages (1024/256 = 4) with each page having 256 registers. With two Hex digits, 256 registers can be numbered from 00 to FFH; 1024 registers can be numbered with four digits from 0000 to 03FF. If we examine the high-order digits of 1K-byte memory, we find that they range from 00 to 03 representing four pages (00, 01, 02, and 03). In 8-bit microprocessor systems, this page concept is used frequently. In 16- and 32-bit microprocessor systems, the page concept (256 registers) defined here is not applicable; it is defined differently, based on the microprocessor used in a system.

So far we have been using the term *addresses* or *address range* for a given memory chip. The term *memory map* is used generally for the entire address ranges of the memory chips in a given system. The relationship of a row of houses to the road map is similar to the relationship of memory addresses to the memory map. However, these terms are also used synonymously.

Illustrate the memory address range of the chip with 256 bytes of memory, shown in Figure 2.10(a), and explain how the range can be changed by modifying the hardware of the Chip Select $\overline{CS}$ line in Figure 2.10(b).

Example 2.1

Figure 2.10(a) shows a memory chip with 256 registers with eight I/O lines; the memory size of the chip is expressed as 256×8. It has eight address lines (A_7–A_0), one Chip Select signal ($\overline{CS}$) (active low), and two control signals Read ($\overline{RD}$) and Write ($\overline{WR}$). The eight address lines (A_7–A_0) of the microprocessor are required to identify 256 memory registers. The remaining eight lines (A_{15}–A_8) are connected to the Chip Select ($\overline{CS}$) line through inverters and the NAND gate. The memory chip is enabled or selected when $\overline{CS}$ goes low. Therefore, to select the chip, the address lines A_{15}–A_8 should be at logic 0, which will cause the output of the NAND gate to go low. No other logic levels on the lines A_{15}–A_8 can select the chip. Once the chip is selected (enabled), the remaining address lines A_7–A_0 can assume any combination from 00H to FFH and identify any of the 256 memory registers through the decoder. Therefore, the memory addresses of the chip in Figure 2.10(a) will range from 0000H to 00FFH, as shown below.

Solution

$$\begin{array}{cccccccc cccccccc}
A_{15} & A_{14} & A_{13} & A_{12} & A_{11} & A_{10} & A_9 & A_8 & A_7 & A_6 & A_5 & A_4 & A_3 & A_2 & A_1 & A_0 \\
0 & 0 & 0 & 0 & 0 & 0 & 0 & 0 & 0 & 0 & 0 & 0 & 0 & 0 & 0 & 0 = 0000H
\end{array}$$

Chip Enable or Chip Select   1 1 1 1 1 1 1 1 = 00FFH

Register Select

The address lines A15–A8, which are used to select the chip, must have fixed logic levels, and these lines are called high-order address lines. The address lines A_7–A_0, which

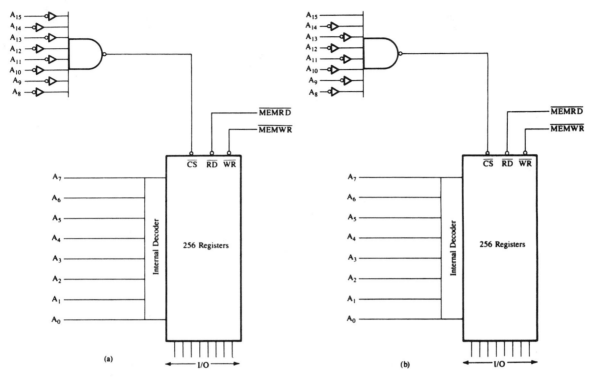

FIGURE 2.10
Memory Maps: 256 Bytes of Memory

are used to select a register, are called low-order address lines, and they can be assigned logic levels from all 0s to all 1s and any in-between combination. For example, when the address lines A_7–A_0 are all 0s, the register number 0 is selected, and when they are all 1s, the register number 255 (FFH) is selected. The Chip Select addresses are determined by the hardware (the inverters and NAND gate); therefore, the memory addresses of the chip can be changed by modifying the hardware. For example, if the inverter on line A_{15} is removed, as shown in Figure 2.10(b), the address required on A_{15}–A_8 to enable the chip will be as follows:

$$\begin{array}{cccccccc} A_{15} & A_{14} & A_{13} & A_{12} & A_{11} & A_{10} & A_9 & A_8 \\ 1 & 0 & 0 & 0 & 0 & 0 & 0 & 0 \end{array} = 80\text{H}$$

The memory address range in Figure 2.10(b) will be 8000H to 80FFH.

The memory chips in Figures 2.10(a) and (b) are the same chips. However, by changing the hardware of the Chip Select logic, the location of the memory in the map can be changed, and memory can be assigned addresses in various locations over the entire map of 0000 to FFFFH.

After reviewing the example and the previous explanation of the memory map, we can summarize the following points.

1. In a numbering system, the number of digits used determines the maximum addressing capacity of the system. Sixteen address lines (16 bits) of the 8085 microprocessor can address 65,536 memory registers; this is similar to three decimal digits providing the postal addresses for one thousand houses.

2. For a given memory chip, the number of address lines required to identify the registers is determined by the number of registers in the chip. The remaining address lines can be used for selecting the chip.

3. The address lines that are used to select registers in memory, called low-order address lines, can be assigned any logic levels (0 or 1) depending on the register being selected. The address lines that are used to select a chip, called high-order address lines, must have a fixed logic for a given address range.

4. The memory address range of a given chip can be changed by changing the hardware of the Chip Select ($\overline{CS}$) line. This line is also known as the Chip Enable ($\overline{CE}$) line.

2.23   Memory Address Range of a 1K Memory Chip

In the previous example, the high-order and the low-order address lines were equally divided, eight each of the 16 address lines. However, if a chip includes more than 256 registers (e.g., 512 or 1024 registers), the number of low-order address lines will be higher than that of the high-order address lines, as shown in the following example.

Explain the memory address range of 1K (1024 × 8) memory shown in Figure 2.11 and explain the changes in the addresses if the hardware of the $\overline{CS}$ line is modified.

**Example
2.2**

The memory chip has 1024 registers; therefore 10 address lines (A_9–A_0) are required to identify the registers. The remaining six address lines (A_{15}–A_{10}) of the microprocessor are used for the Chip Select ($\overline{CS}$) signal. In Figure 2.11, the memory chip is enabled when the address lines A_{15}–A_{10} are at logic 0. The address lines A_9–A_0 can assume any address of the 1024 registers, starting from all 0s to all 1s, as shown next.

Solution

$$
\begin{array}{cccccc}
A_{15} & A_{14} & A_{13} & A_{12} & A_{11} & A_{10}
\end{array}
\qquad
\begin{array}{cccccccccc}
A_9 & A_8 & A_7 & A_6 & A_5 & A_4 & A_3 & A_2 & A_1 & A_0
\end{array}
$$

A_{15} A_{14} A_{13} A_{12} A_{11} A_{10}	A_9 A_8 A_7 A_6 A_5 A_4 A_3 A_2 A_1 A_0	
0 0 0 0 0 0	0 0 0 0 0 0 0 0 0 0	= 0000H
Chip Select Logic	1 1 1 1 1 1 1 1 1 1	= 03FFH

The memory addresses range from 0000H to 03FFH.

By combining the high-order and the low-order address lines, we can specify the complete memory address range of a given chip. As explained in the previous example, the memory addresses of the 1K chip in Figure 2.11 can be changed to any other lo-

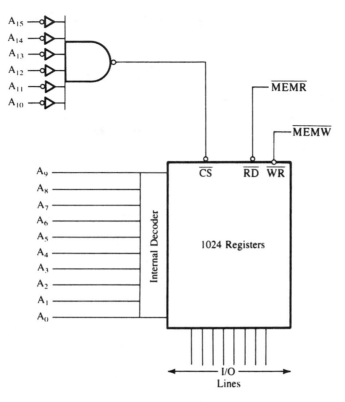

FIGURE 2.11
Memory Address Range: 1024 Bytes of Memory

cation by changing the hardware of the $\overline{\text{CS}}$ line. For example, if A_{15} is connected to the NAND gate without an inverter, the memory addresses will range from 8000H to 83FFH.

The preceding discussion concerning the memory addresses is equally applicable to the Read-Only memory (ROM). The ROM is in many ways organized the same as the R/W memory. The primary difference between the organization of the two memories is in the control signals; the ROM requires only the Read signal from the MPU.

2.24 Memory Address Lines

In the last two examples, the address lines of the memory chips were given. The number of address lines necessary for a given chip can be obtained from data sheets. However, we need to know the relationship between the number of registers in a memory chip and the number address lines. For a chip with 256 registers, we need 256 binary numbers to identify each register. Each address line can assume only two logic states (0 and 1); therefore, we need to find the power of 2 that will give us 256 combinations. The problem can be restated as follows: Find x where $2^x = 256$. By taking the log of both sides, we get:

$$\log 2^x = \log 256 \rightarrow x \log 2 = \log 256$$
$$x = \log 256/\log 2 = 8$$

Here x represents the number of address lines needed to obtain 256 binary numbers.

Calculate the address lines required for an 8K-byte ($1024 \times 8 = 8192$ registers) memory chip.

Example 2.3

Number of address lines $x = \log 8192/\log 2 = 13$ address lines

Solution

2.25 Memory Word Size

Memory devices (chips) are available in various word sizes (1, 4, and 8) and the size of a memory chip is generally specified in terms of the total number of bits it can store. On the other hand, the memory size in a given system is generally specified in terms of bytes. Therefore, it is necessary to design a byte-size memory word. For example, a memory chip of size 1024×4 has 1024 registers and each register can store four bits; thus it can store a total of 4096 ($1024 \times 4 = 4096$) bits. To design 1K-byte (1024×8) memory, we will need two chips; each chip will provide four data lines.

Calculate the number of memory chips needed to design 8K-byte memory if the memory chip size is 1024×1.

Example 2.4

The chip 1024×1 has 1024 (1K) registers and each register can store one bit with one data line. We need eight data lines for byte-size memory. Therefore, eight chips are necessary for 1K-byte memory. For 8K-byte memory, we will need 64 chips. We can arrive at the same answer by dividing 8K-byte by $1K \times 1$ as follows:

Solution

$$8192 \times 8 \div 1024 \times 1 = 64$$

So far we have been concerned primarily with finding necessary address lines and assigning addresses. In the next section, we will examine how the microprocessor communicates with memory using the address lines and control signals.

2.26   Memory and Instruction Fetch

The primary function of memory is to store instructions and data and to provide that information to the MPU whenever the MPU requests it. The MPU requests the information by sending the address of a specific memory register on the address bus and enables the data flow by sending the control signal, as illustrated in the next example.

Example 2.5	The instruction code 0100 1111 (4FH) is stored in memory location 2005H. Illustrate the data flow and list the sequence of events when the instruction code is fetched by the MPU.
Solution	To fetch the instruction located in memory location 2005H, the following steps are performed:

1. The program counter places the 16-bit address 2005H of the memory location on the address bus (Figure 2.12).
2. The control unit sends the Memory Read control signal ($\overline{\text{MEMR}}$, active low) to enable the output buffer of the memory chip.
3. The instruction (4FH) stored in the memory location is placed on the data bus and transferred (copied) to the instruction decoder of the microprocessor.
4. The instruction is decoded and executed according to the binary pattern of the instruction.

Figure 2.12 shows how the 8085 MPU fetches the instruction using the address, the data, and the control buses. Figure 2.12 is similar to Figure 2.2, Memory Read operation, except that Figure 2.12 shows additional details.

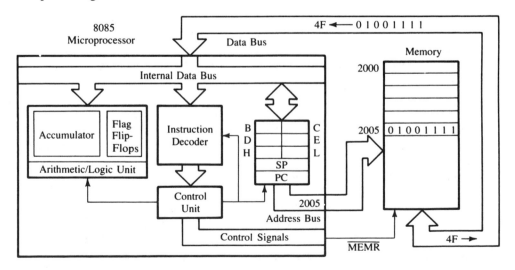

FIGURE 2.12
Instruction Fetch Operation

2.27   Memory Classification

As mentioned earlier, memory can be classified into two groups: prime (system or main) memory and storage memory. The R/WM and ROM are examples of prime memory; this is the memory the microprocessor uses in executing and storing programs. This memory should be able to respond fast enough to keep up with the execution speed of the microprocessor. Therefore, it should be random access memory, meaning that the microprocessor should be able to access information from any register with the same speed (independent of its place in the chip). The size of a memory chip is specified in terms of bits. For example, a 1K memory chip means it can store 1K (1024) bits (not bytes). On the other hand, memory in a system such as a PC is specified in bytes. For example, 4M memory in a PC means it has 4 megabytes of memory.

The other group is the storage memory, such as magnetic disks and tapes (see Figure 2.13). This memory is used to store programs and results after the completion of

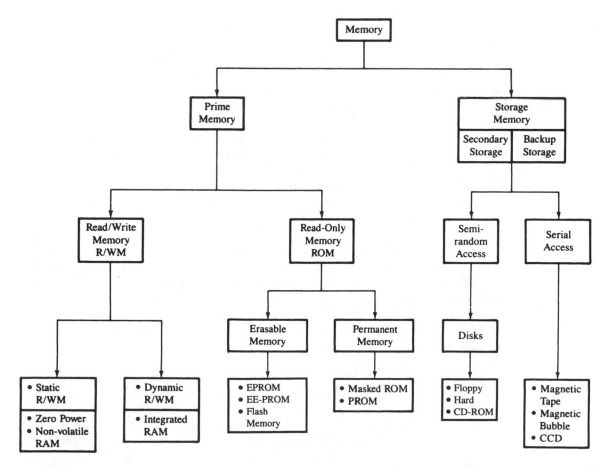

FIGURE 2.13
Memory Classification

program execution. Information stored in these memories is nonvolatile, meaning information remains intact even if the system is turned off. The microprocessor cannot directly execute or process programs stored in these devices; programs need to be copied into the R/W prime memory first. Therefore, the size of the prime memory, such as 512K or 8M (megabytes), determines how large a program the system can process. The size of the storage memory is unlimited; when one disk or tape is full, the next one can be used.

Figure 2.13 shows two groups in storage memory: secondary storage and backup storage. The secondary storage is similar to what you put on a shelf in your study, and the backup is similar to what you store in your attic. The secondary storage and the backup storage include devices such as disks, magnetic tapes, magnetic bubble memory, and charged-coupled devices, as shown in Figure 2.13. The primary features of these devices are high capacity, low cost, and slow access. A disk is similar to a record; the access to the stored information in the disk is semirandom (see Chapter 11 for additional discussion). The remaining devices shown in Figure 2.13 are serial, meaning if information is stored in the middle of the tape, it can be accessed after running half the tape. We will discuss some of these memory storage devices again in Chapter 11. In this chapter, we will focus on various types of prime memory.

Figure 2.13 shows that the prime (system) memory is divided into two main groups: Read/Write memory (R/WM) and Read-Only memory (ROM); each group includes several different types of memory, as discussed below.

R/WM (READ/WRITE MEMORY)

As the name suggests, the microprocessor can write into or read from this memory; it is popularly known as Random Access memory (RAM). It is used primarily for information that is likely to be altered, such as writing programs or receiving data. This memory is volatile, meaning that when the power is turned off, all the contents are destroyed. Two types of R/W memories—static and dynamic—are available; they are described in the following paragraphs.

Static Memory (SRAM)   This memory is made up of flip-flops, and it stores the bit as a voltage. Each memory cell requires six transistors; therefore, the memory chip has low density but high speed. This memory is more expensive and consumes more power than the dynamic memory described in the next paragraph. In high-speed processors (such as Intel 486 and Pentium), SRAM known as cache memory is included on the processor chip. In addition, high-speed cache memory is also included external to the processor to improve the performance of a system.

Dynamic Memory (DRAM)   This memory is made up of MOS transistor gates, and it stores the bit as a charge. The advantages of dynamic memory are that it has high density and low power consumption and is cheaper than static memory. The disadvantage is that the charge (bit information) leaks; therefore, stored information needs to be read and written again every few milliseconds. This is called refreshing the memory, and it requires extra circuitry, adding to the cost of the system. It is generally economical to use dynamic memory when the system memory size is at least 8K; for small systems, the static memory is appropriate. However, in recent years, the processor speed has reached beyond 200 MHz, and 1000 MHz processors are in the design stage. In comparison to the processor speed, the DRAM is too slow. To increase the speed of DRAM various techniques are being used. These techniques have resulted in high-speed memory chips such as EDO (Extended Data Out), SDRAM (Synchronous DRAM), and RDRAM (Rambus DRAM).

ROM (READ-ONLY MEMORY)

The ROM is a nonvolatile memory; it retains stored information even if the power is turned off. This memory is used for programs and data that need not be altered. As the name suggests, the information can be read only, which means once a bit pattern is stored, it is permanent or at least semipermanent. The permanent group includes two types of memory: masked ROM and PROM. The semipermanent group also includes two types of memory: EPROM and EE-PROM, as shown in Figure 2.13. The concept underlying the ROM can be explained with the diodes arranged in a matrix format, as shown in Figure 2.14. The horizontal lines are connected to vertical lines only through the diodes; they are not connected where they appear to cross in the diagram. Each of the eight horizontal rows can be viewed as a register with binary addresses ranging from 000 to 111; information is stored by the diodes in the register as 0s or 1s. The presence of a diode stores 1, and its absence stores 0. When a register is selected, the voltage of that line goes high, and the output lines, where diodes are connected, go high. For example, when the mem-

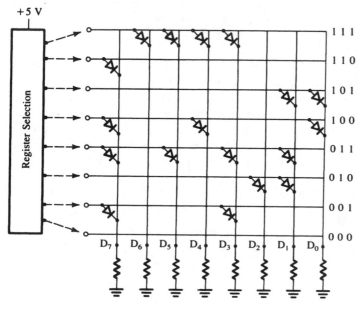

FIGURE 2.14
Functional Representation of ROM Memory Cell

ory register 111 is selected, the data byte 0111 1000 (78H) can be read at the data lines D_7–D_0.

The diode representation is a simplified version of the actual MOSFET memory cell. The manufacturer of the ROM designs the MOSFET matrix according to the information to be stored; therefore, information is permanently recorded in the ROM, as a song is recorded on a record. Five types of ROM—masked ROM, PROM, EPROM, EE-PROM, and Flash Memory—are described in the following paragraphs.

Masked ROM In this ROM, a bit pattern is permanently recorded by the masking and metalization process. Memory manufacturers are generally equipped to do this process. It is an expensive and specialized process, but economical for large production quantities.

PROM (Programmable Read-Only Memory) This memory has nichrome or poly-silicon wires arranged in a matrix; these wires can be functionally viewed as diodes or fuses. This memory can be programmed by the user with a special PROM programmer that selectively burns the fuses according to the bit pattern to be stored. The process is known as "burning the PROM," and the information stored is permanent.

EPROM (Erasable Programmable Read-Only Memory) This memory stores a bit by charging the floating gate of an FET. Information is stored by using an EPROM programmer, which applies high voltages to charge the gate. All the information can be erased by exposing the chip to ultraviolet light through its quartz window, and the chip can be reprogrammed. Because the chip can be reused many times, this memory is ideally suited for product development, experimental projects, and college laboratories. The disadvantages of EPROM are (1) it must be taken out of the circuit to erase it, (2) the entire chip must be erased, and (3) the erasing process takes 15 to 20 minutes.

EE-PROM (Electrically Erasable PROM) This memory is functionally similar to EPROM, except that information can be altered by using electrical signals at the register level rather than erasing all the information. This has an advantage in field and remote control applications. In microprocessor systems, software update is a common occurrence. If EE-PROMs are used in the systems, they can be updated from a central computer by using a remote link via telephone lines. Similarly, in a process control where timing information needs to be changed, it can be changed by sending electrical signals from a central place. This memory also includes a Chip Erase mode, whereby the entire chip can be erased in 10 ms vs. 15 to 20 min. to erase an EPROM. However, this memory is expensive compared to EPROM or flash memory (described in the next paragraph).

Flash Memory This is a variation of EE-PROM that is becoming popular. The major difference between the flash memory and EE-PROM is in the erasure procedure: The EE-PROM can be erased at a register level, but the flash memory must be erased either in its entirety or at the sector (block) level. These memory chips can be erased and programmed at least a million times. The power supply requirement for programming these chips was around 12 V, but now chips are available that can be programmed using a power supply as low as 1.8 V. Therefore, this memory is ideally suited for low-power systems.

In a microprocessor-based product, programs are generally written in ROM, and data that are likely to vary are stored in R/WM. For example, in a microprocessor-controlled oven, programs that run the oven are permanently stored in ROM, and data such as baking period, starting time, and temperature are entered in R/W memory through the keyboard. On the other hand, when microcomputers are used for developing software or for learning purposes, programs are first written in R/W memory, and then stored on a storage memory such as a cassette tape or floppy disk.

ADVANCES IN MEMORY TECHNOLOGY

Memory technology has advanced considerably in recent years. In addition to static and dynamic R/W memory, other options are also available in memory devices. Examples include Zero Power RAM, Nonvolatile RAM, and Integrated RAM.

The Zero Power RAM is a CMOS Read/Write memory with battery backup built internally. It includes lithium cells and voltage-sensing circuitry. When the external power supply voltage falls below 3 V, the power-switching circuitry connects the lithium battery; thus, this memory provides the advantages of R/W and Read-Only memory.

The Nonvolatile RAM is a high-speed static R/W memory array backed up, bit for bit, by EE-PROM array for nonvolatile storage. When the power is about to go off, the contents of R/W memory are quickly stored in the EE-PROM by activating the Store signal on the memory chip, and the stored data can be read into the R/W memory segment when the power is again turned on. This memory chip combines the flexibility of static R/W memory with the nonvolatility of EE-PROM.

The Integrated RAM (iRAM) is a dynamic memory with the refreshed circuitry built on the chip. For the user, it is similar to the static R/W memory. The user can derive the advantages of the dynamic memory without having to build the external refresh circuitry.

INPUT AND OUTPUT (I/O) DEVICES 2.3

Input/output devices are the means through which the MPU communicates with "the outside world." The MPU accepts binary data as input from devices such as keyboards and A/D converters and sends data to output devices such as LEDs or printers. There are two different methods by which I/O devices can be identified: one uses an 8-bit address and the other uses a 16-bit address. These methods are described briefly in the following sections.

2.31 I/Os with 8-Bit Addresses (Peripheral-Mapped I/O)

In this type of I/O, the MPU uses eight address lines to identify an input or an output device; this is known as peripheral-mapped I/O (also known as I/O-mapped I/O). This is an 8-bit numbering system for I/Os used in conjunction with Input and Output instructions. This is also known as I/O space, separate from memory space, which is a 16-bit numbering system. The eight address lines can have 256 (2^8 combinations) addresses; thus, the

MPU can identify 256 input devices and 256 output devices with addresses ranging from 00H to FFH. The input and output devices are differentiated by the control signals; the MPU uses the I/O Read control signal for input devices and the I/O Write control signal for output devices. The entire range of I/O addresses from 00 to FF is known as an I/O map, and individual addresses are referred to as I/O device addresses or I/O port numbers.

If we use LEDs as output or switches as input, we need to resolve two issues: how to assign addresses and how to connect these I/O devices to the data bus. In a bus architecture, these devices cannot be connected directly to the data bus or the address bus; all connections must be made through tri-state interfacing devices so they will be enabled and connected to the buses only when the MPU chooses to communicate with them. In the case of memory, we did not have to be concerned with these problems because of the internal address decoding, Read/Write buffers, and availability of $\overline{CS}$ and control signals of the memory chip. In the case of I/O devices, we need to use external interfacing devices (see Section 2.5).

The steps in communicating with an I/O device are similar to those in communicating with memory and can be summarized as follows:

1. The MPU places an 8-bit address on the address bus, which is decoded by external decode logic (explained in Chapter 4).
2. The MPU sends a control signal (I/O Read or I/O Write) and enables the I/O device.
3. Data are transferred using the data bus.

2.32   I/Os with 16-Bit Addresses (Memory-Mapped I/O)

In this type of I/O, the MPU uses 16 address lines to identify an I/O device; an I/O is connected as if it is a memory register. This is known as memory-mapped I/O. The MPU uses the same control signal (Memory Read or Memory Write) and instructions as those of memory. In some microprocessors, such as the Motorola 6800, all I/Os have 16-bit addresses; I/Os and memory share the same memory map (64K). In memory-mapped I/O, the MPU follows the same steps as if it is accessing a memory register.

The peripheral- and memory-mapped I/O techniques will be discussed in detail in the context of interfacing I/O devices (see Chapter 4).

2.4   EXAMPLE OF A MICROCOMPUTER SYSTEM

On the basis of the discussion in the previous sections, we can expand the microcomputer system shown in Figure 2.1 to include additional details. Figure 2.15 illustrates such a system. It shows the 8085 MPU, two types of memory (EPROM and R/WM), input and output, and the buses linking all peripherals (memory and I/Os) to the MPU.

The address lines A_{15}–A_0 are used to address memory, and the low-order address bus A_7–A_0 is used to identify the input and the output. The data bus D_7–D_0 is bidirectional and common to all the devices. The four control signals generated by the MPU are connected to different peripheral devices, as shown in Figure 2.15.

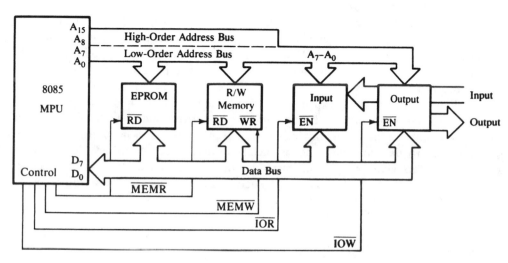

FIGURE 2.15
Example of a Microcomputer System

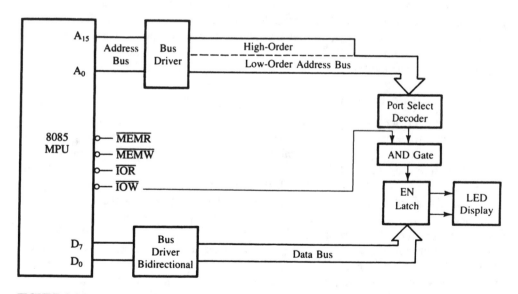

FIGURE 2.16
The Output Section of the Microcomputer System Illustrated in Figure 2.15

The MPU communicates with only one peripheral at a time by enabling the peripheral through its control signal. For example, to send data to the output device, the MPU places the device address (output port number) on the address bus, data on the data bus, and enables the output device using the control signal $\overline{\text{IOW}}$ (I/O Write). The output device latches and displays data if the output device happens to be LEDs. The other peripherals that are not enabled remain in a high impedance state called tri-state (explained later), similar to being disconnected from the system. Figure 2.15 is a simplified block diagram of the system; it does not show such details as data latching and tri-state devices (see Section 2.5).

Figure 2.16 shows an expanded version of the output section and the buses of Figure 2.15. The block diagram includes tri-state bus drivers, a decoder, and a latch. The bus drivers increase the current driving capacity of the buses, the decoder decodes the address to identify the output port, and the latch holds data output for display. These devices are called **interfacing devices.** The interfacing devices are semiconductor chips that are needed to connect peripherals to the bus system. Before we discuss interfacing concepts, we will review these interfacing devices.

2.5 REVIEW: LOGIC DEVICES FOR INTERFACING

Several types of interfacing devices are necessary to interconnect the components of a bus-oriented system. The devices used in today's microcomputer systems are designed using medium-scale integration (MSI) technology. In addition, tri-state logic devices are essential to proper functioning of the bus-oriented system, in which the same bus lines are shared by several components. The concept underlying the tri-state logic, as well as commonly used interfacing devices, will be reviewed in the following section.

2.51 Tri-State Devices

Tri-state logic devices have three states: logic 1, logic 0, and high impedance. The term *Tri-State* is a trademark of National Semiconductor and is used to represent three logic states. A tri-state logic device has a third line called Enable, as shown in Figure 2.17. When this line is activated, the tri-state device functions the same way as ordinary logic devices. When the third line is disabled, the logic device goes into the high impedance state—as if it were disconnected from the system. Ordinarily, current is required to drive a device in logic 0 and logic 1 states. In the high impedance state, practically no current is drawn from the system. Figure 2.17(a) shows a tri-state inverter. When the Enable is high, the circuit functions as an ordinary inverter; when the Enable line is low, the inverter stays in the high impedance state. Figure 2.17(b) also shows a tri-state inverter with active low Enable line—notice the bubble. When the Enable line is high, the inverter stays in the high impedance state.

FIGURE 2.17
Tri-State Inverters with Active High
and Active Low Enable Lines

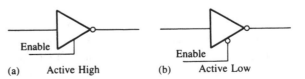

(a)   Active High   (b)   Active Low

In microcomputer systems, peripherals are connected in parallel between the address bus and the data bus. However, because of the tri-state interfacing devices, peripherals do not load the system buses. The microprocessor communicates with one device at a time by enabling the tri-state line of the interfacing device. Tri-state logic is critical to proper functioning of the microcomputer.

2.52 Buffer

The **buffer** is a logic circuit that amplifies the current or power. It has one input line and one output line (a simple buffer is shown in Figure 2.18a). The logic level of the output is the same as that of the input; logic 1 input provides logic 1 output (the opposite of an inverter). The buffer is used primarily to increase the driving capability of a logic circuit. It is also known as a driver.

Figure 2.18b shows a tri-state buffer. When the Enable line is low, the circuit functions as a buffer; otherwise it stays in the high impedance state. The buffer is commonly used to increase the driving capability of the data bus and the address bus.

EXAMPLES OF TRI-STATE BUFFERS

The octal buffer 74LS244 shown in Figure 2.19 is a typical example of a tri-state buffer. It is also known as a **line driver** or **line receiver.** This device is commonly used as a driver for the address bus in a bus-oriented system.

FIGURE 2.18

A Buffer and a Tri-State Buffer

Enable

Active Low

FIGURE 2.19

Logic Diagram of the 74LS244 Octal Buffer

SOURCE: Courtesy of Texas Instruments Incorporated.

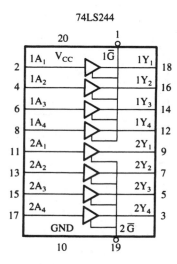

Figure 2.19 shows two groups of four buffers with noninverted tri-state output. The buffers are controlled by two active low Enable lines ($\overline{1G}$ and $\overline{2G}$). Until these lines are enabled, the output of the drivers remains in the high impedance state. Each buffer is capable of sinking 24 mA and sourcing −15 mA of current. The 74LS240 is another example of a tri-state buffer; it has tri-state inverted output.

BIDIRECTIONAL BUFFER

The data bus of a microcomputer system is bidirectional; therefore, it requires a buffer that allows data to flow in both directions. Figure 2.20 shows the logic diagram of the bidirectional buffer 74LS245, also called an **octal bus transceiver.** This is commonly used as a driver for the data bus.

The 74LS245 includes 16 bus drivers, eight for each direction, with tri-state output. The direction of data flow is controlled by the pin DIR. When DIR is high, data flow from the A bus to the B bus; when it is low, data flow from B to A. The schematic also includes an Enable signal ($\overline{G}$), which is active low. The Enable signal and the DIR signal are ANDed to activate the bus lines. The device is designed to sink 24 mA and source −15 mA of current.

2.53 Decoder

The decoder is a logic circuit that identifies each combination of the signals present at its input. For example, if the input to a decoder has two binary lines, the decoder will have four output lines (Figure 2.21). The two lines can assume four combinations of input signals—00, 01, 10, 11—with each combination identified by the output lines 0 to 3. If the

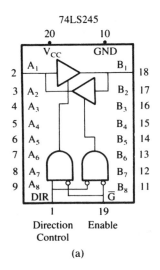

(a)

Function Table		
Enable $\overline{G}$	Direction Control DIR	Operation
L	L	B Data to A Bus
L	H	A Data to B Bus
H	X	Isolation

H = high level, L = low level, X = irrelevant

(b)

FIGURE 2.20
Logic Diagram and Function Table of the 74LS245 Bidirectional Buffer
SOURCE: Courtesy of Texas Instruments Incorporated.

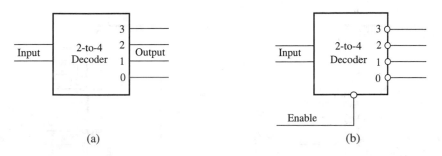

FIGURE 2.21

2-to-4 (1-out-of-4) Decoder Logic Symbol (a) 2-to-4 Decoder with Active Low Output and Enable Line (b)

input is 11_2, the output line 3 will be at logic 1, and the others will remain at logic 0. This is called **decoding.** Figure 2.21(a) shows a symbolic representation for a hypothetical 2:4 decoder. It is also called a 1-out-of-4 decoder. Various types of decoders are available; for example, 3-to-8, 4-to-16 (to decode binary inputs), and 4-to-10 (to decode BCD input). In general, decoders have active low output lines as well as Enable lines, as shown in Figure 2.21(b) and Figure 2.22. The decoders shown in Figures 2.21(b) and 2.22(b) will not function unless they are enabled by a low signal.

A decoder is a commonly used device in interfacing I/O peripherals and memory. In Figure 2.16 the decoder (Port Select Decoder) is used to decode an address bus to identify the output device. Decoders are also built internal to a memory chip to identify individual memory registers.

EXAMPLES OF DECODERS

Figure 2.23 shows the block diagrams of two 3-to-8 decoders, the 74LS138 and the Intel 8205. These are pin-compatible, with slight differences in their switching response and

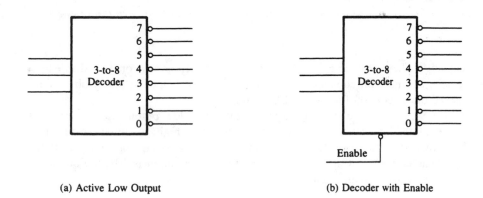

(a) Active Low Output   (b) Decoder with Enable

FIGURE 2.22

3-to-8 (1-out-of-8) Decoder Logic Symbol

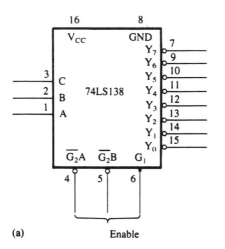

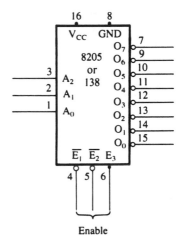

(a)   Enable   Enable

LS138, S138
Function Table

Inputs					Outputs							
Enable		Select										
G_1	G_2	C	B	A	Y_0	Y_1	Y_2	Y_3	Y_4	Y_5	Y_6	Y_7
X	H	X	X	X	H	H	H	H	H	H	H	H
L	X	X	X	X	H	H	H	H	H	H	H	H
H	L	L	L	L	L	H	H	H	H	H	H	H
H	L	L	L	H	H	L	H	H	H	H	H	H
H	L	L	H	L	H	H	L	H	H	H	H	H
H	L	L	H	H	H	H	H	L	H	H	H	H
H	L	H	L	L	H	H	H	H	L	H	H	H
H	L	H	L	H	H	H	H	H	H	L	H	H
H	L	H	H	L	H	H	H	H	H	H	L	H
H	L	H	H	H	H	H	H	H	H	H	H	L

(b)

H = high level, L = low level, X = irrelevant

FIGURE 2.23

Logic Diagrams and Function Table of 3-to-8 Decoders

SOURCE: (a) Courtesy of Texas Instruments Incorporated. (b) Intel corporation, *MCS—80/85 Family User's Manual* (Santa Clara, Calif.: Author, 1979), pp. 6–74.

current capacity. They are also called **1-out-of-8 binary decoders** or **demultiplexers.** There are several other semiconductor manufacturers that use input/output symbols similar to those of the Intel 8205 for their 74LS138 decoder.

    The 74LS138 has three input lines and eight active low output lines. It requires three Enable inputs. Two are active low and one is active high; all three Enable lines should be activated so that the device can function as a decoder. For example, if the 74LS138 is enabled ($\overline{G_2A} = \overline{G_2B} = 0$ and $G_1 = 1$) and if the input is 101, the output Y_5 will go low; others will remain high.

2.54   Encoder

The encoder is a logic circuit that provides the appropriate code (binary, BCD, etc.) as output for each input signal. The process is the reverse of decoding. Figure 2.24 shows an

FIGURE 2.24
Logic Symbols: 8-to-3 Encoder

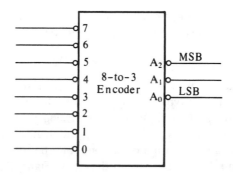

8-to-3 encoder; it has eight active low inputs and three output lines. When the input line 0 goes low, the output is 000; when the input line 5 goes low, the output is 101. However, this encoder is unable to provide an appropriate output code if two or more input lines are activated simultaneously. Encoders called priority encoders can resolve the problem of simultaneous inputs.

Figure 2.25 shows the logic symbol of the 74LS148, an 8-to-3 **priority encoder.** It has eight inputs and one active low enable signal. It has five output signals—three are encoding lines and two are output-enable indicators. The output lines GS and EO can be used to encode more than eight inputs by cascading these devices. When the encoder is enabled and two or more input signals are activated simultaneously, it ignores the low-priority inputs and encodes the highest-priority input. Similarly, the output of this decoder is active low; when input signal 7 is active, the output is 000 rather than 111.

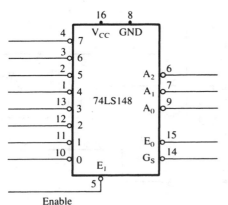

'148, 'LS148
Function Table

	Inputs									Outputs				
E_I	0	1	2	3	4	5	6	7	A_2	A_1	A_0	G_S	E_0
H	X	X	X	X	X	X	X	X	H	H	H	H	H
L	H	H	H	H	H	H	H	H	H	H	H	H	L
L	X	X	X	X	X	X	X	L	L	L	L	L	H
L	X	X	X	X	X	X	L	H	L	L	H	L	H
L	X	X	X	X	X	L	H	H	L	H	L	L	H
L	X	X	X	X	L	H	H	H	L	H	H	L	H
L	X	X	X	L	H	H	H	H	H	L	L	L	H
L	X	X	L	H	H	H	H	H	H	L	H	L	H
L	X	L	H	H	H	H	H	H	H	H	L	L	H
L	L	H	H	H	H	H	H	H	H	H	H	L	H

(a)   (b)

FIGURE 2.25
8-to-3 Priority Encoder—74LS148: Logic Symbol and Function Table
SOURCE: Function table courtesy of Texas Instruments, Incorporated.

Encoders are commonly used with keyboards. For each key pressed, the corresponding binary code is placed on the data bus.

2.55   D Flip-Flops: Latch and Clocked

In its simplest form, a latch is a D flip-flop, as shown in Figure 2.26; it is also called a transparent latch. A typical example of a latch is the 7475 D flip-flop. In this latch, when the enable signal (G) is high, the output changes according to the input D. Figure 2.26(a) shows that the output Q of the 7475 latch changes during T_{12} and T_{34} and data bits are latched at t_2 and t_4. On the other hand, in a positive-edge-triggered flip-flop, the output changes with the positive edge of the clock. Figure 2.26(b) shows the output of the 7474 positive-edge-triggered flip-flop. At the first positive going clock (t_1), the input is low; therefore, the output remains low until the next positive edge (t_3). At t_3, the D input is high; therefore, the output goes high. There is no effect on the output at any other time.

A latch is used commonly to interface output devices. When the MPU sends an output, data are available on the data bus for only a few microseconds; therefore, a latch is used to hold data for display.

EXAMPLES OF LATCHES

A typical example of a transparent latch is the 74LS373 shown in Figure 2.27. This octal latch is used to latch 8-bit data.

The device includes eight D latches with tri-state buffers, and it requires two input signals, Enable (G) and Output Control ($\overline{OC}$). The Enable is an active high signal connected to the clock input of the flip-flop. When this signal goes low, data are latched from

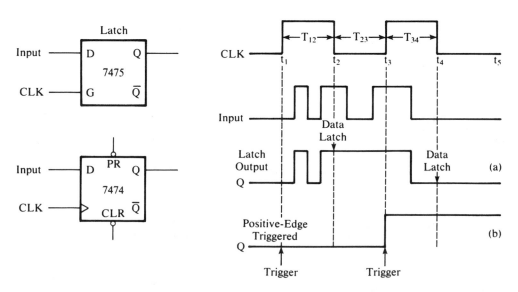

FIGURE 2.26
Output Waveforms of Latch (a) and Positive-Edge-Triggered Flip-Flop (b)

FIGURE 2.27

74LS373 D Latch: Logic Diagram
and Function Table

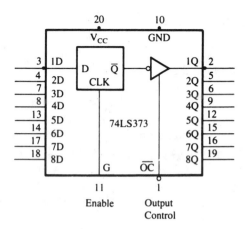

Function Table

Output Control	Enable G	D	Output
L	H	H	H
L	H	L	L
L	L	X	Q_0
H	X	X	Z

the data bus. The Output Control signal is active low, and it enables the tri-state buffers to output data to display devices.

These interfacing devices are discussed briefly here as a review to explain why they are needed and how they are used in microcomputer systems; they will be discussed again with interfacing applications.

SUMMARY

□ The microprocessor (MPU) primarily performs four operations: Memory Read, Memory Write, I/O Read, and I/O Write. For each operation, it generates the appropriate control signal.

□ To communicate with a peripheral (and memory), the MPU identifies the peripheral or the memory location by its address, transfers data, and provides timing signals.

□ **Address Bus**—a group of lines that are used to send a memory address or a device address from the MPU to the memory location or the peripheral. The 8085 microprocessor has 16 address lines.

□ **Data Bus**—a group of bidirectional lines used to transfer data between the MPU and peripherals (or memory). The 8085 microprocessor has eight data lines.

□ **Control Bus**—single lines that are generated by the MPU to provide timing of various operations.

☐ The 8085 microprocessor has six general-purpose 8-bit registers to store data and an accumulator to perform arithmetic and logical operations.

☐ The data conditions, after an arithmetic or logical operation, are indicated by setting or resetting the flip-flops called flags. The 8085 includes five flags: Sign, Zero, Auxiliary Carry, Parity, and Carry.

☐ The 8085 has two 16-bit registers: the program counter and the stack pointer. The program counter is used to sequence the execution of a program, and the stack pointer is used as a memory pointer for the stack memory.

☐ The 8085 can respond to four externally initiated operations: Reset, Interrupt, Ready, and Hold.

☐ Memory is a group of registers, arranged in a sequence, to store bits. The 8085 MPU requires an 8-bit-wide memory word and uses the 16-bit address to select a register called a memory location.

☐ The memory addresses assigned to a memory chip in a system are called the memory map. The assignment of memory addresses is done through the Chip Select logic.

☐ Memory can be classified primarily into two groups: Read/Write memory (R/WM) and Read-Only memory (ROM). The R/W memory is volatile and can be used to read and write information. This is also called the user memory. The ROM is a non-volatile memory and the information written into this memory is permanent.

☐ Input/Output devices or peripherals can be interfaced with the 8085 MPU in two ways, peripheral I/O and memory-mapped I/O. In peripheral I/O, the MPU uses an 8-bit address to identify an I/O, and IN and OUT instructions for data transfer. In memory-mapped I/O, the MPU uses a 16-bit address to identify an I/O and memory-related instructions for data transfer.

☐ To execute an instruction, the MPU places the 16-bit address on the address bus, sends the control signal to enable the memory chip, and fetches the instruction. The instruction is then decoded and executed.

☐ To interconnect peripherals with the 8085 MPU, additional logic circuits, called interfacing devices, are necessary. These circuits include devices such as buffers, decoders, encoders, and latches.

☐ A tri-state logic device has three states: two logic states and one high impedance state. When the device is not enabled, it remains in high impedance and does not draw any current from the system.

QUESTIONS AND PROBLEMS

1. List the four operations commonly performed by the MPU.
2. What is a bus?
3. Specify the function of the address bus and the direction of the information flow on the address bus.
4. How many memory locations can be addressed by a microprocessor with 14 address lines?

5. How many address lines are necessary to address two megabytes (2048K) of memory?

6. Why is the data bus bidirectional?

7. Specify the four control signals commonly used by the 8085 MPU.

8. Specify the control signal and the direction of the data flow on the data bus in a memory-write operation.

9. What is the function of the accumulator?

10. What is a flag?

11. Why are the program counter and the stack pointer 16-bit registers?

12. While executing a program, when the 8085 MPU completes the fetching of the machine code located at the memory address 2057H, what is the content of the program counter?

13. Specify the number of registers and memory cells in a 128×4 memory chip.

14. How many bits are stored by a 256×4 memory chip? Can this chip be specified as 128-byte memory?

15. What is the memory word size required in an 8085 system?

16. If the memory chip size is 2048×8 bits, how many chips are required to make up 16K-byte memory?

17. If the memory chip size is 1024×4 bits, how many chips are required to make up 2K (2048) bytes of memory?

18. If the memory chip size is 256×1 bits, how many chips are required to make up 1K (1024) bytes of memory?

19. What is the function of the $\overline{WR}$ signal on the memory chip?

20. How many address lines are necessary on the chip of 2K (2048) byte memory?

21. The memory map of a 4K (4096) byte memory chip begins at the location 2000H. Specify the address of the last location on the chip and the number of pages in the chip.

22. The memory address of the last location of a 1K byte memory chip is given as FBFFH. Specify the starting address.

23. The memory address of the last location of an 8K byte memory chip is FFFFH. Find the starting address.

24. In Figure 2.10(a), eliminate all the inverters and connect address lines A_8–A_{15} directly to the NAND gate. Identify the memory address range of the chip.

25. In Figure 2.10(b), if the chip is selected and the address lines A_7–A_0 have 01000111, specify the complete address in Hex of the register selected.

26. In Figure 2.11, connect A_{13} to the NAND gate without an inverter, and identify the memory map.

27. Specify the Hex address of the register selected in Question 26 if the address on the address lines A_9–A_0 is 00 1111 1000.

28. How many address lines are used to identify an I/O port in the peripheral I/O and in the memory-mapped I/O methods?

29. What are tri-state devices and why are they essential in a bus-oriented system?

30. In Figure 2.19, if the input to the octal buffer is 4FH and the enable lines $1\overline{G}$ and $2\overline{G}$ are high, what is the output of the buffer?

FIGURE 2.28
Logic Diagram of a 4-to-16
Decoder

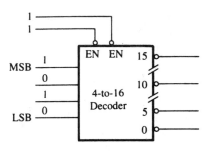

31. In Figure 2.20, if signals $\overline{G}$ (Enable) and DIR are low, specify the direction of data flow.
32. Specify the output line of the 4-to-16 decoder that goes low if the input to the decoder is as shown in Figure 2.28.
33. In Figure 2.29, specify the output line that goes low if the input (including the enable lines) to the 3-to-8 decoder (74LS138) is

$$A_7 \; A_6 \; A_5 \; A_4 \; A_3 \; A_2 \; A_1 \; A_0$$
$$1 \quad 1 \quad 1 \quad 1 \quad 0 \quad 1 \quad 1 \quad 1$$

34. What is the output of the encoder (Figure 2.30) if the key K_6 is pushed? (See Figure 2.25 for the function table.)
35. What is a transparent latch, and why is it necessary to use a latch with output devices such as LEDs?
36. List the high-order, low-order, and don't care address lines in Figure 2.31. How many pages of memory does the chip include?
37. In Figure 2.31, identify the memory addresses, assuming the don't care address line A_{11} at logic 0.

FIGURE 2.29
The 3-to-8 Decoder (74LS138)

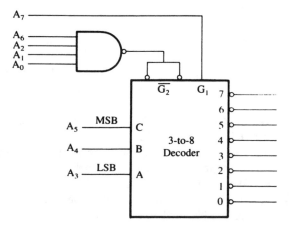

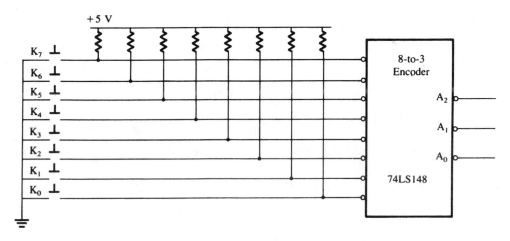

FIGURE 2.30
The 8-to-3 Encoder (74LS148)

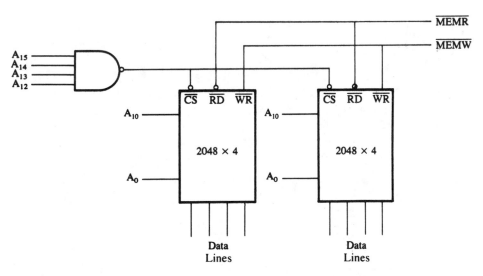

FIGURE 2.31
Memory Schematic

38. Specify the entire memory map (addresses) of the schematic shown in Figure 2.31, and explain the significance of the don't care address line on memory addresses.

8085 Microprocessor Architecture and Memory Interfacing

The 8085 microprocessor is a much improved version of its predecessor, the 8080A. The 8085 includes on its chip most of the logic circuitry for performing computing tasks and for communicating with peripherals. However, eight of its bus lines are **multiplexed;** that is, they are time-shared by the low-order address and data. This chapter discusses the 8085 architecture in detail and illustrates techniques for demultiplexing the bus and generating the necessary control signals.

Later, the chapter describes a typical 8085-based microcomputer designed with general-purpose memory and I/O devices; it also illustrates the bus timing signals in executing an instruction. Then, it examines the requirements of a memory chip based on the timing signals and derives the steps necessary in interfacing memory. In addition, the chapter includes illustrations of a special-purpose device, the 8155 and its interfacing.

OBJECTIVES

□ Recognize the functions of various pins of the 8085 microprocessor.
□ Explain the bus timings in fetching an instruction from memory.

□ Explain how to demultiplex the AD_7–AD_0 bus using a latch.
□ Draw a logic schematic to generate four control signals, using the 8085 IO/$\overline{M}$, $\overline{RD}$, and $\overline{WR}$ signals: (1) $\overline{MEMR}$, (2) $\overline{MEMR}$, (3) $\overline{IOR}$, and (4) $\overline{IOW}$. Explain the functions of these control signals.

☐ List the various internal units that make up the 8085 architecture, and explain their functions in decoding and executing an instruction.

☐ Draw the block diagram of an 8085-based microcomputer.

☐ List the steps performed by the 8085 microprocessor, and identify the contents of buses when an instruction is being executed.

☐ Analyze a memory interfacing circuit, and specify the memory addresses of a given memory device.

☐ Recognize partial decoding and identify foldback (mirror) memory space.

3.1 THE 8085 MPU

The term **microprocessing unit** (MPU) is similar to the term **central processing unit** (CPU) used in traditional computers. We define the MPU as a device or a group of devices (as a unit) that can communicate with peripherals, provide timing signals, direct data flow, and perform computing tasks as specified by the instructions in memory. The unit will have the necessary lines for the address bus, the data bus, and the control signals, and would require only a power supply and a crystal (or equivalent frequency-determining components) to be completely functional.

Using this description, the 8085 microprocessor can almost qualify as an MPU, but with the following two limitations.

1. The low-order address bus of the 8085 microprocessor is **multiplexed** (time-shared) with the data bus. The buses need to be demultiplexed.

2. Appropriate control signals need to be generated to interface memory and I/O with the 8085. (Intel has some specialized memory and I/O devices that do not require such control signals.)

This section shows how to demultiplex the bus and generate the control signals after describing the 8085 microprocessor and illustrates the bus timings.

3.11 The 8085 Microprocessor

The 8085A (commonly known as the 8085) is an 8-bit general-purpose microprocessor capable of addressing 64K of memory. The device has forty pins, requires a +5 V single power supply, and can operate with a 3-MHz single-phase clock. The 8085A-2 version can operate at the maximum frequency of 5 MHz. The 8085 is an enhanced version of its predecessor, the 8080A; its instruction set is upward-compatible with that of the 8080A, meaning that the 8085 instruction set includes all the 8080A instructions plus some additional ones.

Figure 3.1 shows the logic pinout of the 8085 microprocessor. All the signals can be classified into six groups: (1) address bus, (2) data bus, (3) control and status signals, (4) power supply and frequency signals, (5) externally initiated signals, and (6) serial I/O ports.

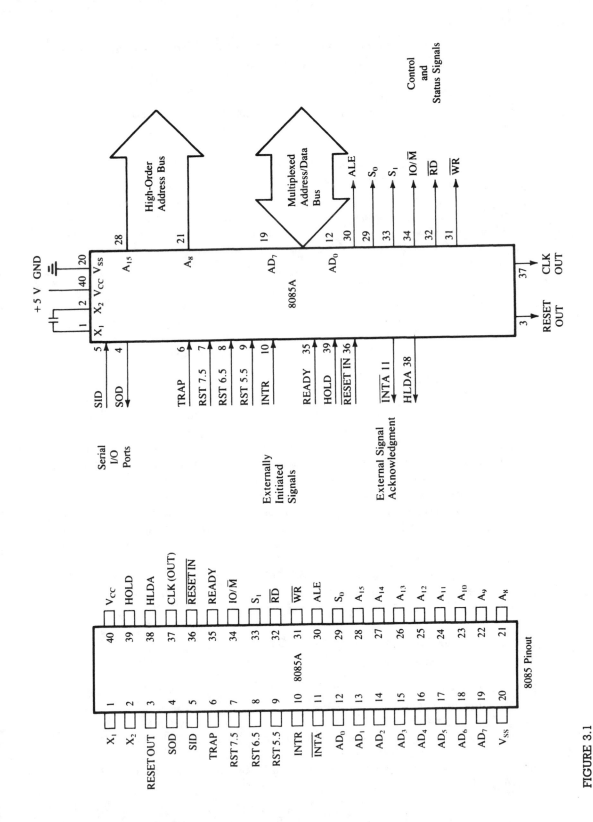

FIGURE 3.1

The 8085 Microprocessor Pinout and Signals

NOTE: The 8085A is commonly known as the 8085.
SOURCE (Pinout): Intel Corporation, *Embedded Microprocessors* (Santa Clara, Calif.: Author, 1994), pp. 1-11.

ADDRESS BUS

The 8085 has eight signal lines, A_{15}–A_8, which are unidirectional and used as the high-order address bus.

MULTIPLEXED ADDRESS/DATA BUS

The signal lines AD_7–AD_0 are bidirectional: they serve a dual purpose. They are used as the low-order address bus as well as the data bus. In executing an instruction, during the earlier part of the cycle, these lines are used as the low-order address bus. During the later part of the cycle, these lines are used as the data bus. (This is also known as multiplexing the bus.) However, the low-order address bus can be separated from these signals by using a latch.

CONTROL AND STATUS SIGNALS

This group of signals includes two control signals ($\overline{RD}$ and $\overline{WR}$), three status signals ($IO/\overline{M}$, S_1, and S_0) to identify the nature of the operation, and one special signal (ALE) to indicate the beginning of the operation. These signals are as follows:

- [ ] ALE—Address Latch Enable: This is a positive going pulse generated every time the 8085 begins an operation (machine cycle); it indicates that the bits on AD_7–AD_0 are address bits. This signal is used primarily to latch the low-order address from the multiplexed bus and generate a separate set of eight address lines, A_7–A_0.
- [ ] RD—Read: This is a Read control signal (active low). This signal indicates that the selected I/O or memory device is to be read and data are available on the data bus.
- [ ] WR—Write: This is a Write control signal (active low). This signal indicates that the data on the data bus are to be written into a selected memory or I/O location.
- [ ] $IO/\overline{M}$: This is a status signal used to differentiate between I/O and memory operations. When it is high, it indicates an I/O operation; when it is low, it indicates a memory operation. This signal is combined with $\overline{RD}$ (Read) and $\overline{WR}$ (Write) to generate I/O and memory control signals.
- [ ] S_1 and S_0: These status signals, similar to $IO/\overline{M}$, can identify various operations, but they are rarely used in small systems. (All the operations and their associated status signals are listed in Table 3.1 for reference.)

POWER SUPPLY AND CLOCK FREQUENCY

The power supply and frequency signals are as follows:

- [ ] V_{CC}: +5 V power supply.
- [ ] V_{SS}: Ground Reference.
- [ ] X_1, X_2: A crystal (or RC, LC network) is connected at these two pins. The frequency is internally divided by two; therefore, to operate a system at 3 MHz, the crystal should have a frequency of 6 MHz.
- [ ] CLK (OUT)—Clock Output: This signal can be used as the system clock for other devices.

TABLE 3.1
8085 Machine Cycle Status and Control Signals

Machine Cycle	Status			Control Signals
	IO/$\overline{\text{M}}$	S_1	S_0	
Opcode Fetch	0	1	1	$\overline{\text{RD}} = 0$
Memory Read	0	1	0	$\overline{\text{RD}} = 0$
Memory Write	0	0	1	$\overline{\text{WR}} = 0$
I/O Read	1	1	0	$\overline{\text{RD}} = 0$
I/O Write	1	0	1	$\overline{\text{WR}} = 0$
Interrupt Acknowledge	1	1	1	$\overline{\text{INTA}} = 0$
Halt	Z	0	0	$\overline{\text{RD}}, \overline{\text{WR}} = \text{Z and } \overline{\text{INTA}} = 1$
Hold	Z	X	X	
Reset	Z	X	X	

NOTE: Z = Tri-state (high impedance)
X = Unspecified

EXTERNALLY INITIATED SIGNALS, INCLUDING INTERRUPTS

The 8085 has five interrupt signals (see Table 3.2) that can be used to interrupt a program execution. One of the signals, INTR (Interrupt Request), is identical to the 8080A microprocessor interrupt signal (INT); the others are enhancements to the 8080A. The microprocessor acknowledges an interrupt request by the $\overline{\text{INTA}}$ (Interrupt Acknowledge) signal. (The interrupt process is discussed in Chapter 12.)

In addition to the interrupts, three pins—RESET, HOLD, and READY—accept the externally initiated signals as inputs. To respond to the HOLD request, the 8085 has one

TABLE 3.2
8085 Interrupts and Externally Initiated Signals

☐ INTR (Input)	Interrupt Request: This is used as a general-purpose interrupt; it is similar to the INT signal of the 8080A.
☐ $\overline{\text{INTA}}$ (Output)	Interrupt Acknowledge: This is used to acknowledge an interrupt.
☐ RST 7.5 (Inputs) RST 6.5 RST 5.5	Restart Interrupts: These are vectored interrupts that transfer the program control to specific memory locations. They have higher priorities than the INTR interrupt. Among these three, the priority order is 7.5, 6.5, and 5.5.
☐ TRAP (Input)	This is a nonmaskable interrupt and has the highest priority.
☐ HOLD (Input)	This signal indicates that a peripheral such as a DMA (Direct Memory Access) controller is requesting the use of the address and data buses.
☐ HLDA (Output)	Hold Acknowledge: This signal acknowledges the HOLD request.
☐ READY (Input)	This signal is used to delay the microprocessor Read or Write cycles until a slow-responding peripheral is ready to send or accept data. When this signal goes low, the microprocessor waits for an integral number of clock cycles until it goes high.

signal called HLDA (Hold Acknowledge). The functions of these signals were previously discussed in Section 2.13. The RESET is again described below, and others are listed in Table 3.2 for reference.

☐ $\overline{\text{RESET}}$ $\overline{\text{IN}}$: When the signal on this pin goes low, the program counter is set to zero, the buses are tri-stated, and the MPU is reset.
☐ RESET OUT: This signal indicates that the MPU is being reset. The signal can be used to reset other devices.

SERIAL I/O PORTS

The 8085 has two signals to implement the serial transmission: SID (Serial Input Data) and SOD (Serial Output Data). They will be discussed in Chapter 16 on serial I/O.

In this chapter, we will focus on the first three groups of signals; others will be discussed in later chapters.

3.12 Microprocessor Communication and Bus Timings

To understand the functions of various signals of the 8085, we should examine the process of communication (reading from and writing into memory) between the microprocessor and memory and the timings of these signals in relation to the system clock. The first step in the communication process is reading from memory or fetching an instruction. This can be easily understood using an analogy of how a package is picked up from your house by a shipping company such as Federal Express. The steps are as follows:

1. A courier gets the address from the office; he or she drives the pickup van, finds the street, and looks for your house number.
2. The courier rings the bell.
3. Somebody in the house opens the door and gives the package to the courier, and the courier returns to the office with the package.
4. The internal office staff disposes the package according to the instructions given by the customer.

Now let us examine the steps in the following example of how the microprocessor fetches or gets a machine code from memory.

Example 3.1

Refer to Example 2.5 in the last chapter (Section 2.26): Illustrate the steps and the timing of data flow when the instruction code 0100 1111 (4FH—MOV C,A), stored in location 2005H, is being fetched.

Solution

To fetch the byte (4FH), the MPU needs to identify the memory location 2005H and enable the data flow from memory. This is called the Fetch cycle. The data flow is shown in Figure 3.2, and the timings are explained below.

Figure 3.3 shows the timing of how a data byte is transferred from memory to the MPU; it shows five different groups of signals in relation to the system clock. The address bus and data bus are shown as two parallel lines. This is a commonly used practice to rep-

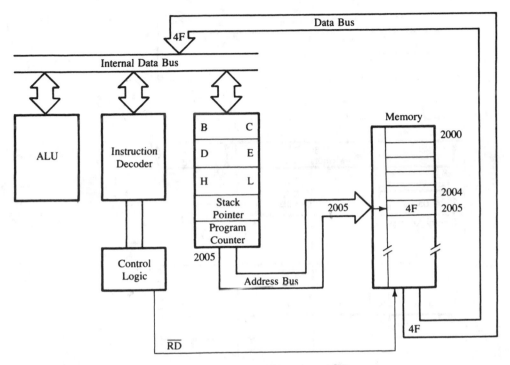

FIGURE 3.2
Data Flow from Memory to the MPU

resent logic levels of groups of lines; some lines are high and others are low. The crossover of the lines indicates that a new byte (information) is placed on the bus, and a dashed straight line indicates the high impedance state. To fetch the byte, the MPU performs the following steps:

Step 1: The program counter places the 16-bit memory address on the address bus (Figure 3.2).

Figure 3.3 shows that at T_1 the high-order memory address 20H is placed on the address lines $A_{15}-A_8$, the low-order memory address 05H is placed on the bus AD_7-AD_0, and the ALE signal goes high. Similarly, the status signal IO/$\overline{M}$ goes low, indicating that this is a memory-related operation. (For the sake of clarity, the other two status signals, S_1 and S_0, are not shown in Figure 3.3; they will be discussed in the next section.)

Step 2: The control unit sends the control signal $\overline{RD}$ to enable the memory chip (Figure 3.2). This is similar to ringing the doorbell in our analogy of a package pickup.

The control signal $\overline{RD}$ is sent out during the clock period T_2, thus enabling the memory chip (Figure 3.3). The $\overline{RD}$ signal is active during two clock periods.

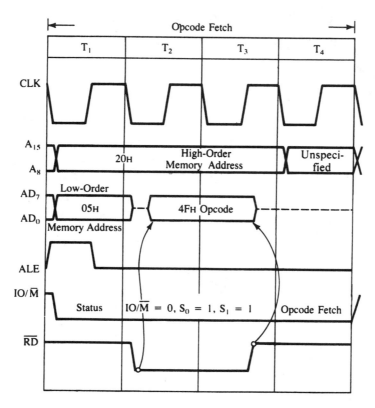

FIGURE 3.3
Timing: Transfer of Byte from Memory to MPU

Step 3: The byte from the memory location is placed on the data bus.

When the memory is enabled, the instruction byte (4FH) is placed on the bus AD_7–AD_0 and transferred to the microprocessor. The $\overline{RD}$ signal causes 4FH to be placed on bus AD_7–AD_0 (shown by the arrow), and when $\overline{RD}$ goes high, it causes the bus to go into high impedance.

Step 4: The byte is placed in the instruction decoder of the microprocessor, and the task is carried out according to the instruction.

The machine code or the byte (4FH) is decoded by the instruction decoder, and the contents of the accumulator are copied into register C. This task is performed during the period T_4 in Figure 3.3.

The above four steps are similar to the steps listed in our analogy of the package pickup.

3.13   Demultiplexing the Bus AD_7–AD_0

The need for demultiplexing the bus AD_7–AD_0 becomes easier to understand after examining Figure 3.3. This figure shows that the address on the high-order bus (20H) remains on the bus for three clock periods. However, the low-order address (05H) is lost after the first clock period. This address needs to be latched and used for identifying the memory address. If the bus AD_7–AD_0 is used to identify the memory location (2005H), the address will change to 204FH after the first clock period.

Figure 3.4 shows a schematic that uses a latch and the ALE signal to demultiplex the bus. The bus AD_7–AD_0 is connected as the input to the latch 74LS373. The ALE signal is connected to the Enable (G) pin of the latch, and the Output control ($\overline{OC}$) signal of the latch is grounded.

Figure 3.3 shows that the ALE goes high during T_1. When the ALE is high, the latch is transparent; this means that the output changes according to input data. During T_1, the output of the latch is 05H. When the ALE goes low, the data byte 05H is latched until the next ALE, and the output of the latch represents the low-order address bus A_7–A_0 after the latching operation.

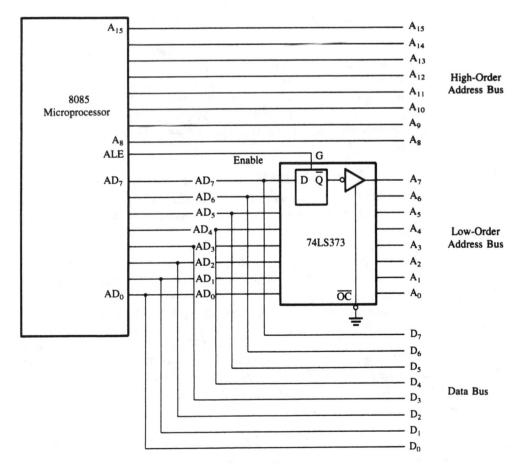

FIGURE 3.4
Schematic of Latching Low-Order Address Bus

Intel has circumvented the problem of demultiplexing the low-order bus by designing special devices such as the 8155 (256 bytes of R/W memory + I/Os), which is compatible with the 8085 multiplexed bus. These devices internally demultiplex the bus using the ALE signal (see Figures 3.18 and 3.19).

After carefully examining Figure 3.3, we can make the following observations:

1. The machine code 4FH (0100 1000) is a one-byte instruction that copies the contents of the accumulator into register C.
2. The 8085 microprocessor requires one external operation—fetching a machine code* from memory location 2005H.
3. The entire operation—fetching, decoding, and executing—requires four clock periods.

Now we can define three terms—instruction cycle, machine cycle, and T-state—and use these terms later for examining timings of various 8085 operations (Section 3.2).

Instruction cycle is defined as the time required to complete the execution of an instruction. The 8085 instruction cycle consists of one to six machine cycles or one to six operations.

Machine cycle is defined as the time required to complete one operation of accessing memory, I/O, or acknowledging an external request. This cycle may consist of three to six T-states. In Figure 3.3, the instruction cycle and the machine cycle are the same.

T-state is defined as one subdivision of the operation performed in one clock period. These subdivisions are internal states synchronized with the system clock, and each T-state is precisely equal to one clock period. The terms T-state and clock period are often used synonymously.

3.14   Generating Control Signals

Figure 3.3 shows the $\overline{RD}$ (Read) as a control signal. Because this signal is used both for reading memory and for reading an input device, it is necessary to generate two different Read signals: one for memory and another for input. Similarly, two separate Write signals must be generated.

Figure 3.5 shows that four different control signals are generated by combining the signals $\overline{RD}$, $\overline{WR}$, and $IO/\overline{M}$. The signal $IO/\overline{M}$ goes low for the memory operation. This signal is ANDed with $\overline{RD}$ and $\overline{WR}$ signals by using the 74LS32 quadruple two-input OR gates, as shown in Figure 3.5. The OR gates are functionally connected as negative NAND gates. When both input signals go low, the outputs of the gates go low and generate $\overline{MEMR}$ (Memory Read) and $\overline{MEMW}$ (Memory Write) control signals. When the $IO/\overline{M}$ signal goes high, it indicates the peripheral I/O operation. Figure 3.5 shows that this signal is complemented using the Hex inverter 74LS04 and ANDed with the $\overline{RD}$ and $\overline{WR}$ signals to generate $\overline{IOR}$ (I/O Read) and $\overline{IOW}$ (I/O Write) control signals.

---

*This code is an operation code (opcode) that instructs the microprocessor to perform the specified task. The term *opcode* is explained in Chapter 5 (Section 5.3).

FIGURE 3.5
Schematic to Generate Read/Write Control Signals for Memory and I/O

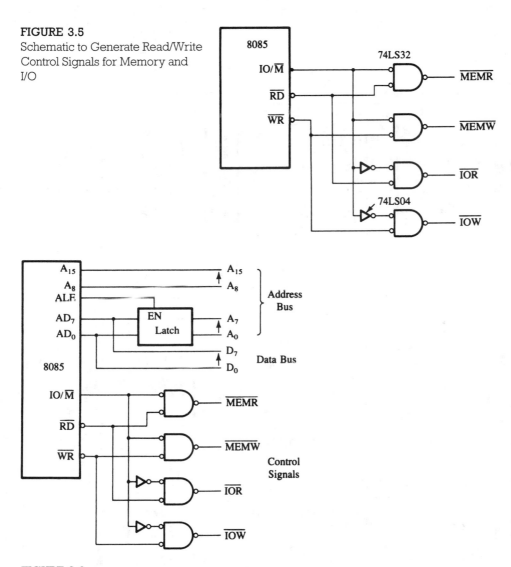

FIGURE 3.6
8085 Demultiplexed Address and Data Bus with Control Signals

To demultiplex the bus and to generate the necessary control signals, the 8085 microprocessor requires a latch and logic gates to build the MPU, as shown in Figure 3.6. This MPU can be interfaced with any memory or I/O.

3.15   A Detailed Look at the 8085 MPU and Its Architecture

Figure 3.7 shows the internal architecture of the 8085 beyond the programmable registers we discussed previously. It includes the ALU (Arithmetic/Logic Unit), Timing and

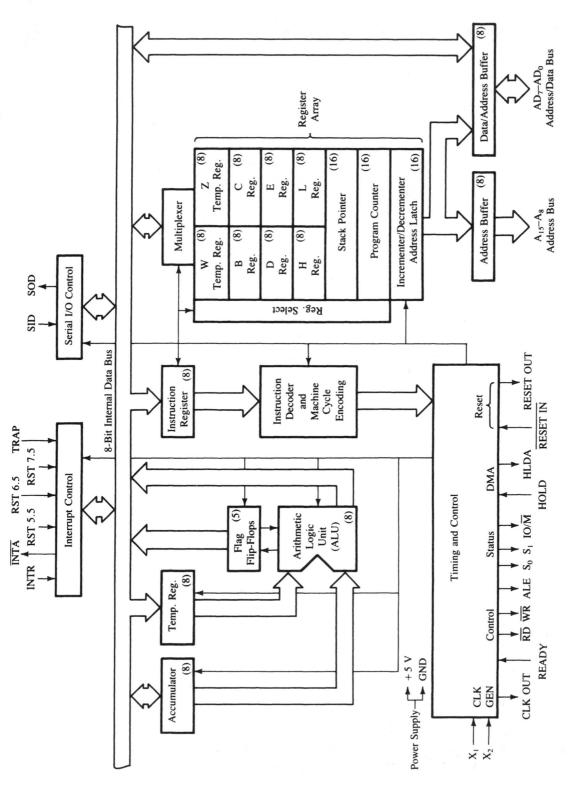

FIGURE 3.7

The 8085A Microprocessor: Functional Block Diagram

NOTE: The 8085A microprocessor is commonly known as the 8085.

SOURCE: Intel Corporation, *Embedded Microprocessors* (Santa Clara, Calif.: Author, 1994), pp. 1-11.

Control Unit, Instruction Register and Decoder, Register Array, Interrupt Control, and Serial I/O Control. We will discuss the first four units below; the last two will be discussed later in the book.

THE ALU

The arithmetic/logic unit performs the computing functions; it includes the accumulator, the temporary register, the arithmetic and logic circuits, and five flags. The temporary register is used to hold data during an arithmetic/logic operation. The result is stored in the accumulator, and the flags (flip-flops) are set or reset according to the result of the operation.

The flags are affected by the arithmetic and logic operations in the ALU. In most of these operations, the result is stored in the accumulator. Therefore, the flags generally reflect data conditions in the accumulator—with some exceptions. The descriptions and conditions of the flags are as follows:

☐ **S—Sign flag:** After the execution of an arithmetic or logic operation, if bit D_7 of the result (usually in the accumulator) is 1, the Sign flag is set. This flag is used with signed numbers. In a given byte, if D_7 is 1, the number will be viewed as a negative number; if it is 0, the number will be considered positive. In arithmetic operations with signed numbers, bit D_7 is reserved for indicating the sign, and the remaining seven bits are used to represent the magnitude of a number. (See Appendix A2 for a discussion of signed numbers.)

☐ **Z—Zero flag:** The Zero flag is set if the ALU operation results in 0, and the flag is reset if the result is not 0. This flag is modified by the results in the accumulator as well as in the other registers.

☐ **AC—Auxiliary Carry flag:** In an arithmetic operation, when a carry is generated by digit D_3 and passed on to digit D_4, the AC flag is set. The flag is used only internally for BCD (binary-coded decimal) operations and is not available for the programmer to change the sequence of a program with a jump instruction.

☐ **P—Parity flag:** After an arithmetic or logical operation, if the result has an even number of 1s, the flag is set. If it has an odd number of 1s, the flag is reset. (For example, the data byte 0000 0011 has even parity even if the magnitude of the number is odd.)

☐ **CY—Carry flag:** If an arithmetic operation results in a carry, the Carry flag is set; otherwise it is reset. The Carry flag also serves as a borrow flag for subtraction.

The bit positions reserved for these flags in the flag register are as follows:

D_7	D_6	D_5	D_4	D_3	D_2	D_1	D_0
S	Z		AC		P		CY

Among the five flags, the AC flag is used internally for BCD arithmetic; the instruction set does not include any conditional jump instructions based on the AC flag. Of the remaining four flags, the Z and CY flags are those most commonly used.

TIMING AND CONTROL UNIT

This unit synchronizes all the microprocessor operations with the clock and generates the control signals necessary for communication between the microprocessor and peripherals.

The control signals are similar to a sync pulse in an oscilloscope. The $\overline{RD}$ and $\overline{WR}$ signals are sync pulses indicating the availability of data on the data bus.

INSTRUCTION REGISTER AND DECODER

The instruction register and the decoder are part of the ALU. When an instruction is fetched from memory, it is loaded in the instruction register. The decoder decodes the instruction and establishes the sequence of events to follow. The instruction register is not programmable and cannot be accessed through any instruction.

REGISTER ARRAY

The programmable registers were discussed in the last chapter. Two additional registers, called temporary registers W and Z, are included in the register array. These registers are used to hold 8-bit data during the execution of some instructions. However, because they are used internally, they are not available to the programmer.

3.16   Decoding and Executing an Instruction

Decoding and executing an instruction after it has been fetched can be illustrated with the example from Section 3.12.

**Example
3.2**

Assume that the accumulator contains data byte 82H, and the instruction MOV C,A (4FH) is fetched. List the steps in decoding and executing the instruction.

Solution

This example is similar to the example in Section 3.12, except that the contents of the accumulator are specified. To decode and execute the instruction, the following steps are performed.

1. The contents of the data bus (4F) are placed in the instruction register and decoded (Figure 3.8).
2. The contents of the accumulator (82H) are transferred to the temporary register in the ALU.
3. The contents of the temporary register are transferred to register C.

3.17   Review of Important Concepts

1. The 8085 microprocessor has a multiplexed bus AD_7–AD_0 used as the lower-order address bus and the data bus.
2. The bus AD_7–AD_0 can be demultiplexed by using a latch and the ALE signal.
3. The 8085 has a status signal $IO/\overline{M}$ and two control signals $\overline{RD}$ and $\overline{WR}$. By ANDing these signals, four control signals can be generated: $\overline{MEMR}$, $\overline{MEMW}$, $\overline{IOR}$, and $\overline{IOW}$.
4. The 8085 MPU
   □ transfers data from memory locations to the microprocessor by using the control signal Memory Read ($\overline{MEMR}$—active low). This is also called **reading from memory.** The term *data* refers to any byte that is placed on the data bus; the byte can be an instruction code, data, or an address.

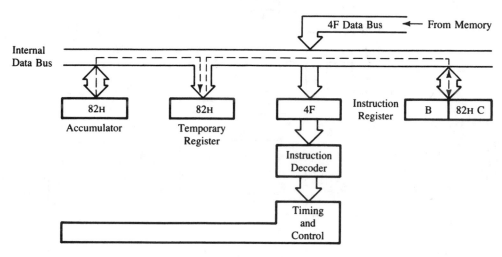

FIGURE 3.8
Instruction Decoding and Execution

☐ transfers data from the microprocessor to memory by using the control signal Memory Write ($\overline{\text{MEMW}}$ active low). This is also called **writing into memory.**

☐ accepts data from input devices by using the control signal I/O Read ($\overline{\text{IOR}}$—active low). This is also known as **reading from an input port.**

☐ sends data to output devices by using the control signal I/O Write ($\overline{\text{IOW}}$—active low). This is also known as **writing to an output port.**

5. To execute an instruction, the MPU

☐ places the memory address of the instruction on the address bus.

☐ indicates the operation status on the status lines.

☐ sends the $\overline{\text{MEMR}}$ control signal to enable the memory, fetches the instruction byte, and places it in the instruction decoder.

☐ executes the instruction.

EXAMPLE OF AN 8085-BASED MICROCOMPUTER 3.2

A general microcomputer system was illustrated in Figure 2.15, in the last chapter. After our discussion of the 8085 microprocessor and the interfacing devices, we can expand the system to include more details, as shown in Figure 3.9. The system includes interfacing devices such as buffers, decoders, and latches.

The 8085 MPU module (Figure 3.9) includes devices such as the 8085 microprocessor, an octal latch, and logic gates, as shown previously in Figure 3.6. The octal latch demultiplexes the bus AD_7–AD_0 using the signal ALE, and the logic gates generate

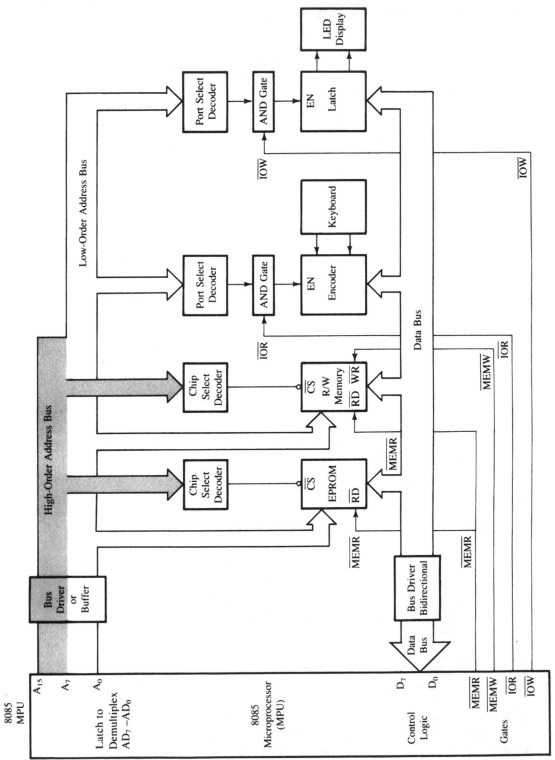

FIGURE 3.9

8085 Single-Board Microcomputer System

NOTE: The bus AD_7–AD_0 is demultiplexed using a latch and memory, and I/O control signals are generated using gates. Figure 3.6 is included in the MPU.

the necessary control signals. Figure 3.9 shows the demultiplexed address bus, the data bus, and the four active low control signals: $\overline{\text{MEMR}}$, $\overline{\text{MEMW}}$, $\overline{\text{IOR}}$, and $\overline{\text{IOW}}$. In addition, to increase the driving capacity of the buses, a unidirectional bus driver is used for the address bus and a bidirectional bus driver is used for the data bus. Now we can examine various operations the 8085 microprocessor performs in terms of machine cycles and T-states.

3.21   The 8085 Machine Cycles and Bus Timings

The 8085 microprocessor is designed to execute 74 different instruction types. Each instruction has two parts: operation code, known as opcode, and operand. The opcode is a command such as Add, and the operand is an object to be operated on, such as a byte or the contents of a register. Some instructions are 1-byte instructions and some are multibyte instructions. To execute an instruction, the 8085 needs to perform various operations such as Memory Read/Write and I/O Read/Write. However, there is no direct relationship between the number of bytes of an instruction and the number of operations the 8085 has to perform (this will be clarified later). In preceding sections, numerous 8085 signals and their functions were described. Now we need to examine these signals in conjunction with execution of individual instructions and their operations. This task may appear overwhelming at the beginning; fortunately, all instructions are divided into a few basic machine cycles and these machine cycles are divided into precise system clock periods.

   Basically, the microprocessor external communication functions can be divided into three categories:

1. Memory Read and Write
2. I/O Read and Write
3. Request Acknowledge

   These functions are further divided into various operations (machine cycles), as shown in Table 3.1. Each instruction consists of one or more of these machine cycles, and each machine cycle is divided into T-states.

   In this section, we will focus on the first three operations listed in Table 3.1—Opcode Fetch, Memory Read, and Memory Write—and examine the signals on various buses in relation to the system clock. In the next section, we will use these timing diagrams to interface memory with the 8085 microprocessor. Similarly, we will discuss timings of other machine cycles in later chapters in the context of their applications. For example, I/O Read/Write machine cycles will be discussed in Chapter 4 on I/O interfacing, and Interrupt Acknowledge will be discussed in Chapter 12, "Interrupts."

3.22   Opcode Fetch Machine Cycle

The first operation in any instruction is Opcode Fetch. The microprocessor needs to get (fetch) this machine code from the memory register where it is stored before the microprocessor can begin to execute the instruction.

   We discussed this operation in Example 3.1. Figure 3.2 shows how the 8085 fetches the machine code, using the address and the data buses and the control signal. Figure 3.3

shows the timing of the Opcode Fetch machine cycle in relation to the system's clock. In this operation, the processor reads a machine code (4FH) from memory. However, to differentiate an opcode from a data byte or an address, this machine cycle is identified as the Opcode Fetch cycle by the status signals (IO/$\overline{\text{M}}$ = 0, S_1 = 1, S_0 = 1); the active low IO/$\overline{\text{M}}$ signal indicates that it is a memory operation, and S_1 and S_0 being high indicate that it is an Opcode Fetch cycle.

This Opcode Fetch cycle is called the M_1 cycle and has four T-states. The 8085 uses the first three states T_1–T_3 to fetch the code and T_4 to decode and execute the opcode. In the 8085 instruction set, some instructions have opcodes with six T-states. When we study the example of the Memory Read machine cycle, discussed in the next section, we may find that these two operations (Opcode Fetch and Memory/Read) are almost identical except that the Memory Read Cycle has three T-states.

3.23   Memory Read Machine Cycle

To illustrate the Memory Read machine cycle, we need to examine the execution of a 2-byte or a 3-byte instruction because in a 1-byte instruction the machine code is an opcode; therefore, the operation is always an Opcode Fetch. The execution of a 2-byte instruction is illustrated in the next example.

Example 3.3

Two machine codes—0011 1110 (3EH) and 0011 0010 (32H)—are stored in memory locations 2000H and 2001H, respectively, as shown below. The first machine code (3EH) represents the opcode to load a data byte in the accumulator, and the second code (32H) represents the data byte to be loaded in the accumulator. Illustrate the bus timings as these machine codes are executed. Calculate the time required to execute the Opcode Fetch and the Memory Read cycles and the entire instruction cycle if the clock frequency is 2 MHz.

Memory Location	Machine Code		Instruction
2000H	0 0 1 1 1 1 1 0	→ 3EH	MVI A,32H ;Load byte 32H
2001H	0 0 1 1 0 0 1 0	→ 32H	; in the accu-
			; mulator

Solution

This instruction consists of two bytes; the first is the opcode and the second is the data byte. The 8085 needs to read these bytes first from memory and thus requires at least two machine cycles. The first machine cycle is Opcode Fetch and the second machine cycle is Memory Read, as shown in Figure 3.10; this instruction requires seven T-states for these two machine cycles. The timings of the machine cycles are described in the following paragraphs.

1. The first machine cycle M_1 (Opcode Fetch) is identical in bus timings with the machine cycle illustrated in Example 3.1, except for the bus contents.

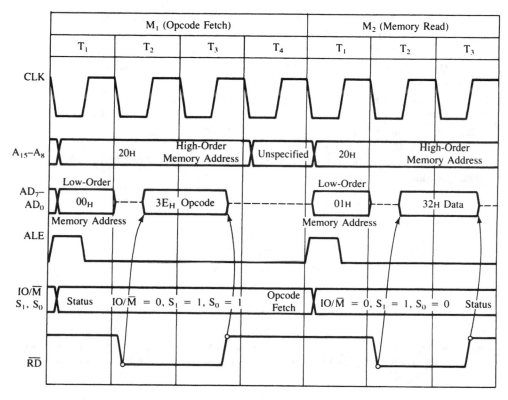

FIGURE 3.10
8085 Timing for Execution of the Instruction MVI A,32H

At T_1, the microprocessor identifies that it is an Opcode Fetch cycle by placing 011 on the status signals* ($IO/\overline{M} = 0$, $S_1 = 1$ and S_0 1). It places the memory address (2000H) from the program counter on the address bus, 20H on A_{15}–A_8, and 00H on AD_7–AD_0 and increments the program counter to 2001H to point to the next machine code. The ALE signal goes high during T_1, which is used to latch the low-order address 00H from the bus AD_7–AD_0. At T_2, the 8085 asserts the $\overline{RD}$ control signal, which enables the memory, and the memory places the byte 3EH from location 2000H on the data bus. Then the 8085 places the opcode in the instruction register and disables the $\overline{RD}$ signal. The fetch cycle is completed in state T_3. During T_4, the 8085 decodes the opcode and finds out that a second byte needs to be read. After the T_3 state, the contents of the bus A_{15}–A_8 are unknown, and the data bus AD_7–AD_0 goes into high impedance.

---

*The status signals S_1 and S_0 can be used to differentiate between various machine cycles. However, they are rarely needed; the necessary control signals can be generated by using $IO/\overline{M}$, $\overline{RD}$, and $\overline{WR}$.

2. After completion of the Opcode Fetch cycle, the 8085 places the address 2001H on the address bus and increments the program counter to the next address 2002H. The second machine cycle M_2 is identified as the Memory Read cycle ($IO/\overline{M} = 0$, $S_1 = 1$, and $S_0 = 0$) and the ALE is asserted. At T_2, the $\overline{RD}$ signal becomes active and enables the memory chip.

3. At the rising edge of T_2, the 8085 activates the data bus as an input bus, memory places the data byte 32H on the data bus, and the 8085 reads and stores the byte in the accumulator during T_3.

The execution times of the Memory Read machine cycle and the instruction cycle are calculated as follows:

- ☐ Clock frequency f = 2 MHz
- ☐ T-state = clock period (1/f) = 0.5 μs
- ☐ Execution time for Opcode Fetch: (4 T) × 0.5 = 2 μs
- ☐ Execution time for Memory Read: (3 T) × 0.5 = 1.5 μs
- ☐ Execution time for Instruction: (7 T) × 0.5 = 3.5 μs

3.24 How to Recognize Machine Cycles

In the last two examples, the number of bytes is the same as the number of machine cycles. However, there is no direct relationship between the number of bytes in an instruction and the number of machine cycles required to execute that instruction. This is illustrated in the following example.

Example 3.4

Explain the machine cycles of the following 3-byte instruction when it is executed.

Opcode	Operand	Bytes	Machine Cycles	T-States	Operation
STA	2065H	3	4	13	This instruction stores (writes) the contents of the accumulator in memory location 2065H

The machine codes are stored in memory locations 2010H, 2011H, and 2012H as follows: the 16-bit address of the operand must be entered in reverse order, the low-order byte first, followed by the high-order byte.

Memory Address	Machine Code	
2010	0011 0010 → 32H	Opcode
2011	0110 0101 → 65H	Low-order address
2012	0010 0000 → 20H	High-order address

This is a 3-byte instruction; however, it has four machine cycles with 13 T-states. The first operation in the execution of an instruction must be an Opcode Fetch. The 8085 requires three T-states for each subsequent operation; thus, nine T-states are required for the remaining machine cycles. Therefore, the Opcode Fetch in this instruction must be a four T-state machine cycle, the same as the one described in the previous example. The two machine cycles following the Opcode Fetch must be Memory Read machine cycles because the microprocessor must read all the machine codes (three bytes) before it can execute the instruction. Now let us examine what the instruction does. It stores (writes) the contents of the accumulator in memory location 2065H; therefore, the last machine cycle must be Memory Write. The execution steps are as follows:

Solution

1. In the first machine cycle, the 8085 places the address 2010H on the address bus and fetches the opcode 32H.
2. The second machine cycle is Memory Read. The processor places the address 2011H and gets the low-order byte 65H.
3. The third machine cycle is also Memory Read; the 8085 gets the high-order byte 20H from memory location 2012H.
4. The last machine cycle is Memory Write. The 8085 places the address 2065H on the address bus, identifies the operation as Memory Write (IO/$\overline{\text{M}}$ = 0, S_1 = 0, and S_0 = 1). It places the contents of the accumulator on the data bus AD_7–AD_0 and asserts the $\overline{\text{WR}}$ signal. During the last T-state, the contents of the data bus are placed in memory location 2065H.

3.25 Review of Important Concepts

1. In each instruction cycle, the first operation is always Opcode Fetch. This cycle can be of four to six T-states duration.
2. The Memory Read cycle requires three T-states and is in many ways similar to the Opcode Fetch cycle. Both use the same control signal ($\overline{\text{RD}}$) and read contents from memory. However, the Opcode Fetch reads opcodes and the Memory Read reads 8-bit data or address; these two machine cycles are differentiated by the status signals.
3. When the status signal IO/$\overline{\text{M}}$ is active low, the 8085 indicates that it is a memory-related operation, and the control signal $\overline{\text{RD}}$ suggests that it is a Read operation. Both signals are necessary to read from memory; the $\overline{\text{MEMR}}$ (Memory Read) control signal is generated by ANDing these two signals (see Section 3.14). The other status signals S_1 and S_0 are generally not needed in simple systems.
4. In the Memory Write cycle, the 8085 writes (stores) data in memory, using the control signal $\overline{\text{WR}}$ and the status signal IO/$\overline{\text{M}}$.
5. In the Memory Read cycle, the 8085 asserts the $\overline{\text{RD}}$ signal to enable memory, and then the addressed memory places data on the data bus; on the other hand, in the Memory Write cycle, the 8085 places the data byte on the data bus and then asserts the $\overline{\text{WR}}$ signal to write into the addressed memory.

6. The Memory Read and Write cycles consist of three T-states. The Memory Read and Write cycles will not be asserted simultaneously—the microprocessor cannot read and write at the same time.

3.3 MEMORY INTERFACING

Memory is an integral part of a microcomputer system, and in this chapter, our focus will be on how to interface a memory chip with the microprocessor. While executing a program, the microprocessor needs to access memory quite frequently to read instruction codes and data stored in memory; the interfacing circuit enables that access. Memory has certain signal requirements to write into and read from its registers. Similarly, the microprocessor initiates a set of signals when it wants to read from and write into memory. The interfacing process involves designing a circuit that will match the memory requirements with the microprocessor signals.

In the following sections, we will examine memory structure and its requirements and the 8085 Memory Read and Write machine cycles. Then we will derive the basic steps necessary to interface memory with the 8085. In the last chapter, we discussed a hypothetical memory chip and the concepts in addressing. In this chapter, we will illustrate memory interfacing, using memory chips such as 2732 EPROM and 6116 static R/W memory, and will discuss address decoding and memory addresses.

3.31 Memory Structure and Its Requirements

As discussed in Chapter 2, Read/Write memory (R/WM) is a group of registers to store binary information. Figure 3.11(a) shows a typical R/W memory chip; it has 2048 registers and each register can store eight bits indicated by eight input and eight output data lines.* The chip has 11 address lines A_{10}–A_0, one Chip Select ($\overline{CS}$), and two control lines: Read ($\overline{RD}$) to enable the output buffer and Write ($\overline{WR}$) to enable the input buffer. Figure 3.11(a) also shows the internal decoder to decode the address lines. Figure 3.11(b) shows the logic diagram of a typical EPROM (Erasable Programmable Read-Only Memory) with 4096 (4K) registers. It has 12 address lines A_{11}–A_0, one Chip Select ($\overline{CS}$), and one Read control signal. This chip must be programmed (written into) before it can be used as a read-only memory. Figure 3.11(b) also shows a quartz window on the chip that is used to expose the chip to ultraviolet rays for erasing the program. Once the chip is programmed, the window is covered with opaque tape to avoid accidental erasing. For interfacing the R/W memory, Figure 3.11(a), and the EPROM, Figure 3.11(b), the process is similar; the only difference is that the EPROM does not require the $\overline{WR}$ signal.

---

*In a typical memory chip, the input and output data lines are not shown separately, but are shown as a group of eight data lines.

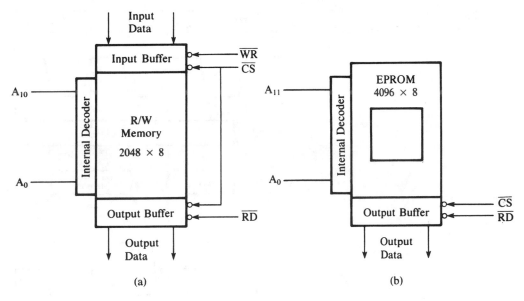

FIGURE 3.11
Typical Memory Chips: R/W Static Memory (a) and EPROM (b)

3.32   Basic Concepts in Memory Interfacing

The primary function of memory interfacing is that the microprocessor should be able to read from and write into a given register of a memory chip. Recall from Chapter 2 that to perform these operations, the microprocessor should

1. be able to select the chip.
2. identify the register.
3. enable the appropriate buffer.

Let us examine the timing diagram of the Memory Read operation (Figure 3.12) to understand how the 8085 can read from memory. Figure 3.12 is the M_2 cycle of Figure 3.10 except that the address bus is demultiplexed. We could also use the M_1 cycle to illustrate these interfacing concepts.

1. The 8085 places a 16-bit address on the address bus, and with this address only one register should be selected. For the memory chip in Figure 3.11(a), only 11 address lines are required to identify 2048 registers. Therefore, we can connect the low-order address lines A_{10}–A_0 of the 8085 address bus to the memory chip. The internal decoder of the memory chip will identify and select the register for the EPROM, Figure 3.11(a).

FIGURE 3.12
Timing of the Memory Read Cycle

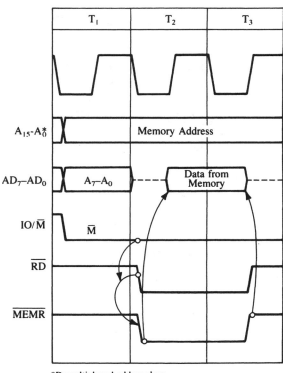

*Demultiplexed address bus

2. The remaining 8085 address lines (A_{15}–A_{11}) should be decoded to generate a Chip Select ($\overline{CS}$) signal unique to that combination of address logic (illustrated in Examples 3.3 and 3.4).

3. The 8085 provides two signals—IO/$\overline{M}$ and $\overline{RD}$—to indicate that it is a memory read operation. The IO/$\overline{M}$ and $\overline{RD}$ can be combined to generate the $\overline{MEMR}$ (Memory Read) control signal that can be used to enable the output buffer by connecting to the memory signal $\overline{RD}$.

4. Figure 3.12 also shows that memory places the data byte from the addressed register during T_2, and that is read by the microprocessor before the end of T_3.

To write into a register, the microprocessor performs similar steps as it reads from a register. Figure 3.13 shows the Memory Write cycle. In the Write operation, the 8085 places the address and data and asserts the IO/$\overline{M}$ signal. After allowing sufficient time for data to become stable, it asserts the Write ($\overline{WR}$) signal. The IO/$\overline{M}$ and $\overline{WR}$ signals can be combined to generate the $\overline{MEMW}$ control signal that enables the input buffer of the memory chip and stores the byte in the selected memory register.

To interface memory with the microprocessor, we can summarize the above steps as follows:

FIGURE 3.13
Timing of the Memory Write Cycle

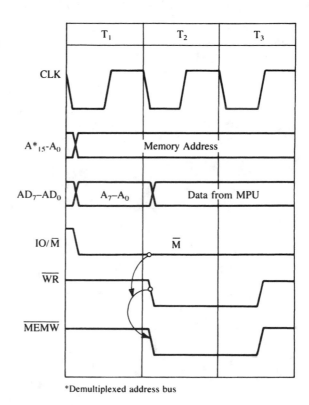

*Demultiplexed address bus

1. Connect the required address lines of the address bus to the address lines of the memory chip.
2. Decode the remaining address lines of the address bus to generate the Chip Select signal, as discussed in the next section (3.33), and connect the signal to select the chip.
3. Generate control signals $\overline{MEMR}$ and $\overline{MEMW}$ by combining $\overline{RD}$ and $\overline{WR}$ signals with $IO/\overline{M}$, and use them to enable appropriate buffers.

3.33 Address Decoding

The process of address decoding should result in identifying a register for a given address. We should be able to generate a unique pulse for a given address. For example, in Figure 3.11(b), 12 address lines (A_{11}–A_0) are connected to the memory chip, and the remaining four address lines (A_{15}–A_{12}) of the 8085 microprocessor must be decoded. Figure 3.14 shows two methods of decoding these lines: one by using a NAND gate and the other by using a 3-to-8 decoder. The output of the NAND goes active and selects the chip only when all address lines A_{15}–A_{12} are at logic 1. We can obtain the same result by using O_7 of the 3-to-8 decoder, which is capable of decoding eight different input addresses. In the decoder circuit, three input lines can have eight different logic combinations from 000 to 111; each input combination can be identified by the corresponding output line if Enable

FIGURE 3.14
Address Decoding Using NAND
Gate (a) and 3-to-8 Decoder (b)

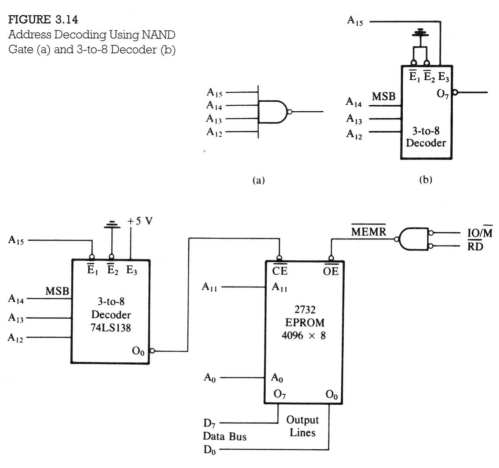

(a)   (b)

FIGURE 3.15
Interfacing the 2732 EPROM

lines are active. In this circuit, the Enable lines $\overline{E}_1$ and $\overline{E}_2$ are enabled by grounding, and A_{15} must be at logic 1 to enable E_3. We will use this address decoding scheme to interface a 4K EPROM and a 2K R/W memory as illustrated in the next two examples.

3.34 Interfacing Circuit

Figure 3.15 shows an interfacing circuit using a 3-to-8 decoder to interface the 2732 EPROM memory chip. It is assumed here that the chip has already been programmed, and we will analyze the interfacing circuit in terms of the same three steps outlined previously:

Step 1: The 8085 address lines $A_{11}-A_0$ are connected to pins $A_{11}-A_0$ of the memory chip to address 4096 registers.

Step 2: The decoder is used to decode four address lines A_{15}–A_{12}. The output O_0 of the decoder is connected to Chip Enable ($\overline{CE}$). The $\overline{CE}$ is asserted only when the address on A_{15}–A_{12} is 0000; A_{15} (low) enables the decoder and the input 000 asserts the output O_0.

Step 3: For this EPROM, we need one control signal: Memory Read ($\overline{MEMR}$), active low. The $\overline{MEMR}$ is connected to $\overline{OE}$ to enable the output buffer; $\overline{OE}$ is the same as $\overline{RD}$ in Figure 3.11.

3.35 Address Decoding and Memory Addresses

We can obtain the address range of this memory chip by analyzing the possible logic levels on the 16 address lines. The logic levels on the address lines A_{15}–A_{12} must be 0000 to assert the Chip Enable, and the address lines A_{11}–A_0 can assume any combinations from all 0s to all 1s. Therefore, the memory address of this chip ranges from 0000H to 0FFFH, as shown below.

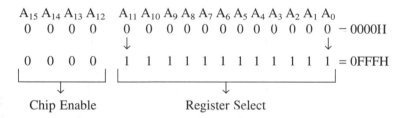

We can verify the memory address range in terms of our analogy of page and line numbers, as discussed in Chapter 2, Section 2.22. The chip's 4096 bytes of memory can be viewed as 16 pages with 256 lines each. The high-order Hex digits range from 00 to 0F, indicating 16 pages—0000H to 00FFH and 0100H to 01FFH, for example.

Now, to examine how an address is decoded and how the microprocessor reads from this memory, let us assume that the 8085 places the address 0FFFH on the address bus. The address 0000 (0H) goes to the decoder, and the output line O_0 of the decoder selects the chip. The remaining address FFFH goes on the address lines of the chip, and the internal decoder of the chip decodes the address and selects the register FFFH. Thus, the address 0FFFH selects the register as shown in Figure 3.16. When the 8085 asserts the RD signal, the output buffer is enabled and the contents of the register 0FFFH are placed on the data bus for the processor to read.

Analyze the interfacing circuit in Figure 3.17 and find its memory address range.

Example 3.5

Figure 3.17 shows the interfacing of the 6116 memory chip with 2048 (2K) registers. The memory chip requires 11 address lines (A_{10}–A_0) to decode 2048 registers. The remaining address lines A_{15}–A_{11} are connected to the decoder. However, in this circuit, the decoder is enabled by the IO/M signal in addition to the address lines A_{15} and A_{14}, and the $\overline{RD}$ and

Solution

FIGURE 3.16
Address Decoding and Reading
from Memory

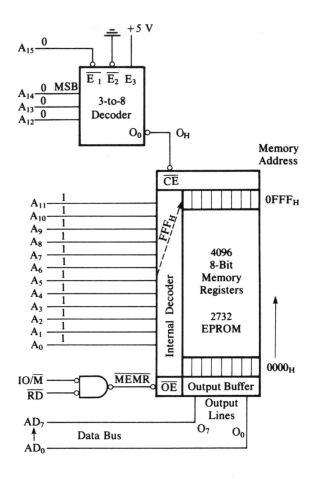

FIGURE 3.17
Interfacing R/W Memory

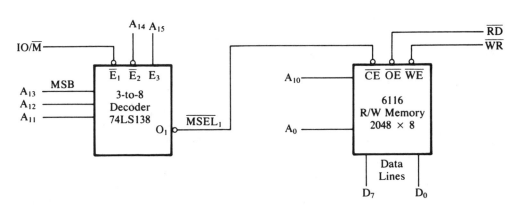

$\overline{\text{WR}}$ signals of the 8085 are directly connected to the memory chip. The signals $\overline{\text{MEMR}}$ and $\overline{\text{MEMW}}$ need not be generated separately; thus, this technique saves two gates. The memory address of this chip ranges from 8800H to 8FFFH, as shown below.

$$A_{15}\ A_{14}\ A_{13}\ A_{12}\ A_{11} \qquad A_{10}\ A_9\ A_8\ A_7\ A_6\ A_5\ A_4\ A_3\ A_2\ A_1\ A_0$$

$$1\quad 0\quad 0\quad 0\quad 1 \qquad 0\ \ 0\ \ 0\ \ 0\ \ 0\ \ 0\ \ 0\ \ 0\ \ 0\ \ 0\ \ 0 = 8800H$$

$$\downarrow \qquad\qquad\qquad\qquad\qquad \downarrow$$

$$\underline{\qquad\qquad\qquad\qquad}|1 \quad 1\ \ 1\ \ 1\ \ 1\ \ 1\ \ 1\ \ 1\ \ 1\ \ 1\ \ 1\ \ 1 = 8FFFH$$

$$\downarrow$$

$$\text{MSEL}_1$$

The output line O_1 of the decoder is connected to $\overline{\text{CE}}$ of the memory chip, and it is identified as MSEL_1 because it is asserted only when $\text{IO}/\overline{\text{M}}$ is low.

The address lines A_{15} (high) and A_{14} (low) enable the decoder, and the input lines to the decoder—A_{13}, A_{12}, and A_{11} (001)—activate the output O_1 ($\overline{\text{MSEL}_1}$) to select the memory chip. The address lines A_{10}–A_0 can assume any logic combination between all 0s to all 1s, as shown above.

INTERFACING THE 8155 MEMORY SEGMENT 3.4

The SDK-85 is a single-board microcomputer designed by Intel and widely used in college laboratories. The system is designed using the 8085 microprocessor and specially compatible devices, such as the 8155/8156.

The 8155/8156 includes multiple devices on the same chip. The 8155 has 256 bytes of R/W memory, two programmable I/O ports, and a timer. The 8156 is identical to the 8155, except that its Chip Enable (CE) signal is active high. The programmable I/O ports of this device is discussed in Chapter 14. The memory section of this chip and its memory addresses in the SDK-85 system will now be discussed.

3.41 Interfacing the 8155 Memory Section

Figure 3.18(a) shows the block diagram of the 8155 memory section. It has eight address lines, one $\overline{\text{CE}}$ (Chip Enable) line, and five lines compatible with the control and status signals of the 8085: $\text{IO}/\overline{\text{M}}$, ALE, $\overline{\text{RD}}$, $\overline{\text{WR}}$, and RESET. These control and status lines are not found in the general-purpose memory devices shown in the previous section. These lines eliminate the need for external demultiplexing of the bus AD_7–AD_0 and for generating separate control signals for memory and I/O.

Figure 3.18(b) also shows the internal structure of the 8155 memory section. The memory section includes 256×8 memory locations and an internal latch to demultiplex the bus lines AD_7–AD_0. The memory section also requires a Chip Enable ($\overline{\text{CE}}$) signal and the Memory Write ($\overline{\text{MEMW}}$) and Memory Read ($\overline{\text{MEMR}}$) control signals, generated internally by combining the $\text{IO}/\overline{\text{M}}$, $\overline{\text{WR}}$, and $\overline{\text{RD}}$ signals.

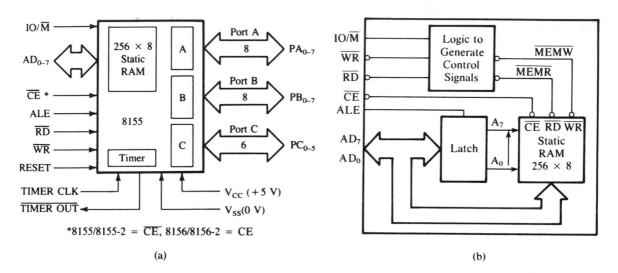

FIGURE 3.18

The 8155 Memory Section: The Block Diagram (a) and the Internal Structure (b)

SOURCE: (a) Intel Corporation, *Embedded Microprocessors* (Santa Clara, Calif.: Author, 1994), pp. 1-31.

FIGURE 3.19

Interfacing the 8155 Memory
Schematic from the SDK-85 System

SOURCE: Intel Corporation, *SDK—85 User's Manual* (Santa Clara, Calif.: Author, 1978), Appendix B.

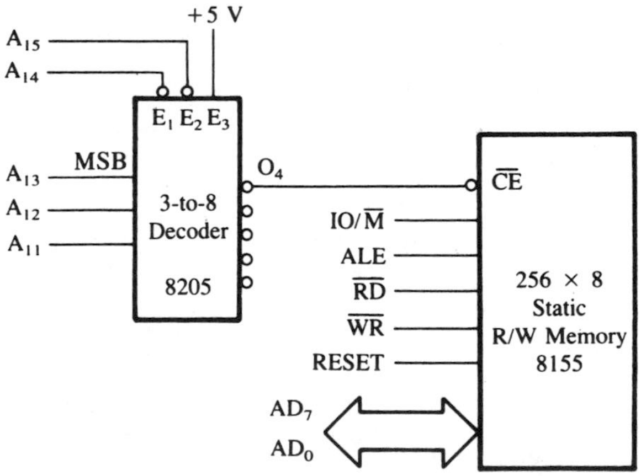

Figure 3.19 shows a schematic of the SDK-85 system of interfacing the 8155 memory section with the 8085. The 8205, a 3-to-8 decoder (identical to the 74LS138 decoder), decodes the address lines A_{15}–A_{11}, and the output line O_4 of the decoder enables the memory chip. The control and the status signals from the 8085 are connected directly to the respective signals on the memory chip. Similarly, the bus lines AD_7–AD_0 are also connected directly to the memory chip to address any one of the 256 memory locations.

Example
3.6

Solution

Explain the decoding logic and the memory address range of the 8155 shown in Figure 3.19.

The interfacing logic shows the 3-to-8 decoder; its output line 4 (O_4) is used to select the 8155. The address lines A_{11} to A_{13} are connected as input to the decoder, and the lines A_{15} and A_{14} are used as active low Enable lines. The third Enable line (active high) is permanently enabled by tying it to +5 V. The address lines A_{10}, A_9, and A_8 are not connected; thus, they are left as don't care lines. The output line O_4 of the decoder goes low when the address lines have the following address:

$$A_{15}\ A_{14}\ A_{13}\ A_{12}\ A_{11}\ A_{10}\ A_9\ A_8$$
$$\ \ 0\ \ \ \ \ 0\ \ \ \ \ 1\ \ \ \ \ 0\ \ \ \ \ 0\ \ \ \ \ 0\ \ \ \ 0\ \ 0\ = 20H \text{ (assuming the don't care lines are at logic 0)}$$

The address lines AD_7–AD_0 can assume any combination of logic levels from all 0s to all 1s. Thus, the memory addresses of the 8155 memory will range from 2000H to 20FFH. In reality, the memory section of this 8155 uses the memory space from 2000H to 27FFH.

In Figure 3.19, the address lines A_{10}, A_9, and A_8 are not connected; thus, they are don't care lines capable of assuming any logic state 0 or 1. Three don't care address lines can be assumed to have any one of the eight combinations from 000 to 111. Thus each combination can generate one set of complete addresses. The address range given by assuming all don't care lines at logic 0 is, by convention, specified as the memory address range of the system or the primary address; the remaining address ranges are known as either foldback memory or mirror memory. In this example, the primary address range is 2000H to 20FFH and the foldback memory range is 2100H to 27FFH as follows.

If A_{10} and A_9 are assumed to be 0 and A_8 is assumed to be 1, the memory address range will be 2100H to 21FFH, as shown below.

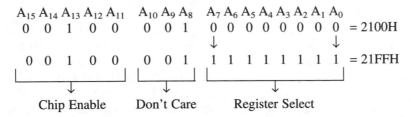

If the the address lines A_{10}, A_9, and A_8 are assumed to be 0 1 0, the address range will be 2200H to 22FFH. Thus we can obtain a total of seven additional address ranges called foldback memory. In reality, these are the same memory registers as 2000H to 20FFH. Attempting to store an instruction in location 2100H (or 2200H or 2700H) is the same as entering the instruction in location 2000H. This results in assigning eight different addresses to the same memory register. To find each address range, we should assume only one combination of the don't care lines at a time; this is particularly important when the don't care lines are not in a particular sequence. (See Problem 34.)

3.42 Absolute vs. Partial Decoding and Multiple Address Ranges

In Figures 3.15 and 3.17, all the high-order address lines were decoded to select the memory chip, and the memory chip is selected only for the specified logic levels on these high-order address lines; no other logic levels can select the chip. This is called absolute decoding, a desirable design practice commonly used in large memory systems.

However, in Figure 3.19, three address lines out of eight (A_{15}–A_8) were not decoded, resulting in multiple addresses; this is called partial decoding. The 8155 chip has 256 memory registers, but it occupies the memory space of 2048 locations, eight times the space of its size. In a small system where the total memory space is not needed, such a technique of partial decoding can be used. The primary advantage of such a technique is in cost saving. In Figure 3.19, we can use the same decoder for multiple size memory chips. This decoder is designed to decode 2K memory chips without generating any multiple address ranges. (See Problem 27.)

3.5 TESTING AND TROUBLESHOOTING MEMORY INTERFACING CIRCUITS

In the previous section, we analyzed the interfacing circuit of the 8155 (Figure 3.19) and determined the address range was 2000H to 20FFH. Assuming that we connected all the wires to an existing working system, we need to test whether we have memory in the given address range, and if we are unable to access memory from 2000H to 20FFH, we need to troubleshoot the interfacing circuit and correct the problem.

3.51 Testing

Testing a memory chip in an existing working system is fairly simple. If we have an input device such as a keyboard, we can load a byte at the memory address 2000H and verify that the byte is stored in that location. Similarly, we can check a few more locations including the last memory address, 20FFH. We can also test the multiple addresses (fold-back memory). If we access locations such as 2100H, 2200H, and 2700H, we will find the same byte in all these locations. If we attempt to load a byte in location 2800H and the byte is not accepted, this is an indication that the memory register does not exist at location 2800H. Now if we find that we are not able to load a byte in location 2000H, we need to troubleshoot the circuit.

3.52 Troubleshooting Microprocessor-Based Systems

To troubleshoot the circuit in Figure 3.19, the first obvious step is a visual inspection. Check the wiring and pin connections. After this preliminary check, most traditional methods used in checking analog circuits (such as an amplifier) are ineffective because the logic levels on the buses are dynamic; they constantly change depending upon the operation being performed at a given instant by the microprocessor. In troubleshooting analog circuits, a commonly used technique is signal injection, whereby a known signal is injected at the input and the output signal is verified against the expected outcome. To use this concept, we need to generate a constant and identifiable signal and check various points in relation to that signal. We can generate such a signal by asking the processor to execute a continuous loop, called diagnostic routine, as discussed below.

3.53 Diagnostic Routine to Generate a Steady Signal

To generate a steady signal, we need to write a continuous loop as shown below.

```
START: MVI A, 55H     ;Load a byte in the accumulator
       STA 2000H      ;Store 55H in memory location 2000H
       JMP START      ;Jump back to beginning and repeat
```

This routine has three instructions. The first instruction loads 55H in the accumulator, and the second instruction stores the byte in location 2000H. The byte 55H has no particular significance. The third instruction is an unconditional Jump instruction that takes the program execution back to the beginning. These instructions are executed continuously. Now we need to examine the machine cycles of these instructions to find an identifiable signal that is repeated at a certain interval. We can analyze the machine cycles in the loop as follows (it will be helpful to have read Chapter 7 to understand the diagnostic routine).

This loop has 30 T-states and nine operations. To execute the loop once, the microprocessor asserts the $\overline{RD}$ signal eight times (the Opcode Fetch is also a Read operation) and the $\overline{WR}$ signal once. Assuming the system clock frequency is 3 MHz, the loop is executed in 9.9 μs, and the $\overline{WR}$ signal is repeated every 9.9 μs that can be observed on a scope. If we sync the scope on the $\overline{WR}$ pulse from the 8085, we can check the output of the decoder and memory control signals $\overline{WR}$ and $\overline{RD}$; some of these signals are shown in Figure 3.20.

When the 8085 asserts the $\overline{WR}$ signal, the address on A_{15}–A_0 must be 2000H, and the output of the decoder must be asserted low. If it is high, it indicates that the address lines A_{15}–A_{11} are improperly connected or that the decoder chip is faulty.

If the decoder output is low, it confirms that the decoding circuit is functioning properly. Now if we check the entire data bus in relation to the $\overline{WR}$ signal, one line at a time, we must read the data 55H. If we check the $\overline{RD}$ signal, it must be high when the $\overline{WR}$ is asserted, and we will observe eight $\overline{RD}$ signals between every two $\overline{WR}$ signals, as shown in Figure 3.20.

Now we can use the $\overline{WR}$ ($\overline{MEMW}$) signal shown in Figure 3.20 as the reference signal. With the second probe, we check the circuit in Figure 3.19 at various points. The potential symptoms and probable conclusions are as follows.

1. **Symptom:** The data byte read from memory is different from the one that is stored (55H).
   The data lines D_7–D_0 are incorrectly connected.
2. **Symptom:** The signals at O_4 of the decoder and $\overline{CE}$ of the memory are low (similar to the $\overline{WR}$ signal).

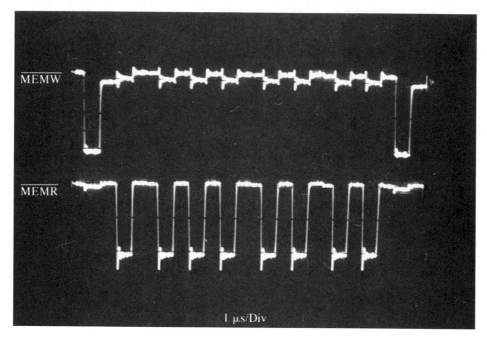

FIGURE 3.20
Timing Signals of Diagnostic Routine

The address lines A_{15}–A_{11} are connected properly and the decoder chip is functioning correctly. Check the $\overline{IO/M}$, ALE, and $\overline{RD}$ signals on the memory chip (assuming that we are checking the $\overline{WR}$ signal at the memory chip).

3. Symptom: The signals at O_4 of the decoder and $\overline{CE}$ of the memory are high.
   Check other outputs of the decoder. If one of the outputs is low, it is an indication that the input address lines of the decoder (A_{13}–A_{10}) are incorrectly connected. If none of the outputs is low, it is an indication that either the address lines A_{15} and A_{14} are not properly connected or the decoder chip is functioning incorrectly.

4. Symptom: The signals at O_4 of the decoder and $\overline{CE}$ of the memory are low similar to the ground signal.
   Either the V_{CC} to the decoder is not connected or the chip is functioning improperly.

3.6   HOW DOES AN 8085-BASED SINGLE-BOARD MICROCOMPUTER WORK?

Hardware is the skeleton of the computer; software is its life. The software (programs) makes the computer live; without it, the hardware is a dead piece of semiconductor material. Single-board microcomputers, such as the one shown in Figure 3.9, the PRIMER

(Appendix B), or the SDK-85, have a program called Key Monitor or Key Executive permanently stored in memory. This program is stored either in EPROM or in ROM, beginning at the memory location 0000H.

When the power is turned on, the monitor program comes alive. Initially, the program counter has a random address. When the system is reset, the program counter in the 8085 is cleared, and it holds the address 0000H. The PRIMER (Appendix B) system includes a "power on" reset circuit, which resets the system and clears the program counter when the system is turned on. The MPU places the address 0000H on the address bus. The instruction code stored in location 0000H is fetched and executed, and the execution continues according to the instructions in the monitor program. The primary functions of the monitor program are as follows:

1. Reading the Hex keyboard and checking for a key closure. Continuing to check the keyboard until a key is pressed.
2. Displaying the Hex equivalent of the key pressed at the output port, such as the seven-segment LEDs.
3. Identifying the key pressed and storing its binary equivalent in memory, if necessary.
4. Transferring the program execution sequence to the user program when the Execute key is pressed.

The programmer enters a program in R/W memory in sequential memory locations by using the data keys (0 to F) and the function key called Enter. When the system is reset, the program counter is cleared, and the monitor program begins to check a key closure again. By using the keyboard, the programmer enters the first memory address where the user program is stored in R/W memory and directs the MPU to execute the program by pressing the Run key. The MPU fetches, decodes, and executes one instruction code at a time and continues to do so until it fetches the Halt instruction.

The key monitor program is a critical element in entering, storing, and executing a program. Until the Execute key is pushed, the monitor program in the EPROM (or ROM) directs all the operations of the MPU. After the Execute key is pushed, the user program directs the MPU to perform the functions written in the program.

SUMMARY

This chapter described the architecture of the 8085 microprocessor and illustrated the techniques for demultiplexing the bus $AD_7–AD_0$ and generating the control signals. The bus timings of three operations—Opcode Fetch, Memory Read, and Memory Write—were examined and were used in discussing the basic concepts in memory interfacing. Several examples of memory interfacing were illustrated, and the concepts of memory addressing, absolute and partial decoding, and multiple addresses were discussed. The important concepts are summarized below.

☐ The 8085 microprocessor signals can be classified in six groups: address bus, data bus, control and status signals, externally initiated signals and their acknowledgment, power and frequency, and serial I/O signals.

☐ The data bus and the low-order address bus are multiplexed; they can be demultiplexed by using the ALE (Address Latch Enable) signal and a latch.

☐ The IO/M is a status signal—when it is high, it indicates an I/O operation; when it is low, it indicates a memory operation.

☐ The $\overline{RD}$ and $\overline{WR}$ are control signals; the $\overline{RD}$ is asserted to read from an external device, and the $\overline{WR}$ is asserted to write into an external device (memory or I/O).

☐ The $\overline{RD}$ and $\overline{WR}$ signals are logically ANDed with the IO/M signal to generate four active low control signals: $\overline{MEMR}$, $\overline{MEMW}$, $\overline{IOR}$, and $\overline{IOW}$.

☐ Each instruction of the 8085 microprocessor can be divided into a few basic operations called machine cycles, and each machine cycle can be divided into T-states.

☐ The frequently used machine cycles are Opcode Fetch, Memory Read and Write, and I/O Read and Write.

☐ When the 8085 performs any of the operations, it asserts the appropriate control signal and status signal.

☐ Most Opcode Fetch operations consist of four T-states, and the subsequent Memory Read or Memory Write cycles require three T-states. Some Opcode Fetch operations require six T-states.

☐ The Opcode Fetch and the Memory Read are operationally similar; the 8085 reads from memory in both machine cycles. However, the 8085 reads opcode during the Opcode Fetch cycle, and it reads 8-bit data during the Memory Read cycle. In the Memory Write cycle, the processor writes data into memory.

☐ The 8085 performs three basic steps in any of these machine cycles: It places the address on the address bus, sends appropriate control signals, and transfers data via data bus.

☐ To read from memory, the address of the register (to be read from) should be placed on the address lines and the $\overline{RD}$ signal must be asserted low to enable the output buffer.

☐ To write into memory, the address of the register (to be written into) should be placed on the address lines, a data byte should be placed on the data lines, and the $\overline{WR}$ signal must be asserted low to enable the input buffer.

☐ To interface a memory chip with the 8085, the necessary low-order address lines of the 8085 address bus are connected to the address lines of the memory chip. The high-order address lines are decoded to generate $\overline{CS}$ signals to enable the chip.

☐ In the absolute decoding technique, all the address lines that are not used for the memory chip to identify a memory register must be decoded; thus, the Chip Select can be asserted by only one address. In the partial decoding technique, some address lines are left don't care. This technique reduces hardware, but generates multiple addresses resulting in foldback memory space.

QUESTIONS AND PROBLEMS

1. Explain the functions of the ALE and IO/$\overline{M}$ signals of the 8085 microprocessor.
2. Explain the need to demultiplex the bus AD_7–AD_0.

3. Figure 3.21 shows the 74LS138 (3-to-8) decoder with the three input signals: IO/$\overline{\text{M}}$, $\overline{\text{RD}}$, and $\overline{\text{WR}}$ from the 8085 microprocessor. Specify and name the valid output signals.

4. Explain why four output signals are invalid or meaningless in Figure 3.21.

5. Identify appropriate control signals that are generated at the output of the 2-to-4 decoder in Figure 3.22.

6. In Figure 3.4, if the 8085 places the address 20H on A_{15}–A_8 and 05H on AD_7–AD_0 and the ALE is high, specify the output of the latch 74LS373.

7. At T_2 (refer to Figure 3.3), the data byte 4FH is placed on AD_7–AD_0. Specify the output of the latch in Figure 3.4. Explain your answer.

8. Specify the crystal frequency required for an 8085 system to operate at 1.1 MHz.

9. List the sequence of events that occurs when the 8085 MPU reads from memory.

10. If the 8085 adds 87H and 79H, specify the contents of the accumulator and the status of the S, Z, and CY flags.

11. If the 8085 has fetched the machine code located at the memory location 205FH, specify the contents of the program counter.

12. If the clock frequency is 5 MHz, how much time is required to execute an instruction of 18 T-states?

13. Assume that memory location 2075H has a data byte 47H. Specify the contents of the address bus A_{15}–AD_8 and the multiplexed bus AD_7–AD_0 when the MPU asserts the $\overline{\text{RD}}$ signal.

14. In the Opcode Fetch cycle, what are the control and status signals asserted by the 8085 to enable the memory buffer?

15. The instruction MOV B,M copies the contents of the memory location in register B. It is a 1-byte instruction with two machine cycles and seven T-states. Identify the second machine cycle and its control signal.

16. The instruction LDA 2050H copies the contents of the memory location 2050H into the accumulator. It is a 3-byte instruction with four machine cycles and 13 T-states. Identify the fourth machine cycle and its control signal.

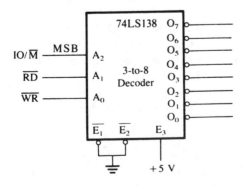

FIGURE 3.21

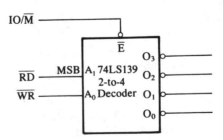

FIGURE 3.22

17. In Question 16, identify the contents of the demultiplexed address bus A_{15}–A_0 and the data bus in the fourth machine cycle when the control signal is asserted.

18. Identify the machine cycles in the following instructions. (You should be able to identify the machine cycles even if you are not familiar with some of the instructions.)

SUB B   ; 1-byte, 4 T-states
            ; Subtract the contents of register B from the accumulator
ADI 47H   ; 2-byte, 7 (4, 3) T-states
            ; Add 47H to the contents of the accumulator
STA 2050H ; 3-byte, 13 (4, 3, 3, 3) T-states
            ; Copies the contents of the accumulator into memory
            ; location 2050H
PUSH B   ; 1-byte, 12 (6, 3, 3) T-states
            ; Copies the contents of the BC register into two stack memory
            ; locations

19. In Figure 3.15, connect the output line O_6 of the decoder to the $\overline{CE}$ of the memory chip instead of O_0, and identify the memory map.

20. In Figure 3.15, connect A_{15} to the active high enable signal E_3 of the decoder, and ground E_1. Identify the memory map of the chip.

21. Identify the actual gate you would use to generate the $\overline{MEMR}$ signal in Figure 3.15.

22. Modify the schematic in Figure 3.15 to eliminate the negative NAND gate and obtain the same memory address range without adding other components.

23. In Figure 3.17, exchange A_{15} and A_{13} and identify the memory map.

24. In Figure 3.17, if we use all the output lines (O_7–O_0) of the decoder to select eight memory chips of the same size as the 6116, what is the total range of the memory map?

25. In the SDK-85 system, the specified map of the 8155 memory is 2000H to 20FFH (see Figure 3.19). If you enter a data byte at the location 2100H, will the system accept the data byte? If it accepts it, where will it store the data byte? Explain your answer.

26. In Figure 3.19, specify the memory address range if output line O_1 of the decoder 8205 is connected to the $\overline{CE}$ signal. Specify the range of the foldback memory.

27. In Figure 3.19 specify the memory address range if output line O_7 of the decoder 8205 is connected to the $\overline{CE}$ signal of a 2K (2048) memory chip.

28. By examining the range of the foldback memory in Figure 3.19, specify the relationship between the range of foldback memory and the number of don't care lines.

29. In Figure 3.23, specify the memory addresses of ROM1, $\overline{ROM2}$, and R/WM1.

30. In Figure 3.23, eliminate the second decoder and connect $\overline{CS_4}$ to $\overline{CE}$ of the R/WM1, and identify its memory map and foldback space.

31. In Figure 3.24, identify the address range of the memory chip.

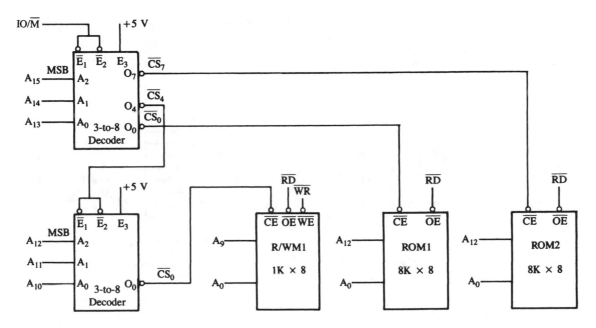

FIGURE 3.23

32. In Figure 3.24, connect Y_1 to $\overline{CE}$ of the memory chip in place of Y_0, and identify the address range of the memory chip.
33. In Figure 3.24, replace the 27128 (16K) memory chip with the 2764 (8K) memory chip. Identify the primary address range and the mirror (foldback) address range of the memory chip for the given decoding circuit.
34. In Question 33, the address line A_{13} was at a don't care logic state. Replace the address line A_{15} by the address line A_{13}, leave A_{15} as don't care, and identify the mirror address range.
35. Refer to the memory schematic of the PRIMER in Appendix B. Identify the address range of the SRAM (U3) if the output Y_1 of the decoder 74HC139 (U9A) is asserted.
36. In the PRIMER schematic, the address line A_{14} is connected to the memory chip as well as to input A of the 2-to-4 decoder. Find the total address range of the memory chip. Explain how this decoding technique enables the designer to use either a 16K or a 32K memory chip.
37. In Figure 3.20, identify the $\overline{MEMR}$ signals of the Opcode Fetch machine cycles.
38. In Figure 3.20, identify the machine cycle and the Hex code read by the processor when it asserts the last $\overline{MEMR}$ signal.

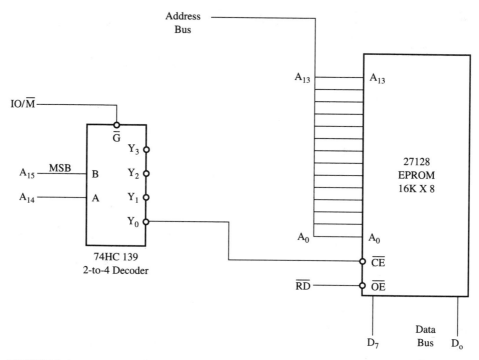

FIGURE 3.24
27128 (16K) EPROM Interfacing Circuit (similar to the circuit in the PRIMER-Appendix B)

Interfacing I/O Devices

The I/O devices, such as keyboards and displays, are the ears and eyes of the MPUs; they are the communication channels to the "outside world." Data can enter (or exit) in groups of eight bits using the entire data bus; this is called the parallel I/O mode. The other method is the serial I/O, whereby one bit is transferred using one data line; typical examples include peripherals such as the CRT terminal and cassette tape. In this chapter, we will focus on interfacing I/O devices in the parallel mode and will discuss the serial mode in Chapter 16.

In the last chapter, we discussed memory interfacing. The 8085 microprocessor uses a 16-bit address bus for identifying and accessing memory registers. This results in a numbering scheme ranging from 0000H to FFFFH, also known as memory space. Similarly, the microprocessor needs to identify I/O devices with a binary number. These I/O devices can be interfaced using addresses (binary numbers) from the memory space; this is called memory-mapped I/O. Another option is to have a separate numbering (addressing) scheme for I/O devices. The 8085 microprocessor has a separate 8-bit addressing scheme (I/O space) for I/O devices; this is called peripheral-mapped I/O (or I/O-mapped I/O), and the I/O space ranges from 00H to FFH. In the 8085-based systems, I/O devices can be inter-

faced using both techniques: peripheral-mapped I/O and memory-mapped I/O. In peripheral-mapped I/O, a device is identified with an 8-bit address and enabled by I/O-related control signals. On the other hand, in memory-mapped I/O, a device is identified with a 16-bit address and enabled by memory-related control signals. The process of data transfer in both is identical. Each device is assigned a binary address, called a device address or port number, through its interfacing circuit. When the microprocessor executes a data transfer instruction for an I/O device, it places the appropriate address on the address bus, sends the control signals, enables the interfacing device, and transfers data. The interfacing device is like a gate for data bits, which is opened by the MPU whenever it intends to transfer data. In peripheral-mapped I/O, data bytes are transferred by using IN/OUT instructions, and in memory-mapped I/O, data bytes are transferred by using memory-related (LDA, STA, etc.) data transfer instructions.

To grasp the essence of interfacing techniques, first we will examine the machine cycles of I/O instructions and find out the timings when I/O data are arriving on the data bus, and then we will latch (or catch) that information. We will derive the basic concepts in interfacing I/O devices from the machine cycles. Then we will illustrate these concepts by interfacing LEDs as an output device and switches as an input device. Specifically, the chapter deals with how the 8085 selects an I/O device, what

hardware chips are necessary, what software instructions are used, and how data are transferred.

OBJECTIVES

☐ Illustrate the 8085 bus contents and control signals when OUT and IN instructions are executed.
☐ Recognize the device (port) address of a peripheral-mapped I/O by analyzing the associated logic circuit.
☐ Illustrate the 8085 bus contents and control signals when memory-related instructions (LDA, STA, etc.) are executed.
☐ Recognize the device (port) address of a memory-mapped I/O by analyzing the associated logic circuit.
☐ Explain the differences between the peripheral-mapped and memory-mapped I/O techniques.
☐ Interface an I/O device to a microcomputer for a specified device address by using logic gates and MSI chips, such as decoders, latches, and buffers.

4.1   BASIC INTERFACING CONCEPTS

The approach to designing an interfacing circuit for an I/O device is determined primarily by the instructions to be used for data transfer. An I/O device can be interfaced with the 8085 microprocessor either as a peripheral I/O or as a memory-mapped I/O. In the peripheral I/O, the instructions IN/OUT are used for data transfer, and the device is identified by an 8-bit address. In the memory-mapped I/O, memory-related instructions are used for data transfer, and the device is identified by a 16-bit address. However, the basic concepts in interfacing I/O devices are similar in both methods. Peripheral I/O is described in the following section, and memory-mapped I/O is described in Section 4.4.

4.11   Peripheral I/O Instructions

The 8085 microprocessor has two instructions for data transfer between the processor and the I/O device: IN and OUT. The instruction IN (Code DB) inputs data from an input device (such as a keyboard) into the accumulator, and the instruction OUT (Code D3) sends the contents of the accumulator to an output device such as an LED display. These are 2-byte instructions, with the second byte specifying the address or the port number of an I/O device. For example, the OUT instruction is described as follows.

Opcode	Operand	Description
OUT	8-bit Port Address:	This is a two-byte instruction with the hexadecimal opcode D3, and the second byte is the port address of an output device. This instruction transfers (copies) data from the accumulator to the output device.

Typically, to display the contents of the accumulator at an output device (such as LEDs) with the address, for example, 01H, the instruction will be written and stored in memory as follows:

Memory Address	Machine Code	Mnemonics		Memory Contents	
2050	D3	OUT 01H	; 2050	$\boxed{1\ 1\ 0\ 1\ 0\ 0\ 1\ 1}$	= D3H
2051	01		; 2051	$\boxed{0\ 0\ 0\ 0\ 0\ 0\ 0\ 1}$	= 01H

(Note: The memory locations 2050H and 2051H are chosen here arbitrarily for the illustration.)

If the output port with the address 01H is designed as an LED display, the instruction OUT will display the contents of the accumulator at the port. The second byte of this OUT instruction can be any of the 256 combinations of eight bits, from 00H to FFH. Therefore, the 8085 can communicate with 256 different output ports with device addresses ranging from 00H to FFH. Similarly, the instruction IN can be used to accept data from 256 different input ports. Now the question remains: How does one assign a device address or a port number to an I/O device from among 256 combinations? The decision is arbitrary and somewhat dependent on available logic chips. To understand a device address, it is necessary to examine how the microprocessor executes IN/OUT instructions.

4.12 I/O Execution

The execution of I/O instructions can best be illustrated using the example of the OUT instruction given in the previous section (4.11). The 8085 executes the OUT instruction in three machine cycles, and it takes ten T-states (clock periods) to complete the execution.

OUT INSTRUCTION (8085)

In the first machine cycle, M_1 (Opcode Fetch, Figure 4.1), the 8085 places the high-order memory address 20H on A_{15}–A_8 and the low-order address 50H on AD_7–AD_0. At the same time, ALE goes high and IO/$\overline{M}$ goes low. The ALE signal indicates the availability of the address on AD_7–AD_0, and it can be used to demultiplex the bus. The IO/$\overline{M}$, being low, indicates that it is a memory-related operation. At T_2, the microprocessor sends the $\overline{RD}$ control signal, which is combined with IO/$\overline{M}$ (externally, see Chapter 3) to generate the $\overline{MEMR}$ signal, and the processor fetches the instruction code D3 using the data bus.

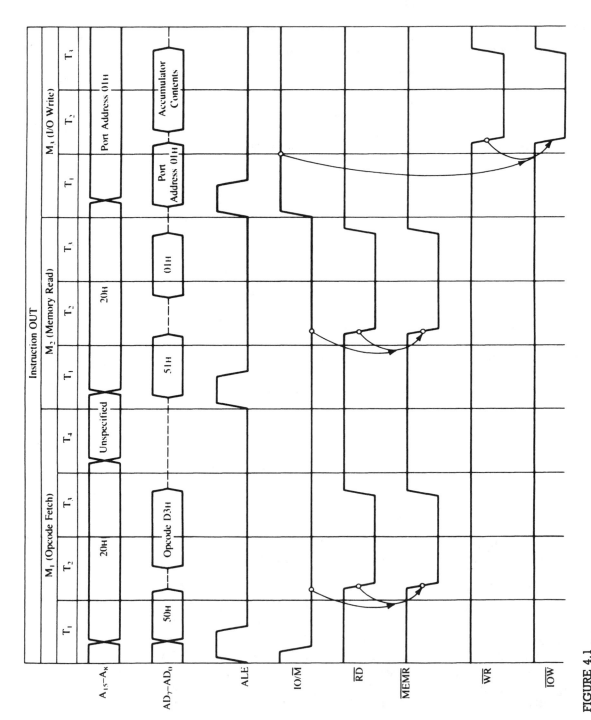

FIGURE 4.1

8085 Timing for Execution of OUT Instruction

When the 8085 decodes the machine code D3, it finds out that the instruction is a 2-byte instruction and that it must read the second byte.

In the second machine cycle, M_2 (Memory Read), the 8085 places the next address, 2051H, on the address bus and gets the device address 01H via the data bus.

In the third machine cycle, M_3 (I/O Write), the 8085 places the device address 01H on the low-order (AD_7–AD_0) as well as the high-order (A_{15}–A_8) address bus. The IO/$\overline{M}$ signal goes high to indicate that it is an I/O operation. At T_2, the accumulator contents are placed on the data bus (AD_7–AD_0), followed by the control signal $\overline{WR}$. By ANDing the IO/$\overline{M}$ and $\overline{WR}$ signals, the $\overline{IOW}$ (see Figure 3.5) signal can be generated to enable an output device.

Figure 4.1 shows the execution timing of the OUT instruction. The information necessary for interfacing an output device is available during T_2 and T_3 of the M_3 cycle. The data byte to be displayed is on the data bus, the 8-bit device address is available on the low-order as well as high-order address bus, and availability of the data byte is indicated by the $\overline{WR}$ control signal. The availability of the device address on both segments of the address bus is redundant information; in peripheral I/O, only one segment of the address bus (low or high) is sufficient for interfacing. The data byte remains on the data bus only for two T-states, then the processor goes on to execute the next instruction. Therefore, the data byte must be latched now, before it is lost, using the device address and the control signal (Section 4.13).

IN INSTRUCTION

The 8085 instruction set includes the instruction IN to read (copy) data from input devices such as switches, keyboards, and A/D data converters. This is a two-byte instruction that reads an input device and places the data in the accumulator. The first byte is the opcode, and the second byte specifies the port address. Thus, the addresses for input devices can range from 00H to FFH. The instruction is described as

IN 8-bit This is a two-byte instruction with the hexadecimal opcode DB, and the second byte is the port address of an input device.

This instruction reads (copies) data from an input device and places the data byte in the accumulator.

To read switch positions, for example, from an input port with the address 84H, the instructions will be written and stored in memory as follows:

Memory Address	Machine Code	Mnemonics		Memory Contents	
2065	DB	IN 84H	; 2065	1 1 0 1 1 0 1 1	= DBH
2066	84		; 2066	1 0 0 0 0 1 0 0	= 84H

(Note: The memory locations 2065H and 66H are selected arbitrarily for the illustration.)

When the microprocessor is asked to execute this instruction, it will first read the machine codes (or bytes) stored at locations 2065H and 2066H, then read the switch positions at port 84H by enabling the interfacing device of the port. The data byte indicating switch positions from the input port will be placed in the accumulator. Figure 4.2 shows the timing of the IN instruction; M_1 and M_2 cycles are identical to that of the OUT instruction. In the M_3 cycle, the 8085 microprocessor places the address of the input port (84H) on the low-order address bus AD_7–AD_0 as well as on the high-order address bus A_{15}–A_8 and asserts the $\overline{RD}$ signal, which is used to generate the I/O Read ($\overline{IOR}$) signal. The $\overline{IOR}$ enables the input port, and the data from the input port are placed on the data bus and transferred into the accumulator.

Machine cycle M_3 (Figure 4.2) is similar to the M_3 cycle of the OUT instruction; the only differences are (1) the control signal is $\overline{RD}$ instead of $\overline{WR}$, and (2) data flow from an input port to the accumulator rather than from the accumulator to an output port.

4.13   Device Selection and Data Transfer

The objective of interfacing an output device is to get information or a result out of the processor and store it or display it. The OUT instruction serves that purpose; during the M_3 cycle of the OUT instruction the processor places that information (accumulator contents) on the data bus. If we connect the data bus to a latch, we can catch that information and display it via LEDs or a printer. Now the questions are: (1) When should we enable the latch to catch that information? and (2) What should be the address of that latch? The answers to both questions can be found in the M_3 cycle (Figure 4.1). The latch should be enabled when IO/M is high and WR is active low. Similarly, the address of an output port is also on the address bus during M_3 (it is 01H in Figure 4.1). Now the task is to generate one pulse by decoding the address bus (A_7–A_0 or A_{15}–A_8) to indicate the presence of the port address we are interested in, generate a timing pulse by combining IO/M and WR signals to indicate that the data byte we are looking for is on the data bus, and use these pulses (by combining them) to enable the latch. These steps are summarized as follows. (For all subsequent discussion, the bus A_7–A_0 is assumed to be the demultiplexed bus AD_7–AD_0.)

1. Decode the address bus to generate a unique pulse corresponding to the device address on the bus; this is called the **device address pulse** or **I/O address pulse.**
2. Combine (AND) the device address pulse with the control signal to generate a device select (I/O select) pulse that is generated only when both signals are asserted.
3. Use the I/O select pulse to activate the interfacing device (I/O port).

The block diagram (Figure 4.3) illustrates these steps for interfacing an I/O device. In Figure 4.3, address lines A_7–A_0 are connected to a decoder, which will generate a unique pulse corresponding to each address on the address lines. This pulse is combined with the control signal to generate a device select pulse, which is used to enable an output latch or an input buffer.

Figure 4.4 shows a practical decoding circuit for the output device with address 01H. Address lines A_7–A_0 are connected to the 8-input NAND gate that functions as a decoder. Line A_0 is connected directly, and lines A_7–A_1 are connected through the inverters. When the address bus carries address 01H, gate G_1 generates a low pulse; otherwise, the

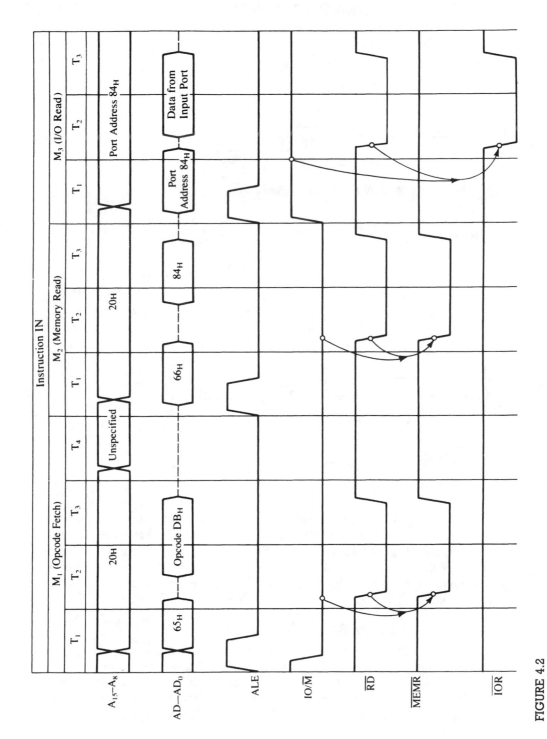

FIGURE 4.2
8085 Timing for Execution of IN Instruction

111

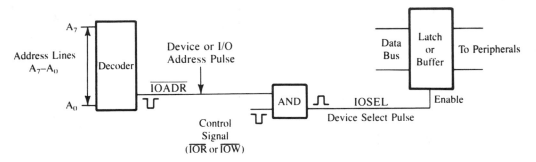

FIGURE 4.3
Block Diagram of I/O Interface

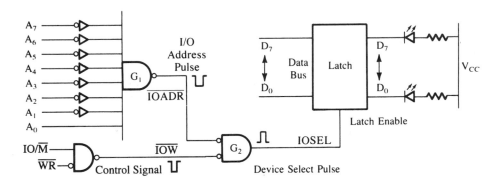

FIGURE 4.4
Decode Logic for LED Output Port
NOTE: To use this circuit with the 8085, the bus AD_7–A_0 must be demultiplexed.

output remains high. Gate G_2 combines the output of G_1 and the control signal $\overline{\text{IOW}}$ to generate an I/O select pulse when both input signals are low. Meanwhile (as was shown in the timing diagram—Figure 4.1, machine cycle M_3), the contents of the accumulator are placed on the data bus and are available on the data bus for a few microseconds and, therefore, must be latched for display. The I/O select pulse clocks the data into the latch for display by the LEDs.

4.14 Absolute vs. Partial Decoding

In Figure 4.4, all eight address lines are decoded to generate one unique output pulse; the device will be selected only with the address, 01H. This is called **absolute decoding** and is a good design practice. However, to minimize the cost, the output port can be selected by decoding some of the address lines, as shown in Figure 4.5; this is called **partial decoding.** As a result, the device has multiple addresses (similar to foldback memory addresses).

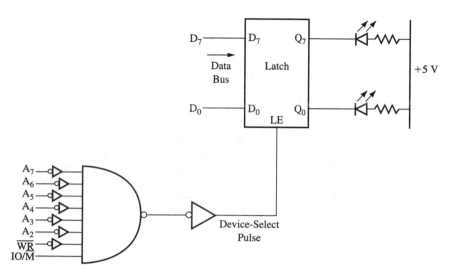

FIGURE 4.5
Partial Decoding: Output Latch with Multiple Addresses

Figure 4.5 is similar to Figure 4.4 except that the address lines A_1 and A_0 are not connected, and they are replaced by $IO/\overline{M}$ and $\overline{WR}$ signals. Because the address lines A_1 and A_0 are at don't care logic level, they can be assumed to be 0 or 1. Thus this output port (latch) can be accessed by the Hex addresses 00, 01, 02, and 03. The partial decoding is a commonly used technique in small systems. Such multiple addresses will not cause any problems, provided these addresses are not assigned to any other output ports.

4.15 Input Interfacing

Figure 4.6 shows an example of interfacing an 8-key input port. The basic concepts behind this circuit are similar to the interfacing concepts explained earlier.

The address lines are decoded by using an 8-input NAND gate. When address lines A_7–A_0 are <u>high</u> (FFH), the output of the NAND gate goes low and is combined with control signal $\overline{IOR}$ in gate G_2. When the MPU executes the instruction (IN FFH), gate G_2 generates the device select pulse that is used to enable the tri-state buffer. Data from the keys are put on the data bus D_7–D_0 and loaded into the accumulator. The circuit for the input port in Figure 4.6 differs from the output port in Figure 4.4 as follows:

1. Control signal $\overline{IOR}$ is used in place of $\overline{IOW}$.
2. The tri-state buffer is used as an interfacing port in place of the latch.
3. In Figure 4.6, data flow from the keys to the accumulator; on the other hand, in Figure 4.4, data flow from the accumulator to the LEDs.

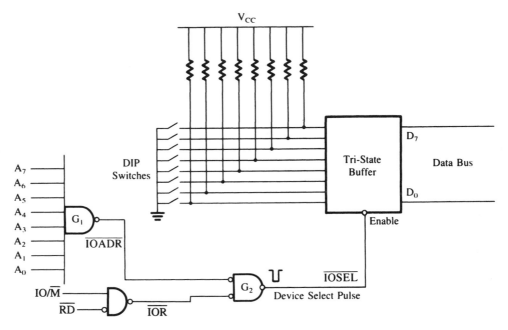

FIGURE 4.6
Decode Logic for a Dip-Switch Input Port

4.16 Interfacing I/Os Using Decoders

Various techniques and circuits can be used to decode an address and interface an I/O device to the microprocessor. However, all of these techniques should follow the three basic steps suggested in Section 4.13. Figures 4.4 and 4.6 illustrate an approach to device selection using an 8-input NAND gate. Figure 4.5 illustrates a technique using minimum hardware; this technique has the disadvantage of having multiple addresses for the same device. Figure 4.7 illustrates another scheme of address decoding. In this circuit, a 3-to-8 decoder and a 4-input NAND gate are used to decode the address bus; the decoding of the address bus is the first step in interfacing I/O devices. The address lines A_2, A_1, and A_0 are used as input to the decoder, and the remaining address lines A_7–A_3 are used to enable the decoder. The address line A_7 is directly connected to E_3 (active high Enable line), and the address lines A_6–A_3 are connected to $\overline{E}_1$ and $\overline{E}_2$ (active low Enable lines) using the NAND gate. The decoder has eight output lines; thus, we can use this circuit to generate eight device address pulses for eight different addresses.

The second step is to combine the decoded address with an appropriate control signal to generate the I/O select pulse. Figure 4.7 shows that the output O_0 of the decoder is logically ANDed in a negative AND gate with the $\overline{\text{IOW}}$ control signal. The output of the gate is the I/O select pulse for an output port. The third step is to use this pulse to enable the output port. Figure 4.7 shows that the I/O select pulse enables the LED latch with the output port address F8H, as shown below (A_7–A_0 is the demultiplexed low-order bus).

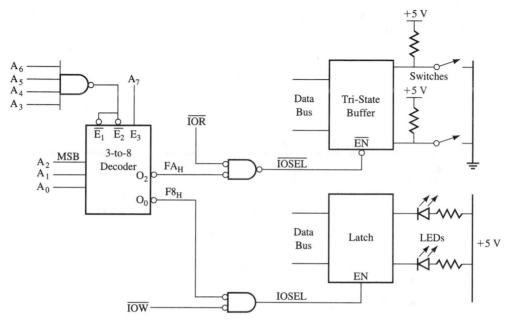

FIGURE 4.7
Address Decoding Using a 3-to-8 Decoder

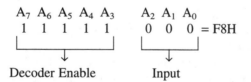

Similarly, the output O_2 of the decoder is combined with the I/O Read ($\overline{\text{IOR}}$) signal, and the I/O select pulse is used to enable the input buffer with the address FAH.

4.17 Review of Important Concepts

In peripheral I/O, the basic concepts and the steps in designing an interfacing circuit can be summarized as follows:

1. When an I/O instruction is executed, the 8085 microprocessor places the device address (port number) on the demultiplexed low-order as well as the high-order address bus.
2. Either the high-order bus (A_{15}–A_8) or the demultiplexed low-order bus (A_7–A_0) can be decoded to generate the pulse corresponding to the device address on the bus.
3. The device address pulse is ANDed with the appropriate control signal ($\overline{\text{IOR}}$ or $\overline{\text{IOW}}$) and, when both signals are asserted, the I/O port is selected.

4. As interfacing devices, a latch is used for an output port and a tri-state buffer is used for an input port.
5. The address bus can be decoded by using either the absolute- or the linear-select decoding technique. The linear-select decoding technique reduces the component cost, but the I/O device ends up with multiple addresses.

4.2   INTERFACING OUTPUT DISPLAYS

This section concerns the analysis and design of practical circuits for data display. The section includes two different types of circuits. The first illustrates the simple display of binary data with LEDs, and the second illustrates the interfacing of seven-segment LEDs.

4.21   Illustration: LED Display for Binary Data

PROBLEM STATEMENT

1. Analyze the interfacing circuit in Figure 4.8(a), identify the address of the output port, and explain the circuit operation.
2. Explain similarities between (a) and (b) in Figure 4.8.
3. Write instructions to display binary data at the port.

CIRCUIT ANALYSIS

Address bus A_7–A_0 is decoded by using an 8-input NAND gate. The output of the NAND gate goes low only when the address lines carry the address FFH. The output of the NAND gate is combined with the microprocessor control signal $\overline{IOW}$ in a NOR gate (connected as negative AND). The output of NOR gate 74LS02 goes high to generate an I/O select pulse when both inputs are low (or both signals are asserted). Meanwhile, the contents of the accumulator have been put on the data bus. The I/O select pulse is used as a clock pulse to activate the D-type latch, and the data are latched and displayed.

In this circuit, the LED cathodes are connected to the $\overline{Q}$ output of the latch. The anodes are connected to +5 V through resistors to limit the current flow through the diodes. When the data line (for example D_0) has 1, the output $\overline{Q}$ is 0 and the corresponding LED is turned on. If the LED anode were connected to Q, its cathode would be connected to the ground. In this configuration, the D flip-flop would not be able to supply the necessary current to the LED.

Figure 4.8(b) uses the 74LS373 octal latch as an interfacing device, and both circuits (a) and (b) are functionally similar. The 74LS373 includes D-latches (flip-flops) followed by tri-state buffers (see Figure 2.27 for details). This device has two control signals: Enable (G) to clock data in the flip-flops and Output Control ($\overline{OC}$) to enable the buffers. In this circuit, the 74LS373 is used as a latch; therefore, the tri-state buffers are enabled by grounding the $\overline{OC}$ signal.

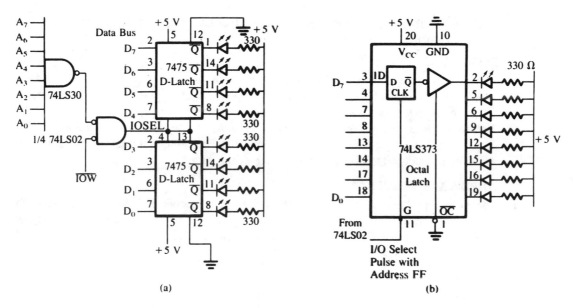

FIGURE 4.8
Interfacing LED Output Port Using the 7475 D-Type Latch (a) and Using the 74LS373 Octal D-Type Latch (b)

PROGRAM

Address (LO)	Machine Code	Mnemonics	Comments
00	3E	MVI A,DATA	;Load accumulator with data
01	DATA*		
02	D3	OUT FFH	;Output accumulator contents ; to port FFH
03	FF		
04	76	HLT	;End of program

PROGRAM DESCRIPTION

Instruction MVI A loads the accumulator with the data you enter, and instruction OUT FFH identifies the LED port as the output device and displays the data.

---

*Enter data you wish to display.

4.22   Illustration: Seven-Segment LED Display as an Output Device

PROBLEM STATEMENT

1. Design a seven-segment LED output port with the device address F5H, using a 74LS138 3-to-8 decoder, a 74LS20 4-input NAND gate, a 74LS02 NOR gate, and a common-anode seven-segment LED.

2. Given $\overline{WR}$ and IO/$\overline{M}$ signals from the 8085, generate the $\overline{IOW}$ control signal.

3. Explain the binary codes required to display 0 to F Hex digits at the seven-segment LED.

4. Write instructions to display digit 7 at the port.

HARDWARE DESCRIPTION

The design problem specifies two MSI chips—the decoder (74LS138) and the latch 74LS373—and a common-anode seven-segment LED. The decoder and the latch have been described in previous sections; the seven-segment LED and its binary code requirement are discussed below.

SEVEN-SEGMENT LED

A seven-segment LED consists of seven light-emitting diode segments and one segment for the decimal point. These LEDs are physically arranged as shown in Figure 4.9(a). To display a number, the necessary segments are lit by sending an appropriate signal for cur-

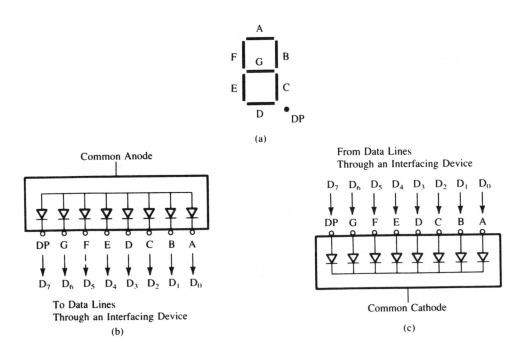

(a)

(b)

(c)

FIGURE 4.9

Seven-Segment LED: LED Segments (a); Common-Anode LED (b); Common-Cathode LED (c)

rent flow through diodes. For example, to display an 8, all segments must be lit. To display 1, segments B and C must be lit. Seven-segment LEDs are available in two types: common cathode and common anode. They can be represented schematically as in Figure 4.9(b) and (c). Current flow in these diodes should be limited to 20 mA.

The seven segments, A through G, are usually connected to data lines D_0 through D_6, respectively. If the decimal-point segment is being used, data line D_7 is connected to DP; otherwise it is left open. The binary code required to display a digit is determined by the type of the seven-segment LED (common cathode or common anode), the connections of the data lines, and the logic required to light the segment. For example, to display digit 7 at the LED in Figure 4.10, the requirements are as follows:

1. It is a common-anode seven-segment LED, and logic 0 is required to turn on a segment.
2. To display digit 7, segments A, B, and C should be turned on.
3. The binary code should be

Data Lines	D_7	D_6	D_5	D_4	D_3	D_2	D_1	D_0	
Bits	X	1	1	1	1	0	0	0	= 78H
Segments	NC	G	F	E	D	C	B	A	

The code for each digit can be determined by examining the connections of the data lines to the segments and the logic requirements.

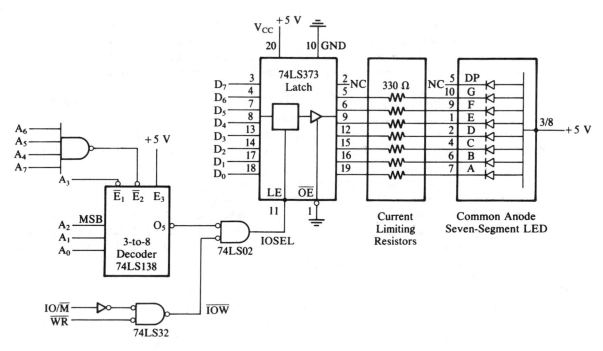

FIGURE 4.10
Interfacing Seven-Segment LED

INTERFACING CIRCUIT AND ITS ANALYSIS

To design an output port with the address F5H, the address lines A_7–A_0 should have the following logic:

$$\begin{array}{cccccccc} A_7 & A_6 & A_5 & A_4 & A_3 & A_2 & A_1 & A_0 \\ 1 & 1 & 1 & 1 & 0 & 1 & 0 & 1 \end{array} = \text{F5H}$$

This can be accomplished by using A_2, A_1, and A_0 as input lines to the decoder. A_3 can be connected to active low enable $\overline{E}_1$, and the remaining address lines can be connected to $\overline{E}_2$ through the 4-input NAND gate. Figure 4.10 shows an output port with the address F5H. The output O_5 of the decoder is logically ANDed with the control signal $\overline{\text{IOW}}$ using the NOR gate (74LS02). The output of the NOR gate is the I/O select pulse that is used to enable the latch (74LS373). The control signal $\overline{\text{IOW}}$ is generated by logically ANDing IO/$\overline{\text{M}}$ and $\overline{\text{WR}}$ signals in the negative NAND gate (physically OR gate 74LS32).

Instructions   The following instructions are necessary to display digit 7 at the output port:

MVI A,78H	;Load seven-segment code in the accumulator
OUT F5H	;Display digit 7 at port F5H
HLT	;End

The first instruction loads 78H in the accumulator; 78H is the binary code necessary to display digit 7 at the common-anode seven-segment LED. The second instruction sends the contents of the accumulator (78H) to the output port F5H. When the 8085 executes the OUT instruction, the digit 7 is displayed at the port as follows:

1. In the third machine cycle M_3 of the OUT instruction (refer to Figure 4.1), the port address F5H is placed on the address bus A_7–A_0 (it is also duplicated on the high-order bus A_{15}–A_8, but we have used the low-order bus for interfacing in this example).
2. The address F5H is decoded by the decoding logic (decoder and 4-input NAND gate), and the output O_5 of the decoder is asserted.
3. During T_2 of the M_3 cycle (see Figure 4.1), the 8085 places the data byte 78H from the accumulator on the data bus and asserts the $\overline{\text{WR}}$ signal.
4. In Figure 4.10, when the $\overline{\text{IOW}}$ signal is asserted, the output of the NOR gate 74LS02 goes high and enables the latch 74LS373. The data byte (78H), which is already on the data bus at the input of the latch, is passed on to the output of the latch and displayed by the seven-segment LED. However, the byte is latched when the $\overline{\text{WR}}$ signal is deasserted during T_3.

Current Requirements   The circuit in Figure 4.10 uses a common-anode seven-segment LED. Each segment requires 10 to 15 mA of current ($I_{D\,max} = 19$ mA) for appropriate illumination. The latch can sink 24 mA when the output is low and can supply approximately 2.6 mA when the output is high. In this circuit, the common-anode LED segments are turned on by zeros on the output of the latch. If common-cathode seven-segment LEDs were used in this circuit, the output of the latch would have to be high to drive the segments. The current supplied would be about 2.6 mA, which is insufficient to make the segments visible.

INTERFACING INPUT DEVICES   4.3

The interfacing of input devices is similar to that of the interfacing output devices, except with some differences in bus signals and circuit components. We will follow the same basic steps described in Section 4.13 and the timing diagram for the execution of the IN instruction shown in Figure 4.2.

4.31   Illustration: Data Input from DIP Switches

In this section, we will analyze the circuit used for interfacing eight DIP switches, as shown in Figure 4.11. The circuit includes the 74LS138 3-to-8 decoder to decode the low-order bus and the tri-state octal buffer (74LS244) to interface the switches to the data bus. The port can be accessed with the address 84H; however, it has multiple addresses, as explained below.

4.32   Hardware

Figure 4.11 shows the 74LS244 tri-state octal buffer used as an interfacing device. The device has two groups of four buffers each, and they are controlled by the active low sig-

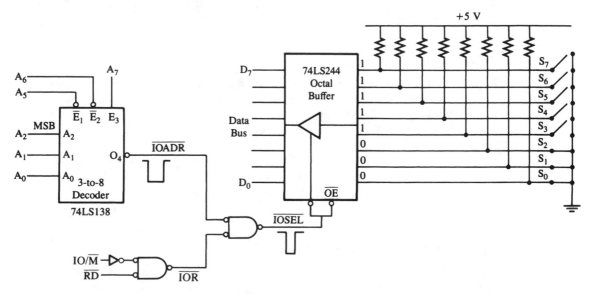

FIGURE 4.11
Interfacing DIP Switches

nals $\overline{OE}$. When $\overline{OE}$ is low, the input data show up on the output lines (connected to the data bus), and when $\overline{OE}$ is high, the output lines assume the high impedance state.

4.33   Interfacing Circuit

Figure 4.11 shows that the low-order address bus, except the lines A_4 and A_3, is connected to the decoder (the 74LS138); the address lines A_4 and A_3 are left in the don't care state. The output line O_4 of the decoder goes low when the address bus has the following address (assume the don't care lines are at logic 0):

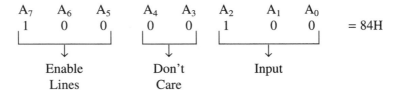

The control signal I/O Read ($\overline{IOR}$) is generated by ANDing the IO/$\overline{M}$ (through an inverter) and $\overline{RD}$ in a negative NAND gate, and the I/O select pulse is generated by ANDing the output of the decoder and the control signal $\overline{IOR}$. When the address is 84H and the control signal $\overline{IOR}$ is asserted, the I/O select pulse enables the tri-state buffer and the logic levels of the switches are placed on the data bus. The 8085, then, begins to read switch positions during T_3 (Figure 4.2) and places the reading in the accumulator. When a switch is closed, it has logic 0, and when it is open, it is tied to +5 V, representing logic 1. Figure 4.11 shows that the switches S_7–S_3 are open and S_2–S_0 are closed; thus, the input reading will be F8H.

4.34   Multiple Port Addresses

In Figure 4.11, the address lines A_4 and A_3 are not used by the decoding circuit; the logic levels on these lines can be 0 or 1. Therefore, this input port can be accessed by four different addresses, as shown below.

A_7	A_6	A_5	A_4	A_3	A_2	A_1	A_0	
1	0	0	0	0	1	0	0	= 84H
			0	1				= 8CH
			1	0				= 94H
			1	1				= 9CH

4.35   Instructions to Read Input Port

To read data from the input port shown in Figure 4.11, the instruction IN 84H can be used. When this instruction is executed, during the M_3 cycle, the 8085 places the address 84H on the low-order bus (as well as on the high-order bus), asserts the $\overline{RD}$ control signal, and reads the switch positions.

MEMORY-MAPPED I/O 4.4

In memory-mapped I/O, the input and output devices are assigned and identified by 16-bit addresses. To transfer data between the MPU and I/O devices, memory-related instructions (such as LDA, STA, etc.)* and memory control signals ($\overline{\text{MEMR}}$ and $\overline{\text{MEMW}}$) are used. The microprocessor communicates with an I/O device as if it were one of the memory locations. The memory-mapped I/O technique is similar in many ways to the peripheral I/O technique. To understand the similarities, it is necessary to review how a data byte is transferred from the 8085 microprocessor to a memory location or vice versa. For example, the following instruction will transfer the contents of the accumulator to the memory location 8000H.

Memory Address	Machine Code	Mnemonics	Comments
2050	32	STA 8000H	;Store contents of accumulator in memory location 8000H
2051	00		
2052	80		

(Note: It is assumed here that the instruction is stored in memory locations 2050H, 51H, and 52H.)

The STA is a three-byte instruction; the first byte is the opcode, and the second and third bytes specify the memory address. However, the 16-bit address 8000H is entered in the reverse order; the low-order byte 00 is stored in location 2051, followed by the high-order address 80H (the reason for the reversed order will be explained in Section 4.6). In this example, if an output device, instead of a memory register, is connected at this address, the accumulator contents will be transferred to the output device. This is called the **memory-mapped I/O technique.**

On the other hand, the instruction LDA (Load Accumulator Direct) transfers the data from a memory location to the accumulator. The instruction LDA is a 3-byte instruction; the second and third bytes specify the memory location. In the memory-mapped I/O technique, an input device (keyboard) is connected instead of a memory. The input device will have the 16-bit address specified by the LDA instruction. When the microprocessor executes the LDA instruction, the accumulator receives data from the input device rather than from a memory location. To use memory-related instructions for data transfer, the control signals Memory Read ($\overline{\text{MEMR}}$) and Memory Write ($\overline{\text{MEMW}}$) should be connected to I/O devices instead of $\overline{\text{IOR}}$ and $\overline{\text{IOW}}$ signals, and the 16-bit address bus (A_{15}–A_0) should be decoded. The hardware details will be described in Section 4.42).

---

*In addition to the instructions STA and LDA, other memory-related data transfer instructions, such as MOV M, LDAX, and STAX, also can be used for memory-mapped I/O. These instructions will not be discussed here to maintain clarity in presenting the concepts of memory-mapped I/O.

4.41   Execution of Memory-Related Data Transfer Instructions

The execution of memory-related data transfer instructions is similar to the execution of
IN or OUT instructions, except that the memory-related instructions have 16-bit addresses. The microprocessor requires four machine cycles (13 T-states) to execute the instruction STA (Figure 4.12). The machine cycle M_4 for the STA instruction is similar to
the machine cycle M_3 for the OUT instruction.

For example, to execute the instruction STA 8000H in the fourth machine cycle
(M_4), the microprocessor places memory address 8000H on the entire address bus
(A_{15}–A_0). The accumulator contents are sent on the data bus, followed by the control signal Memory Write $\overline{MEMW}$ (active low).

On the other hand, in executing the OUT instruction (Figure 4.1), the 8-bit device address is repeated on the low-order address bus (A_0–A_7) as well as on the high-order bus, and
the $\overline{IOW}$ control signal is used. To identify an output device, either the low-order or the high-
order bus can be decoded. In the case of the STA instruction, the entire bus must be decoded.

Device selection and data transfer in memory-mapped I/O require three steps that
are similar to those required in peripheral I/O:

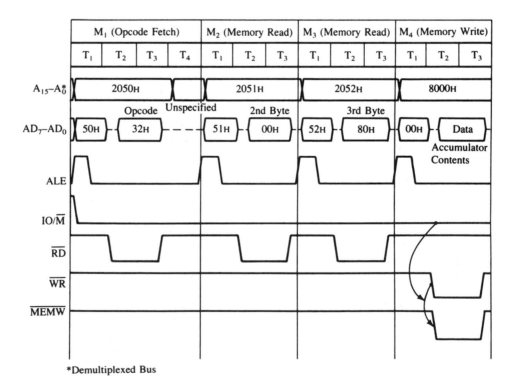

*Demultiplexed Bus

FIGURE 4.12
Timing for Execution of the Instruction: STA 8000H

1. Decode the address bus to generate the device address pulse.
2. AND the control signal with the device address pulse to generate the device select (I/O select) pulse.
3. Use the device select pulse to enable the I/O port.

To interface a memory-mapped input port, we can use the instruction LDA 16-bit, which reads data from an input port with the 16-bit address and places the data in the accumulator. The instruction has four machine cycles; only the fourth machine cycle differs from M_4 in Figure 4.12. The control signal will be $\overline{RD}$ rather than $\overline{WR}$, and data flow from the input port to the microprocessor.

4.42 Illustration: Safety Control System Using Memory-Mapped I/O Technique

Figure 4.13 shows a schematic of interfacing I/O devices using the memory-mapped I/O technique. The circuit includes one input port with eight DIP switches and one output port to control various processes and gates, which are turned on/off by the microprocessor according to the corresponding switch positions. For example, switch S_7 controls the cooling system, and switch S_0 controls the exit gate. All switch inputs are tied high; therefore, when a switch is open (off), it has +5 V, and when a switch is closed (on), it has logic 0. The circuit includes one 3-to-8 decoder, one 8-input NAND gate, and one 4-input NAND gate to decode the address bus. The output O_0 of the decoder is combined with control signal $\overline{MEMW}$ to generate the device select pulse that enables the octal latch. The output O_1 is combined with the control signal $\overline{MEMR}$ to enable the input port. The eight switches are interfaced using a tri-state buffer 74LS244, and the solid state relays controlling various processes are interfaced using an octal latch (74LS373) with tri-state output.

OUTPUT PORT AND ITS ADDRESS

The various process control devices are connected to the data bus through the latch 74LS373 and solid state relays. If an output bit of the 74LS373 is high, it activates the corresponding relay and turns on the process; the process remains on until the bit stays high. Therefore, to control these safety processes, we need to supply an appropriate bit pattern to the latch.

The 74LS373 is a latch followed by a tri-state buffer, as shown in Figure 4.13. The latch and the buffer are controlled independently by the Latch Enable (LE) and Output Enable ($\overline{OE}$). When LE is high, the data enter the latch, and when LE goes low, data are latched. The latched data are available on the output lines of the 74LS373 if the buffer is enabled by $\overline{OE}$ (active low). If $\overline{OE}$ is high, the output lines go into the high impedance state.

Figure 4.13 shows that the $\overline{OE}$ is connected to the ground; thus, the latched data will keep the relays on/off according to the bit pattern. The LE is connected to the device

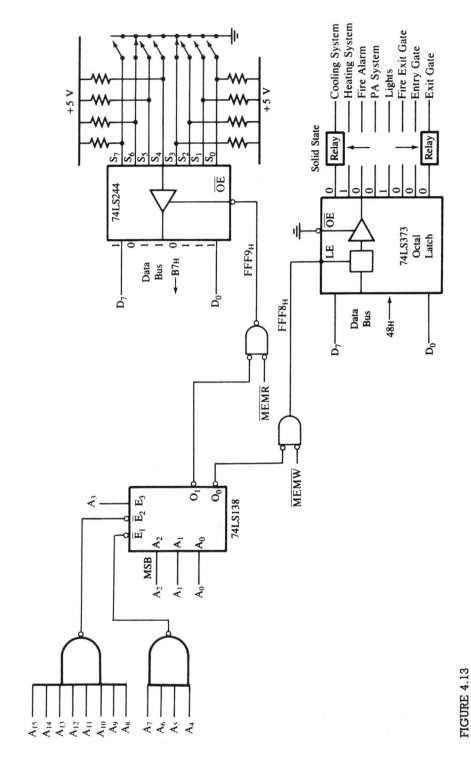

FIGURE 4.13
Memory-Mapped I/O Interfacing

select pulse, which is asserted when the output O_0 of the decoder and the control signal $\overline{\text{MEMW}}$ go low. Therefore, to assert the I/O select pulse, the output port address should be FFF8H, as shown below:

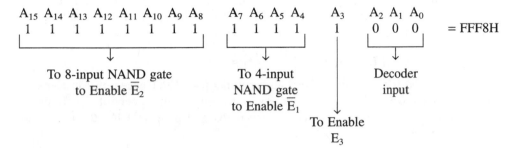

INPUT PORT AND ITS ADDRESS

The DIP switches are interfaced with the 8085 using the tri-state buffer 74LS244. The switches are tied high, and they are turned on by grounding, as shown in Figure 4.13. The switch positions can be read by enabling the signal $\overline{\text{OE}}$, which is asserted when the output O_1 of the decoder and the control signal $\overline{\text{MEMR}}$ go low. Therefore, to read the input port, the port address should be

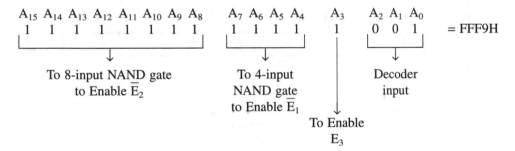

Instructions To control the processes according to switch positions, the microprocessor should read the bit pattern at the input port and send that bit pattern to the output port. The following instructions can accomplish this task:

```
READ: LDA FFF9H      ;Read the switches
      CMA            ;Complement switch reading, convert on-switch (logic 0)
                     ;   into logic 1 to turn on appliances
      STA FFF8H      ;Send switch positions to output port and turn on/off appli-
                     ;   ances
      JMP READ       ;Go back and read again
```

When this program is executed, the first instruction reads the bit pattern 1011 0111 (B7H) at the input port FFF9H and places that reading in the accumulator; this bit pattern represents the on-position of switches S_6 and S_3. The second instruction complements the reading; this instruction is necessary because the on-position has logic 0, and to turn on solid state relays logic 1 is necessary. The third instruction sends the complemented accumulator contents (0100 1000 = 48H) to the output port FFF8H. The 74LS373 latches the data byte 0100 1000 and turns on the heating system and lights. The last instruction, JMP READ, takes the program back to the beginning and repeats the loop continuously. Thus, it monitors the switches continuously.

4.43   Review of Important Concepts

Memory-mapped I/O is in many ways similar to peripheral I/O, except that in memory-mapped I/O, the device has a 16-bit address, and memory-related control signals are required to identify the device. Memory-mapped I/O and peripheral I/O techniques are compared in Table 4.1.

An examination of Table 4.1 shows that the selection of the I/O technique will be determined primarily by the type of application; the advantages seem to balance the disadvantages. In systems in which 64K memory is a requirement, peripheral I/O becomes essential; on the other hand, in control applications in which the number of I/Os exceed the limit (256) and direct data manipulation is preferred, memory-mapped I/O may have an advantage.

TABLE 4.1

Comparison of Memory-Mapped I/O and Peripheral I/O

Characteristics	Memory-Mapped I/O	Peripheral I/O
1. Device address	16-bit	8-bit
2. Control signals for Input/Output	MEMR/MEMW	IOR/IOW
3. Instructions available	Memory-related instructions such as STA; LDA; LDAX; STAX; MOV M,R: ADD M; SUB M; ANA M: etc.	IN and OUT
4. Data transfer	Between any register and I/O	Only between I/O and the accumulator
5. Maximum number of I/Os possible	The memory map (64K) is shared between I/Os and system memory	The I/O map is independent of the memory map; 256 input devices and 256 output devices can be connected
6. Execution speed	13 T-states (STA,LDA) 7 T-states (MOV M,R)	10 T-states
7. Hardware requirements	More hardware is needed to decode 16-bit address	Less hardware is needed to decode 8-bit address
8. Other features	Arithmetic or logical operations can be directly performed with I/O data	Not available

TESTING AND TROUBLESHOOTING I/O INTERFACING CIRCUITS 4.5

In previous sections, we illustrated how to interface an I/O device to a working micro-computer system or add an I/O port as an expansion to the existing system. In Section 4.22, we designed the LED output port with the address F5H. The next step is to test and verify that we can display the digit 7 by sending the code 78H as specified in the design problem. In the first attempt, the most probable outcome will be that nothing is displayed or digit 8 is displayed irrespective of the code sent to the port.

Now we need to troubleshoot the interfacing circuit. The obvious first step is to check the wiring and the pin connections. After this preliminary check, we need to generate a constant and identifiable signal and check various points in relation to that signal. We can generate such a signal by asking the processor to execute a continuous loop, called a diagnostic routine, as discussed in Chapter 3 (Section 3.5).

4.51   Diagnostic Routine and Machine Cycles

We can use the same instructions for the diagnostic routine that we used in the design problem; however, to generate a continuous signal, we need to add a Jump instruction, as shown next.

Instruction	Bytes	T-States	Machine Cycles		
			M_1	M_2	M3
START: MVI A,78H	2	7 (4,3)	Opcode Fetch	Memory Read	
OUT F5H	3	10 (4,3,3)	Opcode Fetch	Memory Read	I/O Write
JMP START	3	10 (4,3,3)	Opcode Fetch	Memory Read	Memory Read

This loop has 27 T-states and eight operations (machine cycles). To execute the loop once, the microprocessor asserts the $\overline{RD}$ signal seven times (the Opcode Fetch is also a Read operation) and the $\overline{WR}$ signal once. Assuming the system clock frequency is 3 MHz, the loop is executed in 8.9 μs, and the $\overline{WR}$ signal is repeated every 8.9 μs that can be observed on a scope. If we sync the scope on the $\overline{WR}$ pulse from the 8085, we can check the output of the decoder, $\overline{IOW}$, and IOSEL signals; some of these signals of a working circuit are shown in Figure 4.14.

When the 8085 asserts the $\overline{WR}$ signal, the port address F5H must be on the address bus A_7–A_0, and the output O_5 of the decoder in Figure 4.9 must be low. Similarly, the $\overline{IOW}$ must be low and the IOSEL (the output of the 74LS02) must be high. Now if we check the data bus in relation to the $\overline{WR}$ signal, one line at a time, we must read the data byte 78H. If the circuit is not properly functioning, we can check various signals in reference to the $\overline{WR}$ signal as suggested below:

1. If IOSEL is low, check $\overline{IOW}$ and O_5 of the decoder.
2. If $\overline{IOW}$ is high, check the input to the OR gate 74LS32. Both should be low.

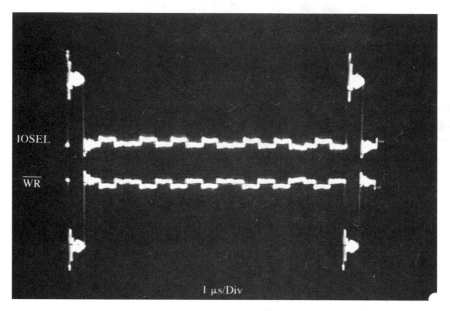

FIGURE 4.14
Timing Signals of Diagnostic Routine

3. If O_5 of the decoder is high, check all the output lines O_0 to O_7 of the decoder. If all of them are high, that means the decoder is not enabled. If one of the outputs of the decoder is low, it suggests that the input address lines are improperly connected.

4. If the decoder is not enabled, check the address lines A_4–A_7; all of them must be high and the address line A_3 must be low.

5. Another possibility is that the port is enabled, but the seven-segment display is wrong. The problem must be with data lines. Try different codes to display other digits. If two data lines are interchanged, you may be able to isolate these two data lines. The final step is to check all the data lines.

4.6   SOME QUESTIONS AND ANSWERS

During the discussion of I/O interfacing, we focused on the basic concepts and avoided some details to maintain the clarity. We will attempt to answer those questions.

1. *Can an input port and an output port have the same port address?*
     Yes. They will be differentiated by control signals. The $\overline{RD}$ is used to enable the input port and the $\overline{WR}$ is used to enable the output port.

2. *How will the port number be affected if we decode the high-order address lines A_{15}–A_8 rather than A_7–A_0?*

The port address will remain the same because the I/O port address is duplicated on both segments of the address bus.

3. *If high-order lines are partially decoded, how can one determine whether it is peripheral I/O or memory-mapped I/O?*

To recognize the type of I/O, examine the control signal. If the control signal is $\overline{\text{IOW}}$ (or $\overline{\text{IOR}}$), it must be a peripheral I/O, and if the control signal is $\overline{\text{MEMW}}$ (or $\overline{\text{MEMR}}$), it must be a memory-mapped I/O (see Problem 14).

4. *In a memory-mapped I/O, how does the microprocessor differentiate between an I/O and memory? Can an I/O have the same address as a memory register?*

In memory-mapped I/O, the microprocessor cannot differentiate between an I/O and memory; it treats an I/O as if it is memory. Therefore, an I/O and memory register cannot have the same address; the entire memory map (64K) of the system has to be shared between memory and I/O.

5. *Why is a 16-bit address (data) stored in memory in the reversed order—the low-order byte first, followed by the high-order byte?*

This has to do with the design of the 8085 microprocessor. The instruction decoder or the associated microprogram is designed to recognize the second byte as the low-order byte in a three-byte instruction.

SUMMARY

In this chapter, we examined the machine cycles of the OUT and IN instructions and derived the basic concepts in interfacing peripheral-mapped I/Os. Similarly, we examined the machine cycles of memory-related data transfer instructions and derived the basic concepts in interfacing memory-mapped I/Os. These concepts were illustrated with various examples of interfacing I/O devices. The interfacing concepts can be summarized as follows.

Peripheral-Mapped I/O

☐ The OUT is a two-byte instruction. It copies (transfers or sends) data from the accumulator to the addressed port.

☐ When the 8085 executes the OUT instruction, in the third machine cycle, it places the output port address on the low-order bus, duplicates the same port address on the high-order bus, places the contents of the accumulator on the data bus, and asserts the control signal $\overline{\text{WR}}$.

☐ A latch is commonly used to interface output devices.

☐ The IN instruction is a two-byte instruction. It copies (transfers or reads) data from an input port and places the data into the accumulator.

☐ When the 8085 executes the IN instruction, in the third machine cycle, it places the input port address on the low-order bus, as well as on the high-order bus, asserts the control signal, $\overline{\text{RD}}$, and transfers data from the port to the accumulator.

☐ A tri-state buffer is commonly used to interface input devices.

☐ To interface an output or an input device, the low-order address bus A_7–A_0 (or high-order bus A_{15}–A_8) needs to be decoded to generate the device address pulse, which must be combined with the control signal $\overline{IOR}$ (or $\overline{IOW}$) to select the device.

Memory-Mapped I/O

☐ Memory-related instructions are used to transfer data.

☐ To interface I/O devices, the entire bus must be decoded to generate the device address pulse, which must be combined with the control signal $\overline{MEMR}$ (or $\overline{MEMW}$) to generate the I/O select pulse. This pulse is used to enable the I/O device and transfer the data.

QUESTIONS AND PROBLEMS

1. Explain why the number of output ports in the peripheral-mapped I/O is restricted to 256 ports.
2. In the peripheral-mapped I/O, can an input port and an output port have the same port address?
3. If an output and input port can have the same 8-bit address, how does the 8085 differentiate between the ports?
4. Specify the two 8085 signals that are used to latch data in an output port.
5. Specify the type of pulse (high or low) required to latch data in the 7475.
6. Are data latched in the 7475 at the leading edge, during the level, or at the trailing edge of the enable (E) signal?
7. In Figure 4.8, explain why the LED cathodes rather than anodes are connected to the latch.
8. Specify the 8085 signals that are used to enable an input port.
9. Explain why a latch is used for an output port, but a tri-state buffer can be used for an input port.
10. What are the control signals necessary in the memory-mapped I/O?
11. Can the microprocessor differentiate whether it is reading from a memory-mapped input port or from memory?
12. Identify the port address in Figure 4.15.
13. In Figure 4.15, if $\overline{OE}$ is connected directly to the $\overline{WR}$ signal and the output of the decoder is connected to the latch enable (through an inverter), can you display a byte at the output port? Explain your answer.
14. In Figure 4.16, can you recognize whether it is the memory-mapped or the peripheral-mapped I/O?
15. In Figure 4.16, what is the port address if all the don't care address lines are assumed to be at logic 0?
16. In Figure 4.10, specify the output port address if the output signal O_1 of the decoder is connected to gate 74LS02 instead of the signal O_5.

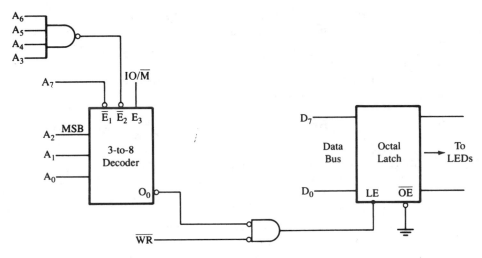

FIGURE 4.15

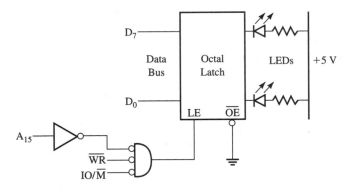

FIGURE 4.16

17. In Figure 4.10, eliminate the gate 74LS32, replace A_7 by the IO/$\overline{M}$ signal, and connect the $\overline{WR}$ signal directly to gate 74LS02. Specify the port addresses if the address line A_7 is left in don't care state.

18. In Question 17, replace A_3 by the IO/$\overline{M}$ signal instead of A_7. Specify the port address assuming all don't care lines (including) A_3 at logic 0. Replace instructions shown in Section 4.22 to display the digit "7" by appropriate instructions.

19. Write instructions to display digit "0" at the output port in Figure 4.10.

20. Write instructions to display letter "H" at the output port in Figure 4.10.

21. In Figure 4.11, eliminate the negative NAND gate that generates the $\overline{IOR}$ signal, replace A_7 by the IO/$\overline{M}$ signal, and connect $\overline{RD}$ directly to the negative NAND gate. Identify all the port addresses.

22. In Figure 4.15, exchange A_7 with IO/$\overline{M}$ signal. Identify the port address assuming any don't care lines at logic 0.

23. In Figure 4.17, identify ports A and B as input or output ports.

24. In Figure 4.17, what are the addresses of ports A and B?

25. In Figure 4.10, explain why the pulse width of the output O_5 of the decoder is larger than the $\overline{IOW}$ signal.

26. The following diagnostic routine can be used to troubleshoot the interfacing circuit of an input port such as in Figure 4.11.

Instruction	Bytes	T-States	Machine Cycles		
			M_1	M_2	M_3
START: IN 84H	2	10 (4,3,3)			
JMP START	3	10 (4,3,3)			

a. Identify the machine cycles.

b. If the system clock is 2 MHz, calculate the time required to execute the routine.

c. Specify the number of times the $\overline{RD}$ signal is asserted if the loop is executed once.

d. If the loop is executed continuously, specify the time between two consecutive $IO/\overline{M}$ signals that are high.

e. Is there a $\overline{WR}$ pulse in the diagnostic routine? If the answer is no, what is the unique identifiable signal that can be used to sync the scope?

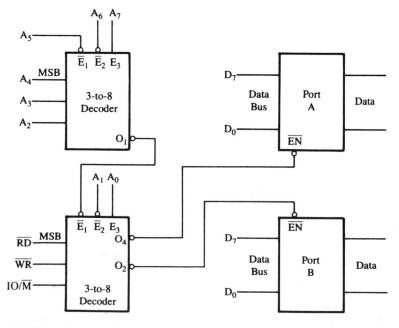

FIGURE 4.17

27. Diagnostic Routine for Figure 4.13:

Instruction	Bytes	T-States	Machine Cycles			
			M_1	M_2	M_3	M_4
START: LDA FFF9H	3	13 (4,3,3,3)				
STA FFF8H	3	13 (4,3,3,3)				
MOV B,A	1	4				
JMP START	3	10 (4,3,3)				

  a. Identify the machine cycles of each instruction.
  b. Specify the contents of the address bus in the fourth machine cycle of the LDA instruction.
  c. Specify the number of $\overline{\text{RD}}$ and $\overline{\text{WR}}$ signals in one loop.
  d. If the system frequency is 2 MHz, calculate the time period between two consecutive $\overline{\text{MEMW}}$ signals.

28. In Figure 4.18, identify the addresses of the input and the output ports.

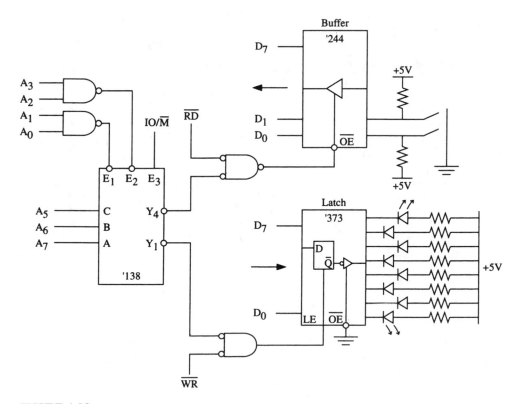

FIGURE 4.18

29. Write instructions to read the input port (Figure 4.18) and continue to read it until both switches are closed (by an operator). When both switches are closed, turn on all the LEDs.

30. In Figure 4.18, write instructions to read the input port and continue to read it until at least one switch is closed. When a switch is closed, turn on the corresponding LED.

31. In Figure 4.18, assume that both switches are normally closed. Write instructions to read the input port and continue to read the port until at least one of the switches is open. Once a switch is opened, turn on the corresponding LED.

32. In Figure 4.19, write instructions to display the number "97" at the common-anode seven-segment LED port.

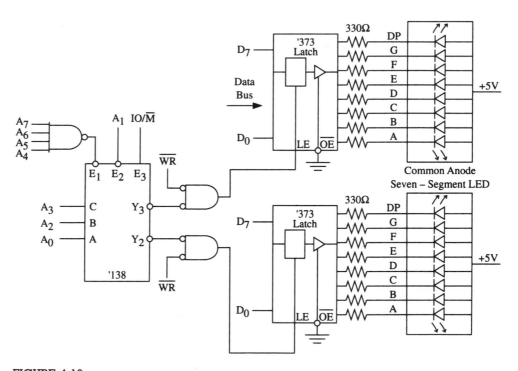

FIGURE 4.19

II
Programming the 8085

Part II of this book is an introduction to assembly language programming for the 8085. It explains commonly used instructions, elementary programming techniques, and their applications.

The content is presented in a format as if for teaching a foreign language. One approach to learning a foreign language is to begin with a few words that can form simple, meaningful, and interactive sentences. After learning a few sentences, the student begins to write a paragraph that can convey an idea in a coherent fashion; then, by sequencing a few paragraphs, begins to compose a letter. Chapters 5 to 11 are arranged in similar fashion—from simple instructions to applications.

Chapter 5 provides an overview of the 8085 instruction set, and Chapters 6 and 7 are concerned primarily with the instructions that occur most frequently. The instructions are not introduced according to the five groups as classified in Chapter 5; instead, a few instructions that can perform a simple task are selected from each group. Chapter 6 includes the discussion of instructions from each group—from data copy to branch instructions. Chapter 7 introduces elementary programming techniques such as looping and indexing. Chapter 8 uses the instructions and techniques presented in these chapters to design software delays and counters.

Chapter 9 introduces the concepts of subroutine and stack, which provide flexibility and variety for program design. Chapter 10 includes applications of the concepts presented in Chapter 9, presenting techniques for writing programs concerned with code conversions, and arithmetic routines. Chapter 11 deals with the uses of assemblers in disk-based software development systems.

PREREQUISITES

The reader is expected to know the following topics:

☐ The 8085 architecture, especially the programming registers.
☐ The concepts related to memory and I/Os.
☐ Logic operations and binary and hexadecimal arithmetic.

Introduction to 8085 Assembly Language Programming

As defined in Chapter 1, an assembly language program is a set of instructions written in the mnemonics of a given microprocessor. These instructions are the commands to the microprocessor to be executed in the given sequence to accomplish a task. To write such programs for the 8085 microprocessor, we should be familiar with the programming model and the instruction set of the microprocessor. This chapter provides such an overview of the 8085 microprocessor.

The 8085 instruction set is classified into five different groups: data transfer, arithmetic, logic, branch, and machine control; each of these groups is illustrated with examples. The chapter also discusses the instruction format and various addressing modes. A simple problem of adding to Hex numbers is used to illustrate writing, assembling, and executing a program. The flowcharting technique and symbols are discussed in the context of the problem. The chapter concludes with a list of selected 8085 instructions.

OBJECTIVES

☐ Explain the various functions of the registers in the 8085 programming model.

☐ Define the term *flag* and explain how the flags are affected.

☐ Explain the terms *operation code* (opcode) and the *operand,* and illustrate these terms by writing instructions.

☐ Classify the instructions in terms of their word size and specify the number of memory registers required to store the instructions in memory.

☐ List the five categories of the 8085 instruction set.

☐ Define and explain the term *addressing mode.*

☐ Write logical steps to solve a simple programming problem.

☐ Draw a flowchart from the logical steps of a given programming problem.

☐ Translate the flowchart into mnemonics and convert the mnemonics into Hex code for a given programming problem.

5.1   THE 8085 PROGRAMMING MODEL

In Chapter 2, we described the 8085 microprocessor registers in reference to the internal data operations. The same information is repeated here briefly to provide the continuity and the context to the instruction set and for readers who prefer to focus initially on the programming aspect of the microprocessor.

5.11   Programming Registers

The 8085 programming model includes six registers, one accumulator, and one flag register, as shown in Figure 5.1. In addition, it has two 16-bit registers: the stack pointer and the program counter. They are described briefly as follows.

REGISTERS

The 8085 has six general-purpose registers to store 8-bit data; these are identified as B, C, D, E, H, and L, as shown in Figure 5.1. They can be combined as register pairs—BC, DE, and HL—to perform some 16-bit operations. The programmer can use these registers to store or copy data into the registers by using data copy instructions.

ACCUMULATOR

The accumulator is an 8-bit register that is part of the arithmetic/logic unit (ALU). This register is used to store 8-bit data and to perform arithmetic and logical operations. The result of an operation is stored in the accumulator. The accumulator is also identified as register A.

FLAGS

The ALU includes five flip-flops, which are set or reset after an operation according to data conditions of the result in the accumulator and other registers. They are called Zero (Z), Carry (CY), Sign (S), Parity (P), and Auxiliary Carry (AC) flags; they are listed in Table 5.1 and their bit positions in the flag register are shown in Figure 5.1(a). The most commonly used flags are Zero, Carry, and Sign. The microprocessor uses these flags to test data conditions.

For example, after an addition of two numbers, if the sum in the accumulator is larger than eight bits, the flip-flop used to indicate a carry—called the Carry flag (CY)—

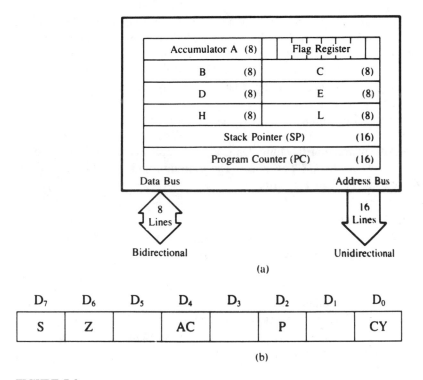

Accumulator A (8)		Flag Register	
B	(8)	C	(8)
D	(8)	E	(8)
H	(8)	L	(8)
Stack Pointer (SP)			(16)
Program Counter (PC)			(16)

Data Bus   Address Bus

8 Lines   16 Lines

Bidirectional   Unidirectional

(a)

D_7	D_6	D_5	D_4	D_3	D_2	D_1	D_0
S	Z		AC		P		CY

(b)

FIGURE 5.1
8085 Programming Model (a) and Flag Register (b)

TABLE 5.1
The 8085 Flags

The following flags are set or reset after the execution of an arithmetic or logic operation; data copy instructions do not affect any flags. See the instruction set (Appendix F) to find how flags are affected by an instruction.

☐ Z—Zero: The Zero flag is set to 1 when the result is zero; otherwise it is reset.
☐ CY—Carry: If an arithmetic operation results in a carry, the CY flag is set; otherwise it is reset.
☐ S—Sign: The Sign flag is set if bit D_7 of the result = 1; otherwise it is reset.
☐ P—Parity: If the result has an even number of 1s, the flag is set; for an odd number of 1s, the flag is reset.
☐ AC—Auxiliary Carry: In an arithmetic operation, when a carry is generated by digit D_3 and passed to digit D_4, the AC flag is set. This flag is used internally for BCD (binary-coded decimal) operations; there is no Jump instruction associated with this flag.

is set to one. When an arithmetic operation results in zero, the flip-flop called the Zero (Z) flag is set to one. Figure 5.1(a) shows an 8-bit register, called the flag register, adjacent to the accumulator. However, it is not used as a register; five bit positions out of eight are used to store the outputs of the five flip-flops. The flags are stored in the 8-bit register so that the programmer can examine these flags (data conditions) by accessing the register through an instruction.

These flags have critical importance in the decision-making process of the microprocessor. The conditions (set or reset) of the flags are tested through software instructions. For example, the instruction JC (Jump On Carry) is implemented to change the sequence of a program when the CY flag is set. The thorough understanding of flags is essential in writing assembly language programs.

PROGRAM COUNTER (PC)

This 16-bit register deals with sequencing the execution of instructions. This register is a memory pointer. Memory locations have 16-bit addresses, and that is why this is a 16-bit register.

The microprocessor uses this register to sequence the execution of the instructions. The function of the program counter is to point to the memory address from which the next byte is to be fetched. When a byte (machine code) is being fetched, the program counter is incremented by one to point to the next memory location.

STACK POINTER (SP)

The stack pointer is also a 16-bit register used as a memory pointer. It points to a memory location in R/W memory, called the stack. The beginning of the stack is defined by loading a 16-bit address in the stack pointer. The stack concept is explained in Chapter 9, "Stack and Subroutines."

This programming model will be used in subsequent chapters to examine how these registers are affected after the execution of an instruction.

5.2   INSTRUCTION CLASSIFICATION

An **instruction** is a binary pattern designed inside a microprocessor to perform a specific function. The entire group of instructions, called the **instruction set,** determines what functions the microprocessor can perform. The 8085 microprocessor includes the instruction set of its predecessor, the 8080A, plus two additional instructions.

5.21   The 8085 Instruction Set

The 8085 instructions can be classified into the following five functional categories: data transfer (copy) operations, arithmetic operations, logical operations, branching operations, and machine-control operations.

DATA TRANSFER (COPY) OPERATIONS

This group of instructions copies data from a location called a source to another location, called a destination, without modifying the contents of the source. In technical manuals, the term *data transfer* is used for this copying function. However, the term *transfer* is misleading; it creates the impression that the contents of a source are destroyed when, in fact, the contents are retained without any modification. The various types of data transfer (copy) are listed below together with examples of each type:

Types	Examples
☐ Between registers	Copy the contents of register B into register D.
☐ Specific data byte to a register or a memory location	Load register B with the data byte 32H.
☐ Between a memory location and a register	From the memory location 2000H to register B.
☐ Between an I/O device and the accumulator	From an input keyboard to the accumulator.

ARITHMETIC OPERATIONS

These instructions perform arithmetic operations such as addition, subtraction, increment, and decrement.

☐ **Addition**—Any 8-bit number, or the contents of a register, or the contents of a memory location can be added to the contents of the accumulator and the sum is stored in the accumulator. No two other 8-bit registers can be added directly (e.g., the contents of register B cannot be added directly to the contents of register C). The instruction DAD is an exception; it adds 16-bit data directly in register pairs.

☐ **Subtraction**—Any 8-bit number, or the contents of a register, or the contents of a memory location can be subtracted from the contents of the accumulator and the results stored in the accumulator. The subtraction is performed in 2's complement, and the results, if negative, are expressed in 2's complement. No two other registers can be subtracted directly.

☐ **Increment/Decrement**—The 8-bit contents of a register or a memory location can be incremented or decremented by 1. Similarly, the 16-bit contents of a register pair (such as BC) can be incremented or decremented by 1. These increment and decrement operations differ from addition and subtraction in an important way; i.e., they can be performed in any one of the registers or in a memory location.

LOGICAL OPERATIONS

These instructions perform various logical operations with the contents of the accumulator.

☐ **AND, OR, Exclusive-OR**—Any 8-bit number, or the contents of a register, or of a memory location can be logically ANDed, ORed, or Exclusive-ORed with the contents of the accumulator. The results are stored in the accumulator.

☐ **Rotate**—Each bit in the accumulator can be shifted either left or right to the next position.

☐ **Compare**—Any 8-bit number, or the contents of a register, or a memory location can be compared for equality, greater than, or less than, with the contents of the accumulator.

☐ **Complement**—The contents of the accumulator can be complemented; all 0s are replaced by 1s and all 1s are replaced by 0s.

BRANCHING OPERATIONS

This group of instructions alters the sequence of program execution either conditionally or unconditionally.

☐ **Jump**—Conditional jumps are an important aspect of the decision-making process in programming. These instructions test for a certain condition (e.g., Zero or Carry flag) and alter the program sequence when the condition is met. In addition, the instruction set includes an instruction called *unconditional jump*.

☐ **Call, Return, and Restart**—These instructions change the sequence of a program either by calling a subroutine or returning from a subroutine. The conditional Call and Return instructions also can test condition flags.

MACHINE CONTROL OPERATIONS

These instructions control machine functions such as Halt, Interrupt, or do nothing.

5.22   Review of the 8085 Operations

The microprocessor operations related to data manipulation can be summarized in four functions:

1. copying data
2. performing arithmetic operations
3. performing logical operations
4. testing for a given condition and altering the program sequence

Some important aspects of the instruction set are noted below:

1. In data transfer, the contents of the source are not destroyed; only the contents of the destination are changed. The data copy instructions do not affect the flags.
2. Arithmetic and logical operations are performed with the contents of the accumulator, and the results are stored in the accumulator (with some exceptions). The flags are affected according to the results.
3. Any register including memory can be used for increment and decrement.
4. A program sequence can be changed either conditionally or by testing for a given data condition.

INSTRUCTION AND DATA FORMAT 5.3

An **instruction** is a command to the microprocessor to perform a given task on specified data. Each instruction has two parts: one is the task to be performed, called the **operation code** (opcode), and the second is the data to be operated on, called the **operand.** The operand (or data) can be specified in various ways. It may include 8-bit (or 16-bit) data, an internal register, a memory location, or an 8-bit (or 16-bit) address. In some instructions, the operand is implicit.

5.31 Instruction Word Size

The 8085 instruction set is classified into the following three groups according to word size:

1. One-word or 1-byte instructions
2. Two-word or 2-byte instructions
3. Three-word or 3-byte instructions

In the 8085, "byte" and "word" are synonymous because it is an 8-bit microprocessor. However, instructions are commonly referred to in terms of bytes rather than words.

ONE-BYTE INSTRUCTIONS

A 1-byte instruction includes the opcode and the operand in the same byte. For example:

Task	Opcode	Operand*	Binary Code	Hex Code
Copy the contents of the accumulator in register C.	MOV	C,A	0100 1111	4FH
Add the contents of register B to the contents of the accumulator.	ADD	B	1000 0000	80H
Invert (complement) each bit in the accumulator.	CMA		0010 1111	2FH

These instructions are 1-byte instructions performing three different tasks. In the first instruction, both operand registers are specified. In the second instruction, the operand B is specified and the accumulator is assumed. Similarly, in the third instruction,

---

*In the operand, the destination register C is shown first, followed by the source register A.

the accumulator is assumed to be the implicit operand. These instructions are stored in 8-bit binary format in memory; each requires one memory location.

TWO-BYTE INSTRUCTIONS

In a 2-byte instruction, the first byte specifies the operation code and the second byte specifies the operand. For example:

Task	Opcode	Operand	Binary Code	Hex Code	
Load an 8-bit data byte in the ac- cumulator.	MVI	A,Data	0011 1110	3E	First Byte
			DATA	Data	Second Byte

Assume the data byte is 32H. The assembly language instruction is written as

Mnemonics	Hex Code
MVI A,32H	3E 32H

This instruction would require two memory locations to store in memory.

THREE-BYTE INSTRUCTIONS

In a 3-byte instruction, the first byte specifies the opcode, and the following two bytes specify the 16-bit address. Note that the second byte is the low-order address and the third byte is the high-order address. For example:

Task	Opcode	Operand	Binary Code	Hex Code	
Transfer the program sequence to the memory location 2085H.	JMP	2085H	1100 0011	C3*	First Byte
			1100 0011	85	Second Byte
			0010 0000	20	Third Byte

This instruction would require three memory locations to store in memory.

These commands are in many ways similar to our everyday conversation. For example, while eating in a restaurant, we may make the following requests and orders:

1. Pass (the) butter.
2. Pass (the) bowl.

---

*The jump location 2085H is stored in memory in the reversed order: 85H first, followed by 20H.

3. (Let us) eat.
4. I will have combination 17 (on the menu).
5. I will have what Susie ordered.

The first request specifies the exact item; it is similar to the instruction for loading a specific data byte in a register. The second request mentions the bowl rather than the contents, even though one is interested in the contents of the bowl. It is similar to the instruction MOV C,A where registers (bowls) are specified rather than data. The third suggestion (let us eat) assumes that one knows what to eat. It is similar to the instruction Complement, which implicitly assumes that the operand is the accumulator. In the fourth sentence, the location of the item on the menu is specified and not the actual item. It is similar to the instruction: Transfer the data byte from the location 2050H. The last order (what Susie ordered) is specified indirectly. It is similar to an instruction that specifies a memory location through the contents of a register pair. (Examples of the last two types of instruction are illustrated in later chapters.)

These various ways of specifying data are called the **addressing modes.** Although microprocessor instructions require one or more words to specify the operands, the notations and conventions used in specifying the operands have very little to do with the operation of the microprocessor. The mnemonic letters used to specify a command are chosen (somewhat arbitrarily) by the manufacturer. When an instruction is stored in memory, it is stored in binary code, the only code the microprocessor is capable of reading and understanding. The conventions used in specifying the instructions are valuable in terms of keeping uniformity in different programs and in writing assemblers. The important point to remember is that the microprocessor neither reads nor understands mnemonics or hexadecimal numbers.

5.32 Opcode Format

To understand operation codes, we need to examine how an instruction is designed into the microprocessor. This information will be useful in reading a user's manual, in which operation codes are specified in binary format and 8-bits are divided in various groups.

In the design of the 8085 microprocessor chip, all operations, registers, and status flags are identified with a specific code. For example, all internal registers are identified as follows:

Code	Registers	Code	Register Pairs
000	B	00	BC
001	C	01	DE
010	D	10	HL
011	E	11	AF OR SP
100	H		
101	L		
111	A		
110	Reserved for Memory-Related operation		

Some of the operation codes are identified as follows:

Function	Operation Code
1. Rotate each bit of the accumulator to the left by one position.	$00000111 = 07H$ (8-bit opcode)
2. Add the contents of a register to the accumulator.	10000 SSS (5-bit opcode—3 bits are reserved for a register)

This instruction is completed by adding the code of the register. For example,

Add	:	10000
Register B	:	000
to A	:	Implicit
Binary Instruction:		$10000\ 000 = 80H$
		Add Reg.B

In assembly language, this is expressed as

Opcode	Operand	Hex Code
ADD	B	80H

3. MOVE (Copy) the content of register Rs (source) to register Rd (destination)

01	DDD	SSS
2-bit Opcode for MOVE	Reg. Rd	Reg. Rs

This instruction is completed by adding the codes of two registers. For example,

Move (copy) the content:		0 1
To register C	:	0 0 1 (DDD)
From register A	:	1 1 1 (SSS)
Binary Instruction	:	$0\ 1\ 0\ 0\ 1\ 1\ 1\ 1 \rightarrow 4FH$
		Opcode Operand

In assembly language, this is expressed as

Opcode	Operand	Hex Code
MOV	C,A	4F

Please note that the first register is the destination and the second register is the source—from A to C—which appears reversed for a general pattern from left to right. Typically, in the 8085 user's manual the data transfer (copy) instruction is shown as follows:

MOV rl, r2*

0	1	D	D	D	S	S	S

5.33 Data Format

The 8085 is an 8-bit microprocessor, and it processes (copy, add, subtract, etc.) only binary numbers. However, the real world operates in decimal numbers and languages of alphabets and characters. Therefore, we need to code binary numbers into different media. Let us examine coding. What is the letter "A"? It is a symbol representing a certain sound in a visual medium that eyes can recognize. Similarly, we can represent or code groups of bits into different media. In 8-bit processor systems, commonly used codes and data formats are ASCII, BCD, signed integers, and unsigned integers. They are explained as follows.

- **ASCII Code**—This is a 7-bit alphanumeric code that represents decimal numbers, English alphabets, and nonprintable characters such as carriage return. Extended ASCII is an 8-bit code. The additional numbers (beyond 7-bit ASCII code) represent graphical characters. This code was discussed in Chapter 1 (Section 1.24).
- **BCD Code**—The term *BCD* stands for binary-coded decimal; it is used for decimal numbers. The decimal numbering system has ten digits, 0 to 9. Therefore, we need only four bits to represent ten digits from 0000 to 1001. The remaining numbers, 1010 (A) to 1111 (F), are considered invalid. An 8-bit register in the 8085 can accommodate two BCD numbers.
- **Signed Integer**—A signed integer is either a positive number or a negative number. In an 8-bit processor, the most significant digit, D_7, is used for the sign; 0 represents the positive sign and 1 represents the negative sign. The remaining seven bits, D_6–D_0, represent the magnitude of an integer. Therefore, the largest positive integer that can be processed by the 8085 at one time is 0111 1111 (7FH); the remaining Hex numbers, 80H to FFH, are considered negative numbers. However, all negative numbers in this microprocessor are represented in 2's complement format (see Appendix A.2 for additional explanation).
- **Unsigned Integers**—An integer without a sign can be represented by all the 8 bits in a microprocessor register. Therefore, the largest number that can be processed at one time is FFH. However, this does not imply that the 8085 microprocessor is limited to handling only 8-bit numbers. Numbers larger than 8 bits (such as 16-bit or 24-bit numbers) are processed by dividing them in groups of 8 bits.

Now let us examine how the microprocessor interprets any number. Let us assume that after performing some operations the result in the accumulator is 0100 0001 (41H). This number can have many interpretations: (1) It is an unsigned number equivalent to 65 in decimal; (2) it is a BCD number representing 41 decimal; (3) it is the ASCII capital let-

---

*In this text, rl is specified as Rd and r2 is specified as Rs to indicate destination and source.

ter "A"; or (4) it is a group of 8 bits where bits D_6 and D_0 turn on and the remaining bits turn off output devices. The processor processes binary bits; it is up to the user to interpret the result. In our example, the number 41H can be displayed on a screen as an ASCII "A" or 41 BCD.

5.4 HOW TO WRITE, ASSEMBLE, AND EXECUTE A SIMPLE PROGRAM

A program is a sequence of instructions written to tell a computer to perform a specific function. The instructions are selected from the instruction set of the microprocessor. To write a program, divide a given problem in small steps in terms of the operations the 8085 can perform, then translate these steps into instructions. Writing a simple program of adding two numbers in the 8085 language is illustrated below.

5.41 Illustrative Program: Adding Two Hexadecimal Numbers

PROBLEM STATEMENT
Write instructions to load the two hexadecimal numbers 32H and 48H in registers A and B, respectively. Add the numbers, and display the sum at the LED output port PORT1.

PROBLEM ANALYSIS
Even though this is a simple problem, it is necessary to divide the problem into small steps to examine the process of writing programs. The wording of the problem provides sufficient clues for the necessary steps. They are as follows:

1. Load the numbers in the registers.
2. Add the numbers.
3. Display the sum at the output port PORT1.

FLOWCHART
The steps listed in the problem analysis and the sequence can be represented in a block diagram, called a flowchart. Figure 5.2 shows such a flowchart representing the above steps. This is a simple flowchart, and the steps are self-explanatory. We will discuss flowcharting in the next chapter.

ASSEMBLY LANGUAGE PROGRAM
To write an assembly language program, we need to translate the blocks shown in the flowchart into 8085 operations and then, subsequently, into mnemonics. By examining the blocks, we can classify them into three types of operations: Blocks 1 and 3 are copy operations; Block 2 is an arithmetic operation; and Block 4 is a machine-control operation. To translate these steps into assembly and machine languages, you should review the instruction set. The translation of each block into mnemonics with comments is shown as follows:

FIGURE 5.2
Flowchart: Adding Two Numbers

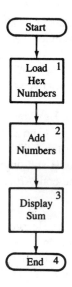

Block 1:	MVI A,32H	Load register A with 32H
	MVI B,48H	Load register B with 48H
Block 2:	ADD B	Add two bytes and save the sum in A
Block 3:	OUT 01H	Display accumulator contents at port 01H
Block 4:	HALT	End

FROM ASSEMBLY LANGUAGE TO HEX CODE

To convert the mnemonics into Hex code, we need to look up the code in the 8085 instruction set; this is called either manual or hand assembly.

Mnemonics	Hex Code	
MVI A,32H	3E	2-byte instruction
	32	
MVI B,48H	06	2-byte instruction
	48	
ADD B	80	1-byte instruction
OUT 01H	D3	2-byte instruction
	01	
HLT	76	1-byte instruction

STORING IN MEMORY AND CONVERTING FROM HEX CODE TO BINARY CODE

To store the program in R/W memory of a single-board microcomputer and display the output, we need to know the memory addresses and the output port address. Let us assume that R/W memory ranges from 2000H to 20FFH, and the system has an LED output port with the address 01H. Now, to enter the program:

1. Reset the system by pushing the RESET key.
2. Enter the first memory address using Hex keys where the program should be stored. Let us assume it is 2000H.
3. Enter each machine code by pushing Hex keys. For example, to enter the first machine code, push the 3, E, and STORE keys. (The STORE key may be labeled differently in different systems.) When you push the STORE key, the program will store the machine code in memory location 2000H and upgrade the memory address to 2001H.
4. Repeat Step 3 until the last machine code, 76H.
5. Reset the system.

Now the question is: How does the Hex code get converted into binary code? The answer lies with the Monitor program stored in Read-Only memory (or EPROM) of the microcomputer system. An important function of the Monitor program is to check the keys and convert Hex code into binary code. The entire process of manual assembly is shown in Figure 5.3.

In this illustrative example, the program will be stored in memory as follows:

Mnemonics	Hex Code	Memory Contents	Memory Address
MVI A,32H	3E	0 0 1 1 1 1 1 0	2000
	32	0 0 1 1 0 0 1 0	2001
MVI B,48H	06	0 0 0 0 0 1 1 0	2002
	48	0 1 0 0 1 0 0 0	2003
ADD B	80	1 0 0 0 0 0 0 0	2004
OUT 01H	D3	1 1 0 1 0 0 1 1	2005
	01	0 0 0 0 0 0 0 1	2006
HLT	76	0 1 1 1 1 1 1 0	2007

This program has eight machine codes and will require eight memory locations to store the program. The critical concept that needs to be emphasized here is that the microprocessor can understand and execute only the binary instructions (or data); everything else (mnemonics, Hex code, comments) is for the convenience of human beings.

EXECUTING THE PROGRAM

To execute the program, we need to tell the microprocessor where the program begins by entering the memory address 2000H. Now, we can push the Execute key (or the key with a similar label) to begin the execution. As soon as the Execute function key is pushed, the microprocessor loads 2000H in the program counter, and the program control is transferred from the Monitor program to our program.

The microprocessor begins to read one machine code at a time, and when it fetches the complete instruction, it executes that instruction. For example, it will fetch the ma-

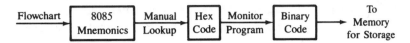

FIGURE 5.3
Manual Assembly Process

chine codes stored in memory locations 2000H and 2001H and execute the instruction MVI A,32H; thus it will load 32H in register A. The ADD instruction will add the two numbers, and the OUT instruction will display the answer 7AH (32H + 48H = 7AH) at the LED port. It continues to execute instructions until it fetches the HLT instruction.

RECOGNIZING THE NUMBER OF BYTES IN AN INSTRUCTION

Students who are introduced to an assembly language for the first time should hand assemble at least a few small programs. Such exercises can clarify the relationship among instruction codes, data, memory registers, and memory addressing. One of the stumbling blocks in hand assembly is in recognizing the number of bytes in a given instruction. The following clues can be used to recognize the number of bytes in an instruction of the 8085 microprocessor.

1. **One-byte instruction**—A mnemonic followed by a letter (or two letters) representing the registers (such as A, B, C, D, E, H, L, M, and SP) is a one-byte instruction. Instructions in which registers are implicit are also one-byte instructions.
   Examples: (a) MOV A, B; (b) DCX SP; (c) RRC
2. **Two-byte instruction**—A mnemonic followed by 8-bit (byte) is a two-byte instruction.
   Examples: (a) MVI A, 8-bit; (b) ADI 8-bit
3. **Three-byte instruction**—A mnemonic followed by 16-bit (also terms such as adr or dble) is a three-byte instruction.
   Examples: (a) LXI B, 16-bit (dble); (b) JNZ 16-bit (adr); (c) CALL 16-bit (adr)

   In writing assembly language programs, we can assign memory addresses in a sequence once we know the number of bytes in a given instruction. For example, a three-byte instruction has three Hex codes and requires three memory locations in a sequence. In hand assembly, omitting a byte inadvertently can have a disastrous effect on program execution, as explained in the next section.

5.42 How Does a Microprocessor Differentiate Between Data and Instruction Code?

The microprocessor is a sequential machine. As soon as a microprocessor-based system is turned on, it begins the execution of the code in memory. The execution continues in a sequence, one code after another (one memory location after another) at the speed of its clock until the system is turned off (or the clock stops). If an unconditional loop is set up in a program, the execution will continue until the system is either reset or turned off.

Now a puzzling question is: How does the microprocessor differentiate between a code and data when both are binary numbers? The answer lies in the fact that the microprocessor interprets the first byte it fetches as an opcode. When the 8085 is reset, its program counter is cleared to 0000H and it fetches the first code from the location 0000H. In the example of the previous section, we tell the processor that our program begins at location 2000H. The first code it fetches is 3EH. When it decodes that code, it knows that it is a two-byte instruction. Therefore, it assumes that the second code, 32H, is a data byte. If we forget to enter 32H and enter the next code, 06H, instead, the 8085 will load 06H in the accumulator, interpret the next code, 48H, as an opcode, and continue the execution in sequence. As a consequence, we may encounter a totally unexpected result.

5.5 OVERVIEW OF THE 8085 INSTRUCTION SET

The 8085 microprocessor instruction set has 74 operation codes that result in 246 instructions. The set includes all the 8080A instructions plus two additional instructions (SIM and RIM, related to serial I/O). It is an overwhelming experience for a beginner to study these instructions. You are strongly advised not to attempt to read all these instructions at one time. However, you should be able to grasp an overview of the set by examining the frequently used instructions listed below.*

The following notations are used in the description of the instructions.

R = 8085 8-bit register   (A, B, C, D, E, H, L)
M = Memory register (location)
Rs = Register source   ⎫
Rd = Register destination ⎬   (A, B, C, D, E, H, L)
Rp = Register pair   (BC, DE, HL, SP)
() = Contents of

1. Data transfer (copy) instructions: From register to register.
                                               Load an 8-bit number in a register.
                                               Between memory and register.
                                               Between I/O and accumulator.
                                               Load 16-bit number in a register pair.

Mnemonics   **Tasks**

MOV Rd,Rs   Copy data from source register Rs into
                                                   destination register Rd.

---

*These instructions are explained and illustrated in the next two chapters. The complete instruction set is explained alphabetically in Appendix F for easy reference; the appendix also includes three lists of instruction summaries arranged according to the functions, hexadecimal sequence of machine codes, and alphabetical order.

MVI R,8-bit	Load 8-bit data in a register.
OUT 8-bit (port address)	Send (write) data byte from the accumulator to an output device.
IN 8-bit (port address)	Accept (read) data byte from an input device and place it in the accumulator.
LXI Rp,16-bit	Load 16-bit in a register pair.
MOV R,M	Copy the data byte from a memory location (source) into a register.
LDAX Rp	Copy the data byte into the accumulator from the memory location indicated by a register pair.
LDA 16-bit	Copy the data byte into the accumulator from the memory location specified by 16-bit address.
MOV M,R	Copy the data byte from register into memory location.
STAX Rp	Copy the data byte from the accumulator into the memory location indicated by a register pair.
STA 16-bit	Copy the data byte from the accumulator in the memory location specified by 16-bit address.

2. Arithmetic instructions:

	Add
	Subtract
	Increment (Add 1)
	Decrement (Subtract 1)
ADD R	Add the contents of a register to the contents of the accumulator.
ADI 8-bit	Add 8-bit data to the contents of the accumulator.
SUB R	Subtract the contents of a register from the contents of the accumulator.
SUI 8-bit	Subtract 8-bit data from the contents of the accumulator.
INR R	Increment the contents of a register.
DCR R	Decrement the contents of a register.
INX Rp	Increment the contents of a register pair.
DCX Rp	Decrement the contents of a register pair.
ADD M	Add the contents of a memory location to the contents of the accumulator.
SUB M	Subtract the contents of a memory location from the contents of the accumultor.
INR M	Increment the contents of a memory location.

DCR M	Decrement the contents of a memory location.
3. Logical instructions:	AND OR X-OR Compare Rotate
ANA R/M	Logically AND the contents of register/memory with the contents of the accumulator.
ANI 8-bit	Logically AND the 8-bit data with the contents of the accumulator.
ORA R/M	Logically OR the contents of register/memory with the contents of the accumulator.
ORI 8-bit	Logically OR the 8-bit data with the contents of the accumulator.
XRA R/M	Exclusive-OR the contents of register/memory with the contents of the accumulator.
XRI 8-bit	Exclusive-OR the 8-bit data with the contents of the accumulator.
CMA	Complement the contents of the accumulator.
RLC	Rotate each bit in the accumulator to the left position.
RAL	Rotate each bit in the accumulator including the carry to the left position.
RRC	Rotate each bit in the accumulator to the right position.
RAR	Rotate each bit in the accumulator including the carry to the right position.
CMP R/M	Compare the contents of register/memory with the contents of the accumulator for less than, equal to, or more than.
CPI 8-bit	Compare 8-bit data with the contents of the accumulator for less than, equal to, or more than.
4. Branch instructions:	Change the program sequence unconditionally. Change the program sequence if specified data conditions are met.
JMP 16-bit address	Change the program sequence to the location specified by the 16-bit address.

JZ 16-bit address	Change the program sequence to the location specified by the 16-bit address if the Zero flag is set.
JNZ 16-bit address	Change the program sequence to the location specified by the 16-bit address if the Zero flag is reset.
JC 16-bit address	Change the program sequence to the location specified by the 16-bit address if the Carry flag is set.
JNC 16-bit address	Change the program sequence to the location specified by the 16-bit address if the Carry flag is reset.
CALL 16-bit address	Change the program sequence to the location of a subroutine.
RET	Return to the calling program after completing the subroutine sequence.

5. Machine control instructions:

| HLT | Stop processing and wait. |
| NOP | Do not perform any operation. |

This set of instructions is a representative sample; it does not include various instructions related to 16-bit data operations, additional Jump instructions, and conditional Call and Return instructions.

SUMMARY

This chapter described the data manipulation functions of the 8085 microprocessor, provided an overview of the instruction set, and illustrated the execution of instructions in relation to the system's clock. The important concepts in this chapter can be summarized as follows.

☐ The 8085 microprocessor operations are classified into five major groups: data transfer (copy), arithmetic, logic, branch, and machine control.

☐ An instruction has two parts: opcode (operation to be performed) and operand (data to be operated on). The operand can be data (8- or 16-bit), address, or register, or it can be implicit. The method of specifying an operand (directly, indirectly, etc.) is called the addressing mode.

☐ The instruction set is classified in three groups according to the word size: 1-, 2-, or 3-byte instructions.

☐ To write an assembly language program, divide the given problem into small steps in terms of the microprocessor operations, translate these steps into assembly language instructions, and then translate them into the 8085 machine code.

QUESTIONS AND PROGRAMMING ASSIGNMENTS

1. List the four categories of 8085 instructions that manipulate data.
2. Define opcode and operand, and specify the opcode and the operand in the instruction MOV H,L.
3. Write the machine code for the instruction MOV H,A if the opcode $= 01_2$, the register code for $H = 100_2$, and the register code for $A = 111_2$.
4. Find the machine codes and the number of bytes of the following instructions. Identify the opcodes and the operands. (Refer to the instruction set on the inside back cover.)

   a. MVI H,47H   **b.** ADI F5H   **c.** SUB C

5. Find the Hex codes for the following instructions, identify the opcodes and operands, and show the order of entering the codes in memory.

   a. STA 2050H   **b.** JNZ 2070H

6. Find the Hex machine code for the following instructions from the instruction set listed on the back cover, and identify the number of bytes of each instruction.

   ```
   MVI B,4FH      ;Load the first byte
   MVI C,78H      ;Load the second byte
   MOV A,C        ;Get ready for addition
   ADD B          ;Add two bytes
   OUT 07H        ;Display the result at port 7
   HLT            ;End of program
   ```

7. If the starting address of the system memory is 2000H, and you were to enter the Hex code for the instructions in Question 6, identify the memory addresses and their corresponding Hex codes.
8. Assemble the following program, starting with the memory address 2020H.

   ```
   MVI A,8FH      ;Load the first byte
   MVI B,68H      ;Load the second byte
   SUB B          ;Subtract the second byte
   ANI 0FH        ;Eliminate D7–D4
   STA 2070H      ;Store D3–D0 in memory location 2070H
   HLT            ;End of program
   ```

9. Assemble the following program, starting at location 2000H.

   ```
   START: IN F2H      ;Read input switches at port F2H
          CMA         ;Set ON switches to logic 1
          ORA A       ;Set Z flag if no switch is ON
          JZ START    ;Go back and read input port if all
                      ;  switches are off
   ```

10. Write logical steps to add the following two Hex numbers. Both the numbers should be saved for future use. Save the sum in the accumulator.

Numbers: A2H and 18H

11. Translate the program in Question 10 into the 8085 assembly language.

12. Data byte 28H is stored in register B and data byte 97H is stored in the accumulator. Show the contents of registers B, C, and the accumulator after the execution of the following two instructions:

MOV A,B
MOV C,A

13. In Question 6, explain the potential results of the program if the code 07H of the out instruction is omitted.

14. In Question 8, explain possible outcomes if the second byte 0FH of the instruction ANI 0FH is omitted.

Introduction to 8085 Instructions

A microcomputer performs a task by reading and executing the set of instructions written in its memory. This set of instructions, written in a sequence, is called a **program.** Each instruction in the program is a command, in binary, to the microprocessor to perform an operation. This chapter introduces 8085 basic instructions, their operations, and their applications.

Chapters 2 and 3 described the architecture of the 8085 microprocessor and Chapter 5 provided an overview of the instruction set and the tasks the 8085 can perform. This chapter is concerned with using instructions within the constraints and capabilities of its registers and the bus system. A few instructions are introduced from each of the five groups (Data Transfer, Arithmetic, Logical, Branch, and Machine Control) and are used to write simple programs to perform specific tasks.

The simple illustrative programs given in this chapter can be entered and executed on the single-board microcomputers used commonly in college laboratories.

OBJECTIVES

☐ Explain the functions of data transfer (copy) instructions and how the contents of the source register and the destination register are affected.

☐ Explain the Input/Output instructions and port addresses.

☐ Explain the functions of the machine control instructions HLT and NOP.

☐ Recognize the addressing modes of the instructions.

☐ Draw a flowchart of a simple program.

☐ Write a program in 8085 mnemonics to illustrate an application of data copy instructions, and translate those mnemonics manually into their Hex codes.

☐ Write a program in the proper format showing memory addresses, Hex machine codes, mnemonics, and comments.

☐ Explain the arithmetic instructions, and recognize the flags that are set or reset for given data conditions.

☐ Write a set of instructions to perform an addition and a subtraction (in 2's complement).

☐ Explain the logic instructions, and recognize the flags that are set or reset for given data conditions.

☐ Write a set of instructions to illustrate logic operations.

☐ Explain the use of logic instructions in masking, setting, and resetting individual bits.

☐ Explain the unconditional and conditional Jump instructions and how flags are used by the conditional Jump instructions to change the sequence of a program.

☐ Write a program to illustrate an application of Jump instructions.

☐ List the important steps in writing and troubleshooting a simple program.

6.1 DATA TRANSFER (COPY) OPERATIONS

One of the primary functions of the microprocessor is copying data, from a register (or I/O or memory) called the source, to another register (or I/O or memory) called the destination. In technical literature, the copying function is frequently labeled as the **data transfer function,** which is somewhat misleading. In fact, the contents of the source are not transferred, but are copied into the destination register without modifying the contents of the source.

Several instructions are used to copy data (as listed in Chapter 5). This section is concerned with the following operations.

MOV	: Move	Copy a data byte.
MVI	: Move Immediate	Load a data byte directly.
OUT	: Output to Port	Send a data byte to an output device.
IN	: Input from Port	Read a data byte from an input device.

The term *copy* is equally valid for input/output functions because the contents of the source are not altered. However, the term *data transfer* is used so commonly to indicate the data copy function that, in this book, these terms are used interchangeably when the meaning is not ambiguous.

In addition to data copy instructions, it is necessary to introduce two machine-control operations to execute programs.

HLT: Halt	Stop processing and wait.
NOP: No Operation	Do not perform any operation.

These operations (opcodes) are explained and illustrated below with examples.

INSTRUCTIONS

The data transfer instructions copy data from a source into a destination without modifying the contents of the source. The previous contents of the destination are replaced by the contents of the source.

Important Note: In the 8085 processor, data transfer instructions do not affect the flags.

Opcode	Operand	Description
MOV	Rd,Rs*	Move ☐ This is a 1-byte instruction ☐ Copies data from source register Rs to destination register Rd
MVI	R,8-bit	Move Immediate ☐ This is a 2-byte instruction ☐ Loads the 8 bits of the second byte into the register specified
OUT	8-bit port address	Output to Port ☐ This is a 2-byte instruction ☐ Sends (copies) the contents of the accumulator (A) to the output port specified in the second byte
IN	8-bit port address	Input from Port ☐ This is a 2-byte instruction ☐ Accepts (reads) data from the input port specified in the second byte, and loads into the accumulator
HLT		Halt ☐ This is a 1-byte instruction ☐ The processor stops executing and enters wait state ☐ The address bus and data bus are placed in high impedance state. No register contents are affected
NOP		No Operation ☐ This is a 1-byte instruction ☐ No operation is performed ☐ Generally used to increase processing time or substitute in place of an instruction. When an error occurs in a program and an instruction needs to be eliminated, it is more convenient to substitute NOP than to reassemble the whole program

*The symbols Rd and Rs are generic terms; they represent any of the 8085 registers: A, B, C, D, E, H, and L.

Example 6.1	Load the accumulator A with the data byte 82H (the letter H indicates hexadecimal number), and save the data in register B.

Instructions   MVI A, 82H,
                     MOV B,A

The first instruction is a 2-byte instruction that loads the accumulator with the data byte 82H, and the second instruction MOV B,A copies the contents of the accumulator in register B without changing the contents of the accumulator.

Example 6.2	Write instructions to read eight ON/OFF switches connected to the input port with the address 00H, and turn on the devices connected to the output port with the address 01H, as shown in Figure 6.1. (I/O port addresses are given in hexadecimal.)

Solution	The input has eight switches that are connected to the data bus through the tri-state buffer. Any one of the switches can be connected to +5 V (logic 1) or to ground (logic 0), and each switch controls the corresponding device at the output port. The microprocessor needs to read the bit pattern on the switches and send the same bit pattern to the output port to turn on the corresponding devices.

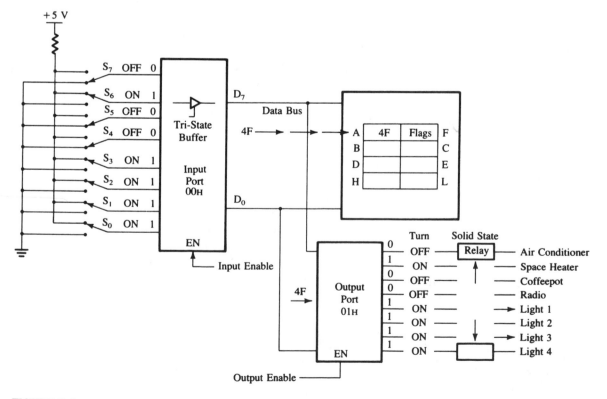

FIGURE 6.1
Reading Data at Input Port and Sending Data to Output Port

Instructions   IN   00H
              OUT   01H
              HLT

When the microprocessor executes the instruction IN 00H, it enables the tri-state buffer. The bit pattern 4FH formed by the switch positions is placed on the data bus and transferred to the accumulator. This is called reading an input port.

When the microprocessor executes the next instruction, OUT 01H, it places the contents of the accumulator on the data bus and enables the output port 01H. (This is also called writing data to an output port.) The output port latches the bit pattern and turns ON/OFF the devices connected to the port according to the bit pattern. In Figure 6.1, the bit pattern 4FH will turn on the devices connected to the output port data lines D_6, D_3, D_2, D_1, and D_0: the space heater and four light bulbs. To turn off some of the devices and turn on other devices, the bit pattern can be modified by changing the switch positions. For example, to turn on the radio and the coffeepot and turn off all other devices, the switches S_4 and S_5 should be on and the others should be off. The microprocessor will read the bit pattern 0011 0000, and this bit pattern will turn on the radio and the coffeepot and turn off other devices.

The preceding explanation raises two questions:

1. What are the second bytes in the instructions IN and OUT?
2. How are they determined?

In answer to the first question, the second bytes are I/O port addresses. Each I/O port is identified with a number or an address similar to the postal address of a house. The second byte has eight bits, meaning $256(2^8)$ combinations; thus 256 input ports and 256 output ports with addresses from 00H to FFH can be connected to the system.

The answer to the second question depends on the logic circuit (called interfacing) used to connect and identify a port by the system designer (see Chapter 4).

6.11 Addressing Modes

The above instructions are commands to the microprocessor to copy 8-bit data from a source into a destination. In these instructions, the source can be a register, an input port, or an 8-bit number (00H to FFH). Similarly, a destination can be a register or an output port. The sources and destination are, in fact, operands. The various formats of specifying the operands are called the addressing modes. The 8085 instruction set has the following addressing modes. (Each mode is followed by an example and by the corresponding piece of restaurant conversation from the analogy discussed in Chapter 5.)

1. Immediate Addressing—MVI R,Data (Pass the butter)
2. Register Addressing—MOV Rd,Rs (Pass the bowl)
3. Direct Addressing—IN/OUT Port# (Combination number 17 on the menu)
4. Indirect Addressing—Illustrated in the next chapter (I will have what Susie has)

The classification of the addressing modes is unimportant, except that it provides some clues in understanding mnemonics. For example, in the case of the MVI opcode, the letter I suggests that the second byte is data and not a register. What is important is to become familiar with the instructions. After you study the examples given in this chapter, you will see a pattern begin to emerge.

6.12 Illustrative Program: Data Transfer—From Register to Output Port

PROBLEM STATEMENT

Load the hexadecimal number 37H in register B, and display the number at the output port labeled PORT1.

PROBLEM ANALYSIS

This problem is similar to the illustrative program discussed in Section 5.41. Even though this is a very simple problem it is necessary to break the problem into small steps and to outline the thinking process in terms of the tasks described in Section 6.1.

STEPS

Step 1: Load register B with a number.
Step 2: Send the number to the output port.

QUESTIONS TO BE ASKED

☐ Is there an instruction to load the register B? YES—MVI B.
☐ Is there an instruction to send the data from register B to the output port? NO. Review the instruction OUT. This instruction sends data from the accumulator to an output port.
☐ The solution appears to be as follows: Copy the number from register B into accumulator A.
☐ Is there an instruction to copy data from one register to another register? YES—MOV Rd,Rs.

FLOWCHART

The thinking process described here and the steps necessary to write the program can be represented in a pictorial format, called a **flowchart.** Figure 6.2 describes the preceding steps in a flowchart.

Flowcharting is an art. The flowchart in Figure 6.2 does not include all the steps described earlier. Although the number of steps that should be represented in a flowchart is ambiguous, not all of them should be included. That would defeat the purpose of the flowchart. It should represent a logical approach and sequence of steps in solving the problem. A flowchart is similar to the block diagram of a hardware system or to the outline of a chapter. Information in each block of the flowchart should be similar to the heading of a paragraph. Generally, a flowchart is used for two purposes: to assist and clarify the thinking process and to communicate the programmer's thoughts or logic to others.

Symbols commonly used in flowcharting are shown in Figure 6.3. Two types of symbols—rectangles and ovals—are already illustrated in Figure 6.2. The diamond is

FIGURE 6.2
Flowchart

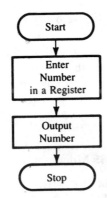

Meaning

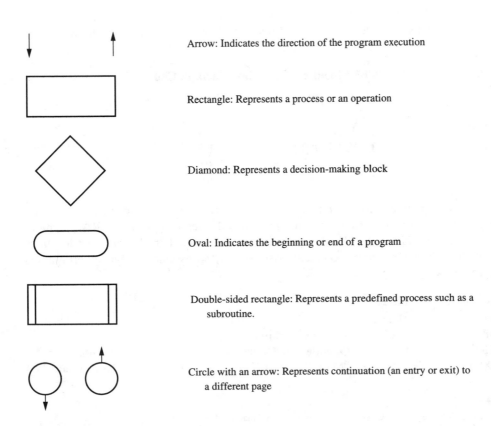

Arrow: Indicates the direction of the program execution

Rectangle: Represents a process or an operation

Diamond: Represents a decision-making block

Oval: Indicates the beginning or end of a program

Double-sided rectangle: Represents a predefined process such as a subroutine.

Circle with an arrow: Represents continuation (an entry or exit) to a different page

FIGURE 6.3
Flowcharting Symbols

used with Jump instructions for decision making (see Figure 6.10), and the double-sided rectangle is used for subroutines (see Chapter 9).

The flowchart in Figure 6.2 includes what steps to do and in what sequence. As a rule, a general flowchart does not include how to perform these steps or what registers are being used. The steps described in the flowchart are translated into an assembly language program in the next section.

ASSEMBLY LANGUAGE PROGRAM

Tasks	8085 Mnemonics
1. Load register B with 37H.	MVI B,37H*
2. Copy the number from B to A.	MOV A,B
3. Send the number to the output—port 01H.	OUT PORT1
4. End of the program.	HLT

TRANSLATION FROM ASSEMBLY LANGUAGE TO MACHINE LANGUAGE

Now, to translate the assembly language program into machine language, look up the hexadecimal machine codes for each instruction in the 8085 instruction set and write each machine code in the sequence, as follows:

8085 Mnemonics	Hex Machine Code
1. MVI B,37H	06
	37
2. MOV A,B	78
3. OUT PORT1	D3
	01
4. HLT	76

This program has six machine codes and will require six bytes of memory to enter the program into your system. If your single-board microcomputer has R/W memory starting at the address 2000H, this program can be entered in the memory locations 2000H to 2005H. The format generally used to write an assembly language program is shown below.

PROGRAM FORMAT

Memory Address (Hex)	Machine Code (Hex)	Instruction Opcode	Operand	Comments
XX00†	06	MVI	B,37H	;Load register B with data 37H
XX01	37			

---

*A number followed by the letter H represents a hexadecimal number.
†Enter high-order address (page number) of your R/W memory in place of XX.

XX02	78	MOV	A,B	;Copy (B) into (A)
XX03	D3	OUT	PORT1	;Display accumulator contents
XX04	PORT1*			; (37H) at Port1
XX05	76	HLT		;End of the program

This program has five columns: Memory Address, Machine Code, Opcode, Operand, and Comments. Each is described in the context of a single-board microcomputer.

Memory Address   These are 16-bit addresses of the user (R/W) memory in the system, where the machine code of the program is stored. The beginning address is shown as XX00; the symbol XX represents the page number of the available R/W memory in the microcomputer, and 00 represents the line number. For example, if the microcomputer has the user memory at 2000H, the symbol XX represents page number 20H; if the user memory begins at 0300H, the symbol XX represents page 03H. Substitute the appropriate page when entering the machine code of a program.

Machine Code   These are the hexadecimal numbers (instruction codes) that are entered (or stored) in the respective memory addresses through the hexadecimal keyboard of the microcomputer. The monitor program, which is stored in Read-Only memory (ROM) of the microcomputer, translates the Hex numbers into binary digits and stores the binary digits in the R/W memory.

If the system has R/W memory with the starting address at 2000H and the output port address 01H, the program will be stored as follows:

Memory Address	Memory Contents		Hex Code
2000	0 0 0 0 0 1 1 0	$\rightarrow$	06
2001	0 0 1 1 0 1 1 1	$\rightarrow$	37
2002	0 1 1 1 1 0 0 0	$\rightarrow$	78
2003	1 1 0 1 0 0 1 1	$\rightarrow$	D3
2004	0 0 0 0 0 0 0 1	$\rightarrow$	01
2005	0 1 1 1 0 1 1 0	$\rightarrow$	76

Opcode (Operation Code)   An instruction is divided into two parts: Opcode and Operand. Opcodes are the abbreviated symbols specified by the manufacturer (Intel) to indicate the type of operation or function that will be performed by the machine code.

Operand   The operand part of an instruction specifies the item to be processed; it can be 8- or 16-bit data, a register, or a memory address.

---

*Enter the output port address of your system. If an output port is not available on your system, see "How to Execute a Program without an Output Port" later in this section.

An instruction, called a mnemonic or mnemonic instruction, is formed by combining an opcode and an operand. The mnemonics are used to write programs in the 8085 assembly language; and then the mnemonics in these programs are translated manually into the binary machine code by looking them up in the instruction set.

Comments The comments are written as a part of the proper documentation of a program to explain or elaborate the purpose of the instructions used. These are separated by a semicolon (;) from the instruction on the same line. They play a critical role in the user's understanding of the logic behind a program. Because the illustrative programs in the early part of this chapter are simple, most of the comments are either redundant or trivial. The purpose of the comments in these programs is to reinforce the meaning of the instructions. In actual usage, the comments should not just describe the operation of an instruction.

HOW TO ENTER AND EXECUTE THE PROGRAM

This program assumes that one output port is available on your microcomputer system. The program cannot be executed without modification if your microcomputer has no independent output ports other than the system display of memory address and data or if it has programmable I/O ports. (See Chapter 14.) To enter the program:*

1. Push the Reset key.
2. Enter the 16-bit memory address of the first machine code of your program. (Substitute the page number of your R/W memory for the letters XX and the output port address for the label PORT1.)
3. Enter and store all the machine codes sequentially, using the hexadecimal keyboard on your system.
4. Reset the system.
5. Enter the memory address where the program begins and push the Execute key.

If the program is properly entered and executed, the data byte 37H will be displayed at the output port.

HOW TO EXECUTE A PROGRAM WITHOUT AN OUTPUT PORT

If your system does not have an output port, either eliminate the instruction OUT PORT1, or substitute NOP (No Operation) in place of the OUT instruction. Assuming your system has R/W memory starting at 2000H, you can enter the program as follows:

Memory Address	Machine Code	Mnemonic Instruction
2000	06	MVI B,37H
2001	37	
2002	78	MOV A,B
2003	00	NOP

---

*Refer to the user's manual of your microcomputer for details.

| 2004 | 00 | NOP |
| 2005 | 76 | HLT |

After you have executed this program, you can find the answer in the accumulator by pushing the Examine Register key (see your user's manual).

The program also can be executed by entering the machine code 76 in location 2003H, thus eliminating the OUT instruction.

6.13 Illustrative Program: Data Transfer to Control Output Devices

PROBLEM STATEMENT

A microcomputer is designed to control various appliances and lights in your house. The system has an output port with the address 01H, and various units are connected to the bits D_7 to D_0 as shown in Figure 6.4. On a cool morning you want to turn on the radio, the coffeepot, and the space heater. Write appropriate instructions for the microcomputer. Assume the R/W memory in your system begins at 2000H.

PROBLEM ANALYSIS

The output port in Figure 6.4 is a latch (D flip-flop). When data bits are sent to the output port they are latched by the D flip-flop. A data bit at logic 1 supplies approximately 5 V as output and can turn on solid-state relays.

To turn on the radio, the coffeepot, and the space heater, set D_6, D_5, and D_4 at logic 1, and the other bits at logic 0:

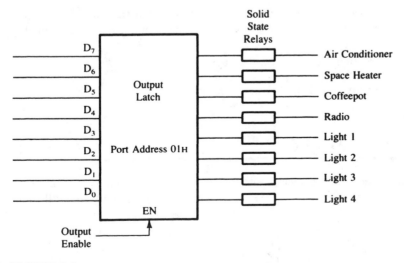

FIGURE 6.4
Output Port to Control Devices

$$D_7 \ D_6 \ D_5 \ D_4 \ D_3 \ D_2 \ D_1 \ D_0$$
$$0 \ \ 1 \ \ 1 \ \ 1 \ \ 0 \ \ 0 \ \ 0 \ \ 0 = 70H$$

The output port requires 70H, and it can be sent to the port by loading the accumulator with 70H.

PROGRAM

Memory Address	Machine Code	Mnemonic Instruction	Comments
HI-LO*			
2000	3E	MVI A,70H	;Load the accumulator with the bit pattern
2001	70		; necessary to turn on the devices
2002	D3	OUT 01H	;Send the bit pattern to the port 01H, and
2003	01†		; turn on the devices
2004	76	HLT	;End of the program

PROGRAM OUTPUT

This program simulates controlling of the devices connected to the output port by displaying 70H on a seven-segment LED display. If your system has individual LEDs, the binary pattern—0111 0000—will be displayed.

6.14   Review of Important Concepts

1. Registers are used to load data directly or to save data bytes.
2. In data transfer (copying), the destination register is modified but the source register retains its data.
3. The 8085 transfers data from an input port to the accumulator (IN) and from the accumulator to an output port (OUT). The instruction OUT cannot send data from any other register.
4. The data copy instructions do not affect the flags.

See Questions and Assignments 1–7 at the end of this chapter.

6.2   ARITHMETIC OPERATIONS

The 8085 microprocessor performs various arithmetic operations, such as addition, subtraction, increment, and decrement. These arithmetic operations have the following mnemonics.

---

*Change the high-order memory address 20 to the appropriate address for your system.
†Substitute the appropriate port address.

ADD : Add Add the contents of a register.*
ADI : Add Immediate Add 8-bit data.
SUB : Subtract Subtract the contents of a register.
SUI : Subtract Immediate Subtract 8-bit data.
INR : Increment Increase the contents of a register by 1.
DCR : Decrement Decrease the contents of a register by 1.

The arithmetic operations Add and Subtract are performed in relation to the contents of the accumulator. However, the Increment or the Decrement operations can be performed in any register. The instructions for these operations are explained below.

INSTRUCTIONS

These arithmetic instructions (except INR and DCR)

1. assume implicitly that the accumulator is one of the operands.
2. modify all the flags according to the data conditions of the result.
3. place the result in the accumulator.
4. do not affect the contents of the operand register.

The instructions INR and DCR

1. affect the contents of the specified register.
2. affect all flags except the CY flag.

The descriptions of the instructions (including INR and DCR) are as follows:

Opcode	Operand	Description
ADD	$R^\dagger$	Add ☐ This is a 1-byte instruction ☐ Adds the contents of register R to the contents of the accumulator
ADI	8-bit	Add Immediate ☐ This is a 2-byte instruction ☐ Adds the second byte to the contents of the accumulator
SUB	$R^\dagger$	Subtract ☐ This is a 1-byte instruction ☐ Subtracts the contents of register R from the contents of the accumulator
SUI	8-bit	Subtract Immediate ☐ This is a 2-byte instruction

---

*Memory-related arithmetic operations are excluded here; they are discussed in Chapter 7.
†R represents any of registers A, B, C, D, E, H, and L.

□ Subtracts the second byte from the contents of the accumulator

INR   R*   Increment
□ This is a 1-byte instruction
□ Increases the contents of register R by 1
Caution: All flags except the CY are affected

DCR   R*   Decrement
□ This is a 1-byte instruction
□ Decreases the contents of register R by 1
Caution: All flags except the CY are affected

6.21 Addition

The 8085 performs addition with 8-bit binary numbers and stores the sum in the accumulator. If the sum is larger than eight bits (FFH), it sets the Carry flag. Addition can be performed either by adding the contents of a source register (B, C, D, E, H, L, or memory) to the contents of the accumulator (ADD) or by adding the second byte directly to the contents of the accumulator (ADI).

Example 6.3

The contents of the accumulator are 93H and the contents of register C are B7H. Add both contents.

Instruction   ADD C

$$\begin{array}{llll}
 & & & \text{CY} \quad D_7 \ D_6 \ D_5 \ D_4 \quad D_3 \ D_2 \ D_1 \ D_0 \\
(A) & : & 93H = & \quad\quad 1 \ \ 0 \ \ 0 \ \ 1 \quad\ 0 \ \ 0 \ \ 1 \ \ 1 \\
 & + & & \\
(C) & : & B7H = & \quad\quad 1 \ \ 0 \ \ 1 \ \ 1 \quad\ 0 \ \ 1 \ \ 1 \ \ 1 \\
\end{array}$$

Carry: 1   1 1   1 1 1

SUM (A)   :   [1] 4AH = [1]   0 1 0 0   1 0 1 0
              CY

Flag Status:[†] S = 0, Z = 0, CY = 1

When the 8085 adds 93H and B7H, the sum is 14AH; it is larger than eight bits. Therefore, the accumulator will have 4AH in binary, and the CY flag will be set. The result in the accumulator (4AH) is not 0, and bit D_7 is not 1; therefore, the Zero and the Sign flags will be reset.

---

*R represents any of registers A, B, C, D, E, H, and L.
[†]The P and AC flags are not shown here. In this chapter, the focus will be on the Sign, Zero, and Carry flags.

Add the number 35H directly to the sum in the previous example when the CY flag is set.

**Example
6.4**

Instruction   ADI 35H

$$
\begin{array}{rl}
& \text{CY} \\
(\text{A}) \quad : \quad 4\text{AH} = & \boxed{1}\ 0\ 1\ 0\ 0\ \ 1\ 0\ 1\ 0 \\
+ & \\
(\text{Data}) \quad : \quad 35\text{H} = & \ \ \ \ \ 0\ 0\ 1\ 1\ \ 0\ 1\ 0\ 1 \\
(\text{A}) \quad : \quad 7\text{FH} = & \boxed{0}\ 0\ 1\ 1\ 1\ \ 1\ 1\ 1\ 1
\end{array}
$$

Flag Status: S = 0, Z = 0, CY = 0

The addition of 4AH and 35H does not generate a carry and will reset the previous Carry flag. Therefore, in adding numbers, it is necessary to count how many times the CY flag is set by using some other programming techniques (see Section 7.32).

Assume the accumulator holds the data byte FFH. Illustrate the differences in the flags set by adding 01H and by incrementing the accumulator contents.

**Example
6.5**

Instruction   ADI 01H

$$
\begin{array}{rl}
& \text{CY} \\
(\text{A}) \quad : \quad \text{FFH} = & \ \ \ 1\ 1\ 1\ 1\ \ 1\ 1\ 1\ 1 \\
+ & \\
(\text{Data}) \quad : \quad 01\text{H} = & \ \ \ 0\ 0\ 0\ 0\ \ 0\ 0\ 0\ 1 \\
& \ \ \ 1\ 1\ 1\ 1\ \ 1\ 1\ 1\ 1 \ \ \text{Carry} \\
(\text{A}) \quad : \quad \boxed{1}\ 00\text{H} = & \boxed{1}\ 1\ 0\ 0\ 0\ \ 0\ 0\ 0\ 0 \\
\text{CY} &
\end{array}
$$

Flag Status: S = 0, Z = 1, CY = 1

After adding 01H to FFH, the sum in the accumulator is 0 with a carry. Therefore, the CY and Z flags are set. The Sign flag is reset because D_7 is 0.

Instruction   INR A
The accumulator contents will be 00H, the same as before. However, the instruction INR will not affect the Carry flag; it will remain in its previous status.

Flag Status: S = 0, Z = 1, CY = NA

FLAG CONCEPTS AND CAUTIONS

As described in the previous chapter, the flags are flip-flops that are set or reset after the execution of arithmetic and logic operations, with some exceptions. In many ways, the flags are like signs on an interstate highway that help drivers find their destinations.

Drivers may see one or more signs at a time. They may take the exit when they find the sign they are looking for, or they may continue along the interstate and ignore the signs.

Similarly, flags are signs of data conditions. After an operation, one or more flags may be set, and they can be used to change the direction of the program sequence by using Jump instructions, which will be described later. However, the programmer should be alert for them to make a decision. If the flags are not appropriate for the tasks, the programmer can ignore them.

Caution #1 In Example 6.3, the CY flag is set, and in Example 6.4, the CY flag is reset. The critical concept here is that if the programmer ignores the flag, it can be lost after the subsequent instructions. However, the flag can be ignored when the programmer is not interested in using it.

Caution #2 In Example 6.5, two flags are set. The programmer may use one or more flags to make decisions or may ignore them if they are irrelevant.

Caution #3 The CY flag has a dual function; it is used as a carry in addition and as a borrow in subtraction.

The importance of flags cannot be emphasized enough, and a thorough understanding of them is critical in writing assembly language programs.

1. Flags are flip-flops in the ALU (arithmetic/logic unit). They are affected (set or reset) by the operations in the ALU; therefore, operations, such as copy, that take place outside the ALU do not affect the flags.
2. The status of the flags is determined by the result of an operation. In most instances, the result is in the accumulator. However, in some operations, such as Increment (INR), results can be in registers other than the accumulator.
3. There is no relationship between a result and the bit positions of the flag register. In Example 6.3, the answer of the addition is 4AH with a carry. The binary answer is as follows:

	Result								Flag Register							
CY	D_7	D_6	D_5	D_4	D_3	D_2	D_1	D_0	D_7	D_6	D_5	D_4	D_3	D_2	D_1	D_0
1	0	1	0	0	1	0	1	0	S	Z						CY

4A

Carry Flag Set to 1 because the answer is larger than eight bits; there is a carry generated out of the last bit D_7. During the addition, bits D_0 through D_6 may generate carries, but these carries do not affect the CY flag.

Misconception #1 Bit D_0 in the result (4AH) corresponds to the bit position of the Carry flag D_0 in the flag register; therefore, the Carry flag is reset.

Misconception #2 In the addition process, bits D_0 of 93H and B7H generate a carry (or other bit additions generate carries); therefore the Carry flag is set.

Zero Flag   Reset to 0 because the answer is not zero. The Zero flag is set only when all eight bits in the result are 0.

Misconception #3   Bit D_6 in the result (4AH) is 1, and it corresponds to bit D_6 (Zero flag position) in the flag register. Therefore, the Z flag is set.

Sign Flag   Reset to 0 because D_7 in the result is 0. The position of the sign flag in the flag register is also D_7. But it is just a coincidence. The microprocessor designer could have chosen bit D_6 for the Sign flag and bit D_7 for the Zero flag in the flag register. The Sign flag is relevant only when we are using signed numbers.

Misconception #4   If the Sign flag is set, the result must be negative.

See Questions and Assignments 9–19 at the end of this chapter.

6.22   Illustrative Program: Arithmetic Operations—Addition and Increment

PROBLEM STATEMENT
Write a program to perform the following functions, and verify the output.

1. Load the number 8BH in register D.
2. Load the number 6FH in register C.
3. Increment the contents of register C by one.
4. Add the contents of registers C and D and display the sum at the output PORT1.

PROGRAM
The illustrative program for arithmetic operations using addition and increment is presented as Figure 6.5 to show the register contents during some of the steps.

PROGRAM DESCRIPTION
1. The first four machine codes load 8BH in register D and 6FH in register C (see Figure 6.5). These are Data Copy instructions. Therefore, no flags will be affected; the flags will remain in their previous status. The status of the flags is shown as X to indicate no change in their status.
2. Instruction INR C adds 1 to 6FH and changes the contents of C to 70H. The result is nonzero and bit D_7 is zero; therefore, the S and Z flags are reset. However, the CY flag is not affected by the INR instruction.
3. To add (C) to (D), the contents of one of the registers must be transferred to the accumulator because the 8085 cannot add two registers directly. Review the ADD instruction. The instruction MOV A,C copies 70H from register C into the accumulator without affecting (C). See the register contents.

Memory Address (H)	Machine Code	Instruction Opcode	Operand	Comments and Register Contents

The first four machine codes load the registers as

Memory Address (H)	Machine Code	Opcode	Operand
HI-LO XX00	16	MVI	D,8BH
01	8B		
02	0E	MVI	C,6FH
03	6F		
04	0C	INR	C
05	79	MOV	A,C
06	82	ADD	D
07	D3	OUT	PORT1
08	PORT #	PORT1	
09	76	HLT	

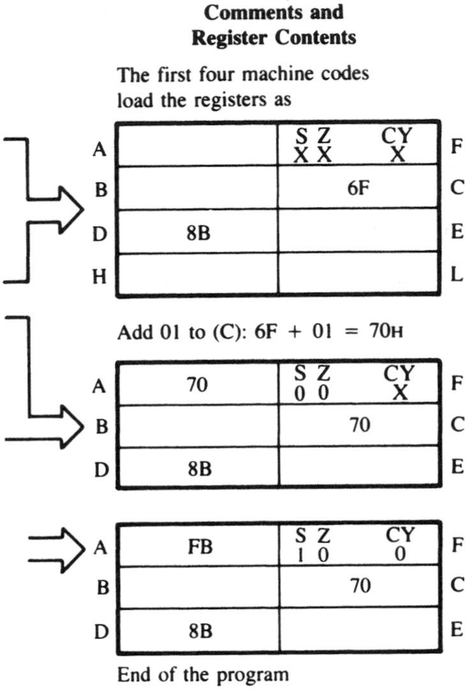

FIGURE 6.5
Illustrative Program for Arithmetic Operations—Using Addition and Increment

4. Instruction ADD D adds (D) to (A), stores the sum in A, and sets the Sign flag as shown below:

$$
\begin{array}{rll}
(A) & : \ 70H = & 0\ 1\ 1\ 1\ \ 0\ 0\ 0\ 0 \\
& + & \\
(D) & : \ 8BH = & 1\ 0\ 0\ 0\ \ 1\ 0\ 1\ 1 \\ \hline
(A) & : \ FBH = & \boxed{0}\ 1\ 1\ 1\ 1\ \ 1\ 0\ 1\ 1 \ \ \text{(see Figure 6.5)} \\
& & \text{CY}
\end{array}
$$

Flag Status: S = 1, Z = 0, CY = 0

5. The sum FBH is displayed by the OUT instruction.

PROGRAM OUTPUT

This program will display FBH at the output port. If an output port is not available, the program can be executed by entering NOP instructions in place of the OUT instruction and the answer FBH can be verified by examining the accumulator A. (Most systems have the Examine-Register operation.) Similarly, the contents of registers C and D and the flags can be verified.

By examining the contents of the registers, the following points can be confirmed:

1. The sum is stored in the accumulator.
2. The contents of the source registers are not changed.
3. The Sign (S) flag is set.

Even though the Sign (S) flag is set, this is not a negative sum. The microprocessor sets the Sign flag whenever an operation results in $D_7 = 1$. The microprocessor cannot recognize whether FBH is a sum, a negative number, or a bit pattern. It is your responsibility to interpret and use the flags. (See "Flag Concepts and Cautions" in Section 6.21.) In this example, the addition is not concerned with the signed numbers. With the signed numbers, bit D_7 is reserved for a sign by the programmer (not by the microprocessor), and no number larger than $+127_{10}$ can be entered.

6.23 Subtraction

The 8085 performs subtraction by using the method of 2's complement. (If you are not familiar with the method of 2's complement, review Appendix A2.)

Subtraction can be performed by using either the instruction SUB to subtract the contents of a source register or the instruction SUI to subtract an 8-bit number from the contents of the accumulator. In either case, the accumulator contents are regarded as the minuend (the number from which to subtract).

The 8085 performs the following steps internally to execute the instruction SUB (or SUI).

Step 1: Converts subtrahend (the number to be subtracted) into its 1's complement.
Step 2: Adds 1 to 1's complement to obtain 2's complement of the subtrahend.
Step 3: Add 2's complement to the minuend (the contents of the accumulator).
Step 4: Complements the Carry flag.

These steps are illustrated in the following example.

Register B has 65H and the accumulator has 97H. Subtract the contents of register B from the contents of the accumulator.

Example
6.6

Instruction SUB B

	Subtrahend (B): 65H =	0 1 1 0 0 1 0 1
Step 1:	1's complement of 65H =	1 0 0 1 1 0 1 0
	(Substitute 0 for 1 and 1 for 0)	
		+
Step 2:	Add 01 to obtain	0 0 0 0 0 0 0 1
	2's complement of 65H =	1 0 0 1 1 0 1 1
		+

To subtract: 97H − 65H,

Add 97H to 2's complement of 65H =   1 0 0 1   0 1 1 1

                                                         1 1   1 1 1   Carry

Step 3:   CY $\boxed{1}$ 0 0 1 1   0 0 1 0

Step 4: Complement Carry $\boxed{0}$ 0 0 1 1   0 0 1 0
Result (A): 32H

Flag Status: S = 0, Z = 0, CY = 0

If the answer is negative, it will be shown in the 2's complement of the actual magnitude. For example, if the above subtraction is performed as 65H − 97H, the answer will be the 2's complement of 32H with the Carry (Borrow) flag set.

6.24   Illustrative Program: Subtraction of Two Unsigned Numbers

PROBLEM STATEMENT

Write a program to do the following:

1. Load the number 30H in register B and 39H in register C.
2. Subtract 39H from 30H.
3. Display the answer at PORT1.

PROGRAM

The illustrative program for subtraction of two unsigned numbers is presented as Figure 6.6 to show the register contents during some of the steps.

PROGRAM DESCRIPTION

1. Registers B and C are loaded with 30H and 39H, respectively. The instruction MOV A,B copies 30H into the accumulator (shown as register contents). This is an essential step because the contents of a register can be subtracted only from the contents of the accumulator and not from any other register.
2. To execute the instruction SUB C the microprocessor performs the following steps internally:

Step 1:   39H = 0 0 1 1   1 0 0 1
                                1's complement of 39H = 1 1 0 0   0 1 1 0
                                                             +
Step 2:   Add 01 = 0 0 0 0   0 0 0 1
                                2's complement of 39H = 1 1 0 0   0 1 1 1
                                                             +
Step 3:   Add 30H to 2's complement of 39H = 0 0 1 1   0 0 0 0
                                               CY $\boxed{0}$ 1 1 1 1   0 1 1 1

Step 4: Complement carry

$\boxed{1}$ $\quad$ 1 1 1 1 $\quad$ 0 1 1 1 = F7H

$\qquad$ Flag Status: S = 1, Z = 0, CY = 1

3. The number F7H is a 2's complement of the magnitude (39H − 30H) = 09H.

4. The instruction OUT displays F7H at PORT1.

PROGRAM OUTPUT

This program will display F7H as the output. In this program, the unsigned numbers were used to perform the subtraction. Now, the question is: How do you recognize that the answer F7H is really a 2's complement of 09H and not a straight binary F7H?

$\qquad$ The answer lies with the Carry flag. If the Carry flag (also known as the Borrow flag in subtraction) is set, the answer is in 2's complement. The Carry flag raises a second question: Why isn't it a positive sum with a carry? The answer is implied by the instruction SUB (it is a subtraction).

$\qquad$ There is no way to differentiate between a straight binary number and 2's complement by examining the answer at the output port. The flags are internal and not easily displayed. However, a programmer can test the Carry flag by using the instruction Jump On Carry (JC) and can find a way to indicate that the answer is in 2's complement. (This is discussed in Branch instructions.)

Memory Address (H)	Machine Code	Instruction Opcode	Instruction Operand	Comments and Register Contents
HI-LO				
XX00	06	MVI	B,30H	Load the minuend in register B
01	30			Load the subtrahend in register C
02	0E	MVI	C,39H	The register contents:
03	39			
04	78	MOV	A,B	

A	30		F
B	30	39	C

Memory Address (H)	Machine Code	Instruction Opcode	Instruction Operand
05	91	SUB	C

A	F7	S Z / 1 0	CY / 1	F
B	30	39		C

06	D3	OUT	PORT1
07	PORT#		
08	76	HLT	

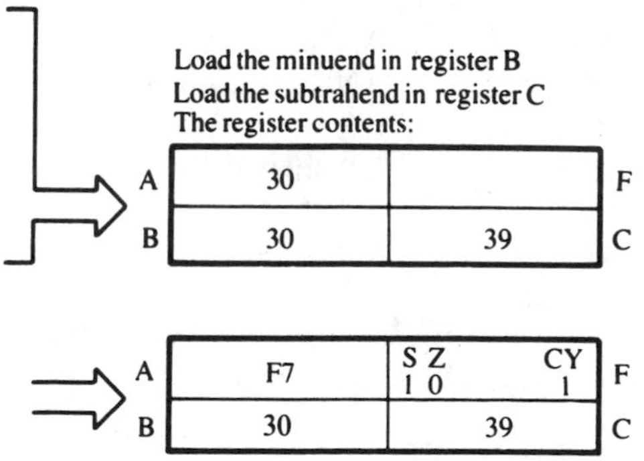

FIGURE 6.6
Illustrative Program for Subtraction of Two Unsigned Numbers

6.25 Review of Important Concepts

1. The arithmetic operations implicitly assume that the contents of the accumulator are one of the operands.

2. The results of the arithmetic operations are stored in the accumulator; thus, the previous contents of the accumulator are altered.

3. The flags are modified to reflect the data conditions of an operation.

4. The contents of the source register are not changed as a result of an arithmetic operation.

5. In the Add operation, if the sum is larger than 8-bit, CY is set.

6. The Subtract operation is performed by using the 2's complement method.

7. If a subtraction results in a negative number, the answer is in 2's complement and CY (the Borrow flag) is set.

8. In unsigned arithmetic operations, the Sign flag (S) should be ignored.

9. The instructions INR (Increment) and DCR (Decrement) are special cases of the arithmetic operations. These instructions can be used for any one of the registers, and they do not affect CY, even if the result is larger than 8-bit. All other flags are affected by the result in the register used (not by the contents of the accumulator).

6.3   LOGIC OPERATIONS

A microprocessor is basically a programmable logic chip. It can perform all the logic functions of the hard-wired logic through its instruction set. The 8085 instruction set includes such logic functions as AND, OR, Ex OR, and NOT (complement). The opcodes of these operations are as follows:*

ANA:	AND	Logically AND the contents of a register.
ANI :	AND Immediate	Logically AND 8-bit data.
ORA:	OR	Logically OR the contents of a register.
ORI :	OR Immediate	Logically OR 8-bit data.
XRA:	X-OR	Exclusive-OR the contents of a register.
XRI :	X-OR Immediate	Exclusive-OR 8-bit data.

All logic operations are performed in relation to the contents of the accumulator. The instructions of these logic operations are described below.

INSTRUCTIONS

The logic instructions

1. implicitly assume that the accumulator is one of the operands.

2. reset (clear) the CY flag. The instruction CMA is an exception; it does not affect any flags.

3. modify the Z, P, and S flags according to the data conditions of the result.

4. place the result in the accumulator.

5. do not affect the contents of the operand register.

---

*Memory-related logic operations are excluded here; they will be discussed in the next chapter.

Opcode	Operand	Description
ANA	R	Logical AND with Accumulator ☐ This is a 1-byte instruction ☐ Logically ANDs the contents of the register R with the contents of the accumulator ☐ 8085: CY is reset and AC is set
ANI	8-bit	AND Immediate with Accumulator ☐ This is a 2-byte instruction ☐ Logically ANDs the second byte with the contents of the accumulator ☐ 8085: CY is reset and AC is set
ORA	R	Logically OR with Accumulator ☐ This is a 1-byte instruction ☐ Logically ORs the contents of the register R with the contents of the accumulator
ORI	8-bit	OR Immediate with Accumulator ☐ This is a 2-byte instruction ☐ Logically ORs the second byte with the contents of the accumulator
XRA	R	Logically Exclusive-OR with Accumulator ☐ This is a 1-byte instruction ☐ Exclusive-ORs the contents of register R with the contents of the accumulator
XRI	8-bit	Exclusive-OR Immediate with Accumulator ☐ This is a 2-byte instruction ☐ Exclusive-ORs the second byte with the contents of the accumulator
CMA		Complement Accumulator ☐ This is a 1-byte instruction that complements the contents of the accumulator ☐ No flags are affected

6.31 Logic AND

The process of performing logic operations through the software instructions is slightly different from the hardwired logic. The AND gate shown in Figure 6.7(a) has two inputs and one output. On the other hand, the instruction ANA simulates eight AND gates, as shown in Figure 6.7(b). For example, assume that register B holds 77H and the accumulator A holds 81H. The result of the instruction ANA B is 01H and is placed in the accumulator replacing the previous contents, as shown in Figure 6.7(b).

Figure 6.7(b) shows that each bit of register B is independently ANDed with each bit of the accumulator, thus simulating eight 2-input AND gates.

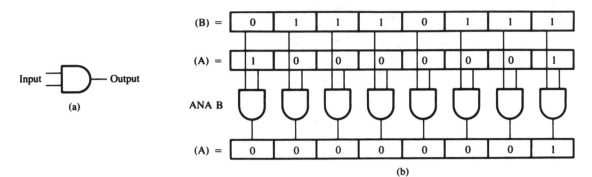

FIGURE 6.7
AND Gate (a) and a Simulated ANA Instruction (b)

6.32   Illustrative Program: Data Masking with Logic AND

PROBLEM STATEMENT

To conserve energy and to avoid an electrical overload on a hot afternoon, implement the following procedures to control the appliances throughout the house (Figure 6.8). Assume that the control switches are located in the kitchen, and they are available to anyone in the house. Write a set of instructions to

1. turn on the air conditioner if switch S_7 of the input port 00H is on.
2. ignore all other switches of the input port even if someone attempts to turn on other appliances.

(To perform this experiment on your single-board microcomputer, simulate the reading of the input port 00H with the instruction MVI A, 8-bit data.)

PROBLEM ANALYSIS

In this problem you are interested in only one switch position, S_7, which is connected to data line D_7. Assume that various persons in the family have turned on the switches of the air conditioner (S_7), the radio (S_4), and the lights (S_3, S_2, S_1, S_0).

   If the microprocessor reads the input port (IN 00H), the accumulator will have data byte 9FH. This can be simulated by using the instruction MVI A,9FH. However, if you are interested in knowing only whether switch S_7 is on, you can mask bits D_6 through D_0 by ANDing the input data with a byte that has 0 in bit positions D_6 through D_0 and 1 in bit position D_7.

$$\begin{array}{cccccccc} D_7 & D_6 & D_5 & D_4 & D_3 & D_2 & D_1 & D_0 \\ 1 & 0 & 0 & 0 & 0 & 0 & 0 & 0 \end{array} = 80H$$

After bits D_6 through D_0 have been masked, the remaining byte can be sent to the output port to simulate turning on the air conditioner.

PROGRAM

Memory Address	Machine Code	Instruction Opcode	Operand	Comments
HI-LO				
XX00	3E	MVI	A,Data	;This instruction simulates the
01	9F			; instruction IN 00H
02	E6	ANI	80H	;Mask all the bits except D_7
03	80			
04	D3	OUT	01H	;Turn on the air conditioner if
05	01			; S_7 is on
06	76	HLT		;End of the program

PROGRAM OUTPUT

The instruction ANI 80H ANDs the accumulator data as follows:

$$(A) = 1\ 0\ 0\ 1\quad 1\ 1\ 1\ 1\quad (9FH)$$

AND

$$(\text{Masking Byte} = 1\ 0\ 0\ 0\quad 0\ 0\ 0\ 0\quad (80H)$$

$$\overline{\hspace{4cm}}$$

$$(A) = 1\ 0\ 0\ 0\quad 0\ 0\ 0\ 0\quad (80H)$$

Flag Status: $S = 1$, $Z = 0$, $CY = 0$

The ANDing operation always resets the CY flag. The result (80H) will be placed in the accumulator and then sent to the ouptput port, and logic 1 of data bit D_7 turns on the air conditioner. In this example, the output (80H) is the same as the masking data byte (80H) because switch S_7 (or data bit D_7) is on. If S_7 is off, the output will be zero.

The masking is a commonly used technique to eliminate unwanted bits in a byte. The masking byte to be logically ANDed is determined by placing 0s in bit positions that are to be masked and by placing 1s in the remaining bit positions.

6.33 OR, Exclusive-OR, and NOT

The instruction ORA (and ORI) simulates logic ORing with eight 2-input OR gates; this process is similar to that of ANDing, explained in the previous section. The instruction XRA (and XRI) performs Exclusive-ORing of eight bits, and the instruction CMA inverts the bits of the accumulator.

Assume register B holds 93H and the accumulator holds 15H. Illustrate the results of the instructions ORA B, XRA B, and CMA.

Example 6.7

1. The instruction ORA B will perform the following operation:

$$(B) = 1\ 0\ 0\ 1\quad 0\ 0\ 1\ 1\quad (93H)$$

OR

$$(A) = 0\ 0\ 0\ 1\quad 0\ 1\ 0\ 1\quad (15H)$$
$$\overline{(A) = 1\ 0\ 0\ 1\quad 0\ 1\ 1\ 1\quad (97H)}$$

Flag Status: $S = 1, Z = 0, CY = 0$

The result 97H will be placed in the accumulator, the CY flag will be reset, and the other flags will be modified to reflect the data conditions in the accumulator.

2. The instruction XRA B will perform the following operation.

$$(B) = 1\ 0\ 0\ 1\quad 0\ 0\ 1\ 1\quad (93H)$$

X-OR

$$(A) = 0\ 0\ 0\ 1\quad 0\ 1\ 0\ 1\quad (15H)$$
$$\overline{(A) = 1\ 0\ 0\ 0\quad 0\ 1\ 1\ 0\quad (86H)}$$

Flag Status: $S = 1, Z = 0, CY = 0$

The result 86H will be placed in the accumulator, and the flags will be modified as shown.

3. The instruction CMA will result in

$$(A) = 0\ 0\ 0\ 1\quad 0\ 1\ 0\ 1\quad (15H)$$

CMA

$$(A) = 1\ 1\ 1\ 0\quad 1\ 0\ 1\ 0\quad (EAH)$$

The result EAH will be placed in the accumulator and no flags will be modified.

6.34   Setting and Resetting Specific Bits

At various times, we may want to set or reset a specific bit without affecting the other bits. OR logic can be used to set the bit, and AND logic can be used to reset the bit.

Example 6.8

In Figure 6.8, keep the radio on (D_4) continuously without affecting the functions of other appliances, even if someone turns off the switch S_4.

Solution

To keep the radio on without affecting the other appliances, the bit D_4 should be set by ORing the reading of the input port with the data byte 10H as follows:

$$\text{IN 00H: } (A) = D_7\ D_6\ D_5\ D_4\quad D_3\ D_2\ D_1\ D_0$$
$$\text{ORI 10H: } \quad\ \ = 0\ \ 0\ \ 0\ \ 1\quad 0\ \ 0\ \ 0\ \ 0$$
$$\overline{(A) = D_7\ D_6\ D_5\ 1\quad D_3\ D_2\ D_1\ D_0}$$

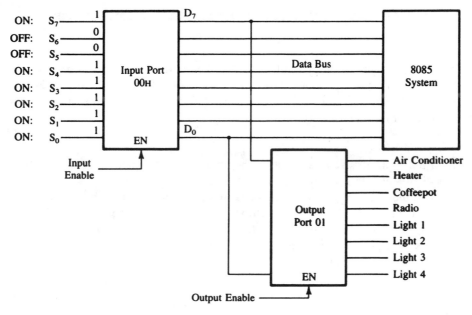

FIGURE 6.8
Input Port to Control Appliances

Flag Status: $CY = 0$; others will depend on data.

The instruction IN reads the switch positions shown as D_7–D_0 and the instruction ORI sets the bit D_4 without affecting any other bits.

In Figure 6.8, assume it is winter, and turn off the air conditioner without affecting the other appliances.

Example 6.9

To turn off the air conditioner, reset bit D_7 by ANDing the reading of the input port with the data byte 7FH as follows:

Solution

$$
\begin{array}{llccccccccc}
\text{IN } 00\text{H:} & (A) = & D_7 & D_6 & D_5 & D_4 & D_3 & D_2 & D_1 & D_0 \\
\text{ANI } 7\text{FH:} & = & 0 & 1 & 1 & 1 & 1 & 1 & 1 & 1 \\
\hline
& & 0 & D_6 & D_5 & D_4 & D_3 & D_2 & D_1 & D_0
\end{array}
$$

Flag Status: $CY = 0$; others will depend on the data bits.

The ANI instruction resets bit D_7 without affecting the other bits.

6.35   Illustrative Program: ORing Data from Two Input Ports

PROBLEM STATEMENT

An additional input port with eight switches and the address 01H (Figure 6.9) is connected to the microcomputer shown in Figure 6.8 to control the same appliances and lights from the bedroom as well as from the kitchen. Write instructions to turn on the devices from any of the input ports.

PROBLEM ANALYSIS

To turn on the appliances from any one of the input ports, the microprocessor needs to read the switches at both ports and logically OR the switch positions.

Assume that the switch positions in one input port (located in the bedroom) correspond to the data byte 91H and the switch positions in the second port (located in the kitchen) correspond to the data byte A8H. The person in the bedroom wants to turn on the air conditioner, the radio, and the bedroom light; and the person in the kitchen wants to

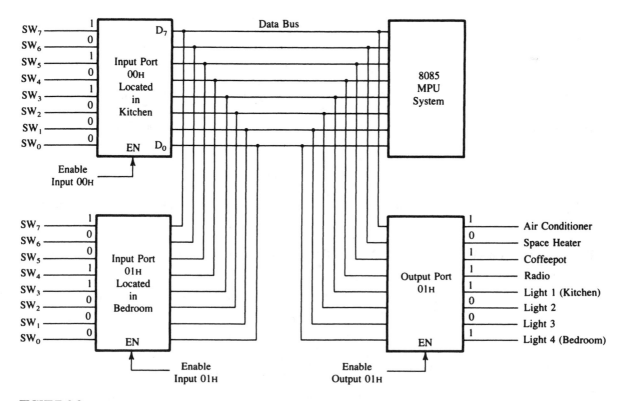

FIGURE 6.9
Two Input Ports to Control Output Devices

turn on the air conditioner, the coffeepot, and the kitchen light. By ORing these two data bytes the microprocessor can turn on the necessary appliances.

To test this program, we must simulate the readings of the input port by loading the data into registers—for example, into B and C.

PROGRAM

Memory Address	Machine Code	Opcode	Operand	Comments
HI-LO				
XX00	06	MVI	B,91H	;This instruction simulates read-
01	91			; ing input port 01H
02	0E	MVI	C,A8H	;This instruction simulates read-
03	A8			; ing input port 00H
04	78	MOV	A,B	;It is necessary to transfer data
				; byte from B to A to OR with
				; C. B and C cannot be ORed
				; directly
05	B1	ORA	C	;Combine the switch positions
				; from registers B and C in the
				; accumulator
06	D3	OUT	PORT1	;Turn on appliances and lights
07	PORT1			
08	76	HLT		;End of the program

PROGRAM OUTPUT

By logically ORing the data bytes in registers B and C

$$(B) \rightarrow (A) = 1\ 0\ 0\ 1\quad 0\ 0\ 0\ 1\ (91H)$$
$$(C) = 1\ 0\ 1\ 0\quad 1\ 0\ 0\ 0\ (A8H)$$
$$\overline{}$$
$$(A) = 1\ 0\ 1\ 1\quad 1\ 0\ 0\ 1\ (B9H)$$

Flag Status: S = 1, Z = 0, CY = 0

Data byte B9H is placed in the accumulator that turns on the air conditioner, radio, coffeepot, and bedroom and kitchen lights.

6.36   Review of Important Concepts

1. Logic operations are performed in relation to the contents of the accumulator.

2. Logic operations simulate eight 2-input gates (or inverters).

3. The Sign, Zero (and Parity) flags are modified to reflect the status of the operation. The Carry flag is reset. However, the NOT operation does not affect any flags.

4. After a logic operation has been performed, the answer is placed in the accumulator replacing the original contents of the accumulator.

5. The logic operations cannot be performed directly with the contents of two registers.

6. The individual bits in the accumulator can be set or reset using logic instructions.

See Questions and Assignments 20–29 at the end of this chapter.

6.4   BRANCH OPERATIONS

The **branch instructions** are the most powerful instructions because they allow the microprocessor to change the sequence of a program, either unconditionally or under certain test conditions. These instructions are the key to the flexibility and versatility of a computer.

The microprocessor is a sequential machine; it executes machine codes from one memory location to the next. Branch instructions instruct the microprocessor to go to a different memory location, and the microprocessor continues executing machine codes from that new location. The address of the new memory location is either specified explicitly or supplied by the microprocessor or by extra hardware. The branch instructions are classified in three categories:

1. Jump instructions

2. Call and Return instructions

3. Restart instructions

This section is concerned with applications of Jump instructions. The Call and Return instructions are associated with the subroutine technique and will be discussed in Chapter 9; Restart instructions are associated with the interrupt technique and will be discussed in Chapter 12.

The Jump instructions specify the memory location explicitly. They are 3-byte instructions: one byte for the operation code, followed by a 16-bit memory address. Jump instructions are classified into two categories: Unconditional Jump and Conditional Jump.

6.41   Unconditional Jump

The 8085 instruction set includes one unconditional Jump instruction. The unconditional Jump instruction enables the programmer to set up continuous loops.

INSTRUCTION

Opcode	Operand	Description
JMP	16-bit	Jump
		☐ This is a 3-byte instruction
		☐ The second and third bytes specify the 16-bit memory address. However, the second byte specifies the low-order and the third byte specifies the high-order memory address

For example, to instruct the microprocessor to go to the memory location 2000H, the mnemonics and the machine code entered will be as follows:

Machine Code	Mnemonics
C3	JMP 2000H
00	
20	

Note the sequence of the machine code. The 16-bit memory address of the jump location is entered in the reverse order, the low-order byte (00H) first, followed by the high-order byte (20H). The 8085 is designed for such a reverse sequence. The jump location can also be specified using a label. While writing a program, you may not know the exact memory location to which a program sequence should be directed. In that case, the memory address can be specified with a label (or a name). This is particularly useful and necessary for an assembler. However, you should not specify both a label and its 16-bit address in a Jump instruction. Furthermore, you cannot use the same label for different memory locations. The next illustrative program shows the use of the Jump instruction.

6.42   Illustrative Program: Unconditional Jump to Set Up a Continuous Loop

PROBLEM STATEMENT
Modify the program in Example 6.2 to read the switch positions continuously and turn on the appliances accordingly.

PROBLEM ANALYSIS
One of the major drawbacks of the program in Example 6.2 is that the program reads switch positions once and then stops. Therefore, if you want to turn on/off different appliances, you have to reset the system and start all over again. This is impractical in real-life situations. However, the unconditional Jump instruction, in place of the HLT instruction, will allow the microcomputer to monitor the switch positions continuously.

PROGRAM

Memory Address	Machine Code	Label	Mnemonics	Comments
2000	DB	START:	IN 00H	;Read input switches
2001	00			
2002	D3		OUT 01H	;Turn on devices according to
2003	01			; switch positions
2004	C3		JMP START	;Go back to beginning and
2005	00			; read the switches again
2006	20			

PROGRAM FORMAT

The program includes one more column called *label*. The memory location 2000H is defined with the label START; therefore, the operand of the Jump instruction can be specified by the label START. The program sets up the endless loop, and the microprocessor monitors the input port continuously. The output will reflect any change in the switch positions.

6.43   Conditional Jumps

Conditional Jump instructions allow the microprocessor to make decisions based on certain conditions indicated by the flags. After logic and arithmetic operations, flip-flops (flags) are set or reset to reflect data conditions. The conditional Jump instructions check the flag conditions and make decisions to change or not to change the sequence of a program.

FLAGS

The 8085 flag register has five flags, one of which (Auxiliary Carry) is used internally. The other four flags used by the Jump instructions are

1. Carry flag
2. Zero flag
3. Sign flag
4. Parity flag

Two Jump instructions are associated with each flag. The sequence of a program can be changed either because the condition is present or because the condition is absent. For example, while adding the numbers you can change the program sequence either because the carry is present (JC = Jump On Carry) or because the carry is absent (JNC = Jump On No Carry).

INSTRUCTIONS

All conditional Jump instructions in the 8085 are 3-byte instructions; the second byte specifies the low-order (line number) memory address, and the third byte specifies the high-order (page number) memory address. The following instructions transfer the program sequence to the memory location specified under the given conditions:

Opcode	Operand	Description
JC	16-bit	Jump On Carry (if result generates carry and CY = 1)
JNC	16-bit	Jump On No Carry (CY = 0)
JZ	16-bit	Jump On Zero (if result is zero and Z = 1)
JNZ	16-bit	Jump On No Zero (Z = 0)
JP	16-bit	Jump On Plus (if $D_7 = 0$, and S = 0)
JM	16-bit	Jump On Minus (if $D_7 = 1$, and S = 1)
JPE	16-bit	Jump On Even Parity (P = 1)
JPO	16-bit	Jump On Odd Parity (P = 0)

All the Jump instructions are listed here for an overview. The Zero and Carry flags and related Jump instructions are used frequently. They are illustrated in the following examples.

6.44 Illustrative Program: Testing of the Carry Flag

PROBLEM STATEMENT
Load the hexadecimal numbers 9BH and A7H in registers D and E, respectively, and add the numbers. If the sum is greater than FFH, display 01H at output PORT0; otherwise, display the sum.

PROBLEM ANALYSIS AND FLOWCHART
The problem can be divided into the following steps:

1. Load the numbers in the registers.
2. Add the numbers.
3. Check the sum.
   Is the sum > FFH?

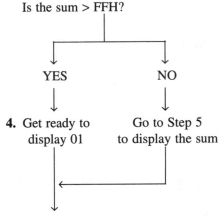

       YES   NO

4. Get ready to   Go to Step 5
   display 01   to display the sum

5. Display.
6. End.

FLOWCHART AND ASSEMBLY LANGUAGE PROGRAM
The six steps listed above can be converted into a flowchart and assembly language program as shown in Figure 6.10.

Step 3 is a decision-making block. In a flowchart, the decision-making process is represented by a diamond shape. It is important to understand how this block is translated into the assembly language program. By examining the block carefully you will notice the following:

1. The question is: Is there a Carry?
2. If the answer is no, change the sequence of the program. In the assembly language this is equivalent to Jump On No Carry—JNC.

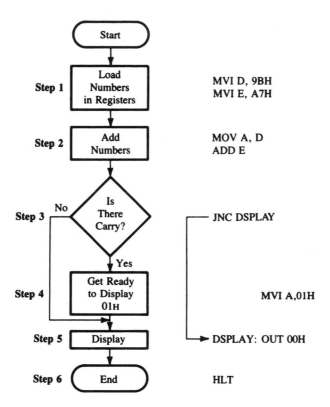

FIGURE 6.10
Flowchart and Assembly Language Program to Test Carry Flag

3. Now the next question is where to change the sequence—to Step 5. At this point the exact location is not known, but it is labeled DSPLAY.
4. The next step in the sequence is 4. Get ready to display byte 01H.
5. After completing the straight line sequence, translate Step 5 and Step 6: Display at the port and halt.

MACHINE CODE WITH MEMORY ADDRESSES

Assuming your R/W memory begins at 2000H, the preceding assembly language program can be translated as follows:

Memory Address	Machine Code	Label	Mnemonics
2000	16	START:	MVI D,9BH
2001	9B		
2002	1E		MVI E,A7H
2003	A7		

2004	7A		MOV A,D
2005	83		ADD E
2006	D2		JNC DSPLAY
2007	X		
2008	X		
2009	3E		MVI A,01H
200A	01		
200B	D3	DSPLAY:	OUT 00H
200C	00		
200D	76		HLT

While translating into the machine code, we leave memory locations 2007H and 2008H blank because the exact location of the transfer is not known. What is known is that two bytes should be reserved for the 16-bit address. After completing the straight line sequence, we know the memory address of the label DSPLAY; i.e., 200BH. This address must be placed in the reversed order as shown:

| 2007 | 0B | Low-order: Line Number |
| 2008 | 20 | High-order: Page Number |

USING THE INSTRUCTION JUMP ON CARRY (JC)

Now the question remains: Can the same problem be solved by using the instruction Jump On Carry (JC)? To use instruction JC, exchange the places of the answers YES and NO to the question: Is there a Carry? The flowchart will be as in Figure 6.11, and it shows that the program sequence is changed if there is a Carry. This flowchart has two end points; thus it will require a few more instructions than that of Figure 6.10. In this particular example, it is unimportant whether to use instruction JC or JNC, but in most cases the choice is made by the logic of a problem.

FIGURE 6.11
Flowchart for the Instruction Jump On Carry

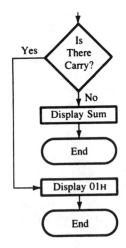

6.45   Review of Important Concepts

1. The Jump instructions change program execution from its sequential order to a different memory location.
2. The Jump instructions can transfer program execution ahead of the sequence (Jump Forward) or behind the sequence (Jump Backward).
3. The unconditional Jump is, generally, used to set up continuous loops.
4. The conditional Jumps are used for the decision-making process based on the data conditions of the result, reflected by the flags.
5. Arithmetic and logic instructions modify the flags according to the data of the result, and the conditional branch instructions use them to make decisions. However, the branch instructions do not affect the flags.

CAUTION

The conditional Jump instructions will not function properly unless the preceding instruction sets the necessary flag. Data Copy instructions do not affect the flags; furthermore, some arithmetic and logic instructions either do not affect the flags or affect only certain flags.

See Questions and Assignments 30–40 at the end of this chapter.

6.5   WRITING ASSEMBLY LANGUAGE PROGRAMS

Communicating with a microcomputer—giving it commands to perform a task and watching it perform them—is exciting. However, one can be uneasy communicating in strange mnemonics and hexadecimal machine codes. This feeling is like the uneasiness one has when beginning to speak a foreign language. How do we learn to communicate with a microcomputer in its assembly language? By using a few mnemonics at a time such as the mnemonics for Read the switches and Display the data. This chapter has introduced a group of basic instructions that can command the 8085 microprocessor to perform simple tasks.

After we know a few instructions, how do we begin to write a program? Any program, no matter how large, begins with mnemonics. And, just as several persons contribute to the construction of a hundred-story building, so the writing of a large program is usually the work of a team. In addition, the 8085 instruction set contains only 74 different instructions, some of them used quite frequently.

In a hundred-story building, most of the rooms are similar. If one knows the basic fundamentals of constructing a room, one can learn how to tie these rooms together in a coherent structure. However, planning and forethought are critical. Before beginning to build a structure, an architectural plan must be drawn. Similarly, to write a program, one needs to draw up a plan of logical thoughts. A given task should be broken down into small units that can be built independently. This is called the **modular design approach.**

6.51 Getting Started

Writing a program is equivalent to giving specific *commands* to the microprocessor in a *sequence* to *perform a task*. The italicized words provide clues to writing a program. Let us examine these terms.

☐ *Perform a Task.* What is the task you are asking it to do?
☐ *Sequence.* What is the sequence you want it to follow?
☐ *Commands.* What are the commands (instruction set) it can understand?

These terms can be translated into steps as follows:

Step 1: Read the problem carefully.
Step 2: Break it down into small steps.
Step 3: Represent these small steps in a possible sequence with a flowchart—a plan of attack.
Step 4: Translate each block of the flowchart into appropriate mnemonic instructions.
Step 5: Translate mnemonics into the machine code.
Step 6: Enter the machine code in memory and execute. Only on rare occasions is a program successfully executed on the first attempt.
Step 7: Start troubleshooting (see Section 6.6, "Debugging a Program").

These steps are illustrated in the next section.

6.52 Illustrative Program: Microprocessor-Controlled Manufacturing Process

PROBLEM STATEMENT

A microcomputer is designed to monitor various processes (conveyer belts) on the floor of a manufacturing plant, presented schematically in Figure 6.12. The microcomputer has two input ports with the addresses F1H and F2H and an output port with the address F3H. Input port F1H has six switches, five of which (corresponding to data lines D_4–D_0) control the conveyer belts through the output port F3H. Switch S_7, corresponding to the data line D_7, is reserved to indicate an emergency on the floor. As a precautionary measure, input port F2H is controlled by the foreman, and its switch, S_7', is also used to indicate an emergency. Output line D_6 of port F3H is connected to the emergency alarm.

Write a program to

1. turn on the five conveyer belts according to the ON/OFF positions of switches S_4–S_0 at port F1H.
2. turn off the conveyer belts and turn on the emergency alarm only when both switches—S_7 from port F1H and S_7' from port F2H—are triggered.
3. monitor the switches continuously.

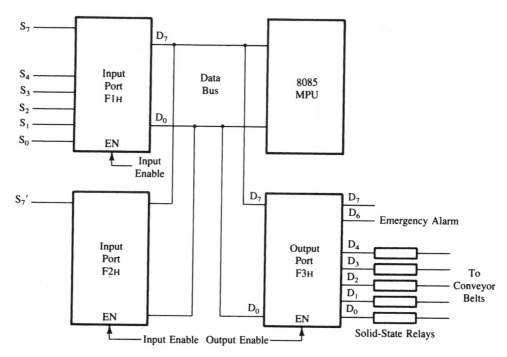

FIGURE 6.12
Input/Output Ports to Control Manufacturing Processes

PROBLEM ANALYSIS

To perform the tasks specified in the problem, the microprocessor needs to

1. read the switch positions.
2. check whether switches S_7 and S_7' from the ports F1H and F2H are on.
3. turn on the emergency signal if both switches are on, and turn off all the conveyer belts.
4. turn on the conveyer belts according to the switch positions S_0 through S_4 at input port F1H if both switches, S_7 and S_7', are not on simultaneously.
5. continue checking the switch positions.

FLOWCHART AND PROGRAM

The five steps listed above can be translated into a flowchart and an assembly language program as shown in Figure 6.13.

6.53 Documentation

A program is similar to a circuit diagram. Its purpose is to communicate to others what the program does and how it does it. Appropriate comments are critical for conveying the logic behind a program. The program as a whole should be self-documented.

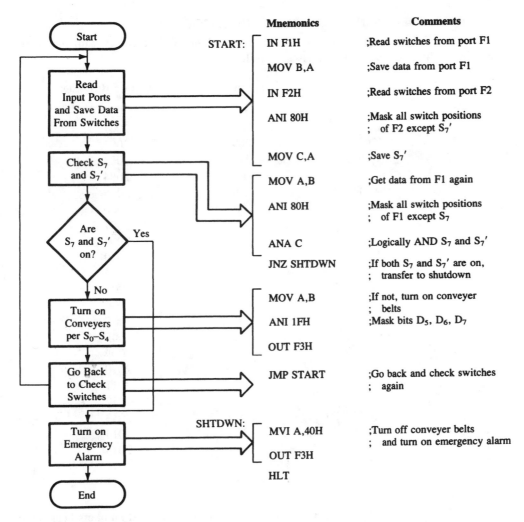

	Mnemonics	Comments
START:	IN F1H	;Read switches from port F1
	MOV B,A	;Save data from port F1
	IN F2H	;Read switches from port F2
	ANI 80H	;Mask all switch positions ; of F2 except S_7'
	MOV C,A	;Save S_7'
	MOV A,B	;Get data from F1 again
	ANI 80H	;Mask all switch positions ; of F1 except S_7
	ANA C	;Logically AND S_7 and S_7'
	JNZ SHTDWN	;If both S_7 and S_7' are on, ; transfer to shutdown
	MOV A,B	;If not, turn on conveyer ; belts
	ANI 1FH	;Mask bits D_5, D_6, D_7
	OUT F3H	
	JMP START	;Go back and check switches ; again
SHTDWN:	MVI A,40H	;Turn off conveyer belts ; and turn on emergency alarm
	OUT F3H	
	HLT	

FIGURE 6.13

Flowchart and Program for Controlling Manufacturing Processes

The comments should explain what is intended; they should not explain the mnemonics. For example, the first comment in Figure 6.13 indicates that the switch positions are being read at the input port and that the reading is saved. There is no point in writing: MOV B,A means transfer the contents of the accumulator to register B. Similarly, in a schematic, one does not write the word *resistor* once a resistor is represented by a symbol. A comment can be omitted if it does not say anything more than repeat a mnemonic.

The labels START and SHTDWN indicate what actions are being taken. These are the landmarks in a program. Avoid "cute" labels. Using cute labels in a program is similar to representing a power supply in a schematic by a picture of the sun or a ground with the digit zero.

FROM ASSEMBLY LANGUAGE TO MACHINE CODE

Illustrative Program 6.52 is translated into its machine code, together with the corresponding suggested memory addresses, as follows:

Mnemonics		Machine Code	Memory Addresses
1. START:	IN F1H	DB	2000
		F1	2001
2.	MOV B,A	78①	2002
3.	IN F2H	DB	2003
		F2	2004
4.	ANI 80H	E6	2005
		80	2006
5.	MOV C,A	4F	2007
6.	MOV A,B	78	2008
7.	ANI 80H	E6②	2009
8.	ANA C	A1	200A
9.	JNZ SHTDWN	C2③	200B
		20	200C
		14	200D
10.	MOV A,B	78	200E
11.	ANI 1FH	E6	200F
		1F	2010
12.	OUT F3H	D3	2011
		F3	2012
13.	JMP START	C3④	2013
14. SHTDWN:	MVI A,40H	3E	2014
		40	2015
15.	OUT F3H	D3⑤	2016
16.	HLT	76	2017

This program includes several errors, indicated by the circled numbers beside the codes. (See Assignment 41 for the debugging of this program.)

PROGRAM EXECUTION

The above machine codes can be loaded in R/W memory, starting with memory address 2000H. The execution of the program can be done two ways. The first is to execute the entire code by pressing the Execute key, and the second is to use the Single-Step key on a single-board computer. The Single-Step key executes one instruction at a time, and by using the Examine Register key, you can observe the contents of the registers and the flags as each instruction is being executed. The Single-Step and Examine Register techniques are discussed in Chapter 7 under the topic "Dynamic Debugging."

DEBUGGING A PROGRAM   6.6

Debugging a program is similar to troubleshooting hardware, but it is much more difficult and cumbersome. It is easy to poke and pinch at the components in a circuit, but, in a program, the result is generally binary: either it works or it does not work. When it does not work, very few clues alert you to what exactly went wrong. Therefore, it is essential to search carefully for the errors in the program logic, machine codes, and execution.

The debugging process can be divided into two parts: static debugging and dynamic debugging.

Static debugging is similar to visual inspection of a circuit board; it is done by a paper-and-pencil check of a flowchart and machine code. **Dynamic debugging** involves observing the output, or register contents, following the execution of each instruction (the single-step technique) or of a group of instructions (the breakpoint technique). Dynamic debugging will be discussed in the next chapter.

6.61 Debugging Machine Code

Translating the assembly language to the machine code is similar to building a circuit from a schematic diagram; the machine code will have errors just as would the circuit board. The following errors are common:

1. Selecting a wrong code.
2. Forgetting the second or third byte of an instruction.
3. Specifying the wrong jump location.
4. Not reversing the order of high and low bytes in a Jump instruction.
5. Writing memory addresses in decimal, thus specifying wrong jump locations.

The program for controlling manufacturing processes listed in Section 6.53 has several of these errors. These errors must be corrected before entering the machine code in the R/W memory of your system.

See Questions and Assignments 41–43 at the end of this chapter.

SOME PUZZLING QUESTIONS AND THEIR ANSWERS   6.7

After one learns something about the microprocessor architecture, memory, I/O, the instruction set, and simple programming, a few questions still remain unanswered. These questions do not fit into any particular discussion. They just lurk in the corners of one's mind to reappear once in a while when one is in a contemplative mood. This section attempts to answer some of these unasked questions.

1. *What happens in a single-board microcomputer when the power is turned on and the Reset key is pushed?*

When the power is turned on, the monitor program stored either in EPROM or ROM comes alive. The Reset key clears the program counter, and the program counter holds the memory address 0000H. Some systems are automatically reset when the power is turned on (called power-on reset).

2. *How does the microprocessor know how and when to start?*

As soon as the Reset key is pushed, the program counter places the memory address 0000H on the address bus, the instruction at that location is fetched, and the execution of the Key Monitor program begins. Therefore, the Key Monitor program is stored on page 00H.

3. *What is a monitor program?*

In a single-board microcomputer with a Hex keyboard, the instructions are entered in R/W memory through the keyboard. The Key Monitor program is a set of instructions that continuously checks whether a key is pressed and stores the binary equivalent of a pressed key in a memory location.

4. *What is an assembler?*

An assembler is a program that translates the mnemonics into their machine code. It is generally not available on a single-board microcomputer.

A program can be entered in mnemonics in a microcomputer equipped with an ASCII keyboard. The assembler will translate mnemonics into the 8085 machine code and assign memory locations to each machine code, thus avoiding the manual assembly and the errors associated with it. Additional instructions can be inserted anywhere in the program, and the assembler will reassign all the new memory locations and jump locations.

5. *How does the microprocessor know what operation to perform first (Read/Write memory or Read/Write I/O)?*

The first operation is always a Fetch instruction.

6. *How does the microprocessor differentiate among a positive number, a negative number, and a bit pattern?*

It does not know the difference. The microprocessor views any data byte as eight binary digits. The programmer is responsible for providing the interpretation.

For example, after an arithmetic or logic operation, if the bits in the accumulator are

$$1\ 1\ 1\ 1\ \ 0\ 0\ 1\ 0 = F2H$$

the Sign flag is set because $D_7 = 1$. This does not mean it is a negative number, even if the Sign flag is set. The Sign flag indicates only that $D_7 = 1$. The eight bits in the accumulator could be a bit pattern, or a positive number larger than 127_{10}, or the 2's complement of a number.

7. *If flags are individual flip-flops, can they be observed on an oscilloscope?*

No, they cannot be observed on an oscilloscope. The flag register is internal to the microprocessor. However, they can be tested through conditional branch instructions, and they can be examined by storing them on the stack memory (see Chapter 9).

8. *If the program counter is always one count ahead of the memory location from which the machine code is being fetched, how does the microprocessor change the sequence of program execution with a Jump instruction?*

When a Jump instruction is fetched, its second and third bytes (a new memory location) are placed in the W and Z registers of the microprocessor. After the execution of the Jump instruction, the contents of the W and Z registers are placed on the address bus to fetch the instruction from a new memory location, and the program counter is loaded by updating the contents of the W and Z registers.

SUMMARY

The instructions from the 8085 instruction set introduced in this chapter are summarized below to provide an overview. After careful examination of these instructions, you will begin to see a pattern emerge from the mnemonics, the number of bytes required for the various instructions, and the tasks the 8085 can perform. Read the notations (Rs) as the contents of the source register, (Rd) as the contents of the destination register, (A) as the contents of the accumulator, and (R) as the contents of the register R.*

Instructions	Tasks	Addressing Mode
Data transfer (Copy) Instructions		
1. MOV Rd,Rs	Copy (Rs) into (Rd).	Register
2. MVI R,8-bit	Load register R with the 8-bit data.	Immediate
3. IN 8-bit port address	Read data from the input port.	Direct
4. OUT 8-bit port address	Write data in the output port.	Direct
Arithmetic Instructions		
1. ADD R	Add (R) to (A).	Register
2. ADI 8-bit	Add 8-bit data to (A).	Immediate
3. SUB R	Subtract (R) from (A).	Register
4. SUI 8-bit	Subtract 8-bit data from (A).	Immediate

*R, Rs, and Rd represent any one of the 8-bit registers—A, B, C, D, E, H, and L.

| 5. INR R | Increment (R). | Register |
| 6. DCR R | Decrement (R). | Register |

Logic Instructions

1. ANA R	Logically AND (R) with (A).	Register
2. ANI 8-bit	Logically AND 8-bit data with (A).	Immediate
3. ORA R	Logically OR (R) with (A).	Register
4. ORI 8-bit	Logically OR 8-bit data with (A).	Immediate
5. XRA R	Logically Exclusive-OR (R) with (A).	Register
6. XRI 8-bit	Logically Exclusive-OR 8-bit data with (A).	Immediate
7. CMA	Complement (A).	

Branch Instructions

1. JMP 16-bit	Jump to 16-bit address unconditionally.	Immediate
2. JC 16-bit	Jump to 16-bit address if the CY flag is set.	Immediate
3. JNC 16-bit	Jump to 16-bit address if the CY flag is reset.	Immediate
4. JZ 16-bit	Jump to 16-bit address if the Zero flag is set.	Immediate
5. JNZ 16-bit	Jump to 16-bit address if the Zero flag is reset.	Immediate
6. JP 16-bit	Jump to 16-bit address if the Sign flag is reset.	Immediate
7. JM 16-bit	Jump to 16-bit address if the Sign flag is set.	Immediate
8. JPE 16-bit	Jump to 16-bit address if the Parity flag is set.	Immediate
9. JPO 16-bit	Jump to 16-bit address if the Parity flag is reset.	Immediate

Machine Control Instructions

| 1. NOP | No operation. |
| 2. HLT | Stop processing and wait. |

The set of instructions listed here is used frequently in writing assembly language programs. The important points to be remembered about these instructions are as follows:

1. The data transfer (copy) instructions copy the contents of the source into the destination without affecting the source contents.
2. The results of the arithmetic and logic operations are usually placed in the accumulator.
3. The conditional Jump instructions are executed according to the flags set after an operation. Not all instructions set the flags; in particular, the data transfer instructions do not set the flags.

See Section 5.5 for an overview of the instruction set and see inside the cover page for a complete instruction set according to the functions.

QUESTIONS AND PROGRAMMING ASSIGNMENTS

Note: To execute the instructions in the following assignments and use the Examine Register key, the HLT instruction must be replaced by an appropriate instruction for your system. In the Intel SDK-85 system, the HLT can be replaced by the RST1 (code CFH) instruction. If the HLT instruction is used, the system must be Reset to exit from the Halt instruction; that clears the registers.

Section 6.1: Data Transfer (Copy) Operations

1. Specify the contents of the registers and the flag status as the following instructions are executed.

	A	B	C	D	S	Z	CY
MVI A,00H							
MVI B,F8H							
MOV C,A							
MOV D,B							
HLT							

2. Assemble the instructions in Assignment 1, enter the code in R/W memory of a single-board microcomputer, and execute the instructions using the single-step key. Before you begin to execute the instructions, record the initial conditions of the registers and the flags using the Examine Register key. Observe the register contents and the flags as you execute each instruction.

3. Write instructions to load the hexadecimal number 65H in register C, and 92H in the accumulator A. Display the number 65H at PORT0 and 92H at PORT1.

4. Write instructions to read the data at input PORT 07H and at PORT 08H. Display the input data from PORT 07H at output PORT 00H, and store the input data from PORT 08H in register B.

5. Specify the output at PORT1 if the following program is executed.

MVI B,82H
MOV A,B
MOV C,A
MVI D,37H
OUT PORT1
HLT

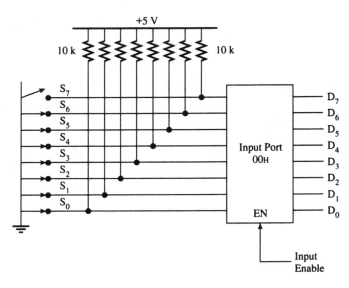

FIGURE 6.14
Input Port with Switches

6. If the switch S_7 of the input PORT0 (see Figure 6.14) connected to the data line D_7 is at logic 1 and other switches are at logic 0, specify the contents of the accumulator when the instruction IN PORT0 is executed.

                              MVI A,9FH
                              IN PORT0
                              MOV B,A
                              OUT PORT1
                              HLT

7. Specify the output at PORT1 and the contents of register B after executing the instructions in Assignment 6.

Section 6.2: Arithmetic Operations

8. Specify the register contents and the flag status as the following instructions are executed. Specify also the output at PORT0.

	A	B	S	Z	CY	
	00	FF	0	1	0	Initial Contents

MVI A,F2H
MVI B,7AH
ADD B
OUT PORT0
HLT

9. What operation can be performed by using the instruction ADD A?

10. What operation can be performed by using the instruction SUB A? Specify the status of Z and CY.

11. Specify the register contents and the flag status as the following instructions are executed.

A	C	S	Z	CY	
XX	XX	0	0	0	Initial Contents

MVI A,5EH
ADI A2H
MOV C,A
HLT

12. Assemble the program in Assignment 11, and enter the code in R/W memory of your system. Reset the system, and examine the contents of the flag register using the Examine Register key. Clear the flags by inserting 00H in the flag register, and execute each instruction using the single-step key. Verify the register contents and the flags as each instruction is being executed.

13. Write a program using the ADI instruction to add the two hexadecimal numbers 3AH and 48H and to display the answer at an output port.

14. Write instructions to
   a. load 00H in the accumulator.
   b. decrement the accumulator.
   c. display the answer.
   Specify the answer you would expect at the output port.

15. The following instructions subtract two unsigned numbers. Specify the contents of register A and the status of the S and CY flags. Explain the significance of the sign flag if it is set.

MVI A,F8H
SUI 69H

16. Specify the register contents and the flag status as the following instructions are executed.

A	B	S	Z	CY
XX	XX	X	X	X

SUB A
MOV B,A
DCR B
INR B
SUI 01H
HLT

17. Assemble the program in Assignment 16, and execute each instruction using the Single-Step key. Verify the register contents and the flags as each instruction is being executed.

18. Write a program to
    a. clear the accumulator.
    b. add 47H (use ADI instruction).
    c. subtract 92H.
    d. add 64H.
    e. display the results after subtracting 92H and after adding 64H.
    Specify the answers you would expect at the output ports.

19. Specify the reason for clearing the accumulator before adding the number 47H directly to the accumulator in Assignment 18.

Section 6.3: Logic Operations

20. What operation can be performed by using the instruction XRA A (Exclusive-OR the contents of the accumulator with itself)? Specify the status of Z and CY.

21. Specify the register contents and the flag status (S, Z, CY) after the instruction ORA A is executed.

> MVI A,A9H
> MVI B,57H
> ADD B
> ORA A

22. Assemble the program in Assignment 21 by adding an End instruction (such as RST1 in Intel's SDK-85 system), and execute the program. Verify the register contents and the flags by using the Examine Register key.

23. When the microprocessor reads an input port, the instruction IN does not set any flag. If the input reading is zero, what logic instruction can be used to set the Zero flag without affecting the contents of the accumulator?

24. Specify the register contents and the flag status as the following instructions are executed.

A	B	S	Z	CY
XX	XX	X	X	X

> XRA A
> MVI B,4AH
> SUI 4FH
> ANA B
> HLT

25. Assemble the instructions in Assignment 24. Execute each instruction using the Single-Step key, and verify the register contents and flags.

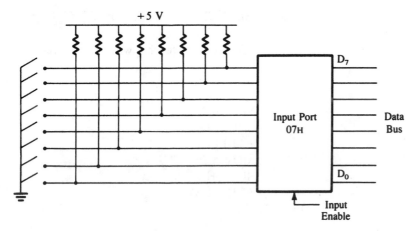

FIGURE 6.15
DIP Switch Input Port with Address 07H

26. Load the data byte A8H in register C. Mask the high-order bits (D_7–D_4), and display the low-order bits (D_3–D_0) at an output port.

27. Load the data byte 8EH in register D and F7H in register E. Mask the high-order bits (D_7–D_4) from both the data bytes, Exclusive-OR the low-order bits (D_3–D_0), and display the answer.

28. Load the bit pattern 91H in register B and 87H in register C. Mask all the bits except D_0 from registers B and C. If D_0 is at logic 1 in both registers, turn on the light connected to the D_0 position of output port 01H; otherwise, turn off the light.

29. Figure 6.15 shows an input port with an 8-key DIP switch. When all switches are off, the microprocessor reads the data FFH. When a switch is turned on (closed), it goes to logic 0 (for all switches ON, the data will be 00H). Write instructions to read the input port and, if all switches are open, set the Zero flag. (Use the instruction CMA to complement the input reading and ORA A to set the Zero flag.)

Section 6.4: Branch Operations

30. What is the output at PORT1 when the following instructions are executed?

```
                MVI A,8FH
                ADI 72H
                JC DSPLAY
                OUT PORT1
                HLT
       DSPLAY:  XRA A
                OUT PORT1
                HLT
```

31. In Question 30, replace the instruction ADI 72H by the instruction SUI 67H, and specify the output.

32. In the following program, explain the range of the bytes that will be displayed at PORT2.

```
                MVI A,BYTE1
                MOV B,A
                SUI 50H
                JC DELETE
                MOV A,B
                SUI 80H
                JC DSPLAY
        DELETE: XRA A
                OUT PORT1
                HLT
        DSPLAY: MOV A,B
                OUT PORT2
                HLT
```

33. Specify the address of the output port, and explain the type of numbers that can be displayed at the output port.

```
                MVI A,BYTE1     ;Get a data byte
                ORA A           ;Set flags
                JP OUTPRT       ;Jump if the byte is positive
                XRA A
        OUTPRT: OUT F2H
                HLT
```

34. In Question 33, if BYTE1 = 92H, what is the output at port F2H?

35. Explain the function of the following program.

```
                MVI A,BYTE1     ;Get a data byte
                ORA A           ;Set flags
                JM OUTPRT
                OUT 01H
                HLT
        OUTPRT: CMA             ;Find 2's complement
                ADI 01H
                OUT 01H
                HLT
```

36. In Question 35, if BYTE1 = A7H, what will be displayed at port 01H?

37. Rewrite the Section 6.44 illustrative program for testing the Carry flag using the instruction JC (Jump On Carry).

38. Write instructions to clear the CY flag, to load number FFH in register B, and increment (B). If the CY flag is set, display 01 at the output port; otherwise, display the contents of register B. Explain your results.

39. Write instructions to clear the CY flag, to load number FFH in register C, and to add 01 to (C). If the CY flag is set, display 01 at an output port; otherwise, display the contents of register C. Explain your results. Are they the same as in Question 38?

40. Write instructions to load two unsigned numbers in register B and register C, respectively. Subtract (C) from (B). If the result is in 2's complement, convert the result in absolute magnitude and display it at PORT1; otherwise, display the positive result. Execute the program with the following sets of data.

$$\begin{aligned} &\text{Set 1:} \quad (B) = 42H, \quad (C) = 69H \\ &\text{Set 2:} \quad (B) = 69H, \quad (C) = 42H \\ &\text{Set 3:} \quad (B) = F8H, \quad (C) = 23H \end{aligned}$$

Section 6.6: Debugging a Program

41. In Section 6.53, a program for controlling manufacturing processes is examined, all the errors are marked as 1 through 5. Rewrite the program with the errors corrected.

42. To test this program, substitute the instructions IN F1H and IN F2H by loading the two data bytes 97H and 85H in registers D and E. Rewrite the program to include other appropriate changes. Enter and execute the program on your system.

43. In the program presented in Section 6.53, assume that an LED indicator is connected to the output line D_7 of the port F3. Modify the program to turn on the LED when switch S_7 from port F1 is turned on, even if switch S_7' is off.

Programming Techniques with Additional Instructions

A computer is at its best, surpassing human capability, when it is asked to repeat such simple tasks as adding thousands of numbers. It does this accurately with electronic speed and without showing any signs of boredom. The programming techniques—such as looping, counting, and indexing—required for repetitive tasks are introduced in this chapter.

Data needed for repetitive tasks generally are stored in the system's R/W memory. The data must be transferred (copied) from memory to the microprocessor for manipulation (processing). The instructions related to data manipulations and data transfer (copy) between memory and the microprocessor are introduced in this chapter, as well as instructions related to 16-bit data and additional logic operations. Applications of these instructions are shown in five illustrative programs. The chapter concludes with a discussion of dynamic debugging techniques.

OBJECTIVES

☐ Draw a flowchart of a conditional loop illustrating indexing and counting.

☐ List the seven blocks of a generalized flowchart illustrating data acquisitions and data processing.

☐ Explain the functions of the 16-bit data transfer instructions LXI and of the arithmetic instructions INX and DCX.

☐ Explain the functions of memory-related data transfer instructions, and illustrate how a memory location is specified using the indirect and the direct addressing modes.

☐ Write a program to illustrate an application of instructions related to memory data transfer and 16-bit data.

☐ Explain the functions of arithmetic instructions related to data in memory: ADD/SUB M. Write a program to perform arithmetic operations that generate carry.

☐ Explain the functions and the differences between the four instructions: RLC, RAL, RRC, and RAR. Write a program to illustrate uses of these instructions.

☐ Explain the functions of the Compare instructions: CMP and CPI and the flags set under various conditions. Write a program to illustrate uses of the Compare instructions.

☐ Explain the term *dynamic debugging* and the debugging techniques: Single Step and Breakpoint.

7.1 PROGRAMMING TECHNIQUES: LOOPING, COUNTING, AND INDEXING

The programming examples illustrated in previous chapters are simple and can be solved manually. However, the computer surpasses manual efficiency when tasks must be repeated, such as adding a hundred numbers or transferring a thousand bytes of data. It is fast and accurate.

The programming technique used to instruct the microprocessor to repeat tasks is called **looping.** A loop is set up by instructing the microprocessor to change the sequence of execution and perform the task again. This process is accomplished by using Jump instructions. In addition, techniques such as counting and indexing (described below) are used in setting up a loop.

Loops can be classified into two groups:

☐ Continuous loop—repeats a task continuously
☐ Conditional loop—repeats a task until certain data conditions are met

They are described in the next two sections.

7.11 Continuous Loop

A continuous loop is set up by using the unconditional Jump instruction shown in the flowchart (Figure 7.1).

FIGURE 7.1
Flowchart of a Continuous Loop

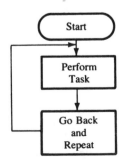

A program with a continuous loop does not stop repeating the tasks until the system is reset. Typical examples of such a program include a continuous counter (see Chapter 8, Section 8.2) or a continuous monitor system.

7.12 Conditional Loop

A conditional loop is set up by the conditional Jump instructions. These instructions check flags (Zero, Carry, etc.) and repeat the specified tasks if the conditions are satisfied. These loops usually include counting and indexing.

CONDITIONAL LOOP AND COUNTER

A counter is a typical application of the conditional loop. For example, how does the microprocessor repeat a task five times? The process is similar to that of a car racer in the Indy 500 going around the track 500 times. How does the racer know when 500 laps have been completed? The racing team manager sets up a counting and flagging method for the racer. This can be symbolically represented as in Figure 7.2(a). A similar approach is needed for the microprocessor to repeat the task five times. The microprocessor needs a counter, and when the counting is completed, it needs a flag. This can be accomplished with the conditional loop, as illustrated in the flowchart in Figure 7.2(b).

FIGURE 7.2
Flowcharts to Indicate Number of
Repetitions Completed

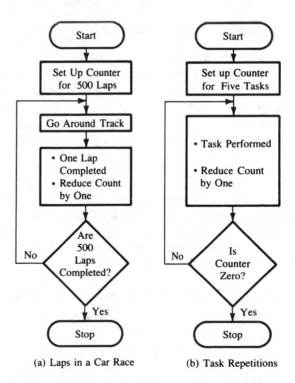

(a) Laps in a Car Race   (b) Task Repetitions

The computer flowchart of Figure 7.2(b) is translated into a program as follows:

1. Counter is set up by loading an appropriate count in a register.
2. Counting is performed by either incrementing or decrementing the counter.
3. Loop is set up by a conditional Jump instruction.
4. End of counting is indicated by a flag.

It is easier to count down to zero than to count up because the Zero flag is set when the register becomes zero. (Counting up requires the Compare instruction, which is introduced later.)

Conditional Loop, Counter, and Indexing   Another type of loop includes indexing along with a counter. (*Indexing* means pointing or referencing objects with sequential numbers. In a library, books are arranged according to numbers, and they are referred to or sorted by numbers. This is called indexing.) Similarly, data bytes are stored in memory locations, and those data bytes are referred to by their memory locations.

Example 7.1

Illustrate the steps necessary to add ten bytes of data stored in memory locations starting at a given location, and display the sum. Draw a flowchart.

Procedure   The microprocessor needs

a. a counter to count 10 data bytes
b. an index or a memory pointer to locate where data bytes are stored
c. to transfer data from a memory location to the microprocessor (ALU)
d. to perform addition
e. registers for temporary storage of partial answers
f. a flag to indicate the completion of the task
g. to store or output the result

These steps can be represented in the form of a flowchart as in Figure 7.3.
This generalized flowchart can be used in solving many problems. Some blocks may have to be expanded with additional loops, or some blocks may need to be interchanged in their positions.

7.13   Review of Important Concepts

1. Programming is a logical approach to instruct the microprocessor to perform operations in a given sequence.
2. The computer is at its best in repeating tasks. It is fast and accurate.
3. Loops are set up by using the looping technique along with counting and indexing.
4. The computer is a versatile and powerful computing tool because of its capability to set up loops and to make decisions based on data conditions.

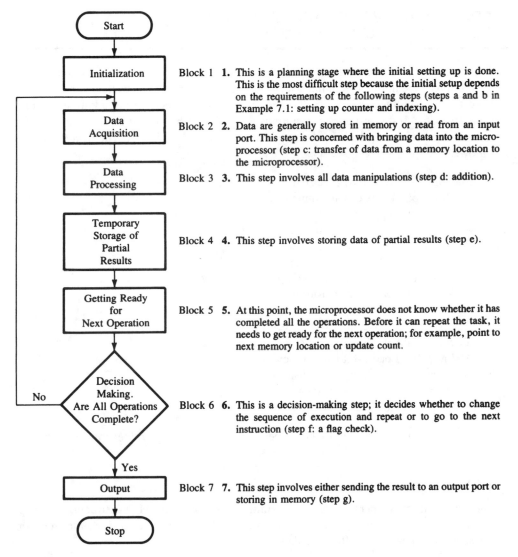

Block 1 **1.** This is a planning stage where the initial setting up is done. This is the most difficult step because the initial setup depends on the requirements of the following steps (steps a and b in Example 7.1: setting up counter and indexing).

Block 2 **2.** Data are generally stored in memory or read from an input port. This step is concerned with bringing data into the microprocessor (step c: transfer of data from a memory location to the microprocessor).

Block 3 **3.** This step involves all data manipulations (step d: addition).

Block 4 **4.** This step involves storing data of partial results (step e).

Block 5 **5.** At this point, the microprocessor does not know whether it has completed all the operations. Before it can repeat the task, it needs to get ready for the next operation; for example, point to next memory location or update count.

Block 6 **6.** This is a decision-making step; it decides whether to change the sequence of execution and repeat or to go to the next instruction (step f: a flag check).

Block 7 **7.** This step involves either sending the result to an output port or storing in memory (step g).

FIGURE 7.3
Generalized Programming Flowchart

LOOKING AHEAD

The programming techniques and the flowcharting introduced in this section will be illustrated with various applications throughout the chapter. Additional instructions necessary for these applications will be introduced first. The primary focus here is to analyze a given programming problem in terms of the basic building blocks of the flowchart shown in Figure 7.3.

See Questions and Assignments 1–4 at the end of the chapter.

7.2   ADDITIONAL DATA TRANSFER AND 16-BIT ARITHMETIC INSTRUCTIONS

The instructions related to the data transfer among microprocessor registers and the I/O instructions were introduced in the last chapter; this section introduces the instructions related to the data transfer between the microprocessor and memory. In addition, instructions for some 16-bit arithmetic operations are included because they are necessary for using the programming techniques introduced earlier in this chapter. The opcodes are as follows:

1. Loading 16-bit data in register pairs
   LXI Rp: Load Register Pair Immediate
2. Data transfer (copy) from memory to the microprocessor
   MOV R,M: Move (from memory to register)
   LDAX B/D: Load Accumulator Indirect
   LDA 16-bit: Load Accumulator Direct
3. Data transfer (copy) from the microprocessor to memory
   MOV M,R: Move (from register to memory)
   STAX B/D: Store Accumulator Indirect
   STA 16-bit: Store Accumulator Direct
4. Loading 8-bit data directly in memory register (location)
   MVI M,8-bit: Load 8-bit data in memory
5. Incrementing/Decrementing Register Pair
   INX Rp: Increment Register Pair
   DCX Rp: Decrement Register Pair

   The instructions related to these operations are illustrated with examples in the following sections.

7.21   16-Bit Data Transfer to Register Pairs (LXI)

The LXI instructions perform functions similar to those of the MVI instructions, except that the LXI instructions load 16-bit data in register pairs and the stack pointer register. These instructions do not affect the flags.

INSTRUCTIONS

Opcode	Operand	
LXI	Rp, 16-bit	Load Register Pair
LXI	B, 16-Bit	☐ This is a 3-byte instruction
LXI	D,16-bit	☐ The second byte is loaded in the low-order register of the register pair (e.g., register C)
LXI	H,16-bit	☐ The third byte is loaded in the high-order register pair (e.g., register B)

LXI SP,16-bit ☐ There are four such instructions in the set as shown. The operands B, D, and H represent BC, DE, and HL registers, and SP represents the stack pointer register

---

Write instructions to load the 16-bit number 2050H in the register pair HL using LXI and MVI opcodes, and explain the difference between the two instructions.

Example 7.2

Instructions Figure 7.4 shows the register contents and the instructions required for Example 7.2.

The LXI instruction is functionally similar to two MVI instructions. The LXI instruction takes three bytes of memory and requires ten clock periods (T-states). On the other hand, two MVI instructions take four bytes of memory and require 14 clock periods (T-states).

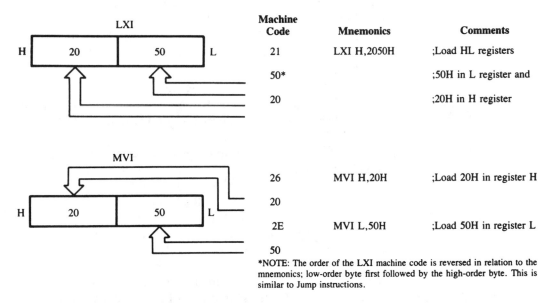

	Machine Code	Mnemonics	Comments
	21	LXI H,2050H	;Load HL registers
	50*		;50H in L register and
	20		;20H in H register
	26	MVI H,20H	;Load 20H in register H
	20		
	2E	MVI L,50H	;Load 50H in register L
	50		

*NOTE: The order of the LXI machine code is reversed in relation to the mnemonics; low-order byte first followed by the high-order byte. This is similar to Jump instructions.

FIGURE 7.4
Instructions and Register Contents for Example 7.2

7.22 Data Transfer (Copy) from Memory to the Microprocessor

The 8085 instruction set includes three types of memory transfer instructions; two use the indirect addressing mode and one uses the direct addressing mode. These instructions do not affect the flags.

1. MOV R,M: Move (from Memory to Register)
☐ This is a 1-byte instruction

☐ It copies the data byte from the memory location into a register
☐ R represents microprocessor registers A, B, C, D, E, H, and L
☐ The memory location is specified by the contents of the HL register
☐ This specification of the memory location is indirect; it is called the indirect addressing mode

2. LDAX B/D: Load Accumulator Indirect

☐ This is a 1-byte instruction

LDAX B ☐ It copies the data byte from the memory location into the accumulator

LDAX D ☐ The instruction set includes two instructions as shown

☐ The memory location is specified by the contents of the registers BC or DE
☐ The addressing mode is indirect

3. LDA 16-bit: Load Accumulator Direct

☐ This is a 3-byte instruction
☐ It copies the data byte from the memory location specified by the 16-bit address in the second and third byte
☐ The second byte is a line number (low-order memory address)
☐ The third byte is a page number (high-order memory address)
☐ The addressing mode is direct

Example 7.3

The memory location 2050H holds the data byte F7H. Write instructions to transfer the data byte to the accumulator using three different opcodes: MOV, LDAX, and LDA.

Solution

Figure 7.5 shows the register contents and the instructions required for Example 7.3. All of these three instructions copy the data byte F7H from the memory location 2050H to the accumulator.

In Figure 7.5(a), register HL is first loaded with the 16-bit number 2050H. The instruction MOV A,M uses the contents of the HL register as a memory pointer to location 2050H; this is the indirect addressing mode. The HL register is used frequently as a memory pointer because any instruction that uses M as an operand can copy from and into any one of the registers.

In Figure 7.5(b), the contents of register BC are used as a memory pointer to location 2050H by the instruction LDAX B. Registers BC and DE can be used as restricted memory pointers to copy the contents of only the accumulator into memory and vice versa; however, they cannot be used to copy the contents of other registers.

Figure 7.5(c) illustrates the direct addressing mode; the instruction LDA specifies the memory address 2050H directly as a part of its operand.

After examining all three methods, you may notice that the indirect addressing mode takes four bytes and the direct addressing mode takes three bytes. The question is: Why not just use the direct addressing mode?

If only one byte is to be transferred, the LDA instruction is more efficient. But for a block of memory transfer, the instruction LDA (three bytes) will have to be repeated for

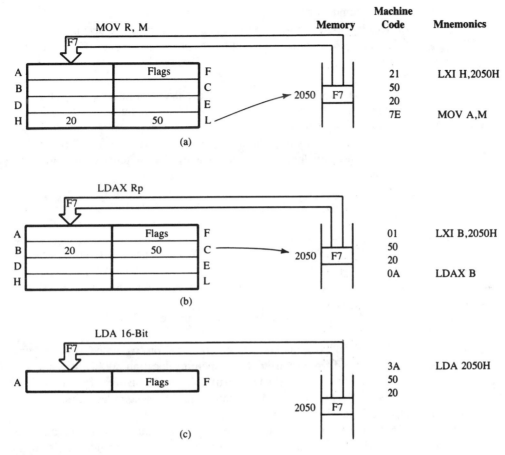

FIGURE 7.5
Instructions and Register Contents for Example 7.3

each memory. On the other hand, a loop can be set up with two other instructions, and the contents of a register pair can be incremented or decremented. This is further illustrated in Section 7.26.

7.23   Data Transfer (Copy) from the Microprocessor to Memory or Directly into Memory

The instructions for copying data from the microprocessor to a memory location are similar to those described in the previous section. These instructions are as follows:

1. MOV M,R: Move (from Register to Memory).

       □ This is a 1-byte instruction that copies data from a register, R, into the memory location specified by the contents of HL registers

2. STAX B/D: Store Accumulator Indirect

               ☐ This is a 1-byte instruction that copies data from the accumulator into

   STAX B   the memory location specified by the contents of either BC or DE reg-

   STAX D   isters

3. STA 16-bit: Store Accumulator Direct

               ☐ This is a 3-byte instruction that copies data from the accumulator into
the memory location specified by the 16-bit operand.

4. MVI M,8-bit: Load 8-bit data in memory

               ☐ This is a two-byte instruction; the second byte specifies 8-bit data

               ☐ The memory location is specified by the contents of the HL reg-
ister

Example 7.4

1. Register B contains 32H. Illustrate the instructions MOV and STAX to copy the contents of register B into memory location 8000H using indirect addressing.
2. The accumulator contains F2H. Copy (A) into memory location 8000H, using direct addressing.
3. Load F2H directly in memory location 8000H using indirect addressing.

Solution

Figure 7.6 shows the register contents and the instructions for Example 7.4. In Figure 7.6(a), the byte 32H is copied from register B into memory location 8000H by using the HL as a memory pointer. However, in Figure 7.6(b), where the DE register is used as a memory pointer, the byte 32H must be copied from B into the accumulator first because the instruction STAX copies only from the accumulator.

    In Figure 7.6(c), the instruction STA copies 32H from the accumulator into the memory location 8000H. The memory address is specified as the operand; this is an illustration of the direct addressing mode. On the other hand, Figure 7.6(d) illustrates how to load a byte directly in memory location by using the HL as a memory pointer.

7.24 Arithmetic Operations Related to 16 Bits or Register Pairs

The instructions related to incrementing/decrementing 16-bit contents in a register pair are introduced below. These instructions do not affect flags.

1. INX Rp: Increment Register Pair

               ☐ This is a 1-byte instruction

   INX B   ☐ It treats the contents of two registers as one 16-bit number and in-

   INX D   creases the contents by 1

   INX H   ☐ The instruction set includes four instructions, as shown

   INX SP

	Machine Code	Mnemonics

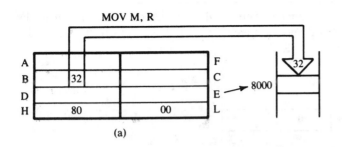

(a)

	21	LXI H,8000H
	00	
	80	
	70	MOV M,B

This instruction copies the contents of the accumulator into memory. Therefore, it is necessary first to copy (B) into A.

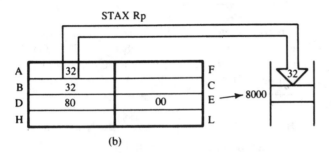

(b)

	11	LXI D,8000H
	00	
	80	
	78	MOV A,B
	12	STAX D

This also requires the transfer of (B) to A.

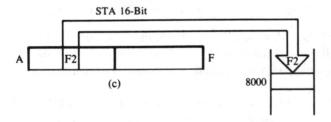

(c)

	32	STA 8000H
	00	
	80	

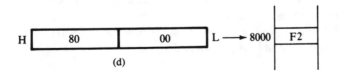

(d)

	21	LXI H,8000H
	00	
	80	
	36	MVI M,F2H
	F2	

FIGURE 7.6
Instructions and Register Contents for Example 7.4

2. DCX Rp: Decrement Register Pair
      □ This is a 1-byte instruction
   DCX B   □ It decreases the 16-bit contents of a register pair by 1
   DCX D   □ The instruction set includes four instructions, as shown
   DCX H
   DCX SP

Example 7.5

Write the instruction to load the number 2050H in the register pair BC. Increment the number using the instruction INX B and illustrate whether the INX B instruction is equivalent to the instructions INR B and INR C.

FIGURE 7.7
Instructions and Register
Contents for Example 7.5

				Machine Code	Mnemonics
B	20	50	C	01	LXI B,2050H
				50	
				20	
B	20	51	C	03	INX B

Solution

Figure 7.7 shows the instructions and register contents for Example 7.5. The instruction INX B views 2050H as one 16-bit number and increases the number to 2051H. On the other hand, the instructions INR B and INR C will increase (B) and (C) separately and the contents of the BC register pair will be 2151H.

7.25 Review of Instructions

In this section, we examined primarily how to copy data from the microprocessor into memory and vice versa. The 8085 instruction provides three methods of copying data between the microprocessor and memory:

1. Indirect addressing using HL as a memory pointer: This is the most flexible and frequently used method. The HL register can be used to copy between any one of the registers and memory. Any instruction with the operand M automatically assumes the HL is the memory pointer.
2. Indirect addressing using BC and DE as memory pointers: The instructions LDAX and STAX use BC and DE as memory pointers. However, this method is restricted to copying from and into the accumulator and cannot be used for other registers. In addition, the mnemonics (LDAX and STAX) are somewhat misleading; therefore, careful attention must be given to their interpretion.
3. Direct addressing using LDA and STA instructions: These instructions include memory address as the operand. This method is also restricted to copying from and into the accumulator.

In addition to the above data copy instructions, we discussed two instructions, INX and DCX, concerning the register pairs. The critical feature of these instructions is that they do not affect the flags.

7.26 Illustrative Program: Block Transfer of Data Bytes

PROBLEM STATEMENT

Sixteen bytes of data are stored in memory locations at XX50H to XX5FH. Transfer the entire block of data to new memory locations starting at XX70H.

Data(H)   37, A2, F2, 82, 57, 5A, 7F, DA, E5, 8B, A7, C2, B8, 10, 19, 98

PROBLEM ANALYSIS

The problem can be analyzed in terms of the blocks suggested in the flowchart (Figure 7.8). The steps are as follows:

The flowchart in Figure 7.8 includes five blocks; these blocks are identified with numbers referring to the blocks in the generalized flowchart in Figure 7.3. This problem is not concerned with data manipulation (processing); therefore, the flowchart does not require Blocks 3 and 4 (data processing and temporary storage of partial results). The problem simply deals with the transferring of the data bytes from one location to another location in memory; therefore, the Store Data Byte block is equivalent to the Output block in the generalized flowchart.

Block 1 is the initialization block; this block sets up two memory pointers and one counter. Block 5 is concerned with updating the memory pointers and the counter. The

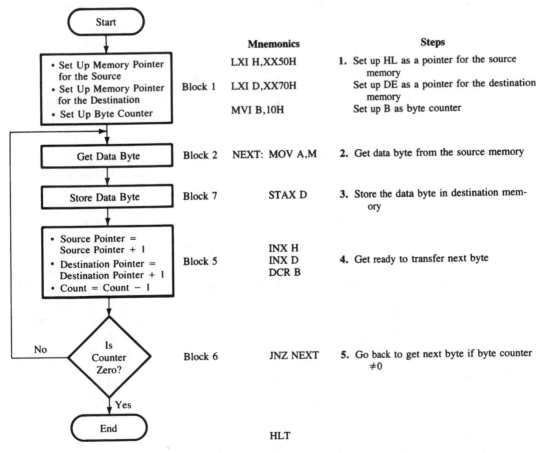

FIGURE 7.8
Flowchart for Block Transfer of Data Bytes

statements shown in the block appear strange if they are read as algebraic equations; however, they are not algebraic equations. The statement Pointer = Pointer + 1 means the new value is obtained by incrementing the previous value by one.

The statements in the flowchart correspond one-to-one with the mnemonics. In large programs, such details in the flowchart are impractical as well as undesirable. However, these details are included here to show the logic flow in writing programs. In Figure 7.8, some of the details can be eliminated very easily from the flowchart. For example, Blocks 2 and 7 can be combined in one statement; such as, Transfer Data Byte from Source to Destination. Similarly, Block 5 can be reduced to one statement; such as, Update memory Pointers and Counter.

PROGRAM

Memory Address HI-LO	Hex Code	Label	Opcode	Operand	Comments
XX00	21	START:	LXI	H,XX50H	;Set up HL as a
01	50				; pointer for source
02	XX				; memory
03	11		LXI	D,XX70H	;Set up DE as
04	70				; a pointer for
05	XX				; destination
06	06		MVI	B,10H	;Set up B to count
07	10				; 16 bytes
08	7E	NEXT:	MOV	A,M	;Get data byte from
09	12		STAX	D	; source memory
09	12		STAX	D	;Store data byte at
0A	23		INX	H	; destination
0A	23		INX	H	;Point HL to next
0B	13		INX	D	; source location
0B	13		INX	D	;Point DE to
0C	05		DCR	B	; next destination
0C	05		DCR	B	;One transfer is
					; complete,
					; decrement count
0D	C2		JNZ	NEXT	;If counter is not 0,
0E	08				; go back to transfer
0F	XX				; next byte
10	76		HLT		;End of program
XX50	37				;Data
↓	↓				
XX5F	98				

FIGURE 7.9
Data Transfer from Memory to
Accumulator (a), Then to New
Memory Location (b)

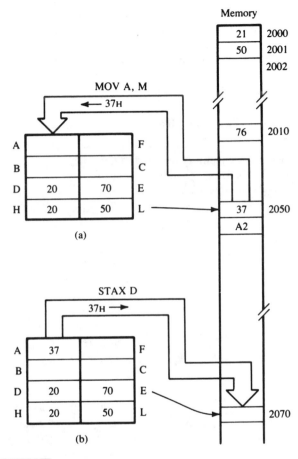

(a)

(b)

PROGRAM EXECUTION AND OUTPUT

To execute the program, substitute the page number of your system's R/W memory in place of XX, enter the program and the data, and execute it. To verify the proper execution, check the memory locations from XX70H to XX7FH.

Let us assume the system R/W user memory starts at 2000H. Figure 7.9(a) shows how the contents of the memory location 2050H are copied into the accumulator by the instruction MOV A,M; the HL register points to location 2050 and instruction MOV A,M copies 37H into A. Figure 7.9(b) shows that the DE register points to the location 2070H and the instruction STAX D copies (A) into the location 2070H.

See Questions and Assignments 5–21 at the end of this chapter.

ARITHMETIC OPERATIONS RELATED TO MEMORY　7.3

In the last chapter, the arithmetic instructions concerning three arithmetic tasks—Add, Subtract, and Increment/Decrement—were introduced. These instructions dealt with microprocessor register contents or numbers. In this chapter, instructions concerning the arithmetic tasks related to memory will be introduced:

ADD M/SUB M: Add/Subtract the contents of a memory location to/from the contents
                     of the accumulator.
INR M/DCR M: Increment/Decrement the contents of a memory location.

7.31   Instructions

The arithmetic instructions referenced to memory perform two tasks: one is to copy a byte from a memory location to the microprocessor, and the other is to perform the arithmetic operation. These instructions (other than INR and DCR) implicitly assume that one of the operands is (A); after an operation, the previous contents of the accumulator are replaced by the result. All flags are modified to reflect the data conditions (see the exceptions: INR and DCR).

Opcode	Operand	
ADD	M	Add Memory ☐ This is a 1-byte instruction ☐ It adds (M) to (A) and stores the result in A ☐ The memory location is specified by the contents of HL register
SUB	M	Subtract Memory ☐ This is a 1-byte instruction ☐ It subtracts (M) from (A) and stores the result in A ☐ The memory location is specified by (HL)
INR	M	This is a 1-byte instruction ☐ It increments the contents of a memory location by 1, not the memory address ☐ The memory location is specified by (HL) ☐ All flags except the Carry flag are affected
DCR	M	This is a 1-byte instruction ☐ It decrements (M) by 1 ☐ The memory location is specified by (HL) ☐ All flags except the Carry flag are affected

Example 7.6

Write instructions to add the contents of the memory location 2040H to (A), and subtract the contents of the memory location 2041H from the first sum. Assume the accumulator has 30H, the memory location 2040H has 68H, and the location 2041H has 7FH.

Solution

Before asking the microprocessor to perform any memory-related operations, we must specify the memory location by loading the HL register pair. In the example illustrated in Figure 7.10, the contents of the HL pair 2040H specify the memory location. The instruction ADD M adds 68H, the contents of memory location 2040H, to the contents of

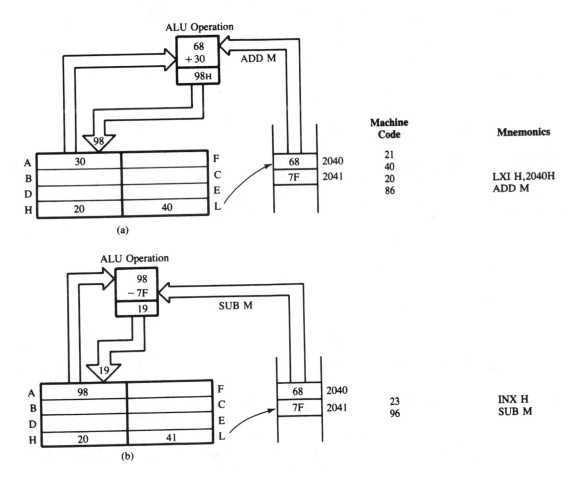

FIGURE 7.10
Register and Memory Contents and Instructions for Example 7.6

the accumulator (30H). The instruction INX H points to the next memory location, 2041H, and the instruction SUB M subtracts the contents (7FH) of memory location 2041H from the previous sum.

Write instructions to

1. load 59H in memory location 2040H, and increment the contents of the memory location.
2. load 90H in memory location 2041H, and decrement the contents of the memory location.

Example 7.7

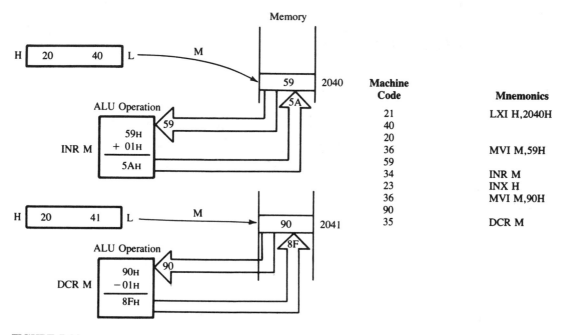

FIGURE 7.11
Register and Memory Contents, and Instructions for Example 7.7

Solution   Figure 7.11 shows register contents and the instructions required for Example 7.7. The instruction MVI M loads 59H in the memory location indicated by (HL). The instruction INR M increases the contents, 59H, of the memory location to 5AH. The instruction INX H increases (HL) to 2041H. The next two instructions load and decrement 90H.

7.32 Illustrative Program: Addition with Carry

PROBLEM STATEMENT
Six bytes of data are stored in memory locations starting at XX50H. Add all the data bytes. Use register B to save any carries generated, while adding the data bytes. Display the entire sum at two output ports, or store the sum at two consecutive memory locations, XX70H and XX71H.

Data(H)   A2, FA, DF, E5, 98, 8B

PROBLEM ANALYSIS
This problem can be analyzed in relation to the general flowchart in Figure 7.3 as follows:

1. Because of the memory-related arithmetic instructions just introduced in this section, two blocks in the general flowchart—data acquisition and data processing—can be combined in one instruction.

2. The fourth block—temporary storage of partial results—is unnecessary because the sum can be stored in the accumulator.

3. The data processing block needs to be expanded to account for carry.

In the first block (Block 1) of the flowchart in Figure 7.12, the accumulator and the carry register (for example, register B) must be cleared in order to use them for arithmetic operations; otherwise, residual data will cause erroneous results.

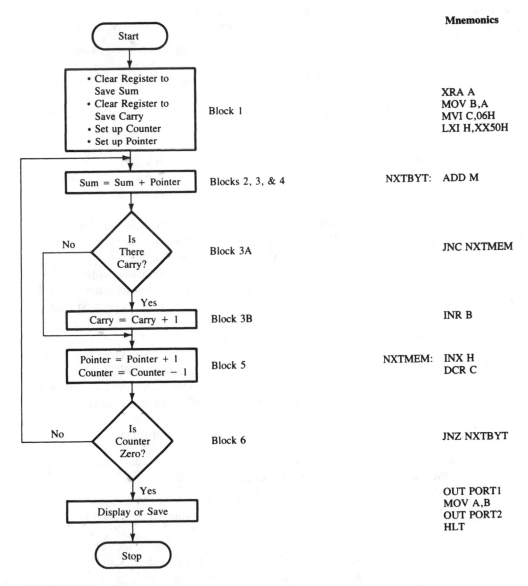

FIGURE 7.12

Flowchart for Addition with Carry

After the addition, it is necessary to check whether that operation has generated a carry (Block 3A). If a carry is generated, the carry register is incremented by one (Block 3B); otherwise, it is bypassed. The instruction ADC (Add with Carry) is inappropriate for this operation. (See Appendix F for the description of the instruction ADC.)

PROGRAM

Memory Address HI-LO	Machine Code	Label	Opcode	Operand	Comments
				Instructions	
XX00	AF		XRA	A	;Clear (A) to save sum
01	47		MOV	B,A	;Clear (B) to save carry
02	0E		MVI	C,06H	;Set up register C as a counter
03	06				
04	21		LXI	H,XX50H	;Set up HL as memory pointer
05	50				
06	XX				
07	86	NXTBYT:	ADD	M	;Add byte from memory
08	D2		JNC	NXTMEM	;If no carry, do not increment
09	0C				; carry register
0A	XX				
0B	04		INR	B	;If carry, save carry bit
0C	23	NXTMEM:	INX	H	;Point to next memory location
0D	0D		DCR	C	;One addition is completed;
					; decrement counter
0E	C2		JNZ	NXTBYT	;If all bytes are not yet added,
0F	07				; go back to get next byte
10	XX				
		;Output Display			
11	D3		OUT	PORT1	;Display low-order byte of the
12	PORT1				; sum at PORT1
13	78		MOV	A,B	;Transfer carry to accumulator
14	D3		OUT	PORT2	;Display carry digits
15	PORT2				
16	76		HLT		;End of program

		;Storing in Memory—Alternative to Output Display			
11	21		LXI	H,XX70H	;Point to the memory
12	70				; location to store answer
13	XX				
14	77		MOV	M,A	;Store low-order byte at XX70H
15	23		INX	H	;Point to location XX71H
16	70		MOV	M,B	;Store carry bits
17	76		HLT		;End of program

50	A2	;Data Bytes
51	FA	
52	DF	
53	E5	
54	98	
55	8B	

PROGRAM DESCRIPTION AND OUTPUT

In this program, register B is used as a carry register, register C as a counter to count six data bytes, and the accumulator to add the data bytes and save the partial sum.

After the completion of the summation, the high-order byte (bits higher than eight bits) of the sum is saved in register B and the low-order byte is in the accumulator. Both are displayed at two different ports, or they can be stored at the memory locations XX70H and 71H.

See Questions and Assignments 22–31 at the end of this chapter.

LOGIC OPERATIONS: ROTATE 7.4

In the last chapter, the logic instructions concerning the four operations AND, OR, Ex-OR, and NOT were introduced. This chapter introduces instructions related to rotating the accumulator bits. The opcodes are as follows:

☐ RLC: Rotate Accumulator Left
☐ RAL: Rotate Accumulator Left Through Carry
☐ RRC: Rotate Accumulator Right
☐ RAR: Rotate Accumulator Right Through Carry

7.41 Instructions

This group has four instructions; two are for rotating left and two are for rotating right. The differences between these instructions are illustrated in the following examples.

1. RLC: Rotate Accumulator Left
☐ Each bit is shifted to the adjacent left position. Bit D_7 becomes D_0.
☐ CY flag is modified according to bit D_7.

Assume the accumulator contents are AAH and CY = 0. Illustrate the accumulator contents after the execution of the RLC instruction twice.

Example 7.8

Figure 7.13 shows the contents of the accumulator and the CY flag after the execution of the RLC instruction twice. The first RLC instruction shifts each bit to the left by one position, places bit D_7 in bit D_0 and sets the CY flag because $D_7 = 1$. The accumulator byte AAH becomes 55H after the first rotation. In the second rotation, the byte is again AAH, and the CY flag is reset because bit D_7 of 55H is 0.

Solution

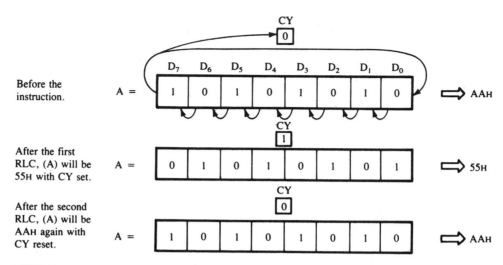

FIGURE 7.13
Accumulator Contents after RLC

2. RAL: Rotate Accumulator Left Through Carry
   □ Each bit is shifted to the adjacent left position. Bit D_7 becomes the carry bit and the carry bit is shifted into D_0.
   □ The Carry flag is modified according to bit D_7.

Example 7.9

Assume the accumulator contents are AAH and CY = 0. Illustrate the accumulator contents after the execution of the instruction RAL twice.

Solution

Figure 7.14 shows the contents of the accumulator and the CY flag after the execution of the RAL instruction twice. The first RAL instruction shifts each bit to the left by one position, places bit D_7 in the CY flag, and the CY bit in bit D_0. This is a 9-bit rotation; CY is assumed to be the ninth bit of the accumulator. The accumulator byte AAH becomes 54H after the first rotation. In the second rotation, the byte becomes A9H, and the CY flag is reset.

Examining these two examples, you may notice that the primary difference between these two instructions is that (1) the instruction RLC rotates through eight bits, and (2) the instruction RAL rotates through nine bits.

3. RRC: Rotate Accumulator Right
   □ Each bit is shifted right to the adjacent position. Bit D_0 becomes D_7.
   □ The Carry flag is modified according to bit D_0.
4. RAR: Rotate Accumulator Right Through Carry
   □ Each bit is shifted right to the adjacent position. Bit D_0 becomes the carry bit, and the carry bit is shifted into D_7.

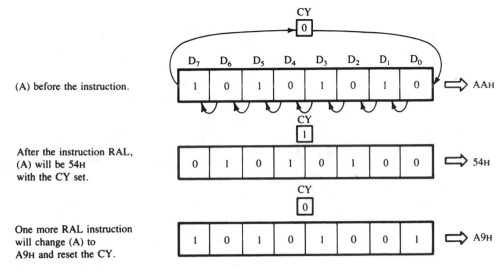

FIGURE 7.14
Accumulator Contents after RAL

Assume the contents of the accumulator are 81H and CY = 0. Illustrate the accumulator contents after the RRC and RAR instructions.

Example 7.10

Figure 7.15 shows the changes in the contents of the accumulator (81H) when the RRC instruction is used and when the RAR instruction is used. The 8-bit rotation of the RRC instruction changes 81H into C0H, and the 9-bit rotation of the RAR instruction changes 81H into 40H.

Solution

APPLICATIONS OF ROTATE INSTRUCTIONS

The rotate instructions are primarily used in arithmetic multiply and divide operations and for serial data transfer.

For example, if (A) is 0000 1000 = 08H,

☐ By rotating 08H right: (A) = 0000 0100 = 04H
   This is equivalent to dividing by 2
☐ By rotating 08H left: (A) = 0001 0000 = 10H
   This is equivalent to multiplying by 2 (10H = 16_{10})

However, these procedures are invalid when logic 1 is rotated left from D_7 to D_0 or vice versa. For example, if 80H is rotated left, it becomes 01H. Applications of serial data transfer are discussed in Chapter 16.

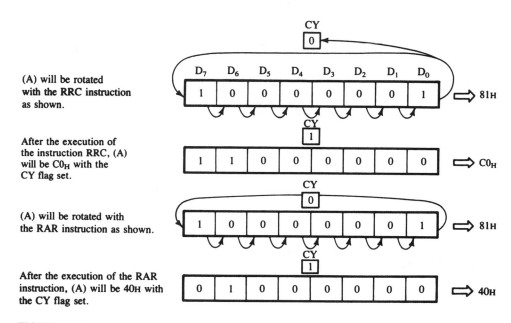

(A) will be rotated with the RRC instruction as shown.

After the execution of the instruction RRC, (A) will be C0H with the CY flag set.

(A) will be rotated with the RAR instruction as shown.

After the execution of the RAR instruction, (A) will be 40H with the CY flag set.

FIGURE 7.15
Rotate Right Instructions

7.42 Illustrative Program: Checking Sign with Rotate Instructions

PROBLEM STATEMENT

A set of ten current readings is stored in memory locations starting at XX60H. The readings are expected to be positive ($<127_{10}$). Write a program to

1. check each reading to determine whether it is positive or negative.
2. reject all negative readings.
3. add all positive readings.
4. output FFH to PORT1 at any time when the sum exceeds eight bits to indicate overload; otherwise, display the sum. If no output port is available in the system, go to step 5.
5. store FFH in the memory location XX70H when the sum exceeds eight bits; otherwise, store the sum.

Data(H)   28, D8, C2, 21, 24, 30, 2F, 19, F2, 9F

PROBLEM ANALYSIS

This problem can be divided into the following steps:

1. Transfer a data byte from the memory location to the microprocessor, and check whether it is a negative number. The sign of the number can be verified by rotating bit D_7 into the Carry position and checking for CY. (See Assignment 38 at the end of this chapter to verify the sign of a number with the Sign flag.)
2. If it is negative, reject the data and get the next data byte.

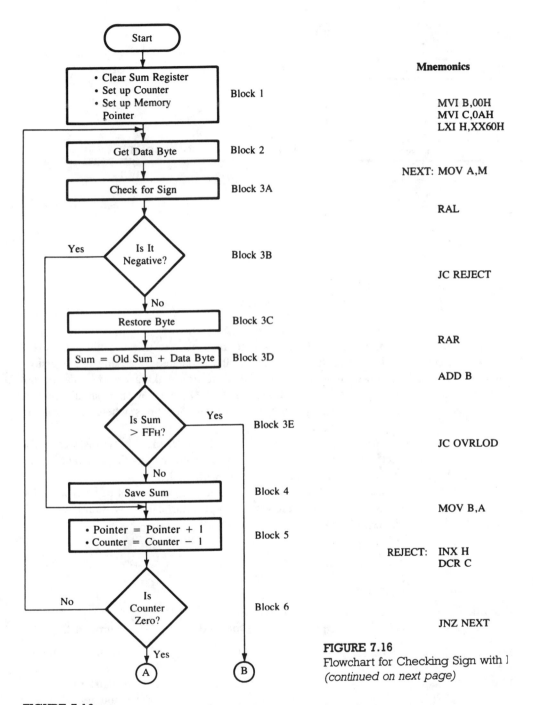

Mnemonics

MVI B,00H
MVI C,0AH
LXI H,XX60H

NEXT: MOV A,M

RAL

JC REJECT

RAR

ADD B

JC OVRLOD

MOV B,A

REJECT: INX H
DCR C

JNZ NEXT

FIGURE 7.16
Flowchart for Checking Sign with]
(continued on next page)

FIGURE 7.16
Flowchart for Checking Sign with Rotate Instructions

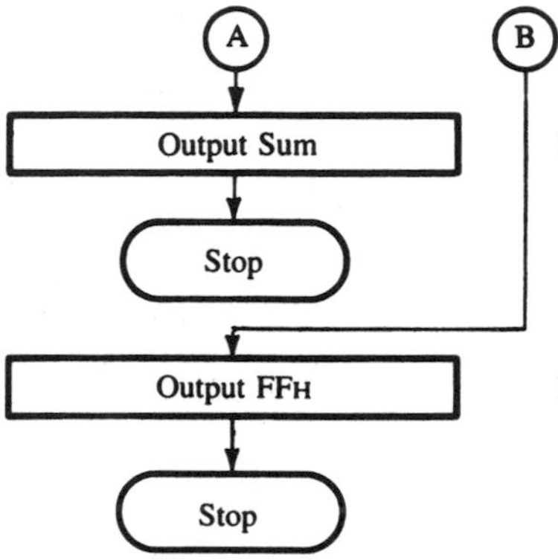

Block 7A

```
         MOV A,B
         OUT PORT1

         HLT
```

Block 7B

```
OVRLOD:  MVI A,FFH
         OUT PORT1
         HLT
```

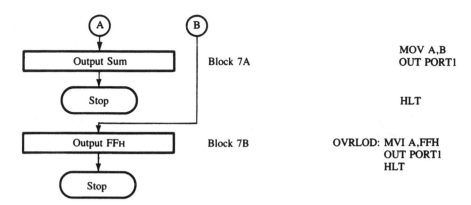

FIGURE 7.16
(continued)

3. If it is positive, add the data byte.
4. Check the sum for a carry and display an appropriate output.

The steps are shown in the flowchart in Figure 7.16.

In this flowchart, Blocks 1, 2, 4, 5, and 6 are similar to those in the generalized flowchart. Block 3—data processing—is substantially expanded by adding two decision-making blocks. This flowchart is first drawn with decision-making answers that do not change the sequence. For example, in Blocks 3B and 3E, the flowchart should first be continued with the answers NO. Then find appropriate locations where the program should be directed to the answers YES.

In this program, the sign of a data byte cannot be checked with the instruction JM (Jump On Minus) because the instruction MOV A,M does not set any flags. (However, the flags can be set by ORing the accumulator contents with itself.)

Similarly, in Block 3C, the instruction RRC (Rotate Right) cannot be used when the instruction RAL is used to rotate left. However, the program can be written using RLC and RRC in both places (3A and 3C).

PROGRAM

Memory Address HI-LO	Machine Code	Label	Instructions Opcode	Operand	Comments
XX00	06		MVI	B,00H	;Clear (B) to save sum
01	00				
02	0E		MVI	C,0AH	;Set up register C
					; as a counter

03	0A				
04	21		LXI	H,XX60H	;Set up HL as memory
					; pointer
05	60				
06	XX				
07	7E	NEXT:	MOV	A,M	;Get byte
08	17		RAL		;Shift D_7 into CY
09	DA		JC	REJECT	;If $D_7 = 1$, reject byte and
0A	12				; go to increment
					; pointer
0B	XX				
0C	1F		RAR		;If byte is positive, re-
					; store it
0D	80		ADD	B	;Add previous sum to (A)
0E	DA		JC	OVRLOD	;If sum >FFH, it is over-
					; load;
0F	1C				; turn on emergency
10	XX				
11	47		MOV	B,A	;Save sum
12	23	REJECT:	INX	H	;Point to next reading
13	0D		DCR	C	;One reading is checked;
					; decrement counter
14	C2		JNZ	NEXT	;If all readings are not
15	07				; checked, go back to
					; transfer next byte
16	XX				
		;Output Display Section			
17	78		MOV	A,B	
18	D3		OUT	PORT1	;Display sum
19	PORT1				
1A	76		HLT		;End of program
1B	00		NOP		;To match Jump location,
					; OVRLOD in memory
					; storage
1C	3E	OVRLOD:	MVI	A,FFH	;It is an overload
1D	FF				
1E	D3		OUT	PORT1	;Display overload signal
1F	PORT1				; at PORT1
20	76		HLT		

		;Storing Result in Memory—Alternative to Output Display			
18	32		STA	XX70H	;Store sum in memory ; XX70H
19	70				
1A	XX				
1B	76				
1C	3E	OVRLOD:	MVI	A,FFH	;Store overload signal in ; memory XX70
1D	FF				
1E	32		STA	XX70H	
1F	70				
20	XX				
21	76		HLT		

XX60	28	;Current Readings
61	D8	
62	C2	
63	21	
64	24	
65	30	
66	2F	
67	19	
68	F2	
69	9F	

PROGRAM DESCRIPTION AND OUTPUT

In this program, register C is used as a counter to count ten bytes. Register B is used to save the sum. The sign of the number is checked by verifying whether D_7 is 1 or 0. If the Carry flag is set to indicate the negative sign, the program rejects the number and goes to Block 5, Getting Ready for Next Operation.

The program should reject the data bytes D8, C2, F2, and 9F, and should add the rest. The answer displayed should be E5.

See Questions and Assignments 32–40 at the end of this chapter.

7.5   LOGIC OPERATIONS: COMPARE

The 8085 instruction set has two types of Compare operations: CMP and CPI.

☐ CMP: Compare with Accumulator
☐ CPI: Compare Immediate (with Accumulator)

The microprocessor compares a data byte (or register/memory contents) with the contents of the accumulator by subtracting the data byte from (A), and indicates whether the data byte is ≥\≤ (A) by modifying the flags. However, the contents are not modified.

7.51 Instructions

1. **CMP R/M**: Compare (Register or Memory) with Accumulator
   - ☐ This is a 1-byte instruction.
   - ☐ It compares the data byte in register or memory with the contents of the accumulator.
   - ☐ If (A) < (R/M), the CY flag is set and the Zero flag is reset.
   - ☐ If (A) = (R/M), the Zero flag is set and the CY flag is reset.
   - ☐ If (A) > (R/M), the CY and Zero flags are reset.
   - ☐ When memory is an operand, its address is specified by (HL).
   - ☐ No contents are modified; however, all remaining flags (S, P, AC) are affected according to the result of the subtraction.
2. **CPI 8-bit**: Compare Immediate with Accumulator
   - ☐ This is a 2-byte instruction, the second byte being 8-bit data.
   - ☐ It compares the second byte with (A).
   - ☐ If (A) < 8-bit data, the CY flag is set and the Zero flag is reset.
   - ☐ If (A) = 8-bit data, the Zero flag is set, and the CY flag is reset.
   - ☐ If (A) > 8-bit data, the CY and Zero flags are reset.
   - ☐ No contents are modified; however, all remaining flags (S, P, AC) are affected according to the result of the subtraction.

Write an instruction to load the accumulator with the data byte 64H, and verify whether the data byte in memory location 2050H is equal to the accumulator contents. If both data bytes are equal, jump to location OUT1.

Example 7.11

Solution

Figure 7.17 illustrates Example 7.11.

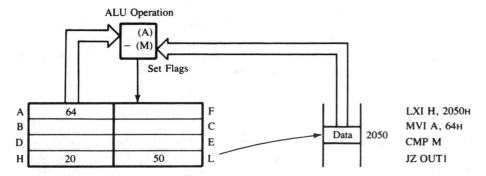

FIGURE 7.17
Compare Instructions

In these instructions, the instruction CMP M selects the memory location pointed out by the HL register (2050H) and compares the contents of that location with (A). If they are equal, the Zero flag is set, and the program jumps to location OUT1.

Let us assume the data byte in memory location 2050H is 9AH. The microprocessor compares 64H with 9AH by subtracting 9AH from 64H as shown below.

$$(A) = 64H \qquad 0110 \ 0100$$
$$+$$
$$\text{2's Complement of 9AH} \qquad \underline{0110 \ 0110}$$
$$CY \qquad 0 \ 1100 \ 1010$$

$$\text{Complement CY} \qquad 1 \ 1100 \ 1010$$

The result CAH will modify the flags as S = 1, Z = 0, and CY = 1; however, the original byte in the accumulator 64H will not be changed.

7.52 Illustrative Program: Use of Compare Instruction to Indicate End of Data String

PROBLEM STATEMENT

A set of current readings is stored in memory locations starting at XX50H. The end of the data string is indicated by the data byte 00H. Add the set of readings. The answer may be larger than FFH. Display the entire sum at PORT1 and PORT2 or store the answer in the memory locations XX70 and XX71H.

Data(H) 32, 52, F2, A5, 00

PROBLEM ANALYSIS

In this problem, the number of data bytes is variable, and the end of the data string is indicated by loading 00H. Therefore, the counter technique will not be useful to indicate the end of the readings. However, by comparing each data byte with 00H, the end of the data can be determined. This is shown in the flowchart in Figure 7.18.

FLOWCHART

The flowchart shows that after a data byte is transferred, it is first checked for 00H (Block 6). This operation is similar to checking for a negative number in the previous problem. However, this is also an exit point. If the byte is zero, the program goes to the output Block 7. The other significant change is in Block 5 where the unconditional Jump brings the sequence back into the program.

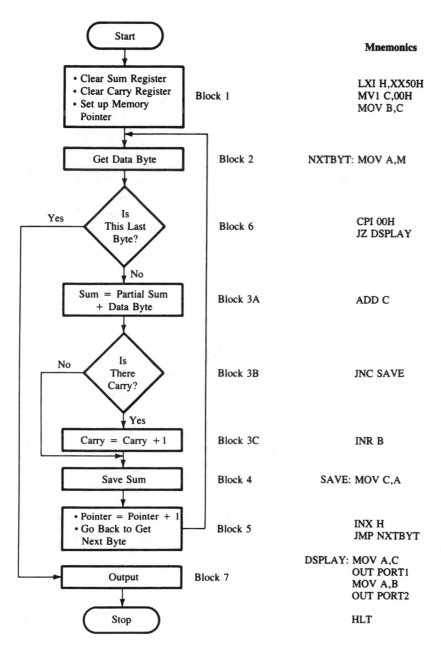

		Mnemonics
• Clear Sum Register • Clear Carry Register • Set up Memory Pointer	Block 1	LXI H,XX50H MV1 C,00H MOV B,C
Get Data Byte	Block 2	NXTBYT: MOV A,M
Is This Last Byte?	Block 6	CPI 00H JZ DSPLAY
Sum = Partial Sum + Data Byte	Block 3A	ADD C
Is There Carry?	Block 3B	JNC SAVE
Carry = Carry + 1	Block 3C	INR B
Save Sum	Block 4	SAVE: MOV C,A
• Pointer = Pointer + 1 • Go Back to Get Next Byte	Block 5	INX H JMP NXTBYT
Output	Block 7	DSPLAY: MOV A,C OUT PORT1 MOV A,B OUT PORT2
Stop		HLT

FIGURE 7.18
Flowchart for Compare Instruction to Check End of Data String

ROGRAM

Memory Address HI-LO	Machine Code	Label	Instruction Opcode	Operand	Comments
XX00	21	START:	LXI	H,XX50H	;Set up HL as memory pointer
01	50				
02	XX				
03	0E		MVI	C,00H	;Clear (C) to save sum
04	00				
05	41		MOV	B,C	;Clear (B) to save carry
06	7E	NXTBYT:	MOV	A,M	;Transfer current reading to (A)
07	FE		CPI	00H	;Is this the last reading?
08	00				
09	CA		JZ	DSPLAY	;If yes; go to display section
0A	16				
0B	XX				
0C	81		ADD	C	;Add previous sum to accumulator
0D	D2		JNC	SAVE	;Skip CY register if there is no
0E	11				; carry
0F	XX				
10	04		INR	B	;Update carry register
11	4F	SAVE:	MOV	C,A	;Save sum
12	23		INX	H	;Point to next reading
13	C3		JMP	NXTBYT	;Go back to get next reading
14	06				
15	XX				
		;Output Display			
16	79	DSPLAY:	MOV	A,C	
17	D3		OUT	PORT1	;Display low-order byte of sum
18	PORT1				; at PORT1
19	78		MOV	A,B	;Transfer carry bits to accumulator
1A	D3		OUT	PORT2	;Display high-order byte of sum
1B	PORT2				; at PORT2
1C	76		HLT		;End of program .

		;Storing Result in Memory—Alternative to Output Display			
16	21	DSPLAY:	LXI	H,XX70	;Point index to XX70H location
17	70				
18	XX				
19	71		MOV	M,C	;Store low-order byte of sum ; in XX70H
1A	23		INX	H	;Point index to XX71H
1B	70		MOV	M,B	;Store high-order byte
1C	76		HLT		;End of program

50	32	;Data—Current Readings	
51	52		
52	F2		
53	A5		
54	00		

PPROGRAM DESCRIPTION AND OUTPUT

This program adds the first four readings, which results in the sum 21BH. The low-order byte 1BH is saved in register C, and the high-order byte 02H is stored in the carry register B. These are each displayed at different output ports by the OUT instruction or stored in the memory locations XX70 and XX71.

See Questions and Assignments 41–56 at the end of the chapter.

7.53   Illustrative Program: Sorting

PROBLEM STATEMENT

A set of three readings is stored in memory starting at XX50H. Sort the readings in ascending order.

Data(H)   87, 56, 42

PROBLEM ANALYSIS

In this problem, three readings should be arranged in ascending order: 42, 56, and 87. The number of readings is kept small to simplify the explanation. However, the programming technique used here, called bubble sort, can be applied to a large set of data. Ordering data in a given sequence such as ascending, descending, or alphabetical order is a common application in programming.

The technique involved is comparing two bytes at a time and placing them in the proper sequence. For example, in our data set, we will compare the first two bytes (87H and 56H), and if the first byte is larger than the second byte, we will exchange their memory locations to arrange them in ascending order; otherwise, we will keep them in the same locations. We will follow the same procedure for the second and third bytes. The number of comparisons necessary is always $N - 1$ where N is the number of data bytes. Therefore, we need a counter of 2 for one complete comparison. Table 7.1 shows the memory locations and their contents after each pass (execution of one cycle). Two exchanges occur in the first pass, and the bytes are sequenced as 56, 42, and 87. In the second pass, only one exchange occurs. The microprocessor should continue these comparisons until no exchanges occur. We need to devise a technique or set up a reminder for the processor to recognize that no exchanges occur in a given pass. This is called setting a flag. In daily life, we write a note to remind ourselves. Similarly, we can use any register to write a binary note to the microprocessor. For example, we can write 1 (or any number such as FFH) in register D when an exchange takes place; otherwise, register D will be cleared.

PROGRAM

START:	LXI H, XX50H	;Set up HL as a memory pointer for bytes
	MVI D, 00	;Clear register D to set up a flag
	MVI C, 02	;Set register C for comparison count
CHECK:	MOV A, M	;Get data byte
	INX H	;Point to next byte
	CMP M	;Compare bytes
	JC NXTBYT	;If (A) < second byte, do not exchange
	MOV B, M	;Get second byte for exchange
	MOV M, A	;Store first byte in second location
	DCX H	;Point to first location
	MOV M, B	;Store second byte in first location
	INX H	;Get ready for next comparison
	MVI D, 01	;Load 1 in D as a reminder for exchange
NXTBYT:	DCR C	;Decrement comparison count
	JNZ CHECK	;If comparison count $\neq$ 0, go back
	MOV A, D	;Get flag bit in A
	RRC	;Place flag bit D_0 in Carry
	JC START	;If flag is 1, exchange occurred
		;Start the next pass
	HLT	;End of sorting

TABLE 7.1
Execution of Sort Program

Memory Locations	Initial Bytes	First Pass	Second Pass	Third Pass
XX50	87	56	42	42
XX51	56	42	56	56
XX52	42	87	87	87
Reg. D	0	1	1	0

PROGRAM DESCRIPTION AND OUTPUT

During the initialization phase, register HL is set up as a pointer where readings are stored; register D is cleared to set up a flag when an exchange occurs; and register C is initialized for a count of two because we have three data bytes to sort. Next is the comparison phase. The programs gets the first byte (87H) in A and compares that with the second byte (56H). In this comparison, the carry flag is reset; therefore, the next set of instructions places 56H in location XX50H and 87H in location XX51H. The flag in register D is set, the comparison counter in register C is decremented by 1, and the program returns to location CHECK for the next comparison. In this second comparison, 87H is compared with 42H, and again the bytes are exchanged. The flag (reminder) in D is set again, but now the comparison counter is zero. Therefore, the program falls through the first loop and checks the status of the flag in D by rotating the flag bit into Carry. Because Carry is set, the program goes back to the beginning and starts the process again from location XX50H. In the second pass, 56H and 42H are exchanged; therefore, the process is repeated a third time. In this pass, every comparison generates a carry. The program jumps to location NXTBYT, and the flag in register D remains zero. Thus, the program places the bytes in ascending order: 42H, 56H, and 87H.

DYNAMIC DEBUGGING **7.6**

After you have completed the steps in the process of static debugging (described in the previous chapter), if the program still does not produce the expected output, you can attempt to debug the program by observing the execution of instructions. This is called **dynamic debugging.**

7.61 Tools for Dynamic Debugging

In a single-board microcomputer, techniques and tools commonly used in dynamic debugging are

☐ Single Step
☐ Register Examine
☐ Breakpoint

Each will be discussed below; the Single-Step and Register Examine keys were discussed briefly in the previous chapter.

SINGLE STEP

The Single-Step key on a keyboard allows you to execute one instruction at a time, and to observe the results following each instruction. Generally, a single-step facility is built with a hard-wired logic circuit. As you push the Single-Step key, you will be able to observe addresses and codes as they are executed. With the single-step technique you will be able to spot

☐ incorrect addresses
☐ incorrect jump locations for loops
☐ incorrect data or missing codes

To use this technique effectively, you will have to reduce loop and delay counts to a minimum number. For example, in a program that transfers 100 bytes, it is meaningless to set the count to 100 and single-step the program 100 times. By reducing the count to two bytes, you will be able to observe the execution of the loop. (If you reduce the count to one byte, you may not be able to observe the execution of the loop.) By single-stepping the program, you will be able to infer the flag status by observing the execution of Jump instructions. The single-step technique is very useful for short programs.

REGISTER EXAMINE

The Register Examine key allows you to examine the contents of the microprocessor register. When appropriate keys are pressed, the monitor program can display the contents of the registers. This technique is used in conjunction either with the single-step or the breakpoint facilities discussed below.

After executing a block of instructions, you can examine the register contents at a critical juncture of the program and compare these contents with the expected outcomes.

BREAKPOINT

In a single-board computer, the breakpoint facility is, generally, a software routine that allows you to execute a program in sections. The breakpoint can be set in your program by using RST instructions. (See "Interrupts," Chapter 12.) When you push the Execute key, your program will be executed until the breakpoint, where the monitor takes over again. The registers can be examined for expected results. If the segment of the program is found satisfactory, a second breakpoint can be set at a subsequent memory address to debug the next segment of the program. With the breakpoint facility you can isolate the segment of the program with errors. Then that segment of the program can be debugged with the single-step facility. The breakpoint technique can be used to check out the timing loop, I/O section, and interrupts. (See Chapter 12 for how to write a breakpoint routine.)

7.62   Common Sources of Errors

Common sources of errors in the instructions and programs illustrated in this chapter are as follows:

1. Failure to clear the accumulator when it is used to add numbers.
2. Failure to clear the carry registers or keep track of a carry.
3. Failure to update a memory pointer or a counter.
4. Failure to set a flag before using a conditional Jump instruction.
5. Inadvertently clearing the flag before using a Jump instruction.
6. Specification of a wrong memory address for a Jump instruction.
7. Use of an improper combination of Rotate instructions.

8. Specifying the Jump instruction on a wrong flag. This is a very common error with the Compare instructions.

See Questions and Assignments 57–59 at the end of the chapter

SUMMARY

In this chapter, programming techniques—such as looping, counting, and indexing—were illustrated using memory-related data transfer instructions, 16-bit arithmetic instructions, and logic instructions. Techniques used commonly in debugging a program—single step, register examine, and breakpoint—were discussed; and common sources of errors were listed.

Review of Instructions

The instructions introduced and illustrated in this chapter are summarized below for an overview.

Instructions	Task	Addressing Mode
Data Transfer (Copy) Instructions		
1. LXI rp,16-bit	Load 16-bit data in a register pair.	Immediate
2. MOV R,M	Copy (M) into (R).	Indirect
3. MOV M,R	Copy (R) into (M).	Indirect
4. LDAX B/D	Copy the contents of the memory, indicated by the register pair, into the accumulator.	Indirect
5. STAX B/D	Copy (A) into the memory, indicated by the register pair.	Indirect
6. LDA 16-bit	Copy (M) into (A), memory specified by the 16-bit address.	Direct
7. STA 16-bit	Copy (A) into memory, specified by the 16-bit address.	Direct
8. MV1 M,8-bit	Load 8-bit data in memory; the memory address is specified by (HL).	Indirect
Arithmetic Instructions		
1. ADD M	Add (M) to (A).	Indirect
2. SUB M	Subtract (M) from (A).	Indirect
3. INR M	Increment the contents of (M) by 1.	Indirect
4. DCR M	Decrement the contents of (M) by 1.	Indirect
Logic Instructions		
1. RLC	Rotate each bit in the accumulator to the left.	

2. RAL	Rotate each bit in the accumulator to the left through the Carry.	
3. RRC	Rotate each bit in the accumulator to the right.	
4. RAR	Rotate each bit in the accumulator to the right through the Carry.	
5. CMP R	Compare (R) with (A).	Register
6. CMP M	Compare (M) with (A).	Indirect
7. CPI 8-bit	Compare 8-bit data with (A).	Immediate

LOOKING AHEAD

The instructions that have been introduced in this and the previous chapter make up the major segment of the instruction set. Applications of these instructions in designing counters and time delays are illustrated in the next chapter. The other group of instructions critical for assembly language programming is related to the subroutine technique, illustrated in Chapter 9.

QUESTIONS AND PROGRAMMING ASSIGNMENTS

If data bytes are given in the programming assignments, enter those data bytes manually in the respective memory locations before executing the programs. If an assignment calls for an output port and your microcomputer does not have an independent output port, store the results in memory. The high-order byte of a memory address is shown as XX. Substitute the high-order byte that is appropriate to the trainer in your laboratory.

Section 7.1

1. You are given a set of 100 resistors with 10 k value and asked to test them for 10 percent tolerance. Reject all the resistors that are outside the tolerance and give the final number of resistors you found within the tolerance. Draw a flowchart.

2. You are given a big grocery list for a party and asked to buy the items starting from number 60. The maximum amount you can spend is $125. Draw a flowchart.

3. In the receiving department of a manufacturing company, auto parts are placed in a sequential order from 0000 to 4050. These parts need to be transferred to the bins on the factory floor, marked with the starting number 8000. You have an intelligent robot that can read the bin numbers and transfer the parts. Draw a flowchart to instruct the robot to transfer 277 parts starting from bin number 1025 in the receiving department to the starting bin number 8060 on the floor.

4. A newly hired librarian was given a set of new books on computers and English literature. The stack of computer books was placed above the English books. The librarian was asked to find the total amount spent on computer books; the prices were marked on the back cover of each book. Draw a flowchart representing the process.

Section 7.2

5. Specify the memory location and its contents after the following instructions are executed.

```
MVI B,F7H
MOV A,B
STA XX75H
HLT
```

6. Show the register contents as each of the following instructions is being executed.

	A	B	C	D	E	H	L

```
MVI C,FFH
LXI H,XX70H
LXI D,XX70H
MOV M,C
LDAX D
HLT
```

7. In Question 6, specify the contents of the accumulator and the memory location XX70H after the execution of the instruction LDAX D.

8. Identify the memory locations that are cleared by the following instructions.

```
MVI B,00H
LXI H,XX75H
MOV M,B
INX H
MOV M,B
HLT
```

9. Specify the contents of registers A, D, and HL after execution of the following instructions.

```
      LXI H,XX90H ;Set up register HL as a memory pointer
      SUB A       ;Clear accumulator
      MVI D,0FH   ;Set up register D as a counter
LOOP: MOV M,A     ;Clear memory
      INX H       ;Point to the next memory location
```

```
DCR D          ;Update counter
JNZ LOOP       ;Repeat until (D) = 0
HLT
```

10. Explain the result after the execution of the program in Question 9.

11. Rewrite the instructions in Question 9 using the register BC as a memory pointer.

12. Explain how many times the following loop will be executed.

```
          LXI B,0007H
LOOP: DCX B
          JNZ LOOP
```

13. Explain how many times the following loop will be executed.

```
          LXI B,0007H
LOOP: DCX B
          MOV A,B
          ORA C
          JNZ LOOP
```

14. The following instructions are intended to clear ten memory locations starting from the memory address 0009H. Explain why a large memory block will be erased or cleared and the program will stay in an infinite loop.

```
          LXI H,0009H
LOOP: MVI M,00H
          DCX H
          JNZ LOOP
          HLT
```

15. The following block of data is stored in the memory locations from XX55H to XX5AH. Transfer the data to the locations XX80H to XX85H in the reverse order (e.g., the data byte 22H should be stored at XX85H and 37H at XX80H).
    Data(H)   22, A5, B2, 99, 7F, 37

16. Data bytes are stored in memory locations from XX50H to XX5FH. To insert an additional five bytes of data, it is necessary to shift the data string by five memory locations. Write a program to store the data string from XX55H to XX64H. Use any sixteen bytes of data to verify your program.

    Hint: This is a block transfer of data bytes with overlapping memory locations. If the data transfer begins at location XX50H, a segment of the data string will be destroyed.

17. A system is designed to monitor the temperature of a furnace. Temperature readings are recorded in 16 bits and stored in memory locations starting at XX60H. The high-order byte is stored first and the low-order byte is stored in the next consecutive memory location. However, the high-order byte of all the temperature readings is constant.

Write a program to transfer low-order readings to consecutive memory locations starting at XX80H and discard the high-order bytes.

Temperature Readings (H) 0581, 0595, 0578, 057A, 0598

18. A string of six data bytes is stored starting from memory location 2050H. The string includes some blanks (bytes with zero value). Write a program to eliminate the blanks from the string. (*Hint:* To check a blank, set the Zero flag by using the ORA. Use two memory pointers: one to get a byte and the other to store the byte.)

    Data(H)   F2, 00, 00, 4A, 98, 00

19. Write a program to add the following five data bytes stored in memory locations starting from XX60H, and display the sum. (The sum does not generate a carry. Use register pair DE as a memory pointer to transfer a byte from memory into a register.)

    Data(H)   1A, 32, 4F, 12, 27

20. Write a program to add the following data bytes stored in memory locations starting at XX60H and display the sum at the output port if the sum does not generate a carry. If a result generates a carry, stop the addition, and display 01H at the output port.

    Data(H)   First Set: 37, A2, 14, 78, 97
                   Second Set: 12, 1B, 39, 42, 07

21. In Assignment 20, modify the program to count the number of data bytes that have been added and display the count at the second output port.

Section 7.3

22. Specify the contents of memory locations XX70H to XX74H after execution of the following instructions.

```
        LXI H,XX70H    ;Set up HL as a memory pointer
        MVI B,05H      ;Set up register B as a counter
        MVI A,01
STORE:  MOV M,A        ;Store (A) in memory
        INR A
        INX H
        DCR B
        JNZ STORE
        HLT
```

23. Identify the contents of the registers, the memory location (XX55H), and the flags as the following instructions are executed.

	A	H	L	S	Z	CY	M XX55H
LXI H,XX55H							
MVI M,8AH							
MVI A,76H							
ADD M							
STA XX55H							
HLT							

24. Assemble and execute the instructions in Question 23, and verify the contents.
25. Identify the contents of the memory location XX65H and the status of the flags S, Z, and CY when the instruction INR M is executed.

> LXI H,XX65H
> MVI M,FFH
> INR M
> HLT

26. Repeat the Illustrative Program: Addition with Carry (Section 7.32) using the DE register as a memory pointer and memory location XX40H as a counter. (*Hints:* Use the instruction MVI M to load the memory location XX40H with a count, and DCR M to decrement the counter.)
27. The temperatures of two furnaces are being monitored by a microcomputer. A set of five readings of the first furnace, recorded by five thermal sensors, is stored at the memory location starting at XX50H. A corresponding set of five readings from the second furnace is stored at the memory location starting at XX60H. Each reading from the first set is expected to be higher than the corresponding reading from the second set. For example, the temperature reading at the location 54H (T_{54}) is expected to be higher than the temperature reading at the location 64H (T_{64}).

   Write a program to check whether each reading from the first set is higher than the corresponding reading from the second set. If all readings from the first set are higher than the corresponding readings from the second set, turn on the bit D_0 of the output PORT1. If any one of the readings of the first set is lower than the corresponding reading of the second set, stop the process and output FF as an emergency signal to the output PORT1.

   Data(H) First Set: 82, 89, 78, 8A, 8F
   Second Set: 71, 78, 79, 82, 7F
28. Repeat Assignment 27 with the following modification. Check whether any two readings are equal, and if they are equal, turn on bit D_7 of PORT1 and continue checking. (*Hint:* Check for the Zero flag when two readings are equal.)

   Data(H) First Set: 80, 85, 8F, 82, 87
   Second Set: 71, 74, 7A, 82, 77
29. A set of eight data bytes are stored in memory locations starting from XX70H. Write a program to add two bytes at a time and store the sum in the same memory locations, low-order sum replacing the first byte and a carry replacing the second byte. If any pair does not generate a carry, the memory location of the second byte should be cleared.

   Data(H) F9, 38, A7, 56, 98, 52, 8F, F2
30. A set of eight data bytes are stored in memory locations starting from XX70H. Write a program to subtract two bytes at a time and store the result in a sequential order in memory locations starting from XX70H.

   Data(H) F9, 38, A7, 56, 98, A2, F4, 67
31. In Assignment 30, if any one of the results of the subtraction is in the 2's complement, it should be discarded.

Section 7.4

32. Specify the contents of the accumulator and the status of the CY flag when the following instructions are executed.

a. MVI A,B7H	**b.** MVI A,B7H
ORA A	ORA A
RLC	RAL

33. Identify the register contents and the flags as the following instructions are being executed.

	A	S	Z	CY

MVI A,80H
ORA A
RAR

34. Specify the contents of the accumulator and the CY flag when the following instructions are executed.

a. MVI A,C5H	**b.** MVI A,A7H
ORA A	ORA A
RAL	RAR
RRC	RAL

35. The accumulator in the following set of instructions contains a BCD number. Explain the function of these instructions.

MVI A,79H
ANI F0H
RRC
RRC
RRC
RRC

36. Calculate the decimal value of the number in the accumulator before and after the Rotate instructions are executed, and explain the mathematical functions performed by the instructions.

a. MVI A,18H	**b.** MVI A,78H
RLC	RRC
	RRC

37. Explain the mathematical function that is performed by the following instructions.

MVI A,07H
RLC
MOV B,A
RLC
RLC
ADD B

38. Repeat the Illustrative Program in Section 7.42 ("Checking Sign with Rotate Instructions") with the following modifications:

a. Check the sign of a number by using the instruction JM (Jump On Minus), instead of the instruction RAL. (*Hint:* Set the flags by using the instruction ORA A.)

b. If the sum of the positive readings exceeds eight bits, continue the addition, save generated carry, and display the total sum at two different ports.

Data(H) 48, 72, 8F, 7F, 6B, F2, 98, 7C, 67, 19

39. In Assignment 38, in addition to modifications a and b, count the number of positive readings in the set and display the count at PORT3.

40. A set of eight data bytes is stored in the memory location starting at XX50H. Check each data byte for bits D_7 and D_0. If D_7 or D_0 is 1, reject the data byte; otherwise, store the data bytes at memory locations starting at XX60H.

Data(H) 80, 52, E8, 78, F2, 67, 35, 62

Section 7.5

41. Identify the contents of the accumulator and the flag status as the following instructions are executed.

	A	S	Z	CY

```
MVI A,7FH
ORA A
CPI A2H
```

42. Identify the register contents and the flag status as the following instructions are executed.

	A	S	Z	CY

```
LXI H,2070H
MVI M,64H
MVI A,8FH
CMP M
```

43. Identify the bytes from the following set that will be displayed at PORT1, assuming one byte is loaded into the accumulator at a time.

Data(H) 58, 32, 7A, 87, F2, D7

```
MVI A,BYTE
MVI B,64H
MVI C,C8H
CMP B
JC REJECT
CMP C
```

```
                    JNC REJECT
                    OUT PORT1
                    HLT
           REJECT:  SUB A
                    OUT PORT1
                    HLT
```

44. In Question 43, identify the range of numbers in decimal that will be displayed at PORT1.
45. The following program reads one data byte at a time. Identify the data bytes from the following set that will transfer the program to location ACCEPT.
    Data(H) 19, 20, 64, 8F, D8, F2

```
                    IN PORT1
                    MVI B,20H
                    CMP B
                    JC REJECT
                    JM REJECT
                    STA 2070H
                    JMP ACCEPT
           REJECT:  JMP INVALID
```

46. In Question 45, identify the range of numbers in decimal that will transfer the program to location INVALID.
47. Repeat the Illustrative Program: Use of Compare Instruction to Indicate the End of a Data String (Section 7.52), but include the following modifications: Clear register D and use CMP D to check a byte in the memory location. If a byte is not zero, add the byte and continue adding; otherwise, go to the output.
48. A set of eight readings is stored in memory starting at location XX50H. Write a program to check whether a byte 40H exists in the set. If it does, stop checking and display its memory location; otherwise output FFH.
    Data(H) 48, 32, F2, 38, 37, 40, 82, 8A
49. Refer to Assignment 48. Write a program to find the highest reading in the set, and display the reading at an output port.
50. Refer to Assignment 48. Write a program to find the lowest reading in the set, and display the reading at the output port.
51. A set of ten bytes are stored in memory starting with the address XX50H. Write a program to check each byte, and save the bytes that are higher than 60_{10} and lower than 100_{10} in memory locations starting from XX60H.
    Data(H) 6F, 28, 5A, 49, C7, 3F, 37, 4B, 78, 64
52. In Assignment 51, in addition to saving the bytes in the given range, display at PORT1 the number of bytes saved.

53. A string of readings is stored in memory locations starting at XX70H, and the end of the string is indicated by the byte 0DH. Write a program to check each byte in the string, and save the bytes in the range of 30H to 39H (both inclusive) in memory locations starting from XX90H.

   Data(H)   35, 2F, 30, 39, 3A, 37, 7F, 31, 0D, 32

54. In Assignment 53, display the number of bytes accepted from the string between 30H and 39H.

55. A bar code scanner scans the boxes being shipped from the loading dock and records all the codes in computer memory; the end of the data is indicated by the byte 00. The code 1010 0011 (A3H) is assigned to 19″ television sets. Write a program to count the number of 19″ television sets that were shipped from the following data set.

   Data(H)   FA, 67, A3, B8, A3, A3, FA, 00

56. Sort the following set of marks scored by ten students in a circuit course in descending order.

   Data(H)   63, 41, 56, 62, 48, 5A, 4F, 4C, 56, 56

Section 7.6

57. The following program adds the number of bytes stored in memory locations starting from XX00H and saves the result in memory. Read the program and answer the questions given below.

```
            LX1 H,XX00H      ;Set up HL as a data pointer
            LX1, D,0000H     ;Set up D as a byte counter
                             ;   and E as a Carry register
     NEXT: ADD M            ;Add byte
            JNC SKIP         ;If the result has no carry, do not
                             ;   increment Carry register
            INR E
     SKIP: DCR D             ;Update byte counter
            JNZ NEXT         ;Go to next byte
            LXI H,XX90H
            MOV M,A          ;Save the result
            INX H
            MOV M,E
            HLT
```

   a. Assuming the byte counter is set up appropriately, specify the number of bytes that are added by the program.

   b. Specify the memory locations where the result is stored.

   c. Identify the two errors in the program.

58. The following program checks a set of six signed numbers and adds the positive numbers. The numbers are stored in memory locations starting from XX60H. The final result is expected to be less than FFH and stored in location XX70H.

Data(H)   First Set: 20, 87, F2, 37, 79, 17
              Second Set: A2, 15, 3F, B7, 47, 9A

```
            MVI B,00H        ;Clear (B) to save result
            MVI C,06H        ;Set up register C to count six numbers
            LXI H,XX60H      ;Set up HL as a memory pointer
     NEXT:  MOV A,M          ;Get a byte
            RAL              ;Place D7 into CY flag
            JC REJECT        ;If D7 = 1, reject the byte
            RAR              ;Restore the byte
            ADD B            ;Add the previous sum
            MOV B,A          ;Save the sum
     REJECT: INX H           ;Next location
            DCR B            ;Update the byte counter
            JNZ NEXT         ;Go back to get the next byte
            STA XX70H        ;Save the result
            HLT
```

   a. Calculate the answers for the given data sets.
   b. Assemble and execute the program for the first data set, and verify the answer.
   c. Execute the program for the second data set, and verify the answer.
   d. Debug the program using the Single-Step and Examine Register techniques if the result is different from the expected answer.

59. The following program adds five bytes stored in memory location starting from 2055H. The result is stored in locations 2060H and 2061H.

Data(H)   First Set: 17, 1B, 21, 7F, 9D
              Second Set: F2, 87, A9, B8, C2

```
2000    21 55 20      LXI H,2055H     ;Set HL as a memory pointer
2003    AF            XRA A           ;Clear accumulator
2004    01 05 00      LXI B,0005H     ;Clear B to save carries and set up C as
                                      ;   a byte counter
2007    8E            ADC M           ;Add byte
2008    D2 0C 20      JNC 200CH       ;Skip CY register if no carry
200B    04            INR B           ;Update Cy register
200C    0D            DCR C           ;Update counter
200D    23            INX H           ;Next memory location
200E    C2 20 08      JNZ 2008H       ;Go back to add next byte
2011    32 60 20      STA 2060H       ;Store low-order sum
2014    78            MOV A,B         ;Get high-order sum
2015    32 61 20      STA 2061        ;Store high-order sum
2018                  HLT
```

a. Calculate the answers for the given data sets.
b. Enter the program and the first data set. Execute the program and verify the answer. Debug the program if necessary.
c. Execute the program for the second data set, and verify the answer.
d. Debug the program using the Single-Step and Examine Register techniques if the result is different from the expected answer.

Counters and Time Delays

This chapter deals with the designing of counters and timing delays through software (programming). Two of the programming techniques discussed in the last chapter—looping and counting—are used to design counters and time delays. The necessary instructions have already been introduced in the previous two chapters.

A counter is designed by loading an appropriate count in a register. A loop is set up to decrement the count for a down-counter* or to increment the count for an up-counter.[†] Similarly, a timing delay is designed by loading a register with a delay count and setting up a loop to decrement the count until zero. The delay is determined by the clock period of the system and the time required to execute the instructions in the loop.

Counters and time delays are important techniques. They are commonly used in applications such as traffic signals, digital clocks, process control, and serial data transfer.

OBJECTIVES

☐ Write instructions to set up time delays, using one register, a register pair, and a loop-within-a-loop technique.

☐ Calculate the time delay in a given loop.
☐ Draw a flowchart for a counter with a delay.
☐ Design an up/down counter for a given delay.
☐ Write a program to turn on/off specific bits at a given interval.

---

*A down-counter counts in descending order.

[†]An up-counter counts in ascending order.

8.1   COUNTERS AND TIME DELAYS

Designing a counter is a frequent programming application. Counters are used primarily to keep track of events; time delays are important in setting up reasonably accurate timing between two events. The process of designing counters and time delays using software instructions is far more flexible and less time consuming than the design process using hardware.

COUNTER

A counter is designed simply by loading an appropriate number into one of the registers and using the INR (Increment by One) or the DCR (Decrement by One) instructions. A loop is established to update the count, and each count is checked to determine whether it has reached the final number; if not, the loop is repeated.

The flowchart shown in Figure 8.1 illustrates these steps. However, this counter has one major drawback; the counting is performed at such high speed that only the last count can be observed. To observe counting, there must be an appropriate time delay between counts.

TIME DELAY

The procedure used to design a specific delay is similar to that used to set up a counter. A register is loaded with a number, depending on the time delay required, and then the register is decremented until it reaches zero by setting up a loop with a conditional Jump instruction. The loop causes the delay, depending upon the clock period of the system, as illustrated in the next sections.

8.11   Time Delay Using One Register

The flowchart in Figure 8.2 shows a time-delay loop. A count is loaded in a register, and the loop is executed until the count reaches zero. The set of instructions necessary to set up the loop is also shown in Figure 8.2.

FIGURE 8.1
Flowchart of a Counter

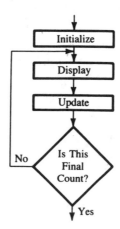

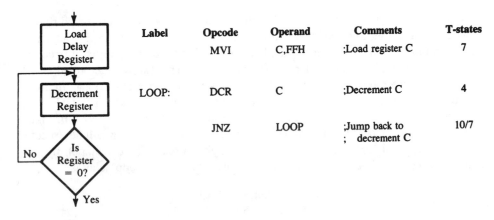

Label	Opcode	Operand	Comments	T-states
	MVI	C,FFH	;Load register C	7
LOOP:	DCR	C	;Decrement C	4
	JNZ	LOOP	;Jump back to ; decrement C	10/7

FIGURE 8.2
Time Delay Loop: Flowchart and Instructions

The last column in Figure 8.2 shows the T-states (clock periods) required by the 8085 microprocessor to execute each instruction. (See Appendix F for the list of the 8085 instructions and their T-states.) The instruction MVI requires seven clock periods. An 8085-based microcomputer with 2 MHz clock frequency will execute the instruction MVI in 3.5 μs as follows:

$$\text{Clock frequency of the system } f = 2 \text{ MHz}$$
$$\text{Clock period } T = 1/f = 1/2 \times 10^{-6} = 0.5 \text{ μs}$$
$$\text{Time to execute MVI} = 7 \text{ T-states} \times 0.5$$
$$= 3.5 \text{ μs}$$

However, if the clock frequency of the system is 1 MHz, the microprocessor will require 7 μs to execute the same instruction. To calculate the time delay in a loop, we must account for the T-states required for each instruction and for the number of times the instructions are executed in the loop.

In Figure 8.2, register C is loaded with the count FFH (255_{10}) by the instruction MVI, which is executed once and takes seven T-states. The next two instructions, DCR and JNZ, form a loop with a total of 14 (4 + 10) T-states. The loop is repeated 255 times until register C = 0.

The time delay in the loop T_L with 2 MHz clock frequency is calculated as

$$\text{TL} = (\text{T} \times \text{Loop T-states} \times \text{N10})$$

where T_L = Time delay in the loop
   T = System clock period
   N_{10} = Equivalent decimal number of the hexadecimal count loaded in the delay
        register
   $T_L = (0.5 \times 10^{-6} \times 14 \times 255)$
      $= 1785$ μs
      ≈ 1.8 ms

In most applications, this approximate calculation of the time delay is considered reasonably accurate. However, to calculate the time delay more accurately, we need to adjust for the execution of the JNZ instruction and add the execution time of the initial instruction.

The T-states for JNZ instruction are shown as 10/7. This can be interpreted as follows: The 8085 microprocessor requires ten T-states to execute a conditional Jump instruction when it jumps or changes the sequence of the program and seven T-states when the program falls through the loop (goes to the instruction following the JNZ). In Figure 8.2, the loop is executed 255 times; in the last cycle, the JNZ instruction will be executed in seven T-states. This difference can be accounted for in the delay calculation by subtracting the execution time of three states. Therefore, the adjusted loop delay is

$$T_{LA} = T_L - (3 \text{ T-states} \times \text{Clock period})$$
$$= 1785.0 \text{ μs} - 1.5 \text{ μs} = 1783.5 \text{ μs}$$

Now the total delay must take into account the execution time of the instructions outside the loop. In the above example, we have only one instruction (MVI C) outside the loop. Therefore, the total delay is

$$\text{Total Delay} = \frac{\text{Time to execute instructions}}{\text{outside loop}} + \frac{\text{Time to execute}}{\text{loop instructions}}$$

$$T_D = T_O + T_{LA}$$
$$= (7 \times 0.5 \text{ μs}) + 1783.5 \text{ μs} = 1787 \text{ μs}$$
$$\approx 1.8 \text{ ms}$$

The difference between the loop delay T_L and these calculations is only 2 μs and can be ignored in most instances.

The time delay can be varied by changing the count FFH; however, to increase the time delay beyond 1.8 ms in a 2 MHz microcomputer system, a register pair or a loop within a loop technique should be used.

8.12   Time Delay Using a Register Pair

The time delay can be considerably increased by setting a loop and using a register pair with a 16-bit number (maximum FFFFH). The 16-bit number is decremented by using the instruction DCX. However, the instruction DCX does not set the Zero flag and, without the test flags, Jump instructions cannot check desired data conditions. Additional techniques, therefore, must be used to set the Zero flag.

The following set of instructions uses a register pair to set up a time delay.

Label	Opcode	Operand	Comments	T-states
	LXI	B,2384H	;Load BC with 16-bit count	10
LOOP:	DCX	B	;Decrement (BC) by one	6
	MOV	A,C	;Place contents of C in A	4
	ORA	B	;OR (B) with (C) to set Zero flag	4
	JNZ	LOOP	;If result ≠ 0, jump back to LOOP	10/7

In this set of instructions, the instruction LXI B,2384H loads register B with the number 23H, and register C with the number 84H. The instruction DCX decrements the entire number by one (e.g., 2384H becomes 2383H). The next two instructions are used only to set the Zero flag; otherwise, they have no function in this problem. The OR instruction sets the Zero flag only when the contents of B and C are simultaneously zero. Therefore, the loop is repeated 2384H times, equal to the count set in the register pair.

TIME DELAY

The time delay in the loop is calculated as in the previous example. The loop includes four instructions: DCX, MOV, ORA, and JNZ, and takes 24 clock periods for execution. The loop is repeated 2384H times, which is converted to decimals as

$$2384H = 2 \times (16)^3 + 3 \times (16)^2 + 8 \times (16)^1 + 4(16^0)$$
$$= 9092_{10}$$

If the clock period of the system = 0.5 μs, the delay in the loop T_L is

$$T_L = (0.5 \times 24 \times 9092_{10})$$
$$\approx 109 \text{ ms (without adjusting for the last cycle)}$$
$$\text{Total Delay } T_D = 109 \text{ ms} + T_O$$
$$\approx 109 \text{ ms (The instruction LXI adds only 5 μs.)}$$

A similar time delay can be achieved by using the technique of two loops, as discussed in the next section.

8.13   Time Delay Using a Loop within a Loop Technique

A time delay similar to that of a register pair can also be achieved by using two loops; one loop inside the other loop, as shown in Figure 8.3(a). For example, register C is used in the inner loop (LOOP1) and register B is used for the outer loop (LOOP2). The following instructions can be used to implement the flowchart shown in Figure 8.3(a).

	MVI B,38H	7T
LOOP2:	MVI C,FFH	7T
LOOP1:	DCR C	4T
	JNZ LOOP1	10/7T
	DCR B	4T
	JNZ LOOP2	10/7T

DELAY CALCULATIONS

The delay in LOOP1 is $T_{L1} = 1783.5$ μs. These calculations are shown in Section 8.11. We can replace LOOP1 by T_{L1}, as shown in Figure 8.3(b). Now we can calculate the delay in LOOP2 as if it is one loop; this loop is executed 56 times because of the count (38H) in register B:

$$T_{L2} = 56(T_{L1} + 21 \text{ T-states} \times 0.5 \text{ μs})$$
$$= 56(1783.5 \text{ μs} + 10.5 \text{ μs})$$
$$= 100.46 \text{ ms}$$

FIGURE 8.3
Flowchart for Time Delay with
Two Loops

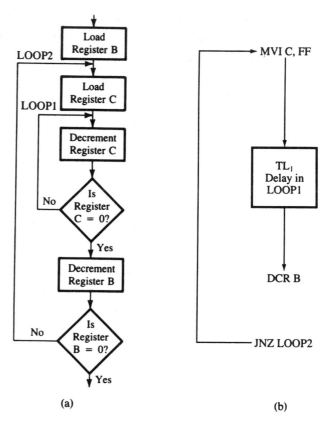

(a)   (b)

The total delay should include the execution time of the first instruction (MVI B,7T); however, the delay outside these loops is insignificant. The time delay can be increased considerably by using register pairs in the above example.

Similarly, the time delay within a loop can be increased by using instructions that will not affect the program except to increase the delay. For example, the instruction NOP (No Operation) can add four T-states in the delay loop. The desired time delay can be obtained by using any or all available registers.

8.14 Additional Techniques for Time Delay

The disadvantages in using software delay techniques for real-time applications in which the demand for time accuracy is high, such as digital clocks, are as follows:

1. The accuracy of the time delay depends on the accuracy of the system's clock.
2. The microprocessor is occupied simply in a waiting loop; otherwise it could be employed to perform other functions.
3. The task of calculating accurate time delays is tedious.

In real-time applications, timers (integrated timer circuits) are commonly used. The Intel 8254 (described in Chapter 15) is a programmable timer chip that can be interfaced with the microprocessor and programmed to provide timings with considerable accuracy. The disadvantages of using the hardware chip include the additional expense and the need for an extra chip in the system.

8.15 Counter Design with Time Delay

To design a counter with a time delay, the techniques illustrated in Figures 8.1 and 8.2 can be combined. The combined flowchart is shown in Figure 8.4.

The blocks shown in the flowchart are similar to those in the generalized flowchart in Figure 7.3. The block numbers shown in Figure 8.4 correspond to the block numbers in the generalized flowchart. Compare Figure 8.4 with Figure 7.3 and note the following points about the counter flowchart (Figure 8.4):

1. The output (or display) block is part of the counting loop.
2. The data processing block is replaced by the time-delay block.
3. The save-the-partial-answer block is eliminated because the count is saved in a counter register, and a register can be incremented or decremented without transferring the count to the accumulator.

The flowchart in Figure 8.4 shows the basic building blocks. However, the sequence can be changed, depending upon the nature of the problem, as shown in Figures 8.5(a) and 8.5(b).

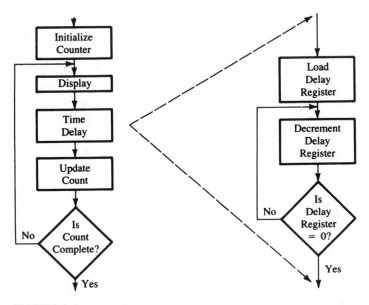

FIGURE 8.4
Flowchart of a Counter with a Time Delay

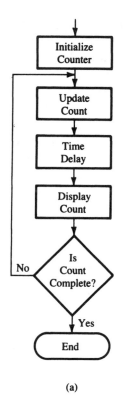

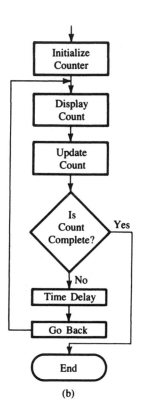

(a)   (b)

FIGURE 8.5
Variations of Counter Flowchart

The flowchart in Figure 8.4 displays the count after initialization. For example, an up-counter starting at zero can be initialized at 00H using the logic shown in Figure 8.4. However, the flowchart in Figure 8.5(a) updates the counter immediately after the initialization. In this case, an up-counter should be initialized at FFH to display the count 00H.

Similarly, the decision-making block differs slightly in these flowcharts. For example, if a counter was counting up to 9, the question in Figure 8.4 would be: Is the count 10? In Figure 8.5(a) the question would be: Is the count 9? The flowchart in Figure 8.5(b) illustrates another way of designing a counter.

8.2   ILLUSTRATIVE PROGRAM: HEXADECIMAL COUNTER

PROBLEM STATEMENT

Write a program to count continuously in hexadecimal from FFH to 00H in a system with a 0.5 μs clock period. Use register C to set up a one millisecond (ms) delay between each count and display the numbers at one of the output ports.

PROBLEM ANALYSIS

The problem has two parts; the first is to set up a continuous down-counter, and the second is to design a given delay between two counts.

1. The hexadecimal counter is set up by loading a register with an appropriate starting number and decrementing it until it becomes zero (shown by the outer loop in the flowchart, Figure 8.6). After zero count, the register goes back to FF because decrementing zero results in a (-1), which is FF in 2's complement.
2. The one millisecond (ms) delay between each count is set up by using the procedure explained previously in Section 8.11—Time Delay Using One Register. Figure 8.2 is identical with the inner loop of the flowchart shown in Figure 8.6. The delay calculations are shown later.

FLOWCHART AND PROGRAM

The flowchart in Figure 8.6 shows the two loops discussed earlier—one for the counter and another for the delay. The counter is initialized in the first block, and the count is displayed

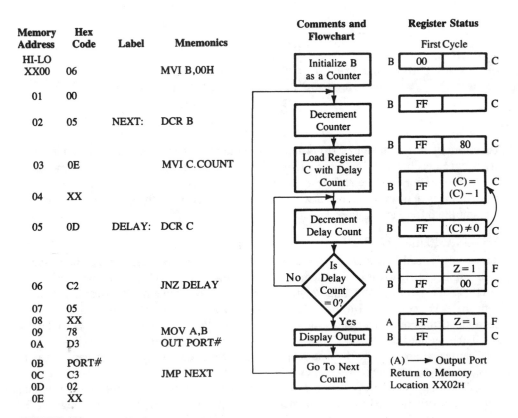

FIGURE 8.6

Program and Flowchart for a Hexadecimal Counter

in the outer loop. The delay is accomplished in the inner loop. This flowchart in Figure 8.6 is similar to that in Figure 8.5(a). You should study the flowchart carefully to differentiate between the counter loop and the delay loop because they may at first appear to be similar.

Delay Calculations   The delay loop includes two instructions: DCR C and JNZ with 14 T-states. Therefore, the time delay T_L in the loop (without accounting for the fact that JNZ requires seven T-states in the last cycle) is

$$T_L = 14 \text{ T-states} \times T \text{ (Clock period)} \times \text{Count}$$
$$= 14 \times (0.5 \times 10^{-6}) \times \text{Count}$$
$$= (7.0 \times 10^{-6}) \times \text{Count}$$

The delay outside the loop includes the following instructions:

DCR B	4T	Delay outside
MVI C,COUNT	7T	the loop: T_O = 35 T-states × T
MOV A,B	4T	= $35 \times (0.5 \times 10^{-6})$
OUT PORT	10T	= 17.5 μs
JMP	10T	
	35 T-states	

$$\text{Total Time Delay } T_D = T_O + T_L$$
$$1 \text{ ms} = 17.5 \times 10^{-6} + (7.0 \times 10^{-6}) \times \text{Count}$$
$$\text{Count} = \frac{1 \times 10^{-3} - 17.5 \times 10^{-6}}{7.0 \times 10^{-6}} \approx 140_{10}$$

Therefore, the delay count 8CH (140_{10}) must be loaded in register C to obtain 1 ms delay between each count. In this problem, the calculation of the delay count does not take into consideration that the JNZ instruction takes seven T-states in the last cycle, instead of 10 T-states. However, the count will remain the same even if the calculations take into account the difference of three T-states.

PROGRAM DESCRIPTION

Register B is used as a counter, and register C is used for delay. Register B initially starts with number 00H; when it is decremented by the instruction DCR, the number becomes FFH. (Verify this by subtracting one from zero in 2's complement.) Register C is loaded with the delay count 8CH to provide a 1 ms delay in the loop. The instruction DCR C decrements the count and the instruction JNZ (Jump On No Zero) checks the Zero flag to see if the number in register C has reached zero. If the number is not zero, instruction JNZ causes a jump back to the instruction labeled DELAY in order to decrement (C) and, thus, the loop is repeated 140_{10} times.

The count is displayed by moving (B) to the accumulator and then to the output port. The instruction JMP causes an unconditional jump for the next count in register B, forming a continuous loop to count from FFH to 00H. After the count reaches zero, (B) is decremented, becoming FFH, and the counting cycle is repeated.

PROGRAM OUTPUT

When the program is executed, the actual output seen may vary according to the device used as the output for the display. The eye cannot see the changes in a count with a 1 ms delay. If the output port has eight LEDs, the LEDs representing the low-order bits will appear to be on continuously, and the LEDs with high-order bits will go on and off according to the count. If the output port is a seven-segment display, all segments will appear to be on; a slight flicker in the display can be noticed. However, the count and the delay can be measured on an oscilloscope.

ILLUSTRATIVE PROGRAM: ZERO-TO-NINE (MODULO TEN)* COUNTER 8.3

PROBLEM STATEMENT

Write a program to count from 0 to 9 with a one-second delay between each count. At the count of 9, the counter should reset itself to 0 and repeat the sequence continuously. Use register pair HL to set up the delay, and display each count at one of the output ports. Assume the clock frequency of the microcomputer is 1 MHz.

Instructions   Review the following instructions.

☐ LXI: Load Register Pair Immediate
☐ DCX: Decrement Register Pair
☐ INX: Increment Register Pair

These instructions manipulate 16-bit data by using registers in pairs (BC, DE, and HL). However, the instructions DCX and INX do not affect flags to reflect the outcome of the operation. Therefore, additional instructions must be used to set the flags.

PROBLEM ANALYSIS

The problem is similar to that in the Illustrative Program for a hexadecimal counter (Section 8.2) except in two respects: the counter is an up-counter (counts *up* to the digit 9), and the delay is too long to use just one register.

1. The counter is set up by loading a register with the appropriate number and incrementing it until it reaches the digit 9. When the counter register reaches the final count, it is reset to zero. This is an additional step compared to the Illustrative Program of Section 8.2 in which the counter resets itself. (Refer to the flowchart in Figure 8.7 to see how each count is checked against the final count.)

---

*The counter goes through ten different states (0 to 9) and is called a *modulo ten* counter.

2. The 1-second delay between each count is set up by using a register pair, as explained in Section 8.12. The delay calculations are shown later.

FLOWCHART AND PROGRAM

The flowchart in Figure 8.7 shows three loops: one for the counter (outer) loop on the left, the second for the delay (inner) loop on the left, and the third to reset the counter (the loop on the right). This flowchart is similar to the flowchart in Figure 8.4. The flowchart indicates that the number is displayed immediately after the initialization; this is different from the flowchart of Figure 8.6, in which the number is displayed at the end of the program.

Memory Address	Hex Code	Label	Mnemonics	T	Comments and Flowchart
HI-LO XX00					Start
01	06	START:	MVI B,00H		Initialize Counter
02	00				
03	D3	DSPLAY:	OUT PORT#	10	Display Output
04	PORT#			$\Big\}$ T_0	
05	21		LXI H,16-Bit	10	Load Delay Register
06	LO*				
07	HI				
08	2B	LOOP:	DCX H	6	• Decrement Delay Register
09	7D		MOV A,L	4	• Set Flags to Check Delay Count
0A	B4		ORA H	4	T_L: 24 T-states
0B	C2		JNZ LOOP	10/7	Is Delay Register = 0? NO
0C	08				
0D	XX†				YES
0E	04		INR B	4	Next Count
0F	78		MOV A,B	4	
10	FE		CPI 0AH	7	T_0
11	0A				Is Count = 0AH? NO
12	C2		JNZ DSPLAY	10/7	
13	03				
14	XX†				YES
15	CA		JZ START		;End of the count,
16	00				start again
17	XX				

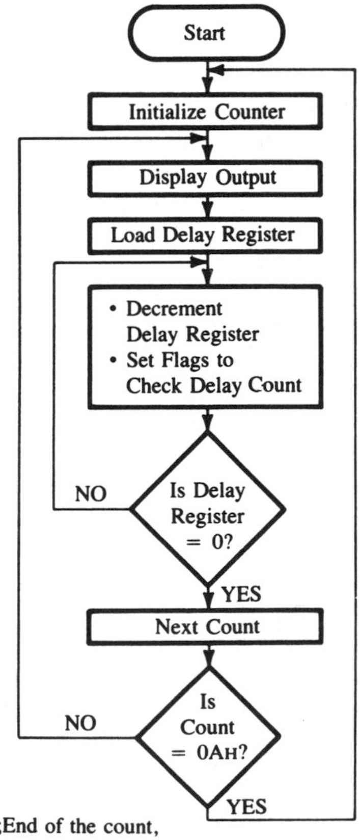

*Enter 16-bit delay count in place of LO and HI, appropriate to the clock period in your system.
†Enter high-order address (page number) of your R/W memory.

FIGURE 8.7
Program and Flowchart for a Zero-to-Nine Counter

The counter is incremented at the end of the program and checked against count 0AH (final count + 1). If the counter has not reached number 0AH, the count is displayed (outer loop on the left); otherwise, it is reset by the loop on the right (Figure 8.7).

PROGRAM DESCRIPTION

Register B is used as a counter, and register pair HL is used for the delay. The significant differences between this program and the Illustrative Program for a hexadecimal counter (Section 8.2) are as follows:

1. Register pair HL contains a 16-bit number that can be manipulated in two ways: first, as a 16-bit number, and second, as two 8-bit numbers. The instruction DCX views the HL register as one register with a 16-bit number. On the other hand, the instructions MOV A,L and ORA H treat the contents of the HL registers as two separate 8-bit numbers.

2. In the delay loop, the sequence is repeated by the instruction JNZ (Jump on No Zero) until the count becomes zero, thus providing a delay of 1 second. However, the instruction DCX does not set the Zero flag. Therefore, the instruction JNZ would be unable to recognize when the count has reached zero, and the program would remain in a continuous loop.

   In this program, the instruction ORA is used to set the Zero flag. The purpose here is to check when the 16-bit number in register pair HL has reached zero. This is accomplished by ORing the contents of register L with the contents of register H. However, the contents of two registers cannot be ORed directly. The instruction MOV A,L loads the accumulator with (L), which is then ORed with (H) by the instruction ORA H. The Zero flag is set only when both registers are zero.

3. The instruction CPI,0AH (Compare Immediate with 0AH) checks the contents of the counter (register B) in every cycle. When register B reaches the number 0AH, the program sequence is redirected to reset the counter without displaying the number 0AH.

4. The time required to reset the counter to zero, indicated by the right-hand loop in the flowchart, is slightly different from the time delay between each count set by the left-hand outer loop.

Delay Calculations The major delay between two counts is provided by the 16-bit number in the delay register HL (the inner loop in the flowchart). This delay is set up by using a register pair, as explained in Section 8.12.

$$\text{Loop Delay } T_L = 24 \text{ T-states} \times T \times \text{Count}$$
$$1 \text{ second} = 24 \times 1.0 \times 10^{-6} \times \text{Count}$$
$$\text{Count} = \frac{1}{24 \times 10^{-6}} = 41666 = \text{A2C2H}$$

The delay count A2C2H in register HL would provide approximately a 1-second delay between two counts. To achieve higher accuracy in the delay, the instructions outside the loop starting from OUT PORT# to JNZ DSPLAY and the difference of three states in the last execution of JNZ must be accounted for in the delay calculations.

The instructions outside the loop are: OUT, LXI, INR, MOV, CPI, and JNZ (DSPLAY). These instructions require 45 T-states; therefore, the delay count is calculated as follows:

$$\text{Total Delay } T_D = T_O + T_L$$
$$1 \text{ second} = (45 \times 1.0 \times 10^{-6}) + (24 \times 1.0 \times 10^{+6} \times \text{Count})$$
$$\text{Count} \approx 41665$$

The difference between the two delay counts calculated above is of very little significance in many applications.

PROGRAM OUTPUT

In an LED output port, each LED will be lit according to the binary count representing 0 to 9. In a seven-segment display, the continuous sequence of the numbers from 0 to 9 can be observed very easily because of the 1-second delay between each count. However, it will be difficult to observe a steady pattern on the oscilloscope (except on a storage scope) because the reset cycle time is slightly different from the delay time. (See Assignment 3 at the end of this chapter to run this program on your system.)

8.4   ILLUSTRATIVE PROGRAM: GENERATING PULSE WAVEFORMS

PROBLEM STATEMENT

Write a program to generate a continuous square wave with the period of 500 μs. Assume the system clock period is 325 ns, and use bit D_0 to output the square wave.

Instructions   The following instructions should be reviewed and the differences between various rotate instructions observed:

☐ RAL: Rotate Accumulator Left Through Carry
☐ RAR: Rotate Accumulator Right Through Carry
☐ RLC: Rotate Accumulator Left
☐ RRC: Rotate Accumulator Right

The first two instructions, RAL and RAR, use the Carry flag as the ninth bit, and the accumulator can be viewed as a 9-bit register. The last two instructions, RLC and RRC, rotate the accumulator contents through eight positions.

In addition to using these instructions, you should review the concept of masking with the ANI instruction.

PROBLEM ANALYSIS

In this problem, the period of the square wave is 500 μs; therefore, the pulse should be on (logic 1) for 250 μs and off (logic 0) for the remaining 250 μs. The alternate pattern of 0/1 bits can be provided by loading the accumulator with the number AAH (1010 1010) and rotating the pattern once through each delay loop. Bit D_0 of the output port is used to provide logic 0 and 1; therefore, all other bits can be masked by ANDing the accumulator with the byte 01H. The delay of 250 μs can be very easily obtained with an 8-bit delay count and one register.

PROGRAM

Memory Address HI-LO	HEX Code	Label	Mnemonics	Comments
XX00	16			
01	AA		MVI D,AA	;Load bit pattern AAH
02	7A	ROTATE:	MOV A,D	;Load bit pattern in A
03	07		RLC	;Change data from AAH to ; 55H and vice versa
04	57		MOV D,A	;Save (A)
05	E6		ANI 01H	;Mask bits D_7–D_1
06	01			
07	D3		OUT PORT1	;Turn on or off the lights
08	PORT1			
09	06		MVI B,COUNT (7T)	;Load delay count for 250 μs
0A	COUNT			
0B	05	DELAY:	DCR B (4T)	;Next count
0C	C2		JNZ DELAY (10/7T)	;Repeat until (B) = 0
0D	0B			
0E	XX			
0F	C3		JMP ROTATE (10T)	;Go back to change logic level
10	02			
11	XX			

PROGRAM DESCRIPTION

Register D is loaded with the bit pattern AAH (1010 1010), and the bit pattern is moved into the accumulator. The bit pattern is rotated left once and saved again in register D. The accumulator contents must be saved because the accumulator is used later in the program.

The next instruction, ANI, ANDs (A) to mask all but bit D_0, as illustrated below.

(A)	→1	0	1	0	1	0	1	0
After RLC	→0	1	0	1	0	1	0	1
AND with 01H	→0	0	0	0	0	0	0	1
Remaining contents	→0	0	0	0	0	0	0	1

This shows that 1 in D_0 provides a high pulse that stays on 250 μs because of the delay. In the next cycle of the loop, bit D_0 is at logic 0 because of the Rotate instruction, and the output pulse stays low for the next 250 μs.

Delay Calculations   In this problem, the pulse width is relatively small (250 μs); therefore, to obtain a reasonably accurate output pulse width, we should account for all the T-states. The total delay should include the delay in the loop and the execution time of the instructions outside the loop.

1. The number of instructions outside the loop is seven; it includes six instructions before the loop beginning at the symbol ROTATE and the last instruction JMP.

$$\text{Delay outside the Loop: } T_O = 46 \text{ T-states} \times 325 \text{ ns} = 14.95 \text{ μs}$$

2. The delay loop includes two instructions (DCR and JNZ) with 14 T-states except for the last cycle, which has 11 T-states.

$$\text{Loop Delay: } T_L = 14 \text{ T-states} \times 325 \text{ ns} \times (\text{Count} - 1) + 11 \text{ T-states} \times 325 \text{ ns}$$
$$= 4.5 \text{ μs} (\text{Count} - 1) + 3.575 \text{ μs}$$

3. The total delay required is 250 μs. Therefore, the count can be calculated as follows:

$$T_D = T_O + T_L$$
$$250 \text{ μs} = 14.95 \text{ μs} + 4.5 \text{ μs} (\text{Count} - 1) + 3.575 \text{ μs}$$
$$\text{Count} = 52.4_{10} = 34\text{H}$$

Program Output   The output of bit D_0 can be observed on an oscilloscope; it should be a square wave with a period of 500 μs.

8.5 DEBUGGING COUNTER AND TIME-DELAY PROGRAMS

The debugging techniques discussed in Chapters 6 and 7 can be used to check errors in a counter program. The following is a list of common errors in programs similar to those illustrated in this chapter.

1. Errors in counting T-states in a delay loop. Typically, the first instruction—to load a delay register—is mistakenly included in the loop.
2. Errors in recognizing how many times a loop is repeated.
3. Failure to convert a delay count from a decimal number into its hexadecimal equivalent.
4. Conversion error in converting a delay count from decimal to hexadecimal number or vice versa.
5. Specifying a wrong Jump location.
6. Failure to set a flag, especially with 16-bit Decrement/Increment instructions.
7. Using a wrong Jump instruction.
8. Failure to display either the first or the last count.
9. Failure to provide a delay between the last and the last-but-one count.

Some of these errors are illustrated in the following program.

8.51   Illustrative Program for Debugging

The following program is designed to count from 100_{10} to 0 in Hex continuously, with a 1-second delay between each count. The delay is set up by using two loops—a loop within a loop. The inner loop is expected to provide approximately 100 ms delay, and it is repeated ten times, using the outer loop to provide a total delay of 1 second. The clock period of the system is 330 ns. The program includes several deliberate errors. Recognize the errors as specified in the following assignment.

		Mnemonics	**T-states**
1.		MVI A,64H	7
2.	DSPLAY:	OUT PORT1	10
3.	LOOP2:	MVI B,10H	7
4.	LOOP1:	LXI D,DELAY	10
5.		DCX D	6
6.		NOP	4
7.		NOP	4
8.		MOV A,D	4
9.		ORA E	4
10.		JNZ LOOP1	10/7
11.		DCR B	4
12.		JZ LOOP2	10/7
13.		DCR A	4
14.		CPI 00H	7
15.		JNZ DSPLAY	10/7

DELAY CALCULATIONS

Delay in LOOP1 = Loop T-states × Count × Clock period (330×10^{-9})

$$100 \text{ ms} = 32 \text{ T} \times \text{Count} \times 330 \times 10^{-9}$$

$$\text{DELAY COUNT} = \frac{100 \times 10^{-3}}{32 \times 330 \times 10^{-9}}$$

$$= 9470$$

This delay calculation ignores the initial T-states in loading the count and the difference of T-states in the last execution of the conditional Jump instruction.

DEBUGGING QUESTIONS

1. Examine LOOP1. Is the label LOOP1 at the appropriate location? What is the effect of the present location on the program?
2. What is the appropriate place for the label LOOP1?
3. Is the delay count accurate?
4. What is the effect of instruction 8 (MOV A,D) on the count?
5. Should instruction 3 be part of LOOP2?
6. Is the byte in register B (instruction 3) accurate?
7. Calculate T-states in the outer loop using the appropriate place for the label LOOP2. (Do not include the T-states of LOOP1.)

8. Is there any need for instruction 14 (CPI)?
9. Is there any need for an additional instruction, such as number 16?
10. What is the effect of instruction 12 (JZ) on the program?
11. Calculate the total delay in LOOP2 inclusive of LOOP1 if the byte in register B = 0AH.
12. Assuming instruction 8 is necessary, make appropriate changes in instructions 1 and 13.
13. Calculate the time delay between the display of two consecutive counts.
14. Will this program display the last count, assuming the other errors are corrected?

SUMMARY

□ Counters and time delays can be designed using software.
□ Time delays are designed simply by loading a count in a register or a register pair and decrementing the count by setting a loop until the count reaches zero. Time delay is determined by the number of T-states in a delay loop, the clock frequency, and the number of times the loop is repeated.
□ Counters are designed using techniques similar to those used for time delays. A counter design generally includes a delay loop.
□ In the 8085 microprocessor, 8-bit registers can be combined as register pairs (B and C, D and E, and H and L) to manipulate 16-bit data. Furthermore, the contents of each register can be examined separately even if registers are being used as register pairs.
□ Sixteen-bit instructions such as DCX and INX do not affect flags; therefore, some other technique must be used to set flags.

QUESTIONS AND PROGRAMMING ASSIGNMENTS*

1. In Figure 8.2, calculate the loop delay T_L if register C contains 00H and the system clock frequency is 3.072 MHz. Adjust the delay calculations to account for seven T-states of the JNZ instruction in the last iteration.
2. In Section 8.12, load register pair BC with 8000H, and calculate the loop delay T_L if the system clock frequency is 3.072 MHz (ignore three T-state difference of the last cycle).
3. In Section 8.12, load register pair BC with 0000H and calculate the total delay T_D if the system clock period is 325 ns (adjust for the last cycle).

---

*The illustrative programs shown in this chapter and these assignments can be verified on a single-board microcomputer with a display output port.

4. Calculate the delay in the following loop, assuming the system clock period is 0.33 μs:

		8085
Label	**Mnemonics**	**T-states**
	LXI B,12FFH	10
DELAY:	DCX B	6
	XTHL	16
	XTHL	16
	NOP	4
	NOP	4
	MOV A,C	4
	ORA B	4
	JNZ DELAY	10/7

5. In Figure 8.3, load register C with 00H and register B with C8H. Calculate the loop delay in LOOP1 and LOOP2 (clock period = 325 ns).

6. Specify the number of times the following loops are executed.

a.		b.		c.	
	MVI A,17H		MVI A,17H		LXI B,1000H
LOOP:	ORA A	LOOP:	RAL	LOOP:	DCX B
	RAL		ORA A		NOP
	JNC LOOP		JNC LOOP		JNZ LOOP

7. Specify the number of times the following loops are executed.

a.		b.		c.	
LOOP:	MVI B,64H		ORA A		MVI A,17H
	NOP		MVI B,64H	LOOP:	ORA A
	DCR B	LOOP:	DCR B		RRC
	JNZ LOOP		JNC LOOP		JNC LOOP

8. Calculate the time delay in the Illustrative Program for a hexadecimal counter (Section 8.2), assuming a count of CFH in register C.

9. Recalculate the delay in the Illustrative Program for a zero-to-nine counter (Section 8.3) using the clock frequency of your system.

10. Calculate the COUNT to obtain a 100 μs loop delay, and express the value in Hex. (Use the clock frequency of your system.)

		T-states
	MVI B,COUNT	
LOOP:	NOP	4
	NOP	4
	DCR B	4
	JNZ LOOP	10/7

11. In Section 8.12, calculate the value of the 16-bit number that should be loaded in register BC to obtain the loop delay of 250 ms if the system clock period is 325 ns. Does the value change if it is calculated with seven T-states for the JNZ instruction in the last cycle?

12. Calculate the 16-bit count to be loaded in register DE to obtain the loop delay of two seconds in LOOP2 (use the clock period of your system and ignore the execution time of the first instruction MVI B).

```
        MVI B,14H      (7)     ;Count for outer loop
LOOP2:  LXI D,16-bit   (10)    ;Count for LOOP1
LOOP1:  DCX D          (6)
        MOV A,D        (4)
        ORA E          (4)
        JNZ LOOP1      (10/7)
        DCR B          (4)
        JNZ LOOP2      (10/7)
```

Use the clock frequency of your system in the following assignments.

13. Write a program to count from 0 to 20H with a delay of 100 ms between each count. After the count 20H, the counter should reset itself and repeat the sequence. Use register pair DE as a delay register. Draw a flowchart and show your calculations to set up the 100 ms delay.

14. Design an up-down counter to count from 0 to 9 and 9 to 0 continuously with a 1.5-second delay between each count, and display the count at one of the output ports. Draw a flowchart and show the delay calculations.

15. Write a program to turn a light on and off every 5 seconds. Use data bit D_7 to operate the light.

16. Write a program to generate a square wave with period of 400 μs. Use bit D_0 to output the square wave.

17. Write a program to generate a rectangular wave with a 200 μs on-period and a 400 μs off-period.

18. A railway crossing signal has two flashing lights run by a microcomputer. One light is connected to data bit D_7 and the second light is connected to data bit D_6. Write a program to turn each signal light alternately on and off at an interval of 1 second.

Stack and Subroutines

The **stack** is a group of memory locations in the R/W memory that is used for temporary storage of binary information during the execution of a program. The starting memory location of the stack is defined in the main program, and space is reserved, usually at the high end of the memory map. The method of information storage resembles a stack of books. The contents of each memory location are, in a sense, "stacked"—one memory location above another—and information is retrieved starting from the top. Hence, this particular group of memory locations is called the *stack*. This chapter introduces the stack instructions in the 8085 set.

The latter part of this chapter deals with the subroutine technique, which is frequently used in programs. A **subroutine** is a group of instructions that performs a subtask (e.g., time delay or arithmetic operation) of repeated occurrence. The subroutine is written as a separate unit, apart from the main program, and the microprocessor transfers the program execution from the main program to the subroutine whenever it is called to perform the task. After the completion of the subroutine task, the microprocessor returns to the main program. The subroutine technique eliminates the need to write a subtask repeatedly; thus, it uses memory more efficiently. Before implementing the subroutine tech-

nique, the stack must be defined; the stack is used to store the memory address of the instruction in the main program that follows the subroutine call.

The stack and the subroutine offer a great deal of flexibility in writing programs. A large software project is usually divided into subtasks called **modules.** These modules are developed independently as subroutines by different programmers. Each programmer can use all the microprocessor registers to write a subroutine without affecting the other parts of the program. At the beginning of the subroutine module, the register contents of the main program are stored on the stack, and these register contents are retrieved before returning to the main program.

This chapter includes two illustrative programs: The first illustrates the use of the stack-related instructions to examine and manipulate the flags; and the second illustrates the subroutine technique in a traffic-signal controller.

OBJECTIVES

☐ Define the stack, the stack pointer (register), and the program counter, and describe their uses.

☐ Explain how information is stored and retrieved from the stack using the instructions PUSH and POP and the stack pointer register.

☐ Demonstrate how the contents of the flag register can be displayed and how a given flag can be set or reset.

☐ Define a subroutine and explain its uses.

☐ Explain the sequence of a program execution when a subroutine is called and executed.

☐ Explain how information is exchanged between the program counter and the stack, and identify the contents of the stack pointer register when a subroutine is called.

☐ List and explain conditional Call and Return instructions.

☐ Illustrate the concepts in the following subroutines: multiple calling, nesting, and common ending.

☐ Compare similarities and differences between PUSH/POP and CALL/RET instructions.

9.1   STACK

The **stack** in an 8085 microcomputer system can be described as a set of memory locations in the R/W memory, specified by a programmer in a main program. These memory locations are used to store binary information (bytes) temporarily during the execution of a program.

The beginning of the stack is defined in the program by using the instruction LXI SP, which loads a 16-bit memory address in the stack pointer register of the microprocessor. Once the stack location is defined, storing of data bytes begins at the memory address that is one less than the address in the stack pointer register. For example, if the stack pointer register is loaded with the memory address 2099H (LXI SP,2099H), the storing of data bytes begins at 2098H and continues in reversed numerical order (decreasing memory addresses such as 2098H, 2097H, etc.). Therefore, as a general practice, the stack is initialized at the highest available memory location to prevent the program from being destroyed by the stack information. The size of the stack is limited only by the available memory.

Data bytes in the register pairs of the microprocessor can be stored on the stack (two at a time) in reverse order (decreasing memory address) by using the instruction PUSH. Data bytes can be transferred from the stack to respective registers by using the instruction POP. The stack pointer register tracks the storage and retrieval of the information. Because two data bytes are being stored at a time, the 16-bit memory address in the stack pointer register is decremented by two; when data bytes are retrieved, the address is incremented by two. An address in the stack pointer register indicates that the next two memory locations (in descending numerical order) can be used for storage.

The stack is shared by the programmer and the microprocessor. The programmer can store and retrieve the contents of a register pair by using PUSH and POP instructions. Similarly, the microprocessor automatically stores the contents of the program counter when a subroutine is called (to be discussed in the next section). The instructions necessary for using the stack are explained below.

INSTRUCTIONS

Opcode	Operand
LXI	SP,16-bit

☐ Load Stack Pointer
☐ Load the stack pointer register with a 16-bit address. The LXI instructions were discussed in Chapter 7

PUSH	Rp

Store Register Pair on Stack
☐ This is a 1-byte instruction

PUSH	B
PUSH	D

☐ It copies the contents of the specified register pair on the stack as described below

PUSH	H
PUSH	PSW

☐ The stack pointer register is decremented, and the contents of the high-order register (e.g., register B) are copied in the location shown by the stack pointer register
☐ The stack pointer register is again decremented, and the contents of the low-order register (e.g., register C) are copied in that location
☐ The operands B, D, and H represent register pairs BC, DE, and HL, respectively
☐ The operand PSW represents Program Status Word, meaning the contents of the accumulator and the flags

POP	Rp

Retrieve Register Pair from Stack
☐ This is a 1-byte instruction

POP	B
POP	D

☐ It copies the contents of the top two memory locations of the stack into the specified register pair

POP	H
POP	PSW

☐ First, the contents of the memory location indicated by the stack pointer register are copied into the low-order register (e.g., register L), and then the stack pointer register is incremented by 1
☐ The contents of the next memory location are copied into the high-order register (e.g., register H), and the stack pointer register is again incremented by 1

All three of these instructions belong to the data transfer (copy) group; thus, the contents of the source are not modified, and no flags are affected.

In the following set of instructions (illustrated in Figure 9.1), the stack pointer is initialized, and the contents of register pair HL are stored on the stack by using the PUSH instruction. Register pair HL is used for the delay counter (actual instructions are not

Example 9.1

284

shown); at the end of the delay counter, the contents of HL are retrieved by using the instruction POP. Assuming the available user memory ranges from 2000H to 20FFH, illustrate the contents of various registers when PUSH and POP instructions are executed.

Solution

In this example, the first instruction—LXI SP,2099H—loads the stack pointer register with the address 2099H (Figure 9.1). This instruction indicates to the microprocessor that memory space is reserved in the R/W memory as the stack and that the locations beginning at 2098H and moving upward can be used for temporary storage. This instruction also suggests that the stack can be initialized anywhere in the memory; however, the stack location should not interfere with a program. The next instruction—LXI H—loads data in the HL register pair, as shown in Figure 9.1.

When instruction PUSH H is executed, the following sequence of data transfer takes place. After the execution, the contents of the stack and the register are as shown in Figure 9.2.

1. The stack pointer is decremented by one to 2098H, and the contents of the H register are copied to memory location 2098H.
2. The stack pointer register is again decremented by one to 2097H, and the contents of the L register are copied to memory location 2097H.
3. The contents of the register pair HL are not destroyed; however, HL is made available for the delay counter.

After the delay counter, the instruction POP H restores the original contents of the register pair HL, as follows. Figure 9.3 illustrates the contents of the stack and the registers following the POP instruction.

1. The contents of the top of the stack location shown by the stack pointer are copied in the L register, and the stack pointer register is incremented by one to 2098H.
2. The contents of the top of the stack (now it is 2098H) are copied in the H register, and the stack pointer is incremented by one.

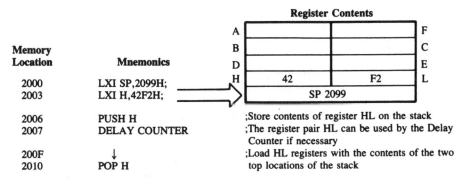

FIGURE 9.1
Instructions and Register Contents in Example 9.1

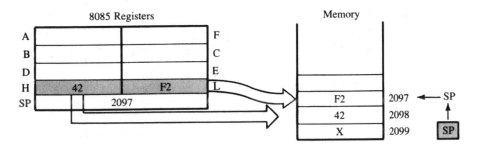

FIGURE 9.2
Contents on the Stack and in the Registers after the PUSH Instruction

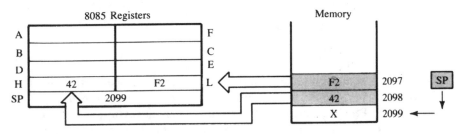

FIGURE 9.3
Contents on the Stack and in the Registers After the POP Instruction

3. The contents of memory locations 2097H and 2098H are not destroyed until some other data bytes are stored in these locations.

The available user memory ranges from 2000H to 23FFH. A program of data transfer and arithmetic operations is stored in memory locations from 2000H to 2050H, and the stack pointer is initialized at location 2400H. Two sets of data are stored, starting at locations 2150H and 2280H (not shown in Figure 9.4). Registers HL and BC are used as memory pointers to the data locations. A segment of the program is shown in Figure 9.4.

**Example
9.2**

1. Explain how the stack pointer can be initialized at one memory location beyond the available user memory.
2. Illustrate the contents of the stack memory and registers when PUSH and POP instructions are executed, and explain how memory pointers are exchanged.
3. Explain the various contents of the user memory.

1. The program initializes the stack pointer register at location 2400H, one location beyond the user memory (Figure 9.4). This procedure is valid because the initialized location is never used for storing information. The instruction PUSH first decrements the stack pointer register, and then stores a data byte.

Solution

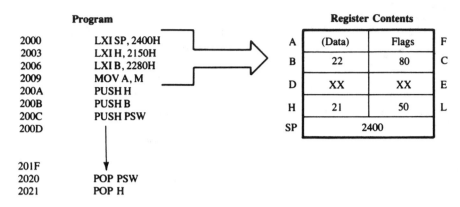

FIGURE 9.4
Instructions and Register Contents in Example 9.2

2. Figure 9.5 shows the contents of the stack pointer register and the contents of the stack locations after the three PUSH instructions are executed. After the execution of the PUSH (H, B, and PSW) instructions, the stack pointer moves upward (decreasing memory locations) as the information is stored. Thus the stack can grow upward in the user memory even to the extent of destroying the program.

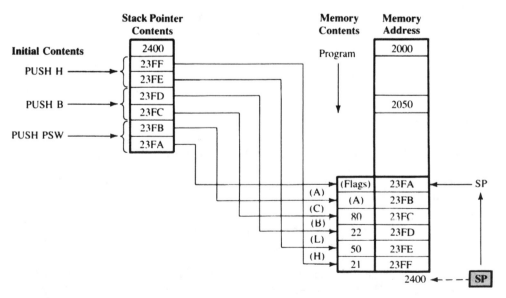

FIGURE 9.5
Stack Contents After the Execution of PUSH Instructions

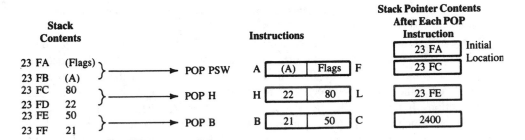

FIGURE 9.6

Register Contents After the Execution of POP Instructions

Figure 9.6 shows how the contents of various register pairs are retrieved. To restore the original contents in the respective registers, follow the sequence Last-In–First-Out (LIFO). In the example, the register contents were pushed on the stack in the order of HL, BC, and PSW. The contents should have been restored in the order of PSW, BC, and HL. However, the order is altered in this example to demonstrate how register contents are exchanged.

The instruction POP PSW copies the contents of the two top locations to the flag register and the accumulator, respectively, and increments the stack pointer by two to 23FCH. The next instruction, POP H, takes the contents of the top two locations (23FC and 23FD), and copies them in registers L and H, respectively, while incrementing the stack pointer by two to 23FEH. The instruction POP B copies the contents of the next two locations in registers C and B, incrementing the stack pointer to 2400H. By reversing the positions of two instructions, POP H and POP B, the contents of the BC pair are exchanged with those of the HL pair. It is important to remember that the instruction POP H does not restore the original contents of the HL

FIGURE 9.7

R/W Memory Contents

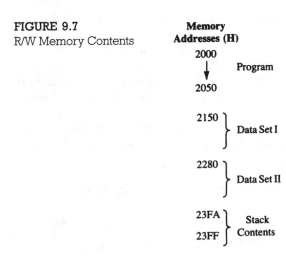

pair; instead it copies the contents of the top two locations shown by the stack pointer in the HL pair.

3. Figure 9.7 shows the sketch of the memory map. The R/W memory includes three types of information. The user program is stored from 2000H to 2050H. The data are stored, starting at locations 2150H and 2280H. The last section of the user memory is initialized as the stack where register contents are stored as necessary, using the PUSH instructions. In this example, the memory locations from 23FFH to 23FAH are used as the stack, which can be extended up to the locations of the second data set.

9.11   Review of Important Concepts

The following points can be summarized from the preceding examples:

1. Memory locations in R/W memory can be employed as temporary storage for information by initializing (loading) a 16-bit address in the stack pointer register; these memory locations are called the stack. The terms *stack* and *stack pointer* appear similar; however, they are not the same. The stack is memory locations in R/W memory; the stack pointer is a 16-bit register in the 8085 microprocessor.

2. Read/Write memory generally is used for three different purposes:
   a. to store programs or instructions;
   b. to store data;
   c. to store information temporarily in defined memory locations called the stack during the execution of the program.

3. The stack space grows upward in the numerically decreasing order of memory addresses.

4. The stack can be initialized anywhere in the user memory map. However, as a general practice, the stack is initialized at the highest user memory location so that it will be less likely to interfere with a program.

5. A programmer can employ the stack to store contents of register pairs by using the instruction PUSH and can restore the contents of register pairs by using the instruction POP. The address in the stack pointer register always points to the top of the stack, and the address is decremented or incremented as information is stored or retrieved.

6. The contents of the stack pointer can be interpreted as the address of the memory location that is already used for storage. The retrieval of bytes begins at the address in the stack pointer; however, the storage begins at the next memory location (in the decreasing order).

7. The storage and retrieval of data bytes on the stacks should follow the LIFO (Last In–First-Out) sequence. Information in stack locations is not destroyed until new information is stored in those locations.

9.12   Illustrative Program: Resetting and Displaying Flags

PROBLEM STATEMENT
Write a program to perform the following functions:

1. Clear all the flags.
2. Load 00H in the accumulator, and demonstrate that the Zero flag is not affected by the data transfer instruction.
3. Logically OR the accumulator with itself to set the Zero flag, and display the flag at PORT1 or store all the flags on the stack.

PROBLEM ANALYSIS

The problem concerns examining the Zero flag after instructions have been executed. There is no direct way of observing the flags; however, they can be stored on the stack by using the instruction PUSH PSW. The contents of the flag register can be retrieved in any one of the registers by using the instruction POP, and the flags can be displayed at an output port. In this example, the result to be displayed is different from the result to be stored on the stack memory. To display only the Zero flag, all other flags should be masked; however, the masking is not necessary to store all the flags.

PROGRAM

Memory Address	Machine Code	Instructions	Comments
XX00	31	LXI SP,XX99H	;Initialize the stack
01	99		
02	XX		
03	2E	MVI L,00H	;Clear L
04	00		
05	E5	PUSH H	;Place (L) on stack
06	F1	POP PSW	;Clear flags
07	3E	MVI A,00H	;Load 00H
08	00		
09	F5	PUSH PSW	;Save flags on stack
0A	E1	POP H	;Retrieve flags in L
0B	7D	MOV A,L	
0C	D3	OUT PORT0	;Display flags
0D	PORT0		
0E	3E	MVI A,00H	;Load 00H again
0F	00		
10	B7	ORA A	;Set flags and reset CY, AC
11	F5	PUSH PSW	;Save flags on stack
12	E1	POP H	;Retrieve flags in L
13	7D	MOV A,L	
14	E6	ANI 40H	;Mask all flags except Z
15	40		
16	D3	OUT PORT1	
17	PORT1		
18	76	HLT	;End of program

Storing in Memory: Alternative to Output Display

XX0A	3E	MVI A,00H	;Load 00H again
0B	00		
0C	B7	ORA A	;Set flags and reset CY and AC
0D	F5	PUSH PSW	;Save flags on stack
0E	76	HLT	;End of program

Program Description The stack pointer register is initialized at XX99H. The instruction MVI L clears (L), and (L) is placed on the stack, which is subsequently placed into the flag register to clear all the flags.

To verify the flags after the execution of the MVI A instruction, the PUSH and POP instructions are used in the same way as these instructions were used to clear the flags, and the flags are displayed at PORT0. Similarly, the Zero flag is displayed at PORT1 after the instructions MVI and ORA.

Program Output Data transfer (copy) instructions do not affect the flags; therefore, no flags should be set after the instruction MVI A, even if (A) is equal to zero. PORT0 should display 00H. However, the instruction ORA will set the Zero and the Parity flags to reflect the data conditions in the accumulator, and it also resets the CY and AC flags. In the flag register, bit D_6 represents the Z flag, and the ANI instruction masks all flags except the Z flag. PORT1 should display 40H as shown below.

$$D_7 \ D_6 \ D_5 \ D_4 \ D_3 \ D_2 \ D_1 \ D_0$$
$$0 \ \ 1 \ \ 0 \ \ 0 \ \ 0 \ \ 1 \ \ 0 \ \ 0 \ = 40H$$
$$S \ \ Z \ \ \ \ \ AC \ \ \ \ P \ \ \ \ CY$$

Storing Output in Memory If output ports are not available, the results can be stored in the stack memory. The machine code (F5) at memory location XX09H saves the flags affected by the instruction MVI A,00H. Then the instructions can be modified starting from memory location XX0AH. The alternative set of instructions is shown above; it sets the flags (using ORA instruction), and saves them on the stack without masking. The result (44H) includes the parity flag. The contents of the stack locations should be as shown in Figure 9.8.

Instructions	Stack Memory	Contents	
	XX99 ◄		Stack Pointer Initialization
XX09 PUSH PSW	XX98	(A)=00H	
	XX97	(F)=00H	
XX0D PUSH PSW	XX96	(A)=00H	
	XX95	(F)=44H	

FIGURE 9.8
Output Stored in Stack Memory

SUBROUTINE 9.2

A **subroutine** is a group of instructions written separately from the main program to perform a function that occurs repeatedly in the main program. For example, if a time delay is required between three successive events, three delays can be written in the main program. To avoid repetition of the same delay instructions, the subroutine technique is used. Delay instructions are written once, separately from the main program, and are called by the main program when needed.

The 8085 microprocessor has two instructions to implement subroutines: CALL (call a subroutine), and RET (return to main program from a subroutine). The CALL instruction is used in the main program to call a subroutine, and the RET instruction is used at the end of the subroutine to return to the main program. When a subroutine is called, the contents of the program counter, which is the address of the instruction following the CALL instruction, is stored on the stack and the program execution is transferred to the subroutine address. When the RET instruction is executed at the end of the subroutine, the memory address stored on the stack is retrieved, and the sequence of execution is resumed in the main program. This sequence of events is illustrated in Example 9.3.

INSTRUCTIONS

Opcode	Operand	
CALL	16-bit memory address of a subroutine	Call Subroutine Unconditionally □ This is a 3-byte instruction that transfers the program sequence to a subroutine address □ Saves the contents of the program counter (the address of the next instruction) on the stack □ Decrements the stack pointer register by two □ Jumps unconditionally to the memory location specified by the second and third bytes. The second byte specifies a line number and the third byte specifies a page number □ This instruction is accompanied by a return instruction in the subroutine
RET		Return from Subroutine Unconditionally □ This is a 1-byte instruction □ Inserts the two bytes from the top of the stack into the program counter and increments the stack pointer register by two □ Unconditionally returns from a subroutine

The conditional Call and Return instructions will be described later in the chapter.

Example 9.3	Illustrate the exchange of information between the stack and the program counter for the following program if the available user memory ranges from 2000H to 23FFH.

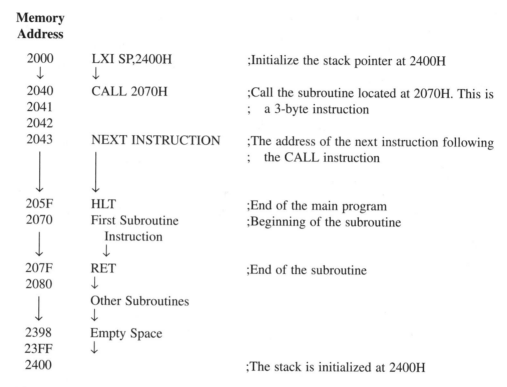

Memory
Address

2000 ↓	LXI SP,2400H ↓	;Initialize the stack pointer at 2400H
2040 2041 2042	CALL 2070H	;Call the subroutine located at 2070H. This is ; a 3-byte instruction
2043	NEXT INSTRUCTION	;The address of the next instruction following ; the CALL instruction
205F	HLT	;End of the main program
2070	First Subroutine Instruction	;Beginning of the subroutine
207F	RET	;End of the subroutine
2080		
	Other Subroutines	
2398	Empty Space	
23FF		
2400		;The stack is initialized at 2400H

Solution   After reviewing the above program note the following points:

1. The available user memory is from 2000H to 23FFH (1024 or 1K bytes); however,
   the stack pointer register is initialized at 2400H, one location beyond the user memory. This allows maximum use of the memory because the actual stack begins at
   23FFH. The stack can expand up to the location 2398H without overlapping with the
   program.
2. The main program is stored at memory locations from 2000H to 205FH.
3. The CALL instruction is located at 2040H to 2042H (3-byte instruction). The next instruction is at 2043H.
4. The subroutine begins at the address 2070H and ends at 207FH.

PROGRAM EXECUTION

The sequence of the program execution and the events in the execution of the CALL and
subroutine are shown in Figures 9.9 and 9.10.

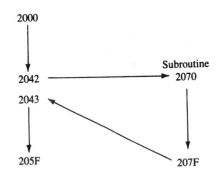

The program execution begins at 2000_H, continues until the end of CALL 2042_H, and transfers to the subroutine at 2070_H. At the end of the subroutine, after executing the RET instruction, it comes back o the main program at 2043H and continues.

FIGURE 9.9
Subroutine Call and Program Transfer

CALL EXECUTION

Memory Address	Machine Code	Mnemonics	Comments
2040	CD	CALL 2070H	;Call subroutine located at the memory
2041	70		; location 2070H
2042	20		
2043	NEXT	INSTRUCTION	

The sequence of events in the execution of the CALL instruction by the 8085 is shown in Figure 9.10. The instruction requires five machine cycles and eighteen T-states. The sequence of events in each machine cycle is as follows.

1. M_1—Opcode Fetch: In this machine cycle, the contents of the program counter (2040H) are placed on the address bus, and the instruction code CD is fetched using the data bus. At the same time, the program counter is upgraded to the next memory address, 2041H. After the instruction is decoded and executed, the stack pointer register is decremented by one to 23FFH.

2. M_2 and M_3—Memory Read: These are two Memory Read cycles during which the 16-bit address (2070H) of the CALL instruction is fetched. The low-order address 70H is fetched first and placed in the internal register Z. The high-order address 20H is fetched next, and placed in register W. During M_3, the program counter is upgraded to 2043H, pointing to the next instruction.

3. M_4 and M_5—Storing of Program Counter: At the beginning of the M_4 cycle, the normal operation of placing the contents of the program counter on the address bus is suspended; instead, the contents of the stack pointer register 23FFH are placed on the address bus. The high-order byte of the program counter (PCH = 20H) is placed on the data bus and stored in the stack location 23FFH. At the same time, the stack pointer register is decremented to 23FEH.

Instruction: CALL 2070H

Machine Cycles	Stack Pointer (SP) 2400	Address Bus (AB)	Program Counter (PCH)(PCL)	Data Bus (DB)	Internal Registers (W)(Z)
M₁ Opcode Fetch	23FF (SP−1)	2040	20 41	CD Opcode	—
M₂ Memory Read		2041	20 42	70 Operand	70
M₃ Memory Read	23FF	2042	20 43	20 Operand	20
M₄ Memory Write	23FE (SP−2)	23FF	20 43	20 (PCH)	
M₅ Memory Write	23FE	23FE	20 43	43 (PCL)	(20)(70)
M₁ Opcode Fetch of Next Instruction		20 70 → 2071 (W)(Z) ◄			(2070) (W)(Z)

Memory Address	Code (H)
2040	CD
2041	70
2042	20

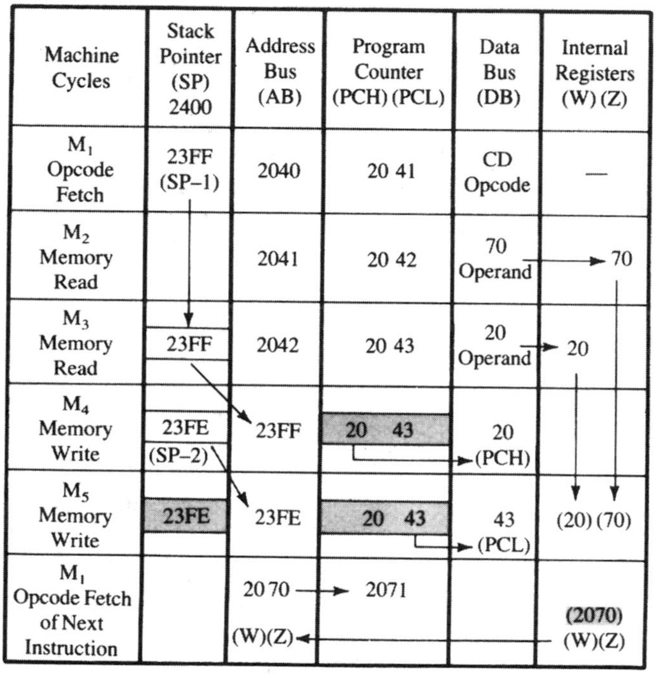

FIGURE 9.10
Data Transfer During the Execution of the CALL Instruction

During machine cycle M_5, the contents of the stack pointer 23FEH are placed on the address bus. The low-order byte of the program counter (PCL = 43H) is placed on the data bus and stored in stack location 23FEH.

4. Next Instruction Cycle: In the next instruction cycle, the program execution sequence is transferred to the CALL location 2070H by placing the contents of W and Z registers (2070H) on the address bus. During M_1 of the next instruction cycle, the program counter is upgraded to location 2071 (W,Z + 1).

In summary: After the CALL instruction is fetched, the 16-bit address (the operand) is read during M_2 and M_3 and stored temporarily in W/Z registers. (Examine the contents of the address bus and the data bus in Figure 9.10.) In the next two cycles, the contents of the program counter are stored on the stack. This is the address where the microprocessor will continue the execution of the program after the completion of the subroutine. Figure 9.11 shows the contents of the program counter, the stack pointer register, and the stack during the execution of the CALL instruction.

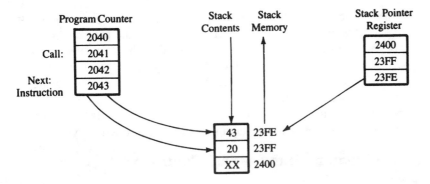

FIGURE 9.11
Contents of the Program Counter, the Stack Pointer, and the Stack During the Execution of the CALL Instruction

RET EXECUTION

At the end of the subroutine, when the instruction RET is executed, the program execution sequence is transferred to the memory location 2043H. The address 2043H was stored in the top two locations of the stack (23FEH and 23FFH) during the CALL instruction. Figure 9.12 shows the sequence of events that occurs as the instruction RET is executed.

Instruction: RET

Memory Address	Code (H)
207F	C9

Contents of Stack Memory	
23FE	43
23FF	20

Machine Cycles	Stack Pointer (23FE)	Address Bus (AB)	Program Counter	Data Bus (DB)	Internal Registers (W) (Z)
M₁ Opcode Fetch	23FE	207F	2080	C9 Opcode	
M₂ Memory Read	23FF	23FE		43 (Stack) →	43
M₃ Memory Read	2400	23FF		20 (Stack–1) →	20
M₁ Opcode Fetch of Next Instruction		2043 (W) (Z) ◄	2044		2043 (W) (Z)

FIGURE 9.12
Data Transfer During the Execution of the RET Instruction

M_1 is a normal Opcode Fetch cycle. However, during M_2 the contents of the stack pointer register are placed on the address bus, rather than those of the program counter. Data byte 43H from the top of the stack is fetched and stored in the Z register, and the stack pointer register is upgraded to the next location, 23FFH. During M_2, the next byte—20H—is copied from the stack and stored in register W, and the stack pointer register is again upgraded to the next location, 2400H.

The program sequence is transferred to location 2043H by placing the contents of the W/Z registers on the address bus at the beginning of the next instruction cycle.

9.21   Illustrative Program: Traffic Signal Controller

PROBLEM STATEMENT

Write a program to provide the given on/off time to three traffic lights (Green, Yellow, and Red) and two pedestrian signs (WALK and DON'T WALK). The signal lights and signs are turned on/off by the data bits of an output port as shown below:

Lights	Data Bits	On Time
1. Green	D_0	15 seconds
2. Yellow	D_2	5 seconds
3. Red	D_4	20 seconds
4. WALK	D_6	15 seconds
5. DON'T WALK	D_7	25 seconds

The traffic and pedestrian flow are in the same direction; the pedestrian should cross the road when the Green light is on.

PROBLEM ANALYSIS

The problem is primarily concerned with providing various time delays for a complete sequence of 40 seconds. The on/off times for the traffic signals and pedestrian signs are as follows:

Time Sequence in Seconds	DON'T WALK D_7	WALK D_6	D_5	Red D_4	D_3	Yellow D_2	D_1	Green D_0		Hex Code
0										
(15) ↓										
15	0	1	0	0	0	0	0	1	=	41H
(5) ↓										
20	1	0	0	0	0	1	0	0	=	84H
(20) ↓										
40	1	0	0	1	0	0	0	0	=	90H

The Green light and the WALK sign can be turned on by sending data byte 41H to the output port. The 15-second delay can be provided by using a 1-second subroutine and a counter with a count of 15_{10}. Similarly, the next two bytes, 84H and 90H, will turn on/off the appropriate lights/signs as shown in the flowchart (Figure 9.13). The necessary time delays are provided by changing the values of the count in the counter.

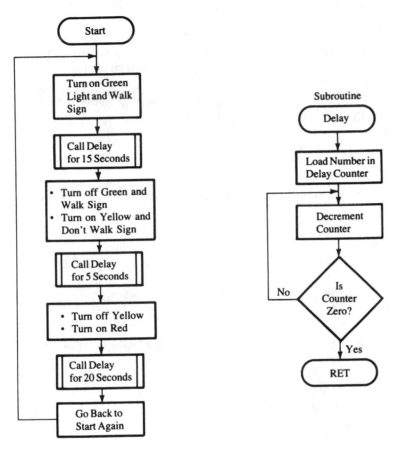

FIGURE 9.13
Flowchart for Traffic Signal Controller

PROGRAM

Memory Address	Code	Mnemonics		Comments
XX00	31			;Initialize stack pointer at location XX99H
01	99		LXI SP,XX99	
02	XX			;High-order address (page) of user memory
03	3E	START:	MVI A,41H	;Load accumulator with the bit pattern for
04	41			; Green light and WALK sign
05	D3		OUT PORT#	;Turn on Green light and WALK sign

06	PORT#		
07	06	MVI B,0FH	;Use B as a counter to count 15seconds.
08	0F		; B is decremented in the subroutine
09	CD	CALL DELAY	;Call delay subroutine located at XX50H
0A	50		
0B	XX		;High-order address (page) of user memory
0C	3E	MVI A,84H	;Load accumulator with the bit pattern for
0D	84		; Yellow light and DON'T WALK
0E	D3	OUT PORT#	;Turn on Yellow light and DON'T WALK
0F	PORT#		; and turn off Green light and WALK
10	06	MVI B,05	;Set up 5-second delay counter
11	05		
12	CD	CALL DELAY	
13	50		
14	XX		;High-order address of user memory
15	3E	MVI A,90H	;Load accumulator with the bit pattern for
16	90		; Red light and DON'T WALK
17	D3	OUT PORT#	;Turn on Red light, keep DON'T WALK on,
18	PORT#		; and turn off Yellow light
19	06	MVI B,14H	;Set up the counter for 20-second delay
1A	14		
1B	CD	CALL DELAY	
1C	50		
1D	XX		
1E	C3	JMP START	;Go back to location START to repeat the
1F	03		; sequence
20	XX		

;**DELAY:** This is a 1-second delay subroutine that provides delay
         ;   according to the parameter specified in register B
         ;*Input:* Number of seconds is specified in register B
         ;*Output:* None
         ;*Registers Modified:* Register B

XX50	D5	DELAY:	PUSH D	;Save contents of DE and accumulator
51	F5		PUSH PSW	
52	11	SECOND:	LXI D,COUNT	;Load register pair DE with a count for
53	LO			; 1-second delay
54	HI			
55	1B	Loop:	DCX D	;Decrement register pair DE
56	7A		MOV A,D	
57	B3		ORA E	;OR (D) and (E) to set Zero flag
58	C2		JNZ LOOP	;Jump to Loop if delay count is not equal to 0
59	55			
5A	XX			

5B	05	DCR B	;End of 1 second delay; decrement the counter
5C	C2	JNZ SECOND	;Is this the end of time needed? If not, go
5D	52		; back to repeat 1-second delay
5E	XX		;High-order memory address of user memory
5F	F1	POP PSW	;Retrieve contents of saved registers
60	D1	POP D	
61	C9	RET	;Return to main program

PROGRAM DESCRIPTION

The stack pointer register is initialized at XX99H so that return addresses can be stored on the stack whenever a CALL instruction is used. As shown in the flowchart this program loads the appropriate bit pattern in the accumulator, sends it to the output port, and calls the delay routine. Register B is loaded in the main program and used in the subroutine to provide appropriate timing.

The DELAY subroutine is similar to the delays discussed in Chapter 8 except it requires the instruction RET at the end of the routine.

This example illustrates the type of subroutine that is called many times from various locations in the main program, as illustrated in Figure 9.14

In this program, the subroutine is called from the locations XX09, 0A, and 0BH. The return address, XX0C, is stored on the stack, and the stack pointer is decremented by two to location XX97H. At the end of the subroutine, the contents of the top two loca-

FIGURE 9.14
Multiple-Calling for a Subroutine

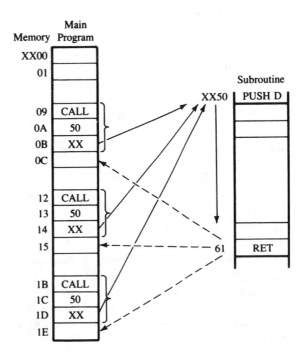

tions of the stack (XX0C) are retrieved, the stack pointer register is incremented by two to the original location (XX99H), and the main program is resumed. This sequence is repeated two more times in the main program, as shown in Figure 9.14. This is called a multiple-calling subroutine.

9.22 Subroutine Documentation and Parameter Passing

In a large program, subroutines are scattered all over the memory map and are called from many locations. Various information is passed between a calling program and a subroutine, a procedure called **parameter passing.** Therefore, it is important to document a subroutine clearly and carefully. The documentation should include at least the following:

1. Functions of the subroutine
2. Input/Output parameters
3. Registers used or modified
4. List of other subroutines called by this subroutine

The delay subroutine in the traffic-signal controller program shows one example of subroutine documentation.

FUNCTIONS OF THE SUBROUTINE

It is important to state clearly and precisely what the subroutine does. A user should understand the function without going through the instructions.

INPUT/OUTPUT PARAMETERS

In the delay subroutine illustrated in Section 9.21, the information concerning the number of seconds is passed from the main program to the subroutine by loading an appropriate count in register B. In this example, a register is used to pass the parameter. The parameters passed to a subroutine are listed as **inputs,** and parameters returned to calling programs are listed as **outputs.**

When many parameters must be passed, R/W memory locations are used to store the parameters, and HL registers are used to point to parameter locations. Similarly, the stack is also used to store and pass parameters.

REGISTERS USED OR MODIFIED

Registers used in a subroutine also may be used by the calling program. Therefore, it is necessary to save the register contents of the calling program on the stack at the beginning of the subroutine and to retrieve the contents before returning from the subroutine.

In the delay subroutine, the contents of registers DE, the accumulator, and the flag register are pushed on the stack because these registers are used in the subroutine. The contents are restored at the end of the routine using the LIFO method. However, the contents of registers that pass parameters should not be saved on the stack because this could cause irrelevant information to be retrieved and passed on to the calling program.

LIST OF SUBROUTINES CALLED

If a subroutine is calling other subroutines, the user should be provided with a list. The user can check what parameters need to be passed to various subroutines and what registers are modified in the process.

RESTART, CONDITIONAL CALL, AND RETURN INSTRUCTIONS 9.3

In addition to the unconditional CALL and RET instructions, the 8085 instruction set includes eight Restart instructions and eight conditional Call and Return instructions.

9.31 Restart (RST) Instructions

RST instructions are 1-byte Call instructions that transfer the program execution to a specific location on page 00H. They are executed the same way as Call instructions. When an RST instruction is executed, the 8085 stores the contents of the program counter (the address of the next instruction) on the top of the stack and transfers the program to the Restart location. These instructions are generally used in conjunction with the interrupt process discussed in Chapter 12. These instructions are listed here to emphasize that they are Call instructions and not necessarily always associated with the interrupts. The list of eight RST instructions is as follows:

RST 0	Call 0000H	RST 4	Call 0020H
RST 1	Call 0008H	RST 5	Call 0028H
RST 2	Call 0010H	RST 6	Call 0030H
RST 3	Call 0018H	RST 7	Call 0038H

9.32 Conditional Call and Return Instructions

The conditional Call and Return instructions are based on four data conditions (flags): Carry, Zero, Sign, and Parity. The conditions are tested by checking the respective flags. In case of a conditional Call instruction, the program is transferred to the subroutine if the condition is met; otherwise, the main program is continued. In case of a conditional Return instruction, the sequence returns to the main program if the condition is met; otherwise, the sequence in the subroutine is continued. If the Call instruction in the main program is conditional, the Return instruction in the subroutine can be conditional or unconditional. The conditional Call and Return instructions are listed for reference.

CONDITIONAL CALL

CC	Call subroutine if Carry flag is set (CY = 1)
CNC	Call subroutine if Carry flag is reset (CY = 0)
CZ	Call subroutine if Zero flag is set (Z = 1)
CNZ	Call subroutine if Zero flag is reset (Z = 0)
CM	Call subroutine if Sign flag is set (S = 1, negative number)
CP	Call subroutine if Sign flag is reset (S = 0, positive number)

CPE   Call subroutine if Parity flag is set (P = 1, even parity)
CPO   Call subroutine if Parity flag is reset (P = 0, odd parity)

CONDITIONAL RETURN

RC   Return if Carry flag is set (CY = 1)
RNC   Return if Carry flag is reset (CY = 0)
RZ   Return if Zero flag is set (Z = 1)
RNZ   Return if Zero flag is reset (Z = 0)
RM   Return if Sign flag is set (S = 1, negative number)
RP   Return if Sign flag is reset (S = 0, positive number)
RPE   Return if Parity flag is set (P = 1, even parity)
RPO   Return if Parity flag is reset (P = 0, odd parity)

9.4 ADVANCED SUBROUTINE CONCEPTS

In Section 9.2, one type of subroutine (multiple-calling of a subroutine by a main program) was illustrated. Other types of subroutine techniques, such as nesting and multiple-ending, are briefly illustrated below.

9.41 Nesting

The programming technique of a subroutine calling another subroutine is called **nesting.** This process is limited only by the number of available stack locations. When a subroutine calls another subroutine, all return addresses are stored on the stack. Nesting is illustrated in Figure 9.15.

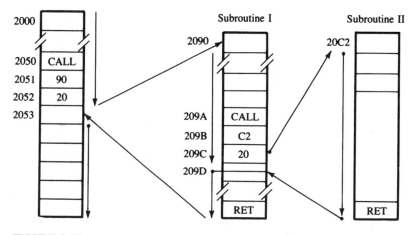

FIGURE 9.15
Nesting of Subroutines

FIGURE 9.16
Multiple-Ending Subroutine

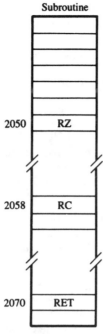

The main program in Figure 9.15 calls the subroutine from location 2050H. The address of the next instruction, 2053H, is placed on the stack, and the program is transferred to the subroutine at 2090H. Subroutine I calls Subroutine II from location 209AH. The address 209DH is placed on the stack, and the program is transferred to Subroutine II. The sequence of execution returns to the main program, as shown in Figure 9.15.

9.42 Multiple-Ending Subroutines

Figure 9.16 illustrates three possible endings to one CALL instruction. The subroutine has two conditional returns (RZ—Return on Zero, and RC—Return on Carry) and one unconditional return (RET). If the Zero flag (Z) is set, the subroutine returns from location 2050H. If the Carry flag (CY) is set, it returns from location 2058H. If neither the Z nor the CY flag is set, it returns from location 2070H. This technique is illustrated in Chapter 10, Section 10.42.

SUMMARY

To implement a subroutine, the following steps are necessary:

1. The stack pointer register must be initialized, preferably at the highest memory location of the R/W memory.

2. The CALL (or conditional Call) instruction should be used in the main program accompanied by the RET (or conditional Return) instruction in the subroutine.

The instructions CALL and RET are similar to the instructions PUSH and POP. The similarities and differences are as follows:

CALL and RET

1. When CALL is executed, the microprocessor automatically stores the 16-bit address of the instruction next to CALL on the stack.

2. When CALL is executed, the stack pointer register is decremented by two.

3. The instruction RET transfers the contents of the top two locations of the stack to the program counter.

4. When the instruction RET is executed, the stack pointer is incremented by two.

5. In addition to the unconditional CALL and RET instructions, there are eight conditional CALL and RETURN instructions.

PUSH and POP

1. The programmer uses the instruction PUSH to save the contents of a register pair on the stack.

2. When PUSH is executed, the stack pointer register is decremented by two.

3. The instruction POP transfers the contents of the top two locations of the stack to the specified register pair.

4. When the instruction POP is executed, the stack pointer is incremented by two.

5. There are no conditional PUSH and POP instructions.

QUESTIONS AND PROGRAMMING ASSIGNMENTS

1. Check the appropriate answer in the following statements.
    a. A stack is
      (1) an 8-bit register in the microprocessor.
      (2) a 16-bit register in the microprocessor.
      (3) a set of memory locations in R/WM reserved for storing information temporarily during the execution of a program.
      (4) a 16-bit memory address stored in the program counter.
    b. A stack pointer is
      (1) a 16-bit register in the microprocessor that indicates the beginning of the stack memory.
      (2) a register that decodes and executes 16-bit arithmetic expressions.
      (3) the first memory location where a subroutine address is stored.
      (4) a register in which flag bits are stored.

c. When a subroutine is called, the address of the instruction following the CALL instruction is stored in/on the
   (1) stack pointer.
   (2) accumulator.
   (3) program counter.
   (4) stack.

d. When the RET instruction at the end of a subroutine is executed,
   (1) the information where the stack is initialized is transferred to the stack pointer.
   (2) the memory address of the RET instruction is transferred to the program counter.
   (3) two data bytes stored in the top two locations of the stack are transferred to the program counter.
   (4) two data bytes stored in the top two locations of the stack are transferred to the stack pointer.

e. Whenever the POP H instruction is executed,
   (1) data bytes in the HL pair are stored on the stack.
   (2) two data bytes at the top of the stack are transferred to the HL register pair.
   (3) two data bytes at the top of the stack are transferred to the program counter.
   (4) two data bytes from the HL register that were previously stored on the stack are transferred back to the HL register.

f. The instruction RST 7 is a:
   (1) restart instruction that begins the execution of a program.
   (2) one-byte call to the memory address 0038H.
   (3) one-byte call to the memory address 0007H.
   (4) hardware interrupt.

2. Read the following program and answer the questions given below.

Line No.	Mnemonics
1	LXI SP,0400H
2	LXI B,2055H
3	LXI H,22FFH
4	LXI D,2090H
5	PUSH H
6	PUSH B
7	MOV A,L
↓	↓
20	POP H

a. What is stored in the stack pointer register after the execution of line 1?

b. What is the memory location of the stack where the first data byte will be stored?

c. What is stored in memory location 03FEH when line 5 (PUSH H) is executed?

d. After the execution of line 6 (PUSH B), what is the address in the stack pointer register, and what is stored in stack memory location 03FDH?

e. Specify the contents of register pair HL after the execution of line 20 (POP H).

3. The following program has a delay subroutine located at location 2060H. Read the program and answer the questions given at the end of the program.

Memory Locations	Mnemonics	
2000	LXI SP,20CDH	;Main Program
2003	LXI H,00008H	
2006	MVI B,0FH	
2008	CALL 2060H	
200B	OUT 01H	
	DCR B	
	CONTD	
2060	PUSH H	:Delay Subroutine
2061	PUSH B	
	MVI B,05H	
	LXI H,COUNT	
	POP B	
	POP H	
	RET	

a. When the execution of the CALL instruction located at 2008H–200AH is completed, list the contents stored at 20CCH and 20CBH, the contents of the program counter, and the contents of the stack pointer register.

b. List the stack locations and their contents after the execution of the instructions PUSH H and PUSH B in the subroutine.

c. List the contents of the stack pointer register after the execution of the instruction PUSH B located at 2061H.

d. List the contents of the stack pointer register after the execution of the instruction RET in the subroutine.

4. Explain the functions of the following routines:

a. LXI SP,209FH
MVI C,00H
PUSH B
POP PSW
RET

b. LXI SP,STACK
PUSH B
PUSH D
POP B
POP D
RET

5. Read the following program and answer the questions.

2000	LXI SP,2100H	DELAY:	2064	PUSH H
2003	LXI B,0000H		2065	PUSH B
2006	PUSH B		2066	LXI B,80FFH
2007	POP PSW	LOOP:	2069	DCX B
2008	LXI H,200BH		206A	MOV A,B
200B	CALL 2064H		206B	ORA C
200E	OUT 01H		206C	JNZ LOOP
2010	HLT		206F	POP B
			2070	RET

   a. What is the status of the flags and the contents of the accumulator after the execution of the POP instruction located at 2007H?

   b. Specify the stack locations and their contents after the execution of the CALL instruction (not the Call subroutine).

   c. What are the contents of the stack pointer register and the program counter after the execution of the CALL instruction?

   d. Specify the memory location where the program returns after the subroutine.

   e. What is the ultimate fate of this program?

6. Write a program to add the two Hex numbers 7A and 46 and to store the sum at memory location XX98H and the flag status at location XX97H.

7. In Assignment 6, display the sum and the flag status at two different output ports.

8. Write a program to meet the following specifications:

   a. Initialize the stack pointer register at XX99H.

   b. Clear the memory locations starting from XX90H to XX9FH.

   c. Load register pairs B, D, and H with data 0237H, 1242H, and 4087H, respectively.

   d. Push the contents of the register pairs B, D, and H on the stack.

   e. Execute the program and verify the memory locations from XX90H to XX9FH.

9. Write a program to clear the initial flags. Load data byte FFH into the accumulator and add 01H to the byte FFH by using the instruction ADI. Mask all the flags except the CY flag and display the CY flag at PORT0 (or store the results on the stack). Repeat the program by replacing the ADI instruction with the INR instruction and the byte 01H with the NOP instruction. Display the flag at PORT1 (or store the results on the stack). Explain the results.

10. Write a 20 ms time delay subroutine using register pair BC. Clear the Z flag without affecting any other flags in the flag register and return to the main program.

11. Write a program to control a railway crossing signal that has two alternately flashing red lights, with a 1-second delay on time for each light.

12. Write a program to simulate a flashing yellow light with 750 ms on time. Use bit D_7 to control the light. (*Hint:* To simulate flashing, the light must be turned off.)

Code Conversion, BCD Arithmetic, and 16-Bit Data Operations

In microcomputer applications, various number systems and codes are used to input data or to display results. The ASCII (American Standard Code for Information Interchange) keyboard is a commonly used input device for disk-based microcomputer systems. Similarly, alphanumeric characters (letters and numbers) are displayed on a CRT (cathode ray tube) terminal using the ASCII code. However, inside the microprocessor, data processing is usually performed in binary. In some instances, arithmetic operations are performed in BCD numbers. Therefore, data must be converted from one code to another code. The programming techniques used for code conversion fall into four general categories:

1. Conversion based on the position of a digit in a number (BCD to binary and vice versa).
2. Conversion based on hardware consideration (binary to seven-segment code using table look-up procedure).
3. Conversion based on sequential order of digits (binary to ASCII and vice versa).
4. Decimal adjustment in BCD arithmetic operations. (This is an adjustment rather than a code conversion.)

This chapter discusses these techniques with various examples written as subroutines. The subroutines are

written to demonstrate industrial practices in writing software, and can be verified on single-board microcomputers. In addition, instructions related to 16-bit data operations are introduced and illustrated.

OBJECTIVES
Write programs and subroutines to

☐ Convert a packed BCD number (0–99) into its binary equivalent.

□ Convert a binary digit (0 to F) into its ASCII Hex code and vice versa.

□ Select an appropriate seven-segment code for a given binary number using the table look-up technique.

□ Convert a binary digit (0 to F) into its ASCII Hex code and vice versa.

□ Decimal-adjust 8-bit BCD addition and subtraction.

□ Perform such arithmetic operations as multiplication and subtraction using 16-bit data related instructions.

□ Demonstrate uses of instructions such as DAD, PCHL, XTHL, and XCHG.

10.1 BCD-TO-BINARY CONVERSION

In most microprocessor-based products, data are entered and displayed in decimal numbers. For example, in an instrumentation laboratory, readings such as voltage and current are maintained in decimal numbers, and data are entered through a decimal keyboard. The system-monitor program of the instrument converts each key into an equivalent 4-bit binary number and stores two BCD numbers in an 8-bit register or a memory location. These numbers are called **packed BCD.** Even if data are entered in decimal digits, it is inefficient to process data in BCD numbers because, in each 4-bit combination, digits A through F are unused. Therefore, BCD numbers are generally converted into binary numbers for data processing.

The conversion of a BCD number into its binary equivalent employs the principle of *positional weighting* in a given number.

For example: $72_{10} = 7 \times 10 + 2$.

The digit 7 represents 70, based on its second position from the right. Therefore, converting 72_{BCD} into its binary equivalent requires multiplying the second digit by 10 and adding the first digit.

Converting a 2-digit BCD number into its binary equivalent requires the following steps:

1. Separate an 8-bit packed BCD number into two 4-bit unpacked BCD digits: BCD_1 and BCD_2.
2. Convert each digit into its binary value according to its position.
3. Add both binary numbers to obtain the binary equivalent of the BCD number.

Example 10.1

Convert 72_{BCD} into its binary equivalent.

Solution

$$72_{10} = 0111\ 0010_{BCD}$$

Step 1: $0111\ 0010 \rightarrow 0000\ 0010$ Unpacked BCD_1
            $\rightarrow 0000\ 0111$ Unpacked BCD_2
Step 2: Multiply BCD_2 by 10 (7×10)
Step 3: Add BCD_1 to the answer in Step 2

The multiplication of BCD_2 by 10 can be performed by various methods. One method is multiplication with repeated addition: add 10 seven times. This technique is illustrated in the next program.

10.11   Illustrative Program: 2-Digit BCD-to-Binary Conversion

PROBLEM STATEMENT

A BCD number between 0 and 99 is stored in an R/W memory location called the **Input Buffer (INBUF)**. Write a main program and a conversion subroutine (BCDBIN) to convert the BCD number into its equivalent binary number. Store the result in a memory location defined as the **Output Buffer (OUTBUF)**.

PROGRAM

```
START:      LXI SP,STACK        ;Initialize stack pointer
            LXI H,INBUF         ;Point HL index to the Input Buffer memory location where BCD
                                ;   number is stored
            LXI B,OUTBUF        ;Point BC index to the Output Buffer memory where binary
                                ;   number will be stored
            MOV A,M             ;Get BCD number
            CALL BCDBIN         ;Call BCD to binary conversion routine
            STAX B              ;Store binary number in the Output Buffer
            HLT                 ;End of program

BCDBIN:     ;Function: This subroutine converts a BCD number into its binary equivalent
            ;Input: A 2-digit packed BCD number in the accumulator
            ;Output: A binary number in the accumulator
            ;No other register contents are destroyed
```

		Example: Assume BCD number is 72:			
PUSH B	;Save BC registers				
PUSH D	;Save DE registers	A	0111	0010	$\rightarrow 72_{10}$
MOV B,A	;Save BCD number	B	0111	0010	$\rightarrow 72_{10}$
ANI OFH	;Mask most significant four bits	A	0000	0010	$\rightarrow 02_{10}$
MOV C,A	;Save unpacked BCD_1 in C	C	0000	0010	$\rightarrow 02_{10}$
MOV A,B	;Get BCD again	A	0111	0010	$\rightarrow 72_{10}$
ANI FOH	;Mask least significant four bits	A	0111	0000	$\rightarrow 70M_{10}$
RRC	;Convert most significant four				
RRC	; bits into unpacked BCD_2				
RRC					
RRC		A	0000	0111	$\rightarrow 07_{10}$
MOV D,A	;Save BCD_2 in D	D	0000	0111	$\rightarrow 07_{10}$
XRA A	;Clear accumulator				

	MVI E,0AH	;Set E as multiplier of 10	E 0000 1010 → 0AH
SUM:	ADD E	;Add 10 until (D) = 0	Add E as many times as (D)
	DCR D	;Reduce BCD₂ by one	
	JNZ SUM	;Is multiplication complete?	After adding E seven times A
		;If not, go back and add again	contains: 0100 0110
	ADD C	;Add BCD₁	C +0000 0010
	POP D	;Retrieve previous contents	A 0100 1000 → 48H
	POP B		
	RET		

The above represents:

- E $0000\ 1010 \rightarrow 0AH$
- Add E as many times as (D)
- After adding E seven times A contains: $0100\ 0110$
- C $+0000\ 0010$
- A $0100\ 1000 \rightarrow 48H$

PROGRAM DESCRIPTION

1. In writing assembly language programs, the use of labels is a common practice. Rather than writing a specific memory location or a port number, a programmer uses such labels as INBUF (Input Buffer) and OUTBUF (Output Buffer). Using labels gives flexibility and ease of documentation.
2. The main program initializes the stack pointer and two memory indexes. It brings the BCD number into the accumulator and passes that parameter to the subroutine.
3. After returning from the subroutine, the main program stores the binary equivalent in the Output Buffer memory.
4. The subroutine saves the contents of the BC and DE registers because these registers are used in the subroutine. Even if this particular main program does not use the DE registers, the subroutine may be called by some other program in which the DE registers are being used. Therefore, it is a good practice to save the registers that are used in the subroutine, unless parameters are passed to the subroutine. The accumulator contents are not saved because that information is passed on to the subroutine.
5. The conversion from BCD to binary is illustrated in the subroutine with the example of 72_{BCD} converted to binary.

The illustrated multiplication routine is easy to understand; however, it is rather long and inefficient. Another method is to multiply BCD₂ by shifting, as illustrated in Assignments 3 and 4 at the end of this chapter.

PROGRAM EXECUTION

To execute the program on a single-board computer, complete the following steps:

1. Assign memory addresses to the instructions in the main program and in the subroutine. Both can be assigned consecutive memory addresses.
2. Define STACK: the stack location with a 16-bit address in the R/W memory (such as 2099H).
3. Define INBUF (Input Buffer) and OUTBUF (Output Buffer): two memory locations in the R/W memory (e.g., 2050H and 2060H).
4. Enter a BCD byte in the Input Buffer (e.g., 2050H).
5. Enter and execute the program.

6. Check the contents of the Output Buffer memory location (2060H) and verify your answer.

See Assignments 1–4 at the end of this chapter.

BINARY-TO-BCD CONVERSION 10.2

In most microprocessor-based products, numbers are displayed in decimal. However, if data processing inside the microprocessor is performed in binary, it is necessary to convert the binary results into their equivalent BCD numbers just before they are displayed. Results are quite often stored in R/W memory locations called the **Output Buffer.**

The conversion of binary to BCD is performed by dividing the number by the powers of ten; the division is performed by the subtraction method.

For example, assume the binary number is

$$1\ 1\ 1\ 1\ \ 1\ 1\ 1\ 1_2\ (FFH) = 255_{10}$$

To represent this number in BCD requires twelve bits or three BCD digits, labeled here as BCD_3 (MSB), BCD_2, and BCD_1 (LSB),

$$= 0\ 0\ 1\ 0\ \ 0\ 1\ 0\ 1\ \ 0\ 1\ 0\ 1$$
$$BCD_3\quad BCD_2\quad BCD_1$$

The conversion can be performed as follows:

	Example		**Quotient**
Step 1: If the number is less than 100, go to Step 2; otherwise, divide by 100 or subtract 100 repeatedly until the remainder is less than 100. The quotient is the most significant BCD digit, BCD_3.	255		
	$-100 =$	155	1
	$-100 =$	55	1
		$BCD_3 =$	2
Step 2: If the number is less than 10, go to Step 3; otherwise divide by 10 repeatedly until the remainder is less than 10. The quotient is BCD_2.	55		
	$-10 =$	45	1
	$-10 =$	35	1
	$-10 =$	25	1
	$-10 =$	15	1
	$-10 =$	05	1
		$BCD_2 =$	5
Step 3: The remainder from Step 2 is BCD_1.		$BCD_1 =$	5

These steps can be converted into a program as illustrated next.

10.21 Illustrative Program: Binary-to-Unpacked-BCD Conversion

PROBLEM STATEMENT

A binary number is stored in memory location BINBYT. Convert the number into BCD, and store each BCD as two unpacked BCD digits in the Output Buffer. To perform this task, write a main program and two subroutines: one to supply the powers of ten, and the other to perform the conversion.

PROGRAM

This program converts an 8-bit binary number into a BCD number; thus it requires 12 bits to represent three BCD digits. The result is stored as three unpacked BCD digits in three Output-Buffer memory locations.

START:	LXI SP,STACK	;Initialize stack pointer
	LXI H,BINBYT	;Point HL index where binary number is stored
	MOV A,M	;Transfer byte
	CALL PWRTEN	;Call subroutine to load powers of 10
	HLT	

PWRTEN: ;this subroutine loads the powers of 10 in register B and calls the binary-to-BCD
    ;   conversion routine
    ;Input: Binary number in the accumulator
    ;Output: Powers of ten and stores BCD_1 in the first Output-Buffer memory
    ;Calls BINBCD routine and modifies register B

	LXI H,OUTBUF	;Point HL index to Output-Buffer memory
	MVI B,64H	;Load 100 in register B
	CALL BINBCD	;Call conversion
	MVI B,0AH	;Load 10 in register B
	CALL BINBCD	
	MOV M,A	;Store BCD_1
	RET	

BINBCD: ;This subroutine converts a binary number into BCD and stores BCD_2 and BCD_3 in the
    ;   Output Buffer.
    ;Input: Binary number in accumulator and powers of 10 in B
    ;Output: BCD_2 and BCD_3 in Output Buffer
    ;Modifies accumulator contents

	MVI M,FFH	;Load buffer with (0 − 1)
NXTBUF:	INR M	;Clear buffer and increment for each subtraction
	SUB B	;Subtract power of 10 from binary number
	JNC NXTBUF	;Is number > power of 10? If yes, add 1 to buffer memory
	ADD B	;If no, add power of 10 to get back remainder

```
          INX H              ;Go to next buffer location
          RET
```

PROGRAM DESCRIPTION

This program illustrates the concepts of the **nested subroutine** and the **multiple-call subroutine.** The main program calls the PWRTEN subroutine; in turn, the PWRTEN calls the BINBCD subroutine twice.

1. The main program transfers the byte to be converted to the accumulator and calls the PWRTEN subroutine.
2. The subroutine PWRTEN supplies the powers of ten by loading register B and the address of the first Output-Buffer memory location, and calls conversion routine BIN-BCD.
3. In the BINBCD conversion routine, the Output-Buffer memory is used as a register. It is incremented for each subtraction loop. This step also can be achieved by using a register in the microprocessor. The BINBCD subroutine is called twice, once after loading register B with 64H (100_{10}), and again after loading register B with 0AH (10_{10}).
4. During the first call of BINBCD, the subroutine clears the Output Buffer, stores BCD_3, and points the HL registers to the next Output-Buffer location. The instruction ADD B is necessary to restore the remainder because one extra subtraction is performed to check the borrow.
5. During the second call of BINBCD, the subroutine again clears the output buffer, stores BCD_2, and points to the next buffer location. BCD_3 is already in the accumulator after the ADD instruction, which is stored in the third Output-Buffer memory by the instruction MOV M,A in the PWRTEN subroutine.

This is an efficient subroutine; it combines the functions of storing the answer and finding a quotient. However, two subroutines are required, and the second subroutine is called twice for a conversion.

See Assignments 5–8 at the end of this chapter.

BCD-TO-SEVEN-SEGMENT-LED CODE CONVERSION 10.3

When a BCD number is to be displayed by a seven-segment LED, it is necessary to convert the BCD number to its seven-segment code. The code is determined by hardware considerations such as common-cathode or common-anode LED; the code has no direct relationship to binary numbers. Therefore, to display a BCD digit at a seven-segment LED, the **table look-up technique** is used.

In the table look-up technique, the codes of the digits to be displayed are stored sequentially in memory. The conversion program locates the code of a digit based on its magnitude and transfers the code to the MPU to send out to a display port. The table look-up technique is illustrated in the next program.

10.31　Illustrative Program: BCD-to-Common-Cathode-LED Code Conversion

PROBLEM STATEMENT

A set of three packed BCD numbers (six digits) representing time and temperature are stored in memory locations starting at XX50H. The seven-segment codes of the digits 0 to 9 for a common-cathode LED are stored in memory locations starting at XX70H, and the Output-Buffer memory is reserved at XX90H.

　　　　Write a main program and two subroutines, called UNPAK and LEDCOD, to un-pack the BCD numbers and select an appropriate seven-segment code for each digit. The codes should be stored in the Output-Buffer memory.

PROGRAM

	LXI SP,STACK	;Initialize stack pointer
	LXI H,XX50H	;Point HL where BCD digits are stored
	MVI D,03H	;Number of digits to be converted is placed in D
	CALL UNPAK	;Call subroutine to unpack BCD numbers
	HLT	;End of conversion
UNPAK:	;This subroutine unpacks the BCD number in two single digits	
	;Input: Starting memory address of the packed BCD numbers in HL registers	
	;　　　　Number of BCDs to be converted in register D	
	;Output: Unpacked BCD into accumulator and output	
	;　　　　Buffer address in BC	
	;Calls subroutine LEDCOD	
	LXI B,BUFFER	;Point BC index to the buffer memory
NXTBCD:	MOV A,M	;Get packed BCD number
	ANI F0H	;Masked BCD_1
	RRC	;Rotate four times to place BCD_2 as unpacked single-digit BCD
	RRC	
	RRC	
	RRC	
	CALL LEDCOD	;Find seven-segment code
	INX B	;Point to next buffer location
	MOV A,M	;Get BCD number again
	ANI 0FH	;Separate BCD_1
	CALL LEDCOD	
	INX B	
	INX H	;Point to next BCD
	DCR D	;One conversion complete, reduce BCD count
	JNZ NXTBCD	;If all BCDs are not yet converted, go back to convert next BCD
	RET	

```
LEDCOD:       ;This subroutine converts an unpacked BCD into its seven-segment-LED code
              ;Input: An unpacked BCD in accumulator
              ;Memory address of the buffer in BC register
              ;Output: Stores seven-segment code in the output buffer
              PUSH H            ;Save HL contents of the caller
              LXI H,CODE        ;Point index to beginning of seven-segment code
              ADD L             ;Add BCD digit to starting address of the code
              MOV L,A           ;Point HL to appropriate code
              MOV A,M           ;Get seven-segment code
              STAX B            ;Store code in buffer
              POP H
              RET
CODE:         3F                :Digit 0: Common-cathode codes
              06                ;Digit 1
              5B                ;Digit 2
              4F                ;Digit 3
              66                ;Digit 4
              6D                ;Digit 5
              7D                ;Digit 6
              07                ;Digit 7
              7F                ;Digit 8
              6F                ;Digit 9
              00                ;Invalid Digit
```

PROGRAM DESCRIPTION/OUTPUT

1. The main program initializes the stack pointer, the HL register as a pointer for BCD digits, and the counter for the number of digits; then it calls the UNPAK subroutine.
2. The UNPAK subroutine transfers a BCD number into the accumulator and unpacks it into two BCD digits by using the instructions ANI and RRC. This subroutine also supplies the address of the buffer memory to the next subroutine, LEDCOD. The subroutine is repeated until counter D becomes zero.
3. The LEDCOD subroutine saves the memory address of the BCD number and points the HL register to the beginning address of the code.
4. The instruction ADD L adds the BCD digit in the accumulator to the starting address of the code. After storing the sum in register L, the HL register points to the seven-segment code of that BCD digit.
5. The code is transferred to the accumulator and stored in the buffer.

This illustrative program uses the technique of the nested subroutine (one subroutine calling another). Parameters are passed from one subroutine to another; therefore, you should be careful in using Push instructions to store register contents on the stack. In addition, the LEDCOD subroutine does not account for a situation if by adding the register L a carry is generated. (See Assignment 12.)

See Assignments 9–12 at the end of this chapter.

10.4 BINARY-TO-ASCII AND ASCII-TO-BINARY CODE CONVERSION

The American Standard Code for Information Interchange (known as ASCII) is used commonly in data communication. It is a seven-bit code, and its 128 (2^7) combinations are assigned different alphanumeric characters (see Appendix E). For example, the hexadecimal numbers 30H to 39H represent 0 to 9 ASCII decimal numbers, and 41H to 5AH represent capital letters A through Z; in this code, bit D_7 is zero. In serial data communication, bit D_7 can be used for parity checking (see Chapter 16, Serial I/O and Data Communication).

The ASCII keyboard is a standard input device for entering programs in a microcomputer. When an ASCII character is entered, the microprocessor receives the binary equivalent of the ASCII Hex number. For example, when the ASCII key for digit 9 is pressed, the microprocessor receives the binary equivalent of 39H, which must be converted to the binary 1001 for arithmetic operations. Similarly, to display digit 9 at the terminal, the microprocessor must send out the ASCII Hex code (39H). These conversions are done through software, as in the following illustrative program.

10.41 Illustrative Program: Binary-to-ASCII Hex Code Conversion

PROBLEM STATEMENT

An 8-bit binary number (e.g., 9FH) is stored in memory location XX50H.

1. Write a program to
   a. Transfer the byte to the accumulator.
   b. Separate the two nibbles (as 09 and 0F).
   c. Call the subroutine to convert each nibble into ASCII Hex code.
   d. Store the codes in memory locations XX60H and XX61H.
2. Write a subroutine to convert a binary digit (0 to F) into ASCII Hex code.

MAIN PROGRAM

```
LXI SP,STACK     ;Initialize stack pointer
LXI H,XX50H      ;Point index where binary number is stored
LXI D,XX60H      ;Point index where ASCII code is to be stored
MOV A,M          ;Transfer byte
MOV B,A          ;Save byte
RRC              ;Shift high-order nibble to the position of low-order
RRC              ;  nibble
```

```
              RRC
              RRC
              CALL ASCII      ;Call conversion routine
              STAX D          ;Store first ASCII Hex in XX60H
              INX D           ;Point to next memory location, get ready to store
                             ;   next byte
              MOV A,B         ;Get number again for second digit
              CALL ASCII
              STAX D
              HLT

ASCII:        ;This subroutine converts a binary digit between 0 and F to ASCII Hex
              ;   code
              ;Input: Single binary number 0 to F in the accumulator
              ;Output: ASCII Hex code in the accumulator
              ANI 0FH         ;Mask high-order nibble
              CPI 0AH         ;Is digit less than 10₁₀?
              JC CODE         ;If digit is less than 10₁₀, go to CODE to add 30H
              ADI 07H         ;Add 7H to obtain code for digits from A to F
CODE:         ADI 30H         ;Add base number 30H
              RET
```

PROGRAM DESCRIPTION

1. The main program transfers the binary data byte from the memory location to the accumulator.
2. It shifts the high-order nibble into the low-order nibble, calls the conversion subroutine, and stores the converted value in the memory.
3. It retrieves the byte again and repeats the conversion process for the low-order nibble.

   In this program, the masking instruction ANI is used once in the subroutine rather than twice in the main program as illustrated in the program for BCD-to-Common-Cathode-LED Code Conversion (Section 10.31).

10.42 Illustrative Program: ASCII Hex-to-Binary Conversion

PROBLEM STATEMENT

Write a subroutine to convert an ASCII Hex number into its binary equivalent. A calling program places the ASCII number in the accumulator, and the subroutine should pass the conversion back to the accumulator.

SUBROUTINE

```
ASCBIN:       ;This subroutine converts an ASCII Hex number into its binary
              :   equivalent
              ;Input: ASCII Hex number in the accumulator
```

;Output: Binary equivalent in the accumulator

```
SUI 30H      ;Subtract 0 bias from the number
CPI 0AH      ;Check whether number is between 0 and 9
RC           ;If yes, return to main program
SUI 07H      ;If not, subtract 7 to find number between A and F
RET
```

PROGRAM DESCRIPTION

This subroutine subtracts the ASCII weighting digits from the number. This process is exactly opposite to that of the Illustrative Program that converted binary into ASCII Hex (Section 10.41). However, this program uses two return instructions, an illustration of the multiple-ending subroutine.

See Assignments 13 and 14 at the end of this chapter.

10.5 BCD ADDITION

In some applications, input/output data are presented in decimal numbers, and the speed of data processing is unimportant. In such applications, it may be convenient to perform arithmetic operations directly in BCD numbers. However, the addition of two BCD numbers may not represent an appropriate BCD value. For example, the addition of 34_{BCD} and 26_{BCD} results in 5AH, as shown below:

$$
\begin{array}{rl}
34_{10} = & 0\ 0\ 1\ 1\quad 0\ 1\ 0\ 0_{BCD} \\
+\ 26_{10} = & 0\ 0\ 1\ 0\quad 0\ 1\ 1\ 0_{BCD} \\
\hline
60_{10} = & 0\ 1\ 0\ 1\quad 1\ 0\ 1\ 0 \rightarrow 5AH
\end{array}
$$

The microprocessor cannot recognize BCD numbers; it adds any two numbers in binary. In BCD addition, any number larger than 9 (from A to F) is invalid and needs to be adjusted by adding 6 in binary. For example, after 9, the next BCD number is 10; however, in Hex it is A. The Hex number A can be adjusted as a BCD number by adding 6 in binary. The BCD adjustment in an 8-bit binary register can be shown as follows:

$$
\begin{array}{rl}
A = & 0\ 0\ 0\ 0\quad 1\ 0\ 1\ 0 \\
+\ 6 = & 0\ 0\ 0\ 0\quad 0\ 1\ 1\ 0 \\
\hline
& 0\ 0\ 0\ 1\quad 0\ 0\ 0\ 0 \rightarrow 10_{BCD}
\end{array}
$$

Any BCD sum can be adjusted to proper BCD value by adding 6 when the sum exceeds 9. In case of packed BCD, both BCD_1 and BCD_2 need to be adjusted; if a carry is generated by adding 6 to BCD_1, the carry should be added to BCD_2, as shown in the following example.

Add two packed BCD numbers: 77 and 48.	**Example 10.2**

Addition:

$$
\begin{array}{r}
77 = 0\ 1\ 1\ 1\quad 0\ 1\ 1\ 1 \\
+\ 48 = 0\ 1\ 0\ 0\quad 1\ 0\ 0\ 0 \\
\hline
125 = 1\ 0\ 1\ 1\quad 1\ 1\ 1\ 1
\end{array}
$$

Solution

The value of the least significant four bits is larger than 9. Add 6.

$$
\begin{array}{r}
+0\ 1\ 1\ 0 \\
\hline
\mathrm{CY}\boxed{1}\quad 0\ 1\ 0\ 1
\end{array}
$$

The value of the most significant four bits is larger than 9. Add 6 and the carry from the previous adjustment.

$$
\begin{array}{r}
+0\ 1\ 1\ 0 \\
\hline
\mathrm{CY}\boxed{1}\quad 0\ 0\ 1\ 0\quad 0\ 1\ 0\ 1
\end{array}
$$

In this example, the carry is generated after the adjustment of the least significant four bits for the BCD digit and is again added to the adjustment of the most significant four bits.

A special instruction called DAA (Decimal Adjust Accumulator) performs the function of adjusting a BCD sum in the 8085 instruction set. This instruction uses the Auxiliary Carry flip-flop (AC) to sense that the value of the least four bits is larger than 9 and adjusts the bits to the BCD value. Similarly, it uses the Carry flag (CY) to adjust the most significant four bits. However, the AC flag is used internally by the microprocessor; this flag is not available to the programmer through any Jump instruction.

INSTRUCTION

DAA: Decimal Adjust Accumulator
  □ This is a 1-byte instruction
  □ It adjusts an 8-bit number in the accumulator to form two BCD numbers by using the process described above
  □ It uses the AC and the CY flags to perform the adjustment
  □ All flags are affected

It must be emphasized that instruction DAA

□ adjusts a BCD sum.
□ does not convert a binary number into BCD numbers.
□ works only with addition when BCD numbers are used; does not work with subtraction.

10.51 Illustrative Program: Addition of Unsigned BCD Numbers

PROBLEM STATEMENT

A set of ten packed BCD numbers is stored in the memory location starting at XX50H.

1. Write a program with a subroutine to add these numbers in BCD. If a carry is generated, save it in register B, and adjust it for BCD. The final sum will be less than 9999_{BCD}.

2. Write a second subroutine to unpack the BCD sum stored in registers A and B, and store them in the output-buffer memory starting at XX60H. The most significant digit (BCD_4) should be stored at XX60H and the least significant digit (BCD_1) at XX63H.

PROGRAM

```
START:      LXI SP,STACK        ;Initialize stack pointer
            LXI H,XX50H         ;Point index to XX50H
            MVI C,COUNT         ;Load register C with the count of BCD numbers to be added
            XRA A               ;Clear accumulator
            MOV B,A             ;Clear register B to save carry
NXTBCD:     CALL BCDADD         ;Call subroutine to add BCD numbers
            INX H               ;Point to next memory location
            DCR C               ;One addition of BCD number is complete, decrement the counter
            JNZ NXTBCD          ;If all numbers are added go to next step, otherwise go back
            LXI H,XX63H         ;Point index to store BCD₁ first
            CALL UNPAK          ;Unpack the BCD stored in the accumulator
            MOV A,B             ;Get ready to store high-order BCD—BCD₃ and BCD₄
            CALL UNPAK          ;Unpack and store BCD₃ and BCD₄ at XX61H and XX60
            HLT

BCDADD:     ;This subroutine adds the BCD number from the memory to the accumulator and decimal-
            ;   adjusts it. If the sum is larger than eight bits, it saves the carry and decimal-adjusts the
            ;   carry sum
            ;Input: The memory address in HL register where the BCD number is stored
            ;Output: Decimal-adjusted BCD number in the accumulator and the carry in register B

            ADD M       ;Add packed BCD byte and adjust it for BCD sum
            DAA
            RNC         ;If no carry, go back to next BCD
            MOV D,A     ;If carry is generated, save the sum from the accumulator
            MOV A,B     ;Transfer CY sum from register B and add 01
            ADI 01H
            DAA         ;Decimal-adjust BCD from B
            MOV B,A     ;Save adjusted BCD in B
            MOV A,D     ;Place BCD₁ and BCD₂ in the accumulator
            RET

UNPAK:      ;This subroutine unpacks the BCD in the accumulator and the carry register and stores
            ;   them in the output buffer
            ;Input: BCD number in the accumulator, and the buffer address in HL registers
            ;Output: Unpacked BCD in the output buffer

            MOV D,A     ;Save BCD number
            ANI 0FH     ;Mask high-order BCD
```

```
MOV M,A      ;Store low-order BCD
DCX H        ;Point to next memory location
MOV A,D      ;Get BCD again
ANI F0H      ;Mask low-order BCD
RRC          ;Convert the most significant four bits into unpacked BCD
RRC
RRC
RRC
MOV M,A      ;Store high-order BCD
DCX H        ;Point to the next memory location
RET
```

PROGRAM DESCRIPTION

1. The expected maximum sum is 9090, which requires two registers. The main program clears the accumulator to save BCD_1 and BCD_2, clears register B to save BCD_3 and BCD_4, and calls the subroutine to add the numbers. The BCD bytes are added until the counter C becomes zero.
2. The BCDADD subroutine is an illustration of the multiple-ending subroutine. It adds a byte, decimal-adjusts the accumulator and, if there is no carry, returns the program execution to the main program. If there is a carry, it adds 01 to the carry register B by transferring the contents to the accumulator and decimal-adjusting the contents. The final sum is stored in registers A and B.
3. The main program calls the UNPAK subroutine, which takes the BCD number from the accumulator (e.g., 57_{BCD}), unpacks it into two separate BCDs (e.g., 05_{BCD} and 07_{BCD}), and stores them in the output buffer. When a subroutine stores a BCD number in memory, it decrements the index because BCD_1 is stored first.

See Assignments 15–16 at the end of this chapter.

BCD SUBTRACTION 10.6

When subtracting two BCD numbers, the instruction DAA cannot be used to decimal-adjust the result of two packed BCD numbers; the instruction applies only to addition. Therefore, it is necessary to devise a procedure to subtract two BCD numbers. Two BCD numbers can be subtracted by using the procedure of 100's complement (also known as 10's complement), similar to 2's complement. The 100's complement of a subtrahend can be added to a minuend as illustrated:

For example, $82 - 48$ ($= 34$) can be performed as follows:

$$
\begin{array}{lrl}
\text{100's complement of subtrahend} & 52 & (100 - 48 = 52) \\
\text{Add minuend} & +\,82 & \\
\hline
& 1/34 &
\end{array}
$$

The sum is 34 if the carry is ignored. This is similar to subtraction by 2's complement. However, in an 8-bit microprocessor, it is not a simple process to find 100's complement of a subtrahend (100_{BCD} requires twelve bits). Therefore, in writing a program, 100's complement is obtained by finding 99's complement and adding 01.

10.61 Illustrative Program: Subtraction of Two Packed BCD Numbers

PROBLEM STATEMENT

Write a subroutine to subtract one packed BCD number from another BCD number. The minuend is placed in register B, and the subtrahend is placed is register C by the calling program. Return the answer into the accumulator.

SUBROUTINE

```
SUBBCD:     ;This subroutine subtracts two BCD numbers and adjusts the result to
            ;   BCD values by using the 100's complement method
            ;Input: A minuend in register B and a subtrahend in register C
            ;Output: The result is placed in the accumulator
            MVI A,99H
            SUB C           ;Find 99's complement of subtrahend
            INR A           ;Find 100's complement of subtrahend
            ADD B           ;Add minuend to 100's complement of subtrahend
            DAA             ;Adjust for BCD
            RET
```

See Assignments 17 and 18 at the end of this chapter.

10.7 INTRODUCTION TO ADVANCED INSTRUCTIONS AND APPLICATIONS

The instructions discussed in the last several chapters deal primarily with 8-bit data (except LXI). However, in some instances data larger than eight bits must be manipulated, especially in arithmetic manipulations and stack operations. Even if the 8085 is an 8-bit microprocessor, its architecture allows specific combinations of two 8-bit registers to form 16-bit registers. Several instructions in the instruction set are available to manipulate 16-bit data. These instructions will be introduced in this section.

10.71 16-Bit Data Transfer (Copy) and Data Exchange Group

LHLD: Load HL registers direct
            ☐ This is a 3-byte instruction
            ☐ The second and third bytes specify a memory location (the second byte is a line number and the third byte is a page number)

- □ Transfers the contents of the specified memory location to L register
- □ Transfers the contents of the next memory location to H register

SHLD: Store HL registers direct
- □ This is a 3-byte instruction
- □ The second and third bytes specify a memory location (the second byte is a line number and the third byte is a page number)
- □ Stores the contents of L register in the specified memory location
- □ Stores the contents of H register in the next memory location

XCHG: Exchange the contents of HL and DE
- □ This is a 1-byte instruction
- □ The contents of H register are exchanged with the contents of D register, and the contents of L register are exchanged with the contents of E register

---

Memory locations 2050H and 2051H contain 3FH and 42H, respectively, and register pair DE contains 856FH. Write instructions to exchange the contents of DE with the contents of the memory locations.

**Example
10.3**

                                                                   Memory

Before Instructions: D [85   6F] E   [3F] 2050
                                                                [42] 2051

Instructions

Machine Code	Mnemonics								
2A	LHLD 2050H							3F	2050
50								42	2051
20		H	42	3F	L				
EB	XCHG	D	42	3F	E			3F	2050
		H	85	6F	L			42	2051
22	SHLD 2050H							6F	2050
50								85	2051
20		H	85	6F	L				

10.72 Arithmetic Group

Operation: Addition with Carry

      ADC R   □ These instructions add the contents of the operand, the
      ADC M   carry, and the accumulator. All flags are affected
      ACI 8-bit

Example 10.4	Registers BC contain 2793H, and registers DE contain 3182H. Write instructions to add these two 16-bit numbers, and place the sum in memory locations 2050H and 2051H.

Before instructions: B | 27 | 93 | C
                        D | 31 | 82 | E

Instructions

MOV A,C	A	93		F		93H
ADD E	A	15	CY = 1	F	+	82H
MOV L,A	H		15	L		1/15H
MOV A,B						27H
ADC D						+ 31H
MOV H,A	H	59	15	L		59H
SHLD 2050H						

Operation: Subtraction with Carry
            SBB R   □ These instructions subtract the contents of the operand and
            SBB M   the borrow from the contents of the accumulator
            SBI 8-bit

Example 10.5	Registers BC contain 8538H and registers DE contain 62A5H. Write instructions to subtract the contents of DE from the contents of BC, and place the result in BC.

Instructions

MOV A,C	(B) 85	38 (C)
SUB E		—
MOV C,A	(D) 62	A5 (E)
MOV A,B	−1	1/93
SBB D	(B) 22	93 (C)
MOV B,A		

Operation: Double Register ADD
            DAD Rp   Add register pair to register HL
                        □ This is a 1-byte instruction
            DAD B   □ Adds the contents of the operand (register pair or stack
            DAD D   pointer) to the contents of HL registers
            DAD H   □ The result is placed in HL registers
            DAD SP   □ The Carry flag is altered to reflect the result of the 16-bit
                           addition. No other flags are affected
                        □ The instruction set includes four instructions

Write instructions to display the contents of the stack pointer register at output ports.

Example
10.6

Instructions

```
LXI H,0000H    ;Clear HL
DAD SP         ;Place the stack pointer contents in HL
MOV A,H        ;Place high-order address of the stack pointer in the accumulator
OUT PORT1
MOV A,L        ;Place low-order address of the stack pointer in the accumulator
OUT PORT2
```

The instruction DAD SP adds the contents of the stack pointer register to the HL register pair, which is already cleared. This is the only instruction in the 8085 that enables the programmer to examine the contents of the stack pointer register.

10.73 Instructions Related to the Stack Pointer and the Program Counter

XTHL: Exchange Top of the Stack with H and L
   ☐ The contents of L are exchanged with the contents of the memory location shown by the stack pointer, and the contents of H are exchanged with the contents of memory location of the stack pointer +1.

Write a subroutine to set the Zero flag and check whether the instruction JZ (Jump on Zero) functions properly, without modifying any register contents other than flags.

Example
10.7

Subroutine

```
CHECK:     PUSH H
           MVI L,FFH      ;Set all bits in L to logic 1
           PUSH PSW       ;Save flags on the top of the stack
           XTHL           ;Set all bits in the top stack location
           POP PSW        ;Set Zero flag
           JZ NOEROR
           JMP ERROR
NOEROR:    POP H
           RET
```

The instruction PUSH PSW places the flags in the top location of the stack, and the instruction XTHL changes all the bits in that location to logic 1. The instruction POP PSW sets all the flags. If the instruction JZ is functioning properly, the routine returns to the calling program; otherwise, it goes to the ERROR routine (not shown). This example shows that the flags can be examined, and they can be set or reset to check malfunctions in the instructions.

SPHL: Copy H and L registers into the Stack Pointer Register
   □ The contents of H specify the high-order byte and the contents of L specify the low-order byte
   □ The contents of HL registers are not affected

This instruction can be used to load a new address in the stack pointer register. (For an example, see SPHL in the instruction set, Appendix F.)

PCHL: Copy H and L registers into the Program Counter
   □ The contents of H specify the high-order byte and the contents of L specify the low-order byte

---

Example 10.8

Assume that the HL registers hold address 2075H. Transfer the program to location 2075H.

Solution

The program can be transferred to location 2075H by using Jump instructions. However, PCHL is a 1-byte instruction that can perform the same function as the Jump instruction. (For an illustration, see the instruction PCHL in the instruction set, Appendix F.)

This instruction is commonly used in monitor programs to transfer the program control from the monitor program to the user's program (see Chapter 17, Section 17.5).

10.74   Miscellaneous Instruction

CMC: Complement the Carry Flag (CY)
   If the Carry flag is 1, it is reset; and if it is 0, it is set
STC: Set the Carry Flag

These instructions are used in bit manipulation, usually in conjunction with rotate instructions. (See the instruction set in Appendix F for examples.)

10.8 MULTIPLICATION

Multiplication can be performed by repeated addition; this technique is used in BCD-to-binary conversion. It is, however, an inefficient technique for a large multiplier. A more efficient technique can be devised by following the model of manual multiplication of decimal numbers. For example,

$$
\begin{array}{r}
108 \\
\times\ \ 15 \\
\hline
\end{array}
$$

Step 1:	$(108 \times 5) =$	540
Step 2: Shift left and add	$(108 \times 1) =$	+ 108
		1620

In this example, the multiplier multiplies each digit of the multiplicand, starting from the farthest right, and adds the product by shifting to the left. The same process can be applied in binary multiplication.

10.81 Illustrative Program: Multiplication of Two 8-Bit Unsigned Numbers

PROBLEM STATEMENT

A multiplicand is stored in memory location XX50H and a multiplier is stored in location XX51H. Write a main program to

1. transfer the two numbers from memory locations to the HL registers.
2. store the product in the Output Buffer at XX90H.

Write a subroutine to

1. multiply two unsigned numbers placed in registers H and L.
2. return the result into the HL pair.

MAIN PROGRAM

```
        LXI SP,STACK
        LHLD XX50H      ;Place contents of XX50 in L register and contents of
                        ;   XX51 in H register
        XCHG            ;Place multiplier in D and multiplicand in E
        CALL MLTPLY     ;Multiply the two numbers
        SHLD XX90H      ;Store the product in locations XX90 and 91H
        HLT
```

Subroutine

```
        MLTPLY: This subroutine multiplies two 8-bit unsigned numbers
        ;Input: Multiplicand in register E and multiplier in register D
        ;Output: Results in HL register

MLTPLY:     MOV A,D         ;Transfer multiplier to accumulator
            MVI D,00H       ;Clear D to use in DAD instruction
            LXI H,0000H     ;Clear HL
            MVI B,08H       ;Set up register B to count eight rotations
NXTBIT:     RAR             ;Check if multiplier bit is 1
            JNC NOADD       ;If not, skip adding multiplicand.
            DAD D           ;If multiplier is 1, add multiplicand to HL and place
                            ;   partial result in HL
```

```
NOADD:      XCHG            ;Place multiplicand in HL
            DAD H           ;And shift left
            XCHG            ;Retrieve shifted multiplicand
            DCR B           ;One operation is complete, decrement counter
            JNZ NXTBIT      ;Go back to next bit
            RET
```

PROGRAM DESCRIPTION

1. The objective of the main program is to demonstrate use of the instructions LHLD, SHLD, and XCHG. The main program transfers the two bytes (multiplier and multiplicand) from memory locations to the HL registers by using the instruction LHLD, places them in the DE register by the instruction XCHG, and places the result in the Output Buffer by the instruction SHLD.
2. The multiplier routine follows the format—add and shift to the left—illustrated at the beginning of Section 10.8. The routine places the multiplier in the accumulator and rotates it eight times until the counter (B) becomes zero. The reason for clearing D is to use the instruction DAD to add register pairs.
3. After each rotation, when a multiplier bit is 1, the instruction DAD D performs the addition, and DAD H shifts bits to the left. When a bit is 0, the subroutine skips the instruction DAD D and just shifts the bits.

10.9   SUBTRACTION WITH CARRY

The instruction set includes several instructions specifying arithmetic operations with carry (for example, add with carry or subtract with carry, Section 10.72). Descriptions of these instructions convey an impression that these instructions can be used to add (or subtract) 8-bit numbers when the addition generates carries. In fact, in these instructions when a carry is generated, it is added to bit D_0 of the accumulator in the next operation. Therefore, these instructions are used primarily in 16-bit addition and subtraction, as shown in the next program.

10.91   Illustrative Program: 16-Bit Subtraction

PROBLEM STATEMENT

A set of five 16-bit readings of the current consumption of industrial control units is monitored by meters and stored at memory locations starting at XX50H. The low-order byte is stored first (e.g., at XX50H), followed by the high-order byte (e.g., at XX51H). The corresponding maximum limits for each control unit are stored starting at XX90H. Subtract each reading from its specified limit, and store the difference in place of the readings. If any reading exceeds the maximum limit, call the indicator routine and continue checking.

MAIN PROGRAM

```
             LXI D, 2050H      ;Point index to readings
             LXI H, 2080H      ;Point index to maximum limits
             MVI B,05H         ;Set up B as a counter
NEXT: CALL SBTRAC             ;Point to next location
             INX D             ;Point to next location
             INX H
             DCR B
             JNZ NEXT
             HLT
```

Subroutine
;SBTRAC: This subroutine subtracts two 16-bit numbers
;Input: The contents of registers DE point to reading locations
;   The contents of registers HL point to maximum limits
;Output: The results are placed in reading locations, thus destroying the initial readings
;The comment section illustrates one example, assuming the following data:

Memory Contents

The first current reading = 6790H   2050 = 90H   LSB
                                      2051 = 67H   MSB
Maximum limit = 7000 H   2090 = 00H   LSB
                                      2091 = 70H   MSB

```
                                 ;Illustrative Example
SBTRAC:     MOV A,M;(A)          ;(A)  = 00H LSB of maximum limit
            XCHG                 ;(HL) = 2050H
            SUB M                ;(A)  = 0000 0000   2's complement of 90H
                                 ;(M)  = 0111 0000   Borrow flag is set to
                                      1 0111 0000   indicate the result is in
                                                     2's complement.
            MOV M,A              ;Store at 2050H
            XCHG                 ;(HL) = 2090H
            INX H                ;(HL) = 2091H
            INX D                ;(DE) = 2051H
            MOV A,M              ;(A)  = 70H   MSB of the maximum limit
            XCHG                 ;(HL) = 2051H
            SBB M                ;(A)  = 0111  0000   (70H)
                                 ;(M)  = 1001  1001   2's complement of 67H
                                 ;(CY) =           1   Borrow flag
            CC INDIKET           ;Call Indicate subroutine if reading is higher than the
                                 ;   maximum limit
            MOV M,A
            RET
```

PROGRAM DESCRIPTION

This is a 16-bit subtraction routine that subtracts one byte at a time. The low-order bytes are subtracted by using the instruction SUB M. If a borrow is generated, it is accounted for by using the instruction SBB M (Subtract with Carry) for high-order bytes. In the illustrative example, the first subtraction (00H – 90H) generates a borrow that is subtracted from high-order bytes. The instruction XCHG changes the index pointer alternately between the set of readings and the maximum limits.

SUMMARY

The following code conversion and 16-bit arithmetic techniques were illustrated in this chapter:

☐ BCD-to-binary (Section 10.1)
☐ Binary-to-BCD (Section 10.2)
☐ BCD to seven-segment-LED code (Section 10.3)
☐ Binary-to-ASCII code and ASCII-to-binary (Section 10.4)
☐ BCD addition and subtraction (Sections 10.5 and 10.6)
☐ Multiplication of two 8-bit unsigned numbers and 16-bit subtraction with carry (Section 10.7)

Review of Instructions

The instructions introduced and illustrated in this section are summarized below.

16-Bit Data Transfer (Copy) and Data Exchange Instructions

☐ LHLD 16-bit   Load HL registers direct
☐ SHLD 16-bit   Store HL registers direct
☐ XCHG   Exchange the contents of HL with DE
☐ XTHL   Exchange the top of the stack with HL
☐ SPHL   Copy HL registers into the stack pointer
☐ PCHL   Copy HL registers into the program counter

Arithmetic Instructions Used in 16-Bit Operations

Addition: The following instructions add the contents of the operand, the carry, and the accumulator.

ADC R   Add register contents with carry
ADC M   Add memory contents with carry
ACI 8-bit   Add immediate 8-bit data with carry

Subtraction: The following instructions subtract the contents of the operand and the borrow from the contents of the accumulator.

SBB R	Subtract register contents with borrow
SBB M	Subtract memory contents with borrow
SBI 8-bit	Subtract immediate 8-bit data with borrow

LOOKING AHEAD

This chapter included various types of code conversion techniques and illustrated applications of more advanced instructions. The illustrative programs were written as independent modules, similar to the circuit boards of an electronic system. Now, what is needed is to link these modules to perform a specific task.

However, single-board microcomputer systems arc unsuitable for writing and coding programs with more than 50 instructions. Coding becomes cumbersome, modifications become tedious, and troubleshooting becomes next to impossible. To write a large program, therefore, it is necessary to have access to an assembler and a disk-based system, which will be discussed in the next chapter.

QUESTIONS AND PROGRAMMING ASSIGNMENTS

Section 10.1: BCD-to-Binary Conversion

1. Rewrite the BCDBIN subroutine to include storing results in the Output Buffer. Eliminate unnecessary PUSH and POP instructions.
2. Modify the program (Section 10.11) to convert a set of numbers of 2-digit BCD numbers into their binary equivalents and store them in the Output Buffer. The number of BCD digits in the set is specified by the main program in register D and passed on as a parameter to the subroutine.
3. Rewrite the multiplication section using the RLC (Rotate Left) instruction. *Hints:* Rotating left once is equivalent to multiplying by two. To multiply a digit by ten, rotate left three times and add the result of the first rotation (times 10 = times 8 + times 2).
4. In Assignment 3, multiplication is performed by rotating the high-order digit to the left. However, in the BCDBIN subroutine, just before the multiplication section, the high-order digit BCD_2 is shifted right to place it in the low-order position. Rewrite the subroutine to combine the two operations. *Hint:* Rotating BCD_2 right once from the high-order position is equivalent to multiplying it by eight.

Section 10.2: Binary-to-BCD Conversion

5. Assume the STACK is defined as 20B8H in the illustrative program to convert binary to unpacked BCD (Section 10.21). Specify the stack addresses and their (symbolic) contents when the BINBCD subroutine is called the second time.

6. Rewrite the main program to supply the powers of ten in registers B and C and to store converted BCD numbers in the Output Buffer. Modify the BINBCD subroutine to accommodate the changes in the main program and eliminate the PWRTEN subroutine.

7. Rewrite the program to convert a given number of binary data bytes into their BCD equivalents, and store them as unpacked BCDs in the Output Buffer. The number of data bytes is specified in register D in the main program. The converted numbers should be stored in groups of three consecutive memory locations. If the number is not large enough to occupy all three locations, zeros should be loaded in those locations.

8. A set of ten BCD readings is stored in the Input Buffer. Convert the numbers into binary and add the numbers. Store the sum in the Output Buffer; the sum can be larger than FFH.

Section 10.3: BCD-to-Seven-Segment-LED Code Conversion

9. List the common-cathode and the common-anode seven-segment-LED look-up table to include hexadecimal digits from 0 to F. (See Figure 4.9 for the diagram.)

10. A set of data is stored as unpacked (single-digit) BCD numbers in memory from XX50H to XX5FH. Write a program to look up the common-cathode-LED code for each reading and store the code from XX55H to XX64H (initial data can be eliminated).

11. Design a counter to count continuously from 00H to FFH with 500 ms delay between each count. Display the count at PORT1 and PORT2 (one digit per port) with a common-anode seven-segment-LED code.

12. Modify the LEDCOD subroutine to account for a carry when the instruction ADD L is executed. For example, if the starting address of the CODE table is 02F9, the codes of digits larger than six will be stored on page 03H. The subroutine given in the illustrative program to convert BCD to LED code (Section 10.31) does not account for such a situation. *Hint:* Check for the carry (CY) after the addition, and increment H whenever a carry is generated.

Section 10.4: Binary-to-ASCII Code Conversion

13. A set of ASCII Hex digits is stored in the Input-Buffer memory. Write a program to convert these numbers into binary. Add these numbers in binary, and store the result in the Output-Buffer memory.

14. Extend the program in Assignment 13 to convert the result from binary to ASCII Hex code.

Section 10.5:   BCD Addition

15.   Write a counter program to count continuously from 0 to 99 in BCD with a delay of 750 ms between each count. Display the count at an output port.

16.   Modify the illustrative program for the addition of unsigned numbers (Section 10.51) to convert the unpacked BCD digits located at XX60 to XX63 into ASCII characters, and store them in the Output Buffer.

Section 10.6:   BCD Subtraction

17.   Design a down-counter to count from 99 to 0 in BCD with 500 ms delay between each count. Display the count at an output port. *Hint:* Check for the low-order digit; when it reaches zero, adjust the next digit to nine.

18.   Write a program to subtract a 2-digit BCD number from another 2-digit BCD number; the numbers are stored in two consecutive memory locations. Rather than using the 100's complement method, subtract one BCD digit at a time, and decimal-adjust each digit after subtraction. (The instruction DAA cannot be used in subtraction.) Display the result at an output port. Verify that if the subtrahend is larger than the minuend, the result will be negative and will be displayed as the 100's complement.

Section 10.7:   Advanced Instructions and Applications

19.   A set of 16-bit readings is stored in memory locations starting at 2050H. Each reading occupies two memory locations: the low-order byte is stored first, followed by the high-order byte. The number of readings stored is specified by the contents of register B. Write a program to add all the readings and store the sum in the Output-Buffer memory. (The maximum limit of a sum is 24 bits.)

20.   In Assignment 19, save the contents of the stack pointer from the main program, point the stack pointer to location 2050H, and transfer the readings to registers by using the POP instruction. Add the readings as in Assignment 19; however, retrieve the original contents of the stack pointer after the addition is completed.

21.   Assume that the monitor program stores a memory address in the DE registers. When a Hex key is pressed to enter a new memory address, the keyboard subroutine places the 4-bit code of the key pressed in the accumulator. Write a subroutine to shift out the most significant four bits of the old address and to insert the new code from the accumulator as the least significant four bits in register E.

   Hint: Place the memory address in the HL registers, and use the instruction DAD H four times; this will shift all bits to the left by four positions and will clear the least significant four bits.

22.   A pair of 32-bit readings is stored in groups of four consecutive memory locations; the memory location with the lowest memory address in each group contains the least significant byte. Write a program to add these readings; if a carry is generated, call an error routine.

Software Development Systems and Assemblers

A **software development system** is a computer that enables the user to develop programs (**software**) with the assistance of other programs. The development process includes writing, modifying, testing, and debugging of the user programs. Programs such as Editor, Assembler, Loader (or Linker), and Debugger enable the user to write programs in mnemonics, translate mnemonics into binary code, and debug the binary code. All the activities of the computer—hardware and software—are directed by another program, called the **operating system** of the computer.

This chapter describes a microprocessor-based software development system, its hardware, and related programs. It also describes widely used operating systems and assemblers and illustrates the use of an assembler to write assembly language programs.

OBJECTIVES

☐ Describe the components of a software development system.

☐ List various types of floppy disks, and explain how information is stored on the disk.

☐ Define the operating system of a microcomputer, and explain its function.

☐ Explain the functions of programs such as Editor, Assembler, Loader (or Linker), and Debugger.

☐ List the advantages of an assembler over manual assembly.

☐ List assembler directives, and explain their functions.

☐ Write assembly language programs with appropriate directives.

11.1   MICROPROCESSOR-BASED SOFTWARE DEVELOPMENT SYSTEMS

A **software development system** is simply a computer that enables the user to write, modify, debug, and test programs. In a microprocessor-based development system, a microcomputer is used to develop software for a particular microprocessor. Generally, the microcomputer has a large R/W memory (typically, 8M to 64M), disk storage, and a video terminal with an ASCII keyboard. The system includes programs that enable the user to develop software in either assembly language or high-level languages. This text will focus on developing programs in the **8085 assembly language.**

Conceptually, this type of microcomputer is similar to a single-board microcomputer except that it has features that can assist in developing large programs. Programs are accessed and stored under a file name (title), and they are written by using other programs such as text editors and assemblers. The system (I/Os, files, programs, etc.) is managed by a program called the **operating system.** The hardware and software features of a typical software development system are described in the next sections.

11.11   System Hardware and Storage Memory

Figure 11.1 shows a typical software development system; it includes an ASCII keyboard, a CRT terminal, an MPU board with 8M to 64M R/W memory and disk controllers, and disk drives. The disk controller is an interfacing circuit through which the MPU can access a disk and provide Read/Write control signals. The disk drives have Read/Write ele-

FIGURE 11.1
Typical software development system
SOURCE: Courtesy of International Business Machines

ments, which are responsible for reading and writing data on the disk. At present, most systems are generally equipped with a 3½-inch disk, a hard disk, and CD-ROM (described later). A high-density 3½-inch disk stores 1.44M bytes of information. The storage capacity of a typical hard disk in a PC (personal computer) is 2.2 G (giga) bytes or higher.

FLOPPY DISK

Figure 11.2(a) shows a typical 3½-inch disk housed in a hard plastic coating. It is made of a thin magnetic material (iron oxide) that can store logic 0s and 1s in the form of magnetic directions. The surface of the disk is divided into a number of concentric tracks, and each track is divided into sectors, as shown in Figure 11.2(b). Data are stored on concentric circular tracks on both sides (known as doubled-sided). The large hole in the center (on the back side, not shown in the figure) is locked by the disk drive when it spins the disk at a constant speed (approximately 300 rpm). The oblong segment, shown as media, is the read/write (R/W) segment; this is the only segment of the surface that comes in contact with the R/W head of the controller. A piece of metal called the shutter normally covers this recording area. When the disk is inserted into a disk drive, the shutter slides over, and the recording surface of the disk is exposed to the R/W head. At the edge of the disk, there is a notch called the *write protect notch*. If the disk notch is open, data cannot be written on the disk; the disk is "write protected."

Each sector and track (Figure 11.2b) is assigned a binary address using a program (FORMAT); this is called *formatting* a disk. The MPU can access any information on the disk with the sector and the track addresses; however, the access is semirandom. To go from one track to another track, the access is random. Once the track is found, the sector is located serially by counting the sectors. Once data bytes are located, they are transferred to the system's R/W memory. These various functions in data transfer between a floppy disk and the system are performed by the disk controller and controlled by the operating system, which is also known as the Disk Operating System (DOS) and is described in Section 11.2.

HARD DISK

Another type of storage memory used with computers is called a hard disk. The hard disk is similar to the floppy disk except that the magnetic material is coated on a rigid aluminum base and enclosed in a sealed container. In general, the disk is permanently fastened in a dust-free drive mechanism. While it is highly precise and reliable, the hard disk requires sophisticated controller circuitry. The hard disk is more stable mechanically than the floppy disk; therefore, it can be spun at a higher rate (1000 to 3600 rpm) resulting in a faster (by almost ten times) readout rate than that of the floppy disk. Hard disks are available in various sizes and their storage capacity is quite large—in the order of gigabytes.

CD-ROM (COMPACT DISK READ-ONLY MEMORY)

A CD-ROM is an optical disk that uses a laser beam to store digital information that can be read with a laser diode. The disk is immune to dust and mechanical wear because of

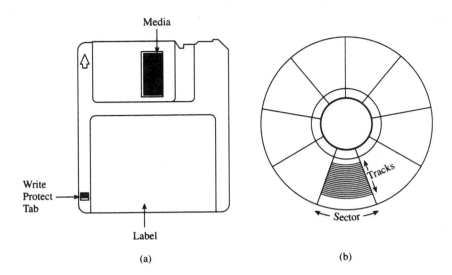

FIGURE 11.2

A Typical 3½-Inch Floppy Disk (a), Its Sectors and Tracks (b)

its optical nature. This is the same technology used in audio CD recording. The CD-ROM comes in various sizes (3½-inch to 14-inch) and stores a huge amount of data, from hundreds of megabytes to several gigabytes. Because of its huge storage capacity, the disk can store text, graphics, and audio information, and because it is removable, information can be very easily distributed widely as audio CDs. CD-ROM technology enables the computer to combine and display various types of information using its video screen and attached speakers; this has come to be known as multimedia.

OPERATING SYSTEMS AND PROGRAMMING TOOLS 11.2

The **operating system** of a computer is a group of programs that manages or oversees all the operations of the computer. The computer transfers information constantly among peripherals such as a floppy disk, printer, keyboard, and video monitor. It also stores user programs under file names on a disk. (A **file** is defined as related records stored as a single entity.) The operating system is responsible primarily for managing the files on the disk and the communication between the computer and its peripherals. The functional relationship between the operating system and the computer's various subsystems is shown in Chapter 1, Figure 1.5. From the user's point of view, the operating system is an access or a pathway to run application programs; it is a user interface with the computer.

11.21   Operating Systems

Each computer has its own operating system. We focus briefly here on four operating systems, known respectively as MS-DOS (Microsoft Disk Operating System), OS/2 (Operating System 2), Windows 95 (98), and UNIX. In the 1970s when most microcomputers were designed using 8-bit microprocessors, the CP/M (Control Program/Monitor) operating system was in common use. The early 1980s began the era of PCs (personal computers) based on 16-bit microprocessors (such as the Intel 8088/8086) with the addressing capability of 1M-byte memory. The CP/M was replaced by MS-DOS in PC-compatible microcomputers. MS-DOS and CP/M are in many ways similar, except that MS-DOS is designed to handle 16-bit microprocessors and 640K memory, and the CP/M was designed for 8-bit microprocessors and 64K memory. In the 1990s, 32-bit processors with gigabytes of memory addressing capacity are being used widely in microcomputers, and therefore, MS-DOS is being replaced by newer operating systems such as Microsoft Windows 95 (Windows 98), IBM OS/2 (Operating System/2), and UNIX (and UNIX compatibles).

MS-DOS OPERATING SYSTEM

In 16-bit microcomputers, such as IBM PC, XT, and AT, MS-DOS was so widely used that it became the industry standard. Initially, when it was installed on the IBM PC, it was known as PC-DOS; the terms MS-DOS and PC-DOS are interchangeable. MS-DOS is designed to handle 16-bit data word and 640K system memory and disks with quad (high)-density disk format with memory capacity of 720K and 1,200K. Similarly, it can support a hard disk and includes a hierarchical file directory.

OS/2 (OPERATING SYSTEM 2)

OS/2 (and its various versions) is a 32-bit single-user operating system designed by IBM to exploit various powerful features of recent 32-bit (and 64-bit) microprocessors. It has removed the memory limitation of DOS (640K bytes); it can assign 512M bytes of virtual memory space (using storage space from a disk drive) to each application program. Furthermore, it is compatible with DOS and Windows application programs. A few of the important features of OS/2 are as follows. It supports:

☐ multitasking
☐ telecommunications
☐ multimedia

It is a multitasking operating system, meaning the user can run multiple applications concurrently. For example, if the user is using a word processor, OS/2 can send or receive a fax, or send a message (e-mail) electronically in the background. In addition to 32-bit computing capability, it is designed to support various telecommunications features such as electronic mail, fax, and telephone voice mail. Furthermore, it is also well suited for the multimedia applications of CD-ROM.

WINDOWS 95 (98)

Windows 95 is a 32-bit single-user operating system designed by Microsoft, and it is by far the most widely used operating system in personal computers. It has a graphical interface and it supports the use of a mouse, icons, and menus. It is also a multitasking operating system, similar to OS/2. It includes many of the features of OS/2 described above. Windows 98 is an upgraded version of Windows 95.

UNIX OPERATING SYSTEM

This is a multiuser, multitasking operating system. Initially, it was designed for minicomputers, but is now used on various machines ranging from powerful microcomputers to supercomputers. It is independent of any particular hardware structure. It is widely used in engineering, scientific, and research environments and is not limited by any memory constraints. It is well suited for networking and graphical environments.

Solaris is a Unix-based operating system designed by Sun Microsystems and is widely used in high-end microcomputers such as workstations and network servers. In the network environment, Solaris is capable of handling 64 computers and is being upgraded to handle 128 computers. One of the major problems with Unix is that there are too many versions in the market; however, several companies have begun to accept Solaris as a standard operating system for high-end microcomputers.

11.22 Tools for Developing Assembly Language Programs

In addition to the operating system of a computer, various programs called utility programs are necessary to develop assembly language programs. These programs can be classified in two categories: (1) file-management utilities and (2) program-development utilities. The file-management utilities are programs that enable the user to perform such functions as copying, printing, erasing, and renaming files. In MS-DOS, some programs such as COPY (Copy), DEL (Delete), and DIR (Directory) are part of the internal commands, meaning these programs are loaded into system memory along with the operating system. Other programs, such as PRINT (Print), FORMAT (Formatting Disk), and MODE (to set up printer options) are part of the external commands, meaning these programs are stored on the disk under file names and copied into system memory whenever they are needed. The program-development utilities enable the user to write, assemble, and test assembly language programs; they include programs such as Editor, Assembler, Linker (or Loader), and Debugger. The descriptions of various programs and the assembly process described below may vary in details depending upon an operating system and its utility programs.

EDITOR

The Editor is a program that allows the user to enter, modify, and store a group of instructions or text under a file name. The Editor programs can be classified in two groups: line editors and full-screen editors. Line editors, such as EDIT in MS-DOS, work with and manage one line at a time. Full-screen editors (also known as word processors), such as MSWord and WordPerfect, manage the full screen or a paragraph at a time. To write text, the user must call the Editor under control of the operating system. As soon as the Editor program is transferred from the disk to the system memory, the program control is transferred from the operating system to the Editor program. The Editor has its own commands, and the user can enter and modify text by using those commands. Some Editor programs, such as WordPerfect, are very easy to use. At the completion of writing a program, the exit command of the Editor program will save the program on the disk under the file name and will transfer the program control to the operating system. This file is known as a source file or a source program. If the source file is intended to be a program in the 8085 assembly language, the user should follow the syntax of the assembly language and the rules of the assembler that are described next.

The Editor program is not concerned with whether one is writing a letter or an assembly language program. The full-screen editors are convenient for writing either line- or paragraph-oriented text; they automatically adjust lines as words are typed, and the text can be modified or erased with ease.

ASSEMBLER

The Assembler is a program that translates source code or mnemonics into the binary code, called object code, of the microprocessor and generates a file called the Object file. This function is similar to manual assembly, whereby the user looks up the code for each mnemonic in the listing. In addition to translating mnemonics, the Assembler performs various functions, such as error checking and memory allocations. The Assembler is described in more detail in Section 11.3.

LOADER

The Loader (or Linker) is a program that takes the object file generated by the Assembler program and generates a file in binary code called the COM file or the EXE file. The COM (or EXE) file is the only executable file—i.e., the only file that can be executed by the microcomputer. To execute the program, the COM file is called under the control of the operating system and executed. In different assemblers, the COM file may be labeled by other names.

DEBUGGER

The Debugger is a program that allows the user to test and debug the object file. The user can employ this program to perform the following functions:

☐ Make changes in the object code.
☐ Examine and modify the contents of memory.
☐ Set breakpoints, execute a segment of the program, and display register contents after the execution.
☐ Trace the execution of the specified segment of the program, and display the register and memory contents after the execution of each instruction.
☐ Disassemble a section of the program; i.e., convert the object code into the source code or mnemonics.

11.23   MS-DOS and Cross-Assemblers

MS-DOS is an operating system designed primarily for 16-bit microprocessors. Now the question is: Why are we discussing it in the context of an 8-bit microprocessor, such as the 8085? The answer lies with the widespread use of IBM PCs or their compatibles on college campuses. We can use PCs to develop (assemble) 8085 assembly language programs by using a program called Cross-Assembler. The computers that operate under MS-DOS, such as IBM PCs, are designed around the 16-bit microprocessor Intel 8088 or compatibles. The mnemonics of the 8088 microprocessor are different from those of the 8085 microprocessor; thus, we need a program that can translate the 8085 mnemonics, but operate under the 8088 microprocessor. Such a program is called a **cross-assembler.** For example, the 8085 cross-assembler from 2500 AD Software Inc. has two programs: one is an assembler named X8085 and the other is a linker with the file name LINK. After assembling a program, the Hex file (described later) can be directly transferred to R/W memory of your 8085 single-board microcomputer by using a download program. Thus, programs and/or hardware-related laboratory experiments can be easily performed.

Writing and assembling a program using a cross-assembler such as X8085 on the PC is described as follows.

Step 1:   **Call an Editor** program (such as EDIT in MS-DOS or WordPerfect) and write an assembly language program in 8085 mnemonics. This is called a **Source** file. The format of the program is similar to the handwritten programs written in earlier chapters except that it should include **assembler directives** (described in Section 11.3). The name of this file should be limited to eight characters and the extension to three characters. Some assemblers require that the extension be ASM. As an example, the Source file can be named as LAB1.ASM. **Save** this file as LAB1.ASM and exit from the Editor program. If you use a word processor to write this program, the file must be saved in **ASCII format** (check various options of saving a file in your word processor).

Step 2:   **Call a cross-assembler** such as X8085 and assemble the Source file LAB1.ASM. The X8085 generates an intermediate binary file called an *object file,* such as LAB1.OBJ, and provides a list of errors. **Go back** and repeat Step 1 to correct errors. **Repeat Step 2** and reassemble the program. Repeat Steps 1 and 2 until the cross-assembler gives a message of zero errors. The cross-assembler also generates a List (LAB1.LST) file that includes memory addresses, machine codes in Hex, labels, and comments. This file is used primarily for documentation.

Step 3:   **Call a Link** program and use the intermediate file LAB1.OBJ to generate either an executable file such as LAB1.COM or Hex file LAB1.HEX. However, the executable COM file in 8085 machine code is rather meaningless—the file cannot be executed on the PC because the machine codes of the PC are different from that of the 8085 processor. Assuming you have a single-board microcomputer and a program (called Download) that has the capability of transferring the machine code from PC into R/W memory of the microcomputer, you can use the Hex file and copy the machine codes in proper memory locations of the single-board microcomputer. This is similar to entering or storing a program in the single-board memory by hand.

Step 4: Execute the program on the single-board machine. If you are unable to observe the expected output on a display of the single-board, you must conclude that the program has logic errors (not assembly syntax errors). You can debug the program by using techniques such as single-step and/or break point described in Chapter 7 (Section 7.6). Once you find an error, you must **go back to Step 1 and repeat Steps 1 through 4.** (If errors are simple and obvious, and no additional memory locations are needed, the machine code can be corrected in the single-board microcomputer.)

After the end of the assembly process, you will have the following files on your PC disk.

☐ **ASM file** This is the Source file written by the user using an Editor. The filename can be one to eight characters long with an extension of a maximum of three characters. The filename and the extension are separated by a dot. As described before, the filename can be LAB1.ASM; the extension ASM suggests that this is an assembly language file.
☐ **OBJ file** This is the intermediate binary file generated by the cross-assembler.
☐ **LST file** This is the list file generated by the assembler program for documentation purposes. It contains memory locations, Hex code, mnemonics, and comments.
☐ **HEX file** This is generated by the Link program and contains program code in hexadecimal notations. This file can be used for debugging the program and to transfer files from one system to another.
☐ **COM file** This is the executable file generated by the Link program, and it contains binary code. However, this file cannot be executed on the PC. This type of a file can be executed if you were to write an assembly language program for the microprocessor in the PC. However, in such a situation, you would not use a cross-assembler; you would use the assembler for the PC microprocessor.

A summary of steps in writing, assembling, and executing a program using a cross-assembler is shown in the flowchart (Figure 11.3).

ASSEMBLERS AND CROSS-ASSEMBLERS 11.3

The assembler,* as previously described, is a program that translates assembly language mnemonics or source code into binary executable code. This translation requires that the source program be written strictly according to the specified syntax of the assembler. The assembly language source program includes three types of statements:

1. The program statements in 8085 mnemonics that are to be translated into binary code.
2. Comments that are reproduced as part of the program documentation.
3. Directives to the assembler that specify items such as starting memory locations, label definitions, and required memory spaces for data.

---

*The following description is equally applicable to cross-assemblers.

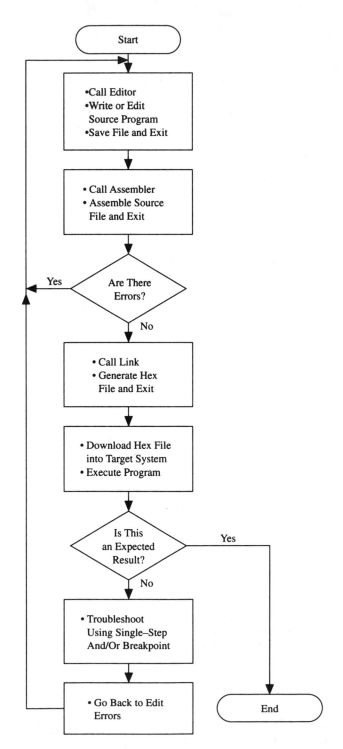

FIGURE 11.3
Flowchart: Program Assembly and Execution

The first two types of statements have been used throughout Part II of this book; the format of these statements as they appear in an assembly language source program is identical to the format used here. The third type—directive statements—and their functions will be described in the next sections.

11.31 Assembly Language Format

A typical assembly language programming statement is divided into four parts called **fields:** *label, operation code* (opcode), *operand,* and *comments.* These fields are separated by *delimiters* as shown in Table 11.1.

The assembler statements have a free-field format, which means that any number of blanks can be left between the fields. As mentioned before, comments are optional but are generally included for good documentation. Similarly, a label for an instruction is also optional, but its use greatly facilitates specifying jump locations. As an example, a typical assembly language statement is written as follows:

Label	Opcode	Operand	Comments
START:	LXI	SP,20FFH	;Initialize stack pointer

Delimiters include the colon following START, the space following LXI, the comma following SP, and the semicolon preceding the comment.

TABLE 11.1
Typical Delimiters Used in
Assembler Statements

Delimiter	Placement
1. Colon	After label (optional)
2. Space	Between an opcode and an operand
3. Comma	Between two operands
4. Semicolon	Before the beginning of a comment

11.32 Assembler and Cross-Assembler Directives

The **assembler directives** are the instructions to the assembler concerning the program being assembled; they also are called *pseudo instructions* or *pseudo opcodes.* These instructions are neither translated into machine code nor assigned any memory locations in the object file. Some of the important assembler directives for an assembler or a cross-assembler are listed and described here.

Assembler Directives	Example	Description
1. ORG (Origin)	ORG 2000H	The next block of instructions should be stored in memory locations starting at 2000H.
2. END	END	End of assembly. The HLT instruction suggests the end of a program, but that does not necessarily mean it is the end of the assembly.

3. EQU (Equate)	PORT1 EQU 01H	The value of the term PORT1 is equal to 01H. Generally, this means the PORT1 has the port address 01H.
	INBUF EQU 2099H	The value of the term INBUF is 2099H. This may be the memory location used as Input Buffer.
	STACK EQU INBUF+1	The equate can be expressed by using the label of another equate. This example defines the stack as next location of INBUF.
4. DB (Define Byte)	DATA: DB A2H,9FH	Initializes an area byte by byte. Assembled bytes of data are stored in successive memory locations until all values are stored. This is a convenient way of writing a data string. The label is optional.
5. DW (Define Word)	DW 2050H	Initializes an area two bytes at a time.
6. DS (Define Storage)	OUTBUF: DS 4	Reserves a specified number of memory locations. In this example, four memory locations are reserved for OUTBUF.

11.33   Cross-Assembler Format

As mentioned earlier, a cross-assembler is a program that can be used to translate 8085 mnemonics by a computer that has a microprocessor other than the 8085. The format of a cross-assembler is almost identical to that of an assembler with a few variations. Programmers who write these programs generally tend to follow a similar format. Occasionally, personal preferences of these programmers or some added features may cause some variations. For example, the cross-assembler from 2500AD Software Inc. has pseudo opcodes similar to those in the previous section. The directives such as LWORD (Long Word: 32-bit) or BLKW (Block Word for storage) can be cited as additional features. From the user point of view, however, the process of assembling code is almost identical. An example of how to write a program by using a cross-assembler is illustrated in Section 11.4.

11.34   Advantages of the Assembler/Cross-Assembler

The **assembler** is a tool for developing programs with the assistance of the computer. Assemblers are absolutely essential for writing industry-standard software; manual assembly is too difficult for programs with more than 50 instructions. The assembler performs many functions in addition to translating mnemonics, and it has several advantages over manual assembly. The salient features of the assembler are as follows:

1. The assembler translates mnemonics into binary code with speed and accuracy, thus eliminating human errors in looking up the codes.
2. The assembler assigns appropriate values to the symbols used in a program. This facilitates specifying jump locations.
3. It is easy to insert or delete instructions in a program; the assembler can reassemble the entire program quickly with new memory locations and modified addresses for jump locations. This avoids rewriting the program manually.
4. The assembler checks syntax errors, such as wrong labels and expressions, and provides error messages. However, it cannot check logic errors in a program.
5. The assembler can reserve memory locations for data or results.
6. The assembler can provide files for documentation.
7. A Debugger program can be used in conjunction with the assembler to test and debug an assembly language program.

The above comments, except number 7, are equally applicable to cross-assemblers.

WRITING PROGRAMS USING A CROSS-ASSEMBLER* 11.4

This section deals primarily with writing programs using a cross-assembler written for IBM PCs. The following examples are taken from previous chapters in which the programs were assembled manually. An assembler source program is identical to a program the user writes with paper and pencil, except that the assembler source program includes assembler directives.

11.41 Illustrative Program: Unconditional Jump to Set Up a Loop

The following program is taken from Chapter 6 (Section 6.42). Its source program is rewritten here for the assembler.

SOURCE PROGRAM
;This program monitors the switch positions of the input port and turns on/off devices
;   connected to the output port.

PORT0	EQU 00H	;Input port address
PORT1	EQU 01H	;Output port address
	ORG 2000H	;Start assembling the program from location 2000H
START:	IN PORT0	;Read input switches
	OUT PORT1	;Turn on devices
	JMP START	;Go back and read switches again
	END	

*The terms *assembler* and *cross-assembler* are used synonymously.

This program illustrates the following assembler directives:

☐ ORG The object code will be stored starting at the location 2000H.
☐ EQU The program defines two equates, PORT0 and PORT1. In this program it
would have been easier to write port addresses directly with the instructions.
The equates are essential in development projects in which hardware and soft-
ware design are done concurrently. In such a situation, equates are convenient.
Equates are also useful in long programs because it is easy to change or define
port addresses by defining equates.
☐ Label The program illustrates one label: START
☐ END The end of assembly.

TWO-PASS ASSEMBLER

To assemble the program, the assembler scans through the program twice; this is known
as a **two-pass assembler.** In the first pass, the first memory location is determined from
the ORG statement, and the counter known as the location counter is initialized. Then the
assembler scans each instruction and records locations in the address column of the first
byte of each instruction; the location counter keeps track of the bytes in the program. The
assembler also generates a symbol table during the first pass. When it comes across a label, it records the label and its location. In the second pass, each instruction is examined,
and mnemonics and labels are replaced by their machine codes in Hex notation as shown
below.

Pass 1

Address Hex	Machine Code Hex	Label Opcode	Operand	Symbol Table
2000		START: IN	PORT0	PORT0 00H
2002		OUT	PORT1	PORT1 01H
2004		JMP	START	START 2000H

Pass 2

2000	DB00	
2002	D301	
2004	C30020	

ASSEMBLED LIST FILE

The file lists the memory addresses of the first byte of each instruction.

;This program monitors the switch positions of the input port and turns on/off devices
;   connected to the output port

```
0000 =           PORT0 EQU 00H          ;Input port address
0001 =           PORT1 EQU 01H          ;Output port address
2000                   ORG 2000H        ;Start assembling program from location
                                        ;   2000H
2000 DB00        START: IN PORT0        ;Read input switches
2002 D301              OUT PORT1        ;Turn on devices
2004 C30020            JMP START        ;Go back and read switches again
2007                   END
```

11.42   Illustrative Program: Addition with Carry

The following program is from Chapter 7 (Section 7.32). The program adds six bytes of data stored in memory locations starting at 2050H and stores the sum in the Output Buffer memory in two consecutive memory locations.

Data(H)   A2,FA,DF,E5,98,8B.

SOURCE PROGRAM

```
;Addition with Carry
INBUF        EQU 2050H              ;Input Buffer location
COUNTR       EQU 06H                ;Number of bytes to add
             ORG 2000H
             XRA A                  ;Clear accumulator
             MOV B,A                ;Set up B for carry
             MVI C,COUNTR           ;Set up C to count bytes
             LXI H,INBUF            ;Point to data address
NXTBYT:      ADD M
             JNC NXTMEM
             INR B                  ;If there is carry, add 1
NXTMEM:      INX H                  ;Point to next data byte
             DCR C                  ;One is added, decrement counter
             JNZ NXTBYT             ;Get next byte if all bytes not yet added
             LXI H,OUTBUF           ;Point index to output buffer
             MOV M,A                ;Store low-order byte of the sum
             INX H
             MOV M,B                ;Store high-order byte of the sum
             HLT
OUTBUF:      DS 2                   ;Reserve two memory locations
             ORG 2050H              ;Assemble next instructions starting at 2050H
DATA:        DB 0A2H,0FAH,0DFH,0E5H,98H
             DB 8BH
             END
```

This program illustrates two more assembler directives—DB (Define Byte) and DS (Define Storage)—and the use of the ORG statement to store data starting at location 2050H. A data byte or an address that begins with Hex digits (A through F) should be preceded by 0; otherwise, the assembler cannot interpret it as a Hex number (see the data string above). The assembled program is shown below.

PRINT FILE
```
;Addition with Carry
2050 =                  INBUF      EQU 2050H         ;Input Buffer location
0006 =                  COUNTR     EQU 06H           ;Number of bytes to add
2000                               ORG 2000H
2000 AF                            XRA A             ;Clear accumulator
2001 47                            MOV B,A           ;Set up B for carry
2002 0E06                          MVI C,COUNTR      ;Set up C to count bytes
2004 215020                        LXI H,INBUF       ;Point to data address
2007 86                 NXTBYT     ADD M
2008 D20C20                        JNC NXTMEM
200B 04                            INR B             ;If there is carry, add 1
200C 23                 NXTMEM:    INX H             ;Point to next data byte
200D 0D                            DCR C             ;Decrement counter
200E C20720                        JNZ NXTBYT        ;Get next byte if all bytes
                                                     ;   not yet added
2011 211820                        LXI H,OUTBUF      ;Point index to Output
2014 77                            MOV M,A           ;Store low-order byte
2015 23                            INX H
2016 70                            MOV M,B           ;Store high-order byte
2017 76                            HLT
2018                    OUTBUF:    DS 2              ;Reserve two memory
                                                     ;   locations
2050                               ORG 2050H         ;Assemble next instructions
                                                     ;   starting at 2050H
2050 A2FADFE598         DATA:      DB 0A2H,0FAH,0DFH,0E5H,98H
2055 8B                            DB 8BH
2056                               END
```

The list file shows the memory addresses of the first byte of each instruction; this is a typical printout of an assembly language program. The file also shows that two memory locations (2018H and 2019H) are reserved for OUTBUF, and six locations are used to store data starting from 2050H.

ERROR MESSAGES

In addition to translating the mnemonics into object code, the assembler also gives error messages. The two types of error messages are terminal error messages and source program error messages. In the first case, the assembler is not able to complete the assembly. In the second case, the assembler is able to complete the assembly, but it lists the errors.

11.43 Illustration of a List File Using a Cross-Assembler

The following program, which adds two Hex bytes, was assembled on an IBM PC-compatible microcomputer by using the cross-assembler from 2500AD Software Inc. The source program shown below was written using an Editor.

SOURCE PROGRAM

;This program adds two Hex bytes and stores the sum in memory

```
            ORG 2000H           ;Begin assembly at 2000H
OUTBUF      EQU 2050H           ;Address to store sum
START:      MVI B,32H           ;Load first byte
            MVI C,0A2H          ;Load second byte
            MOV A,C
            ADD B               ;Add two bytes
            STA OUTBUF          ;Store sum in buffer
            HLT                 ;End of program
            END                 ;End of assembly
```

The source program was assembled by calling the X8085, the file name of the cross-assembler. This cross-assembler generates two files: object (.OBJ) and list (.LST) files. The object file is used as an input to the program called LINK1 to generate the executable file (.TSE). The object file can be assembled at any starting memory location. The LINK1 program can relocate a file at a desired location or combine several files and assign appropriate memory locations. However, the 8085 processor executable file, also known as a COM file, cannot be executed on the PC. The LINK2 program can generate a Hex file that can be used to download the machine code to a single-board computer. The list file and the messages generated by the cross-assembler are shown below.

List File:

2500AD 8085 CROSS-ASSEMBLER-VERSION 3.01a

INPUT FILENAME: ADDHEX.ASM
OUTPUT FILENAME: ADDHEX.OBJ


```
1              ;This program adds two Hex bytes
2              ;   and stores the sum in memory
3
4    2000                          ORG 2000H       ;Begin assembly at 2000H
5
6          50 20      OUTBUF:      EQU 2050H       ;Address to store sum
```

```
  7
  8    2000    06 32       START:    MVI B,32H       ;Load first byte
  9
 10    2002    0E A2                 MVI C,0A2H      ;Load second byte
 11
 12    2004    79                    Mov A,C
 13
 14    2005    80                    ADD B           ;Add two bytes
 15
 16    2006    32 50 20              STA OUTBUF      ;Store sum in buffer
 17
 18    2009    76                    HLT             ;End of program
 19
 20    200A                          END             ;End of assembly
```

******************** SYMBOLIC REFERENCE TABLE ********************

OUTBUF   =2050   START   2000

LINES ASSEMBLED: 20   ASSEMBLY ERRORS: 0

Precautions in Writing Programs Assembler and cross-assembler programs are available from various software companies; for the most part, they follow similar formats. However, we suggest the following precautions in writing assembly language programs.

1. Some assemblers do not allow free format, meaning the unnecessary spaces are not tolerated.
2. The letter following a number specifies the type of a number. A hexadecimal number is followed by the letter H, an octal by the letter O or Q, a binary by the letter B. A number without a letter is interpreted as a decimal number.
3. Any Hex number that begins with A through F must be preceded by zero; otherwise, the assembler interprets the number as a label and gives an error message because it cannot find the label.
4. Some assemblers must have a colon after a label.
5. In some assemblers, Equate statements must begin in column 1.

SUMMARY

A software development system and an assembler are essential tools for writing large assembly language programs. These tools facilitate the writing, assembling, testing, and debugging of assembly language programs.

A disk-based microcomputer, its operating system, and assembler programs can serve as a development system. All the operations of the computer are managed and directed by the operating system of the computer. The Assembler and other utility programs assist the user in developing software. The Editor allows the user to enter text; the Assembler translates mnemonics into machine code and provides error messages. The Debugger assists in debugging the program. The 8085 code can be assembled by using a program called a cross-assembler on a system that has a processor other than the 8085.

The program assembled using the Assembler or a cross-assembler is in many ways similar to that of the hand assembly program except that the program written for the Assembler includes assembler directives concerning how to assemble the program. The Assembler has many advantages over manual assembly; without the Assembler, it would be extremely difficult to develop industry-standard software.

QUESTIONS AND PROGRAMMING ASSIGNMENTS

1. Check the appropriate answer in the following statements:
   a. The process of accessing information on a floppy disk is
      (1) random.
      (2) serial.
      (3) semirandom.
   b. The operating system of a computer is defined as
      (1) hardware that operates the floppy disk.
      (2) a program that manages files on the disk.
      (3) a group of programs that manages and directs hardware and software in the system.
   c. The Editor is
      (1) an assembly language program that reads and writes information on the disk.
      (2) a high-level language program that allows the user to write programs.
      (3) a program that allows the user to write, modify, and store text in the computer system.
   d. The Assembler is
      (1) a compiler that translates statements from high-level language into assembly language.
      (2) a program that translates mnemonics into binary code.
      (3) an operating system that manages all the programs in the system.
   e. A file is
      (1) a group of related records stored as a single entity.
      (2) a program that transfers information between the system and the floppy disk.
      (3) a program that stores data.

    f. A disk controller is
       (1) a program that manages the files on the disk.
       (2) a circuit that interfaces the disk with the microcomputer system.
       (3) a mechanism that controls the spinning of the disk.

2. Write a print file for the following program with the starting location 0100H, and list the errors in the source file. Assume that the subroutine BCDBIN is written separately.

```
                ORG 0100
                LXI SP,STACK        ;Initialize stack pointer
                LXI H,INBUF         ;Output buffer
                LXI B,OUTBUF        ;Output buffer
                MVI D,0AH           ;Initialize counter
NEXT:           MOV A,M             ;Get byte
                CALL BCDBIN         ;Call BCD to binary routine
                STAX B              ;Store result
                DCR D
                JNZ NEXT
INBUF:          DW
OUTBUF:         DW
                HLT
```

3. Rewrite the illustrative program in Section 10.31 for a BCD to common-cathode-LED code conversion to assemble it with an assembler or a cross-assembler.

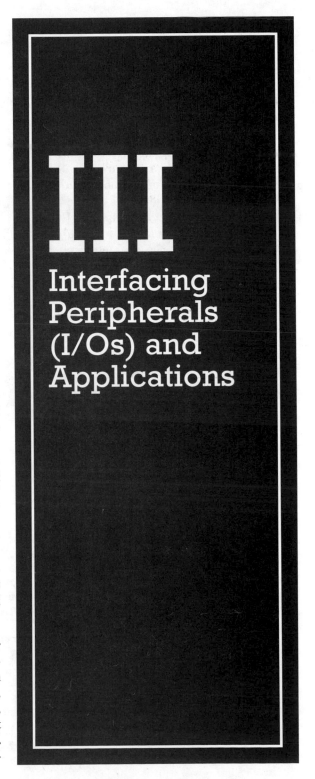

III
Interfacing Peripherals (I/Os) and Applications

Part III of this book is concerned with the interfacing of peripherals (I/Os) and design processes of microcomputer-based systems. The primary objectives of Part III are

1. To examine the concepts and processes of data transfer, such as interrupts, Direct Memory Access (DMA), and serial I/O, between the microprocessor and peripherals.
2. To apply the concepts of data transfer for interfacing peripherals with the microprocessor.
3. To synthesize the concepts of microprocessor architecture, software, and interfacing by designing a simple microprocessor-based system and by discussing the process of troubleshooting.

The primary function of the **microprocessor** is to accept data from input devices such as keyboards and A/D converters, read instructions from memory, process data according to the instructions, and send the results to output devices such as LEDs, printers, and video monitors. These input and output devices are called either **peripherals** or **I/Os.** Designing logic circuits (hardware) and writing instructions (software) to enable the microprocessor to communicate with these peripherals is called **interfacing,** and the logic circuits are called **I/O ports** or **interfacing devices.**

The microprocessor (or more precisely, MPU) communicates with the peripherals in either of two formats: **asynchronous** or **synchronous.** Similarly, it transfers data in either of two modes: **parallel I/O** or **serial I/O.** The 8085 identifies peripherals either as **memory-mapped I/O** or **peripheral I/O** on the basis of their interfacing logic circuits (Chapter 4). Data transfer between the microprocessor and its peripherals can take place under various conditions, as shown in the chart. The modes, the techniques, the instructions, and the conditions of data transfer are briefly described in the following paragraphs and summarized in the chart.

FORMATS OF DATA TRANSFER: SYNCHRONOUS AND ASYNCHRONOUS

Synchronous means at the same time; the transmitter and receiver are synchronized with the same clock. *Asynchronous* means at irregular intervals. The synchronous format is used in high-speed data transmission and the asynchronous format is used for low-speed data transmission. Data transfer between the microprocessor and the peripherals is primarily asynchronous.

MODES OF DATA TRANSFER: PARALLEL AND SERIAL

The microprocessor receives (or transmits) binary data in either of two modes: parallel or serial. In the parallel mode, the entire word (4-bit, 8-bit, or 16-bit) is transferred at one time. In the 8085, an 8-bit word is transferred simultaneously over the eight data lines. The devices commonly used for parallel data transfer are keyboards, seven-segment LEDs, data converters, and memory.

In the serial mode, data are transferred one bit at a time over a single line between the microprocessor and a peripheral. For data transmission from the microprocessor to a peripheral, a word is converted into a stream of eight bits; this is called

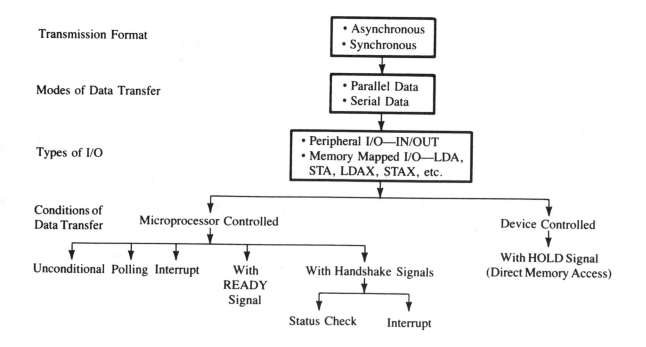

parallel-to-serial conversion. For reception, a stream of eight bits is converted into a parallel word; this is called serial-to-parallel conversion. The serial I/O mode is commonly used with peripherals such as CRT terminals, printers (also used in parallel I/O mode), cassette tapes, and modems for telephone lines.

TYPES OF I/O: PERIPHERAL AND MEMORY-MAPPED

In peripheral I/O, a peripheral is identified with an 8-bit address. The 8085 has two instructions—IN and OUT—to implement data transfer between the microprocessor and peripherals. These are 2-byte instructions; the second byte specifies the address or the port number of a peripheral. The instruction IN transfers (copies) data from an input device to the accumulator, and the instruction OUT transfers data from the accumulator to an output device.

In memory-mapped I/O, a peripheral is connected as if it were a memory location, and it is identified with a 16-bit address. Data transfer is implemented by using memory-related instructions such as STA; LDA; MOV M,R; and MOV R,M.

CONDITIONS OF DATA TRANSFER

The process of data transfer between the microprocessor and the peripherals is controlled either by the microprocessor or by the peripherals, as shown in the chart. Data transfer is generally implemented under the microprocessor control when the peripheral response is slow relative to that of the microprocessor.

MICROPROCESSOR-CONTROLLED DATA TRANSFER

Most peripherals respond slowly in comparison with the speed of the microprocessor. Therefore, it is necessary to set up conditions for data transfer so that data will not be lost during the transfer. Microprocessor-controlled data transfer can take place under five different conditions: unconditional, polling (also known as status check), interrupt, with READY signal, and with handshake signals. These conditions are described briefly.

Unconditional Data Transfer In this form of data transfer, the microprocessor assumes that a peripheral is always available. For example, to display data at an LED port, the microprocessor simply enables the port, transfers data, and goes on to execute the next instruction.

Data Transfer with Polling (Status Check) In this form of data transfer, the microprocessor is kept in a loop to check whether data are available; this is called polling. For example, to read data from an input keyboard in a single-board microcomputer, the microprocessor can keep polling the port until a key is pressed.

Data Transfer with Interrupt In this condition, when a peripheral is ready to transfer data, it sends an interrupt signal to the microprocessor. The microprocessor stops the execution of the program, accepts the data from the peripheral, and then returns to the program. In the interrupt technique, the processor is free to perform other tasks rather than being held in a polling loop.

Data Transfer with READY Signal When peripheral response time is slower than the execution time of the microprocessor, the READY signal can be used to add T-states, thus extending the execution time. This process provides sufficient time for the peripheral to complete the data transfer. The technique is commonly used in a system with slow memory chips.

Data Transfer with Handshake Signals In this data transfer, signals are exchanged between the microprocessor and a peripheral prior to actual data transfer; these signals are called handshake signals. The function of handshake signals is to ensure the readiness of the peripheral and to synchronize the timing of the data transfer. For example, when an A/D converter is used as an input device, the microprocessor needs to wait because of the slow conversion time of the converter. At the end of the conversion, the A/D converter sends the Data Ready (DR), also known as End of Conversion, signal to the microprocessor. Upon receiving the DR signal, the microprocessor reads the data and acknowledges by sending a signal to the converter that the data have been read. During the conversion period, the microprocessor keeps checking the DR signal; this technique is called the status check with handshake signals. This status check method is functionally similar to the polling method and achieves the same results.

Rather than using the handshake signals for the status check, the signals can be used to implement data transfer with interrupt. In the above example of the A/D converter, the DR signal can be used to interrupt the microprocessor.

Handshake signals prevent the microprocessor from reading the same data more than once, from a slow device, and from writing new data before the device has accepted the previous data.

PERIPHERAL-CONTROLLED DATA TRANSFER

The last category of data transfer shown in the chart is device-controlled I/O. This type of data transfer is employed when the peripheral is much faster than the microprocessor. For example, in the case of Direct Memory Access (DMA), the DMA controller sends a HOLD signal to the microprocessor, the microprocessor releases its data bus and the address bus to the DMA controller, and data are transferred at high speed without the intervention of the microprocessor.

CHAPTER TOPICS

In Chapter 4, we discussed the basic concepts of unconditional data transfer in parallel I/O. It included examples of interfacing simple devices, such as LEDs and switches. The remaining processes of data transfer shown in the chart are discussed in Chapters 12 through 16.

Chapter 12 deals primarily with interrupts. It includes the 8085 interrupts with several examples.

Chapter 13 is concerned with the interfacing of data converters. After reviewing the basic concepts underlying data converters, the chapter presents examples of interfacing data converters.

Chapter 14 deals with programmable interface devices commonly used in small microprocessor-based systems. These devices can be set up to perform I/O tasks by writing instructions in their control registers, thus the title programmable devices. The chapter includes several illustrations of interfacing using programmable devices from the Intel family, such as the 8155/8156 (Memory with I/O and Timer) and the 8279 (Programmable Keyboard Display Interface).

Chapter 15 is an extension of the topics examined in Chapter 14. It includes general-purpose programmable interface devices, such as the 8255A (Programmable Peripheral Interface), the 8254 (Timer), the 8259A (Interrupt Controller), and the 8237 (DMA Controller).

Chapter 16 deals with serial I/O. It includes discussion of the software approach and the hardware approach to serial I/O. The hardware approach is illustrated with the example of interfacing a terminal using the 8251A (Programmable Communication Interface).

Chapter 17 is concerned with the process of designing a microprocessor-based product. The primary objective of this chapter is to synthesize the concepts, using both hardware and software, discussed in all previous chapters. The chapter includes several interfacing projects and a design project.

Chapter 18 introduces concepts in 16-bit microprocessors using the Intel 8088/86 microprocessor family and suggests the trends in microprocessor technology. It also describes other 8-bit microprocessors and microcontrollers such as the Zilog Z80, Intel 8051, and Motorola 68HC11. The chapter concludes with descriptions of the latest 32- and 64-bit microprocessors.

PREREQUISITES

☐ Basic concepts of microprocessor architecture, memory, and I/Os (see Part I).
☐ Familiarity with the 8085 instruction set and programming techniques (see Part II).

Interrupts

The interrupt I/O is a process of data transfer whereby an external device or a peripheral can inform the processor that it is ready for communication and it requests attention. The process is initiated by an external device and is asynchronous, meaning that it can be initiated at any time without reference to the system clock. However, the response to an interrupt request is directed or controlled by the microprocessor.

The interrupt requests are classified in two categories: maskable interrupt and nonmaskable interrupt. The 8085 microprocessor includes four maskable interrupts and one nonmaskable interrupt. Among the four maskable interrupts, one is nonvectored, which requires external hardware to supply a Call location to restart the execution. The other three are vectored to specific locations (explained later in this chapter).

The microprocessor can ignore or delay a maskable interrupt request if it is performing some critical task; however, it has to respond to a nonmaskable request immediately. In many ways, the maskable interrupt is like a telephone, which can be kept off the hook if one is not interested in receiving any messages. The nonmaskable interrupt is like a smoke detector, which should be attended to if set off.

The interrupt process allows the microprocessor to respond to these external requests for attention or service on a demand basis and leaves the microprocessor free to perform other tasks. On the other hand, in the polled or the status check I/O, the microprocessor remains in a loop, doing nothing, until the device is ready for data transfer.

This chapter first describes the nonvectored interrupt process. It includes a discussion of how multiple interrupts are implemented with one interrupt line and how priorities are determined. The remaining three vectored interrupts and the nonmaskable interrupt are described later in the chapter. Two examples of the interrupt I/O are illustrated: a clock timer with the 60 Hz power line as the interrupting source, and a software breakpoint routine. The chapter also includes a brief explanation of the interrupt controller, the 8259, and the I/O process called Direct Memory Access (DMA).

OBJECTIVES

☐ Explain an interrupt process and the difference between a nonmaskable and a maskable interrupt.
☐ Explain the instructions EI, DI, and RST and their functions in the 8085 interrupt process.

☐ List the eight steps to initiate and implement the 8085 interrupt.
☐ Design and implement an interrupt with a given RST instruction.
☐ Explain how to connect multiple interrupts with the INTR interrupt line and how to determine their priorities using logic circuits.
☐ List the 8085 vectored interrupts, nonmaskable interrupt and their vectored memory locations.
☐ Explain the instructions SIM and RIM, and illustrate how to use them for the 8085 interrupts.
☐ Explain how to use an RST instruction to implement a software breakpoint.
☐ Explain features of the programmable interrupt controller, the 8259A, and the Direct Memory Access (DMA) data transfer.

12.1 THE 8085 INTERRUPT

The 8085 interrupt process is controlled by the Interrupt Enable flip-flop, which is internal to the processor and can be set or reset by using software instructions. If the flip-flop is enabled and the input to the interrupt signal INTR (pin 10) goes high, the microprocessor is interrupted. This is a maskable interrupt and can be disabled. The 8085 has a nonmaskable and three additional vectored interrupt signals as well. The best way to describe the 8085 interrupt process is to compare it to a telephone with a blinking light instead of a ring.

Assume that you are reading an interesting novel at your desk, where there is a telephone. For you to receive and respond to a telephone call, the following steps should occur:

1. The telephone system should be enabled, meaning that the receiver should be on the hook.
2. You should glance at the light at certain intervals to check whether someone is calling.
3. If you see a blinking light, you should pick up the receiver, say hello, and wait for a response. Once you pick up the phone, the line is busy, and no more calls can be received until you replace the receiver.
4. Assuming that the caller is your roommate, the request may be: It is going to rain today. Will you please shut all the windows in my room?
5. You insert a bookmark on the page you are reading.
6. You replace the receiver on the hook.
7. You shut your roommate's windows.
8. You go back to your book, find your mark, and start reading again.

Steps 6 and 7 may be interchanged, depending on the urgency of the request. If the request is critical and you do not want to be interrupted while attending to the request, you are likely to attend to the request first, then put the receiver back on the hook. The 8085 interrupt process can be described in terms of those eight steps.

Step 1: The interrupt process should be enabled by writing the instruction EI in the main program. This is similar to keeping the phone receiver on the hook. The instruction EI sets the Interrupt Enable flip-flop. The instruction DI resets the flip-flop and disables the interrupt process.

Instruction   EI (Enable Interrupt)

☐ This is a 1-byte instruction.
☐ The instruction sets the Interrupt Enable flip-flop and enables the interrupt process.
☐ System reset or an interrupt disables the interrupt process.

Instruction   DI (Disable Interrupt)

☐ This is a 1-byte instruction.
☐ The instruction resets the Interrupt Enable flip-flop and disables the interrupt.
☐ It should be included in a program segment where an interrupt from an outside source cannot be tolerated.

Step 2: When the microprocessor is executing a program, it checks the INTR line during the execution of each instruction.

Step 3: If the line INTR is high and the interrupt is enabled, the microprocessor completes the current instruction, disables the Interrupt Enable flip-flop and sends a signal called INTA—Interrupt Acknowledge (active low). The processor cannot accept any interrupt requests until the interrupt flip-flop is enabled again.

Step 4: The signal INTA is used to insert a restart (RST) instruction (or a Call instruction) through *external hardware*. The RST instruction is a 1-byte call instruction (explained below) that transfers the program control to a specific memory location on page 00H and restarts the execution at that memory location after executing Step 5.

Step 5: When the microprocessor receives an RST instruction (or a Call instruction), it saves the memory address of the next instruction on the stack. This is similar to inserting a bookmark. The program is transferred to the CALL location.

Step 6: Assuming that the task to be performed is written as a subroutine at the specified location, the processor performs the task. This subroutine is known as a service routine.

Step 7: The service routine should include the instruction EI to enable the interrupt again. This is similar to putting the receiver back on the hook.

Step 8: At the end of the subroutine, the RET instruction retrieves the memory address where the program was interrupted and continues the execution. This is similar to finding the page where you were interrupted by the phone call and continuing to read.

We will elaborate further on the restart instructions, additional hardware mentioned in Step 4, and multiple interrupts.

12.11   RST (Restart) Instructions

The 8085 instruction set includes eight RST (Restart) instructions listed in Section 9.3. These are 1-byte Call instructions that transfer the program execution to a specific location on page 00H, as listed in Table 12.1. The RST instructions are executed in a similar way to that of Call instructions. The address in the program counter (meaning the address of the next instruction to an RST instruction) is stored on the stack before the program execution is transferred to the RST call location. When the processor encounters a Return instruction in the subroutine associated with the RST instruction, the program returns to the address that was stored on the stack. In case of a hardware interrupt, we will use an RST instruction to restart the program execution.

To implement Step 4 in the interrupt process, insert one of these instructions in the microprocessor by using external hardware and the signal $\overline{INTA}$ (Interrupt Acknowledge), as shown in Figure 12.1.

In Figure 12.1, the instruction RST 5 is built using resistors and a tri-state buffer. Figure 12.2 shows the timing of the 8085 Interrupt Acknowledge machine cycle. In response to the INTR (Interrupt Request) high signal, the 8085 sends the $\overline{INTA}$ (Interrupt Acknowledge) low signal, which is used to enable the buffer, and the RST instruction is placed on the data bus during M_1. During M_1, the program counter holds the memory address of the next instruction, which should be stored on the stack so that the program can continue after the service routine. During M_2, the address of the stack pointer minus one $(SP - 1)$ location is placed on the address bus, and the high-order address of the program counter is stored on the stack. During M_3, the low-order address of the program counter is stored in the next location $(SP - 2)$ of the stack.

The machine cycle M_1 of the Interrupt Acknowledge is identical with the Opcode Fetch cycle, with two exceptions. The $\overline{INTA}$ signal is sent out instead of the $\overline{RD}$ signal,

TABLE 12.1
Restart Instructions

Mnemonics	Binary Code								Hex Code	Call Location in Hex
	D_7	D_6	D_5	D_4	D_3	D_2	D_1	D_0		
RST 0	1	1	0	0	0	1	1	1	C7	0000
RST 1	1	1	0	0	1	1	1	1	CF	0008
RST 2	1	1	0	1	0	1	1	1	D7	0010
RST 3	1	1	0	1	1	1	1	1	DF	0018
RST 4	1	1	1	0	0	1	1	1	E7	0020
RST 5	1	1	1	0	1	1	1	1	EF	0028
RST 6	1	1	1	1	0	1	1	1	F7	0030
RST 7	1	1	1	1	1	1	1	1	FF	0038

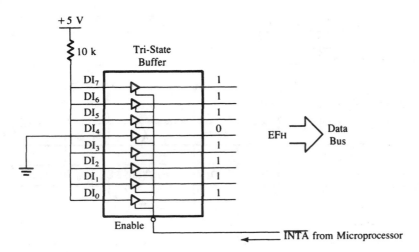

FIGURE 12.1
A Circuit to Implement the Instruction RST 5

and the status lines ($IO/\overline{M}$, S_0 and S_1) are 1 1 1 instead of 0 1 1 (see Figure 12.2). During M_1, the RST 5 is decoded, a 1-byte Call instruction to location 0028H. The machine cycles M_2 and M_3 are Memory Write cycles that store the contents of the program counter on the stack, and then a new instruction cycle begins.

In this next instruction cycle, the program is transferred to location 0028H. The service routine is written somewhere else in memory, and the Jump instruction is written at 0028H to specify the address of the service routine. All these steps are illustrated in the following example.

12.12 Illustration: An Implementation of the 8085 Interrupt

PROBLEM STATEMENT

1. Write a main program to count continuously in binary with a one-second delay between each count.
2. Write a service routine at XX70H to flash FFH five times when the program is interrupted, with some appropriate delay between each flash.

MAIN PROGRAM

Memory Address	Label	Mnemonics	Comments
XX00		LXI SP,XX99H	;Initialize stack pointer
03		EI	;Enable interrupt process
04		MVI A,00H	;Initialize counter
06	NXTCNT:	OUT PORT1	;Display count
08		MVI C,01H	;Parameter for 1-second delay

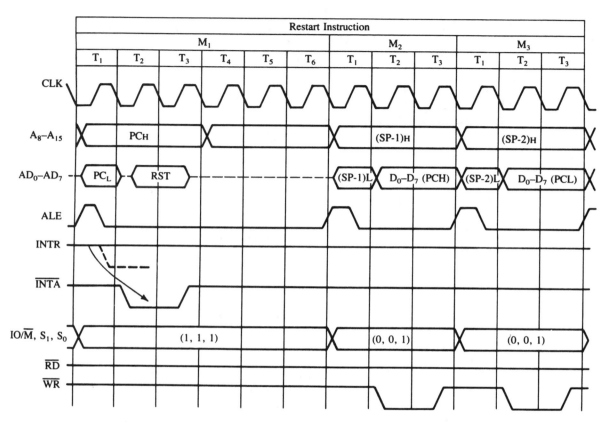

FIGURE 12.2
8085 Timing of the Interrupt Acknowledge Machine Cycle and Execution of an RST Instruction

SOURCE: Intel Corporation, *MCS 80/85 Student Study Guide* (Santa Clara, Calif.: Author, 1979), pp. 2–90.

0A	CALL DELAY	;Wait one second
0D	INR A	;Next count
0E	JMP NXTCNT	;Continue

Delay Routine   Use delay subroutine illustrated in Chapter 9, Section 9.21.

Service Routine

Memory Address	Label	Mnemonics	Comments
XX70	SERV:	PUSH B	;Save contents
71		PUSH PSW	
72		MVI B,0AH	;Load register B for five flashes and ; five blanks

74		MVI A,00H	;Load 00 to blank display
76	FLASH:	OUT PORT1	
78		MVI C,01H	;Parameter for 1-second delay
7A		CALL DELAY	
7D		CMA	;Complement display count
7E		DCR B	;Reduce count
7F		JNZ FLASH	
82		POP PSW	
83		POP B	
84		EI	;Enable interrupt process
85		RET	;Service is complete; go back to
			; main program

DESCRIPTION OF THE INTERRUPT PROCESS

1. The main program initializes the stack pointer at XX99H and enables the interrupts. The program will count continuously from 00H to FFH, with a delay of one second between each count.
2. To interrupt the processor, push the switch. The INTR line goes high.
3. Assuming the switch is pushed when the processor is executing the instruction OUT at memory location XX06H, the following sequence of events occurs:*
   a. The microprocessor completes the execution of the instruction OUT.
   b. It senses that the line INTR is high, and that the interrupt is enabled.
   c. The microprocessor disables the interrupt, stops execution, and sends out a control signal INTA (Interrupt Acknowledge).
   d. The INTA (active low) enables the tri-state buffer, and the instruction EFH is placed on the data bus.
   e. The microprocessor saves the address XX08H of the next instruction (MVI C,01H) on the stack at locations XX98H and XX97H, and the program is transferred to memory location 0028H. The locations 0028-29-2AH should have the following Jump instruction to transfer the program to the service routine.

JMP XX70H

   (However, you do not have access to write at 0028H in the monitor program. See the next section.)
4. The program jumps to the service routine at XX70H.
5. The service routine saves the registers that are being used in the subroutine and loads the count ten in register B to output five flashes and also five blanks.
6. The service routine enables the interrupt before returning to the main program.
7. When the service routine executes the RET instruction, the microprocessor retrieves the memory address XX08H from the top of the stack and continues the binary counting.

---

*It is assumed here that the Hold signal is inactive; Hold has a higher priority than any interrupt.

TESTING INTERRUPT ON A SINGLE-BOARD COMPUTER SYSTEM

Step 3e in the above description assumes that you are designing the system and have access to locations in EPROM or ROM on page 00H. In reality, you have no direct access to restart locations if the system has already been designed. Then how do you transfer the program control from a restart location to the service routine?

In single-board microcomputers, some restart locations are usually reserved for users, and the system designer provides a Jump instruction at a restart location to jump somewhere in R/W memory. For example, in Intel's SDK-85 system, R/W memory begins at page 20H, and you may find the following instruction in the monitor program at memory location 0028H:

> 0028   JMP 20C2H

If instruction RST 5 is inserted as shown in Figure 12.3, it transfers the program to location 0028H, and the monitor transfers the program from 0028H to location 20C2H. To implement the interrupt shown in Figure 12.3, you need to store the Jump instruction as shown below:

> 20C2   C3   JMP SERV
> 20C3   70
> 20C4   20

This instruction will transfer the program to the service routine located at 2070H.

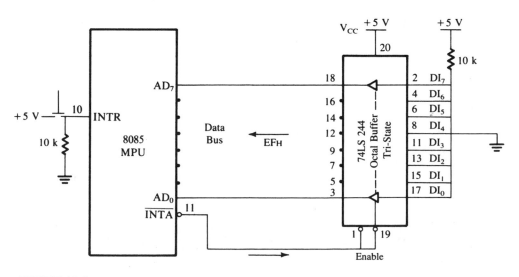

FIGURE 12.3
Schematic to Implement the 8085 Interrupt

ISSUES IN IMPLEMENTING INTERRUPTS

In the above illustration, some questions remain unanswered:

1. *Is there a minimum pulse width required for the INTR signal?*

The microprocessor checks INTR, one clock period before the last T-state of an instruction cycle. In the 8085, the Call instructions require 18 T-states; therefore, the INTR pulse should be high at least for 17.5 T-states. In a system with 3 MHz clock frequency (such as the SDK-85 system), the input pulse to INTR should be at least 5.8 μs long.

2. *How long can the INTR pulse stay high?*

The INTR pulse can remain high until the interrupt flip-flop is set by the EI instruction in the service routine. If it remains high after the execution of the EI instruction, the processor will be interrupted again, as if it were a new interrupt. In Figure 12.3, the manual push button will keep the INTR high for more than 20 ms; however, the service routine has a delay of 1 second, and the EI instruction is executed at the end of the service routine.

3. *Can the microprocessor be interrupted again before the completion of the first interrupt service routine?*

The answer to this question is determined by the programmer. After the first interrupt, the interrupt process is automatically disabled. In the Illustrative Program in Section 12.12, the service routine enables the interrupt at the end of the service routine; in this case, the microprocessor cannot be interrupted before the completion of this routine. If instruction EI is written at the beginning of the routine, the microprocessor can be interrupted again during the service routine. (See Experimental Assignment 1 at the end of this chapter.)

12.13 Multiple Interrupts and Priorities

In the previous section, we illustrated how to implement the interrupt for one peripheral using one line (INTR). Now we will expand our discussion to include how to use INTR for multiple peripherals and how to determine priorities among these peripherals when two or more of the peripherals request interrupt service simultaneously.

The schematic in Figure 12.4 implements multiple interrupting devices using an 8-to-3 priority encoder that determines the priorities among interrupting devices. If you examine the instruction code for eight RST instructions, you will notice that bits D_5, D_4, and D_3 change in a binary sequence and that the others are always at logic 1 (see Table 12.1).

The encoder provides appropriate combinations on its output lines A_0, A_1, and A_2, which are connected to data lines D_3, D_4, and D_5 through a tri-state buffer. The eight inputs to the encoder are connected to eight different interrupting devices.

When an interrupting device requests service, one of the input lines goes low, which makes line E_0 high and interrupts the microprocessor. When the interrupt is acknowledged and the signal $\overline{INTA}$ enables the tri-state buffer, the code corresponding to the input is placed on lines D_5, D_4, and D_3. For example, if the interrupting device on line I_5

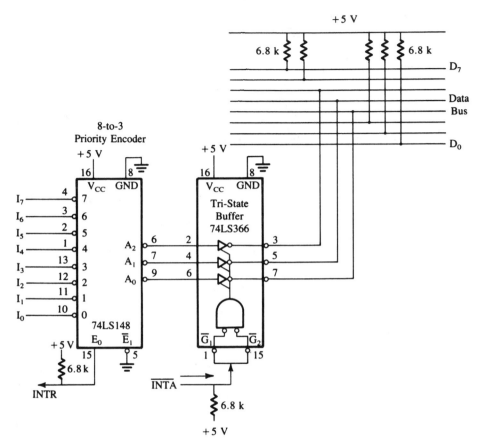

FIGURE 12.4
Multiple Interrupts Using a Priority Encoder

goes low, the output of the encoder will be 010. This code is inverted by the buffer 74LS366 and combined with other high data lines. Thus, the instruction 1110 1111 (EFH) is placed on the data bus. This is instruction RST 5. Similarly, any one of the RST instructions can be generated and placed on the data bus. If there are simultaneous requests, the priorities are determined by the encoder; it responds to the higher-level input, ignoring the lower-level input. One of the drawbacks of this scheme is that the interrupting device connected to the input I_7 always has the highest priority. The interrupt scheme shown in Figure 12.4 also can be implemented by using a special device called a Priority Interrupt Controller—8214. This device includes a status register and a priority comparator in addition to an 8-to-3 priority encoder. Today, however, this device is being replaced by a more versatile one called a Programmable Interrupt Controller—8259A (described briefly later in this chapter).

8085 VECTORED INTERRUPTS 12.2

The 8085 has five interrupt inputs (Figure 12.5). One is called INTR (discussed in the previous section), three are called RST 5.5, 6.5, and 7.5, respectively, and the fifth is called TRAP, a nonmaskable interrupt. These last four (RSTs and TRAP) are automatically vectored (transferred) to specific locations on memory page 00H without any external hardware. They do not require the $\overline{\text{INTA}}$ signal or an input port; the necessary hardware is already implemented inside the 8085. These interrupts and their call locations are as follows:

Interrupts		Call Locations
1. TRAP	$\longrightarrow$	0024H
2. RST 7.5	$\longrightarrow$	003CH
3. RST 6.5	$\longrightarrow$	0034H
4. RST 5.5	$\longrightarrow$	002CH

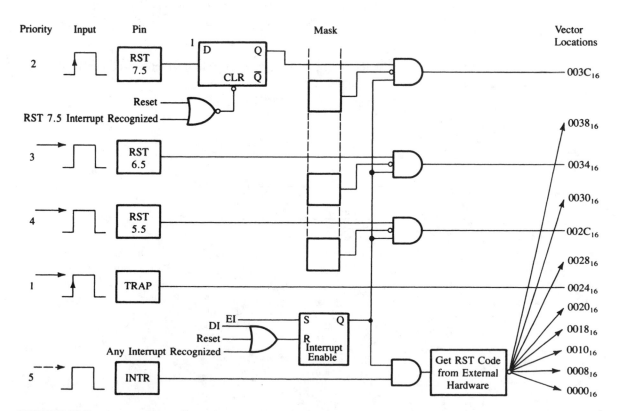

FIGURE 12.5

The 8085 Interrupts and Vector Locations

SOURCE: Intel Corporation, *MCS 80/85 Student Study Guide* (Santa Clara, Calif.: Author, 1979).

The TRAP has the highest priority, followed by RST 7.5, 6.5, 5.5, and INTR, in that order; however, the TRAP has a lower priority than the Hold signal used for DMA (Section 12.42).

12.21   TRAP

TRAP, a nonmaskable interrupt known as NMI, is analogous to the smoke detector described earlier. It has the highest priority among the interrupt signals, it need not be enabled, and it cannot be disabled. It is level- and edge-sensitive, meaning that the input should go high and stay high to be acknowledged. It cannot be acknowledged again until it makes a transition from high to low to high.

Figure 12.5 shows that when this interrupt is triggered, the program control is transferred to location 0024H without any external hardware or the interrupt enable instruction EI. TRAP is generally used for such critical events as power failure and emergency shut-off.

12.22   RST 7.5, 6.5, and 5.5

These maskable interrupts (shown in Figure 12.5) are enabled under program control with two instructions: EI (Enable Interrupt) described earlier, and SIM (Set Interrupt Mask) described below:

Instruction   SIM: Set Interrupt Mask. This is a 1-byte instruction and can be used for three different functions (Figure 12.6).

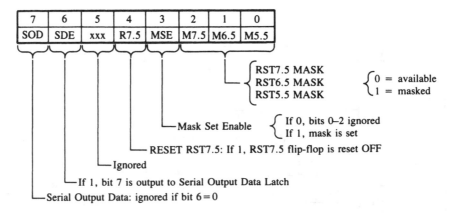

FIGURE 12.6

Interpretation of the Accumulator Bit Pattern for the SIM Instruction

SOURCE: Intel Corporation, *Assembly Language Programming Manual* (Santa Clara, Calif.: Author, 1979), pp. 3–59.

☐ One function is to set mask for RST 7.5, 6.5, and 5.5 interrupts. This instruction reads the content of the accumulator and enables or disables the interrupts according to the content of the accumulator. Bit D_3 is a control bit and should $= 1$ for bits D_0, D_1, and D_2 to be effective. Logic 0 on D_0, D_1, and D_2 will enable the corresponding interrupts, and logic 1 will disable the interrupts.

☐ The second function is to reset RST 7.5 flip-flop (Figure 12.6). Bit D_4 is additional control for RST 7.5. If $D_4 = 1$, RST 7.5 is reset. This is used to override (or ignore) RST 7.5 without servicing it.

☐ The third function is to implement serial I/O (discussed in Chapter 16). Bits D_7 and D_6 of the accumulator are used for serial I/O and do not affect the interrupts. Bit $D_6 = 1$ enables the serial I/O and bit D_7 is used to transmit (output) bits.

Here we are concerned with RST 7.5, 6.5, and 5.5 interrupts and not with serial I/O.

The mnemonic SIM is confusing. The wording—Set Interrupt Mask—implies that the instruction masks the interrupts. However, the instruction must be executed in order to use the interrupts. The process required to enable these interrupts can be likened to a switchboard controlling three telephone extensions in a company. Let us assume these phone extensions are assigned to the president (RST 7.5), the vice president (RST 6.5), and the manager (RST 5.5), in that priority, and are monitored by their receptionist according to the instructions given. The protocols of placing a telephone call to one of the executives and of interrupting the microprocessor using RST 7.5, 6.5, and 5.5 can be compared as follows:

Placing a Telephone Call	**Interrupting the 8085 (Figure 12.6)**
1. The switchboard is functional and all telephone lines are open.	1. The interrupt process is enabled. The instruction EI sets the Interrupt Enable flip-flop, and one of the inputs to the AND gates is set to logic 1 (Figure 12.5). These AND gates activate the program transfer to various vectored locations.
2. All executives leave instructions on the receptionist's desk as to whether they wish to receive any phone calls.	2. An appropriate bit pattern is loaded into the accumulator.
3. The receptionist reads the instructions.	3. If bit $D_3 = 1$, the respective interrupts are enabled according to bits D_2–D_0.
4. The receptionist is on duty and sends calls through for whoever is available.	4. RST 7.5, 6.5, and 5.5 are being monitored.
5. The receptionist is busy typing. Phone calls can be received directly according to previous instructions.	5. If bit $D_3 = 0$, bits D_2–D_0 have no effect on previous conditions.
6. No calls for the president now. Call back later.	6. Bit $D_4 = 1$; this resets RST 7.5.

This analogy can be extended to the interrupt INTR, which is viewed as one telephone line shared by eight engineers with a switchboard operator (external hardware) who rings the appropriate extension.

The entire interrupt process (except TRAP) is disabled by resetting the Interrupt Enable flip-flop (Figure 12.5). The flip-flop can be reset in one of the three ways:

☐ Instruction DI
☐ System Reset
☐ Recognition of an Interrupt Request

Figure 12.5 shows that these three signals are ORed and the output of the OR gate is used to reset the flip-flop.

TRIGGERING LEVELS

These interrupts are sensitive to different types of triggering as listed below:

☐ **RST 7.5**   This is positive-edge sensitive and can be triggered with a short pulse. The request is stored internally by the D flip-flop (Figure 12.5) until the microprocessor responds to the request or until it is cleared by Reset or by bit D_4 in the SIM instruction.

☐ **RST 6.5** and **RST 5.5**   These interrupts are level-sensitive, meaning that the triggering level should be on until the microprocessor completes the execution of the current instruction. If the microprocessor is unable to respond to these requests immediately, they should be stored or held by external hardware.

Example 12.1

Enable all the interrupts in an 8085 system.

Instructions

```
EI              ;Enable interrupts
MVI A,08H       ;Load bit pattern to enable RST 7.5, 6.5, and 5.5
SIM             ;Enable RST 7.5, 6.5, and 5.5
```

Bit $D_3 = 1$ in the accumulator makes the instruction SIM functional, and bits D_2, D_1, and $D_0 = 0$ enable the interrupts 7.5, 6.5, and 5.5.

Example 12.2

Reset the 7.5 interrupt from Example 12.1.

Instructions

```
MVI A,18H       ;Set D₄ = 1
SIM             ;Reset 7.5 interrupt flip-flop
```

PENDING INTERRUPTS

Because there are several interrupt lines, when one interrupt request is being served, other interrupt requests may occur and remain pending. The 8085 has an additional instruction called RIM (Read Interrupt Mask) to sense these pending interrupts.

Instruction RIM: Read Interrupt Mask. This is a 1-byte instruction that can be used for the following functions.

☐ To read interrupt masks. This instruction loads the accumulator with 8 bits indicating the current status of the interrupt masks (Figure 12.7).
☐ To identify pending interrupts. Bits D_4, D_5, and D_6 (Figure 12.7) identify the pending interrupts.
☐ To receive serial data. Bit D_7 (Figure 12.7) is used to receive serial data.

Assuming the microprocessor is completing an RST 7.5 interrupt request, check to see if RST 6.5 is pending. If it is pending, enable RST 6.5 without affecting any other interrupts; otherwise, return to the main program.

Example 12.3

Instructions

```
            RIM            ;Read interrupt mask
            MOV B,A        ;Save mask information
            ANI 20H        ;Check whether RST 6.5 is pending
            JNZ NEXT
            EI
            RET            ;RST 6.5 is not pending, return to main program
    NEXT:   MOV A,B        ;Get bit pattern; RST 6.5 is pending
            ANI 0DH        ;Enables RST 6.5 by setting D₁ = 0
            ORI 08H        ;Enable SIM by setting D₃ = 1
            SIM
            JMP SERV       ;Jump to service routine for RST 6.5
```

Correction for LaTeX subscripts in comments:
- ANI 0DH   ;Enables RST 6.5 by setting $D_1 = 0$
- ORI 08H   ;Enable SIM by setting $D_3 = 1$

The RIM instruction loads the accumulator with the following information:

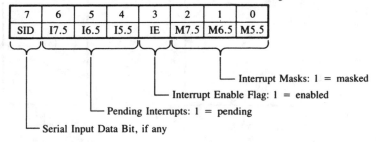

7	6	5	4	3	2	1	0
SID	I7.5	I6.5	I5.5	IE	M7.5	M6.5	M5.5

Interrupt Masks: 1 = masked
Interrupt Enable Flag: 1 = enabled
Pending Interrupts: 1 = pending
Serial Input Data Bit, if any

FIGURE 12.7
Interpretation of the Accumulator Bit Pattern for the RIM Instruction
SOURCE: Intel Corporation, *Assembly Language Programming Manual* (Santa Clara, Calif.: Author, 1979), pp. 3–49.

The instruction RIM checks for a pending interrupt. Instruction ANI 20H masks all the bits except D_5 to check pending RST 6.5. If $D_5 = 0$, the program control is transfered to the main program. $D_5 = 1$ indicates that RST 6.5 is pending. Instruction ANI 0DH sets $D_1 = 0$ (RST 6.5 bit for SIM), instruction ORI sets $D_3 = 1$ (this is necessary for SIM to be effective), and instruction SIM enables RST 6.5 without affecting any other interrupts. The JMP instruction transfers the program to the service routine (SERV) written for RST 6.5.

12.23   Illustration: Interrupt-Driven Clock

PROBLEM STATEMENT

Design a 1-minute timer using a 60 Hz power line as an interrupting source. The output ports should display minutes and seconds in BCD. At the end of the minute, the output ports should continue displaying one minute and zero seconds.

HARDWARE DESCRIPTION

This 1-minute timer is designed with a 60 Hz AC line. The circuit (Figure 12.8) uses a step-down transformer, the 74121 monostable multivibrator, and interrupt pin RST 6.5. After the interrupt, program control is transferred to memory location 0034H in the monitor program.

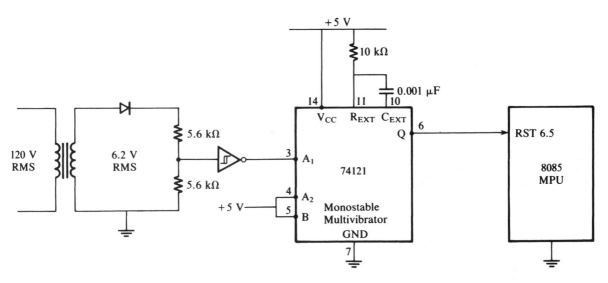

FIGURE 12.8
Schematic of Interrupt-Driven Timer Clock

The AC line with 60 Hz frequency has a period of 16.6 ms; that means it can provide a pulse every sixtieth of a second with 8.3 ms pulse width, which is too long for the interrupt. The interrupt flip-flop is enabled again within 6 μs in the timer service routine; therefore, the pulse should be turned off before the EI instruction in the service routine is executed. The 74121 monostable multivibrator is used to provide appropriate pulse width. Another option is to use the 7474 positive-edge triggered flip-flop (see Problem 4 at the end of this chapter).

Monitor Program

```
0034          JMP RWM         ;This is RST 6.5; go to location in user memory
                              ;   to give Restart access to the user
```

Main Program

```
              LXI SP,STACK    ;Initialize stack pointer
              RIM             ;Read mask
              ORI 08H         ;Bit pattern to enable RST 6.5
              SIM             ;Enable RST 6.5
              LXI B,0000H     ;Set up register B for minutes and register C for
                              ;   seconds
              MVI D,3CH       ;Set up register D to count 60₁₀ interrupts
              EI              ;Enable interrupts
DSPLAY:       MOV A,B
              OUT PORT1       ;Display minutes at PORT1
              MOV A,C
              OUT PORT2       ;Display seconds at PORT2
              JMP DSPLAY

RWM:          JMP TIMER       ;This is RST 6.5 vector location 0034H; go to
                              ;   TIMER routine to upgrade the clock
```

Interrupt Service Routine

```
;Section I
TIMER:        DCR D           ;One interrupt occurred; reduce count by 1
              EI              ;Enable interrupts
              RNZ             ;Has 1 second elapsed? If not, return
;Section II
              DI              ;No other interrupts allowed
              MVI D,3CH       ;1 second is complete; load register D again to count
                              ;   60 interrupts
              MOV A,C
              ADD 01H         ;Increment "Second" register
              DAA             ;Decimal-adjust "Seconds"
              MOV C,A         ;Save "BCD" seconds
```

```
              CPI 60H
              EI
              RNZ                    ;Is time = 60 seconds? If not, return
;Section III
              DI                     ;Disable interrupts
              MVI C,00H              ;60 seconds complete, clear "Second" register
              INR B                  ;Increment "Minutes"
              RET                    ;1 minute elapsed
```

PROGRAM DESCRIPTION

The main program clears registers B and C to store minutes and seconds, respectively; enables the interrupts; sets up register D to count 60 interrupts; and displays the starting time in minutes (00) and seconds (00). Instruction SIM enables RST 6.5 according to the bit pattern in the accumulator.

When the first pulse interrupts the processor, program control is transferred to memory location 0034H, as mentioned earlier. (Check location 0034H in your monitor program; you may find a Jump instruction to transfer the control to a memory location in R/W memory. Write a Jump instruction at that location to locate the service routine labeled TIMER.)

In the service routine (Section I), register D is decremented every second, the interrupt is enabled, and the program is returned to the main routine. This is repeated 60 times. After the sixtieth interrupt, counter D goes to zero and the program enters Section II. In this section, counter D is reloaded, the "second" register is incremented and adjusted for BCD, and the program is returned to the main routine. In this section, instruction DI is used as a precaution to avoid any interrupts from other sources. For the next 60 interrupts, the program remains in Section I. When Section II is repeated 60 times, the program goes to Section III, where the "minute" register is incremented and the "second" register is cleared. To avoid further interrupts, the interrupt is disabled and the program is returned to the main routine where one minute and zero seconds are displayed continuously.

In this particular program, the service routine does not save any register contents by using PUSH instructions before starting the service routine. However, in most service routines, register contents must be saved because the interrupt is asynchronous and can occur at any time.

12.3 RESTART AS SOFTWARE INSTRUCTIONS

External hardware is necessary to insert an RST instruction when an interrupt is requested to INTR. However, the fact that RST is a software instruction is quite often overlooked or misunderstood. RST instructions are commonly used to set up software breakpoints as a debugging technique. A breakpoint is a Restart (RST) instruction in a program where the execution of the program stops temporarily and program control is transferred to the RST location. The program should be transferred from the RST location to the breakpoint service routine to allow the user to examine register or memory contents when specified keys are pressed. After the breakpoint routine, the program should return to executing the main program at the breakpoint. The breakpoint procedure allows the user to test programs in segments. For example, if RST 6 is written in a program, the program execution is transferred to location 0030H; it is equivalent to a 1-byte call instruction. This can be used to write a software breakpoint routine, as illustrated next.

12.31 Illustrative Program: Implementation of Breakpoint Technique

PROBLEM STATEMENT

Implement a breakpoint facility at RST 5 for user. When the user writes RST 5 in the program, the program should

1. be interrupted at the instruction RST 5.
2. display the accumulator content and the flags when Hex key A (1010_2) is pressed.
3. exit the breakpoint routine and continue execution when the Zero key (0000_2) is pressed.

Assume that when a keyboard routine (KBRD) is called, it returns the binary key code of the key pressed in the accumulator.

PROBLEM ANALYSIS

The breakpoint routine should display the accumulator contents and the flags when a user writes the RST instruction in a program. The technique used to display register contents after executing a segment of the user's program is as follows:

1. Store the register contents on the stack.
2. Assign (arbitrarily) a key from the keyboard for the accumulator display. (In this problem Hex key A from the keyboard is assigned to display the accumulator contents.)
3. Wait for the key to be pressed, and retrieve the contents from the stack by manipulating the stack pointer when the key is pressed.
4. Assign a key to return to the user's program. (In this problem, it is the Zero key.)

This approach assumes that a keyboard subroutine can be called from your monitor program and that the codes associated with each key are known.

BREAKPOINT SUBROUTINE

;BRKPNT: This is a breakpoint subroutine; it can be implemented with the instruction
;   RST 5. It displays the accumulator and the flags when the A key is pressed
;   and returns to the calling program when the Zero key is pressed
;Input: None
;Output: None
;Does not modify any register contents
;Calls: KBRD subroutine. The KBRD is a keyboard subroutine that checks a key
           pressed.

```
;       The routine identifies the key and places its binary code in the accumulator
BRKPNT:     PUSH PSW        ;Save registers
            PUSH B
            PUSH D
            PUSH H
KYCHK:      CALL KBRD        ;Check for a key
            CPI 0AH          ;Is it key A?
            JNZ RETKY        ;If not, check Zero key
            LXI H,0007H      ;Load stack pointer displacement count; see program
                             ;  description
            DAD SP           ;Place memory address in HL, where (A) is stored
            MOV A,M
            OUT PORT1        ;Display accumulator contents
            DCX H            ;Point HL to the location of the flags
            MOV A,M
            OUT PORT2        ;Display flags
            JMP KYCHK        ;Go back and check next key
RETKY:      CPI 00H          ;Is it Zero key?
            JNZ KYCHK        ;If not, go and check key program
            POP H            ;Retrieve registers
            POP D            ;
            POP B
            POP PSW
            RET
```

PROGRAM DESCRIPTION

The breakpoint routine saves all the registers on the stack, and the address in the stack pointer is decremented accordingly. (In this particular problem, registers BC and DE need not be saved. These registers are saved here for the assignments given at the end of the chapter.) The accumulator contents are stored in the seventh memory location from the top of the stack, and the flags in the sixth memory location.

When key A is pressed, the HL register adds seven to the stack pointer contents and places that address in the HL register (DAD SP), without modifying the contents of the stack pointer. This is an important point; if the stack pointer is varied, appropriate contents may not be retrieved with POP and RET instructions.

The subroutine displays the accumulator and the flags at the two output ports and returns to the main program.

12.4 ADDITIONAL I/O CONCEPTS AND PROCESSES

The 8085 interrupt I/O, described earlier, is limited because of its single interrupt pin and hardware requirements to determine interrupt priorities. To circumvent these limitations, a programmable interrupt controller such as the 8259A is used to implement and extend the capability of the 8085 interrupt. Another I/O process, Direct Memory Access (DMA), is commonly used for high-speed data transfer. This I/O process is implemented also by using a programmable device such as DMA controller 8257. A programmable device is generally a multifunction chip, and the microprocessor can specify and/or modify its functions by writing appropriate bits in the control register of the device. The concepts of programmable devices are discussed in Chapter 14. The intent here is to maintain the continuity in the discussion of the interrupt I/O and to introduce the concept of Direct Memory Access (DMA). The interrupt controller and the DMA process will be described briefly in the next sections—see Chapter 15 for further details.

12.41 Programmable Interrupt Controller: The 8259A

The 8259A is a programmable interrupt-managing device, specifically designed for use with the interrupt signals (INTR/INT) of the 8085 microprocessor. The primary features of the 8259A are as follows:

1. It manages eight interrupt requests.
2. It can vector an interrupt request anywhere in the memory map through program control without additional hardware for restart instructions. However, all eight requests are spaced at the interval of either four locations or eight locations.
3. It can solve eight levels of interrupt priorities in a variety of modes.
4. With additional 8259A devices, the priority scheme can be expanded to 64 levels.

One of the major limitations of the 8085 interrupt scheme is that all requests are vectored to memory locations on page 00H, which is reserved for ROM or EPROM, and access to these locations is difficult after a system has been designed. In addition, the process of determining priorities is limited, and extra hardware is required to insert Restart instructions. The 8259A overcomes these limitations and provides many more flexible options. It can be employed with such 16-bit Intel microprocessors as the 8086/8088 as well.

The 8259A block diagram (Figure 12.9) includes control logic, registers for interrupt requests, priority resolver, cascade logic, and data bus. The registers manage interrupt requests; the priority resolver determines their priority. The cascade logic is used to connect additional 8259A devices.

8259A INTERRUPT OPERATION

Implementing interrupts in the simplest format without cascading requires two specific instructions. The instructions are written by the MPU in the device registers (explained in Chapter 15). The first instruction specifies features such as mode and/or memory space between two consecutive interrupt levels. The second instruction specifies high-order memory address. After these instructions have been written, the following sequence of events should occur:

1. One or more interrupt request lines go high requesting the service.
2. The 8259A resolves the priorities and sends an INT signal to the MPU.

Block Diagram

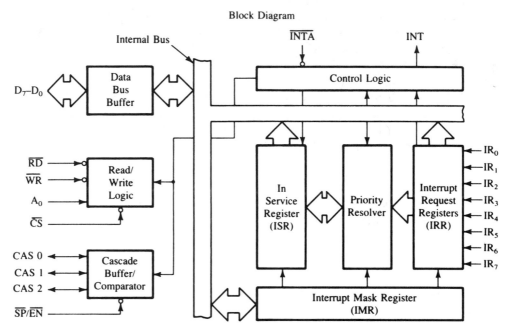

FIGURE 12.9
The 8259A Block Diagram
SOURCE: Intel Corporation, *Peripheral Components* (Santa Clara, Calif.: Author, 1993), pp. 3–171.

3. The MPU acknowledges the interrupt by sending $\overline{\text{INTA}}$.
4. After the $\overline{\text{INTA}}$ has been received, the opcode for the CALL instruction (CDH) is placed on the data bus.
5. Because of the CALL instruction, the MPU sends two more $\overline{\text{INTA}}$ signals.
6. At the first $\overline{\text{INTA}}$, the 8259A places the low-order 8-bit address on the data bus and, at the second $\overline{\text{INTA}}$, it places the high-order 8-bit address of the interrupt vector. This completes the 3-byte CALL instruction.
7. The program sequence of the MPU is transferred to the memory location specified by the CALL instruction.

The 8259A includes additional features such as reading the status and changing the interrupt mode during a program execution.

12.42 Direct Memory Access (DMA)

The Direct Memory Access (DMA) is a process of communication or data transfer controlled by an external peripheral. In situations in which the microprocessor-controlled data transfer is too slow, the DMA is generally used; e.g., data transfer between a floppy disk and R/W memory of the system.

The 8085 microprocessor has two pins available for this type of I/O communication: HOLD (Hold) and HLDA (Hold Acknowledge). Conceptually, this is an im-

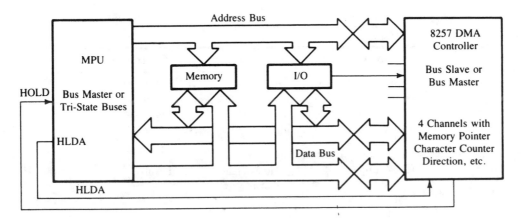

FIGURE 12.10

DMA Data Transfer

SOURCE: Intel Corporation, *MCS 80/85 Student Study Guide* (Santa Clara, Calif.: Author, 1979), pp. 2–21.

portant I/O technique; it introduces two new signals available on the 8085—HOLD and HLDA.

☐ HOLD—Hold. This is an active high input signal to the 8085 from another master requesting the use of the address and data buses. After receiving the HOLD request, the MPU relinquishes the buses in the following machine cycle. All buses are tri-stated and a Hold Acknowledge (HLDA) signal is sent out. The MPU regains the control of buses after HOLD goes low.

☐ HLDA—Hold Acknowledge. This is an active high output signal indicating that the MPU is relinquishing the control of the buses.

Typically, an external peripheral such as a DMA controller sends a request—a high signal—to the HOLD pin (Figure 12.10). The processor completes the execution of the current machine cycle; floats (high impedance state) the address, the data, and the control lines; and sends the Hold Acknowledge (HLDA) signal. The DMA controller takes control of the buses and transfers data directly between source and destination, thus by-passing the microprocessor. At the end of data transfer, the controller terminates the request by sending a low signal to the HOLD pin, and the microprocessor regains control of the buses. Typically, DMA controllers are programmable LSI chips. One such chip, the Intel 8237, is described in Chapter 15.

SUMMARY

The 8085 interrupts and their requirements are listed in Table 12.2 in the order of their priority; TRAP has the highest priority and INTR has the lowest priority. It must be emphasized that an interrupt can be recognized only if the HOLD signal is inactive. Some of the important features of the 8085 interrupts are summarized in Table 12.2.

TABLE 12.2
Summary of Interrupts in 8085

Interrupts	Type	Instructions	Hardware	Trigger	Vector
TRAP	Nonmaskable	☐ Independent of EI and DI	No external hardware	Level- and Edge-sensitive	0024H
RST 7.5	Maskable	☐ Controlled by EI and DI ☐ Unmasked by SIM	No external hardware	Edge-sensitive	003CH
RST 6.5	Maskable	☐ Controlled by EI and DI ☐ Unmasked by SIM	No external hardware	Level-sensitive	0034H
RST 5.5	Maskable	☐ Controlled by EI and DI ☐ Unmasked by SIM	No external hardware	Level-sensitive	002CH
INTR 8085	Maskable	☐ Controlled by EI and DI	RST code from external hardware	Level-sensitive	038H ↑ 0000H

☐ The interrupt is an asynchronous process of communication with the microprocessor, initiated by an external peripheral.

☐ The 8085 has a maskable interrupt that can be enabled or disabled using the instructions EI and DI, respectively. This interrupt is labeled here as an 8085 nonvectored interrupt.

☐ The 8085 has eight RST instructions that are equivalent to 1-byte Calls to specific locations on memory page 00H.

☐ For the nonvectored interrupt, the RST instructions (0 to 7) are implemented using external hardware and the INTA signal.

☐ The 8085 has four additional interrupt inputs, one nonmaskable and three maskable. These three interrupts are implemented without any external hardware and are known as RST 7.5, 6.5, and 5.5.

☐ The instruction SIM is necessary to implement the interrupts 7.5, 6.5, and 5.5.

☐ The instruction RIM can be used to check whether any interrupt requests are pending.

☐ The RST instructions are software commands and can be used in a program to jump to their vectored locations on memory page 00H.

☐ A programmable interrupt controller such as the 8259A is used commonly to implement and extend the capability of the 8085 interrupt.

☐ The Direct Memory Access (DMA) is a process of high-speed data transfer under the control of external devices such as a DMA controller.

QUESTIONS, PROBLEMS, AND PROGRAMMING ASSIGNMENTS

1. Check whether the following statements are true or false.
   a. If the 8085 microprocessor is interrupted while executing a 3-byte instruction (assuming the interrupt is enabled), the processor will acknowledge the interrupt request immediately, even before the completion of the instruction.   (T/F)
   b. When an 8085 system is Reset, all the interrupts including the TRAP are disabled.   (T/F)
   c. When the 8085 microprocessor acknowledges an interrupt, it disables the interrupt system (except TRAP).   (T/F)
   d. Instruction EI (Enable Interrupt) is necessary to implement the TRAP interrupt, but external hardware and the SIM instruction are unnecessary.   (T/F)
   e. If instruction RST 4 is written in a program, the program will jump to location 0020H without any external hardware.   (T/F)
   f. If a DMA request is sent to the microprocessor with a high signal to the HOLD pin, the microprocessor acknowledges the request after completing the present cycle.   (T/F)
   g. Instruction RIM is used to disable the interrupts 7.5, 6.5, and 5.5.   (T/F)
   h. The execution of instructions MVI A,10H, and SIM will enable all three interrupts (7.5, 6.5, and 5.5).   (T/F)
2. a. Identify the RST instruction in Figure 12.11.
   b. Specify the restart memory location when the microprocessor is interrupted.
   c. If the instruction in the monitor program at 0030 is JMP 20BFH and the service routine is written at 2075H, what instruction is necessary (at 20BFH) to locate the service routine?
3. The main program is stored beginning at 0100H. The main program (at 0120H) has called the subroutine at 0150H, and when the microprocessor is executing the

FIGURE 12.11
Schematic for an Interrupt

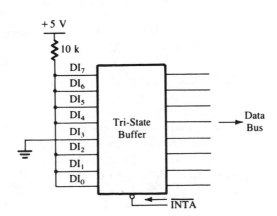

instruction at location 0151 (LXI), it is interrupted. Read the program, then answer the questions that follow:

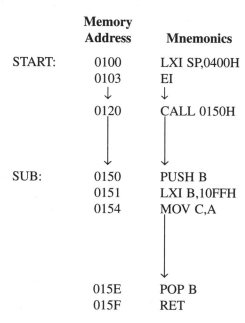

	Memory Address	Mnemonics
START:	0100	LXI SP,0400H
	0103	EI
	↓	↓
	0120	CALL 0150H
SUB:	0150	PUSH B
	0151	LXI B,10FFH
	0154	MOV C,A
	015E	POP B
	015F	RET

    a. Specify the contents of stack location 03FFH after the CALL instruction.
    b. Specify the stack locations where the contents of registers B and C are stored.
    c. When the program is interrupted, what is the memory address stored on the stack?

4. Redraw the schematic in Figure 12.8 with the following changes, and modify the service routine accordingly.
    a. Replace monostable multivibrator 74121 by positive-edge trigger flip-flop 7474.
    b. Design an output port, and use bit D_7 to clear the flip-flop.
    c. Modify Section I of the timer routine to clear the flip-flop at an appropriate place.

5. Answer the following questions in reference to Figure 12.4.
    a. What is the instruction placed on the data bus when input line I_6 of the encoder goes low, thus requesting the interrupt service?
    b. If three input lines (I_2, I_4, and I_5) go low simultaneously, explain how the priority is determined among the three requests, and specify the instruction that is placed on the data bus.

6. A program is stored in memory from 2000H to 205FH. To check the first segment of the program up to location 2025H, a breakpoint routine call is inserted at location 2026H. (Refer to the Illustrative Program in Section 12.31 for the breakpoint subroutine.) If the stack pointer is initialized at 2099H, answer the following questions.

   a. Specify the contents of memory locations 2098H and 2097H.
   b. Specify the memory locations where the accumulator contents and the flags are stored when the microprocessor executes instruction PUSH PSW in the breakpoint routine.
   c. Specify the memory locations where HL register contents are stored after executing the instruction PUSH H.
   d. Specify the contents of the stack pointer when the breakpoint routine returns from the KBRD routine.
   e. What address is placed in the program counter when instruction RET is executed?
7. Modify the breakpoint routine in Section 12.31 to display the memory location where the breakpoint is inserted in a program (for example, location 2026H in Question 6a).
8. Modify the breakpoint routine to display the contents of the BC, DE, and HL registers when the user pushes the Hex keys 1, 2, and 3. (The respective Hex codes are 01, 02, and 03.)

EXPERIMENTAL ASSIGNMENTS

1. a. Build the circuit shown in Figure 12.3. Enter and execute the program given in the illustration.
   b. Verify the interrupt process by pushing the Interrupt key.
   c. Replace the instruction EI at location XX03 by the NOP instruction. Push the Interrupt key and verify whether an interrupt request can be accepted by the microprocessor.
   d. Interrupt the processor; when the processor is in the middle of the service routine, push the Interrupt key again. Explain why the processor does not accept any interrupts during the service routine.
   e. In the routine SERV, write instruction EI at the beginning of the service routine. Push the Interrupt key, and explain your observation. (You may notice interesting results because the manual key keeps INTR high too long.)
2. a. Build the circuit shown in Figure 12.8 and implement the interrupt-driven clock.
   b. Rewrite the program to simulate a 5-minute egg timer.
   c. Modify the program in Experimental Assignment 2a to flash FF with some appropriate delay to indicate the completion of the 5-minute period.

Interfacing Data Converters

The microprocessor is a logic device; it processes digital signals that are binary and discontinuous. On the other hand, the real-world physical quantities such as temperature and pressure are continuous. These are represented by equivalent electrical quantities called analog signals. Even though an analog signal may represent a real physical parameter with accuracy, it is difficult to process or store the analog signal for later use without introducing considerable error. Therefore, in microprocessor-based industrial products, it is necessary to translate an analog signal into a digital signal. The electronic circuit that translates an analog signal into a digital signal is called an analog-to-digital (A/D) converter (ADC). Similarly, a digital signal needs to be translated into an analog signal to represent a physical quantity (e.g., to regulate a machine). This translator is called a digital-to-analog (D/A) converter (DAC). Both A/D and D/A are also known as data converters and are now available as integrated circuits.

This chapter focuses on interfacing data converters with the 8085 microprocessor. First, D/A converters are discussed, including the basic concepts in the conversion process and their interfacing applications. A/D converters are discussed later because some A/D conversion techniques include D/A converters in the conversion process. The chapter in-

cludes illustrations of successive approximation converters.

OBJECTIVES

- ☐ Explain the functions of data converters.
- ☐ Explain the basic circuit of a D/A converter and define the terms *resolution* and *settling time*.
- ☐ Calculate the analog output of a D/A converter for a given digital input signal.

☐ Design a circuit to interface an 8-bit D/A converter with the 8085 microprocessor, and verify the analog output for a digital signal.

☐ Interface a 10- or 12-bit D/A converter with an 8-bit microprocessor.

☐ Explain the basic concepts underlying the successive-approximation A/D converter.

☐ Interface an 8-bit A/D converter with the 8085, using status check and interrupt.

13.1 DIGITAL-TO-ANALOG (D/A) CONVERTERS

Digital-to-Analog converters can be broadly classified in three categories: **current output, voltage output,** and **multiplying type.** The current output DAC, as the name suggests, provides current as the output signal. The voltage output DAC internally converts the current signal into the voltage signal. The voltage output DAC is slower than the current output DAC because of the delay in converting the current signal into the voltage signal. However, in many applications, it is necessary to convert current into voltage by using an external operational amplifier. The multiplying DAC is similar to the other two types except its output represents the product of the input signal and the reference source (Figure 13.3 will explain the reference source), and the product is linear over a broad range. Conceptually, there is not much difference between these three types; any DAC can be viewed as a multiplying DAC.

D/A converters are available as **integrated circuits.** Some are specially designed to be compatible with the microprocessor. Typical applications include digital voltmeters, peak detectors, panel meters, programmable gain and attenuation, and stepping motor drive.

13.11 Basic Concepts

Figure 13.1(a) shows a block diagram of a 3-bit D/A converter; it has three digital input lines (D_2, D_1, and D_0) and one output line for the analog signal. The three input lines can assume eight ($2^3 = 8$) input combinations from 000 to 111, D_2 being the most significant bit (MSB) and D_0 being the least significant bit (LSB). If the input ranges from 0 to 1 V, it can be divided into eight equal parts ($\frac{1}{8}$ V); each successive input is $\frac{1}{8}$ V higher than the previous combination, as shown in Figure 13.1(b).

The following points can be summarized from the graph:

1. The 3-bit D/A converter has eight possible combinations. If a converter has n input lines, it can have 2^n input combinations.
2. If the full-scale analog voltage is 1 V, the smallest unit or the LSB (001_2) is equivalent to $\frac{1}{2^n}$ of 1 V. This is defined as resolution. In this example, the LSB = $\frac{1}{8}$ V.
3. The MSB represents half of the full-scale value. In this example, the MSB (100_2) = $\frac{1}{2}$ V.
4. For the maximum input signal (111_2), the output signal is equal to the value of the full-scale input signal minus the value of the 1 LSB input signal. In this example, the maximum input signal (111_2) represents $\frac{7}{8}$ V.

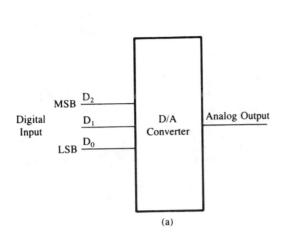

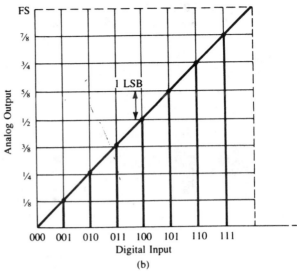

FIGURE 13.1
A 3-Bit D/A Converter: Block Diagram (a) and Digital Input vs. Analog Output (b)

Calculate the values of the LSB, MSB, and full-scale output for an 8-bit DAC for the 0 to 10 V range.

Example 13.1

1. LSB = $\frac{1}{2^8}$ = $\frac{1}{256}$
   For 10 V, LSB = 10 V/256 = 39 mV
2. MSB = $\frac{1}{2}$ full scale = 5 V
3. Full-Scale Output = (Full-Scale Value − 1 LSB)
   $$= 10\ V - 0.039\ V$$
   $$= 9.961\ V$$

Solution

13.12 D/A Converter Circuits

Input signals representing appropriate binary values can be simulated by an operational amplifier with a summing network, as shown in Figure 13.2.

The input resistors R_1, R_2, and R_3 are selected in binary weighted proportion; each has double the value of the previous resistor. If all three inputs are 1 V, the total output current is

$$I_O = I_T = I_1 + I_2 + I_3$$
$$= \frac{V_{in}}{R_1} + \frac{V_{in}}{R_2} + \frac{V_{in}}{R_3}$$
$$= \frac{V_{in}}{1\ k}\left(\frac{1}{2} + \frac{1}{4} + \frac{1}{8}\right)$$
$$= 0.875\ mA$$

FIGURE 13.2
Summing Amplifier with Binary
Weighted Input Resistors

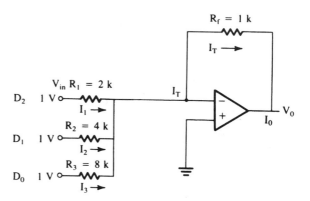

The voltage output is

$$V_O = -R_f I_T$$
$$= -(1\ k)(0.875\ mA)$$
$$= -0.875\ V$$
$$= |7/8\ V|$$

This example shows that for the input = 111_2, the output is equal to either $7/8$ mA or $7/8$ V, representing the D/A conversion process.

Now we can redraw Figure 13.2 as shown in Figure 13.3(a), where input voltage V_{in} is replaced by V_{Ref}, which can be turned on or off by the switches. The output current I_O can be generalized for any number of bits as

$$I_O = \frac{V_{Ref}}{R}\left(\frac{A_1}{2} + \frac{A_2}{4} + \cdots + \frac{A_n}{2^n}\right)$$

where A_1 to A_n can be zero or one.

Figure 13.3(b) shows an illustration of the transitorized switch for input D_0; when bit D_0 is high, it will drive the transistor into saturation, and the current is determined by the resistor R_3 with appropriate binary weighting. When bit D_0 is low, the transistor is turned off. The switching speed of the transistor, shown in Figure 13.3(b), determines the settling time of a D/A converter, which is defined as the time necessary for the output to stabilize within $\pm 1/2$ LSB of its final value. The accuracy of the output is dependent on the tolerance of resistor values, and it is generally specified in terms of relative accuracy (also known as linearity)—the difference between the actual output and the expected fraction for a given digital input. The accuracy of the converter is also specified in terms of monotonicity, which guarantees that analog output increases in magnitude with increasing digital code. Most commercial D/A converters are specified as monotonic; this limits the output error within $\pm 1/2$ LSB at each digital input.

The following points can be inferred from the above example:

1. A D/A converter circuit requires three elements: resistor network with appropriate weighting, switches, and a reference source.

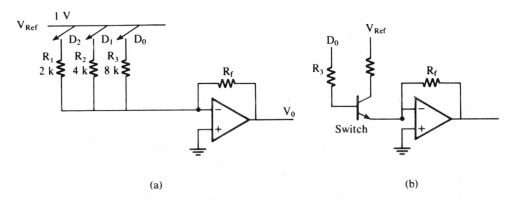

FIGURE 13.3
Simulated D/A Converter (a) and Transister Switch to Turn On/Off Bit D_0 (b)

2. The output can be a current signal or converted into a voltage signal using an operational amplifier.
3. The time required for conversion, called settling time, is dependent on the response time of the switches and the output amplifier (for a voltage output DAC).

R/2R LADDER NETWORK

One major drawback of designing a DAC as shown in Figure 13.3 is the requirement for various precision resistors. The R/2R ladder network shown in Figure 13.4 uses only two resistor values. The resistors are connected in such a way that for any number of inputs, the total current I_T is in binary proportion. The R/2R ladder network (or a similar network called an inverted ladder) is used commonly in designing integrated D/A converters. The interfacing of 8-bit and 10-bit integrated D/A converters is illustrated in the next two sections.

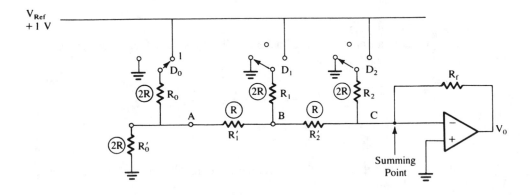

FIGURE 13.4
R/2R Ladder Network

13.13   Illustration: Interfacing an 8-Bit D/A Converter with the 8085

PROBLEM STATEMENT

1. Design an output port with the address FFH to interface the 1408 D/A converter that is calibrated for a 0 to 10 V range; refer to Figure 13.5(a).
2. Write a program to generate a continuous ramp waveform.
3. Explain the operation of the 1408 in Figure 13.5(b), which is calibrated for a bipolar range ±5 V. Calculate the output V_O if the input is 10000000_2.

HARDWARE DESCRIPTION

The circuit shown in Figure 13.5(a) includes an 8-input NAND gate and a NOR gate (negative AND) as the address decoding logic, the 74LS373 as a latch, and an industry-standard 1408 D/A converter. The address lines A_7–A_0 are decoded using the 8-input NAND gate, and the output of the NAND gate is combined with the control signal $\overline{IOW}$. When the microprocessor sends the address FFH, the output of the negative AND gate enables the latch, and the data bits are placed on the input lines of the converter for conversion.

The 1408 is an 8-bit D/A converter compatible with TTL and CMOS logic, with the settling time around 300 ns. It has eight input data lines A_1 (MSB) through A_8 (LSB); the convention of labeling MSB to LSB is opposite to that of what is normally used for the data bus in the microprocessor. It requires 2 mA reference current for full-scale input and two power supplies $V_{CC} = +5$ V and $V_{EE} = -15$ V (V_{EE} can range from -5 V to -15 V).

The total reference current source is determined by the resistor R_{14} and the voltage V_{Ref}. The resistor R_{15} is generally equal to R_{14} to match the input impedance of the reference source. The output I_O is calculated as follows:

$$I_O = \frac{V_{Ref}}{R_{14}}\left(\frac{A_1}{2} + \frac{A_2}{4} + \frac{A_3}{8} + \frac{A_4}{16} + \frac{A_5}{32} + \frac{A_6}{64} + \frac{A_7}{128} + \frac{A_8}{256}\right)$$

where inputs A_1 through $A_8 = 0$ or 1.

This formula is an application of the generalized formula for the current I_O. For full-scale input (D_7 through $D_0 = 1$),

$$I_O = \frac{5\text{ V}}{2.5\text{ k}}\left(\frac{1}{2} + \frac{1}{4} + \frac{1}{8} + \frac{1}{16} + \frac{1}{32} + \frac{1}{64} + \frac{1}{128} + \frac{1}{256}\right)$$
$$= 2\text{ mA }(255/256)$$
$$= 1.992\text{ mA}$$

The output is 1 LSB less than the full-scale reference source of 2 mA. The output voltage V_O for the full-scale input is

$$V_O = 2\text{ mA }(255/256) \times 5\text{ k}$$
$$= 9.961\text{ V}$$

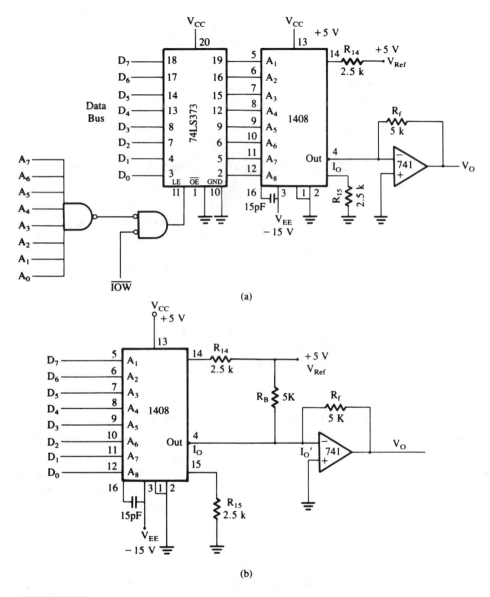

FIGURE 13.5
Interfacing the 1408 D/A Converter: Voltage Output in Unipolar Range (a) and in Bipolar Range (b)

PROGRAM

To generate a continuous waveform, the instructions are as follows:

```
                MVI A,00H          ;Load accumulator with the first input
DTOA:           OUT FFH            ;Output to DAC
                MVI B,COUNT        ;Set up register B for delay
DELAY:          DCR B
                JNZ DELAY
                INR A              ;Next input
                JMP DTOA           ;Go back to output
```

Program Description   This program outputs 00 to FF continuously to the D/A converter. The analog output of the DAC starts at 0 and increases up to 10 V (approximately)
as a ramp. When the accumulator contents go to 0, the next cycle begins; thus the ramp
signal is generated continuously. The ramp output of the DAC can be observed on an oscilloscope with an external sync.

   The delay in the program is necessary for two reasons:

1. The time needed for a microprocessor to execute an output loop is likely to be less
   than the settling time of the DAC.
2. The slope of the ramp can be varied by changing the delay.

OPERATING THE D/A CONVERTER IN A BIPOLAR RANGE

The 1408 in Figure 13.5(b) is calibrated for the bipolar range from −5 V to +5 V by
adding the resistor R_B (5.0 k) between the reference voltage V_{Ref} and the output pin 4.
The resistor R_B supplies 1 mA (V_{Ref}/R_B) current to the output in the opposite direction of
the current generated by the input signal. Therefore, the output current for the bipolar operation I_O' is

$$I_O' = I_O - \frac{V_{Ref}}{R_B}$$

$$= \frac{V_{Ref}}{R_{14}}\left(\frac{A_1}{2} + \frac{A_2}{4} + \frac{A_3}{8} + \frac{A_4}{16} + \frac{A_5}{32} + \frac{A_6}{64} + \frac{A_7}{128} + \frac{A_8}{256}\right) - \frac{V_{Ref}}{R_B}$$

When the input signal is equal to zero, the output V_O is

$$V_O = I_O'R_f$$

$$= \left(I_O - \frac{V_{Ref}}{R_B}\right)R_f$$

$$= \left(0 - \frac{5\ V}{5\ k}\right)(5\ k) \qquad (I_O = 0 \text{ for input} = 0)$$

$$= -5\ V$$

When the input = 1000 0000, output V_O is

$$V_O = \left(I_O - \frac{V_{Ref}}{R_B}\right)R_f$$

$$= \left(\frac{V_{Ref}}{R_{14}} \times \frac{A_1}{2} - \frac{V_{Ref}}{R_B}\right)R_f \qquad (A_2 - A_8 = 0)$$

$$= \left(\frac{5\ V}{2.5\ k} \times \frac{1}{2} - \frac{5\ V}{5\ k}\right)5\ k$$

$$= (1\ mA - 1\ mA)5\ k = 0\ V$$

13.14 Microprocessor-Compatible D/A Converters

In response to the growing need for interfacing data converters with the microprocessor, specially designed microprocessor-compatible D/A converters are now available. These D/A converters generally include a latch on the chip (see Figure 13.6), thus eliminating the need for an external latch as in Figure 13.5.

Figure 13.6 shows the block diagram of Analog Devices AD558, which includes a latch and an output op amp internal to the chip. It can be operated with one power supply voltage between +4.5 V to +16.5 V. To interface the AD558 with the microprocessor, two signals are required: Chip Select ($\overline{CS}$) and Chip Enable ($\overline{CE}$).

Figure 13.6 shows one example of interfacing the AD558 with the 8085. The address line A_7 through an inverter is used for the Chip Select, which assigns the port address 80H (assuming all other address lines are at logic 0) to the DAC port. The control signal $\overline{IOW}$ is used for the Chip Enable. The program shown in Section 13.13 can be used to generate ramp waveforms.

Figure 13.6(b) shows the timing of latching data in relation to the control signals; Figure 13.6(c) shows the truth table of the control logic. When both signals $\overline{CS}$ and $\overline{CE}$ are at logic 0, the latch is transparent, meaning the input is transferred to the DAC section. When either $\overline{CS}$ or $\overline{CE}$ goes to logic 1, the input is latched in the register and held until both control signals go to logic 0.

13.15 Interfacing a 10-Bit D/A Converter

In many D/A converter applications, 10- or 12-bit resolution is required. However, the 8-bit microprocessor has only eight data lines. Therefore, to transfer ten bits, the data bus is time-shared by using two output ports: one for the first eight bits and the second for the remaining two bits. A disadvantage of this method is that the DAC output assumes an intermediate value between the two output operations. This difficulty can be circumvented by using a double-buffered DAC such as Analog Devices AD7522, as shown in Figure 13.7(a).

The AD7522 is a CMOS 10-bit D/A converter with an input buffer and a holding register. The ten bits are loaded into the input register in two steps using two output ports. The low-order eight bits are loaded with the control line LBS, and the remaining two bits are loaded with the control line HBS. Then all ten bits are switched into a holding register for the conversion by enabling the line LDAC. The last operation (enabling the line LDAC) can be combined with the loading of the second byte as shown in Figure 13.7(a); otherwise, this operation will require three output ports.

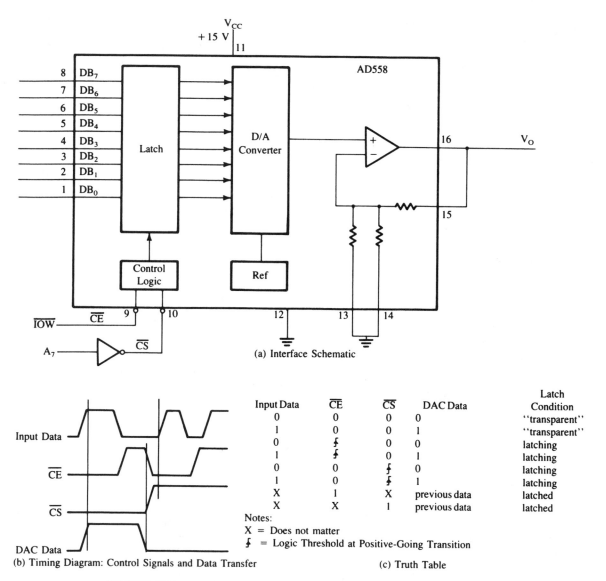

(a) Interface Schematic

Input Data	$\overline{CE}$	$\overline{CS}$	DAC Data	Latch Condition
0	0	0	0	"transparent"
1	0	0	1	"transparent"
0	∫	0	0	latching
1	∫	0	1	latching
0	0	∫	0	latching
1	0	∫	1	latching
X	1	X	previous data	latched
X	X	1	previous data	latched

Notes:
X = Does not matter
∫ = Logic Threshold at Positive-Going Transition

(b) Timing Diagram: Control Signals and Data Transfer   (c) Truth Table

FIGURE 13.6
Interfacing the AD558 (Microprocessor-Compatible D/A Converter) with the 8085.
SOURCE: (b) and (c): Analog Devices, Inc., *Data Converter Reference Manual* (Norwood, Mass.: Author, 1992), pp. 2, 52.

HARDWARE DESCRIPTION

Figure 13.7(a) shows a schematic of interfacing the AD7522 with the 8085. It is a memory-mapped I/O with multiple addresses. The attempt here is to minimize the chip count.

The three input signals to the decoder are address line A_0 and two control signals $IO/\overline{M}$ and $\overline{WR}$. To enable the line LBS, the input to the decoder should be 000; this re-

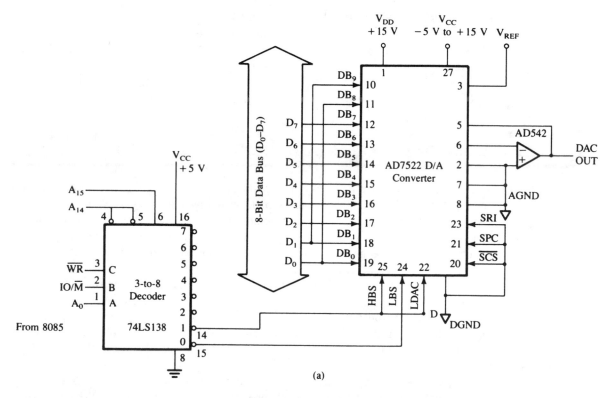

FIGURE 13.7

Interfacing 10-Bit DAC (AD7522) with the 8085 (a) and a Timing Diagram for Loading the Input Data (b)

SOURCE: Analog Devices, Inc., *Data Acquisition Components and Subsystems* (Norwood, Mass.: Author, 1980), pp. 9–73.

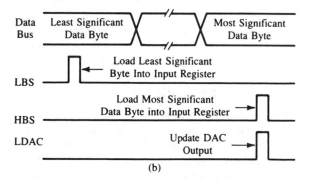

sults in the port address 8000H. ($A_{15} = 1$, $A_{14} = 0$, and the lines A_{13} to A_1 are assumed at logic 0.) When a data byte is sent to the port address 8000H in a memory-mapped I/O, the $\overline{WR}$ and IO/$\overline{M}$ signals go low along with A_0, and the line LBS is enabled. Similarly, the address 8001H enables the lines HBS and LDAC. Figure 13.7(b) shows the timing diagram for loading input data into the converters.

The following instructions illustrate how to load the maximum input of ten bits (all 1s) into the D/A converter.

```
LXI B,03FFH      ;Load ten bits at logic 1 in BC register
LXI H,8000H      ;Load HL with port address for low-order 8-bits
MOV M,C          ;Load eight bits (D7–D0) in the DAC
INX H            ;Point to port address 8001H
MOV M,B          ;Load two bits (D9 and D8) and switch all ten bits for conversion
HLT
```

13.2 ANALOG-TO-DIGITAL (A/D) CONVERTERS

The A/D conversion is a quantizing process whereby an analog signal is represented by equivalent binary states; this is opposite to the D/A conversion process. Analog-to-Digital converters can be classified into two general groups based on the conversion technique. One technique involves comparing a given analog signal with the internally generated equivalent signal. This group includes successive-approximation, counter, and flash-type converters. The second technique involves changing an analog signal into time or frequency and comparing these new parameters to known values. This group includes integrator converters and voltage-to-frequency converters. The trade-off between the two techniques is based on accuracy vs. speed. The successive-approximation and the flash type are faster but generally less accurate than the integrator and the voltage-to-frequency type converters. Furthermore, the flash type is expensive and difficult to design for high accuracy.

The successive-approximation A/D converters are used in applications such as data loggers and instrumentation, where conversion speed is important. On the other hand, integrating-type converters are used in applications such as digital meters, panel meters, and monitoring systems, where the conversion accuracy is critical. The most commonly used A/D converters—the successive-approximation—is discussed in this section with several interfacing examples.

13.21 Basic Concepts

Figure 13.8(a) shows a block diagram of a 3-bit A/D converter. It has one input line for an analog signal and three output lines for digital signals. Figure 13.8(b) shows the graph of the analog input voltage (0 to 1 V) and the corresponding digital output signal. It shows eight (2^3) discrete output states from 000_2 to 111_2, each step being 1/8 V apart. This is defined as the resolution of the converter. The LSB, the MSB, and the full-scale output are calculated the same way as in D/A converters.

In A/D conversion, another critical parameter is **conversion time.** This is defined as the total time required to convert an analog signal into its digital output and is determined by the conversion technique used and by the propagation delay in various circuits.

13.22 Successive-Approximation A/D Converter

Figure 13.9(a) shows the block diagram of a successive approximation A/D converter. It includes three major elements: the D/A converter, the successive approximation register

FIGURE 13.8

A 3-Bit A/D Converter: Block
Diagram (a) and Analog Input vs.
Digital Output (b)

SOURCE: Analog Devices, Inc., *Integrated
Circuit Converters, Data Acquisition Systems,
and Analog Signal Conditioning Components*
(Norwood, Mass.: Author, 1979), pp. 1-18.

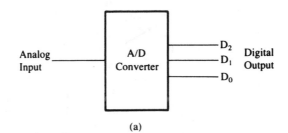

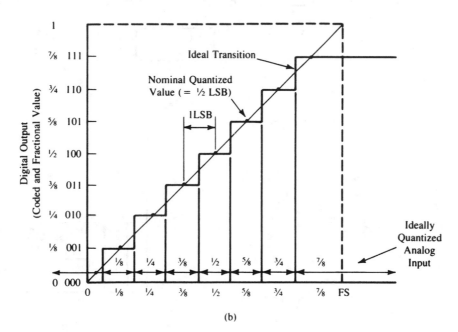

(SAR), and the comparator. The conversion technique involves comparing the output of the D/A converter V_O with the analog input signal V_{in}. The digital input to the DAC is generated using the successive-approximation method (explained below). When the DAC output matches the analog signal, the input to the DAC is the equivalent digital signal.

The successive-approximation method of generating input to the DAC is similar to weighing an unknown material (e.g., less than 1 gram) on a chemical balance with a set of such fractional weights as ½ g, ¼ g, ⅛ g, etc. The weighing procedure begins with the heaviest weight (½ g), and subsequent weights (in decreasing order) are added until the balance is tipped. The weight that tips the balance is removed, and the process is continued until the smallest weight is used. In the case of a 4-bit A/D converter, bit D_3 is turned on first and the output of the DAC is compared with an analog signal. If the comparator changes the state, indicating that the output generated by D_3 is larger than the analog signal, bit D_3 is turned off in the SAR and bit D_2 is turned on. The process continues until the input reaches bit D_0. Figure 13.9(b) illustrates a 4-bit conversion process. When bit D_3

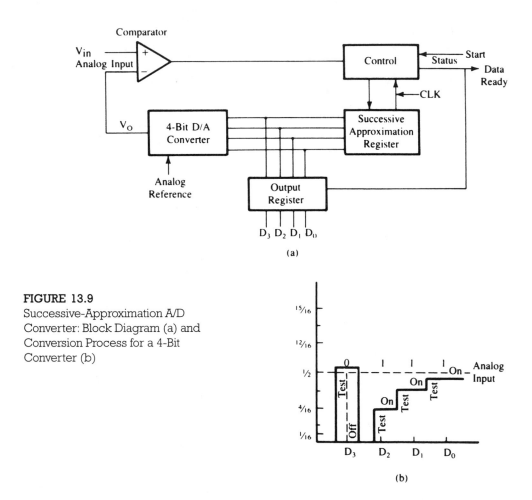

FIGURE 13.9

Successive-Approximation A/D
Converter: Block Diagram (a) and
Conversion Process for a 4-Bit
Converter (b)

is turned on, the output exceeds the analog signal and, therefore, bit D_3 is turned off.
When the next three successive bits are turned on, the output becomes approximately
equal to the analog signal.

The successive-approximation conversion process can be accomplished through either the software or hardware approach. In the software approach, an A/D converter is designed using a D/A converter, and the microprocessor plays the role of the counter and
the SAR. For the hardware approach, a complete ADC, including a tri-state buffer, is now
available as an integrated circuit on a chip. The interfacing of both types of A/D converters is illustrated in the next section.

13.23   Interfacing 8-Bit A/D Converters

As an integrated circuit, the A/D converter includes all three elements—SAR, DAC, and
comparator—on a chip (Figure 13.10). In addition, it has a tri-state output buffer.

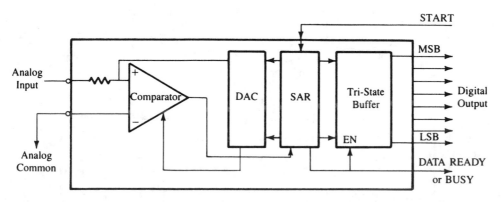

FIGURE 13.10
Block Diagram of a Typical Successive-Approximation A/D Converter as an Integrated Circuit

Typically, it has two control lines, START (or CONVERT) and DATA READY (or BUSY); they are TTL-compatible and can be active low or high depending upon the design.

A pulse to the START pin begins the conversion process and disables the tri-state output buffer. At the end of the conversion period, DATA READY becomes active and the digital output is made available at the output buffer. To interface an A/D converter with the microprocessor, the microprocessor should

1. send a pulse to the START pin. This can be derived from a control signal such as Write ($\overline{WR}$).
2. wait until the end of the conversion. The end of the conversion period can be verified either by status checking (polling) or by using the interrupt.
3. read the digital signal at an input port.

Two examples of interfacing A/D converters are discussed in the following sections. The first example demonstrates the interfacing of a typical A/D converter using the status check approach, and the second example demonstrates the interfacing of the National Semiconductor ADC0801 using the interrupt. An example of using handshake signals is given in Chapter 14.

INTERFACING AN 8-BIT A/D CONVERTER USING STATUS CHECK

Figure 13.11 shows a schematic of interfacing a typical A/D converter using status check. The A/D converter has one input line for analog signal and eight output lines for converted digital signals. Typically, the analog signal can range from 0 to 10 V, or ±5 V. In addition, the converter shows two lines $\overline{START}$ and $\overline{DR}$ (Data Ready), both active low (the active logic level of these lines can be either low or high depending upon the design of a particular converter). When an active low pulse is sent to the $\overline{START}$ pin, the DA

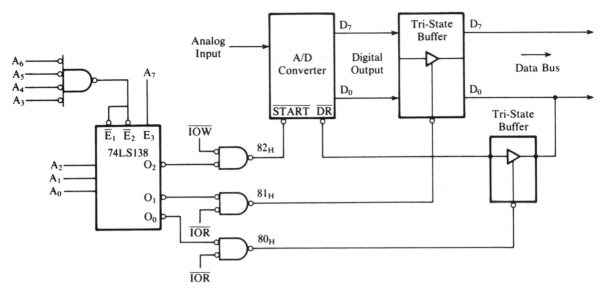

FIGURE 13.11
Interfacing an A/D Converter Using the Status Check

goes high and the output lines go into the high impedance state. The rising edge of the START pulse initiates the conversion. When the conversion is complete, the DR goes low, and the data are made available on the output lines that can be read by the microprocessor. To interface this converter, we need one output port to send a START pulse and two input ports: one to check the status of the DR line and the other to read the output of the converter.

In Figure 13.11, the address decoding is performed by using the 3-to-8 decoder (74LS138), the 4-input NAND gate, and inverters. Three output lines of the decoder are combined with appropriate control signals (IOW and IOR) to assign three port addresses from 80H to 82H. The output port 82H is used to send a START pulse by writing the OUT instruction; in this case, we are interested in getting a pulse from the microprocessor, and the contents of the accumulator are irrelevant to start the conversion. However, for some converters the IOW pulse from the microprocessor may not be long enough to start the conversion. When the conversion begins, the DR (Data Ready) goes high and stays high until the conversion is completed. The status of the DR line is monitored by connecting the line to bit D_0 of the data bus through a tri-state buffer with the input port address 80H. The processor will continue to read port 80H until bit D_0 goes low. When the DR goes active, the data are available on the output lines of the converter, and the processor can access that data by reading the input port 81H. The subroutine instructions, to initiate the conversion and to read output data, and the flowchart are shown in Figure 13.12.

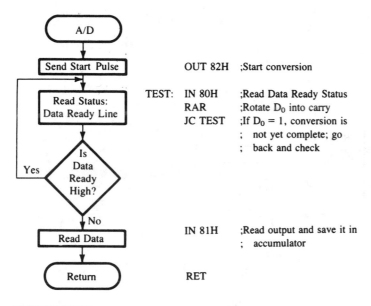

FIGURE 13.12
Flowchart of A/D Conversion Process

13.24 Illustration: Interfacing an 8-Bit A/D Converter Using the Interrupt

PROBLEM STATEMENT

1. Interface the National Semiconductor ADC0801 converter with the 8085 MPU using memory-mapped I/O and the interrupt RST 6.5.
2. Write an interrupt routine to read the output data of the converter, store it in memory, and continue to collect data for the specified number of times.

HARDWARE DESCRIPTION

The ADC0801 is a CMOS 8-bit successive-approximation A/D converter housed in a 20-pin DIP package. The input voltage can range from 0 to 5 V and operates with a single power supply of +5 V. It has two inputs $V_{IN}(+)$ and $V_{IN}(-)$ for the differential analog signal. When the analog signal is single-ended positive, $V_{IN}(+)$ is used as the input and the $V_{IN}(-)$ pin is grounded; when the signal is single-ended negative, $V_{IN}(-)$ is used as the input and the $V_{IN}(+)$ pin is grounded. The converter requires a clock at CLK IN; the frequency range can be from 100 kHz to 800 kHz. The user has two options: either to connect an external clock at CLK IN or to use the built-in internal clock by connecting a resistor and a capacitor externally at pins 19 and 4, respectively, as shown in Figure 13.13. The frequency is calculated by using the formula $f = 1/1.1(RC)$. Typically, the clock frequency is designed for 640 kHz to provide 100 μs of conversion time.

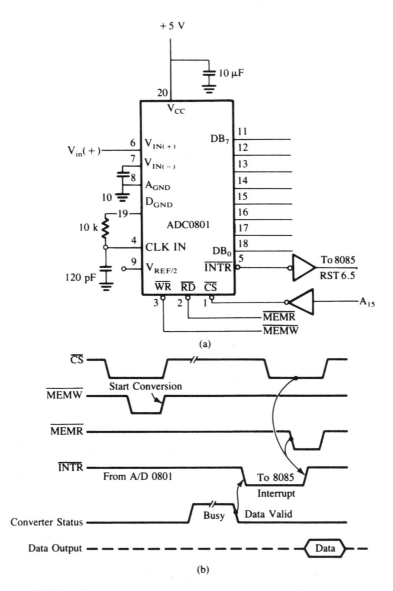

FIGURE 13.13

The ADC 0801 Using the Interrupt: Interface (a) and Timing Diagram for Reading Data from A/D Converter (b)

SOURCE: (b): National Semiconductor, *Data Conversion/Acquisition Data Book* (Santa Clara, Calif.: Author, 1980), pp. 5–25.

The ADC0801 is designed to be microprocessor-compatible. In our previous example of Figure 13.11, we needed external ports to access data and monitor the Data Ready signal; however, in this converter, the necessary control logic is built inside the chip. It has three control signals—$\overline{CS}$, $\overline{WR}$, and $\overline{RD}$—that are used for interfacing. To start conversion, the $\overline{CS}$ and $\overline{WR}$ signals are asserted low. When $\overline{WR}$ goes low, the internal SAR

(Successive-Approximation Register) is reset, and the output lines go into the high impedance state. When $\overline{WR}$ makes the transition from low to high, the conversion begins. When the conversion is completed, the $\overline{INTR}$ is asserted low and the data are placed on the output lines. The $\overline{INTR}$ signal can be used to interrupt the processor. When the processor reads the data by asserting $\overline{RD}$, the $\overline{INTR}$ is reset.

When V_{CC} is +5 V, the input voltage can range from 0 V to 5 V and the corresponding output will be from 00 to FFH. However, the full-scale output can be restricted to the lower range of inputs by using pin 9 ($V_{REF/2}$). For example, if we connect a 0.5 V dc source at pin 9, we can obtain full-scale output FFH for a 1 V input signal (this is twice the voltage of pin 9).

The ADC0801 can be operated in a continuous mode by connecting $\overline{WR}$ to $\overline{INTR}$ and grounding $\overline{CS}$ and $\overline{RD}$. However, our focus here is to use the interrupt RST 6.5 to collect data from the converter.

INTERFACING CIRCUIT

Figure 13.13(a) shows the interfacing of the ADC0801 with the 8085 MPU, using the interrupt. Address line A_{15} with an inverter is used for Chip Select ($\overline{CS}$), and the control signals $\overline{MEMR}$ and $\overline{MEMW}$ are connected to $\overline{RD}$ and $\overline{WR}$ signals, respectively. This is a memory-mapped port with the address 8000H (assuming all don't care lines at logic 0).

The conversion is initiated when the $\overline{CS}$ and $\overline{WR}$ signals go low; see Figure 13.13(b). At the end of the conversion, the $\overline{INTR}$ signal goes low and is used to interrupt the MPU through an inverter. When the service routine reads the data byte, the $\overline{RD}$ signal causes the $\overline{INTR}$ to go high, as shown in the timing diagram, Figure 13.13(b). This chip includes the control logic to set $\overline{INTR}$ at the end of a conversion and to reset it when data are read; by including this logic on the converter chip, extra components necessary for interfacing are eliminated.

To implement data transfer using the interrupt, as shown in Figure 13.13(a), the main program should initialize the stack, enable the microprocessor interrupts (EI), unmask the RST 6.5 (SIM), and initiate a conversion by writing to port 8000H. In addition, the main program should include the initialization of the memory pointer for storing data and the counter to count the readings. At the end of the conversion, the microprocessor is interrupted by using RST 6.5, which transfers the program control to location 0034H and then to the service routine.

SERVICE ROUTINE

LDA 8000H	;Read data
MOV M,A	;Store data in memory
INX H	;Next memory location
DCR B	;Next count
STA 8000H	;Start next conversion
EI	;Enable interrupt again
RNZ	;Go back to main if counter ≠ 0
HLT	;End

The service routine reads the output data by using the instruction LDA, stores the byte in memory, and updates the memory pointer and the counter. The routine assumes that the information concerning the memory pointer (HL) and the counter (B) is supplied by the main program. The memory pointer specifies the location where the data should be stored, and the counter specifies the number of bytes to be collected. The STA instruction starts the next conversion by asserting the $\overline{\text{MEMW}}$ signal; this instruction should not be interpreted to mean that it is storing the contents of the accumulator in the converter. Then the service routine sets the interrupt flip-flop for subsequent interrupts and returns to the main program if the counter is not zero. When the counter goes to zero, the program completes the data collection.

SUMMARY

☐ D/A converters transform a digital signal into an equivalent analog signal, and A/D converters transform an analog signal into an equivalent digital signal.

☐ Resolution of a converter determines the degree of accuracy in conversion. It is equal to $\frac{1}{2^n}$ (n is the number of bits).

☐ Settling time (for DAC) and conversion time (for ADC) are important parameters in selecting data converters; they suggest the speed of data conversion.

☐ D/A converters are classified into three categories according to their output function: current, voltage, and multiplying.

☐ The successive-approximation A/D converter is a high-speed converter and uses a D/A converter to compare an analog signal with an internally generated signal.

☐ A successive-approximation A/D converter is available as an integrated circuit on a chip or can be designed by using software and a D/A converter.

☐ The integrating-type A/D converter is a high-accuracy converter. It is based on the principle of converting an analog signal into a time period and measuring the time period with a digital counter.

☐ A/D converters can serve as input devices and D/A converters can serve as output devices to microprocessor-based systems.

☐ A/D converters can be interfaced with the microprocessor by using techniques such as status check and interrupt.

☐ When 12-bit (or 16-bit) converters are interfaced with an 8-bit microprocessor, data are transferred in two (or three) stages using multiple ports.

QUESTIONS, PROBLEMS, AND PROGRAMMING ASSIGNMENTS

1. Calculate the output current for the 1408, Figure 13.5(a), if the input is 82H and the converter is calibrated for a 0 to 2 mA current range.
2. What is the voltage output in Problem 1?
3. Figure 13.5(b) shows a circuit that is calibrated for −5 V to +5 V. Calculate the output voltage if the input is 45H.

4. What changes are necessary in the ramp program, Figure 13.5(a), to limit the peak voltage to 7.5 V?

5. Modify the program of Figure 13.5(a) to generate a square wave with the amplitude of 5 V and a 1 kHz frequency.

6. Calculate the resolution of a 12-bit D/A converter.

7. A 12-bit D/A converter is calibrated over the range 0 to 10 V. Calculate the outputs if the input is 01H and 82H.

8. Calculate the analog voltages corresponding to the LSB and the MSB for a 12-bit A/D converter calibrated for a 0 to 5 V range.

9. In Figure 13.11, assume that the data line D_7 is connected to the line $(\overline{DR})$, instead of D_0, to check the status, and make the necessary changes in the instructions.

10. Change the circuit in Figure 13.11 from the peripheral I/O to the memory-mapped I/O, and assign the port addresses as 8000H to 8002H. Assume that the unused address lines are at logic 0. Rewrite the instructions.

11. Modify the circuit in Figure 13.11 to interface a 12-bit A/D converter.

12. Write a main program for the service routine in Section 13.24.

EXPERIMENTAL ASSIGNMENTS

1. **a.** Connect the circuit as shown in Figure 13.5(a) and calibrate the D/A converter for 0 to 10 V.
   b. Write a DELAY routine for a 100 µs delay, and enter the program.
   c. Measure the frequency and the slope of the ramp on an oscilloscope.
   d. Change the DELAY routine to 500 µs. Measure the frequency and the slope of the ramp.
   e. Modify the program to limit the maximum peak voltage of the ramp to 5 V.
   f. Modify the program to generate a triangular waveform.
   g. Modify the program as suggested below:
       (1) Store some random data at locations XX50H to XX5FH.
       (2) Change instructions to call out the data in sequence. Display on an oscilloscope.
       (3) Continue displaying the data to observe a stable pattern.
       (4) Eliminate the DELAY routine and observe the output.

2. **a.** Modify the interfacing circuit in Figure 13.13 to change memory-mapped I/O to peripheral I/O and assign the port address F8H to the A/D converter by using a 3-to-8 decoder and a NAND gate.
   b. Calculate the clock frequency and measure at CLK IN.
   c. Record digital output readings for five different V_{IN} values from 0 V to +5 V, and store them in memory. Rewrite the interrupt routine to record one reading at a time.
   d. Repeat step c by connecting +1 V dc at pin 9.

Programmable
Interface Devices:
▪ 8155 I/O and Timer
▪ 8279 Keyboard/Display
Interface

14

A programmable interface device is designed to perform various input/output functions. Such a device can be set up to perform specific functions by writing an instruction (or instructions) in its internal register, called the control register. Furthermore, functions can be changed anytime during execution of the program by writing a new instruction in the control register. These devices are flexible, versatile, and economical; they are widely used in microprocessor-based products.

In Chapter 4, we used simple integrated circuits, such as latches and tri-state buffers for I/O functions. However, they are limited in their capabilities; each device can perform one function, and they are hard-wired.

In a programmable device, on the other hand, functions are determined through software instructions. A programmable interface device can be viewed as multiple I/O devices, but it also performs many other functions, such as time delays, counting, and interrupts. In fact, it consists of many devices on a single chip, interconnected through a common bus. This is a hardware approach through software control to performing the I/O functions discussed earlier. This approach, a trade-off between hardware and software, should reduce programming.

This chapter describes two programmable devices: the 8155/8156 I/Os and timer, and the 8279 Keyboard/Display Interface. The 8155 is specifically designed to be compatible with the 8085 and the 8279 is a general-purpose device. The Intel family of support devices includes several other general-purpose devices that will be discussed in the next chapter.

This chapter first describes the basic concepts underlying these programmable devices. On the basis of these concepts the 8155 and the 8279 are discussed in the context of a single-board microcomputer. The 8155 is a multipurpose chip. Its memory section has been discussed already in Chapter 3; only the programmable I/Os and the timer are discussed in this chapter.

OBJECTIVES

☐ List the elements and characteristics of a typical programmable device.
☐ Explain the functions of handshake signals.
☐ Explain the block diagram of the 8155 I/O section and timer.

☐ Design an interfacing circuit for the 8155 I/O ports and the timer, and write initialization instructions.
☐ Set up the 8155 I/O ports in the handshake mode and write initialization instructions.
☐ Set up the 8155 timer to generate a pulse after a given time delay or a continuous waveform.
☐ Explain the block diagram of the 8279 Keyboard/Display Interface and its operation.
☐ Write instructions to initialize the 8279 in a given mode.

14.1   BASIC CONCEPTS IN PROGRAMMABLE DEVICES

In Chapter 4, we discussed the interfacing of simple input (switches) and output (LEDs) devices. In the illustrations, we assumed that the I/O devices were always ready for data transfer. In fact, the assumption may not be valid in many data transfer situations. The MPU needs to check whether a peripheral is ready before it reads from or writes into a device because the execution speed of the microprocessor is much faster than the response of a peripheral such as a printer. For example, when the MPU sends data bytes (characters) to a printer, the microprocessor can execute the instructions to transfer a byte in microseconds; on the other hand, the printer can take 10 to 25 ms to print a character. After transferring a character to the printer, the MPU should wait until the printer is ready for the next character; otherwise data will be lost. To prevent the loss of data or the MPU reading the same data more than once, signals are exchanged between the MPU and a peripheral prior to actual data transfer; these signals are called handshake signals. To provide such signals in the illustrations of Chapter 4, we need to build additional logic circuitry.

In Chapter 13, we interfaced the A/D converter (ADC0801) using the interrupt I/O; however, the interrupt signal was generated by the internal logic of the data converter. Many peripherals may not have that capability; such signals may have to be provided by the interfacing circuitry. In some applications, data flow is bidirectional (such as data transfer between two computers). In such a situation, the interfacing device should be capable of handling bidirectional data flow. On the basis of the above discussion and the illustrations of Chapters 4 and 13, we can summarize the requirements for a programmable interfacing device as follows:

The device should include

1. input and output registers (a group of latches to hold data).
2. tri-state buffers.
3. capability for bidirectional data flow.
4. handshake and interrupt signals.

5. control logic.
6. chip select logic.
7. interrupt control logic.

To understand the programmability of such a device, we will, in the next section, illustrate a simple example of building a programmable device using a transceiver (bidirectional buffer).

14.11 Making the 74LS245 Transceiver Programmable

The 74LS245 is a bidirectional tri-state octal buffer, and the direction of the data flow is determined by the signal DIR. Figure 14.1 shows the logic diagram of the 74LS245; it shows one buffer (rather than eight) in each direction. The buffer is enabled when $\overline{G}$ is active low; however, the direction of the data flow is determined by the DIR signal. When the DIR is high, data flow from A to B, and when it is low, data flow from B to A. In fact, this is a hard-wired programmable device; the direction of the data flow is programmed through DIR. However, we are interested in a device that can be programmed by writing an instruction through the MPU. This can be accomplished by adding a register called the control register, as shown in Figure 14.2, and by connecting the DIR signal to bit D_0 of the control register. When $D_0 = 1$, data flow from A to B as output, and when $D_0 = 0$, data flow in the opposite direction as input.

Now the question is: How would the MPU write into the control register? The same way it would with any other I/O port—through a port address. Figure 14.2 shows that the address lines A_7–A_1 are used to select the chip through a NAND gate, and A_0 is used to differentiate between the control register and the transceiver. When A_0 is high, the control register is enabled; when A_0 is low, the transceiver is enabled. Thus, the MPU could access the control register through the port address FFH, and the transceiver through FEH. To set up the transceiver as an output device, the control word would be 01H, and to set it up as an input device the control word would be 00H.

FIGURE 14.1

Logic Symbol of 74LS245
Bidirectional Buffer

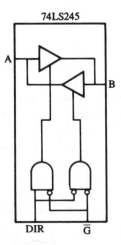

74LS245

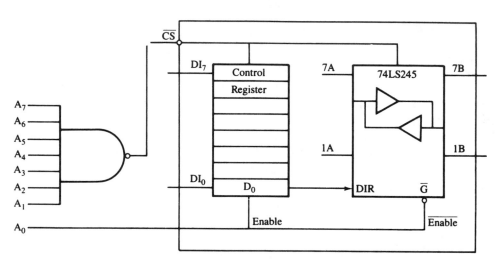

FIGURE 14.2
Making 74LS245 Programmable

Example 14.1   Write instructions to initialize the hypothetical chip (Figure 14.2) as an output buffer.

Solution   **Instructions**

MVI A,01H	;Set $D_0 = 1$; D_1 through D_7 are don't care lines
OUT FFH	;Write in the control register
MVI A,BYTE1	;Load data byte
OUT FEH	;Send data out

In the last example, we used the 74LS245 as a bidirectional buffer. However, in microprocessor applications, we often need registers that can be used as I/O ports. We can build a bidirectional latch with an input and an output buffer, and by controlling the enable signals of the latch and buffers, we can program it to function as an input port or an output port. Figure 14.3 shows a data register (representing eight latches in each direction); the I/O mode of this register is determined by bit D_0 in the control register. If bit D_0 is 0, it functions as an output port; if D_0 is 1, it functions as an input port; thus, by programming bit D_0, the device can function as an input port or an output port. When the device is programmed as an output device, the MPU can write to the port by using the $\overline{WR}$ control signal that enables the tri-state buffer (not shown) and send out a byte. When bit $D_0 = 1$, the input latch is enabled, the output latch is disabled, and the MPU can read by using the $\overline{RD}$ signal.

14.12   Programmable Device with a Status Register

Figure 14.3 shows an additional input register called the status register. In the discussion of interfacing A/D converters in Chapter 13, we needed to build an external input port (Figure 13.11) to monitor the status of the Data Ready signal. In a programmable device, we can build such a register internally that can monitor the data lines of the data register as shown in Figure 14.3. Now we have three registers that can be accessed as ports. The decode logic for Chip Select is similar to that in Figure 14.2, with FFH as the port address for the control and the status registers and FEH as the port address for the data register. The control and the status registers are differentiated by the $\overline{WR}$ and $\overline{RD}$ control signals even if they have the same port address. The port addresses and the functions of these registers can be summarized as follows:

$$\text{Control register (output only)} = \text{FFH } (A_1 \text{ and } A_0 = 1)$$
$$\text{Status register (input only)} = \text{FFH } (A_1 \text{ and } A_0 = 1)$$
$$\text{Data register (input or output)} = \text{FEH } (A_1 = 1, A_0 = 0)$$

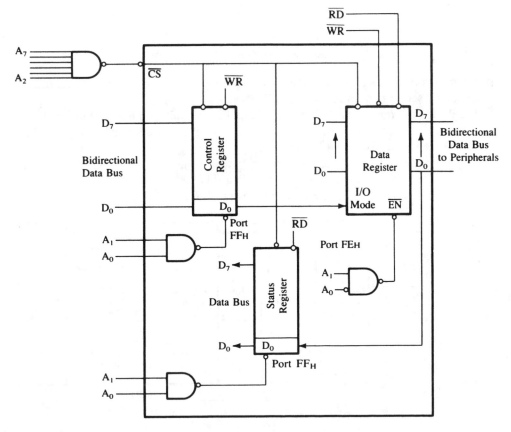

FIGURE 14.3
A Hypothetical Programmable Device with a Status Register

Thus we can build additional ports, and other bits in the control register can be used to define the functions of these registers. Similarly, we can add an interrupt logic and handshake signals.

14.13 Programmable Devices with Handshake Signals

The MPU and peripherals operate at different speeds; therefore, signals are exchanged prior to data transfer between the fast-responding MPU and slow-responding peripherals such as printers and data converters. These signals are called **handshake signals.** The exchange of handshake signals prevents the MPU from writing over the previous data before a peripheral has had a chance to accept it or from reading the same data before a peripheral has had time to send the next data byte. These signals are generally provided by programmable devices. Figure 14.4(a) shows a programmable device in the input mode, with two handshake signals (STB and IBF) and one interrupt signal (INTR). Now the MPU has two ways of finding out whether a peripheral is ready: either by checking the status of a handshake signal or through the interrupt technique, as explained below.

DATA INPUT WITH HANDSHAKE

The steps in data input from a peripheral such as a keyboard are as follows:

1. A peripheral strobes or places a data byte in the input port and informs the interfacing device by sending handshake signal STB (Strobe).
2. The device informs the peripheral that its input port is full—do not send the next byte until this one has been read. This message is conveyed to the peripheral by sending handshake signal IBF (Input Buffer Full).
3. The MPU keeps checking the status until a byte is available. Or the interfacing device informs the MPU, by sending an interrupt, that it has a byte to be read.
4. The MPU reads the byte by sending control signal $\overline{RD}$.

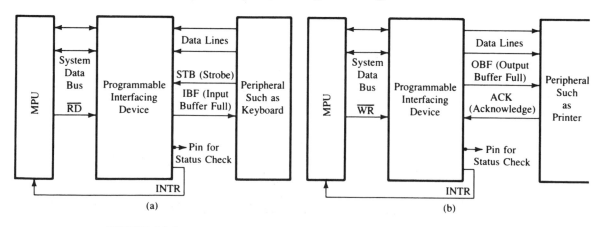

FIGURE 14.4
Interfacing Device with Handshake Signals for Data Input (a) and Data Output (b)

DATA OUTPUT WITH HANDSHAKE

Figure 14.4(b) shows the programmable device in the output mode using the same two handshake signals, except that they are labeled differently. The steps in data output to a peripheral such as a printer are as follows:

1. The MPU writes a byte into the output port of the programmable device by sending control signal $\overline{WR}$.
2. The device informs the peripheral, by sending handshake signal OBF (Output Buffer Full), that a byte is on the way.
3. The peripheral acknowledges the byte by sending back the ACK (Acknowledge) signal to the device.
4. The device interrupts the MPU to ask for the next byte, or the MPU finds out that the byte has been acknowledged through the status check.

Examination of Figure 14.4 shows the following similarities among the handshake signals:

1. Handshake signals ACK and STB are input signals to the device and perform similar functions, although they are called by different names.
2. Handshake signals OBF and IBF are output signals from the device and perform similar functions (Buffer Full).

The active levels and labels of these signals are arbitrary and vary considerably from device to device. For example, in the 8155 (R/W Memory with I/O) both input signals are called $\overline{STB}$ (Strobe); in the 8255 (Programmable Peripheral Interface) one input signal is called $\overline{STB}$ (Strobe) and the other is called $\overline{ACK}$ (Acknowledge).

14.14   Review of Important Concepts

The above examples suggest that a programmable I/O device is likely to have the following elements:

1. A control register in which the MPU can write an instruction
2. A status register that can be read by the MPU
3. I/O devices or registers
4. Control logic
5. Chip Select logic
6. Bidirectional data bus
7. Handshake signals and Interrupt logic

A programmable I/O device is programmed by writing a specific word, called the **control word,** according to the internal logic; its status can be verified by reading the status register. This I/O device can be expanded to include elements such as multiple I/O ports, counters, and parallel-to-serial registers. Two programmable devices commonly used in 8085 single-board systems—the 8155 and the 8279—are described in detail in the next sections.

14.2 THE 8155: MULTIPURPOSE PROGRAMMABLE DEVICE

The 8155 is a multipurpose programmable device specifically designed to be compatible with the 8085 microprocessor. The ALE, IO/M, RD, and WR signals from the 8085 can be connected directly to the device; this eliminates the need for external demultiplexing of the low-order bus AD_7-AD_0 and generation of the control signals such as MEMR, MEMW, IOR, and IOW.

The 8155 includes 256 bytes of R/W memory, three I/O ports, and a timer. The programmable I/O sections of this device are illustrated in the following sections.

14.21 The 8155 Programmable I/O Ports and Timer

The 8155 is a device with two sections: the first is 256 bytes of R/W memory, and the second is a programmable I/O. Functionally, these two sections can be viewed as two independent chips. The I/O section includes two 8-bit parallel I/O ports (A and B), one 6-bit port (C), and a timer (Figure 14.5). All the ports can be configured simply as input/output ports. Ports A and B also can be programmed in the handshake mode, each port using three signals as handshake signals from port C. The timer is a 14-bit down-counter and has four modes. Pins PA, PB, and PC, shown in Figure 14.5, correspond to ports A, B, and C.

CONTROL LOGIC

The control logic of the 8155 is specifically designed to eliminate the need for externally demultiplexing lines AD_7-AD_0 and generating separate control signals for memory and I/O. Figure 14.5 shows five control signals; all except the Chip Enable (CE) are input signals directly generated by the 8085.

☐ CE—Chip Enable: This is a master Chip Select signal connected to the decoded high-order bus.
☐ IO/M—When this signal is low, the memory section is selected, and when it is high, the I/O section (including timer) is selected.
☐ ALE—Address Latch Enable: This signal latches the low-order address AD_7-AD_0, CE, and IO/M into the chip.
☐ RD and WR—These are control signals to read from and write into the chip registers and memory.
☐ RESET—This is connected to RESET OUT of the 8085 and this resets the chip and initializes I/O ports as input.

THE 8155 I/O PORTS

The I/O section of the 8155 includes a control register, three I/O ports, and two registers for the timer (Figure 14.6). The expanded block diagram of the I/O section (Figure 14.6) represents a typical programmable I/O, as discussed in Section 14.12. In that section, two address lines plus the Chip Select logic were used to determine port addresses. The 8155 I/O section requires three address lines—AD_2 to AD_0 (A_2-A_0 internally)—and the Chip Enable logic to specify one of the seven registers. In addition, two control signals, RD and WR, are necessary to read from and write into these I/O registers.

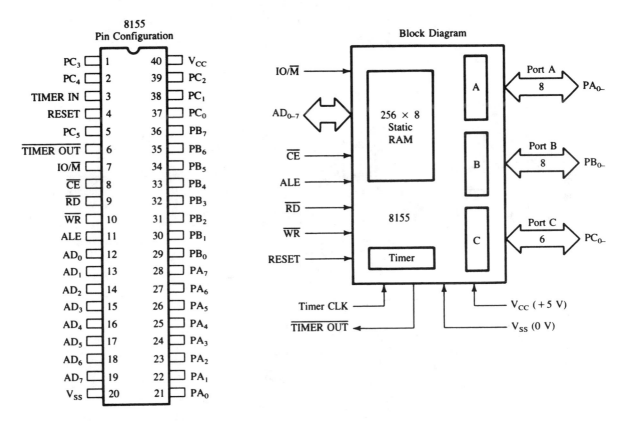

FIGURE 14.5
8155 Pin Configuration and Block Diagram
Source: Intel Corporation, *Embedded Microprocessors* (Santa Clara, Calif.: Author, 1994), pp. 1–31.

To communicate with peripherals through the 8155 the following steps are necessary:

1. Determine the addresses (port numbers of the registers and I/Os) based on the Chip Enable logic and address lines AD_0, AD_1, and AD_2.
2. Write a control word in the control register to specify I/O functions of the ports and the timer characteristics.
3. Write I/O instructions to port addresses to communicate with peripherals.
4. Read the status register, if necessary, to verify the status of the I/O ports and the timer. In simple applications, this step is not necessary.

CHIP ENABLE LOGIC AND PORT ADDRESSES

Address lines AD_2–AD_0, also shown as A_2–A_0 after internal demultiplexing, select one of the registers, as shown in Figure 14.6(b). Address lines A_3–A_7 are don't care lines; however, the logic levels on the corresponding high-order lines, A_{11}–A_{15}, will be duplicated on lines A_3–A_7, as explained in the next example.

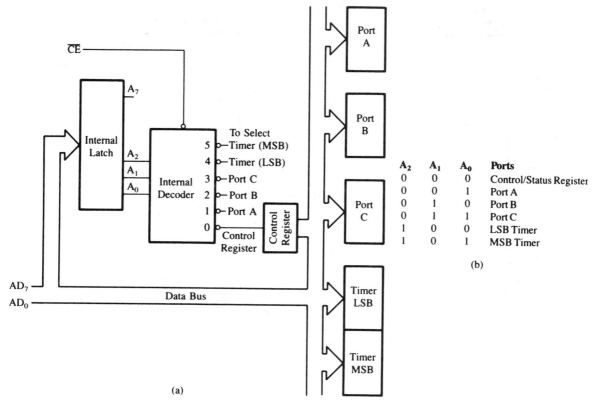

FIGURE 14.6
Expanded Block Diagram of the 8155 (a) and Its I/O Address: Selection (b)

Example 14.2

Determine the addresses of the control/status register, I/O ports, and timer registers in Figure 14.7.

Solution

To select the chip, the output line O_4 of the 8205* (3-to-8) decoder (Figure 14.7) should go low. Therefore, the logic levels of A_{15}–A_{11} should be as follows:

A_{15}	A_{14}	A_{13}	A_{12}	A_{11}
0	0	1	0	0

Enable lines of the 8205 (↑ for A_{15}, A_{14})

Input logic to activate the output line O_4 of the 8205 (↑ for A_{13}, A_{12}, A_{11})

---

*The 8205 is a 3-to-8 decoder similar to the 74LS138 except with higher current capacity and speed.

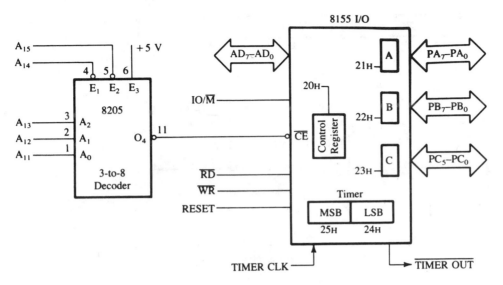

FIGURE 14.7
Interfacing 8155 I/O Ports (Schematic from the SDK-85 System)

By combining five high-order address lines with three low-order address lines (A_2–A_0), the port numbers in Figure 14.7 will range from 20H to 25H, as shown below.

A_{15}	A_{14}	A_{13}	A_{12}	A_{11}	AD_2	AD_1	AD_0	= Addresses	Ports
0	0	1	0	0	0	0	0	= 20H	Control or status register
					0	0	1	= 21H	Port A
		(2H)			0	1	0	= 22H	Port B
					0	1	1	= 23H	Port C
		⇓			1	0	0	= 24H	Timer (LSB)
A_7	A_6	A_5	A_4	A_3	1	0	1	= 25H	Timer (MSB)

This raises a question: How is it possible to combine five high-order address lines with three low-order address lines to generate a port address? To find an answer to this question, examine the execution of either the IN or the OUT instruction. When these instructions are executed, the high-order and low-order address buses carry the same information. In this case, the logic levels required on lines A_{15}–A_{11} for the Chip Enable are also duplicated on the address lines from A_7 through A_3, as shown above.

CONTROL WORD

The I/O ports and the timer can be configured by writing a control word in the control register. The control register bits are defined as shown in Figure 14.8.

In this control word, outputs are defined with logic 1 and inputs with logic 0. The first two LSBs, D_0 and D_1, determine I/O functions of ports A and B; and the MSBs, D_7

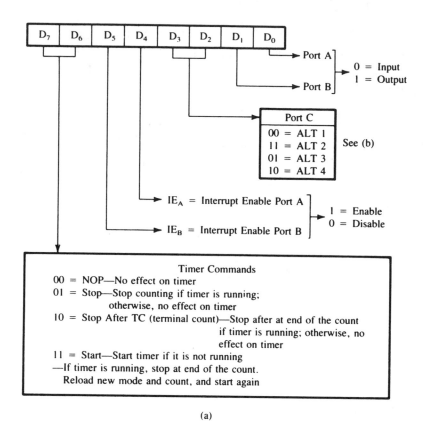

(a)

Table: ALT 1–ALT 4: Port C Bit Assignments, Defined by Bits D_3 and D_2 in the Control Register

ALT	D_3	D_2	PC_5	PC_4	PC_3	PC_2	PC_1	PC_0
ALT 1	0	0	I	I	I	I	I	I
ALT 2	1	1	O	O	O	O	O	O
ALT 3	0	1	O	O	O	$\overline{STB}_A$	BF_A	$INTR_A$
ALT 4	1	0	$\overline{STB}_b$	BF_B	$INTR_B$	$\overline{STB}_A$	BF_A	$INTR_A$

I = Input, STB = Strobe, INTR = Interrupt Request
O = Output, BF = Buffer Full, Subscript A = Port A
B = Port B

(b)

FIGURE 14.8
Control Word Definition in the 8155 (a) and Table of Port C Bit Assignments (b)

and D_6, determine timer functions. Bits D_2 and D_3 determine the functions of port C; their combination specifies one of the four alternatives, from simple I/O to interrupt I/O, as shown in Figure 14.8(b). Bits D_4 and D_5 are used only in the interrupt mode to enable or disable internal flip-flops of the 8155. These bits do not have any effect on the Interrupt Enable flip-flop (INTE) of the MPU.

The next section shows an application of the 8155 to design two output ports. An application of the 8155 in the handshake mode is illustrated later.

14.22   Illustration: Interfacing Seven-Segment-LED Output Ports Using the 8155

PROBLEM STATEMENT

1. Design two seven-segment-LED displays using ports A and B of the 8155.
2. Write initialization instructions and display data bytes at each port.

HARDWARE DESCRIPTION

Figure 14.9 shows two seven-segment output ports: port A with the Hewlett-Packard HP 5082/7340, and port B with the 9370 Hex decoders and common-anode seven-segment LEDs. The HP 5082 includes an internal decoder/driver. Both are functionally similar; however, a seven-segment display with an internal built-in decoder/driver is more expensive.

The decode logic is the same as that used in the previous discussion; therefore, the port addresses are as follows:

$$\text{Control Register} = 20\text{H}$$
$$\text{Port A} \quad = 21\text{H}$$
$$\text{Port B} \quad = 22\text{H}$$

CONTROL WORD

To configure ports A and B as outputs, the control word is as follows:

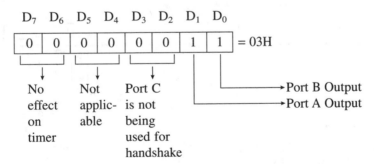

PROGRAM

```
MCI A,03H        ;Initialize ports A and B as output ports
OUT 20H
MVI A,BYTE1
OUT 21H          ;Display BYTE1 at port A
MVI A,BYTE2
OUT 22H          ;Display BYTE2 at port B
HLT
```

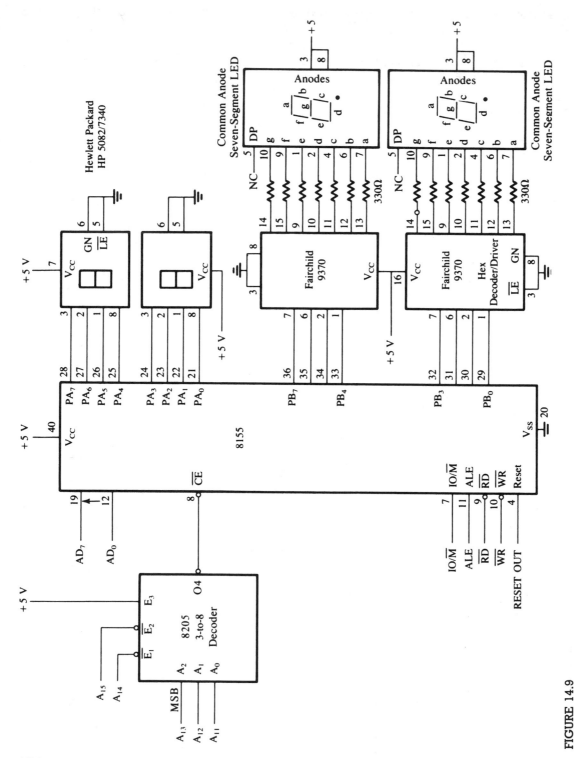

FIGURE 14.9
Interfacing 8155 I/O Ports with Seven-Segment LEDs

PROGRAM DESCRIPTION

The instruction MVI A,03H initializes ports A and B as simple output ports, and the following instructions display data BYTE1 and data BYTE2 at ports A and B, respectively.

14.23 The 8155 Timer

The timer section of the 8155 has two 8-bit registers; 14 bits are used for the counter, two bits for the timer mode, and it requires a clock as an input. This 14-bit down-counter provides output in four different modes, as described below.

Figure 14.10(a) shows two registers for a 14-bit count, one for LSB (low significant byte) and one for MSB (most significant byte). The most significant bits M_2 and M_1 are used to specify the timer mode. To operate the timer, a 14-bit count and mode bits are loaded in the registers. An appropriate control word starts the counter, which decrements the count by two at each clock pulse. The timer outputs vary according to the mode specified; see Figure 14.10(b).

The timer can be stopped either in the midst of counting or at the end of a count (applicable to Modes 1 and 3). In addition, the actual count at a given moment can be obtained by reading the status register. These details will be described later.

14.24   Illustration: Designing a Square-Wave Generator Using the 8155 Timer

PROBLEM STATEMENT

Design a square-wave generator with a pulse width of 100 μs by using the 8155 timer. Set up the timer in Mode 1 if the clock frequency is 3 MHz. Use the same decode logic and the port addresses as in Example 14.2 (Figure 14.7).

PROBLEM ANALYSIS

Timer Count   The pulse width required is 100 μs; therefore, the count should be calculated for the period of 200 μs. The timer output stays high for only half the count.

$$\text{Clock Period} = \frac{1}{f} = \frac{1}{3} \times 10^6 = 330 \text{ ns}$$
$$\text{Timer Count} = \frac{\text{Pulse Period}}{\text{Clock Period}} = \frac{200 \times 10^{-6}}{330 \times 10^{-9}} = 606$$
$$\text{Count} \qquad\quad = 025\text{EH}$$

Assuming the same decode logic for the 8155 Chip Enable line as in Example 14.2, the port addresses for the timer registers are

$$\text{Timer LSB} = 24\text{H}$$
$$\text{Timer MSB} = 25\text{H}$$

Loading Format

M_2	M_1	T_{13}	T_{12}	T_{11}	T_{10}	T_9	T_8

MSB

T_7	T_6	T_5	T_4	T_3	T_2	T_1	T_0

LSB

(a)

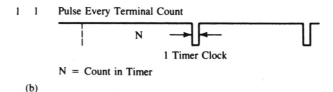

Modes	M_2 M_1	Timer Output
Mode 0: In this mode, the timer output remains high for half the count and goes low for the remaining count, thus providing a single square wave. The pulse width is determined by the count and the clock frequency.	0 0	Single Square Wave Cycle
Mode 1: In this mode, the initial timer count is automatically reloaded at the end of each count, thus providing a continuous square wave.	0 1	Square Wave
Mode 2: In this mode, a single clock pulse is provided at the end of the count.	1 0	Single Pulse Upon Terminal Count
Mode 3: This is similar to Mode 2, except the initial count is reloaded to provide a continuous wave form.	1 1	Pulse Every Terminal Count

N = Count in Timer

(b)

FIGURE 14.10
Timer Loading Format (a) and Modes (b)

The least significant byte, 5EH (of the count 025EH), should be loaded in the timer register with address 24H. The most significant byte is determined as follows:

$$
\begin{array}{cccccccc}
M_2 & M_1 & T_{13} & T_{12} & T_{11} & T_{10} & T_9 & T_8 \\
0 & 1 & 0 & 0 & 0 & 0 & 1 & 0 = 42H
\end{array}
$$

Timer Mode 1   MSB

Therefore, 42H should be loaded in the timer register with the address 25H.

Control Word   Assuming the same configuration for ports A and B as before, only bits D_7 and D_6 should be set to 1 to start the counter (see control word definition in Figure 14.8).

Therefore, Control Word: 1100 0011 = C3H

Initialization Instructions

MVI A,5EH	;LSB of the count
OUT 24H	;Load the LSB timer register
MVI A,42H	;MSB of the count
OUT 25H	;Load the MSB timer register
MVI A,C3H	
OUT 20H	;Start the timer
HLT	

14.25   The 8155 I/O Ports in Handshake Mode

In the handshake mode, data transfer occurs between the MPU and peripherals using control signals called handshake signals. Two I/O ports of the 8155, A and B, can be configured in the handshake mode; each uses three signals from port C as control signals (Figure 14.11). Another alternative (ALT 3 in the table in Figure 14.8) available in the 8155 is to configure port A in the handshake mode with three control signals from port C, configure port B as simple I/O, and configure the remaining three bits of port C as outputs. The details of configuring ports A and B in the handshake mode by using the pins of port C are given below.

CONTROL SIGNALS IN HANDSHAKE MODE

When both ports A and B are configured in the handshake mode, port A uses the lower three signals of port C (PC_0, PC_1, and PC_2), and port B uses the upper three signals (PC_3, PC_4, and PC_5), as shown in Figure 14.11. The functions of these signals are as follows:

☐ **$\overline{STB}$ (Strobe Input):** This is an input handshake signal from a peripheral to the 8155. The low on this signal informs the 8155 that data are strobed into the input port.
☐ **BF (Buffer Full):** This is an active high signal, indicating the presence of a data byte in the port.

FIGURE 14.11
8155 with Handshake Mode

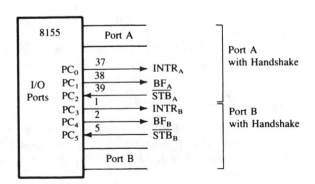

☐ **INTR (Interrupt Request):** This signal is generated by the rising edge of the $\overline{STB}$ signal if the interrupt flip-flop (INTE) is enabled. This signal can be used to interrupt the MPU.

☐ **INTE (Interrupt Enable):** This is an internal flip-flop used to enable or disable the interrupt capability of the 8155. The interrupts for port A and port B are controlled by bits D_4 and D_5, respectively, in the control register.

These control signals can be used to implement either interrupt I/O or status check I/O.

INPUT

Figure 14.12(a) shows the sequence of events and timing in data input to the 8155; they can be described as follows:

1. An external peripheral places data in the input port and informs the 8155 by causing the $\overline{STB}$ signal to go low.
2. The falling edge of the $\overline{STB}$ sets signal BF (Buffer Full) high, informing the peripheral to wait.
3. When the $\overline{STB}$ goes high, the rising edge of the $\overline{STB}$ can generate signal INTR if the internal interrupt flip-flop INTE is set. The interrupt flip-flops are set or reset by the control word.
4. The last step is to transfer data from the 8155 input port to the MPU. This can be done either by interrupting the MPU with the INTR signal or by checking the status of signal BF. The MPU can check the status by reading the status register (described later). When the MPU reads data, the INTR and BF signals are reset. When the BF signal goes low, it informs the peripheral that the port is empty, and the device is ready for the next byte.

OUTPUT

The sequence of events and timing in data output from the 8155 port to a peripheral are as follows; see Figure 14.12(b):

1. When the output is empty, the MPU writes a byte in the port.
2. The falling edge of the $\overline{WR}$ signal resets the INTR signal and the rising edge sets the BF (Buffer Full) signal high, informing the peripheral that a byte is available in the port.
3. After receiving the data byte, the peripheral acknowledges by sending the $\overline{STB}$ signal (active low).
4. The $\overline{STB}$ signal resets the BF signal low and generates the interrupt request by setting INTR high. Now the MPU can be informed by the interrupt signal to send the next byte, or the MPU can sense that the port is empty through a status check.

STATUS WORD

The MPU can read the status register to check the status of the ports or the timer. The control register and the status register have the same port address; they are differentiated only by the $\overline{RD}$ and $\overline{WR}$ signals. The status register bits are defined in Figure 14.13.

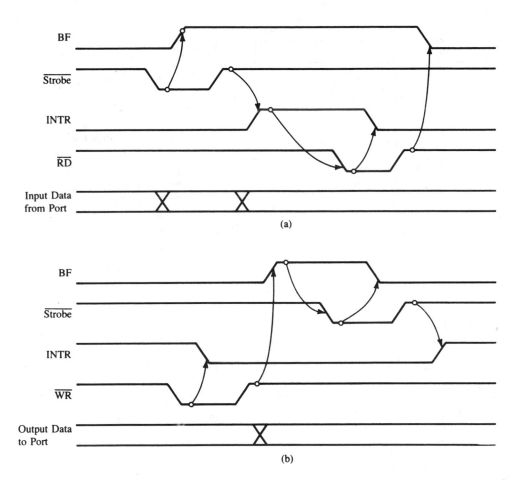

FIGURE 14.12

Timing Waveforms of the 8155 I/O Ports with Handshake: Input Mode (a) and Output Mode (b)

Source: Intel Corporation, *Embedded Microprocessors* (Santa Clara, Calif.: Author, 1994), pp. 1–43.

14.26   Illustration: Interfacing I/O Ports in Handshake Mode Using the 8155

PROBLEM STATEMENT

Design an interfacing circuit using the 8155 to read and display from an A/D converter to meet the following requirements:

1. Set up port A in the handshake mode to read data from an A/D converter.
2. Set up port B as an output port to display data at seven-segment LEDs.
3. Use line PC$_3$ from port C to initiate a conversion.

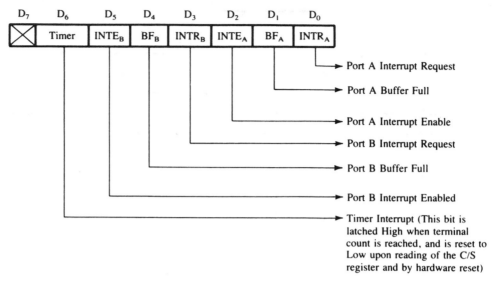

FIGURE 14.13
Status Word Definition

Source: Intel Corporation, *MCS—80/85 Family User's Manual* (Santa Clara, Calif.: Author, 1979), pp. 6–20.

4. Use the same decode logic as in Example 14.2 (Figure 14.7) to assign I/O port addresses.

5. Use the 8155 timer to record the conversion time.

PROBLEM ANALYSIS

Figure 14.14 shows an interfacing circuit that uses the 8155 I/O ports as follows:

1. Port A is configured as an input port in the handshake mode for reading data from the A/D converter.

2. Port B is configured as a simple output port for seven-segment LEDs.

3. The upper half of port C is a simple output port, and bit PC_3 is being used to start conversion.

4. The lower half of port C provides handshake signals for port A. Bit PC_2 is being used as a strobe ($\overline{STB}$) to inform the 8155 that the conversion is complete and that the output of the converter has been placed in port A.

INPUT WITH STATUS CHECK

The circuit shows that the INTR signal (bit PC_0) is not being used. This suggests that port A is configured for status check and not for interrupt I/O. Therefore, the control word (see

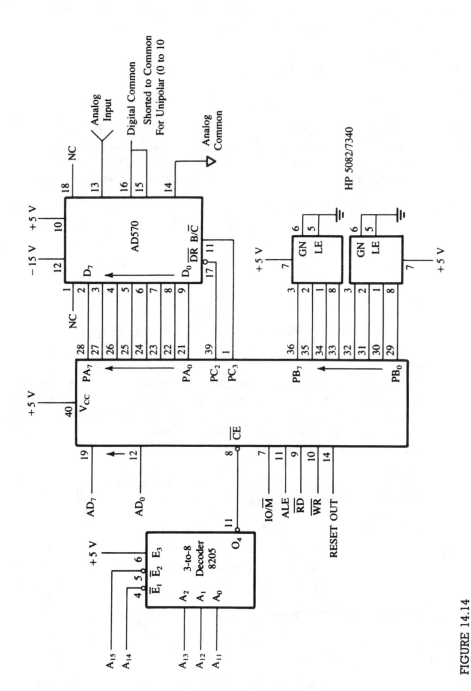

FIGURE 14.14
Interfacing the A/D Converter AD570 in the Handshake Mode

431

Figure 14.8) required to set up the ports as specified above and the masking byte to check the Data Ready ($\overline{\text{DR}}$) line are as follows:

Control Word

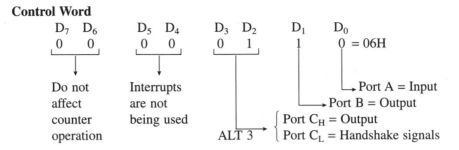

$$
\begin{array}{cccccccc}
D_7 & D_6 & D_5 & D_4 & D_3 & D_2 & D_1 & D_0 \\
0 & 0 & 0 & 0 & 0 & 1 & 1 & 0 = 06H
\end{array}
$$

Do not affect counter operation

Interrupts are not being used

ALT 3

Port A = Input
Port B = Output
Port C_H = Output
Port C_L = Handshake signals

Status Word   The MPU needs to check bit D_1 of the status register to verify the end of conversion and the availability of data in port A. The status word will have the following information (see Figure 14.13):

D_7	D_6	D_5	D_4	D_3	D_2	D_1	D_0
X	X	X	X	X	X	BF_A	X
Timer	$INTE_B$	BF_B	$INTR_B$	$INTE_A$			$INTR_A$

When the status word is masked with byte 02H, the availability of a data byte in port A can be verified.

8155 TIMER

The timer can be used to calculate the conversion time. When a conversion begins, the timer should be started with a known count, and at the end of conversion the timer should be stopped. To calculate the count, the reading must be divided by two because at each clock cycle the count is decremented twice. The difference between the two counts multiplied by the clock period of the timer should provide a fairly accurate reading of the conversion time.

To start the timer, set bits D_7 and D_6 of the control register to 1 without affecting the other bits of the register. Therefore, to start the timer, the control word should be

$$
= 1 \; 1 \quad 0 \; 0 \; 0 \; 1 \; 1 \; 0 \qquad = C6H
$$

Start Timer

I/O Assignments

To stop the timer, the control word should be

$$
= 0 \; 1 \quad 0 \; 0 \; 0 \; 1 \; 1 \; 0 \qquad = 46H
$$

Stop Timer

I/O Assignments

PORT ADDRESSES

The decode logic is the same as in Example 14.2; therefore, the I/O port addresses range from 20H (control register) to 25H (Timer—MSB).

PROGRAM

	MVI A,06H	;Control word for I/O ports
	OUT 20H	;Set up ports as specified
	MVI A,00H	;Load 0000H in the timer registers
	OUT 24H	
	OUT 25H	
	MVI A,08H	;Byte to set $PC_3 = 1$
	OUT 23H	;Send START pulse
	MVI A,C6H	;Control word to start timer
	OUT 20H	;Start timer
	MVI A,00H	;Byte to set $PC_3 = 0$
	OUT 23H	;Start conversion
STATUS:	IN 20H	;Read status register
	ANI 02H	;Check status of $\overline{DR}$
	JZ STATUS	;If $BF_A = 0$, wait in the loop until a data byte is available
	MVI A,46H	;Byte to stop counter
	OUT 20H	;Stop counter
	IN 21H	;Read A/D converter output
	OUT 22H	;Display data at port B
	IN 24H	;Read LSB of timer count
	MOV L,A	;Save timer count in register L
	IN 25H	;Read MSB of timer count
	ANI 3FH	;Delete D_7 and D_6 from the MSB; they represent
		; timer mode
	MOV H,A	;Save MSB timer count in H
	LHLD RWM	;Store timer count from HL register in R/W memory
		; locations
	HLT	

PROGRAM DESCRIPTION

The comments are self-explanatory; however, some explanation is needed for the timer count, start conversion (convert) pulse, and status check.

The program loads 0000H in the timer register, and after the first decrement, the count becomes 3FFFH. This is a 14-bit counter, with bits D_{15} and D_{14} reserved to specify the mode. However, in this particular problem, the counter mode is irrelevant. This program assumes that the A/D conversion time is less than the time period given by the maximum count. The difference between the initial count and the final count will provide the necessary value to calculate the conversion time. The program does not perform this subtraction; it just stores the final count in two consecutive memory locations labeled as RWM.

The second item needing explanation is the start conversion (convert) pulse. This is an active high pulse provided by turning on and off bit PC_3 in port C.

Finally, instruction IN 20H reads the status register, and instructions ANI 02 and JZ check whether the buffer in port A (BF_A) is full. The program stays in the loop until the BF_A goes high, indicating the availability of data.

INTERRUPT I/O

This example illustrates all the important I/O operations of the 8155 except the interrupt I/O in the handshake mode. To implement the interrupt I/O in the above example, the IN-TR_A—the output bit PC_0—should be connected to a vectored interrupt such as RST 6.5 and the control word should be changed accordingly (see Problem 14 at the end of this chapter).

14.27 Interfacing I/O Devices with Multiple Addresses

Figure 14.15 shows a decoding technique using the 74139, a 2-to-4 decoder. This device has two 2-to-4 decoders inside; one is used for interfacing I/O ports and the second is used for interfacing memory (not shown here). In this section, we will focus on I/O interfacing.

In Figure 14.15, inputs to the decoder are the address lines A_7 and A_6, and the decoder is enabled by the $IO/\overline{M}$ signal through an inverter. When the processor asserts $IO/\overline{M}$

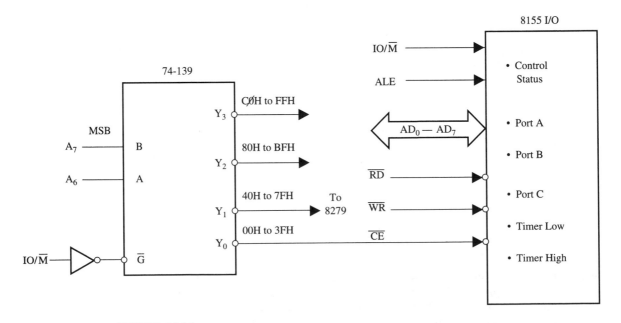

FIGURE 14.15
Interfacing 8155 I/O Ports and 8279 with Multiple Addresses (This figure is extracted from the PRIMER schematic in Appendix B.)

high to access an I/O port, the decoder is enabled. The address lines A_5-A_0 are not decoded in this schematic; some of them are connected to programmable I/O devices such as the 8155 and 8279.

ADDRESSES FOR 8155 I/O PORTS

8155 I/O Addresses

A_7	A_6	A_5	A_4	A_3	A_2	A_1	A_0	
0	0	X	X	X	0	0	0	= 00 Control or status register
					0	0	1	= 01 Port A
					0	1	0	= 02 Port B
					0	1	1	= 01 Port C
					1	0	0	= 04 Timer LSB
					1	0	1	= 05 Timer MSB

The 8155 I/O ports are accessed by the output line Y_0 of the decoder; therefore, the address lines A_7 and A_6 must be 0. In these addresses we are assuming that the don't care address lines A_5, A_4, and A_3 are at logic 0. However, these three lines can assume eight different combinations. If we assume these lines to be at logic 1, the addresses will range from 38H to 3DH.

8279 I/O Addresses   In this keyboard/display device (discussed in the next section), the address line A_0 is connected to 8279; when A_0 is 0, it accesses the internal data port and when it is 1, it accesses the internal command port. The output line Y_1 of the decoder accesses 8279; therefore, the address lines A_7 and A_6 must be 0 and 1, respectively. Therefore, the range of port addresses of the 8279 is as follows:

8279 I/O Addresses

A7	A6	A5	A4	A3	A2	A1	A0	
0	1	X	X	X	X	X	0	= 40H Command/status Port
							1	= 41H Data Port

If we assume the don't care address lines A_5 through A_1 at 0, the possible logic combination, these addresses can range from 40H to 7FH.

Multiple I/O Addresses   The 8279 needs only two addresses, but because of the five don't care lines, it occupies the space of 64 I/O addresses. In the case of the 8155, three addresses lines are don't care; therefore, it has eight sets of address ranges for its I/O ports and the timer. The advantage of such a technique is in saving costs and space on a printed circuit board. By using one two-input decoder, we can decode only two address lines. To decode the remaining lines, we need additional chips and space on the board. In small systems, the space of the printed circuit, called real estate, is of prime importance. The price we pay for this technique is the assignment of multiple addresses. However, if we do not need all the I/O space in a given system, wasting I/O addresses for a given device to reduce cost is of no consequence.

14.3 THE 8279 PROGRAMMABLE KEYBOARD/DISPLAY INTERFACE

The 8279 is a hardware approach to interfacing a matrix keyboard and a multiplexed display. The software approach to interfacing a matrix keyboard and a multiplexed display of seven-segment LEDs is illustrated in Chapter 17. The disadvantage of the software approach is that the microprocessor is occupied for a considerable amount of time in checking the keyboard and refreshing the display. The 8279 relieves the processor from these two tasks. The disadvantage of using the 8279 is the cost. The trade-offs between the hardware approach and the software approach are the production cost vs. the processor time and the software development cost.

The 8279 (Figure 14.16) is a 40-pin device with two major segments: keyboard and display. The keyboard segment can be connected to a 64-contact key matrix. Keyboard entries are debounced and stored in the internal FIFO (First-In–First-Out) memory; an interrupt signal is generated with each entry. The display segment can provide a 16-character scanned display interface with such devices as LEDs. This segment has 16×8 R/W memory (RAM), which can be used to read/write information for display purposes. The display can be set up in either right-entry or left-entry format.

14.31 Block Diagram of the 8279

The block diagram (Figure 14.17) shows four major sections of the 8279: keyboard, scan, display, and MPU interface. The functions of these sections are described below.

KEYBOARD SECTION

This section has eight lines (RL_0–RL_7) that can be connected to eight columns of a keyboard, plus two additional lines: Shift and CNTL/STB (Control/Strobe). The status of the SHIFT key and the Control key can be stored along with a key closure. The keys are automatically debounced, and the keyboard can operate in two modes: two-key lockout or N-key rollover. In the two-key lockout mode, if two keys are pressed almost simultaneously, only the first key is recognized. In the N-key rollover mode, simultaneous keys are recognized and their codes are stored in the internal buffer; it can also be set up so that no key is recognized until only one key remains pressed.

The keyboard section also includes 8×8 FIFO (First-In–First-Out) RAM. The FIFO RAM consists of eight registers that can store eight keyboard entries; each is then read in the order of entries. The status logic keeps track of the number of entries and provides an IRQ (Interrupt Request) signal when the FIFO is not empty.

SCAN SECTION

The scan section has a scan counter and four scan lines (SL_0–SL_3). These four scan lines can be decoded using a 4-to-16 decoder to generate 16 lines for scanning. These lines can be connected to the rows of a matrix keyboard and the digit drivers of a multiplexed display.

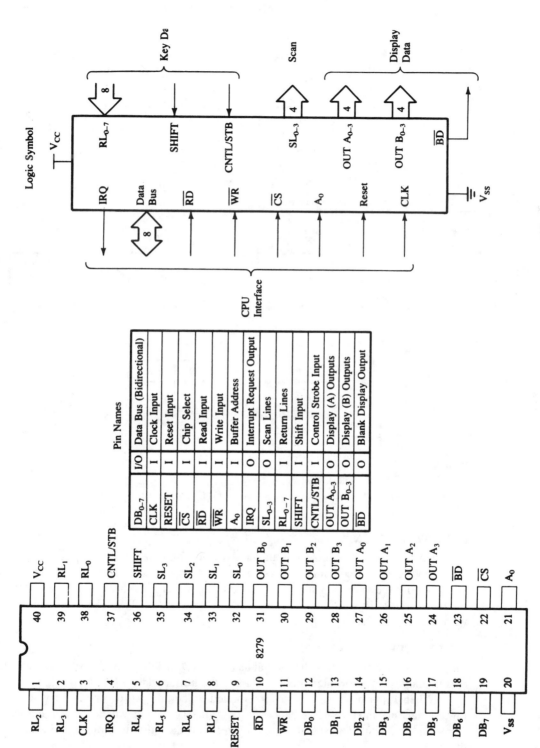

FIGURE 14.16
The 8279 Logic Pinout

437

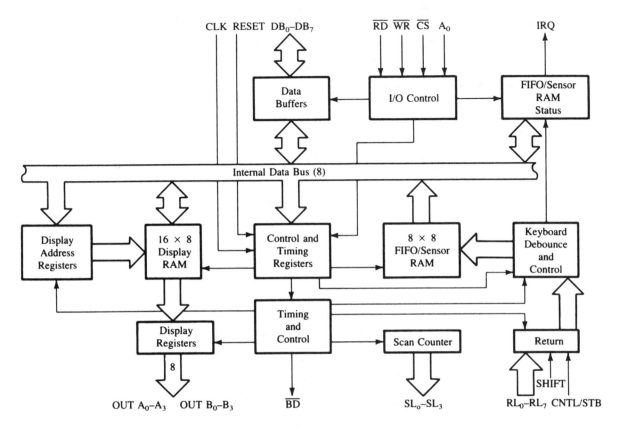

FIGURE 14.17

The 8729 Logic Block Diagram

Source: Intel Corporation, *Peripheral Components* (Santa Clara, Calif.: Author, 1993), pp. 3–218.

DISPLAY SECTION

The display section has eight output lines divided into two groups A_0–A_3 and B_0–B_3. These lines can be used, either as a group of eight lines or as two groups of four, in conjunction with the scan lines for a multiplexed display. The display can be blanked by using the $\overline{BD}$ line. This section includes 16×8 display RAM. The MPU can read from or write into any of these registers.

MPU INTERFACE SECTION

This section includes eight bidirectional data lines (DB_0–DB_7), one Interrupt Request line (IRQ), and six lines for interfacing, including the buffer address line (A_0).

When A_0 is high, signals are interpreted as control words or status; when A_0 is low, signals are interpreted as data. The IRQ line goes high whenever data entries are stored in the FIFO. This signal is used to interrupt the MPU to indicate the availability of data.

14.32   Programming the 8279

The 8279 is a complex device that can accept eight different commands to perform various functions.*

The initialization commands can specify

1. left or right entry and key rollover.
2. clock frequency prescaler.
3. starting address and incrementing mode of the FIFO RAM.
4. RAM address to read and write data and incrementing mode.
5. blanking format.

To illustrate important command words, the next section will illustrate the keyboard/display circuit shown in Figure 14.18.

14.33   Illustration: Keyboard/Display Interfacing Using the 8279

Figure 14.18 shows the keyboard/display circuit from the Primer, a single-board system shown in Appendix B.

1. Explain the functions of various components in the circuit.
2. Explain the decoding logic, and identify the port addresses of the 8279 registers.
3. Explain the initialization instructions given later.

CIRCUIT DESCRIPTION

Figure 14.18 shows the following components:

☐ The 8279 Programmable Keyboard/Display Interface
☐ A matrix keyboard with 20 keys: one matrix of 16 keys (4×4) and the other with four keys (1×4)
☐ Six seven-segment LEDs in groups of two
☐ A 3-to-8 decoder to generate additional scan lines
☐ A 2981A, a current driver for anodes and transistors as current drivers for cathodes

Lines RL_0–RL_3 (Return Lines) of the 8279 are connected to the columns of the matrix keyboard, and the output lines (A_0–A_3 and B_0–B_3) are connected to drive the LED segments. The three scan lines are connected to the decoder, the 74HC138, to generate eight decoded signals. In this circuit, six output lines of the decoder are connected as digit drivers to turn on six seven-segment LEDs; two output lines are unused. The 8279 has four scan lines that can be decoded to generate 16 output lines to drive 16 displays. The data lines of the 8279 are connected to the data bus of the 8085, and the IRQ (Interrupt Request) is connected to the RST 5.5 of the system.

Four signals—$\overline{RD}$, $\overline{WR}$, CLK, and RESET OUT—are connected directly from the 8085. The system has a 3.072 MHz clock; when the 8279 is reset, the clock prescaler is

---

*For a complete description of the 8279 refer to Intel's *MCS-80/85 Family User's Manual*.

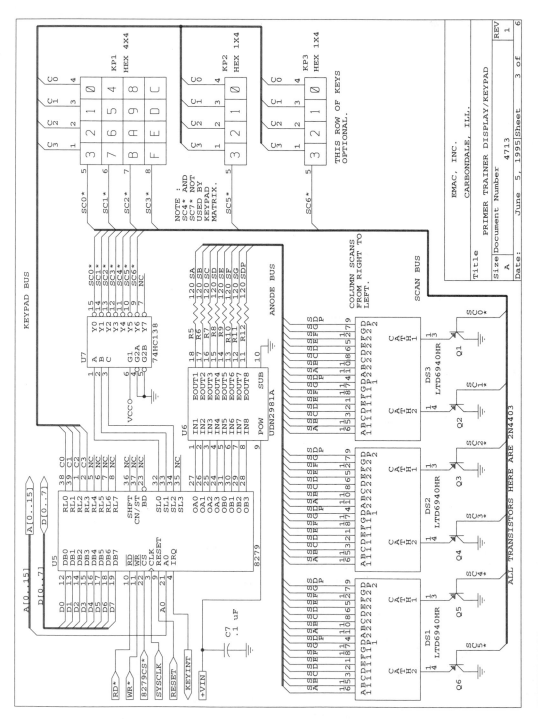

FIGURE 14.18
Keyboard/display circuit
SOURCE: Courtesy of EMAC Inc.

440

set to 31. This divides the clock frequency by 31 to provide the scan frequency of approximately 100 kHz. The RESET signal also sets the 8279 in the mode of 16-character display with two-key lockout keyboard.

After the initialization of the 8279, the respective codes are sent to the display RAM to display any characters. The 8279 takes over the task of displaying the characters by outputting the codes and digit strobes. To read the keyboard, the 8279 scans the columns; if a key closure is detected, it debounces the key. If a key closure is valid, it loads the key code into the FIFO, and the IRQ line goes high to interrupt the system.

DECODING LOGIC AND PORT ADDRESSES

The port addresses of the 8279 registers are determined by two signals: $\overline{CS}$ and A_0. The $\overline{CS}$ of the 8279 is connected to the Y_1 signal generated by the address decoding circuit in Figure 14.15 and A_0 of the 8279 is directly connected to the address line A_0 of the 8085 processor. For commands and status, A_0 should be high and for data transfer A_0 should be low. The addresses of the Data port and the Command/Status port are 40H and 41H, respectively, as discussed in Section 14.27 (Figure 14.15).

INITIALIZATION INSTRUCTIONS

In Figure 14.18, we need to initialize the 8279 to read a key and display the key at the seven-segment display. The definitions of various control words necessary for initialization are given in Appendix D. The RESET signal sets the clock prescaler to 31. This prescaler divides the system clock (3.072 MHz) to set the scan frequency at approximately 100 kHz. When the 8279 detects a key closure, the IRQ signal interrupts the 8085, using RST 5.5. The interrupt service routine sends the command word to read from the keyboard (Command Port 41H), reads the character data from the keyboard (Data Port 40H), and stores it in the system's R/W memory location, the input buffer (IBUFF). The following instructions illustrate the initialization and the interrupt service routine.

Initialization

Keyboard/Display Mode

```
MVI A, 00000000  ;Set mode: Left entry, 8-character,   Code: | 0  0  0  D  D  K  K  K |
                 ;2-key lockout, encoded scan
                 ;  keyboard
OUT 41H          ;Send to Command Port
```

Program Clock

```
MVI A, 00111111  ;Divide clock by 31—this is    Code: | 0  0  1  P  P  P  P  P |
                 ;  unnecessary
OUT 41H          ;  shown here for illustration
```

Clear

```
MVI A, 11000001  ;Clear Display RAM and FIFO status   Code: | 1  1  0  C  C  C  C  C |
OUT 41H
```

Write Display RAM

MVI A, 10000000 ;Set up 8279 for display   Code: | 1 0 0 A A A A A |

Interrupt Routine to Read a Keyboard

PUSH H   ;Save registers
PUSH PSW

Read FIFO RAM

MVI A, 01000000 ;Control word to read FIFO RAM   Code: | 0 1 0 A X A A A |

OUT 41H

IN 40H   ;Read Data

Data Format

ANI 00111111   ;Mask CNTL & SHFT keys

D_7	D_6	D_5 D_4 D_3	D_2 D_1 D_0
CNTL	SHFT	ROW	COL

STA IBUFF   ;Store in R/W memory
POP PSW   ;Restore registers
POP H
RET

    The six seven-segment LED display is divided into two segments: left four LEDs for memory address and right two LEDs for data value. The software instructions should determine whether it is a four-digit or two-digit display and send the control word to write into the display RAM.

    For example, to display a 4-digit memory address, the control word instructions are as follows:

Write Display RAM Control Word

MVI A,90H   ;Control word to write | 1 | 0 | 0 | A_1 | A | A | A | A |
                    ;   starting at first RAM
                    ;   location
OUT 41H
MVI A,CODE   ;Load seven-segment code
OUT 40H   ;Output the code

    To display a 2-digit data value, the control instructions are as follows:

MVI A,94H   ;Control word to display data
OUT 41H

    In this example, the control word 94H points to the fifth memory location in the display RAM; the first four locations are reserved for memory addresses.

SUMMARY

This chapter was concerned with the basic concepts (such as control register, control logic, chip select logic, and handshake signals) underlying a programmable device. The characteristics of a programmable device were discussed using the bidirectional buffer 74LS245, and the important concepts related to a programmable device were reviewed in Section 14.14. On the basis of these concepts, two programmable devices—the 8155 (R/W memory with I/O) and the 8279 (keyboard/display interface)—were discussed.

The 8155 is a multipurpose device that includes memory, timer, and I/O ports. Interfacing applications of the I/O ports in various modes (including handshake) and the timer were illustrated with examples. Similarly, interfacing and initialization of the 8279 (keyboard/display interface) was illustrated, using the circuit from the Primer, a single-board system shown in Appendix B.

QUESTIONS, PROBLEMS, AND PROGRAMMING ASSIGNMENTS

1. List the internal components generally found in a programmable device.
2. In a programmable device, how does the MPU differentiate between the control register and the status register if both registers have the same port address?
3. Explain the functions of handshake signals.
4. Explain the difference between setting the 8155 I/O ports in ALT 1 and ALT 3.
5. Specify the handshake signals for port B of the 8155 if port B is connected as an input port in the interrupt mode. Explain the function of each handshake signal.
6. Port B of the 8155 is set up in the handshake mode, and the reading of the status word is 20H. Is the port set up for status check or interrupt I/O?
7. List the major components of the 8279 keyboard/display interface, and explain their functions.
8. In Figure 14.7, specify all the port addresses if the output line 7 of the decoder is connected to $\overline{CE}$.
9. In Figure 14.7, assume that the decoder is eliminated and address line A_{15} is connected to $\overline{CE}$ through an inverter. Specify the addresses of ports, A, B, and C, assuming all don't care lines are at logic 0.
10. Can any port be accessed with port address FDH in Problem 10?
11. Write the instructions to set up the 8155 timer in Mode 3 with count 3FF8H.
12. In Problem 11, specify the output if the clock frequency is 3 MHz and the count is 3080H.
13. Calculate the count for the 8155 timer to obtain the square wave of the 500 µs period if the clock frequency is 3.072 MHz.
14. Modify the circuit in Figure 14.14 to connect the A/D converter with interrupt I/O. Use RST 6.5 for the interrupt.

15. In Problem 14, write the instructions in the main program to enable the RST 6.5 interrupt. Write a service routine to read a data byte, store it in memory location XX70H, and start the next conversion. (Ignore all the specifications related to the timer in the illustration.)

EXPERIMENTAL ASSIGNMENTS

1. **a.** Connect the A/D converter AD570 as an input (Figure 14.14).
   b. Set up port A as an input port in ALT 3 with interrupt I/O. Use RST 6.5 for the interrupt.
   c. Write a main program to

   □ initialize port A as an input and port C for handshake signals.
   □ start conversion.
   □ display data at an output port.
   □ set up a continuous loop for displaying data.

   d. Write a service routine to

   □ read a data byte.
   □ start conversion for the next reading.

   e. Record data for various analog signals.
2. **a.** Set up the 8155 timer as shown in Section 14.24.
   b. Enter and execute the given program.
   c. Measure the square-wave output on an oscilloscope.
   d. Calculate the frequency and the pulse width of the square wave if bits T_{13}–T_0 all = 0.
   e. Load the count from step (d), start the counter, and measure the frequency and the pulse width of the output.

General-Purpose Programmable Peripheral Devices

This chapter is an extension of Chapter 14, except that the programmable devices discussed in this chapter are designed for general-purpose use. This chapter describes several programmable devices from the Intel family: the 8255A Peripheral Interface, the 8254 Interval Timer, the 8259A Interrupt Controller, and the 8237 DMA controller.

The 8255A and the 8254, two widely used general-purpose programmable devices, can be compatible with any microprocessor. The 8255A includes three programmable ports, one of which can be used for bidirectional data transfer. This is an important additional feature in comparison with the 8155 I/O ports discussed in the last chapter. The 8254 timer is similar to the 8155 timer, except that it has three 16-bit independent timers with various modes.

The next two devices—the 8259A Interrupt Controller and the 8237 DMA controller—were introduced briefly in Chapter 12. These devices illustrate the implementation of interrupts and of Direct Memory Access by using programmable devices. If you are not familiar with the concepts underlying programmable devices and handshake signals, you are strongly advised to read Section 14.1 before reading this chapter.

OBJECTIVES

- [ ] List the elements of the 8255A Programmable Peripheral Interface (PPI) and explain its various operating modes.
- [ ] Set up the 8255A I/O ports in the simple I/O and Bit Set/Reset (BSR) mode.
- [ ] Design an interfacing circuit to set up the 8255A in the handshake mode (Mode 1) and

write instructions to transfer data under status check I/O and interrupt I/O.

☐ List operating modes of the 8254 timer and write instructions to set up the timer in the various modes.

☐ Explain the functions of the 8259A interrupt controller and its operation in the fully nested mode.

☐ Explain the process of the Direct Memory Access (DMA) and the functions of various elements of the 8237.

15.1 THE 8255A PROGRAMMABLE PERIPHERAL INTERFACE

The 8255A is a widely used, programmable, parallel I/O device. It can be programmed to transfer data under various conditions, from simple I/O to interrupt I/O. It is flexible, versatile, and economical (when multiple I/O ports are required), but somewhat complex. It is an important general-purpose I/O device that can be used with almost any microprocessor.

The 8255A has 24 I/O pins that can be grouped primarily in two 8-bit parallel ports: A and B, with the remaining eight bits as port C. The eight bits of port C can be used as individual bits or be grouped in two 4-bit ports: C_{UPPER} (C_U) and C_{LOWER} (C_L), as in Figure 15.1(a). The functions of these ports are defined by writing a control word in the control register.

Figure 15.1(b) shows all the functions of the 8255A, classified according to two modes: the Bit Set/Reset (BSR) mode and the I/O mode. The BSR mode is used to set or reset the bits in port C. The I/O mode is further divided into three modes: Mode 0, Mode 1, and Mode 2. In Mode 0, all ports function as simple I/O ports. Mode 1 is a handshake mode whereby ports A and/or B use bits from port C as handshake signals. In the handshake mode, two types of I/O data transfer can be implemented: status check and interrupt. In Mode 2, port A can be set up for bidirectional data transfer using handshake signals from port C, and port B can be set up either in Mode 0 or Mode 1.

15.11 Block Diagram of the 8255A

The block diagram in Figure 15.2(a) shows two 8-bit ports (A and B), two 4-bit ports (C_U and C_L), the data bus buffer, and control logic. Figure 15.2(b) shows a simplified but expanded version of the internal structure, including a control register. This block diagram includes all the elements of a programmable device; port C performs functions similar to that of the status register in addition to providing handshake signals.

CONTROL LOGIC

The control section has six lines. Their functions and connections are as follows:

☐ $\overline{RD}$ **(Read):** This control signal enables the Read operation. When the signal is low, the MPU reads data from a selected I/O port of the 8255A.

☐ $\overline{WR}$ **(Write):** This control signal enables the Write operation. When the signal goes low, the MPU writes into a selected I/O port or the control register.

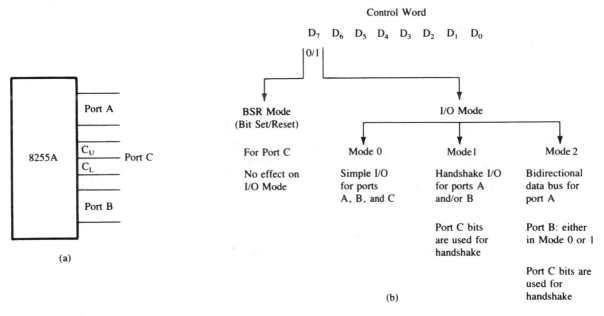

FIGURE 15.1
8255A I/O Ports (a) and Their Modes (b)

□ **RESET (Reset):** This is an active high signal; it clears the control register and sets all ports in the input mode.

□ **CS, A_0, and A_1:** These are device select signals. $\overline{CS}$ is connected to a decoded address, and A_0 and A_1 are generally connected to MPU address lines A_0 and A_1, respectively.

The $\overline{CS}$ signal is the master Chip Select, and A_0 and A_1 specify one of the I/O ports or the control register as given below:

$\overline{CS}$	A_1	A_0	Selected
0	0	0	Port A
0	0	1	Port B
0	1	0	Port C
0	1	1	Control Register
1	X	X	8255A is not selected.

As an example, the port addresses in Figure 15.3(a) are determined by the $\overline{CS}$, A_0, and A_1 lines. The $\overline{CS}$ line goes low when $A_7 = 1$ and A_6 through A_2 are at logic 0. When these signals are combined with A_0 and A_1, the port addresses range from 80H to 83H, as shown in Figure 15.3(b).

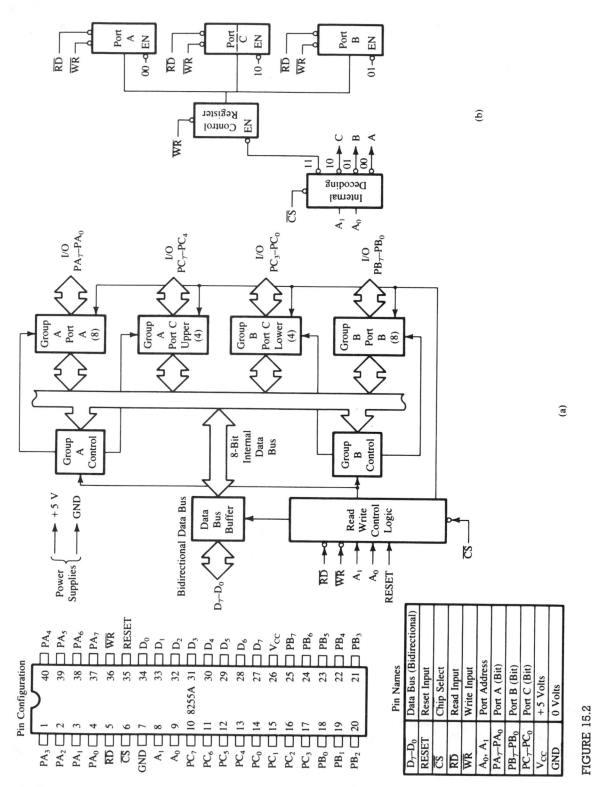

FIGURE 15.2

8255A Block Diagram (a) and an Expanded Version of the Control Logic and I/O Ports (b)

SOURCE: A: Intel Corporation, *Peripheral Components* (Santa Clara, Calif.: Author, 1993), p. 3–100.

448

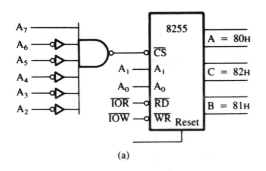

FIGURE 15.3
8255A Chip Select Logic (a) and I/O Port Addresses (b)

CONTROL WORD

Figure 15.2(b) shows a register called the **control register.** The contents of this register, called the **control word,** specify an I/O function for each port. This register can be accessed to write a control word when A_0 and A_1 are at logic 1, as mentioned previously. The register is not accessible for a Read operation.

Bit D_7 of the control register specifies either the I/O function or the Bit Set/Reset function, as classified in Figure 15.1(b). If bit $D_7 = 1$, bits D_6–D_0 determine I/O functions in various modes, as shown in Figure 15.4. If bit $D_7 = 0$, port C operates in the Bit Set/Reset (BSR) mode. The BSR control word does not affect the functions of ports A and B (the BSR mode will be described later).

To communicate with peripherals through the 8255A, three steps are necessary:

1. Determine the addresses of ports A, B, and C and of the control register according to the Chip Select logic and address lines A_0 and A_1.
2. Write a control word in the control register.
3. Write I/O instructions to communicate with peripherals through ports A, B, and C.

Examples of the various modes are given in the next section.

15.12 Mode 0: Simple Input or Output

In this mode, ports A and B are used as two simple 8-bit I/O ports and port C as two 4-bit ports. Each port (or half-port, in case of C) can be programmed to function as simply an input port or an output port. The input/output features in Mode 0 are as follows:

1. Outputs are latched.
2. Inputs are not latched.
3. Ports do not have handshake or interrupt capability.

---

1. Identify the port addresses in Figure 15.5.
2. Identify the Mode 0 control word to configure port A and port C_U as output ports and port B and port C_L as input ports.

Example
15.1

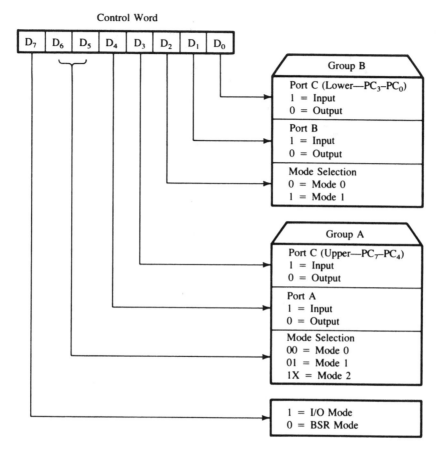

FIGURE 15.4
8255A Control Word Format for I/O Mode
SOURCE: Adapted from Intel Corporation, *Peripheral Components* (Santa Clara, Calif.: Author, 1993), p. 3–104.

3. Write a program to read the DIP switches and display the reading from port B at port A and from port C_L at port C_U.

Solution

1. Port Addresses This is a memory-mapped I/O; when the address line A_{15} is high, the Chip Select line is enabled. Assuming all don't care lines are at logic 0, the port addresses are as follows:

$$
\begin{array}{lll}
\text{Port A} & = & 8000\text{H } (A_1 = 0, A_0 = 0) \\
\text{Port B} & = & 8001\text{H } (A_1 = 0, A_0 = 1) \\
\text{Port C} & = & 8002\text{H } (A_1 = 1, A_0 = 0) \\
\text{Control Register} & = & 8003\text{H } (A_1 = 1, A_0 = 1)
\end{array}
$$

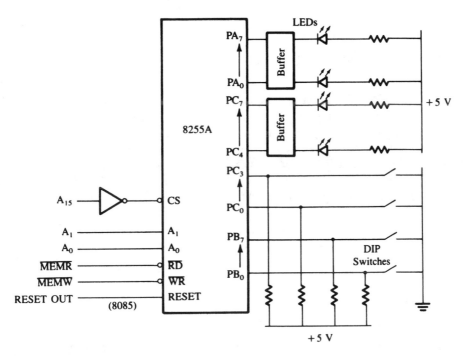

FIGURE 15.5
Interfacing 8255A I/O Ports in Mode 0

2. Control Word

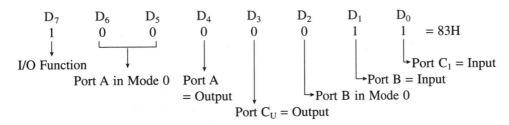

3. Program

MVI A,83H	;Load accumulator with the control word
STA 8003H	;Write word in the control register to initialize the ports
LDA 8001H	;Read switches at port B
STA 8000H	;Display the reading at port A
LDA 8002H	;Read switches at port C
ANI 0FH	;Mask the upper four bits of port C; these bits are not input data
RLC	;Rotate and place data in the upper half of the accumulator
RLC	

```
RLC
RLC
STA 8002H        ;Display data at port C_U
HLT
```

Program Description The circuit is designed for memory-mapped I/O; therefore, the instructions are written as if all the 8255A ports are memory locations.

The ports are initialized by placing the control word 83H in the control register. The instructions STA and LDA are equivalent to the instructions OUT and IN, respectively.

In this example, the low four bits of port C are configured as input and the high four bits are configured as output; even though port C has one address for both halves C_U and C_L (8002H), Read and Write operations are differentiated by the control signals $\overline{MEMR}$ and $\overline{MEMW}$. When the MPU reads port C (e.g., LDA 8002H), it receives eight bits in the accumulator. However, the high-order bits (D_7–D_4) must be ignored because the input data bits are in PC_3–PC_0. To display these bits at the upper half of port C, bits (PC_3–PC_0) must be shifted to PC_7–PC_4.

15.13   BSR (Bit Set/Reset) Mode

The BSR mode is concerned only with the eight bits of port C, which can be set or reset by writing an appropriate control word in the control register. A control word with bit D_7 = 0 is recognized as a BSR control word, and it does not alter any previously transmitted control word with bit D_7 = 1; thus the I/O operations of ports A and B are not affected by a BSR control word. In the BSR mode, individual bits of port C can be used for applications such as an on/off switch.

BSR CONTROL WORD

This control word, when written in the control register, sets or resets one bit at a time, as specified in Figure 15.6.

FIGURE 15.6
8255A Control Word Format in the
BSR Mode

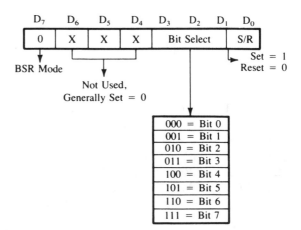

Write a BSR control word subroutine to set bits PC_7 and PC_3 and reset them after 10 ms. Use the schematic in Figure 15.3 and assume that a delay subroutine is available.

Example 15.2

Solution

BSR CONTROL WORDS

		D_7	D_6	D_5	D_4	D_3	D_2	D_1	D_0		
To set bit PC_7	=	0	0	0	0	1	1	1	1	=	0FH
To reset bit PC_7	=	0	0	0	0	1	1	1	0	=	0EH
To set bit PC_3	=	0	0	0	0	0	1	1	1	=	07H
To reset bit PC_3	=	0	0	0	0	0	1	1	0	=	06H

PORT ADDRESS

Control register address = 83H; refer to Figure 15.3(b).

SUBROUTINE

```
BSR:    MVI A,0FH      ;Load byte in accumulator to set PC7
        OUT 83H        ;Set PC7 = 1
        MVI A,07H      ;Load byte in accumulator to set PC3
        OUT 83H        ;Set PC3 = 1
        CALL DELAY     ;This is a 10-ms delay
        MVI A,06H      ;Load accumulator with the byte to reset PC3
        OUT 83H        ;Reset PC7
        MVI A,0EH      ;Load accumulator with the byte to reset PC7
        OUT 83H        ;Rest PC7
        RET
```

From an analysis of the above routine, the following points can be noted:

1. To set/reset bits in port C, a control word is written in the control register and not in port C.
2. A BSR control word affects only one bit in port C.
3. The BSR control word does not affect the I/O mode.

15.14 Illustration: Interfacing A/D Converter Using the 8255A in Mode 0 and BSR Mode

PROBLEM STATEMENT

Design an interfacing circuit to read data from an A/D converter, using the 8255A in the memory-mapped I/O.

1. Set up port A to read data.
2. Set up bit PC_0 to start conversion and bit PC_7 to read the ready status of the converter.

PROBLEM ANALYSIS

The Chip Select logic in Figure 15.7 is the same as in Figure 15.5; therefore, the assigned port addresses range from 8000H for port A to 8003H for the control register. The control signals $\overline{MEMR}$ and $\overline{MEMW}$ specify the memory-mapped I/O.

MODE 0: CONTROL WORD

The configuration of the ports is specified as follows:

☐ Port A: As an input port.
☐ Port C_L: As an output port because bit PC_0 is used to start conversion.
☐ Port C_U: As an input port to read the status at PC_7.
☐ Port B: not used.

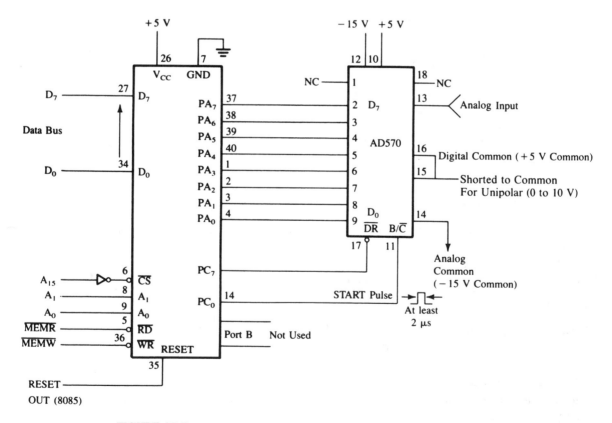

FIGURE 15.7

Schematic: Interfacing the A/D Converter AD570 Using the 8255A in Mode 0 and BSR Mode

Therefore, the control word necessary to meet the requirements is as follows:

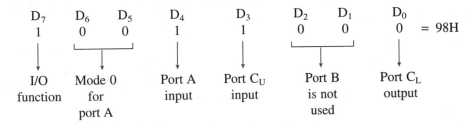

BSR CONTROL WORD FOR START PULSE

Bit PC_0 is used as a START pulse. To set and reset PC_0, the BSR control word is as follows (refer to Figure 15.6):

D_7	D_6	D_5	D_4	D_3	D_2	D_1	D_0	
0	0	0	0	0	0	0	1/0	= 01H to set
								= 00H to reset

BSR mode   Don't care   Bit 0   1 = set   0 = reset

SUBROUTINE

```
A2D:    LXI H,8003H    ;Point the index to control register
        MVI A,98H      ;Load the mode control word
        MOV M,A        ;Set up ports A and C_U as inputs
        MVI A,01H      ;Load BSR control word to set PC_0
        MOV M,A        ;Turn on the START pulse
        CALL DELAY     ;Wait
        MVI A,00H      ;Load BSR control word to reset PC_0
        MOV M,A        ;Start conversion
        DCX H          ;Memory pointer to port C
READ:   MOV A,M        ;Read port C
        RAL            ;Place PC_7 in the carry
        JC READ        ;Wait in the loop until the end of conversion
        LDA 8000H      ;Read A/D converter
        RET
```

PROGRAM DESCRIPTION

Instruction MOV M,A initializes the 8255A ports by placing the control word in the control register. To provide a START pulse to the converter, bit PC_0 is set to 1; it is turned off after the appropriate delay. The end of conversion is checked by verifying the status of line PC_7. When PC_7 goes low, instruction LDA 8000H reads and places data in the accumulator.

15.15   Mode 1: Input or Output with Handshake

In Mode 1, handshake signals are exchanged between the MPU and peripherals prior to data transfer. The features of this mode include the following:

1. Two ports (A and B) function as 8-bit I/O ports. They can be configured either as input or output ports.
2. Each port uses three lines from port C as handshake signals. The remaining two lines of port C can be used for simple I/O functions.
3. Input and output data are latched.
4. Interrupt logic is supported.

   In the 8255A, the specific lines from port C used for handshake signals vary according to the I/O function of a port. Therefore, input and output functions in Mode 1 are discussed separately.

MODE 1: INPUT CONTROL SIGNALS

Figure 15.8(a) shows the associated control signals used for handshaking when ports A and B are configured as input ports. Port A uses the upper three signals: PC_3, PC_4, and

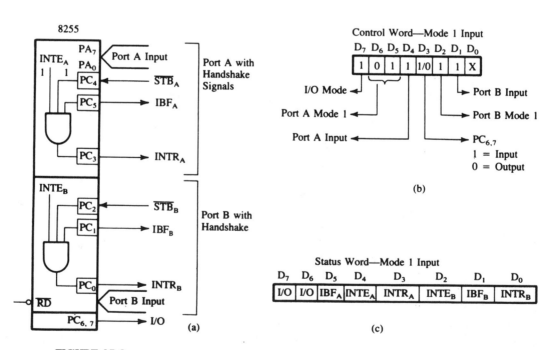

FIGURE 15.8
8255A Mode 1: Input Configuration
SOURCE: Adapted from Intel Corporation, *Peripheral Components* (Santa Clara, Calif.: Author, 1993), p. 3–110.

PC$_5$. Port B uses the lower three signals: PC$_2$, PC$_1$, and PC$_0$. The functions of these signals are as follows:

- **$\overline{\text{STB}}$ (Strobe Input):** This signal (active low) is generated by a peripheral device to indicate that it has transmitted a byte of data. The 8255A, in response to $\overline{\text{STB}}$, generates IBF and INTR, as shown in Figure 15.9.
- **IBF (Input Buffer Full):** This signal is an acknowledgment by the 8255A to indicate that the input latch has received the data byte. This is reset when the MPU reads the data (Figure 15.9).
- **INTR (Interrupt Request):** This is an output signal that may be used to interrupt the MPU. This signal is generated if $\overline{\text{STB}}$, IBF, and INTE (Internal flip-flop) are all at logic 1. This is reset by the falling edge of the $\overline{\text{RD}}$ signal (Figure 15.9).
- **INTE (Interrupt Enable):** This is an internal flip-flop used to enable or disable the generation of the INTR signal. The two flip-flops INTE$_A$ and INTE$_B$ are set/reset using the BSR mode. The INTE$_A$ is enabled or disabled through PC$_4$, and INTE$_B$ is enabled or disabled through PC$_2$.

CONTROL AND STATUS WORDS

Figure 15.8(b) uses control words derived from Figure 15.4 to set up port A and port B as input ports in Mode 1. Similarly, Figure 15.8(c) also shows the status word, which will be placed in the accumulator if port C is read.

PROGRAMMING THE 8255A IN MODE 1

The 8255A can be programmed to function using either status check I/O or interrupt I/O. Figure 15.10(a) shows a flowchart for the status check I/O. In this flowchart, the MPU continues to check data status through the IBF line until it goes high. This is a simplified flowchart; however, it does not show how to handle data transfer if two ports

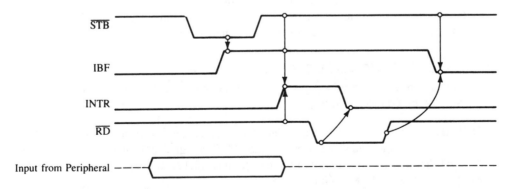

FIGURE 15.9

8255A Mode 1: Timing Waveforms for Strobed Input (with Handshake)

SOURCE: Adapted from Intel Corporation, *Peripheral Components* (Santa Clara, Calif.: Author, 1993), p. 3–110.

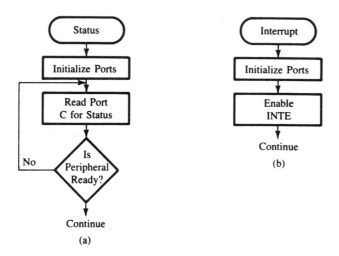

FIGURE 15.10
Flowcharts: Status Check I/O (a) and Interrupt I/O (b)

are being used. The technique is similar to that of Mode 0 combined with the BSR mode. The disadvantage of the status check I/O with handshake is that the MPU is tied up in the loop.

The flowchart in Figure 15.10(b) shows the steps required for the interrupt I/O, assuming that vectored interrupts are available. The confusing step in the interrupt I/O is to set INTE either for port A or port B. Figure 15.8(a) shows that the $\overline{STB}$ signal is connected to pin PC_4 and the $INTE_A$ is also controlled by the pin PC_4. (In port B, pin PC_2 is used for the same purposes.) However, the $INTE_A$ is set or reset in the BSR mode and the BSR control word has no effect when ports A and B are set in Mode 1.

In case the INTR line is used to implement the interrupt, it may be necessary to read the status of $INTR_A$ and $INTR_B$ to identify the port requesting an interrupt service and to determine the priority through software, if necessary.

MODE 1: OUTPUT CONTROL SIGNALS

Figure 15.11 shows the control signals when ports A and B are configured as output ports. These signals are defined as follows:

☐ **OBF (Output Buffer Full):** This is an output signal that goes low when the MPU writes data into the output latch of the 8255A. This signal indicates to an output peripheral that new data are ready to be read (Figure 15.12). It goes high again after the 8255A receives an $\overline{ACK}$ from the peripheral.

☐ **ACK (Acknowledge):** This is an input signal from a peripheral that must output a low when the peripheral receives the data from the 8255A ports (Figure 15.12).

☐ **INTR (Interrupt Request):** This is an output signal, and it is set by the rising edge of the ACK signal. This signal can be used to interrupt the MPU to request the next data

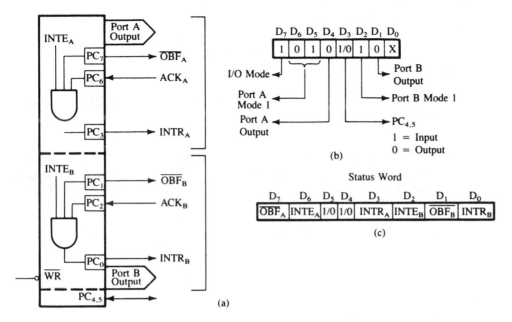

FIGURE 15.11
8255A Mode 1: Output Configuration
SOURCE: Adapted from Intel Corporation, *Peripheral Components* (Santa Clara, Calif.: Author, 1993), p. 3–111.

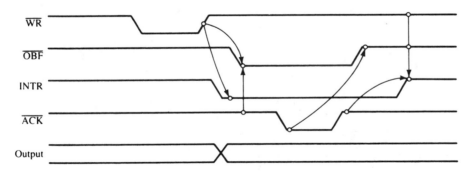

FIGURE 15.12
8255 Mode 1: Timing Waveforms for Strobed (with Handshake) Output
SOURCE: Adapted from Intel Corporation, *Peripheral Components* (Santa Clara, Calif.: Author, 1993), p. 3–111.

byte for output. The INTR is set when $\overline{\text{OBF}}$, $\overline{\text{ACK}}$, and INTE are all one (Figure 15.12) and reset by the falling edge of $\overline{\text{WR}}$.

☐ **INTE (Interrupt Enable):** This is an internal flip-flop to a port and needs to be set to generate the INTR signal. The two flip-flops INTE_A and INTE_B are controlled by bits PC_6 and PC_2, respectively, through the BSR mode.

☐ **$PC_{4,5}$:** These two lines can be set up either as input or output.

CONTROL AND STATUS WORDS

Figure 15.11(b) shows the control word used to set up ports A and B as output ports in Mode 1. Similarly, Figure 15.11(c) also shows the status word, which will be placed in the accumulator if port C is read.

15.16 Illustration: An Application of the 8255A in the Handshake Mode (Mode 1)

PROBLEM STATEMENT

Figure 15.13 shows an interfacing circuit using the 8255A in Mode 1. Port A is designed as the input port for a keyboard with interrupt I/O, and port B is designed as the output port for a printer with status check I/O.

1. Find port addresses by analyzing the decode logic.
2. Determine the control word to set up port A as input and port B as output in Mode 1.
3. Determine the BSR word to enable $INTE_A$ (port A).
4. Determine the masking byte to verify the $\overline{OBF}_B$ line in the status check I/O (port B).
5. Write initialization instructions and a printer subroutine to output characters that are stored in memory.

1. Port Addresses The 8255A is connected as peripheral I/O. When the address lines A_7–A_2 are all 1, the output of the NAND gate goes low and selects the 8255A. The individual ports are selected as follows:

$$
\begin{aligned}
\text{Port A} \quad &= \text{FCH } (A_1 = 0, A_0 = 0)\\
\text{Port B} \quad &= \text{FDH } (A_1 = 0, A_0 = 1)\\
\text{Port C} \quad &= \text{FEH } (A_1 = 1, A_0 = 0)\\
\text{Control Register} &= \text{FFH } (A_1 = 1, A_0 = 1)
\end{aligned}
$$

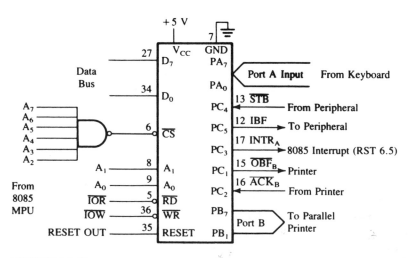

FIGURE 15.13
Interfacing the 8255A in Mode 1 (Strobed Input/Output)

2. Control Word to Set Up Port A as Input and Port B as Output in Mode 1

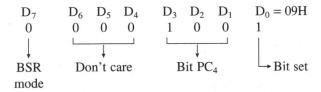

D_7 D_6 D_5 D_4 D_3 D_2 D_1 D_0 = B4H
1   0   1   1   0   1   0   0

| I/O function | Port A in Mode 1 | Port A as input | PC$_{6,7}$ as don't care | Port B as output | Bit D_0 is don't care |

Port B in Mode 1

In the above control word, all bits are self-explanatory, and bits D_3 and D_0 are in a don't care logic state. To generate interrupt signal INTR$_A$, flip-flop INTE$_A$ should be set to 1, which can be accomplished by using the BSR Mode to set PC$_4$.

The output to the printer (port B) is status-controlled. Therefore, the status of line $\overline{OBF_B}$ can be checked by reading bit D_1 of port C$_L$.

3. BSR Word to Set INTE$_A$ To set the Interrupt Enable flip-flop of port A (INTE$_A$), bit PC$_4$ should be 1.

D_7 D_6 D_5 D_4 D_3 D_2 D_1 D_0 = 09H
0   0   0   0   1   0   0   1

BSR mode   Don't care   Bit PC$_4$ Bit set

4. Status Word to Check $\overline{OBF_B}$

D_7	D_6	D_5	D_4	D_3	D_2	$\underline{D_1}$	D_0
X	X	X	X	X	X	$\overline{OBF_B}$	X

Masking byte: 02H

5. Initialization Program

```
          MVI A,B4H        ;Word to initialize port A as input,
                           ;   port B as output in Mode 1
          OUT FFH
          MVI A,09H        ;Set INTE_A (PC_4)
          OUT FFH          ;Using BSR Mode
          EI               ;Enable interrupts
          CALL PRINT
          ;Continue other tasks
```

Print Subroutine

```
PRINT:    LXI H,MEM        ;Point index to location of stored characters
          MVI B,COUNT      ;Number of characters to be printed
NEXT:     MOV A,M          ;Get character from memory
          MOV C,A          ;Save character
```

```
STATUS:     IN FEH          ;Read port C for status of OBF
            ANI 02H         ;Mask all bits except D₁
            JZ STATUS       ;If it is low, the printer is not ready; wait in the loop
            MOV A,C
            OUT FDH         ;Send a character to port B
            INX H           ;Point to the next character
            DCR B
            JNZ NEXT
            RET
```

PROGRAM DESCRIPTION

This I/O design using the 8255A in Mode 1 allows two operations: outputting to the printer and data entry through the keyboard. The printer interfacing is designed with the status check and the keyboard interfacing with the interrupt.

In the PRINT subroutine, the character is placed in the accumulator, and the status is read by the instruction IN FEH. Initially, port B is empty, bit PC_1 ($\overline{OBF_B}$) is high, and the instruction OUT FDH sends the first character to port B. The rising edge of the $\overline{WR}$ signal sets signal $\overline{OBF}$ low, indicating the presence of a data byte in port B, which is sent out to the printer (Figure 15.12). After receiving a character, the printer sends back an acknowledge signal ($\overline{ACK}$), which in turn sets $\overline{OBF_B}$ high, indicating that port B is ready for the next character, and the PRINT subroutine continues.

If a key is pressed during the PRINT, a data byte is transmitted to port A and the $\overline{STB_A}$ goes low, which sets IBF_A high. The initialization routine should set the $INTE_A$ flip-flop. When the $\overline{STB_A}$ goes high, all the conditions (i.e., $IBF_A = 1$, $INTE_A = 1$) to generate $INTR_A$ are met. This signal, which is connected to the RST 6.5, interrupts the MPU, and the program control is transferred to the service routine. This service routine would read the contents of port A, enable the interrupts, and return to the PRINT routine (the interrupt service routine is not shown here).

15.17   Mode 2: Bidirectional Data Transfer

This mode is used primarily in applications such as data transfer between two computers or floppy disk controller interface. In this mode, port A can be configured as the bidirectional port and port B either in Mode 0 or Mode 1. Port A uses five signals from port C as handshake signals for data transfer. The remaining three signals from port C can be used either as simple I/O or as handshake for port B. Figure 15.14 shows two configurations of Mode 2. This mode is illustrated in Section 15.3.

15.2   ILLUSTRATION: INTERFACING KEYBOARD AND SEVEN-SEGMENT DISPLAY

This illustration is concerned with interfacing a pushbutton keyboard and a seven-segment LED display using the 8255A. The emphasis in this illustration is not particularly on the features of 8255A but on how to integrate hardware and software. When a key

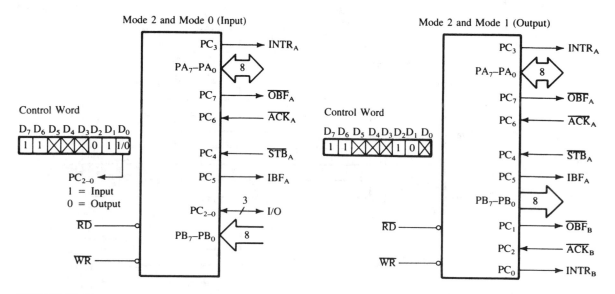

FIGURE 15.14

8255A Mode 2: Bidirectional Input/Output

SOURCE: Intel Corporation, *Peripheral Components* (Santa Clara, Calif.: Author, 1993), p. 3–113.

is pressed, the binary reading of the key has almost no relationship to what it represents. Similarly, to display a number at a seven-segment LED, the binary value of the number needs to be converted into the seven-segment code, which is primarily decided by the hardware consideration. This illustration demonstrates how the microprocessor monitors the changes in hardware reading and converts into appropriate binary reading using its instruction set.

15.21   Problem Statement

A pushbutton keyboard is connected to port A and a seven-segment LED is connected to port B of the 8255A, as shown in Figure 15.15. Port A should be configured as an input port and port B as an output port; this is a simple I/O configuration in Mode 0 without the use of handshake signals or the interrupt.

Write a program to monitor the keyboard to sense a key pressed and display the number of the key at the seven-segment LED. For example, when the key K_7 is pressed, the digit 7 should be displayed at port B.

15.22   Problem Analysis

In this problem, the address decoding circuit is the same as in Figure 15.13; therefore, the port addresses range from FCH to FFH. The keyboard circuit shown in Figure 15.15 is similar to that in Figure 4.11 except that the DIP switches are replaced by pushbutton keys and the buffer is replaced by port A of the 8255A. When a pushbutton key is pressed, it bounces (makes and breaks contact) a few times before it makes a firm contact. To

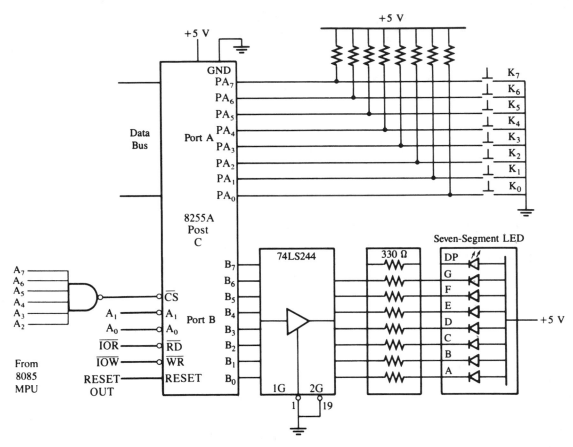

FIGURE 15.15
Interfacing a Keyboard and a Seven-Segment LED

prevent multiple readings of the same key, it is necessary to debounce the key. The hardware solution to this problem is to use a key debounce circuit (cross-coupled NAND or NOR gates), and the software solution is to wait for 10 to 20 ms until the key is settled and then check the key again. The display circuit in Figure 15.15 uses a common-cathode seven-segment LED, connected to port B of the 8255A. To display a digit, it is necessary to turn to the appropriate segments of the LED. The appropriate binary code can be obtained by using the table look-up technique, described in Section 15.24. The programming of this problem can be divided into the following categories:

1. Check if a key is pressed.
2. Debounce the key.
3. Identify and encode the key in appropriate binary format.
4. Obtain the seven-segment code and display it.

The instructions for these steps can be written in separate modules, as shown in the next section.

15.23 Keyboard

The keys K_7–K_0 are tied high through 10 k resistors, and when a key is pressed, the corresponding line is grounded. When all keys are open and if the 8085 reads port A, the reading on the data bus will be FFH. When any key is pressed, the reading will be less than FFH. For example, if K_7 is pressed, the output of port A will be 0111 1111 (7FH). This reading should be encoded into the binary equivalent of the digit 7 (0111) by using software routines. The subroutines KYCHK and KYCODE accomplish the tasks of checking a key pressed and encoding the key in appropriate binary format.

```
KYCHK:      ;This subroutine first checks whether all keys are open.
            ;   Then, it checks for a key closure, debounces the key, and places
            ;   the reading in the accumulator. See Figure 15.16 for flowchart.
            IN PORTA            ;Read keyboard
            CPI 0FFH            ;Are all keys open?
            JNZ KYCHIK          ;If not, wait in loop
            CALL DBONCE         ;If yes, wait 20 ms
KYPUSH:     IN PORTA            ;Read keyboard
            CPI 0FFH            ;Is key pressed?
            JZ KYPUSH           ;If not, wait in loop
            CALL DBONCE         ;If yes, wait 20 ms
            CMA                 ;Set 1 for key closur
            ORA A               ;Set 0 flag for an error
            JZ KYPUSH           ;It is error, check again
            RET
```

PROGRAM DESCRIPTION

This subroutine is based on hardware; when all keys are open the keyboard reading is FFH, and when a key is pressed, the reading is less than FFH. The routine begins with the loop to check whether all keys are open, and it stays in the loop until all keys are open (Figure 15.16). This prevents reading the same key repeatedly if someone were to hold the key for a long time. When the routine finds that a key has been released, it waits for 20 ms for a key debounce.

The loop starting at KYPUSH (Figure 15.16) checks whether a key is pressed. When a key is pressed, the reading is less than FFH; thus, the compare instruction does not set the Z flag and the program goes to the next instruction for a key debounce. The CMA instruction complements the accumulator reading; thus, the reading of the key pressed is set to 1, and other bits are set to 0. The next two instructions check for an error. If it is a momentary contact (false alarm), all bits will be 0s. The ORA instruction sets the Z flag, and the Jump instruction takes the program back to checking keys.

FIGURE 15.16
Flowchart: Key Check Subroutine

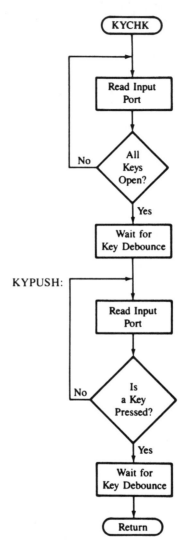

KYCODE: ;This routine converts (encodes) the binary hardware reading of the key
        ;   pressed into appropriate binary format according to the number of the
        ;   key.

```
         MVI C,08H      ;Set code encounter
NEXT:    DCR C          ;Adjust key code
         RAL            ;Place MSB in CY
         JNC NEXT       ;If bit = 0, go back to check next bit
         MOV A,C        ;Place key code in the accumulator
         RET
```

PROGRAM DESCRIPTION

Conceptually, this is an important routine; it establishes the relationship between the hardware and the number of a key. For example, if key K_7 is pressed, the reading from the routine KYCHEK in the accumulator will be 1000 0000 (the reading is already complemented). The KYCODE routine sets register C for the count of eight and immediately decrements the count to seven. The instruction RAL places bit D_7 in the CY flag, and the next instruction checks for the CY flag. If it is set, the key K_7 must be pressed, and the key code (digit 7) is in register C. If CY = 0, the program loops back to check the next bit (D_6). The loop is repeated until 1 is found in CY, and at every iteration of the loop the key code in register C is adjusted for the next key. If more than one key is pressed, this routine ignores the low-order key. Finally, the subroutine places the key code in the accumulator and returns.

KEY DEBOUNCE

When a mechanical pushbutton key, shown in Figure 15.17(a), is pressed or released, the metal contacts of the key momentarily bounce before giving a steady-state reading, as shown in Figure 15.17(b). Therefore, it is necessary that the bouncing of the key should not be read as an input. The key bounce can be eliminated from input data by the **key-debounce technique,** using either hardware or software.

Figure 15.17(c) shows a key debounce circuit. In this circuit, the outputs of the NAND gates do not change even if the key is released from position A_1. The outputs change when the key makes a contact with position B_1. When the key is connected to A_1, A_1 goes low. If one of the inputs to gate G_1 is low, the output O_1 becomes 1, which makes B_2 high. Because line B_1 is already high, the output of O_2 goes low, which makes A_2 low. When the key connection is released from A_1, it goes high, but because A_2 is low the output doesn't change. When the key makes contact with B_1, the outputs change. This means when the key goes from one contact (+5 V) to another contact (ground), the output does not change during the transition period, thus eliminating multiple readings.

In the software technique, when a key closure is found, the microprocessor waits for 10 to 20 ms before it accepts the key as an input. The delay routine is as follows:

```
DBONCE: ;This is a 20 ms delay routine
        ;The delay COUNT should be calculated based on system frequency
        ;This does not destroy any register contents
        ;Input and Output = None
        PUSH B          ;Save register contents
        PUSH PSW
        LXI B,COUNT     ;Load delay count
LOOP:   DCX B           ;Next count
        MOV A,C
        ORA B           ;Set Z flag if (BC) = 0
        JNZ LOOP
        POP PSW         ;Restore register contents
        POP BC
        RET
```

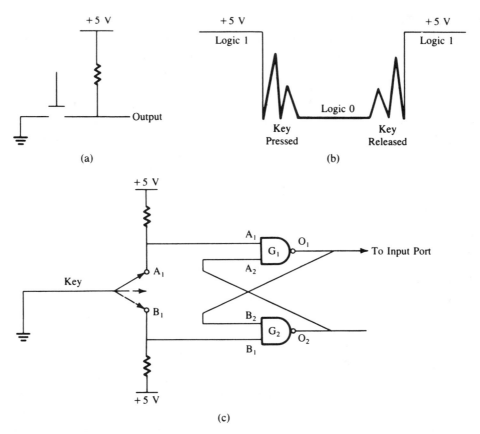

FIGURE 15.17
Pushbutton Key (a), Key Bounce (b), and Key Debounce Circuit Using NAND Gates (c)

PROGRAM DESCRIPTION

This is a simple delay routine similar to the delay routines discussed in Chapter 8. The first instruction loads the BC register with a 16-bit number, and the loop is repeated until BC = 0. In this routine, the 16-bit number (COUNT) should be calculated on the basis of the clock frequency of the system and the T-states in the loop (see Chapter 8 for details).

15.24   Seven-Segment Display

Figure 15.15 shows that a common-anode seven-segment LED is connected to port B through the driver 74LS244. The driver is necessary to increase the current capacity of port B; each LED segment requires 15–20 mA of current. The code for each Hex digit from 0 to F can be determined by examining the connections of the data lines to the segments and the logic requirements.

The driver 74LS244 (Figure 15.15) is an octal noninverting driver with tri-state output and current sinking capacity of 24 mA. It has two active low enable lines ($1\overline{G}$ and

$\overline{2G}$), and the driver is permanently enabled by grounding these lines. In this circuit, this driver functions simply as a current amplifier; whatever logic is at port B will be at the output of the driver.

To display the number of the key pressed, a routine is necessary that will send an appropriate code to port B. The routine KYCODE supplies the binary number of the key pressed; however, there is no relationship between the binary value of a digit and its seven-segment code. Therefore, the table look-up technique (refer to Chapter 10, Section 10.3) will have to be used to find the code for the digit supplied by the KYCODE; this is shown in the next routine, DSPLAY.

```
DSPLAY:     ;This routine takes the binary number and converts into its common-
            ;   anode seven-segment LED code. The codes are stored in memory
            ;   sequentially, starting from the address CACODE
            ;Input: Binary number in accumulator
            ;Output: None
            ;Modifies contents of HL and A
            LXI H,CACODE    ;Load starting address of code table in HL
            ADD L           ;Add digit to low-order address in L
            MOV L,A         ;Place code address in L
            MOV A,M         ;Get code from memory
            OUT PORTB       ;Send code to port B
            RET
CACODE:     ;Common-anode seven-segment codes are stored sequentially in memory
            DB 40H,79H,24H,30H,19H,12H ;Codes for digits from 0 to 5
            DB 02H,78H,00H,18H,08H,03H ;Codes for digits from 6 to B
            DB 46H,21H,06H,0EH          ;Codes from digits from C to F
```

PROGRAM DESCRIPTION

In this routine the HL register is used as a memory pointer to code location. The digit to be displayed is in the accumulator, supplied by the routine KYCODE, and the seven-segment code is stored sequentially in memory, starting from location CACODE. The basic concept in this routine is to modify the memory pointer by adding the value of the digit to the base address and get the code location. For example, let us assume that the starting address of CACODE is 2050H and the digit 7 is in the accumulator. The code for digit 0 is in location 2050H; consequently, the code for digit 7 is in location 2057H. Therefore, to display digit 7, the routine adds the contents of the accumulator (7) to the low-order byte 50H in register L, resulting in the sum 57H. By transferring 57H in register L, the memory pointer in HL is modified to 2057H. Thus, the code for digit 7 is obtained by using this memory pointer.

15.25 Main Program

Now to monitor the keyboard and display the key pressed, we need to initialize the 8255A ports and combine the software modules discussed below:

```
KYBORD:   ;This program initializes the 8255A ports; port A and port B in Mode 0
          ;   and then calls the subroutine modules discussed
          ;   previously to monitor the keyboard
PORTA     EQU FCH              ;Port A address
PORTB     EQU FDH              ;Port B address
CNTRL     EQU FFH              ;Control register
CNWORD    EQU 90H              ;Mode 0 control word, port A input and port B output
STACK     EQU 20AFH            ;Beginning stack address
          LXI SP,STACK
PPI:      MVI A,CNWORD
          OUT CNTRL            ;Set up port A in Mode 1
NEXTKY:   CALL KYCHK           ;Check if a key is pressed
          CALL KYCODE          ;Encode the key
          CALL DSPLAY          ;Display key pressed
          JMP NEXTKY           ;Check the next key pressed
```

PROGRAM DESCRIPTION

This is the main program, which involves the initialization of the 8255A and the stack pointer. The port addresses defined here are from Figure 15.13, and the address of STACK (20AFH) is shown as an illustration; it has no specific significance. Because the problem is divided into small modules, the main program consists primarily of calling these modules.

15.26 Comments and Alternative Approaches

The interfacing of the pushbutton keyboard and seven-segment display is a simplified illustration of industrial applications. The illustration is deliberately kept simple to emphasize the conceptual framework between hardware and software. However, as an application, it has several limitations, as follows:

1. The method of connecting the keyboard demands the number of I/O ports be in proportion to the number of keys; only eight keys can be connected to an 8-bit port. Generally, keys are connected in a matrix format (discussed in Chapter 17). For example, in the matrix format, 16 keys can be connected to one 8-bit port or 64 keys can be connected to two 8-bit ports.
2. The method of connecting a seven-segment LED needs excessive hardware, one port per seven-segment LED and a driver. Furthermore, it consumes a large amount of current (100 to 150 mA per display). To minimize hardware and power consumption, the technique of multiplexing is generally used (discussed in Chapter 17).

In this illustration, the approach is primarily software. For example, in the keyboard, the debouncing and encoding are performed by using instructions. However, nowadays, interfacing chips are available commercially that can sense a key closure, debounce the key, and encode the key. In addition, the chip can generate an interrupt signal when a key is pressed. Similarly, in the seven-segment display, the table look-up can be replaced by a decoder/driver. However, the hardware approach increases unit price. On the other hand, the software approach involves considerable labor (programming and debugging) cost. The choice is generally determined by the production volume and the total unit price.

ILLUSTRATION: BIDIRECTIONAL DATA TRANSFER BETWEEN TWO MICROCOMPUTERS 15.3

Advances in the VLSI technology make it economical to design dedicated microcomputers to perform or monitor specific tasks where the speed of data processing is less important. These dedicated microcomputers are usually controlled by a high-speed computer; this approach is called distributed data processing. The high-speed computer is known as a master and the dedicated computer is called a slave. Parallel I/O with handshake is used to transfer data between a master and a slave, and the data transfer is bidirectional. The bidirectional communication between two microcomputers can be accomplished using the 8255 PPI in Mode 2, as shown in the next illustration.

15.31 Problem Statement

Design an interfacing circuit to set up bidirectional data communication in the master-slave format between two 8085A microcomputers. Use the 8255A as the interfacing device with the master and a tri-state buffer with the slave microcomputer. Write necessary software to transfer a block of data from the master to the slave.

15.32 Problem Analysis

Figure 15.18 shows a block diagram to set up the bidirectional communication between the master and the slave MPUs. The block diagram shows two bidirectional data buses interconnected through the 8255A, which serves as a peripheral device of the master MPU. Port A of the 8255A is used for bidirectional data transfer, and four signals from port C are used for handshaking. The communication process is similar to that of Mode 1 of the 8255A. When the master MPU writes a data byte in the 8255A, the $\overline{OBF}$ signal goes low to inform the slave that a byte is available, and the slave acknowledges when it reads the

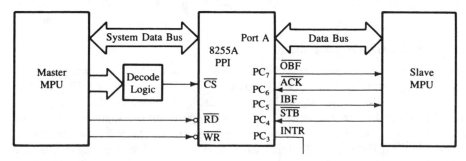

FIGURE 15.18
Block Diagram of Bidirectional Communication between Two Computers Using the 8255A

byte. Similarly, two other handshake signals are used when the slave transfers a data byte to the master.

The master requires I/O ports to read and write data and to check the status of hand-shake signals. Similarly, the slave MPU requires I/O ports to perform Read and Write operations. Therefore, it is necessary to analyze carefully these I/O functions between the MPUs. Data transfer can be accomplished either by status check or interrupt. The speed of handling data is of more importance to the master MPU than to the slave MPU. Therefore, the master MPU is generally set up in the interrupt mode and the slave MPU in the status check mode. However, for this example, both MPUs are set up under the status check mode; the interrupt mode is left as an assignment. The data transfer operations between the two MPUs under the status check I/O are listed in the following sections.

DATA TRANSFER FROM MASTER MPU TO SLAVE MPU

1. The master MPU reads the status of $\overline{OBF}$ to verify whether the previous byte has been read by the slave MPU. This is an input function for the master MPU.
2. The master writes data into port A and the 8255A informs the slave by causing the signal $\overline{OBF}$ to go low. This is an output function for the master MPU.
3. The slave checks the $\overline{OBF}$ signal from the master for data availability. This is an input function for the slave MPU.
4. The slave MPU reads data from port A and acknowledges the reading at the same time by making the signal $\overline{ACK}$ low. This is an input function for the slave MPU.

DATA TRANSFER FROM SLAVE TO MASTER MPU

5. The slave MPU checks the handshake signal IBF (Input Buffer Full) to find out whether port A is available (empty) to transfer a data byte. This is an input function for the slave MPU.
6. The slave MPU places a data byte on the data bus and informs the 8255A by enabling the $\overline{STB}$ (Strobe) signal. This is an output function for the slave MPU.
7. The 8255A causes the IBF (Input Buffer Full) to go high, and the master MPU reads the signal to find out whether a data byte is available. This is an input function for the master MPU.
8. Finally, the master reads the data byte. This is an input function for the master MPU.

This analysis leads to certain hardware requirements that are discussed in the next section.

15.33   Hardware Description

In the first four steps described in the previous section, the master MPU performs one input and one output operation, and the slave MPU performs two input operations. They use two handshake signals: OBF (Output Buffer Full) and ACK (Acknowledge). Steps 5 through 8 are mirror images of the first four steps. The slave MPU performs one input and one output operation, and the master MPU performs two input operations. They use two additional handshake signals: IBF (Input Buffer Full) and $\overline{STB}$ (Strobe). These steps suggest that the master MPU and the slave MPU require three input ports and one output port each. However,

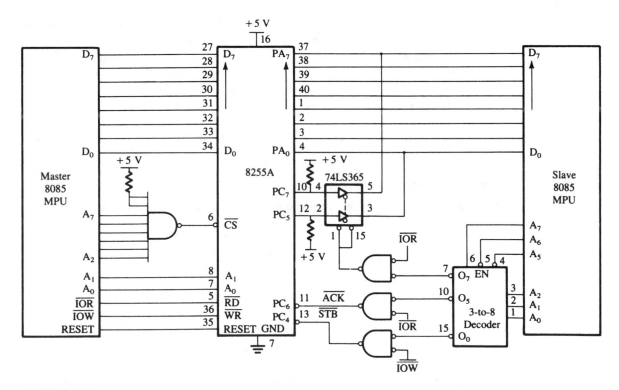

FIGURE 15.19

Schematic: Bidirectional Communication between the Master and Slave MPUs

SOURCE: Adapted from Peter Rony, "Interfacing Fundamentals: Bidirectional I/Os Using Two Semaphores." Reprinted with permission from the April 1981 issue of *Computer Design,* copyright 1981 Computer Design Publishing Company.

if port A is a bidirectional port and port C is a status port, they will meet all the Read/Write requirements of the master MPU. Additional ports need to be designed for the slave MPU.

Figure 15.19 shows the complete schematic of the necessary ports and their decoding logic. The address bus of the master MPU is decoded by using an 8-input NAND gate, and the 8255A is selected when all lines are high, thus assigning the following port addresses:

$$\begin{array}{ll}
\text{Control Register} = \text{FFH} & (A_1 \text{ and } A_0 = 1) \\
\text{Port A} = \text{FCH} & (A_1 \text{ and } A_0 = 0) \\
\text{Port C} = \text{FEH} & (A_1 = 1, A_0 = 0)
\end{array}$$

Port A is configured in Mode 2 using the four signals from port C as shown in the schematic. The $\overline{\text{INTR}}$ signals are unnecessary and, therefore, are not shown. The master MPU checks the $\overline{\text{ACK}}$ and the $\overline{\text{STB}}$ signals by reading the status bits of $\overline{\text{OBF}}$ and IBF in port C.

The other two handshake signals—$\overline{OBF}$ and IBF—are tied, respectively, to bits D_7 and D_0 of the slave data bus through a tri-state buffer so that they can be read by the slave MPU. The decode logic for three input ports and one output port is generated by using the 74LS138 (3-to-8) decoder. Assuming the don't care address lines (A_4 and A_3) are at logic 0, the eight output lines of the decoder can be enabled with the addresses from 80H to 87H. Two output lines of the decoder are combined with the $\overline{IOR}$ control signal to generate two input device select pulses (85H and 87H). Input device select pulse 87H is used to read status on the data lines D_7 and D_0. The decoder line with address 80H is combined with the $\overline{IOW}$ signal to generate the $\overline{STB}$ signal.

CONTROL WORD—MODE 2

To set up the 8255A in the bidirectional mode (Mode 2), the bits of the control word are defined as follows:

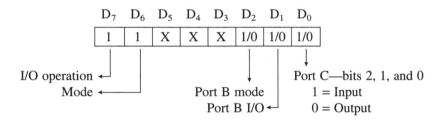

Examination of the control word definition shows that bits D_2 to D_0 are irrelevant in this example because port B and the remaining bits of port C are not being used. Therefore, the required control word is C0H.

STATUS WORD—MODE 2

The status of the I/O operation in Mode 2 can be verified by reading the contents of port C. The status word format is as follows:

D_7	D_6	D_5	D_4	D_3	D_2	D_1	D_0
$\overline{OBF_A}$	$INTE_1$	IBF_A	$INTE_2$	$INTR_A$	X	X	X

for port B

The status of the signal $\overline{OBF}$ can be checked by rotating bit D_7 into the Carry, and the status of the signal IBF can be checked by ANDing the status word with data byte 20H.

READ AND WRITE OPERATIONS OF THE SLAVE MPU

A data byte can be read by the slave MPU from port A simply by sending an active low device select pulse to the $\overline{ACK}$ signal; there is no need to build an input port. Similarly, a data byte can be written by the slave MPU into port A by causing the $\overline{STB}$ signal to go low.

Master Program

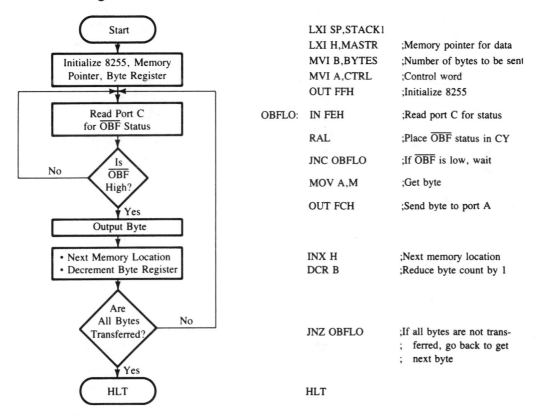

	LXI SP,STACK1	
	LXI H,MASTR	;Memory pointer for data
	MVI B,BYTES	;Number of bytes to be sent
	MVI A,CTRL	;Control word
	OUT FFH	;Initialize 8255
OBFLO:	IN FEH	;Read port C for status
	RAL	;Place $\overline{OBF}$ status in CY
	JNC OBFLO	;If $\overline{OBF}$ is low, wait
	MOV A,M	;Get byte
	OUT FCH	;Send byte to port A
	INX H	;Next memory location
	DCR B	;Reduce byte count by 1
	JNZ OBFLO	;If all bytes are not trans- ; ferred, go back to get ; next byte
	HLT	

FIGURE 15.20
Flowchart of Master Program

15.34 Program

PROGRAM COMMENTS

1. The flowcharts in Figure 15.20 and Figure 15.21 show that both programs check for the status line $\overline{OBF}$. The master program waits until the $\overline{OBF}$ goes high, then writes a byte in port A. On the other hand, the slave program waits until the $\overline{OBF}$ goes low, then reads data.
2. When the master MPU writes a data byte, it is latched by port A, and the data byte is placed on the slave data bus when the $\overline{ACK}$ goes low. The timing diagram in Figure 15.22 shows that when the slave MPU reads data, the $\overline{ACK}$ signal goes low, and the falling edge of the $\overline{ACK}$ signal sets the $\overline{OBF}$ signal high for the master MPU to write the next byte.
3. The programs given above can transfer a block of data from the master MPU to the slave MPU but not vice versa. To transfer a block of data from the slave MPU to the master MPU, additional instructions are necessary (see Problem 7 at the end of this chapter).

Slave Program

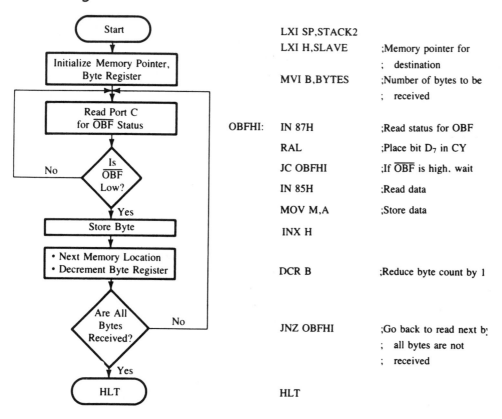

	LXI SP,STACK2	
	LXI H,SLAVE	;Memory pointer for
		; destination
	MVI B,BYTES	;Number of bytes to be
		; received
OBFHI:	IN 87H	;Read status for OBF
	RAL	;Place bit D_7 in CY
	JC OBFHI	;If $\overline{OBF}$ is high. wait
	IN 85H	;Read data
	MOV M,A	;Store data
	INX H	
	DCR B	;Reduce byte count by 1
	JNZ OBFHI	;Go back to read next b;
		; all bytes are not
		; received
	HLT	

FIGURE 15.21
Flowchart of Slave Program

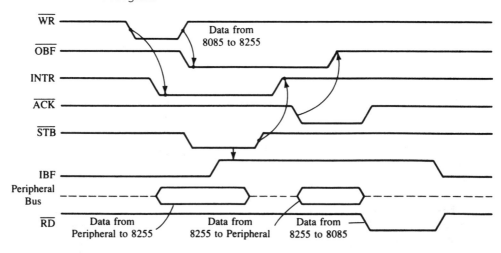

FIGURE 15.22
Timing Diagram Mode 2

SOURCE: Adapted from Intel Corporation, *Peripheral Components* (Santa Clara, Calif.: Author, 1993), p. 3–113.

These instructions should monitor the IBF signal in both programs and Read/Write operations should be interchanged. The master MPU should wait until IBF goes high to read a data byte, and the slave MPU should wait until IBF goes low to write a byte.

4. The timing diagram in Figure 15.22 shows an INTR signal to implement data transfer using the interrupt. However, the signal is irrelevant in this illustration; it is given in Figure 15.22 for Problem 8 at the end of this chapter.

THE 8254 (8253) PROGRAMMABLE INTERVAL TIMER 15.4

The 8254 programmable interval timer/counter is functionally similar to the software-designed counters and timers described in Chapter 8. It generates accurate time delays and can be used for applications such as a real-time clock, an event counter, a digital one-shot, a square-wave generator, and a complex waveform generator.

The 8254 includes three identical 16-bit counters that can operate independently in any one of the six modes (to be described later). It is packaged in a 24-pin DIP and requires a single +5 V power supply. To operate a counter, a 16-bit count is loaded in its register and, on command, begins to decrement the count until it reaches 0. At the end of the count, it generates a pulse that can be used to interrupt the MPU. The counter can count either in binary or BCD. In addition, a count can be read by the MPU while the counter is decrementing.

The 8254 is an upgraded version of the 8253, and they are pin-compatible. The features of these two devices are almost identical except that

☐ the 8254 can operate with higher clock frequency range (DC to 8 MHz and 10 MHz for 8254-2), and the 8253 can operate with clock frequency from DC to 2 MHz.

☐ the 8254 includes a Status Read-Back Command that can latch the count and the status of the counters.

15.41   Block Diagram of the 8254

Figure 15.23 is the block diagram of the 8254; it includes three counters (0, 1, and 2), a data bus buffer, Read/Write control logic, and a control register. Each counter has two input signals—Clock (CLK) and GATE—and one output signal—OUT.

DATA BUS BUFFER
This tri-state, 8-bit, bidirectional buffer is connected to the data bus of the MPU.

CONTROL LOGIC
The control section has five signals: $\overline{RD}$ (Read), $\overline{WR}$ (Write), $\overline{CS}$ (Chip Select), and the address lines A_0 and A_1. In the peripheral I/O mode, the $\overline{RD}$ and $\overline{WR}$ signals are connected to $\overline{IOR}$ and $\overline{IOW}$, respectively. In memory-mapped I/O, these are connected to $\overline{MEMR}$ (Memory Read) and $\overline{MEMW}$ (Memory Write). Address lines A_0 and A_1 of the

Pin Configuration

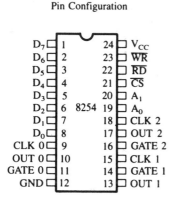

D$_7$–D$_0$	Data Bus (8 Bit)
CLK N	Counter Clock Inputs
GATE N	Counter Gate Inputs
OUT N	Counter Outputs
$\overline{RD}$	Read Counter
$\overline{WR}$	Write Command or Data
$\overline{CS}$	Chip Select
A$_0$–A$_1$	Counter Select
V$_{CC}$	+ 5 Volts
GND	Ground

Block Diagram

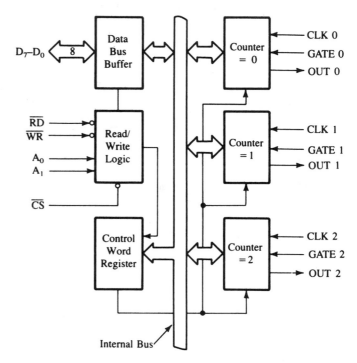

FIGURE 15.23
8254 Block Diagram
SOURCE: Intel Corporation, *Peripheral Components* (Santa Clara, Calif.: Author, 1993), p. 3–62.

MPU are usually connected to lines A$_0$ and A$_1$ of the 8254, and $\overline{CS}$ is tied to a decoded address.

The control word register and counters are selected according to the signals on lines A$_0$ and A$_1$, as shown below:

A$_1$	A$_0$	Selection
0	0	Counter 0
0	1	Counter 1
1	0	Counter 2
1	1	Control Register

CONTROL WORD REGISTER

This register is accessed when lines A$_0$ and A$_1$ are at logic 1. It is used to write a command word which specifies the counter to be used, its mode, and either a Read or a Write operation. The control word format is shown in Figure 15.24.

	D7	D6	D5	D4	D3	D2	D1	D0
	SC1	SC0	RW1	RW0	M2	M1	M0	BCD

SC—Select Counter:

SC1 SC0

0	0	Select Counter 0
0	1	Select Counter 1
1	0	Select Counter 2
1	1	Read-Back Command (See Read Operations)

M—MODE:

M2 M1 M0

0	0	0	Mode 0
0	0	1	Mode 1
X	1	0	Mode 2
X	1	1	Mode 3
1	0	0	Mode 4
1	0	1	Mode 5

RW—Read/Write:

RW1 RW0

0	0	Counter Latch Command (see Read Operations)
0	1	Read/Write least significant byte only.
1	0	Read/Write most significant byte only.
1	1	Read/Write least significant byte first, then most significant byte.

BCD:

0	Binary Counter 16-bits
1	Binary Coded Decimal (BCD) Counter (4 Decades)

Note: Don't Care Bits (X) Should Be 0 to Ensure Compatability with Future Intel Products.

FIGURE 15.24
8254 Control Word Format
SOURCE: Intel Corporation, *Peripheral Components* (Santa Clara, Calif.: Author, 1993), p. 3–67.

MODE

The 8254 can operate in six different modes, and the gate of a counter is used either to disable or enable counting, as shown in Figure 15.25. However, to maintain clarity, only one mode (Mode 0) is illustrated first, and details of the remaining modes are discussed in Section 15.44.

In Mode 0, after the count is written and if the gate is high, the count is decremented every clock cycle. When the count reaches zero, the output goes high and remains high until a new count or mode word is loaded.

15.42 Programming the 8254

The 8254 can be programmed to provide various types of output (see Section 15.44, Figure 15.27) through Write operations, or to check a count while counting through Read operations. The details of these operations are given below.

Modes \ Signal Status	Low or Going Low	Rising	High
0	Disables counting	—	Enables counting
1	—	(1) Initiates counting (2) Resets output after next clock	—
2	(1) Disables counting (2) Sets output immediately high	(1) Reloads counter (2) Initiates counting	Enables counting
3	(1) Disables counting (2) Sets output immediately high	Initiates counting	Enables counting
4	Disables counting	—	Enables counting
5	—	Initiates counting	—

FIGURE 15.25

Gate Settings of a Counter

SOURCE: Intel Corporation, *Peripheral Components* (Santa Clara, Calif.: Author, 1993), p. 3–78.

WRITE OPERATIONS

To initialize a counter, the following steps are necessary.

1. Write a control word into the control register.
2. Load the low-order byte of a count in the counter register.
3. Load the high-order byte of a count in the counter register.

With a clock and an appropriate gate signal to one of the counters, the above steps should start the counter and provide appropriate output according to the control word.

READ OPERATIONS

In some applications, especially in event counters, it is necessary to read the value of the count in progress. This can be done by either of two methods. One method involves reading a count after inhibiting (stopping) the counter to be read. The second method involves reading a count while the count is in progress (known as reading on the fly).

In the first method, counting is stopped (or inhibited) by controlling the gate input or the clock input of the selected counter, and two I/O read operations are performed by the MPU. The first I/O operation reads the low-order byte, and the second I/O operation reads the high-order byte.

In the second method, an appropriate control word is written into the control register to latch a count in the output latch, and two I/O Read operations are performed by the MPU. These Read/Write operations are illustrated below.

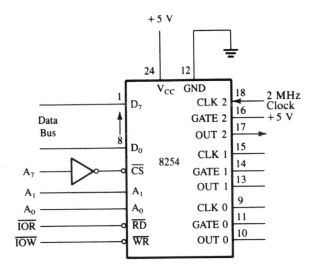

FIGURE 15.26
Schematic: Interfacing the 8254

15.43 Illustration: The 8254 as a Counter

PROBLEM STATEMENT

1. Identify the port addresses of the control register and counter 2 in Figure 15.26.
2. Write a subroutine to initialize counter 2 in Mode 0 with a count of $50,000_{10}$. The subroutine should also include reading counts on the fly; when the count reaches zero, it should return to the main program.
3. Write a main program to display seconds by calling the subroutine as many times as necessary.

1. Port Addresses The Chip Select is enabled when $A_7 = 1$ (see Figure 15.26), and the control register is selected when A_1 and $A_0 = 1$. Similarly, counter 2 is selected when $A_1 = 1$ and $A_0 = 0$. Assuming that the unused address lines A_6 to A_2 are at logic 0, the port addresses will be as follows:

$$\text{Control Register} = 83\text{H}$$
$$\text{Counter 2} = 82\text{H}$$

2. Subroutine Counter To initialize the 8254 for counter 2 in Mode 0, the following control word is necessary (see Figure 15.24):

Control Word (Load Operation)

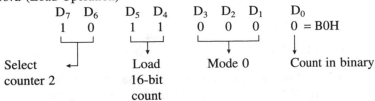

Control Word for Latching: Bits D_5 and D_4 should be 0 = 80H

Subroutine

```
COUNTER:    MVI A,B0H        ;Control word to initialize counter 2
            OUT 83H          ;Write in the control register
            MVI A,LOBYTE     ;Low-order byte of the count 50000
            OUT 82H          ;Load counter 2 with the low-order byte
            MVI A,HIBYTE     ;High-order byte of the count 50000
            OUT 82H          ;Load counter 2 with the high-order byte
READ:       MVI A,80H        ;Control word to latch a count
            OUT 83H          ;Write in the control register
            IN 82H           ;Read low-order byte
            MOV D,A          ;Store low-order byte in register D
            IN 82H           ;Read high-order byte
            ORA D            ;OR low- and high-order bytes to set Z flag
            JNZ READ         ;If counter ≠ 0, go back to read next count
            RET
```

Subroutine Description   The subroutine has two segments. In the first segment, counter 2 is initialized by writing a control word in the control register and a 16-bit count specified as LOBYTE and HIBYTE in the counter register. The hexadecimal value equivalent to 50000_{10} must be calculated.

In the second segment (beginning at READ), a control word is written into the control register to sample a count, and the 16-bit count is read by performing two input operations. The reading of the counter is repeated until the counter reaches 0; the Zero flag is checked by ORing the low- and high-order bytes.

3. Main Program   The subroutine COUNTER provides 25 ms (50000×0.5 μs Clock) delay; if this routine is called 40 times, the total delay will be one second.

Program

```
            LXI SP,STACK     ;Initialize stack pointer
            MVI B,00H        ;Clear register B to save number of seconds
SECOND:     MVI C,28H        ;Set up register C to count $40_{10}$
WAIT:       CALL COUNTER     ;Wait for 25 ms
            DCR C
            JNZ WAIT         ;Is this one second? If not, go back and wait
            MVI A,B
            ADI 01           ;Add one second
            DAA
            OUT PORT1
            MOV B.A          ;Save seconds
            JMP SECOND       ;Go back and start counting the next second
```

Program Description   The main program initializes the stack pointer; loads register C with the count of 28H (40_{10}) and sets up the WAIT loop. The loop calls the COUNTER subroutine 40 times to generate a one-second delay. At the end of the loop, it increments the seconds in register B, decimal-adjusts the byte, and displays seconds. The sequence is repeated until register B reaches 99_{BCD}. After the 99th second, register B is cleared and the clock sequence is repeated.

This program is just to demonstrate the Read and Write operations of the 8254; this clock design does not take into account the errors caused by the delay in executing the program instructions. A better way of designing a real-time clock is to interrupt the MPU at the end of a count (see Problem 15 at the end of this chapter).

15.44 Modes

As mentioned earlier, the 8254 can operate in six different modes; we already illustrated Mode 0 in Section 15.23. Now we will describe briefly various modes of the 8254 including Mode 0.

MODE 0: INTERRUPT ON TERMINAL COUNT

In this mode, initially the OUT is low. Once a count is loaded in the register, the counter is decremented every cycle, and when the count reaches zero, the OUT goes high. This can be used as an interrupt. The OUT remains high until a new count or a command word is loaded. Figure 15.27 also shows that the counting (m = 5) is temporarily stopped when the Gate is disabled (G = 0), and continued again when the Gate is at logic 1.

MODE 1: HARDWARE-RETRIGGERABLE ONE-SHOT

In this mode, the OUT is initially high. When the Gate is triggered, the OUT goes low, and at the end of the count, the OUT goes high again, thus generating a one-shot pulse (Figure 15.27, Mode 1).

MODE 2: RATE GENERATOR

This mode is used to generate a pulse equal to the clock period at a given interval. When a count is loaded, the OUT stays high until the count reaches 1, and then the OUT goes low for one clock period. The count is reloaded automatically, and the pulse is generated continuously. The count = 1 is illegal in this mode.

Write instructions to generate a pulse every 50 μs from Counter 0 (refer to Figure 15.26). **Example 15.3**

To generate a pulse every 50 μs, from Counter 0, it should be initialized in Mode 2 (refer to Figure 15.27 for modes), and Gate 0 should be high. **Solution**

Control Word (Refer to Figure 15.24):

D_7	D_6	D_5	D_4	D_3	D_2	D_1	D_0	
0	0	0	1	0	1	0	0	= 14H

Select Counter 0   Load 8-bit Count   Mode 2   Binary Count

Count: In Mode 2, the count is decremented every clock period, and at the last count, the counter generates a pulse equivalent to the clock period of the timer. Here the clock frequency of the 8254 is 2 MHz (0.5 μs clock period), and the pulse should be generated every 50 μs. Therefore, the count is calculated as follows:

$$\text{Count} = \frac{50 \times 10^{-6}}{0.5 \times 10^{-6}} = 100 = 64H$$

In this example, the frequency of the pulse is 20 kHz (1/50 μs). This count can also be calculated by dividing the clock frequency by the frequency of the pulse (2 MHz/ 20 kHz = 100).

Instructions:

```
PULSE:    MVI A, 00010100B      ;Control word Mode 2 and Counter 0
          OUT 83H               ;Write in 8254 control register
          MVI A,64H             ;Low-order byte of the count
          OUT 80H               ;Load Counter 0 with low-order byte
          HALT
```

MODE 3: SQUARE-WAVE GENERATOR

In this mode, when a count is loaded, the OUT is high. The count is decremented by two at every clock cycle, and when it reaches zero, the OUT goes low, and the count is reloaded again. This is repeated continuously; thus, a continuous square wave with period equal to the period of the count is generated. In other words, the frequency of the square wave is equal to the frequency of the clock divided by the count. If the count (N) is odd, the pulse stays high for $(N + 1)/2$ clock cycles and stays low for $(N - 1)/2$ clock cycles.

Example 15.4

Write instructions to generate a 1 kHz square wave from Counter 1 (refer to Figure 15.26). Assume the gate of Counter 1 is tied to +5 V through a 10 k resistor. Explain the significance of connecting the gate to +5 V.

Solution

To generate a square wave from Counter 1, it should be initialized in Mode 3 (refer to Figure 15.27 for modes).

Control Word (Refer to Figure 15.24):

D_7	D_6	D_5	D_4	D_3	D_2	D_1	D_0	
0	1	1	1	0	1	1	0	= 76H

Select Counter 1   Load 16-bit Count   Mode 3   Binary Count

Mode 0: Interupt on Terminal Count

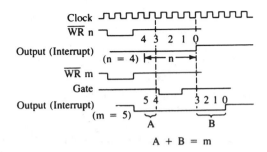

Mode 1: Programmable One-Shot

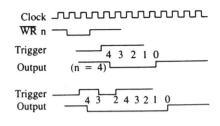

Mode 2: Rate Generator Clock

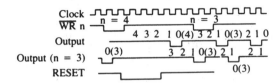

Mode 3: Square Wave Generator

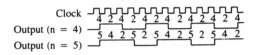

Mode 4: Software Triggered Strobe

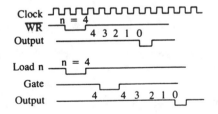

Mode 5: Hardware Triggered Strobe

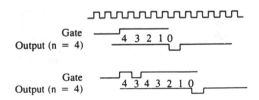

FIGURE 15.27
Six Modes of the 8254

SOURCE: Intel Corporation, *Peripheral Components* (Santa Clara, Calif.: Author, 1993), pp. 3–72–3–75 (adapted).

Count: In Mode 3, the count is decremented by two for every clock period. If the count is N, N/2 clock pulses provide the upper half of the square wave. The count is loaded again and N/2 clock pulses provide the lower half.

In this example, the clock frequency of the 8254 is 2 Mhz (0.5 µs clock period), and the square wave frequency is 1 kHz (1 ms clock period). Therefore, we need a count for 1 ms delay.

$$\text{Count} = \frac{1 \times 10^{-3}}{0.5 \times 10^{-6}} = 2000 = 07\text{D0H}$$

This count can also be calculated by dividing the clock frequency by the square wave frequency (2 MHz/1 kHz = 2000).

Instructions:

SQWAVE: MVI A, 01110110B   ;Control word Mode 3 and Counter 1
            OUT 83H   ;Write in 8254 control register
            MVI A, D0H   ;Low-order byte of the count
            OUT 81H   ;Load Counter 1 with low-order byte
            MVI A, 07H   ;High-order byte of the count
            OUT 81H   ;Load Counter 1 with high-order byte
            HALT

To run Counter 1, the gate of that counter must be tied high; otherwise the counter action is inhibited.

MODE 4: SOFTWARE-TRIGGERED STROBE

In this mode, the OUT is initially high; it goes low for one clock period at the end of the count. The count must be reloaded for subsequent outputs.

MODE 5: HARDWARE-TRIGGERED STROBE

This mode is similar to Mode 4, except that it is triggered by the rising pulse at the gate. Initially, the OUT is low, and when the Gate pulse is triggered from low to high, the count begins. At the end of the count, the OUT goes low for one clock period.

READ-BACK COMMAND

The Read-Back Command in the 8254 allows the user to read the count and the status of the counter; this command is not available in the 8253. The format of the command is shown in Figure 15.28(a).

The command is written in the control register, and the count of the specified counter(s) can be latched if $\overline{COUNT}$ (bit D_5) is 0. A counter or a combination of counters is specified by keeping the respective CNT bits (D_1, D_2, and D_3) high. For example, the control word 1 1 0 1 0 1 1 0 (D6H) written in the control register will latch the counts of Counter 0 and Counter 1, and these counts can be obtained by reading respective counter port addresses. The latched counts are held until they are read or the counters are reprogrammed. The Read-Back Command eliminates the need of writing separate counter-latch commands for different counters.

The status of the counter(s) can be read if $\overline{STATUS}$ bit (D_4) of the Read-Back Command is low. Figure 15.28(b) shows the format of the status byte.

Example 15.5

Write a subroutine to generate an interrupt every 1 sec. Refer to Figure 15.26 for counter addresses.

Solution

The clock frequency shown in Figure 15.26 is 2 MHz; thus, a count is decremented every 0.5 μs. To obtain a delay of one second, the count should be $(1 \text{ sec}/0.5 \times 10^{-6})$ 2 Meg.; this count is too large for one 16-bit counter. We can divide this count, as an example,

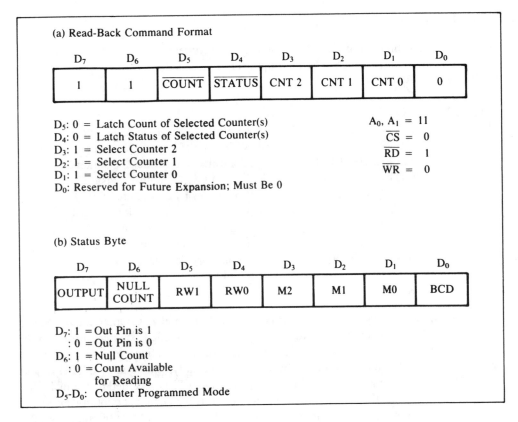

FIGURE 15.28

Read-Back Command Format (a) and Status Byte (b)

SOURCE: Intel Corporation, *Peripheral Components* (Santa Clara, Calif.: Author, 1993), p. 3–69.

50,000 for Counter 1 and 40 for second Counter 2 (50 k × 40 = 2 Meg.). If we set up Counter 1 in Mode 2 with 50,000 as a count, it will generate a pulse every 25 millisec. The output of Counter 1 can be used as a clock input to Counter 2. That means the count in Counter 2 will be decremented every 25 millisec. If Counter 2 is also set up in Mode 2 with the count = 40, the output pulse from Counter 2 is generated every second that can be used to interrupt the processor.

Control Word (Refer to Figure 15.24):

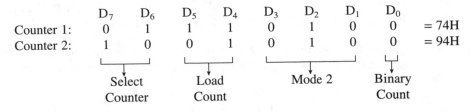

Instructions:

```
;The following subroutine is an initialization for 8254 timer.
;It uses two counters to generate an interrupt every one second.
CNT1LO      EQU 50H
CNT1HI      EQU C3H
COUNT2      EQU 40
SECOND:     MVI A, 01110100B        ;Control word: Mode 2, Counter 1
            OUT 83H                 ;Write in 8254 control register
            MVI A, 10010100B        ;Control word: Mode 2, Counter 2
            OUT 83H                 ;Write in 8254 control register
            MVI A, CNT1LO           ;Low-order byte of count 50,000
            OUT 81H                 ;Load Counter 1 with low-order byte
            MVI A, CNT1HI           ;High-order byte of count 50,000
            OUT 81H                 ;Load Counter 1 with high-order byte
            MVI A, COUNT2           ;Count for Counter 2
            OUT 82H                 ;Load Counter 2
            RET
```

This subroutine sends two control words in a sequence to initialize Counter 1 and Counter 2. The 8254 differentiates these words according to the specified counter in the control words. We could have initialized these counters in a different sequence; for example, the control word for Counter 1 followed by its count. Initially, the equates CNT1LO and CNT1HI are defined in Hex numbers and the equate COUNT2 is specified in decimal. This is based on the assumption that an assembler will convert the decimal count in Hex equivalent. However, this procedure is not feasible for the 16-bit count of 50,000; therefore, the equivalent Hex count (C350H) is loaded into two registers as 50H (low) and C3H (high).

Once this subroutine is called, the output of Counter 2 will generate a pulse every 1 second because the counts are reloaded automatically at the end of each count. This pulse can be used to interrupt the processor, and the processor can update the clock every second. This is a hardware dependent timing and more accurate than the technique used in Section 15.43.

15.5   THE 8259A PROGRAMMABLE INTERRUPT CONTROLLER

The 8259A is a programmable interrupt controller designed to work with Intel microprocessors 8085, 8086, and 8088. The 8259A interrupt controller can

1. manage eight interrupts according to the instructions written into its control registers. This is equivalent to providing eight interrupt pins on the processor in place of one INTR (8085) pin.

2. vector an interrupt request anywhere in the memory map. However, all eight interrupts are spaced at the interval of either four or eight locations. This eliminates the major drawback of the 8085 interrupts in which all interrupts are vectored to memory locations on page 00H.

3. resolve eight levels of interrupt priorities in a variety of modes, such as fully nested mode, automatic rotation mode, and specific rotation mode (to be explained later).

4. mask each interrupt request individually.

5. read the status of pending interrupts, in-service interrupts, and masked interrupts.

6. be set up to accept either the level-triggered or the edge-triggered interrupt request.

7. be expanded to 64 priority levels by cascading additional 8259As.

8. be set up to work with either the 8085 microprocessor mode or the 8086/8088 microprocessor mode.

The 8259A is upward-compatible with its predecessor, the 8259. The main difference between the two is that the 8259A can be used with Intel's 8086/88 16-bit microprocessors. It also includes additional features such as the level-triggered mode, buffered mode, and automatic-end-of-interrupt mode. To simplify the explanation of the 8259A, illustrative examples will not include the cascade mode or the 8086/88 mode and will be limited to modes commonly used with the 8085.

15.51 Block Diagram of the 8259A

Figure 15.29 shows the internal block diagram of the 8259A. It includes eight blocks: control logic, Read/Write logic, data bus buffer, three registers (IRR, ISR, and IMR), priority resolver, and cascade buffer. This diagram shows all the elements of a programmable device, plus additional blocks. The functions of some of these blocks need explanation, which is given below.

READ/WRITE LOGIC

This is a typical Read/Write control logic. When the address line A_0 is at logic 0, the controller is selected to write a command or read a status. The Chip Select logic and A_0 determine the port address of the controller.

CONTROL LOGIC

This block has two pins: INT (Interrupt) as an output, and $\overline{INTA}$ (Interrupt Acknowledge) as an input. The INT is connected to the interrupt pin of the MPU. Whenever a valid interrupt is asserted, this signal goes high. The $\overline{INTA}$ is the Interrupt Acknowledge signal from the MPU.

INTERRUPT REGISTERS AND PRIORITY RESOLVER

The Interrupt Request Register (IRR) has eight input lines (IR_0–IR_7) for interrupts. When these lines go high, the requests are stored in the register. The In-Service Register (ISR) stores all the levels that are currently being serviced, and the Interrupt Mask Register (IMR) stores the masking bits of the interrupt lines to be masked. The Priority Resolver (PR) examines these three registers and determines whether INT should be sent to the MPU.

Pin Configuration

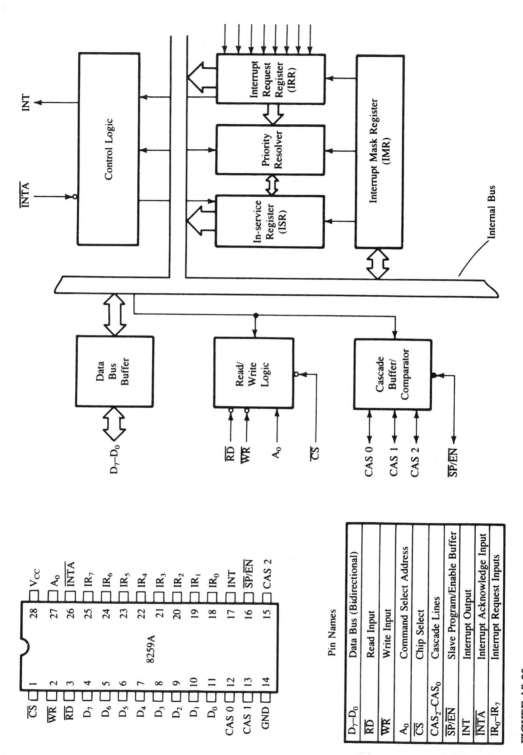

Block Diagram

Pin Names

D_7-D_0	Data Bus (Bidirectional)
$\overline{RD}$	Read Input
$\overline{WR}$	Write Input
A_0	Command Select Address
$\overline{CS}$	Chip Select
CAS_2-CAS_0	Cascade Lines
$\overline{SP}/\overline{EN}$	Slave Program/Enable Buffer
INT	Interrupt Output
$\overline{INTA}$	Interrupt Acknowledge Input
IR_0-IR_7	Interrupt Request Inputs

FIGURE 15.29

The 8259A Block Diagram

SOURCE: Intel Corporation, *Peripheral Components* (Santa Clara, Calif.: Author, 1993), pp. 3–171, 3–172.

CASCADE BUFFER/COMPARATOR

This block is used to expand the number of interrupt levels by cascading two or more 8259As. To simplify the discussion, this block will not be mentioned again.

15.52 Interrupt Operation

To implement interrupts, the Interrupt Enable flip-flop in the microprocessor should be enabled by writing the EI instruction, and the 8259A should be initialized by writing control words in the control register. The 8259A requires two types of control words: Initialization Command Words (ICWs) and Operational Command Words (OCWs). The ICWs are used to set up the proper conditions and specify RST vector addresses. The OCWs are used to perform functions such as masking interrupts, setting up status-read operations, etc. After the 8259A is initialized, the following sequence of events occurs when one or more interrupt request lines go high:

1. The IRR stores the requests.
2. The priority resolver checks three registers: the IRR for interrupt requests, the IMR for masking bits, and the ISR for the interrupt request being served. It resolves the priority and sets the INT high when appropriate.
3. The MPU acknowledges the interrupt by sending $\overline{\text{INTA}}$.
4. After the $\overline{\text{INTA}}$ is received, the appropriate priority bit in the ISR is set to indicate which interrupt level is being served, and the corresponding bit in the IRR is reset to indicate that the request is accepted. Then, the opcode for the CALL instruction is placed on the data bus.
5. When the MPU decodes the CALL instruction, it places two more $\overline{\text{INTA}}$ signals on the data bus.
6. When the 8259A receives the second $\overline{\text{INTA}}$, it places the low-order byte of the CALL address on the data bus. At the third $\overline{\text{INTA}}$, it places the high-order byte on the data bus. The CALL address is the vector memory location for the interrupt; this address is placed in the control register during the initialization.
7. During the third $\overline{\text{INTA}}$ pulse, the ISR bit is reset either automatically (Automatic-End-of-Interrupt—AEOI) or by a command word that must be issued at the end of the service routine (End-of-Interrupt—EOI). This option is determined by the initialization command word (ICW).
8. The program sequence is transferred to the memory location specified by the CALL instruction.

15.53 Priority Modes and Other Features

Many types of priority modes are available under software control in the 8259A, and they can be changed dynamically during the program by writing appropriate command words. Commonly used priority modes are discussed below.

1. **Fully Nested Mode** This is a general-purpose mode in which all IRs (Interrupt Requests) are arranged from highest to lowest, with IR_0 as the highest and IR_7 as the lowest.

In addition, any IR can be assigned the highest priority in this mode; the priority sequence will then begin at that IR. In the example below, IR_4 has the highest priority, and IR_3 has the lowest priority:

IR_0	IR_1	IR_2	IR_3	IR_4	IR_5	IR_6	IR_7
4	5	6	7	0	1	2	3
			↑	↑			
			Lowest	Highest			
			priority	priority			

2. Automatic Rotation Mode   In this mode, a device, after being serviced, receives the lowest priority. Assuming that the IR_2 has just been serviced, it will receive the seventh priority, as shown below:

IR_0	IR_1	IR_2	IR_3	IR_4	IR_5	IR_6	IR_7
1	6	7	0	1	2	3	4

3. Specific Rotation Mode   This mode is similar to the automatic rotation mode, except that the user can select any IR for the lowest priority, thus fixing all other priorities.

END OF INTERRUPT

After the completion of an interrupt service, the corresponding ISR bit needs to be reset to update the information in the ISR. This is called the End-of-Interrupt (EOI) command. It can be issued in three formats:

1. Nonspecific EOI Command   When this command is sent to the 8259A, it resets the highest priority ISR bit.

2. Specific EOI Command   This command specifies which ISR bit to reset.

3. Automatic EOI   In this mode, no command is necessary. During the third $\overline{\text{INTA}}$, the ISR bit is reset. The major drawback with this mode is that the ISR does not have information on which IR is being serviced. Thus, any IR can interrupt the service routine, irrespective of its priority, if the Interrupt Enable flip-flop is set.

ADDITIONAL FEATURES OF THE 8259A

The 8259A is a complex device with various modes of operation. These modes are listed below for reference; the user should refer to the *8085 User's Manual* for details.

☐ Interrupt Triggering: The 8259A can accept an interrupt request with either the edge-triggered mode or the level-triggered mode. The mode is determined by the initialization instructions.

☐ Interrupt Status: The status of the three interrupt registers (IRR, ISR, and IMR) can be read, and this status information can be used to make the interrupt process versatile.

☐ Poll Method: The 8259A can be set up to function in a polled environment. The MPU polls the 8259A rather than each peripheral.

15.54   Programming the 8259A

As mentioned before, the 8259A requires two types of command words: Initialization Command Words (ICWs) and Operational Command Words (OCWs). The 8259A can be initialized with four ICWs; the first two are essential, and the other two are optional based on the modes being used. These words must be issued in a given sequence. Once initialized, the 8259A can be set up to operate in various modes by using three different OCWs; however, they no longer need be issued in a specific sequence.

Figure 15.30 shows the bit specification of the first two ICWs. The ICW_1, shown in Figure 15.30(a), specifies

1. single or multiple 8259As in the system.
2. 4- or 8-bit interval between the interrupt vector locations.
3. the address bits A_7–A_5 of the CALL instruction; the rest are supplied by the 8259A, as shown in Figure 15.30(b).

The ICW_2 of Figure 15.30(c) specifies the high-order byte of the CALL instruction.

EXAMPLES

Figure 15.31 shows the schematic of an interrupt-driven system using the 8259A. Four sources are connected to the IR lines of the 8259A: Emergency Signal, Keyboard, A/D Converter, and Printer. Of these, the Emergency Signal has the highest priority and the Printer has the lowest priority.

Explain the following initialization instructions in reference to Figure 15.31.

**Example
15.6**

Initialization Instructions

```
DI
MVI A,76H     ;ICW₁
OUT 80H       ;Initialize 8259A
MVI 20H       ;ICW₂
OUT 81H       ;Initialize 8259A
```

1. The DI instruction disables the interrupts so that the initialization process will not be interrupted.

Solution

2. The command word 76H specifies the following parameters; see Figure 15.30(a):

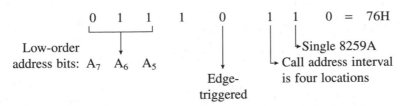

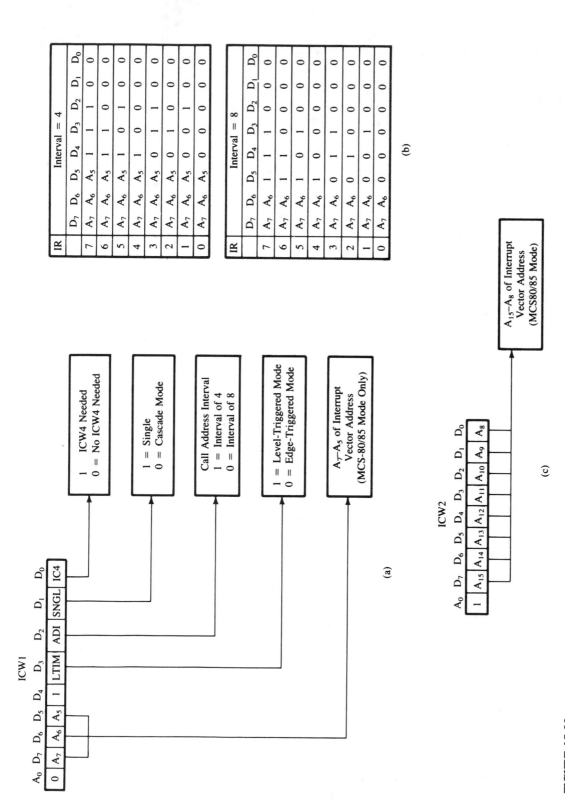

FIGURE 15.30

Initialization Command Words for the 8259A

SOURCE: Intel Corporation, *Peripheral Components* (Santa Clara, Calif.: Author, 1993), p. 3–181.

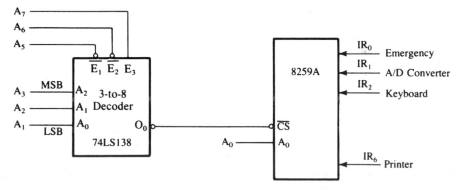

FIGURE 15.31
Schematic of an Interrupt System Using the 8259A

Low-Order Byte of the IR_0 Call Address; see Figure 15.22(b):

A_7	A_6	A_5	A_4	A_3	A_2	A_1	A_0	
0	1	1	0	0	0	0	0	= 60H

The address bits A_4–A_0 are supplied by the 8259A. The subsequent addresses are four locations apart (e.g., IR_1 = 64H).

3. The port address of the 8259A for ICW_1 is 80H; A_0 should be at logic 0, and the other bits are determined by the decoder.

4. Command word ICW_2 is 20H, which specifies the high-order byte of the Call address.

5. The port address of ICW_2 is 81H; A_0 should be at logic 1.

Explain the interrupt process in the fully nested mode relative to Figure 15.32. Assume that the 8259A is initialized with the same instructions as in the previous example.

Example 15.7

Figure 15.32 shows that the interrupts are enabled by the main program, and the 8259A is initialized. After the initialization, the 8259A is set in the fully nested mode by default, unless a different Operational Command Word (OCW) is issued. During the main program, the printer has made a request. Because the interrupts are enabled, the program is transferred first to the vectored location 2078H for IR_6, and then to the service routine. The ISR_6 bit is also set to indicate that IR_6 is being serviced.

Solution

During the IR_6 service routine, the keyboard makes a request (IR_2). Even though IR_2 has a higher priority than the IR_6, the request is not acknowledged until the IR_6 service routine enables the interrupts through the EI instruction. When IR_2 is acknowledged, bit ISR_2 is set, and the program is vectored to the location 2068H and then to the service routine.

At the end of the IR_2 service routine, the instruction EOI (End-of-Interrupt) informs the 8259A that the service has been completed, and the highest ISR bit (ISR_2) has been reset. The program returns to the IR_6 service routine, completes the service, sends the EOI, and then returns to the main program. The EOI in this routine resets the ISR_6 bit.

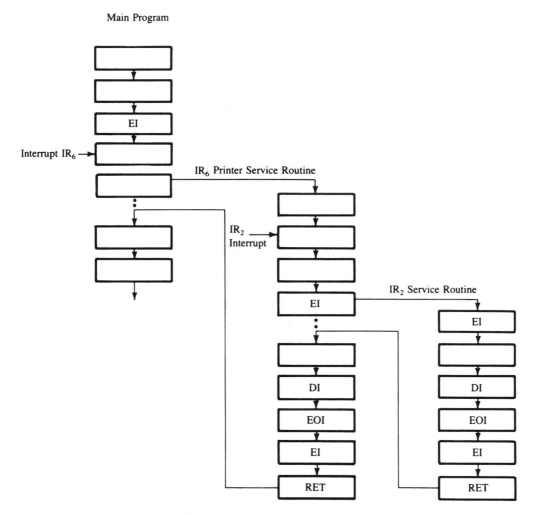

FIGURE 15.32

Interrupt Process: Fully Nested Mode

SOURCE: John Beaston, *Using the Programmable Interrupt Controller,* Intel Application Note AP-31 (Santa Clara, Calif.: Intel Corporation, May 1978), p. 10.

The format for the nonspecific EOI command is as follows:

A_0	D_7	D_6	D_5	D_4	D_3	D_2	D_1	D_0	
0	0	0	1	0	0	0	0	0	= 20H with port address 80H

The nonspecific EOI command can be used in the fully nested mode because it always resets the bit of the highest priority; however, in other priority modes, it may reset the wrong bit. It is always safe to use a specific EOI command; the format is as follows:

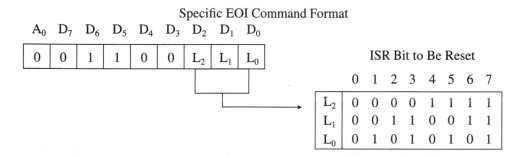

Specific EOI Command Format

DIRECT MEMORY ACCESS (DMA) AND THE 8237
DMA CONTROLLER 15.6

Direct Memory Access is an I/O technique commonly used for high-speed data transfer; for example, data transfer between system memory and a floppy disk. In status check I/O and interrupt I/O, data transfer is relatively slow because each instruction needs to be fetched and executed. In DMA, the MPU releases the control of the buses to a device called a DMA controller. The controller manages data transfer between memory and a peripheral under its control, thus bypassing the MPU. Conceptually, this is an important I/O technique; it introduces two new signals available on the 8085—HOLD and HLDA (Hold Acknowledge).

☐ HOLD—This is an active high input signal to the 8085 from another master requesting the use of the address and data buses. After receiving the Hold request, the MPU relinquishes the buses in the following machine cycle. All buses are tri-stated and the Hold Acknowledge (HLDA) signal is sent out. The MPU regains the control of the buses after HOLD goes low.
☐ HLDA (Hold Acknowledge)—This is an active high output signal indicating that the MPU is relinquishing control of the buses.
    A DMA controller uses these signals as if it were a peripheral requesting the MPU for the control of the buses. The MPU communicates with the controller by using the Chip Select line, buses, and control signals. However, once the controller has gained control, it plays the role of a processor for data transfer. To perform this function the DMA controller should have

1. a data bus,
2. an address bus,
3. Read/Write control signals, and
4. control signals to disable its role as a peripheral and to enable its role as a processor.

This process is called switching from the slave mode to the master mode.
    For all practical purposes, the DMA controller is a processor capable only of copying data at high speed from one location to another location. As an illustration, a programmable DMA controller, the Intel 8237, is described below.

15.61   The 8237 DMA Controller

The 8237 is a programmable Direct Memory Access controller (DMA) housed in a 40-pin package. It has four independent channels with each channel capable of transferring 64K bytes. It must interface with two types of devices: the MPU and peripherals such as floppy disks. As mentioned earlier, the DMA plays two roles in a given system: It is an I/O to the microprocessor (slave mode) and it is a data transfer processor to peripherals such as floppy disks (master mode). Many of its signals that are input in the I/O mode become outputs in the processor mode. It also needs additional signal lines to communicate with the addresses of 64K data bytes, and these signals must be generated externally by using latches and buffers. The 8237 is a complex device. To maintain clarity, the following discussion is divided into five segments: DMA channels and interfacing, DMA signals, system interface, programming, and DMA execution. The specification details of the 8237 are included in Appendix D.

DMA CHANNELS AND INTERFACING

Figure 15.33 shows a logical pin out and internal registers of the 8237. It also shows the interface with the 8085 using a 3-to-8 decoder.

The 8237 has four independent channels, CH0 to CH3. Internally, two 16-bit registers are associated with each channel: One is used to load a starting address of the byte to be copied and the second is used to load a count of the number of bytes to be copied. Figure 15.33 shows eight such registers that can be accessed by the MPU. The addresses of these registers are determined by four address lines, A_3 to A_0, and the chip select ($\overline{CS}$) signal. Address 0000 on lines A_3–A_0 selects CH0 Memory Address Register (MAR) and address 0001 selects the next register, CH0 Count. Similarly, all the remaining registers are selected in sequential order. The last eight registers are used to write commands or read status as shown. In Figure 15.33, the MPU accesses the DMA controller by asserting the signal Y_0 of the decoder. Therefore, the addresses of these internal registers range from 00 to 0FH as follows:

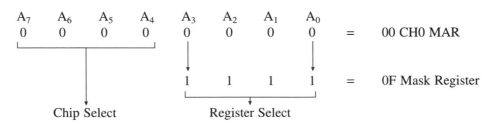

DMA SIGNALS

In Figure 15.33, signals are divided into two groups: (1) one group of signals shown on the left of the 8237 is used for interfacing with the MPU; (2) the second group shown on the right-hand side of the 8237 is for communicating with peripherals. Some of these signals are bidirectional and their functions are determined by the DMA mode of operation (I/O or processor). The signals that are necessary to understand the DMA operation are explained as follows; the remaining signals are listed in Appendix D.

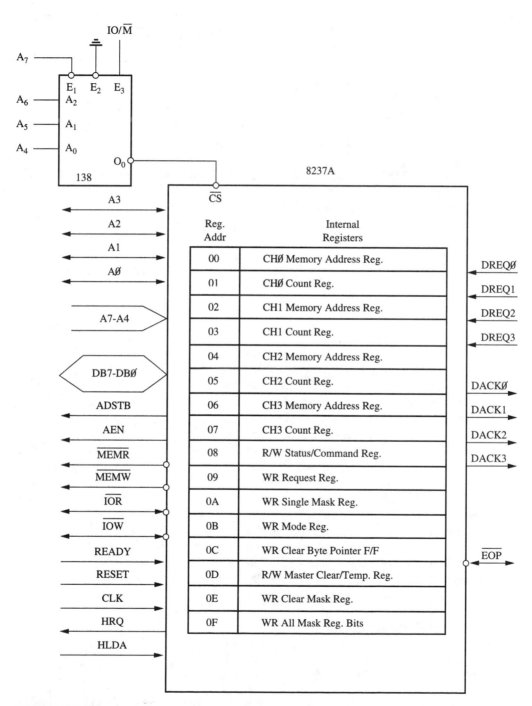

FIGURE 15.33
8237A—DMA Controller with Internal Registers

☐ **DREQ0–DREQ3—DMA Request:** These are four independent, asynchronous input signals to the DMA channels from peripherals such as floppy disks and the hard disk. To obtain DMA service, a request is generated by activating the DREQ line of the channel.

☐ **DACK0–DACK3—DMA Acknowledge:** These are output lines to inform the individual peripherals that a DMA is granted. DREQ and DACK are equivalent to handshake signals in I/O devices.

☐ **AEN and ADSTB—Address Enable and Address Strobe:** These are active high output signals that are used to latch a high-order address byte to generate a 16-bit address.

☐ **MEMR and MEMR—Memory Read and Memory Write:** These are output signals used during the DMA cycle to write and read from memory.

☐ **A_3–A_0 and A_7–A_4—Address:** A_3–A_0 are bidirectional address lines. They are used as inputs to access control registers as shown in the previous section. During the DMA cycle, these lines are used as output lines to generate a low-order address that is combined with the remaining address lines A_7–A_4.

☐ **HRQ and HLDA—Hold Request and Hold Acknowledge:** HRQ is an output signal used to request the MPU control of the system bus. After receiving the HRQ, the MPU completes the bus cycle in process and issues the HLDA signal.

SYSTEM INTERFACE

The DMA is used to transfer data bytes between I/O (such as a floppy disk) and system memory (or from memory to memory) at high speed. It includes eight data lines, four control signals (IOR, IOW, MEMR, and MEMW), and eight address lines (A_7–A_0). However, it needs 16 address lines to access 64K bytes. Therefore, an additional eight lines must be generated as shown in Figure 15.34.

When a transfer begins, the DMA places the low-order byte on the address bus and the high-order byte on the data bus and asserts AEN (Address Enable) and ADSTB (Address Strobe). These two signals are used to latch the high-order byte from the data bus; thus, it places the 16-bit address on the system bus. After the transfer of the first byte, the latch is updated when the lower byte generates a carry (or borrow). Figure 15.34 shows two latches: one latch (373 #1) to latch a high-order address from the data bus by using the AEN and ADSTB signals, and the second latch (373 #2) to demultiplex the 8085 bus and generate the low-order address bus by using the ALE (Address Latch Enable from the 8085) signal. The AEN signal is connected to the OE signal of the second latch to disable the low-order address bus from the 8085 when the first latch is enabled to latch the high-order byte of the address.

PROGRAMMING THE 8237

To implement the DMA transfer, the 8237 should be initialized by writing into various control registers discussed earlier in the DMA channels and interfacing section. To initialize the 8237, the following steps are necessary.

1. Write a control word in the Mode register that selects the channel and specifies the type of transfer (Read, Write, or Verify) and the DMA mode (block, single-byte, etc.).

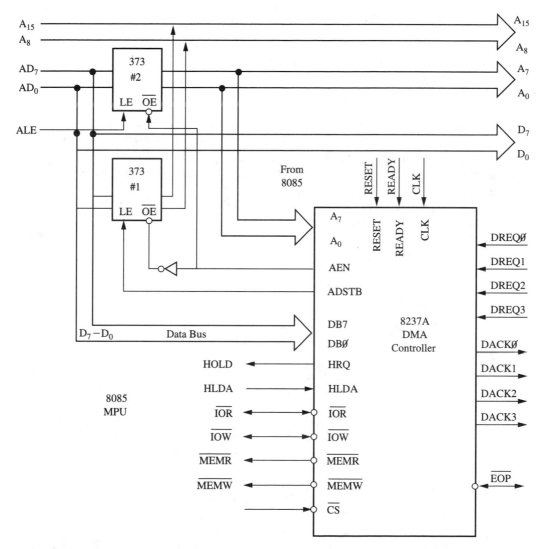

FIGURE 15.34
Interfacing 8237A—DMA Controller with the 8085

2. Write a control word in the Command register that specifies parameters such as priority among four channels, DREQ and DACK active levels, and timing, and enables the 8237.
3. Write the starting address of the data block to be transferred in the channel Memory Address Register (MAR).
4. Write the count (the number of the bytes in the data block) in the channel Count register.

These steps are illustrated in the following example.

Example 15.8

Write initialization instructions for the DMA controller in Figure 15.33 to meet the following specifications. Use the same port (register) addresses (00 to 0FH) as in Figure 15.33.

1. Disable the DMA controller and begin writing initialization instructions.
2. Initialize Channel #3 (CH3) to transfer 1K of bytes from the system memory to the floppy disk assigned to CH3.
3. The starting address of the data block is 4075H and subsequent data bytes have memory addresses in increasing order.
4. The Command parameters should be: normal timing, fixed priority, late write. DREQ and DACK are both active low.
5. Set up the demand mode whereby the DMA can complete the data transfer without any interruption.

Solution

The initialization instructions to set up the DMA controller are as follows (for a detailed explanation see Appendix D):

```
MVI A, 00000100B   ;Command:    0    0    0    0    0    1    0    0
                                                      └┬┘
                                                   Disable DMA

OUT 08H            ;Send to Command Reg.
MVI A, 00000111B   ;Mode:    0  0      0      0      0 1    1 1
                            └─┬─┘    └┬┘    └┬┘    └─┬─┘  └─┬─┘
                            Demand  Incre-  Disable  Write  Ch 3
                            Mode     ment    Auto
                                     Addr.   Load

OUT 0BH            ;Send to Mode Reg.
MVI A, 75H         ;Low-order byte of starting address
OUT 06H            ;Output to CH3 Memory Address Reg.
MVI A, 40H         ;High-order byte of starting address
OUT 06H            ;Output to CH3 Memory Address Reg.
MVI A, FFH         ;Low-order byte of the count 03FFH
OUT 07H            ;Output to CH3 Count Reg.
MVI A, 03H         ;High-order byte of the count 03FFH
OUT 07H            ;Output to CH3 Count Reg.
MVI A, 10000000B   ;Command:  1    0    0    0    0    0  0    0
                            └┬┘  └┬┘  └┬┘  └┬┘       └┬┘  └┬┘
                            DACK Late Fixed Normal  DMA  Disable
                            DREQ Write Priority Time Enable  Mem
                            High                              to
                                                             Mem

OUT 08H            ;Send to Command Reg.
```

DMA EXECUTION

The process of data transfer from the peripheral to the system memory under the DMA controller can be classified under two modes: the slave mode and the master mode.

Slave Mode In the slave mode, the DMA controller is treated as a peripheral, using the following steps:

1. The MPU selects the DMA controller through Chip Select.
2. The MPU writes the control words as illustrated in Example 15.8 in channel registers and command/status registers by using control signals $\overline{\text{IOW}}$ and $\overline{\text{IOR}}$.

In this mode, the output signals of the 8237, such as A_7–A_4, $\overline{\text{MEMW}}$, and $\overline{\text{MEMR}}$, are in tri-state.

Master Mode After the initialization, the 8237 in master mode keeps checking for a DMA request, and the steps in data transfer can be listed as follows:

1. When the peripheral is ready for data transfer, it sends a high signal to DRQ.
2. When the DRQ has been received and the channel enabled, the control logic sets HRQ (Hold Request) high. (HRQ is connected to the HOLD signal of the 8085.)
3. In the next cycle, the MPU relinquishes the buses and sends the HLDA (Hold Acknowledge) signal to the 8237.
4. After receiving the HLDA signal, the DMA asserts AEN (Address Enable) signal high. The high AEN signal disables 373 Latch #2, thus disconnecting the demultiplexed bus A_7–A_0 of the MPU and enables 373 Latch #1 through an inverter. Next, the DMA asserts ADSTB (Address Strobe) high that is connected to Latch Enable (LE) of 373 Latch #1 and places the contents of the data bus, which is a high-order byte of the starting address, on A_{15}–A_8. At the same time, the DMA also outputs the low order address A_7–A_0 on the low-order address bus.
5. When the entire address A_{15}–A_0 is available on the address bus, the DMA sends DACK to the peripheral.
6. The DMA controller continues the data transfer by asserting the necessary control signals ($\overline{\text{IOR}}$, $\overline{\text{IOW}}$, $\overline{\text{MEMR}}$, or $\overline{\text{MEMW}}$) until DACK remains high.
7. At the end of the data transfer, the DMA asserts $\overline{\text{EOP}}$ (End of Process) signal low that can be used to inform the peripheral that the data transfer is complete. The DMA data transfer can also be terminated by sending a low signal to $\overline{\text{EOP}}$ from outside.

SUMMARY

In this chapter, four programmable devices were described: the 8255A (PPI), the 8254 (timer), the 8259A (interrupt controller), and the 8237 (DMA controller). These are general-purpose devices, and each is designed to serve different purposes in the I/O communication process. The common element among them is that the functions of these devices can be programmed by writing instructions in their control registers. Applications of these devices were demonstrated with illustrations and examples.

QUESTIONS, PROBLEMS, AND PROGRAMMING ASSIGNMENTS

1. List the operating modes of the 8255A Programmable Peripheral Interface.
2. Specify the handshake signals and their functions if port A of the 8255A is set up as an output port in Mode 1.
3. Specify the bit of a control word for the 8255, which differentiates between the I/O mode and the BSR mode.
4. Specify the two control words that are necessary to set bit PC_6 (assume that the other ports are not being used).
5. Port A of the 8255A is set up in Mode 1, and the status word is read as 18H. Is there an error in the status word?
6. List the necessary conditions to generate INTR when port A of the 8255A is set up as an output port in Mode 1.
7. Write necessary software to transfer 100 bytes of data from the slave MPU to the master MPU, using the status check I/O (see Figure 15.19).
8. Connect the INTR signal to RST 6.5 (8085 system) to interrupt the master MPU when data transfer is required. Modify the master program to implement data transfer under the interrupt I/O mode.
   Hints: The main program should enable $INTE_1$ and $INTE_2$, using the BSR mode. The Interrupt service routine should verify whether it is a Read or a Write request.
9. Specify the conditions to start the timer 8254.
10. List the major components of the 8259A interrupt controller and explain their functions.
11. Explain how the 8237 DMA controller transfers 64K bytes of data per channel with eight address lines.
12. Write initialization instructions for the 8255A to set up

    □ port A as an output port in Mode 0.
    □ port B as an output port in Mode 1 for interrupt I/O.
    □ port C_U as an output port in Mode 0.

13. In Figure 15.5, connect the system address lines A_9 and A_8 to A_1 and A_0 lines of the 8255A, respectively. Specify the port addresses.
14. Set up the 8254 as a square-wave generator with a 1 ms period, if the input frequency to the 8254 is 1 MHz.
15. Design a five-minute clock (timer) using the 8254 and the interrupt technique. Display minutes and seconds.
16. Write initialization instructions for the 8259A Interrupt Controller to meet the following specifications:

    □ Interrupt vector address: 2090H
    □ Call address interval of eight bits
    □ Nested mode

EXPERIMENTAL ASSIGNMENTS

1. **a.** Connect the circuit as shown in Figure 15.7.
    b. Write initialization instructions, and store binary readings in memory buffer for five different analog signals.
    c. Modify the circuit to record the data, using the interrupt RST 6.5.
    d. Modify the initialization instructions.
    e. Write the service routine to record the data.
    f. Store data in memory for five different analog signals.
2. **a.** Set up the 8254 timer as shown in Figure 15.26.
    b. Write instructions to obtain a square wave with the period of 500 μs.
    c. Enter the instructions and execute the program.
    d. Measure the square wave output on an oscilloscope.
    e. Calculate the pulse width if the count is 8000H.
    f. Load the count in step e, start the counter, and measure the pulse width of the output.

Serial I/O and Data Communication

The 8085 microprocessor is a parallel device; it transfers eight bits of data simultaneously over eight data lines. This is the **parallel I/O mode** discussed in previous chapters. However, in many situations, the parallel I/O mode is either impractical or impossible. For example, parallel data communication over a long distance can become very expensive. Similarly, devices such as a CRT terminal or a cassette tape are not designed for parallel I/O. In these situations, the serial I/O mode is used, whereby one bit at a time is transferred over a single line.

In serial transmission (from the MPU to a peripheral), an 8-bit parallel word should be converted into a stream of eight serial bits; this is known as **parallel-to-serial conversion.** After the conversion, one bit at a time is transmitted over a single line at a given rate called the baud (bits per second). On the other hand, in serial reception, the MPU receives a stream of eight bits, and they should be converted into an 8-bit parallel word; this is known as **serial-to-parallel conversion.** In addition to the conversion, information such as the beginning and the end of transmission and error check is necessary in serial transmission. This process raises several questions about the serial I/O mode.

☐ In serial I/O, how does the MPU identify a peripheral and what are the conditions of data transfer: unconditional, using the status check, or using the interrupt?

☐ What are the codes for alphanumeric data?

☐ What are the requirements of transmission: synchronization, speed, error check, etc.?

☐ What are the standards for interfacing various types of equipment?

☐ What are the trade-offs between software and hardware approaches in implementing serial I/O?

These questions concern basic concepts in serial data communication and are discussed in this chapter. The illustrations include both software- and hardware-controlled serial I/O, as well as uses of the SOD (Serial Output Data) and SID (Serial Input Data) pins—specially designed signals for serial I/O in the 8085.

OBJECTIVES

☐ Explain how data transfer occurs in the serial I/O mode and how it differs from the parallel I/O mode.

☐ Explain the terms *synchronous* and *asynchronous transmission, simplex, half* and *full duplex trans-*

mission; ASCII code; baud (rate); and *parity check.*

☐ Explain how data bits are transmitted (or received) in the asynchronous format, and calculate the delay required between two successive bits for a given baud.

☐ Explain the RS-232C serial I/O standard, and compare it with the RS-422A and 423A standards.

☐ Write instructions to transmit and receive data using the serial I/O lines (SID and SOD) in an 8085 system.

☐ Explain the block diagram and the functions of each block of the Intel 8251 USART (Programmable Communication Interface).

☐ Design an interfacing circuit using the 8251, and write initialization instructions to set up data communication between a microcomputer and a serial peripheral.

16.1   BASIC CONCEPTS IN SERIAL I/O

The basic concepts concerning the serial I/O mode can be classified into the following categories as discussed in the next sections.

1. Interfacing requirements
2. Alphanumeric codes
3. Transmission format
4. Error checks in data communication
5. Data communication over telephone lines
6. Standards in serial I/O
7. Software vs. programmable hardware approaches

16.11   Interfacing Requirements

The interfacing requirements for a serial I/O peripheral are the same as for a parallel I/O device. The microprocessor identifies the peripheral through a port address and enables it using the Read or Write control signal. The primary difference between parallel I/O and serial I/O is in the number of lines used for data transfer—the parallel I/O uses the entire data bus and the serial I/O uses one data line. Figure 16.1 shows a typical configuration of serial I/O transmission; the MPU selects the peripheral through Chip Select and uses the control signals Read to receive data and Write to transmit data. The address decoding can be either peripheral I/O or memory-mapped I/O. Similarly, a serial peripheral can be interfaced under either program control (status check) or interrupt control.

FIGURE 16.1
Block Diagram: Serial I/O
Interfacing

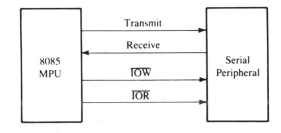

16.12 Alphanumeric Codes

A computer is a binary machine; to communicate with the computer in alphabetic letters and decimal numbers, translation codes are necessary. The commonly used code is known as ASCII, the American Standard Code for Information Interchange. It is a 7-bit code with 128 (2^7) combinations, and each combination from 00H to 7FH is assigned to a letter, a decimal number, a symbol, or a machine command (see Appendix E). For example, hexadecimals 30H to 39H represent numerals 0 to 9; 41H to 5AH represent capital letters A through Z; 21H to 2FH represent various symbols; and the initial codes 00H to 1FH represent machine commands such as Carriage Return (CR) or Line Feed (LF). Devices that use ASCII characters include ASCII terminals, teletype machines (TTY), and printers. When key 9 is pressed on an ASCII terminal, the computer receives 39H in binary, and the system programs (as shown in Chapter 10) translate ASCII characters into appropriate binary or BCD numbers.

This topic was discussed briefly in Chapter 1 and repeated here because of its direct relevance to serial communication.

16.13 Transmission Format

A transmission format is concerned with issues such as synchronization, direction of data flow, speed, errors, and medium of transmission (telephone lines, for example). These topics are described briefly below.

SYNCHRONOUS VS. ASYNCHRONOUS TRANSMISSION

Serial communication occurs in either synchronous or asynchronous format. In the synchronous format, a receiver and a transmitter are synchronized; a block of characters is transmitted along with the synchronization information, as in Figure 16.2(a). This format is generally used for high-speed transmission (more than 20 k bits/second).

The asynchronous format is character-oriented. Each character carries the information of the Start and the Stop bits, shown in Figure 16.2(b). When no data are being transmitted, a receiver stays high at logic 1, called Mark; logic 0 is called Space. Transmission begins with one Start bit (low), followed by a character and one or two Stop bits (high). This is also known as **framing.** Figure 16.2(b) shows the transmission of 11 bits for an ASCII character in the asynchronous format: one Start bit, eight character bits, and two Stop bits. The format shown in Figure 16.2(b) is similar to Morse code, but the dots and dashes are replaced by logic 0s and 1s. The asynchronous format is generally used in low-speed transmission (less than 20 k bits/second).

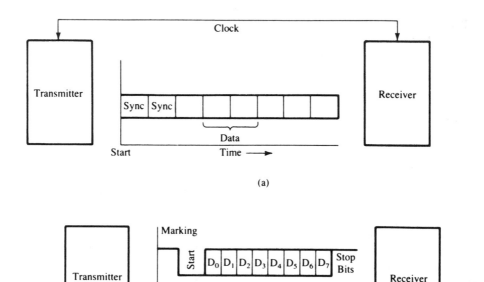

FIGURE 16.2
Transmission Format: Synchronous (a) and Asynchronous (b)

SIMPLEX AND DUPLEX TRANSMISSION

Serial communication also can be classified according to the direction and simultaneity of data flow.

In **simplex transmission,** data are transmitted in only one direction. A typical example is transmission from a microcomputer to a printer.

In **duplex transmission,** data flow in both directions. However, if the transmission goes one way at a time, it is called **half duplex;** if it goes both ways simultaneously, it is called **full duplex.** Generally, transmission between two computers or between a computer and a terminal is full duplex.

RATE OF TRANSMISSION (BAUD)

In parallel I/O, data bits are transferred when a control signal enables the interfacing device; the transfer takes place in less than three T-states. However, in serial I/O, one bit is sent out at a time; therefore, how long the bit stays on or off is determined by the speed at which the bits are transmitted. Furthermore, the receiver should be set up to receive the bits at the same rate as the transmission; otherwise, the receiver may not be able to differentiate between two consecutive 0s or 1s.

The rate at which the bits are transmitted—bits/second—is called a baud; technically, however, it is defined as the number of signal changes/second. Each piece of equip-

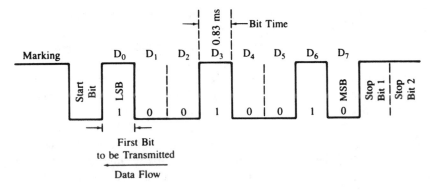

FIGURE 16.3
Serial Bit Format for ASCII Character I at 1200 Baud

ment has its own baud requirement. For example, a teletype (TTY) generally runs at 110 baud. However, in most terminals and printers, the baud is adjustable, typically, in the range of 50 to 9600 baud. Figure 16.3 shows how the ASCII character I (49H) will be transmitted with 1200 baud with the framing information of one Start and two Stop bits. The transmission begins with an active low Start bit, followed by the LSB-bit D_0. The bit time—the delay between any two successive bits—is 0.83 ms; this is determined by the baud as follows:

$$1200 \text{ bits} = 1 \text{ second}$$
$$\text{For 1 bit} = 1/1200 = 0.83 \text{ ms}$$

Therefore, to transmit one character, a parallel byte (49H) should be converted into a stream of 11 bits by adding framing bits (one Start and two Stop bits), and each bit must be transmitted at the interval of 0.83 ms. This can be implemented either through software or through programmable hardware chips. To receive a character in the serial mode, the process is reversed—one bit at a time is received and the bits are converted into a parallel word.

16.14 Error Checks in Data Communication

During transmission, various types of errors can occur. For example, data bits may change because of noise or can be misunderstood by the receiver because of differences in receiver and transmitter clocks. These errors need to be checked; therefore, additional information for error checking is sent during the transmission. The receiver can check the received data against the error check information, and if an error is detected, the receiver can request the retransmission of that data segment. Three methods are generally in common practice; they are parity check, checksum, and cyclic redundancy check.

PARITY CHECK

This is used to check each character by counting the number of 1s in the character; in the ASCII code transmission, bit D_7 is used to transmit parity check information. The technique is based on the principle that, in a given system, each character is transmitted with either an even number of 1s or an odd number of 1s.

In an even parity system, when a character has an odd number of 1s, the bit D_7 is set to 1 and an even number of 1s is transmitted. For example, the code for the character I is 49H (0100 1001), with three 1s. When the character I is transmitted in an even parity system, the bit D_7 is set to 1, making the code C9H (1100 1001). On the other hand, in an odd parity system, the character I is transmitted by keeping bit $D_7 = 0$; thus, the code remains 49H.

In the 8085 system, the parity check is easy to implement and detect because the 8085 has the parity flag, and this flag can be used to check parity information in each character. However, the parity check cannot detect multiple errors in any given character.

CHECKSUM

The checksum technique is used when blocks of data are transferred. It involves adding all the bytes in a block without carries. Then, the 2's complement of the sum (negative of the sum) is transmitted as the last byte. The receiver adds all the bytes, including the 2's complement of the sum; thus, the result should be zero if there is no error in the block (refer to Section 16.44 for an illustration of the checksum technique).

CYCLIC REDUNDANCY CHECK (CRC)

This technique is commonly used when data are transferred from and to a floppy disk and in a synchronous data communication. The technique is based on mathematical relationships of polynomials. A stream of data can be represented as a polynomial that is divided by a constant polynomial, and the remainder, unique to that set of bits, is generated. The remainder is sent out as a check for errors. The receiver checks the remainder to detect an error in the transmission. This is a somewhat complex technique and will not be discussed here.

16.15 Data Communication over Telephone Lines

The serial I/O technique can be used to send data over long distance through telephone lines. However, telephone lines are designed to handle voice; the bandwidth of telephone lines ranges from 300 Hz to 3300 Hz. The digital signal with rise time in nanoseconds requires a bandwidth of several megahertz. Therefore, data bits should be converted into audio tones; this is accomplished through modems.

A modem (Modulator/Demodulator) is a circuit that translates digital data into audio tone frequencies for transmission over the telephone lines and converts audio frequencies into digital data for reception. At the present time, modems are available that can transfer data at rates of 300–2400 bps (bits per second). Generally, two types of modulation techniques are used: frequency-shift keying (FSK) for low-speed modems and phase-shift keying (PSK) for high-speed modems.

Computers can exchange information over telephone lines by using two modems— one on each side (Figure 16.4). A calling computer (or a terminal), also known as an originator, contacts a receiving computer (also known as answering) through a telephone number, and a communication link is established after control signals have been exchanged between computers and modems.

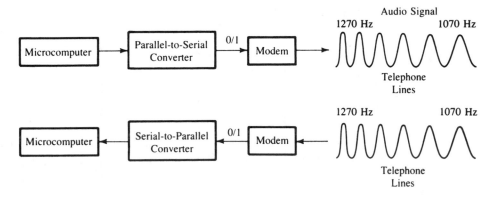

FIGURE 16.4
Communication over Telephone Lines Using Modems

A typical process of communication for a 300 bps modem is shown in Figure 16.4. A parallel word is converted into serial bits; in turn, the originator modem generates two audio frequencies—1070 Hz for logic 0 (Space) and 1270 Hz for logic 1 (Mark). These audio frequencies are transmitted over telephone lines. At the answering end, audio frequencies are converted back into 0s and 1s, and serial bits are converted into a parallel word that can be read by the computer. When the answering-end computer needs to transmit, it transmits on 2025 Hz (Space) and 2225 Hz (Mark).

16.16 Standards in Serial I/O

The serial I/O technique is commonly used to interface terminals, printers, and modems. These peripherals and computers are designed and manufactured by various manufacturers. Therefore, a common understanding must exist, among various manufacturing and user groups, that can ensure compatibility among different equipment. When this understanding is defined and generally accepted in industry (and by users), it is known as a standard. A standard is normally defined by a professional organization (such as IEEE—Institute of Electrical and Electronics Engineers); however, occasionally, a widespread practice can become a de facto standard. A standard may include such items as assignment of pin positions for signals, voltage levels, speed of data transfer, length of cables, and mechanical specifications.

In serial I/O, data can be transmitted as either current or voltage. Typically, 20 mA (or 60 mA) current loops are used in teletype equipment. When a teletype is marking or at logic 1, current flows; when it is at logic 0 (or Space), the current flow is interrupted. The advantage of the current loop method is that signals are relatively noise-free and are suitable for transmission over a distance.

When data are transmitted as voltage, the commonly used standard is known as RS-232C. It is defined in reference to Data Terminal Equipment (DTE) and Data Communication Equipment (DCE)—terminal and modem—as shown in Figure 16.5(a); however, its voltage levels are not compatible with TTL logic levels. The rate of data transmission in RS-232C is restricted to a maximum of 20 kbaud and a distance of 50 ft.

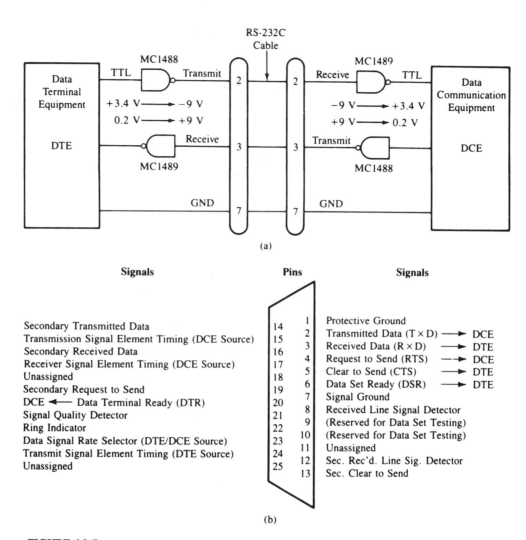

FIGURE 16.5

Minimum Configuration of RS-232C Signals and Voltage Levels (a) and RS-232C Signal
Definitions and Pin Assignments (b)

SOURCE: Courtesy of Electronic Industries Association.

For high-speed data transmission, two new standards—RS-422A and RS-423A—have
been developed in recent years; however, they are not yet widely used.

To appreciate the difficulties and confusion in this standard, one has to examine its
historical background. The RS-232 standard was developed during the initial days of
computer timesharing, long before the existence of TTL logic, and its primary focus was
to have compatibility between a terminal and a modem. However, the same standard is
now being used for communications between computers and peripherals, and the roles of
a data terminal and a modem have become ambiguous. Should a computer be considered

a terminal or a modem? The answer is that it can be either. Therefore, the lines used for transmission and reception will differ, depending on the manufacturer's role-definition of its equipment.

RS-232C

Figure 16.5(b) shows the RS-232C 25-pin connector and its signals. The signals are divided into four groups: data signals, control signals, timing signals, and grounds. For data lines, the voltage level +3 V to +15 V is defined as logic 0; from −3 V to −15 V is defined as logic 1 (normally, voltage levels are ±12 V). This is negative true logic. However, other signals (control and timing) are compatible with the TTL level. Because of incompatibility of the data lines with the TTL logic, voltage translators, called line drivers and line receivers, are required to interface TTL logic with the RS-232 signals, as shown in Figure 16.5(a). The line driver, MC1488, converts logic 1 into approximately −9 V and logic 0 into +9 V, as shown in Figure 16.5(a). Before it is received by the DCE, it is again converted by the line receiver, MC1489, into TTL-compatible logic. This raises the question: If the received signal is to be converted back to the TTL level, what is the reason, in the first place, to convert the transmitted signal to the higher level? The primary reason is that the standard was defined before the TTL levels came into existence; before 1960, most equipment was designed to handle higher voltages. The other reason is that this standard provides a higher level of noise margin—from −3 V to +3 V.

The minimum interface between a computer and a peripheral requires three lines: pins 2, 3, and 7, as shown in Figure 16.5(a). These lines are defined in relation to the DTE; the terminal transmits on pin 2 and receives on pin 3. On the other hand, the DCE transmits on pin 3 and receives on pin 2. Now the dilemma is: How does a manufacturer define the role of its equipment? For example, the user may connect its microcomputer to a serial printer configured as a DTE. Therefore, to remain compatible with the defined signals of the RS-232C, the RS-232 cable must be reconfigured as shown in Figure 16.6(b). In Figure 16.6, the microcomputer is defined as a DTE, and it can be connected to the modem, defined as a DCE, without any modification in the RS-232 cable, as shown in Figure 16.6(a). However, when it is connected to the printer, the transmit and the re-

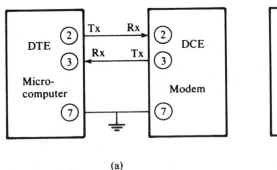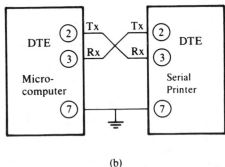

(a)   (b)

FIGURE 16.6
RS-232C Connections: DTE to DCE (a) and DTE to DTE (b)

ceive lines must be crossed as shown in Figure 16.6(b); this is known as a null-modem connection.

Typically, data transmission with a handshake requires eight lines, listed in Table 16.1. Specific functions of handshake lines differ in different peripherals and, therefore, should be referred to in the manufacturers' manuals.

For high-speed transmission, the standards RS-422A and RS-423A are used. These standards use differential amplifiers to reject noise levels and can transmit data at higher speed with longer cable. The RS-422A allows a maximum speed of 10 kbaud for a 40-ft distance and 100 kbaud for 4000 ft. The RS-423A is limited to 100 kbaud for a 30-ft distance and 10 kbaud for 300 ft. See Table 16.2 for comparison of the three standards: RS-232C, RS-422A, and RS-423A.

TABLE 16.1
RS-232C Signals Used with Handshake Data Communication

Pin No.	Signals[a]		Functions
2	Transmitted Data	TxD	Output; transmits data from DTE to DCE
3	Received Data	RxD	Input; DTE receives data from DCE
4	Request to Send	RTS	General-purpose output from DTE
5	Clear to Send	CTS	General-purpose input to DTE; can be used as a handshake signal
6	Data Set Ready	DSR	General-purpose input to DTE; can be used to indicate that DCE is ready
7	Signal Ground	GND	Common reference between DTE and DCE
8	Data Carrier Detect	DCD	Generally used by DTE to disable data reception
20	Data Terminal Ready	DTR	Output; generally used to indicate that DTE is ready

[a]Signals are referenced to DTE.

TABLE 16.2
Comparison of Serial I/O Standards

Specifications	RS-232C	RS-422A	RS-423A
Speed	20 kbaud	10 Mbaud at 40 ft 100 kbaud at 4000 ft	100 kbaud at 30 ft 1 kbaud at 4000 ft
Distance	50 ft	4000 ft	4000 ft
Logic 0	> +3 to +25 V	B > A[a]	+4 to +6 V
Logic 1	< −3 to −25 V	B < A	−4 to −6 V
Receiver Input			
Voltage	±15 V	±7 V	±12 V

[a]B and A are differential input to the op amp.

16.17   Review of Serial I/O Concepts and Approaches to Implementation

The serial data transmission can be implemented through either software or programmable I/O devices. Conceptually, the software and the hardware approaches are similar. In the asynchronous data transmission, the steps can be summarized as follows:

1. Inform the receiver of the beginning and the end of the transmission and the parity check.
2. Convert a parallel word into a stream of serial bits.
3. Transmit one bit at a time with appropriate time delay, using one data line of an output port. The time delay is determined by the speed of the transmission.

In data reception, the above process is reversed. The receiver needs to

1. Recognize the beginning of the transmission.
2. Receive serial bits, one at a time, and convert them into a parallel byte.
3. Check for errors and recognize the end of the transmission.

In the software approach, the speed of transmission is set up by using an appropriate delay between the transmission of two consecutive bits, and the entire word is converted into a serial stream by rotating the byte and outputting one bit at a time, using one of the data lines of an output port. The software provides the time delay between the two consecutive bits and adds framing bits and the parity bit; this is discussed in the next section.

In the hardware approach, the above functions are performed by a programmable device (chip). The device contains a parallel-to-serial register and 1-bit output port for transmission and a serial-to-parallel register and 1-bit input port for reception. The rate of transmission and reception is determined by the clock. The programmable chip also includes a control register that can be programmed to add framing and error-check information and to specify the number of bits to be transferred. The microprocessor transfers a parallel byte using the data bus, and the programmable chip performs the remaining functions for serial I/O.

The software approach is suitable for slow-speed asynchronous data communication where timing requirements are not critical. The approach is simple and inexpensive. The hardware approach is suitable for both asynchronous and synchronous formats. The approach is flexible, and chips can be programmed to accommodate changing requirements. In industrial and commercial products, the hardware approach has become almost universal. This chapter includes detailed discussion and illustrations of a widely used serial I/O device: the Intel 8251. However, to understand the basic concepts in serial I/O, the software approach is more suitable than the hardware approach; thus, the software approach is described here prior to discussion of programmable serial I/O devices. We will limit our discussion to the asynchronous communication mode commonly used in the microcomputer. Synchronous communication, being a specialized technique, will not be discussed here.

TABLE 16.3
Summary of Synchronous and Asynchronous Serial Transmission

Format	Synchronous	Asynchronous
Data Format	Groups of characters	One character at a time
Speed	High (20 k bits/s or higher)	20 k bits/s or lower
Framing Information	Sync characters are sent with each group	Start and Stop bits with every character
Implementation	Hardware	Hardware or software
Data Direction	Simplex, Half and Full Duplex	Simplex, Half and Full Duplex

The basic concepts concerning serial I/O, discussed in the previous sections, are summarized in Table 16.3.

16.2 SOFTWARE-CONTROLLED ASYNCHRONOUS SERIAL I/O

In the software-controlled asynchronous serial mode, the program should perform the following tasks, as discussed in the previous section:

1. Output a Start bit.
2. Convert the character to be sent in a stream of serial bits with appropriate delay.
3. Add parity information if necessary.
4. Output one or two Stop bits.

Figure 16.7 shows the accumulator with the code for the ASCII character I, and it is converted into a stream of 11 bits; it includes one Start bit and two Stop bits. After the Start bit, the character bits are transmitted with bit D_0 first and bit D_7 last; in ASCII characters, bit D_7 can be used to add parity information. The bit time—the delay between two successive bits—is determined by the transmission baud (rate). Figure 16.7 shows the transmission with 1200 baud; the delay between the two consecutive bits is 0.833 ms.

Data reception in the serial mode involves the reverse process: receiving one bit at a time and forming an 8-bit parallel word. The receiving program should continue to read the input port until it receives the Start bit and, then, begin to count character bits with appropriate delay.

16.21 Serial Data Transmission

Figure 16.8(a) shows a flowchart to transmit an ASCII character; it can be explained in the context of the block diagram shown in Figure 16.7(a). When no character is being transmitted, the transmit line of the output port stays high—in Mark position. The transmission begins with the Start bit, active low. The initialization block of the flowchart includes setting up a counter to count eight character bits; Start and Stop bits are sent out

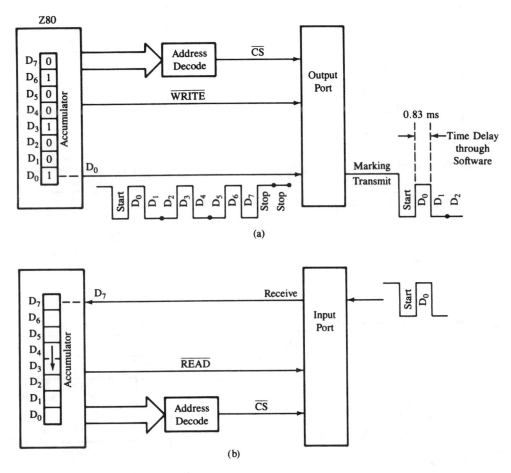

FIGURE 16.7

Serial Data Transmission (a) and Serial Data Reception under Software Control (b)

separately. The program waits for bit time—0.833 ms for 1200 baud—and begins to send one character-bit at a time over the data line D_0 at the interval of 0.833 ms. To get ready for the next bit, the program rotates the bits—for example, D_1 into D_0. It repeats the loop eight times and finally sends out two Stop bits. Assuming that the character is being sent out to a printer, the printer waits until it receives all the bits serially, forms a character, and prints it during the Stop bits. The Stop bits perform two functions: they allow sufficient time for the printer to print the character, and they leave the transmit line in Mark position at the end of the character.

16.22 Serial Data Reception

In serial data reception, the program begins with reading the input port. When no character is being received, the input line stays high. The program stays in the loop and contin-

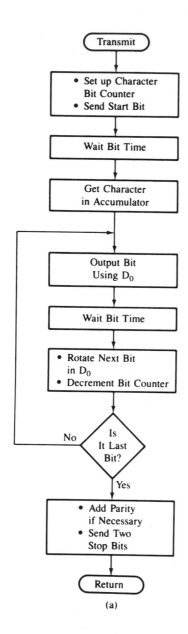

(a)

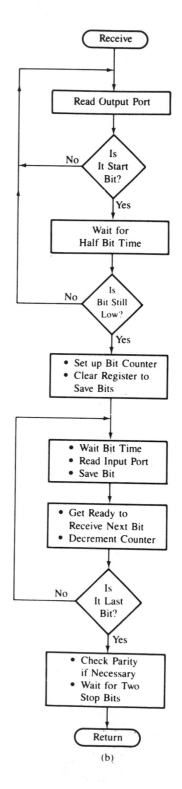

(b)

FIGURE 16.8

Flowcharts: Transmission (a) and Reception (b) of an ASCII Character

ues to read the port until the Start bit (active low) is received, as shown in the flowchart of Figure 16.8(b).

When the Start bit is received, it waits for half the bit time and samples the character bits in the middle of the pulse rather than at the beginning to avoid errors in transition. In the next block, it initializes the counter to count eight bits and clears a register to save the partial readings. The program reads the input port at the interval of bit time until it reads all the character bits and ignores the last two bits by just waiting for bit times. The character reception begins also with the LSB; that means the microprocessor will receive bit D_0 first. In Figure 16.7(a), the data line D_7 is used for the reception. The line D_7 provides some programming convenience for serial-to-parallel conversion; the word can be formed by shifting bits to the right whenever a bit is read, and eventually the LSB will reach its proper position.

THE 8085—SERIAL I/O LINES: SOD AND SID 16.3

The 8085 microprocessor has two pins specially designed for software-controlled serial I/O. One is called SOD (Serial Output Data), and the other is called SID (Serial Input Data). Data transfer is controlled through two instructions: SIM and RIM. Instructions SIM and RIM are used for two different processes: interrupt and serial I/O. These instructions have already been discussed in the context of the 8085 interrupt process (Chapter 12); now the focus is on their uses in serial I/O.

SERIAL OUTPUT DATA (SOD)

The instruction SIM is necessary to output data serially from the SOD line. It can be interpreted for serial output as in Figure 16.9.

Instructions

MVI A,80H	;Set D_7 in the accumulator $= 1$
RAR	;Set $D_6 = 1$ and bring Carry into D_7
SIM	;Output D_7

In this set of instructions, the serial output line is enabled by rotating 1 into bit position D_6; the instruction SIM outputs the Carry bit through bit position D_7.

FIGURE 16.9
Interpretation of Accumulator
Contents by the SIM Instruction

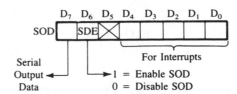

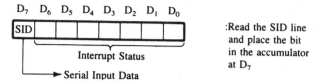

:Read the SID line
and place the bit
in the accumulator
at D_7

FIGURE 16.10
Interpretation of Accumulator Contents After the RIM Instruction

SERIAL INPUT DATA (SID)

Instruction RIM is used to input serial data through the SID line. Instruction RIM can be interpreted for serial I/O as in Figure 16.10.

In the context of serial I/O, instruction RIM is similar to instruction IN, except RIM reads only one bit and places it in the accumulator at D_7.

The SID and SOD lines in the 8085 eliminate the need for an input port and an output port in the software-controlled serial I/O. Essentially, SID is a 1-bit input port and SOD is a 1-bit output port. Similarly, instruction RIM is equivalent to a 1-bit IN instruction and instruction SIM is equivalent to a conditional 1-bit OUT instruction. The software necessary to implement serial I/O using SID and SOD lines is conceptually similar to that illustrated in Section 16.2.

16.31 Illustration: Data Transmission Using the SOD Line

PROBLEM STATEMENT

Write a subroutine to transmit an ASCII character, stored in register B, using the SOD line as a 1-bit output port.

SUBROUTINE*

The following subroutine SODATA (Serial Output Data) follows the flowchart given in Figure 16.8(a), except that it sets the counter for 11 bits and repeats the loop 11 times. The Start and the Stop bits are included with the character, and the Stop bits are set up by using the instruction STC (Set Carry):

```
                ;Input: An ASCII character without parity check in register B
                ;Output: None
SODATA:    MVI C,0BH         ;Set up counter C to count eleven bits
           XRA A             ;Reset Carry to 0
NXTBIT:    MVI A,80H         ;Set D7 = 1 in the accumulator
           RAR               ;Bring Carry in D7 and set D6 = 1
           SIM               ;Output D7
           CALL BITTIME      ;Wait for 9.1 ms
           STC               ;Set Carry = 1
           MOV A,B           ;Place ASCII character in the accumulator
```

---

*This subroutine is adapted from John Wharton, *Using the 8085 Serial Lines,* Application Note AP 29 (Santa Clara, Calif.: Intel Corporation, 1977).

RAR	;Place ASCII D_0 in the Carry, shift 1 in D_7,
	; and continue shifting for each loop
MOV B,A	;Save
DCR C	;One bit transmitted, decrement counter
JNZ NXTBIT	;If all bits are not transmitted, go back
RET	

Subroutine Description   This is an efficient subroutine based on the same serial I/O concepts discussed earlier. However, it uses some programming tricks to output data. To understand these tricks, it is necessary to examine the role of the following instructions, illustrated with an example of the ASCII letter G (47H = 0100 0111):

1. MVI A,10000000 (80H)
   RAR
   When D_7 is rotated into D_6, the SOD line is enabled for each loop and the contents of the Carry are placed in D_7.
2. STC
   MOV A,B
   RAR
   Instruction STC places 1 into the Carry, and instruction MOV A,B places the ASCII character in the accumulator. Instruction RAR brings 1 from the Carry into D_7 and places the ASCII bit into the Carry.
3. In the second iteration, the first RAR (described in step 1) places the ASCII bit from the Carry into D_7, and 1 from 80H into D_6. Instruction SIM outputs the ASCII bit from bit D_7.
4. The logic 1s, set by instruction STC and saved in register B, are shifted right by one position in every iteration. In the ninth iteration, when ASCII D_7 is sent out, register B will have all 1s from D_0 to D_7. In the last two iterations, logic 1s are sent out as Stop bits.
5. The contents of the accumulator after execution of these instructions are as follows (assume ASCII letter G is being transmitted):

Input from Main
Program: (B) = 47H   0   1   0   0   0   1   1   1   (B)
Subroutine
SODATA:

		CY	D_7	D_6	D_5	D_4	D_3	D_2	D_1	D_0	
XRA		0	0	0	0	0	0	0	0	0	(A)
MVI A,80H		0	1	0	0	0	0	0	0	0	(A)
RAR	D_0	0	0	1	0	0	0	0	0	0	(A)
SIM	$\longrightarrow$		0			SOD					
			- - - - - -	$\rightarrow$OUTPUT							
STC		1									
		$\downarrow$									
MOV A,B		1	0	1	0	0	0	1	1	1	(A)
RAR		1	1	0	1	0	0	0	1	1	(A)

16.32 Illustration: Data Reception Using the SID Line

PROBLEM STATEMENT

Write a subroutine to receive an ASCII character using the 8085 SID line.

SUBROUTINE*

The subroutine shown in Figure 16.11, SIDATA (Serial Input Data), is conceptually similar to the flowchart in Figure 16.8(b).

SUBROUTINE DESCRIPTION

The flowchart of this subroutine is slightly different from the flowchart in Figure 16.8(b). The difference is in the procedure of saving a bit and forming a parallel word.

After the first bit (D_0) is read, it is saved in the Carry with instruction RAL. The bit is stored and shifted right by using the instruction RAR. This procedure is repeated eight times. In the ninth iteration, the subroutine returns to the main program, ignoring the last two Stop bits.

16.4   HARDWARE-CONTROLLED SERIAL I/O USING PROGRAMMABLE CHIPS

The hardware approach to serial I/O incorporates the same basic principles and requirements necessary for the software approach. The functions performed separately under software control must be combined in one chip. These functions and requirements for the software approach are summarized below.

1. An input port and an output port are required for interfacing.
2. In data transmission, the MPU converts a parallel word into a stream of serial bits.
3. In data reception, the MPU converts serial bits into a parallel word.
4. Data transfer is synchronized between the MPU and slow-responding peripherals through time delays.

An integrated circuit called **USART** (Universal Synchronous/Asynchronous Receiver/Transmitter) incorporates all the features described above on the chip, as well as many more sophisticated features used for serial data communication. It is a programmable device; i.e., its functions and specifications for serial I/O can be determined by writing instructions in its internal registers. The Intel 8251A USART is a device widely used in serial I/O; the device and its applications are described in the next section.

---

*This subroutine is adapted from John Wharton, *Using the 8085 Serial Lines,* Application Note AP 29 (Santa Clara, Calif.: Intel Corporation, 1977).

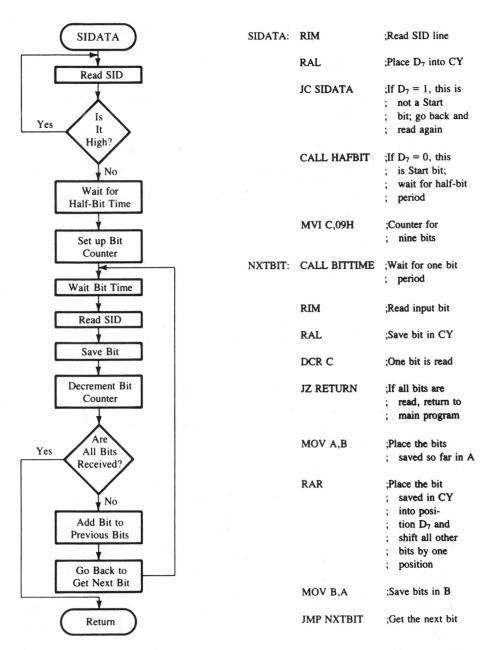

SIDATA:	RIM	;Read SID line
	RAL	;Place D_7 into CY
	JC SIDATA	;If D_7 = 1, this is ; not a Start ; bit; go back and ; read again
	CALL HAFBIT	;If D_7 = 0, this ; is Start bit; ; wait for half-bit ; period
	MVI C,09H	;Counter for ; nine bits
NXTBIT:	CALL BITTIME	;Wait for one bit ; period
	RIM	;Read input bit
	RAL	;Save bit in CY
	DCR C	;One bit is read
	JZ RETURN	;If all bits are ; read, return to ; main program
	MOV A,B	;Place the bits ; saved so far in A
	RAR	;Place the bit ; saved in CY ; into posi- ; tion D_7 and ; shift all other ; bits by one ; position
	MOV B,A	;Save bits in B
	JMP NXTBIT	;Get the next bit

FIGURE 16.11
Flowchart of Data Reception Using SID Line

16.41   The 8251A Progammable Communication Interface

The 8251A is a programmable chip designed for synchronous and asynchronous serial data communication, packaged in a 28-pin DIP. The 8251A is the enhanced version of its predecessor, the 8251, and is compatible with the 8251. Figure 16.12 shows the block diagram of the 8251A. It includes five sections: Read/Write Control Logic, Transmitter, Receiver, Data Bus Buffer, and Modem Control.

The control logic interfaces the chip with the MPU, determines the functions of the chip according to the control word in its register (to be explained below), and monitors the data flow. The transmitter section converts a parallel word received from the MPU into serial bits and transmits them over the TxD line to a peripheral. The receiver section receives serial bits from a peripheral, converts them into a parallel word, and transfers the word to the MPU. The modem control is used to establish data communication through modems over telephone lines. The 8251A is a complex device, capable of performing various functions. For the sake of clarity, this chapter focuses only on the asynchronous mode of serial I/O and excludes any discussion of the synchronous mode and the modem control. The asynchronous mode is often used for data communication between the MPU and serial peripherals such as terminals and floppy disks.

Figure 16.13 shows an expanded version of the 8251A block diagram. The block diagram shows all the elements of a programmable chip; it includes the interfacing signals, the control register, and the status register. The functions of various blocks are described below.

READ/WRITE CONTROL LOGIC AND REGISTERS

This section includes R/W control logic, six input signals, control logic, and three buffer registers: data register, control register, and status register. The input signals to the control logic are as follows.

Input Signals

☐ $\overline{\text{CS}}$—**Chip Select:** When this signal goes low, the 8251A is selected by the MPU for communication. This is usually connected to a decoded address bus.

☐ **C/$\overline{\text{D}}$—Control/Data:** When this signal is high, the control register or the status register is addressed; when it is low, the data buffer is addressed. The control register and the status register are differentiated by $\overline{\text{WR}}$ and $\overline{\text{RD}}$ signals, respectively.

☐ $\overline{\text{WR}}$—**Write:** When this signal goes low, the MPU either writes in the control register or sends output to the data buffer. This is connected to $\overline{\text{IOW}}$ or $\overline{\text{MEMW}}$.

☐ $\overline{\text{RD}}$—**Read:** When this signal goes low, the MPU either reads a status from the status register or accepts (inputs) data from the data buffer. This is connected to either $\overline{\text{IOR}}$ or $\overline{\text{MEMR}}$.

☐ **RESET—Reset:** A high on this input resets the 8251A and forces it into the idle mode.

☐ **CLK-Clock:** This is the clock input, usually connected to the system clock. This clock does not control either the transmission or the reception rate. The clock is necessary for communication with the microprocessor.

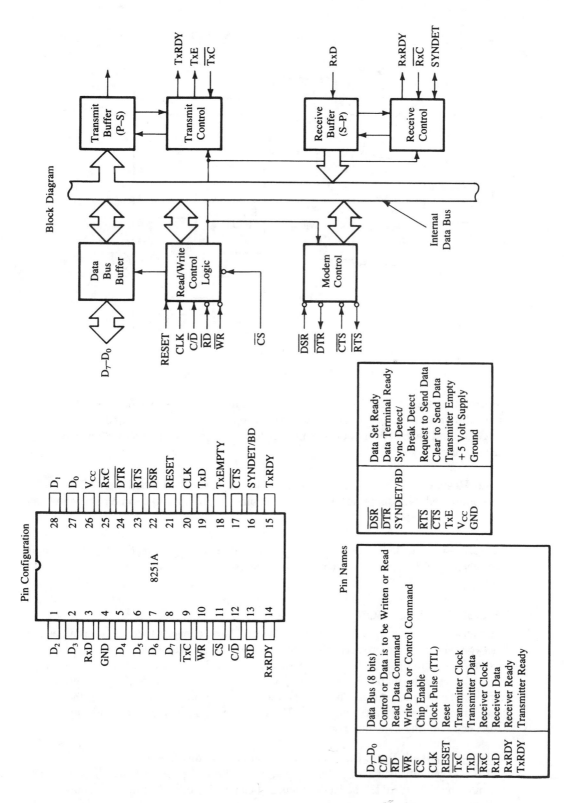

Block Diagram

Internal Data Bus

Pin Configuration

8251A

D_2	1		28	D_1
D_3	2		27	D_0
RxD	3		26	V_{CC}
GND	4		25	$\overline{RxC}$
D_4	5		24	$\overline{DTR}$
D_5	6		23	$\overline{RTS}$
D_6	7		22	$\overline{DSR}$
D_7	8		21	RESET
$\overline{TxC}$	9		20	CLK
$\overline{WR}$	10		19	TxD
$\overline{CS}$	11		18	TxEMPTY
$C/\overline{D}$	12		17	$\overline{CTS}$
$\overline{RD}$	13		16	SYNDET/BD
RxRDY	14		15	TxRDY

Pin Names

D_7-D_0	Data Bus (8 bits)
$C/\overline{D}$	Control or Data is to be Written or Read
$\overline{RD}$	Read Data Command
$\overline{WR}$	Write Data or Control Command
$\overline{CS}$	Chip Enable
CLK	Clock Pulse (TTL)
RESET	Reset
$\overline{TxC}$	Transmitter Clock
TxD	Transmitter Data
$\overline{RxC}$	Receiver Clock
RxD	Receiver Data
RxRDY	Receiver Ready
TxRDY	Transmitter Ready

$\overline{DSR}$	Data Set Ready
$\overline{DTR}$	Data Terminal Ready
SYNDET/BD	Sync Detect/ Break Detect
$\overline{RTS}$	Request to Send Data
$\overline{CTS}$	Clear to Send Data
TxE	Transmitter Empty
V_{CC}	+5 Volt Supply
GND	Ground

FIGURE 16.12

The 8251A Block Diagram, Pin Configuration, and Description

SOURCE: Intel Corporation, *Connectivity* (Santa Clara, Calif.: Author, 1993), p. 2–2.

527

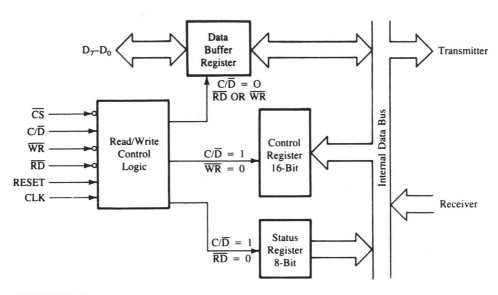

FIGURE 16.13
The 8251A: Expanded Block Diagram of Control Logic and Registers

Control Register This 16-bit register for a control word consists of two independent bytes: the first byte is called the **mode instruction** (word) and the second byte is called the **command instruction** (word). This register can be accessed as an output port when the C/$\overline{D}$ pin is high.

Status Register This input register checks the ready status of a peripheral. This register is addressed as an input port when the C/$\overline{D}$ pin is high; it has the same port address as the control register.

Data Buffer This bidirectional register can be addressed as an input port and an output port when the C/$\overline{D}$ pin is low. Table 16.4 summarizes all the interfacing and control signals.

TRANSMITTER SECTION

The transmitter accepts parallel data from the MPU and converts them into serial data. It has two registers: a buffer register to hold eight bits and an output register to convert eight bits into a stream of serial bits (Figure 16.14). The MPU writes a byte in the buffer register; whenever the output register is empty, the contents of the buffer register are transferred to the output register. This section transmits data on the TxD pin with the appropriate framing bits (Start and Stop). Three output signals and one input signal are associated with the transmitter section.

☐ **TxD—Transmit Data:** Serial bits are transmitted on this line.
☐ **TxC—Transmitter Clock:** This input signal controls the rate at which bits are transmitted by the USART. The clock frequency can be 1, 16, or 64 times the baud.

TABLE 16.4
Summary of Control Signals for the 8251A

$\overline{CS}$	C/$\overline{D}$	$\overline{RD}$	$\overline{WR}$	Function
0	1	1	0	MPU writes instructions in the control register
0	1	0	1	MPU reads status from the status register
0	0	1	0	MPU outputs data to the Data Buffer
0	0	0	1	MPU accepts data from the Data Buffer
1	X	X	X	USART is not selected

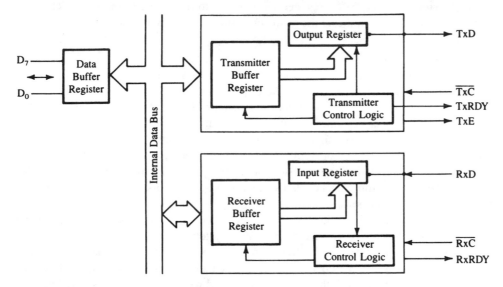

FIGURE 16.14
The 8251A: Expanded Block Diagram of Transmitter and Receiver Sections

- ☐ **TxRDY—Transmitter Ready:** This is an output signal. When it is high, it indicates that the buffer register is empty and the USART is ready to accept a byte. It can be used either to interrupt the MPU or to indicate the status. This signal is reset when a data byte is loaded into the buffer.
- ☐ **TxE—Transmitter Empty:** This is an output signal. Logic 1 on this line indicates that the output register is empty. This signal is reset when a byte is transferred from the buffer to the output registers.

RECEIVER SECTION

The receiver accepts serial data on the RxD line from a peripheral and converts them into parallel data. The section has two registers: the receiver input register and the buffer register (Figure 16.14).

When the RxD line goes low, the control logic assumes it is a Start bit, waits for half a bit time, and samples the line again. If the line is still low, the input register accepts the following bits, forms a character, and loads it into the buffer register. Subsequently, the parallel byte is transferred to the MPU when requested. In the asynchronous mode, two input signals and one output signal are necessary, as described below.

☐ **RxD—Receive Data:** Bits are received serially on this line and converted into a parallel byte in the receiver input register.

☐ **RxC—Receiver Clock:** This is a clock signal that controls the rate at which bits are received by the USART. In the asynchronous mode, the clock can be set to 1, 16, or 64 times the baud.

☐ **RxRDY—Receiver Ready:** This is an output signal. It goes high when the USART has a character in the buffer register and is ready to transfer it to the MPU. This line can be used either to indicate the status or to interrupt the MPU.

INITIALIZING THE 8251A

To implement serial communication, the MPU must inform the 8251A of all details, such as mode, baud, Stop bits, parity, etc. Therefore, prior to data transfer, a set of control words must be loaded into the 16-bit control register of the 8251A. In addition, the MPU must check the readiness of a peripheral by reading the status register. The control words are divided into two formats: mode words and command words. The mode word specifies the general characteristics of operation (such as baud, parity, number of Stop bits), the command word enables data transmission and/or reception, and the status word provides the information concerning register status and transmission errors. Figure 16.15 shows the definitions of these words.

To initialize the 8251A in the asynchronous mode, a certain sequence of control words must be followed. After a Reset operation (system Reset or through instruction), a mode word must be written in the control register followed by a command word. Any control word written into the control register immediately after a mode word will be interpreted as a command word; that means a command word can be changed anytime during the operation. However, the 8251A should be reset prior to writing a new mode word, and it can be reset by using the Internal Reset bit (D_6) in the command word.

16.42   Illustration: Interfacing an RS-232 Terminal Using the 8251A

PROBLEM STATEMENT

1. Identify the port addresses of the control register, the status register, and the data register in Figure 16.16.
2. Explain the RS-232 signals and the operations of the line driver (MC 1488) and the line receiver (MC 1489) shown in Figure 16.16.
3. Specify the initialization instructions and the status word to transmit characters with the following parameters if the transmitter clock frequency (TxC) is 153.6 kHz.
   ☐ Asynchronous mode with 9600 baud
   ☐ Character length = seven bits and two Stop bits
   ☐ No parity check

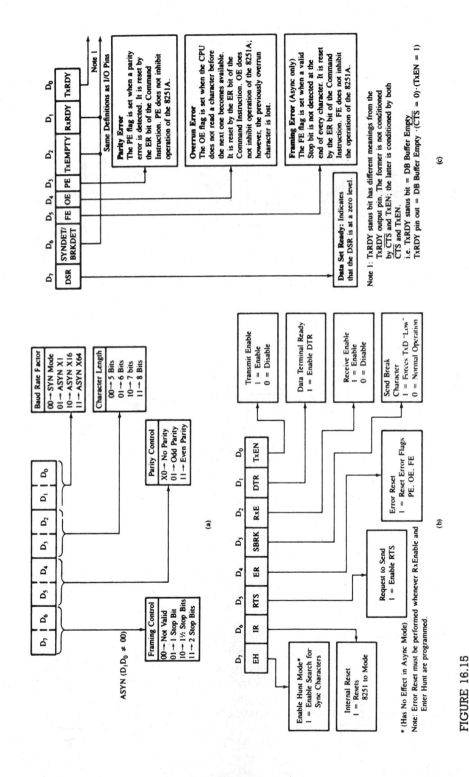

FIGURE 16.15

Mode Word Format (a), Command Word Format (b), and Status Word Format (c)

SOURCE: (a) and (c): Intel Corporation, *Connectivity* (Santa Clara, Calif.: Author, 1993), p. 2–8.

531

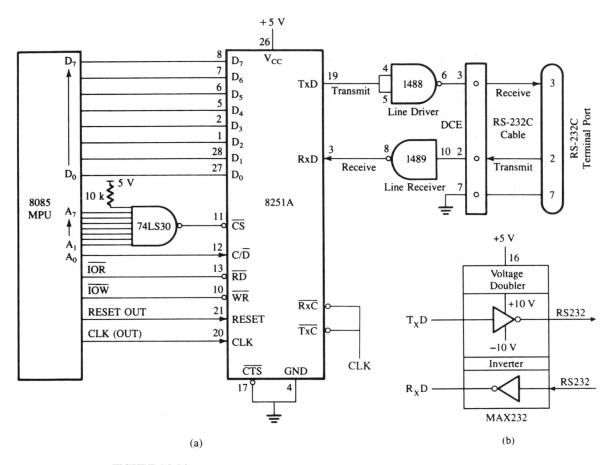

FIGURE 16.16
Schematic of Interfacing an RS-232 Terminal with an 8085 System Using the 8251A (a) and
MAX 232 Logic Diagram (b)

4. Write instructions to initialize the 8251A, to read the status word, and set up a loop
until the transmitter (TxRDY) is ready.

1. PORT ADDRESSES

a. The Chip Select line of the 8251A is enabled when the address lines A_7 through A_1 are
at logic 1. To select the control register or the status register, the C/$\overline{D}$ line should be
high, which means that address line A_0 should be 1. Therefore, the port address of the
control register and the status register = FFH.

    The control register is an output port and the status register is an input port; they
are identified by $\overline{WR}$ and $\overline{RD}$ signals, even if their port addresses are the same.

b. The data register is selected when the C/$\overline{\text{D}}$ line goes low; thus, A_0 should be low. The port address of the data register = FEH. The register is bidirectional, and the same address is used to receive or transmit data. The input and output functions are identified by $\overline{\text{RD}}$ and $\overline{\text{WR}}$ signals.

2. RS-232C SIGNALS, LINE DRIVERS, AND LINE RECEIVERS

Figure 16.16 shows that three RS-232 signals—TxD, RxD, and Ground—are being used for serial communication between the CRT terminal and the 8085 system. The terminal transmits data on pin 2 and receives on pin 3; on the other hand, the 8085 system receives on pin 2 and transmits on pin 3 using the 8251A. Therefore, the terminal is connected as the DTE and the system plays the role of the DCE; the 8251A is part of the 8085 system.

Data transmitted over the TxD line (pin 19 of the 8251A) are at the TTL logic level. These bits are converted to RS-232 voltage levels and negative logic by line driver MC 1488. Data received by the 8251A over the RxD line (pin 3) should be at the TTL logic level. Therefore, the RS-232 signals at pin 2 of the connector are converted to the positive TTL logic level by line receiver MC 1489. The line driver and receiver are described here briefly.

Line Driver: MC 1488 This is a quad line driver that converts TTL input levels to a maximum +15 V_{DC} output signal. Typically, it is used with a ±12 V power supply. For logic 0 input (< 0.8 V_{DC}) the output is around +10 V, and for logic 1 input (> +2.4 V_{DC}) the output is around −10 V; thus, the positive true logic is converted into negative true logic for RS-232C signals. The internal circuit of the MC 1488 functions much like a comparator. For an input lower than the threshold voltage, the output approaches positive power supply voltage, and for an input higher than the threshold voltage, the output approaches negative power supply voltage.

Line Receiver: MC 1489 This is a quad line receiver that converts high voltage signals (+15 V) into TTL logic levels. Output voltages usually range from 0.2 V (low) to 4.0 V (high). The internal circuit functions as an on/off transistor. When the transistor base has a negative input voltage, the transistor is turned off and the collector voltage (the output of the MC 1489) is high. When the transistor base has a positive input voltage, the transistor is driven into saturation to 0.2 V.

RS-232 Drivers/Receivers: MAX 232 One of the drawbacks of the line driver MC 1488 is that it requires additional voltages ±12 V from a power supply. Typically, such voltages are unavailable in systems compatible with the TTL logic. This difficulty can be resolved by using a specialized integrated device, MAX 232 (Figure 16.16(b)). It includes drivers and receivers (two each), and it generates ±10 V internally by using a voltage doubler and inverter circuits. The MAX 232 can replace the 1488 and 1489 and eliminate the need for ±12 V power supplies.

3. INITIALIZATION

The control words necessary for the given specifications are as follows:

Mode Word   Refer to Figure 16.15(a).

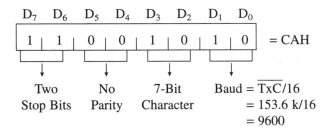

In a mode word, bits D_1 and D_0 can specify a baud factor that divides the clock frequency (TxC) to provide three different transmission rates. In this illustration, TxC is sixteen times the specified baud.

Command Word   Refer to Figure 16.15(b).

$$
\begin{array}{cccccccc}
D_7 & D_6 & D_5 & D_4 & D_3 & D_2 & D_1 & D_0 \\
\hline
X & 0 & X & 1 & X & 0 & X & 1
\end{array} = 11H
$$

	Error	Receive	Transmit
	Reset	Disable	Enable

In this command word, bit D_0 enables the transmitter, bit D_4 ignores any errors, and bit D_6 prevents reset of the 8251A; all other bits are don't care. In this illustration, bit D_4 also can assume don't care logic level.

Status Word   Refer to Figure 16.15(c).

$$
\begin{array}{cccccccc}
D_7 & D_6 & D_5 & D_4 & D_3 & D_2 & D_1 & D_0 \\
\hline
X & X & X & X & X & X & X & 1
\end{array} = 01H
$$

Transmitter
Ready

The MPU should check bit D_0 before transferring a character to the 8251A; it indicates the status of the pin TxRDY. When a byte is transferred from the transmitter buffer to the output register (see Figure 16.14 for the block diagram), bit D_0 is set to 1, and it is reset when the MPU loads the next byte in the buffer. Bit D_2 (TxEMPTY) indicates the status of the output register; this bit usually is not used except in applications such as half duplex mode.

4. INITIALIZATION INSTRUCTIONS

SETUP: MVI A,CAH   ;Load the mode word
              OUT FFH   ;Write mode word in control register
              MVI A,11H   ;Load the command word to enable transmitter
              OUT FFH   ;Enable the transmitter
STATUS: IN FFH   ;Read status register
              ANI 01H   ;Mask all bits except D_0
              JZ STATUS   ;If $D_0 = 0$, the transmitter buffer is full; go back
                               ;   and wait

     Figure 16.17 shows the contents of various registers after the initialization. Mode word CAH and command word 11H are loaded in the control register by the OUT FFH instructions. The MPU checks the transmitter status by reading the status register and examining bit D_0 (Transmitter Ready).

     When a character is transferred from the buffer register to the output (parallel-to-serial) register, bit D_0 is set to 1, indicating to the MPU that the transmitter is ready to accept the next character.

16.43 Illustration: Data Transmission to a CRT Terminal Using the 8251A in the Status Check Mode

PROBLEM STATEMENT

Write a program including the initialization of the USART—the 8251A—to transmit a message from an 8085 single-board microcomputer to a CRT terminal (Figure 16.16). The requirements are as follows:

1. A message is stored as ASCII characters (without parity) in memory locations starting at XX70H.
2. The message specifies the number of characters to be transmitted as the first byte (excluding the first byte) and concludes with the characters for the Carriage Return and the Line Feed.

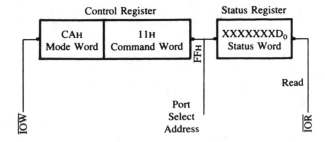

FIGURE 16.17
Control Register Contents After Initialization

3. The initialization instructions are the same as in the previous illustration (Figure 16.16).

4. The program should check status before it transmits a character. See Figure 16.18.

PROGRAM DESCRIPTION

According to the problem statement, the first character of the message specifies the number of characters to be transmitted. Therefore, the instruction MOV C,M loads the first character (in this case 08H) in register C and sets that register as the counter.

Before the initialization of the 8251A, the dummy mode word and the reset command are sent to the control register. Initially, the control register may have any random command; therefore, it is a good practice to reset the 8251A. However, it expects the first instruction as a mode word followed by a command word. Therefore, the reset command is sent after sending three dummy mode words, which are recommended to avoid problems when it is turned on.

16.44   Illustration: Transferring (Downloading) a File from the Microcomputer to the 8085 Single-Board Computer Using the Intel Hex Format

The communication between computers is a common occurrence. For example, to interface an A/D converter with a single-board computer, a program is generally written using an assembler or a cross-assembler on a software development system such as an IBM PC. Then, to test the program and the interfacing hardware, it must be transferred from the IBM PC to R/W memory of the single-board computer. The commonly used format for the Intel family of microprocessors is known as the Intel Hex format. For example, the Hex file of the ADDHEX program illustrated in Chapter 11 (Section 11.43) will appear on a CRT or a printer as follows. However, these are sent as ASCII characters. For example, the colon (:) will be sent as 3AH and zero will be sent as 30H.

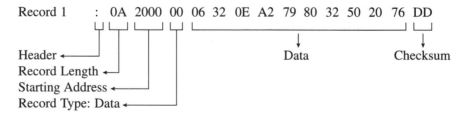

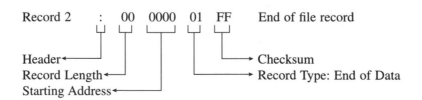

PROGRAM

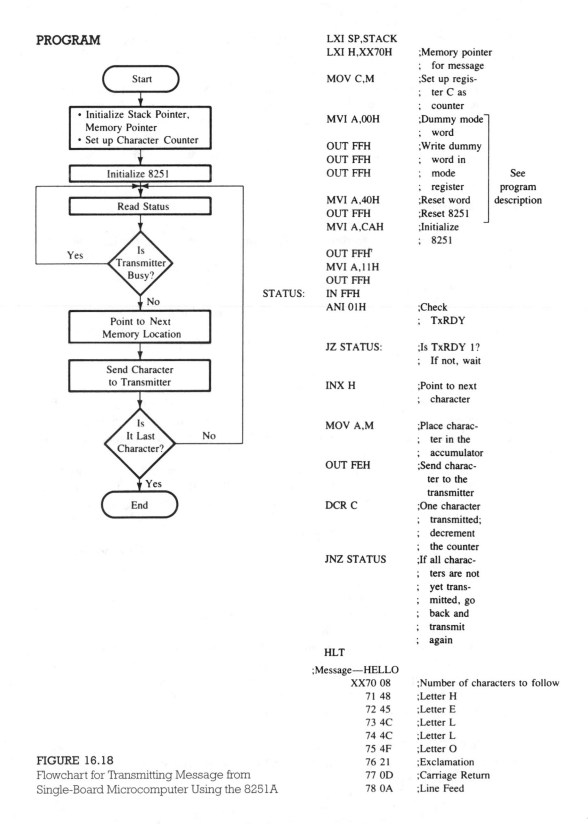

```
        LXI SP,STACK
        LXI H,XX70H     ;Memory pointer
                        ;  for message
        MOV C,M         ;Set up regis-
                        ;  ter C as
                        ;  counter
        MVI A,00H       ;Dummy mode
                        ;  word
        OUT FFH         ;Write dummy
        OUT FFH         ;  word in
        OUT FFH         ;  mode
                        ;  register
        MVI A,40H       ;Reset word
        OUT FFH         ;Reset 8251
        MVI A,CAH       ;Initialize
                        ;  8251
        OUT FFH'
        MVI A,11H
        OUT FFH
STATUS: IN FFH
        ANI 01H         ;Check
                        ;  TxRDY
        JZ STATUS:      ;Is TxRDY 1?
                        ;  If not, wait
        INX H           ;Point to next
                        ;  character
        MOV A,M         ;Place charac-
                        ;  ter in the
                        ;  accumulator
        OUT FEH         ;Send charac-
                        ;  ter to the
                        ;  transmitter
        DCR C           ;One character
                        ;  transmitted;
                        ;  decrement
                        ;  the counter
        JNZ STATUS      ;If all charac-
                        ;  ters are not
                        ;  yet trans-
                        ;  mitted, go
                        ;  back and
                        ;  transmit
                        ;  again
        HLT
;Message—HELLO
        XX70 08         ;Number of characters to follow
        71 48           ;Letter H
        72 45           ;Letter E
        73 4C           ;Letter L
        74 4C           ;Letter L
        75 4F           ;Letter O
        76 21           ;Exclamation
        77 0D           ;Carriage Return
        78 0A           ;Line Feed
```

See program description

FIGURE 16.18
Flowchart for Transmitting Message from
Single-Board Microcomputer Using the 8251A

To transfer this program in the R/W memory of the single-board computer using the serial I/O, we need an interfacing circuit, as shown in Figure 16.16, and a program that can recognize the header of a file, receive the data bytes, and store them in memory locations with the given starting address. The following illustration discusses the Intel Hex format and the receiver program that must be stored permanently in EPROM or ROM of the single-board computer.

PROBLEM STATEMENT

1. Explain the Intel Hex format for the program shown above.
2. Draw flowcharts for a receiver (download) program and explain the function of the DTR signal.

1. INTEL HEX FORMAT

A program or a file is divided into records, and each record has the format shown in the example of the ADDHEX program. Each record has six segments in ASCII characters: the header, the record length, the starting address, the record type, the data, and the checksum. These are described below.

☐ Header: The colon is the first byte used to indicate the beginning of the record.
☐ Record Length: The number of data bytes in the record; it can be from 00 to 10H. In the above example, the record length in Record 1 is 0AH to indicate that the number of bytes in the record is ten, and in Record 2, the length is zero to indicate the last record in the file.
☐ Starting Address: This is the memory address where the data in that record are to be stored. In the example, the starting address is 2000H. If there were additional records, the starting memory address of each record would be automatically calculated and shown. Record 2 is the end of the file, and it does not have any data bytes. Therefore, the memory address shown is 0000H.
☐ Record Type: This includes two types of records: 00 means normal data and 01 means end of file record. Record 1 shows 00 to indicate normal data, and Record 2 shows 01 to indicate the end of file record.
☐ Data: These are the Hex bytes of the mnemonics in the program. In the example, we have 10 data bytes; the maximum number of bytes allowed in a record is 16 (10H). There are no data bytes in Record 2.
☐ Checksum: This is the 2's complement of the sum of all the bytes in the record, excluding the header. In Record 1, the sum of all the bytes, from 0AH to 76H, is 323H. The checksum ignores the carry digit 3 and takes the 2's complement of 23H, which is DDH. The receiver program adds all the bytes, including the checksum, and the result should be zero for an error-free transmission.

2. DESIGN OF DOWNLOAD PROGRAM

After examining the Intel Hex format, the tasks of the receiver program can be divided into the following segments.

☐ Initialize the 8251 to receive a file.

☐ Check for the header character.

☐ Read two ASCII characters at a time.

☐ Convert the ASCII characters into binary values, and combine them in a byte.

☐ Extract the information concerning the byte count and the memory address. Add the Hex values in the checksum counter.

☐ Check for the record type. If it is the end of file, display the successful transfer.

☐ If the record type is data, store the bytes in memory, update the memory pointer and the byte counter, and add the Hex values in the checksum counter.

☐ After reading all the data bytes, add the checksum to the value of the checksum received (the last two ASCII characters in the record). If the result is zero, display end of data transfer; if it is other than zero, display error message.

The flowchart in Figure 16.19 shows the above steps. The program is divided into five subroutines, which are explained below:

1. RDASKY—Figure 16.20(a). This subroutine enables the 8251 receiver, checks the status, and reads an ASCII character. Before it returns to the main program, it deactivates the $\overline{DTR}$ signal (Figure 16.21) so that no character can be sent until it is ready to read again. This allows the program to perform other functions, such as ASCII conversion and checksum.

2. ASCBIN—Figure 16.20(c). This subroutine converts an ASCII character into binary value and adds the value to the checksum counter. (Refer to Problem 9.)

3. HEXBYTE—Figure 16.20(b). This subroutine reads two ASCII characters by calling RDASKY, converts the characters into binary by calling ASCBIN, and combines the binary values to form a byte. The binary value of the first ASCII character is saved as high-order four bits and of the second character as low-order four bits.

4. CHKSUM. This subroutine reads the last two ASCII characters by calling the routine HEXBYTE and compares the byte with the checksum counter. If the result is not zero, it calls the error message.

5. ERROR. This subroutine displays an error message. The details of this message are dependent on a particular single-board computer and its display output ports.

Figure 16.21 shows that the $\overline{DTR}$ signal is connected to pin 20 of the RS-232 cable; this signal is used as a handshake signal. Before reading an ASCII character, the $\overline{DTR}$ is enabled, and after reading the ASCII character, the $\overline{DTR}$ signal is set high by sending $D_1 = 1$ in the command word. This prevents the transmission of subsequent characters and allows the program to perform ASCII-to-binary conversion and the checksum procedure.

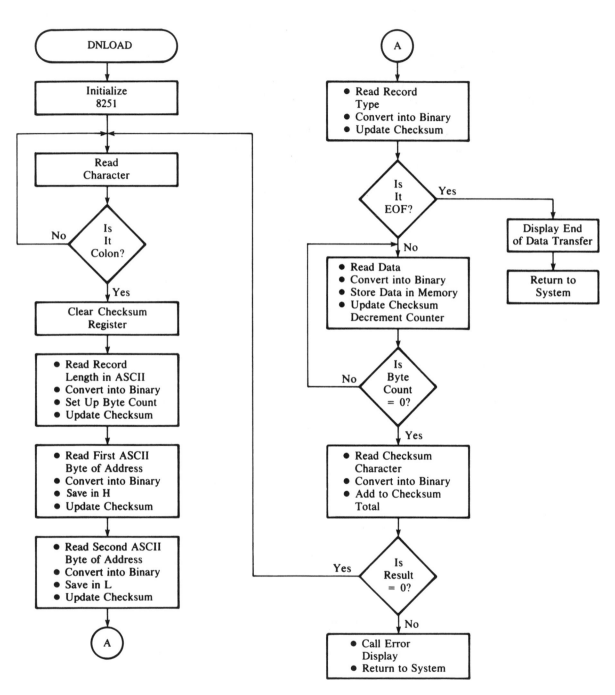

FIGURE 16.19
Flowchart: Download—Main Program

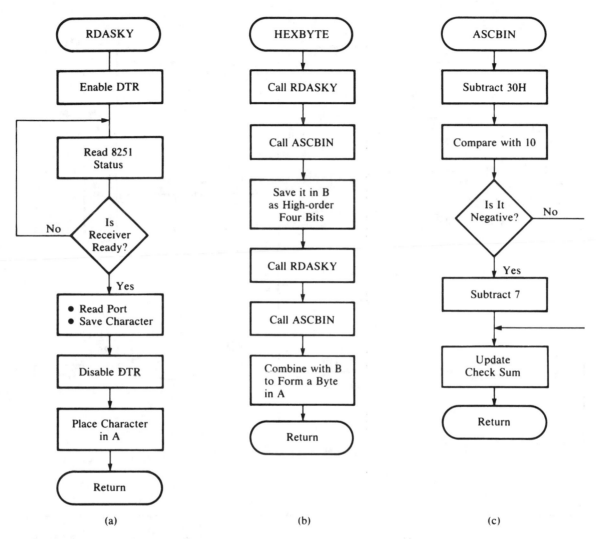

FIGURE 16.20
Flowcharts: Download Program Subroutines

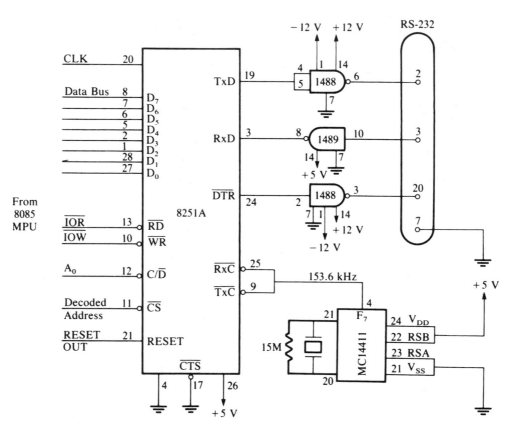

FIGURE 16.21
Using $\overline{DTR}$ as a Handshake Signal

SUMMARY

In this chapter, we discussed the technique of serial I/O for data communication. In this technique, one bit is transferred over one line rather than using eight data lines to transfer a byte. The serial I/O technique is necessary for certain types of equipment and media, such as magnetic tapes and telephone lines. The rate of data transfer in serial I/O is determined by the time delay between two successive bits. Therefore, a host of issues, such as synchronization of data transfer between the transmitter and the receiver, and error check, need to be resolved. The serial I/O data transfer can be implemented using software techniques; however, programmable devices called USART (Universal Synchronous/Asynchronous Receiver/Transmitter) are commonly used in industrial and commercial products. The basic concepts involved in the serial I/O can be summarized as follows:

1. In serial I/O communication, a word is transmitted one bit at a time over a single line by converting a parallel word into a stream of serial bits. On the other hand, a word is received by converting a stream of bits into a parallel word.
2. Serial data communication can be either synchronous or asynchronous. The synchronous mode is used for high-speed and the asynchronous mode is used for low-speed data communication.
3. The MPU identifies a serial peripheral through a decoded address and an appropriate control signal. Data transfer occurs under various conditions—unconditional, status check, and interrupt—depending upon logic design.
4. In software-controlled serial transmission, the MPU converts a parallel word into serial bits by using time delays and transmits bits over one data line of an output port.
5. In software-controlled serial reception, the MPU converts a serial word into a parallel word by using time delays and receives bits over one data line.
6. In the 8085, two specially designed 1-bit ports are provided for serial I/O: SOD and SID. Data transfer is controlled through two instructions: SIM and RIM.
7. The 8251A is a programmable serial I/O chip known as USART, which can perform all the functions of serial I/O.

DEFINITION OF TERMS

- ☐ ASCII (American Standard Code for Information Interchange). A 7-bit alphanumeric code commonly used in computers.
- ☐ EBCDIC (Extended Binary Coded Decimal Interchange Code). An 8-bit alphanumeric code used primarily in large IBM computers.
- ☐ Asynchronous Serial Data Transmission. In this format, the transmitter is not synchronized with the receiver by the same master clock. A transmitted character includes information about the starting and ending of the character.
- ☐ Synchronous Serial Data Transmission. In this format, the transmitter is synchronized with the receiver on the same frequency.
- ☐ Simplex Transmission. One-way data communication.
- ☐ Duplex Transmission. Two-way data communication. Full duplex is simultaneous in both directions and half duplex is one direction at a time.
- ☐ Baud (Rate). The number of signal changes per second. In serial I/O, it is equal to bits per second, the rate of data transmission.
- ☐ Current Loop. The transmission of serial data bits as current signals.
- ☐ RS-232C. A data communications standard that defines voltage signals in reference to data terminal equipment and data communication equipment.
- ☐ RS-422A and 423A. Data communication standards for high-speed data transmission.
- ☐ SID (Serial Input Data Line). A specially designed 1-bit input port in the 8085 to receive serial data.

☐ SOD (Serial Output Data). A specially designed 1-bit output port in the 8085 to transmit serial data.

☐ USART (Universal Synchronous/Asynchronous Receiver/Transmitter). A programmable chip designed for synchronous/asynchronous serial data communication.

QUESTIONS, PROBLEMS, AND PROGRAMMING ASSIGNMENTS

1. Check whether the following statements are true or false.
   a. Serial data communication cannot be implemented using the memory-mapped I/O technique. (T/F)
   b. ASCII is an 8-bit binary code that represents 256 different characters. (T/F)
   c. In the synchronous serial I/O format, all eight bits are sent simultaneously. (T/F)
   d. In the half duplex transmission mode, data flow is bidirectional between the MPU and a serial peripheral, but not simultaneously. (T/F)
   e. In serial transmission from the MPU to a peripheral, bit D_0 is transmitted first after the Start bit. (T/F)
   f. In serial reception from a peripheral to the MPU, bit D_7 of a character is received first after the Start bit. (T/F)
   g. In a system with the odd parity check, the letter A is transmitted with the code C1H. (T/F)
   h. In a system with the even parity check, the letter M is transmitted with the code 4CH. (T/F)
   i. The delay between the successive bits for 9600 baud rate is approximately 0.1 ms. (T/F)
   j. RS-232C is a 25-pin serial I/O voltage standard compatible with TTL logic. (T/F)
   k. To enable the 8085 serial output data line (SOD), bits D_7 and D_6 of the accumulator should be at logic 1. (T/F)
   l. The instruction RIM is equivalent to the instruction IN with one input data line (D_7). (T/F)
   m. The 8085 SID and SOD lines receive and transmit characters starting from bit D_7 after the Start bit. (T/F)
2. Write delay loops for BITTIME and half BITTIME for 1200 baud if the system frequency is 3 MHz.
3. Sketch the serial output waveform for the ASCII character A when it is transmitted with 9600 baud and even parity.
4. Sketch the serial output waveform for the ASCII sign + when it is transmitted with 2400 baud and odd parity.
5. Write a program to transmit letters A to Z from the MPU to the terminal in Figure 16.16.

6. Write a subroutine to accept a letter from the CRT terminal (Figure 16.16).
7. Write a program to receive ASCII characters from the terminal to the 8085 system (Figure 16.16) and to transmit the same characters back to the terminal to display on the CRT.
8. Specify the control word and the command word for data communication having the following specifications:
   a. asynchronous mode
   b. 1200 baud ($\overline{\text{TxC}} = \overline{\text{RxC}} = 76.8$ kHz)
   c. 8-bit character
   d. even parity
   e. one Stop bit
9. An ASCII character is supplied in the accumulator. Write the subroutine ASCBIN (Section 16.44) to convert the ASCII character into its binary value; if the character is not within Hex values 0 to F, call the Error routine; otherwise, update the checksum counter and return to the main program. (Refer to Section 10.42. Modify the illustrative program in this section to check whether the binary value is within the range.)
10. Write the subroutine HEXBYTE (Section 16.44) that reads two ASCII characters by calling the subroutine RDASKY, converts them into binary values by calling ASCBIN, and combines the binary values in a byte.

EXPERIMENTAL ASSIGNMENTS

1. a. Connect the circuit shown in Figure 16.16 to receive and transmit ASCII characters.
   b. Connect the clock with 4.8 kHz frequency to $\overline{\text{RxC}}$ and $\overline{\text{TxC}}$ pins of the 8251A. (Use a pulse generator or set up the 8155 timer for 4.8 kHz.)
   c. Write initialization instructions to set up the 8251A for the asynchronous mode, 300 baud, and 7-bit character with no parity.
   d. Write a program to receive a character from the terminal and echo the character back to the terminal for display. Use the status check technique.
2. Write a download program to transfer a file from a computer system, such as an IBM PC, to the single-board system shown in Figure 16.16.

Microprocessor Applications

Microcomputer systems based on the 8085 micro-processor were introduced at the beginning of the book, an overview of the instruction set was given in Chapter 5, and Chapters 6 through 10 were devoted to programming techniques. Various types of I/O data transfer and interfacing concepts were later discussed in detail in Chapters 12 through 16. This chapter is concerned with integrating or synthesizing all the concepts of the microprocessor architecture, software, and interfacing discussed previously, by designing a microprocessor system.

Designing a single-board microcomputer is the best possible choice, since it can incorporate all the important concepts related to the microprocessor. Furthermore, it allows expansion to include various types of interfacing. The chapter begins with interfacing applications commonly used in industrial environments. These applications include such examples as the scanned LED displays, the matrix keyboard, and memory. Later these examples are used as components for a system design that deals primarily with designing a single-board microcomputer. The chapter also includes troubleshooting techniques using an in-circuit emulator, a logic analyzer, and between an in-circuit emulator and a logic analyzer.

OBJECTIVES

- ☐ Illustrate the interfacing of scanned display, and list the advantages.
- ☐ Illustrate the interfacing of a matrix keyboard using software.
- ☐ Illustrate the interfacing of a matrix keyboard using a keyboard encoder.
- ☐ Draw a system block diagram of a single-board microcomputer based on the 8085 microprocessor

and its family of programmable interfacing devices.

☐ Draw flowcharts to illustrate the software design of a key monitor program and related submodules.

☐ List the primary features of an in-circuit emulator and explain its applications in troubleshooting microprocessor-based systems.

☐ Explain the functions of a logic analyzer as a troubleshooting instrument.

17.1 INTERFACING SCANNED MULTIPLEXED DISPLAYS AND LIQUID CRYSTAL DISPLAYS

In Chapter 15, when we illustrated an interfacing of a seven-segment LED using the 8255A (Figure 15.15), we needed one I/O port and a driver per LED to display one Hex digit. To display several digits with this technique, additional hardware is required in proportion to the number of digits to be displayed—which can be costly. The number of hardware chips needed for multiple-digit display can be minimized by using the technique called **multiplexing,** whereby the data lines and output ports are time-shared by various seven-segment LEDs. Similarly, Liquid Crystal Displays (LCDs) are also used commonly in low-power consumption systems. The interfacing of scanned displays and LCDs is discussed in the next sections.

17.11 Scanned Multiplexed Displays

The basic circuit for multiplexed display is illustrated in Figure 17.1. The circuit has two output ports: one port (P_A) to drive LED segments, and a second port (P_B) to turn on the corresponding cathodes. The output data lines of port P_A are connected to seven segments of each LED, and the output lines of port P_B are connected to the cathodes of each LED. The code of the first digit to be displayed at LED-1 is sent on the data lines by outputting to port P_A. The corresponding seven-segment LED is turned on by sending a bit to the cathode through port P_B. Next, LED-2 is turned on and LED-1 is turned off. Each seven-segment LED is turned on and off sequentially. The cycle is repeated fast enough that the display appears stable. This is also known as scanned display.

In a common-cathode seven-segment LED, all segments are driven by the output lines, which should supply at least 10 mA to 15 mA of current to each segment. The cathode should sink seven or eight times that current. I/O ports of programmable devices (such as the 8255A) are limited in current capacity; therefore, additional transistors or ICs, called **segment** and **digit drivers,** are required, as shown in Figure 17.1 and illustrated in the next problem.

PROBLEM STATEMENT

1. Design a six seven-segment LED display using the technique of multiplexing.

2. Write a program to display the message uP-rdy (microprocessor-ready).

INTERFACING CIRCUIT

Figure 17.2 shows the address-decoding network using the 74LS138, 3-to-8 decoder to assign I/O ports. The address lines A_7–A_2 are connected to the decoder and the remain-

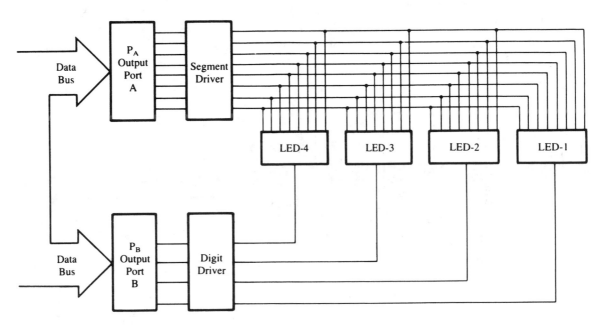

FIGURE 17.1
Block Diagram of Multiplexed Output Display

FIGURE 17.2
Address Decoding

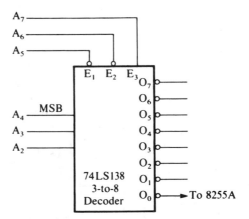

ing two address lines A_1 and A_0 will be connected directly to the 8255A. This decoder provides the capability of eight I/O ports. We will use the output line O_0 of this decoder for the LED scanned display, and we can use the remaining output lines of the decoder for other I/Os. The 8255A is selected when the output line O_0 of the decoder (Figure 17.2) is asserted. Therefore, the port addresses of the 8255A (Figure 17.3) are as follows (refer to Chapter 15, Section 15.11).

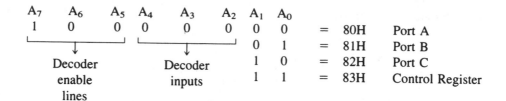

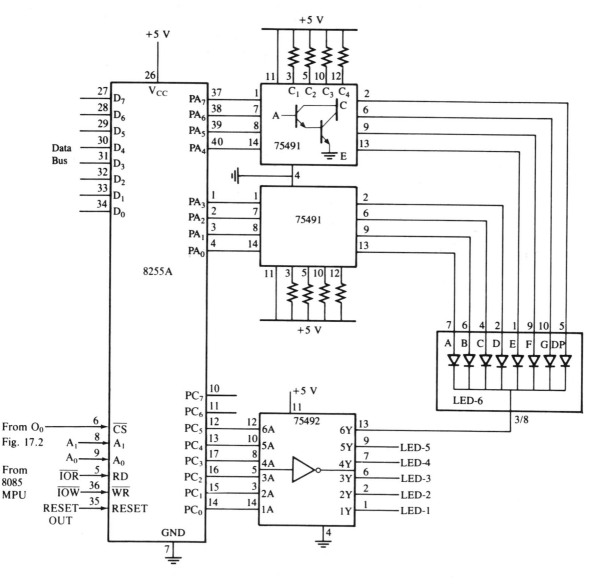

FIGURE 17.3
Schematic: Scanned Multiplexed Display

Figure 17.3 shows the schematic of a scanned display; it has six common-cathode seven-segment LEDs, one 8255A, and two drivers. Ports A and C are used as output ports: port A with the address 80H for segment codes and port C with the address 82H for digits to turn on/off LEDs. The SN 75491 and the SN 75492 are used as the segment code driver and the digit driver, respectively, to increase the current capacity in the circuit.

SN 75491—SEGMENT DRIVER

The SN 75491 is a quad device that has four Darlington pair transistors in a package. To drive eight data lines, we need two devices, as shown in Figure 17.3. It can source or sink 50 mA current (approx. 12.5 mA/pair). Pin A, the base of the transistor, is connected to one of the data lines of the output port and emitter E is connected to one of the LED segments, as shown in Figure 17.4(a). Similarly, the base of the driver is tied to +5 V through a resistor to turn on the driver properly.

SN 75492—DIGIT DRIVER

The SN 75492 has six Darlington pairs in a package, and can sink 250 mA of total current. The collector (pin Y) is connected to the common cathode of the LED, and the data lines from the port are connected to the base of the transistor, as shown in Figure 17.4(b), to turn on/off the LEDs.

To display a digit, the seven-segment code for the digit is sent to port A, the corresponding cathode is turned on and off in sequence, and the loop is repeated continuously.

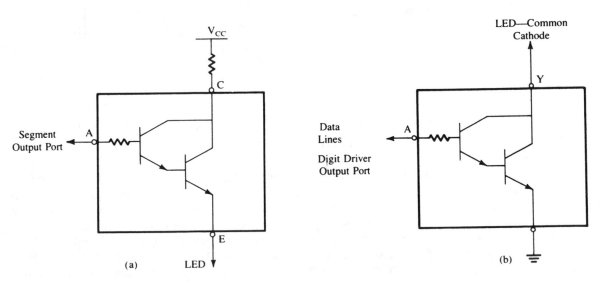

FIGURE 17.4
Simplified Diagram of: SN 75491 (a), SN 75492 (b)
SOURCE: Courtesy of Texas Instruments Incorporated.

PROGRAM

```
;The following program initializes the 8255A ports A and C as output ports
;   and displays a constant message stored at memory location SYSRDY (System
;   Ready). The message has six codes: uP-rdy (microprocessor-ready). The
;   code for the rightmost letter "y" is stored at the first location SYSRDY,
;   and the scanning begins at that location.
```

PORTA EQU 80H	;Port A address—Segment Driver
PORTC EQU 82H	;Port C address—Digit Driver
PORTB EQU 81H	;Port B address, initialized for later use
CONTRL EQU 83H	;Address for Control Register
8255A: MVI A,10000010B	;Control word: Mode 0, ports A and C as output,
	; port B as input for later use
OUT CONTRL	;Initialize ports A and C as outputs
READY: MVI B,00000001B	;Initialize digit code
MVI C,06H	;Initialize counter for six LEDs
LXI H,SYSRDY	;Use HL as memory pointer for message
NEXT: MOV A,M	;Get segment code
OUT PORTA	;Output segment code
MOV A,B	;Get digit code
OUT PORTC	;Turn on one LED
CALL DELAY1	;Wait 1 millisecond
XRA A	;Code to turn off segments
OUT PORTA	;Clear segments
MOV A,B	
RLC	;Shift digit code to turn on next LED
MOV B,A	;Save digit code
INX HL	;Point to next code
DCR C	;Next LED count
JNZ NEXT	
RET	
SYSRDY: DB 6EH, 5EH, 50H, 40H	;Message Codes
73H,1CH	;y d r - P u

Program Description   This routine initializes ports A and C of the 8255A as output ports by sending the word 82H to the control register (refer to Figure 15.4 for the definition of the control word). Port B is not being used in this illustration; however, it is being initialized as an input port to use in the illustration of the matrix keyboard in the next section. The next instruction initializes the scan routine by placing the digit code 00000001 in register B; this code will turn on LED-1 (the first LED at the right). By rotating bit D_0 (logic 1) to the left, the next LED is turned on and the LED presently being displayed is turned off; thus, only one LED is on at a time. Register C is set up as a counter to scan six LEDs, and the HL register is used as a memory pointer to point where the message is stored.

The scanning begins by sending the first code (the last letter "y" in the message) to port A and the LED-1 is turned on by sending the digit code. This LED is kept on for approximately 1 ms by calling the delay routine, and the entire display is turned off by clearing the segment code; this eliminates the flicker and the ghost images. The segment codes are sent in a sequence as they are stored in memory, and the corresponding LED is turned on until the counter reaches zero. To keep the display on, the routine should be called repeatedly.

Comments   In the scanned display, the hardware is minimized. With two output ports, this scheme can scan eight LEDs. In addition, current consumption is considerably reduced. However, the major disadvantage is that the MPU is kept occupied in scanning the display continuously. To relieve the MPU from the continuous scanning task, the Intel 8279 programmable keyboard/display interface device can be used.

This routine is appropriate for a permanent message; however, there are many situations where the routine is expected to scan only four or two LEDs. In such situations, an appropriate parameter must be passed on to register C by the calling program, instead of loading 06H as shown in the routine.

17.12   Liquid Crystal Displays (LCD)

A liquid crystal display (LCD) is commonly used in systems where low power consumption is necessary. Typical examples include watches, calculators, instrument panels, and consumer electronic displays. An LCD display consists of crystal material placed between two plates. The crystal material is arranged in segments or in the form of a dot matrix. The crystal material can pass or block the light that passes through; thus it creates a display. To display a character, certain segments or dots are driven by a square wave pattern, which is supplied by a built-in driver. The driver is interfaced with the processor using signals such as data lines, enable, read/write, and chip select. An interfacing of a commonly available LCD is illustrated in the next example.

PROBLEM STATEMENT

1. Explain the hardware requirements to interface an LCD using the 8255 peripheral chip.
2. Write instructions to initialize the LCD and display the message "Hello!" in the center of the display area (begin writing at address 87H of the first line).

HARDWARE AND INTERFACING CIRCUIT

Figure 17.5(a) shows the interfacing of a 20-character × 2-line dot matrix LCD module using the 8255. The LCD module has a built-in controller driver (Hitachi HD44780)* with a 14-pin dual row header. It can be interfaced with a 4-bit and 8-bit data bus. The pinout includes 8 data lines (DB7–DB0), three control signals (RS-Register Select, R/$\overline{\text{W}}$-Read and $\overline{\text{Write}}$, and E-Enable), and three connections for power supply and ground (V_{ss}, V_{dd}, and V_o). The controller has two 8-bit control registers: an instruction register

---

*Most manufacturers use this type of controller driver with a similar pin configuration.

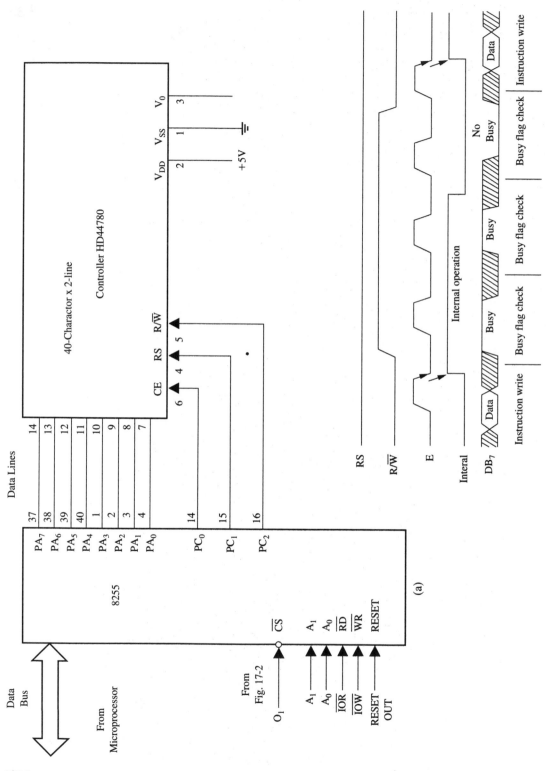

FIGURE 17.5

Interfacing LCD Using 8255 (a); Data Transfer Timing Diagram (b)

(IR) for commands and a data register (DR) for data. The MPU can write commands when RS is low and data when RS is high. When the MPU writes into one of the registers, the controller sets data line DB7 high as a busy flag that can be read by the MPU. When the controller completes its internal operation, it resets the busy flag.

Figure 17.5(b) shows the timing diagram of the read and write operations. It shows that after selection of a register (IR or DR) by RS, the Read/Write signal is asserted, and the enable signal is pulsed low-high-low. At the trailing edge of the enable signal, the controller sets the flag (data line DB7) and executes the internal operation of either writing a command or displaying a character. During the internal operation, the MPU should continue to check the busy flag (DB7) by asserting the enable signal (E) low-high-low until the controller resets the flag. When DB7 is reset, the MPU can write the next command or data (character).

In Figure 17.5, the LCD is interfaced by using the 8255 that is selected by the signal O_1 of the decoder in Figure 17.2. Therefore, the Hex addresses of ports A, B, C, and the control register are 84, 85, 86, and 87, respectively; however, we will use only the symbolic addresses in writing instructions. Port A is set up in I/O mode (Mode 0), the output mode to send a character to the LCD and the input mode to read the busy flag (DB7). Port C is set up in BSR mode; three signal lines of port C (PC_0, PC_1, and PC_2) are used for control signals (EN, RS, and R/$\overline{W}$).

To display a message, we need to initialize the LCD controller by writing into the IR register; the list of commands is given in Appendix D. The list includes various commands such as clearing the display, returning the cursor home (at the beginning), and shifting the cursor automatically after writing a character. After the initialization, we can write an ASCII character into the R/W memory of the LCD called DD RAM (Display Data R/W Memory) or an internally generated character called CG RAM (Character Generator R/W Memory). Each R/W memory location has an address. For example, the address of the first DD RAM of the first line is 80H and of the second line is C0H. This gives the capability of writing in any specific locations and generating signboard-type displays.

PROGRAM
(See Appendix D for LCD command codes.)

```
              MVI A, 10000000B    ;Control word to set up port A as an output in
                                  ;   mode 0
              OUT CNTRL           ;Send to 8255 control port
              MVI A, 00000000B    ;Reset EN signal PC0 low
              OUT CNTRL
INIT:         MVI A, 00110000B    ;Precautionary code if power on conditions are
                                  ;   unmet
              MOV B, A            ;Save the code
              CALL OUTPUT         ;DB7 cannot be checked before the above in-
                                  ;   struction
              CALL CMDOUT         ;Send to port A two more times
```

```
                CALL CMDOUT
                MVI A, 00111000B      ;Function set code (38H) for 8 bits, 2 lines,
                                      ;   5X7 dots
                CALL CMDOUT
                MVI A, 00001000B      ;Code (08H) for display off
                CALL CMDOUT
                MVI A, 00000001B      ;Code (01H) for clear display
                CALL CMDOUT
                MVI A, 00000110B      ;Code (06H) for entry mode set: shift and in-
                                      ;   crement cursor
                CALL CMDOUT
                MVI A, 00001110B      ;Code (0EH) to turn on display, cursor, and
                                      ;   blink
                CALL CMDOUT
DSPLAY:         MVI A, 87H            ;Write to eighth location of DD Ram address
                CALL CMDOUT
                LXI H, MESAGE         ;Point to MESAGE
                MVI B, 06H            ;Set up counter for six characters
NEXT:           MOV A, M             ;Get a character
                CALL DTAOUT
                INX H
                DCR B
                JNZ NEXT
                HLT
MESAGE          DB "HELLO!"
CMDOUT:         MOV B, A             ;Save the command code
                CALL CHKDB7
OUTPUT:         MVI A, 00000010      ;Select command register PC1 = RS = 0
                OUT CNTRL
                MVI A, 00000100B      ;Enable Write. PC2 = R/W = 0
                OUT CNTRL
                MVI A, 00000001B      ;Set EN high
                OUT CNTRL
                MOV A, B             ;Get code
                OUT PORTA            ;Send out on data bus
                MVI A, 00000000B      ;Set EN low
                RET
DTAOUT:         MOV B, A             ;Save data byte
                CALL CHKDB7
                MVI A, 00000011B      ;Select data register PC1 = RS = 1
                OUT CNTRL
                MVI A, 00000100B      ;Enable Write. PC2 = R/W = 0
                OUT CNTRL
                MVI A, 00000001B      ;Set EN high
```

```
                    OUT CNTRL
                    MOV A, B              ;Get data byte
                    OUT PORTA
                    MVI A, 00000000B      ;Set EN low
                    RET
CHKDB7:             MVI A, 10010000B      ;Set up port A as input port
                    OUT CNTRL
                    MVI A, 00000010B      ;Select command register PC₁ = RS = 0
                    OUT CNTRL
                    MVI A, 00000101B      ;Enable Read. PC₂ = R/W = 1
                    OUT CNTRL
READ:               MVI A, 00000001B      ;Set EN high
                    OUT CNTRL
                    IN PORTA              ;Read port A and check DB7
                    MVI A, 00000000B      ;Set EN low
                    RLC
                    JC READ               ;If DB7 = 1, go back and read again
                    MVI A, 10000000B      ;Set up port A as an output again
                    OUT CNTRL
                    RET
```

Program Description The program initializes port A as an output port and resets the EN signal. The LCD has certain power-on conditions; therefore, initial precautionary code 30H is sent in case those power-on conditions are unmet. Before this code is sent out, bit DB7 (flag) cannot be checked. After the first instruction, the program checks bit DB7 before it sends out any code. The first six commands initialize the LCD for a given operation. The command at location DSPLAY loads 87H in the accumulator. The DD RAM address begins at 80H; this command selects the eighth RAM location (approximately near the center of the 20-character line) to write. The loop beginning at NEXT sends out six ASCII characters in a sequence. After each character, the program checks bit DB7 by calling the subroutine CHKDB7, and when the bit is low, it calls the DTAOUT subroutine. The subroutines CMDOUT and DTAOUT are almost identical except for the RS (Register Select) signal; it must be low for commands and high for data. The subroutine CHKDB7 initializes port A as an input port and checks DB7. It continues to check DB7 until it goes low and then reinitializes port A as an output port.

INTERFACING A MATRIX KEYBOARD 17.2

A matrix keyboard is a commonly used input device when more than eight keys are necessary, rather than a row of keys as illustrated in Chapter 15. A matrix keyboard reduces the number of connections, thus the number of interfacing devices required. For example, a keyboard with 16 keys, arranged in a 4×4 (four rows and four columns) matrix as shown in Figure 17.6, requires eight lines from the microprocessor to make all the connections instead of 16 lines if the keys are connected in a linear format. When a key is pressed, it shorts one row and column; otherwise, the row and the column do not have any connections. The interfacing of a matrix keyboard requires two ports: one output port and one input port. Rows are connected to the output port, and the columns are connected to the input port. The schematic in Figure 17.6 shows 8-bit I/O ports; they are capable of interfacing a matrix keyboard as large as 64 keys: eight columns and eight rows.

In a matrix keyboard, the major task is to identify a key that is pressed and decode the key in terms of its binary value. This task can be accomplished through either software or hardware. In this section, we will explore both methods: first we will discuss basic concepts in identifying a key pressed and, then, write subroutines to check, identify, and decode (interpret) the key that is pressed. Finally, we will illustrate how these functions can be replaced by a hardware device, such as the National Semiconductor keyboard encoder MM74C923.

17.21   Problem Statement

1. Interface a 20-key matrix keyboard using ports B and C of the 8255A shown in Figure 17.3.

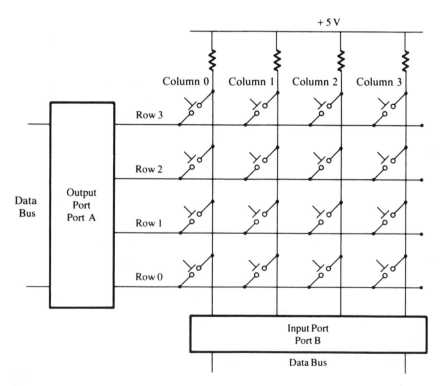

FIGURE 17.6
Matrix Keyboard Connections

2. Use the same address decoding circuit shown in Figure 17.2.

3. Write a keyboard subroutine with a software key debounce to read the keyboard, and return the equivalent binary code of the key pressed in the accumulator.

17.22 Interfacing a Matrix Keyboard

Figure 17.7 shows a matrix keyboard with 20 keys; the keyboard has five rows and four columns. The first 16 keys in a sequence will represent data 0 to F in Hex, and the remaining four will represent various functions such as Store and Execute. The circuit shows that the rows are connected to port C and the columns are connected to port B of the 8255A. To avoid any confusion, let us assume that port C in Figure 17.7 is different from port C in Figure 17.3 even if we are using the same address-decoding circuit. Later, we will discuss how the same I/O lines can be used for scanning both LEDs and keys. Furthermore, we will not repeat the discussion of initialization of the 8255A; we have already set up port B as an input port in Mode 0 in the previous illustration.

In Figure 17.7, the columns and rows make contact only when a key is pressed; otherwise, they remain high (+5 V). When a key is pressed, the key must be identified by its

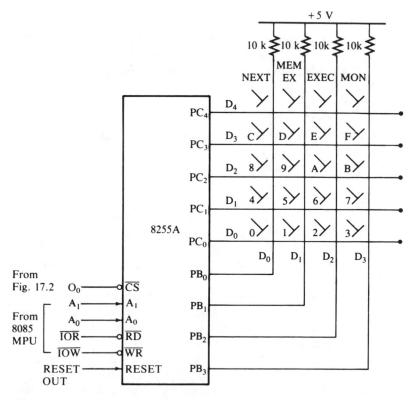

FIGURE 17.7
Interfacing a Matrix Keyboard

column and the row, and the intersection of the column and row must change from high
to low. This can be accomplished as explained in the following steps:

1. Ground all the rows by sending logic 0 through the output port.
2. Check the columns by reading the input port. If no key is pressed, all columns remain
   high. Continue to repeat Steps 1 and 2 until the reading indicates a change.
3. When one of the keys is pressed, the corresponding column goes low; identify and decode the key.

17.23 Program

The matrix keyboard routine is conceptually important: it illustrates how to set up relationships between hardware binary readings and expected codes. For example, when key
0 is pressed, the input reading at port B will be XXXX1110 (D_3–D_0); however, the binary
code for that key must be 00000000. This conversion is performed by the software routines illustrated below. Similarly, when key NEXT is pressed, the input reading for bits
D_3–D_0 will be the same as for the 0 key (1110). Again, the software routines will have to
differentiate between data and function keys.

This matrix keyboard problem can be divided into four steps:

Step 1: Check whether all keys are open.
In this step, the program grounds all the rows by sending 0s to the output port.
It reads the input port to check the key release, and debounces the key release
by waiting for 10 ms. This step is necessary to avoid misinterpretation if a key
is held for a long time.

Step 2: Check a key closure.
In this step the program checks for a key closure by reading the input port. If
all keys are open, the input reading on data lines D_3–D_0 should be 1111, and
if one of the keys is closed, the reading will be less than 1111. (Data lines
D_7–D_4 are not connected; therefore, the data on these lines should be
masked.)

Step 3: Identify the key.
This is a somewhat complex procedure. Once a key closure is found, the key
should be identified by grounding one row at a time and checking each column
for zero. Figure 17.8 (Step 3) shows that two loops are set up: the outer loop
grounds one row at a time, and the inner loop checks each column for zero.

Step 4: Find the binary key code for the key.
The binary key code is identified through the counter procedure. For each row,
the inner loop is repeated four times to check four columns, and for every column check, the counter is incremented. For five rows, the inner loop is repeated
20 times and the counter is incremented from 0 to 13H, thus maintaining the
binary code in the counter. Once the key is identified, the code is transferred
from the counter to the accumulator. The codes 0 to F are used for data keys
and the remaining codes 10H to 13H are assigned various functions, as shown
in Figure 17.7.

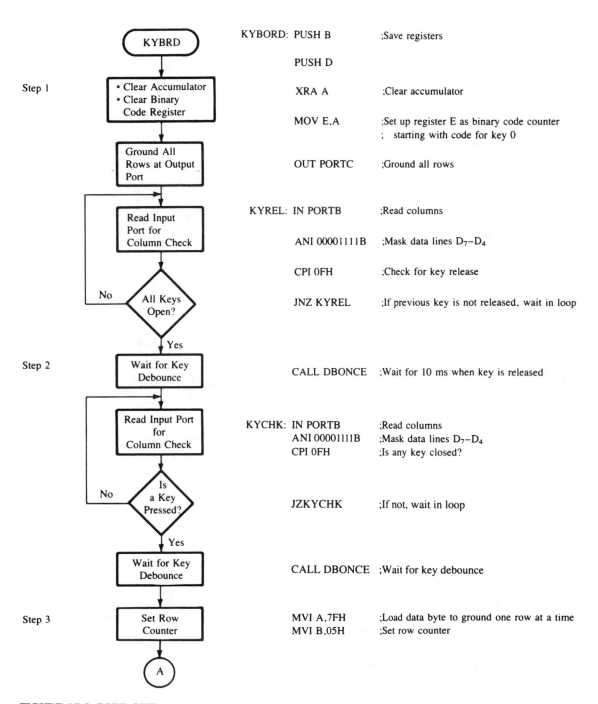

KYBORD:	PUSH B	;Save registers
	PUSH D	
	XRA A	;Clear accumulator
	MOV E,A	;Set up register E as binary code counter ; starting with code for key 0
	OUT PORTC	;Ground all rows
KYREL:	IN PORTB	;Read columns
	ANI 00001111B	;Mask data lines D_7-D_4
	CPI 0FH	;Check for key release
	JNZ KYREL	;If previous key is not released, wait in loop
	CALL DBONCE	;Wait for 10 ms when key is released
KYCHK:	IN PORTB	;Read columns
	ANI 00001111B	;Mask data lines D_7-D_4
	CPI 0FH	;Is any key closed?
	JZKYCHK	;If not, wait in loop
	CALL DBONCE	;Wait for key debounce
	MVI A,7FH	;Load data byte to ground one row at a time
	MVI B,05H	;Set row counter

FIGURE 17.8, PART ONE
Flowchart: Matrix Keyboard Subroutine

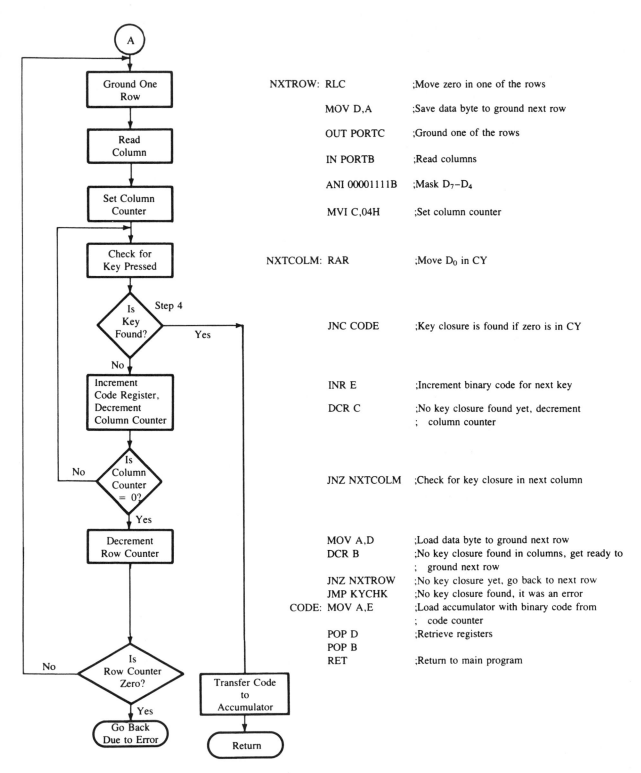

FIGURE 17.8, PART TWO
(continued)

```
DBONCE:   ;This is a 10 ms delay routine, does not destroy any register content
          ;Input: None
          ;Output: None
          PUSH B            ;Save registers
          PUSH PSW
          LXI B,COUNT       ;Load 10 ms delay count
LOOP:     DCX B             ;Repeat loop for delay
          MOV A,C
          ORA B             ;Set zero flag if BC = 0
          JNZ LOOP
          POP PSW
          POP B
          RET
```

FIGURE 17.8, PART THREE
(continued)

KEYBOARD SUBROUTINE
;This subroutine checks a key closure in the keyboard, identifies the key, and supplies the
;   corresponding binary code in the accumulator. It does not modify any register
;   contents.
;Input: None
;Output: Binary key code in the accumulator
;Calls DBONCE, a 10 ms delay subroutine

PROGRAM DESCRIPTION
This keyboard routine saves register contents of the calling program and clears registers
A and E. Register E is used as a binary code counter for the keys; it begins with the code
of the 0 key. The OUT instruction grounds all the rows, and the IN instruction reads the
columns. The ANI instruction masks the data on lines D_7–D_4 because they are not being
used for this keyboard.

The next instruction, CPI 0FH, checks whether the previous key pressed has been
released; this is a precautionary step against someone holding a key for a long time. If all
keys are open, D_3–D_0 will be high, the reading will be 0FH, and the Compare instruction
will set the Zero flag; otherwise, the routine stays in the loop KYREL until all keys are
open. The subroutine DBONCE eliminates the key bounce by waiting for 10 ms.

Once all keys are open, the routine reads the columns to check for a key closure. If
any of the keys is closed, one of the columns will be at logic 0, and the routine will skip
the KYCHK loop. The DBONCE routine will debounce the key closure. At this point, a
key closure is found, but the key is not identified. For example, if the reading on data lines
D_3–D_0 is 1110, any of the keys in Column 0 may have been pressed. Therefore, the next
step is to identify the key.

To identify the key pressed, one row is grounded at a time, beginning at Row 0. The
byte 01111111 (7FH) is loaded into the accumulator and rotated left (RLC) by one position; the byte is converted to 11111110. This byte is sent to port C to ground Row 0.
Then, port B is read, and each column is checked for logic 0 by rotating the reading into

the CY flag. Register C is set up to count four columns, and by rotating the byte to the left four times, each column is checked for logic 0 in the loop labeled NXTCOLM. As each column is being checked, the code counter (register E) is incremented in each iteration. For example, when Row 0 is grounded, four keys, 0 through 3, are checked, and the code counter is incremented from 00 to 03H.

After checking the columns in Row 0, the program loops back to location NXTROW and grounds the next row by sending the code that was previously saved in register D. Register B is set up as a row counter to count five rows. For each row, the loop NXTCOLM is repeated four times; thus, all 20 keys are checked, and for each iteration the code counter is incremented. When the key closure is found and the key is identified, the program jumps to location CODE. The routine copies the key code into the accumulator and returns to the calling program.

17.24 Combining the Matrix Keyboard and the Scanned Multiplexed Display

A display and a keyboard are two devices that are commonly used in microprocessor-based products. To reduce the cost and the chip-count in a product, a matrix keyboard and a scanned display are combined often. For example, in our previous illustrations, we can use port C to scan the display as well as ground the keys. If we use the scanned multiplexed display with the software-driven matrix keyboard, the keyboard subroutine must be coupled with the display; otherwise, the display may go off. For example, when the subroutine is waiting for a key to be pressed, the display cannot be refreshed by turning on and off digits in a sequence at a regular interval. Therefore, the program must alternate between refreshing the display and checking a keyboard to find a key pressed. The time needed for the keyboard subroutine to check a key is relatively short; therefore, it does not affect the display. Another approach is to interface the keyboard using the interrupt technique. In this approach, the program continues to refresh the display until the interrupt signal is received; then, the program checks the keyboard, processes the key, and goes back to the display.

17.25 Hardware Approaches to Matrix Keyboard and Scanned Display

The hardware approach reduces the software and allows the MPU to perform other tasks; however, it may increase the unit cost of the product. One approach is to use a logic device, such as the National Semiconductor MM74C923 keyboard encoder. This keyboard encoder can sense a key closure, debounce the key, provide the binary code of the key, and generate an interrupt. The other approach is to use a programmable device, such as the Intel 8279 keyboard/display interface. The 8279 performs two tasks: one task is to detect and encode a key (this is the same as that of the National Semiconductor keyboard encoder), and the other task is to refresh a scanned display (capable of displaying 16 bytes). We discussed the 8279 in Chapter 14 (Section 14.3); now we will illustrate how to interface a matrix keyboard using the MM74C923 keyboard encoder.

MM74C923 KEYBOARD ENCODER

This is a 20-key encoder with columns and rows (Figure 17.9). The respective columns and rows of a matrix keyboard must be connected to the columns and rows of the encoder. The encoder includes the chip select and the interrupt logic. The decoded address line (I/O Select) is connected to the $\overline{OE}$ signal of the encoder; it does not require an 8255A device or an input buffer. It has five output lines that provide the binary code of a key closure.

Figure 17.10 shows the schematic for interfacing a 20-key matrix keyboard using the encoder. The keyboard is assigned the port address by connecting the output O_5 of the decoder from Figure 17.2. Thus, the keyboard can be accessed by the port addresses 94H to 97H; the address lines A_1 and A_0 are left as don't care.

When a key is pressed, the encoder debounces the key and checks again for a valid key. If a valid key is detected, the encoder generates an interrupt and places the binary code of the key on the output lines. When the MPU acknowledges the interrupt and reads the binary code, the encoder turns off the interrupt. In this interfacing, the keyboard routine is reduced to a few instructions that involve reading the keyboard and storing the code in the input buffer. This technique reduces a considerable software overhead of the MPU. Therefore, the MPU can continue to scan the display until an interrupt request is received. Then, it can process the key and go back to the scanned display routine.

MEMORY DESIGN 17.3

Microprocessor-based products that require inputs from the user generally include two types of memory: Read-Only memory (ROM or EPROM) and R/W memory. ROM is used to store permanent or system programs and R/W memory is used to enter data and parameters and as stack. In memory design, we should be concerned about the size of the memory chips required and their memory maps, future expandability, and access time. Additional consideration is the capability to program Read-Only memories in the laboratory environment. Therefore, during the product development cycle or in instructional laboratories, EPROM is commonly used.

The first consideration of the memory design is the memory size and the memory map. Programs written for instructional projects can be very easily stored in 2K memory, such as the 2716 EPROM. However, the price differences between the 2764 EPROM (8K), the 2732 EPROM (4K), and the 2716 EPROM are negligible; furthermore, the price tends to go down as the usage of a chip in the marketplace increases even if the chip has larger memory size. Therefore, we will use the 2764 EPROM (8K × 8) in our design. The memory map of this EPROM should begin at memory address 0000H because the program counter is cleared to that address whenever the system is reset, and we will assume that the EPROM will be used for the system program. The memory map of EPROM with 8K bytes of memory should be placed in the range from 0000H to 1FFFH, as shown in Figure 17.11. However, there are no such restrictions for the memory map of R/W

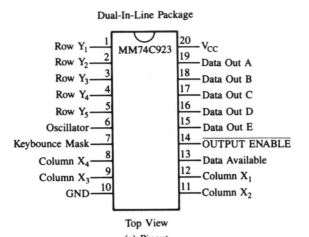

Dual-In-Line Package

Top View

(a) Pinout

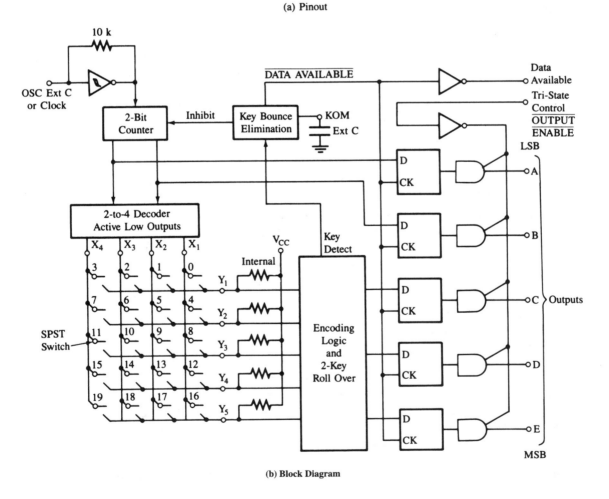

(b) Block Diagram

FIGURE 17.9

Keyboard Encoder MM74C923 (National Semiconductor): Pinout (a) and Block Diagram (b)

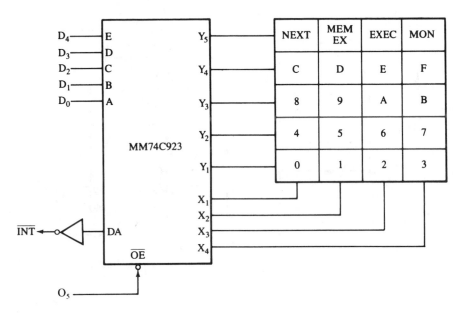

FIGURE 17.10
Interfacing 20-Key Matrix Keyboard Using the MM74C923

memory; it can be mapped anywhere as long as it does not overlap with the map of the EPROM. The other consideration is an appropriate decoding technique for memory devices with different sizes. If we use the same decoding network for devices with different sizes, such as a 3-to-8 decoder, the smaller memory chip will be left with don't care address lines; thus, it will have foldback memory addresses. However, in a small system, the foldback memory is not a serious concern.

The next consideration is future expandability. The 2764 requires a 28-pin socket; however, its pinout is designed in such a way that it is compatible with larger memory chips such as 27128 (16K × 8) and 27256 (32K × 8). However, this type of expansion cannot be easily accomplished by the decoding network shown in Figure 17.11; we will have to use a PROM, which can be reprogrammed to accommodate larger-size memory chips.

The last consideration is the memory access time and whether we need any Wait states in interfacing these memories. In the last decade, the memory access time has improved considerably; memory devices with 200 to 250 ns are commonly available. However, we need to calculate memory response time for a given system frequency to determine the necessity of Wait states.

17.31 EPROM Memory

Figure 17.11 shows the design of EPROM using the 2764 (8192 × 8) and the 74LS138 (3-to-8 decoder). The 13 address lines ($A_{12}-A_0$) of the MPU are directly connected to pins $A_{12}-A_0$ of the 2764 to decode 8192 memory locations. The rest of the address lines

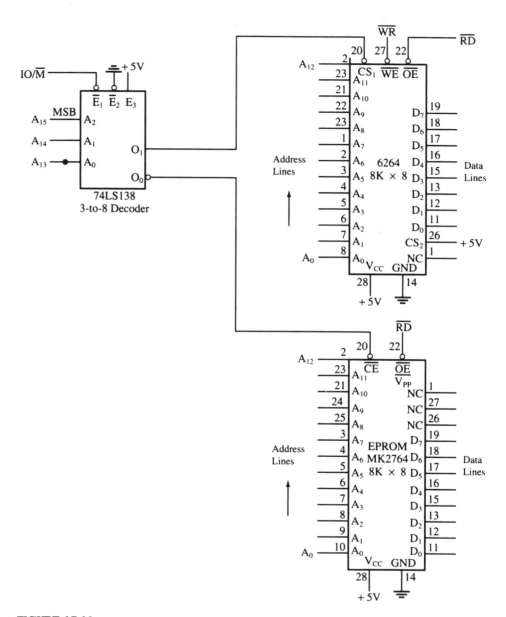

FIGURE 17.11
Schematic: Memory Design

$(A_{15}-A_{13})$ are decoded by the 74LS138; this provides an 8K decoding resolution for each output line of the decoder. The enable lines $\overline{E}_2$ and E_3 of the decoder are permanently enabled, and $\overline{E}_1$ is enabled by the $IO/\overline{M}$ signal of the 8085. Because $\overline{E}_1$ is active low, the decoder is enabled only for memory operations and it is disabled for I/O operations. The $\overline{RD}$ signal of the 8085 can be directly connected to the $\overline{OE}$ of the memory chip, thus eliminating the need to generate a separate $\overline{MEMR}$ signal by combining $\overline{RD}$ and IO/$\overline{M}$.

The memory chip is accessed by connecting O_0 of the decoder to the $\overline{CE}$ of the memory chip. When the address lines A_{15}–A_{13} are at logic 0, and IO/$\overline{M}$ is active low, the decoder output O_0 is asserted to enable the memory chip. The memory address range for this EPROM is 0000H to 1FFFH, as shown below.

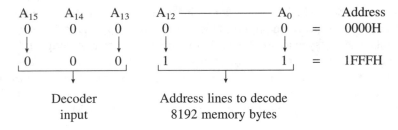

17.32 R/W Memory

The R/W memory can be designed with popular CMOS memory chips, such as 6116 (2048×8) and 6264 (8192×8). If we were to use a 2K memory chip, it would require only 11 address lines to decode the 2048 locations. To use the 3-to-8 decoder from Figure 17.11, two address lines will have to be left don't care, thus generating foldback memory space. Figure 17.11 shows the interfacing of R/W memory using the 6264 (8K) CMOS chip. This memory chip requires 13 address lines A_{12}–A_0, the same as in EPROM. The Chip Select (CS) logic is generated from the same decoder as for EPROM; therefore, the $\overline{RD}$ and $\overline{WR}$ signals are directly connected to the memory chip. When the O_1 line of the decoder goes low, the memory chip is enabled; thus the memory map of this R/W memory ranges from 2000H to 3FFFH, as shown below.

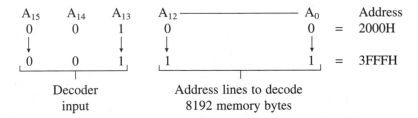

17.33 Interfacing Memory with Wait States

In interfacing memory with the microprocessor, the interfacing circuit must satisfy the timing requirements of the microprocessor and the memory chip. In Chapter 3, we assumed that memory response can match the execution speed of the microprocessor, but this assumption is invalid in some situations. Because of cost considerations, memory chips with slow response time are used occasionally in microprocessor-based systems. Therefore, it is necessary to synchronize the execution speed of the microprocessor with the response time of memory. This can be accomplished by using the READY signal of the 8085 microprocessor.

The READY signal is an active low signal, input to the 8085 as an external request from a slow peripheral, to indicate that the peripheral is not yet ready for data transfer. When this signal is high during the Read or Write cycle, it indicates that I/O or memory is ready to send or receive data. The 8085 samples the READY line during T_2 of each machine cycle, and if the READY line is asserted low, it adds one clock period T^W as a Wait state to its machine cycle. Then it continues to sample the T^W state and adds additional Wait states until the READY signal becomes inactive. During this time, the 8085 extends the time of control signals and preserves the contents of all the buses. Thus, the READY signal can be used to synchronize the response time of any type of peripheral.

Now, to ascertain whether a given memory chip is too slow in comparison with the execution speed of the microprocessor and needs Wait states to synchronize the data transfer, we must examine the timing requirements of the microprocessor and the response time of the memory. When Wait states are not needed, the READY signal must be tied high.

WAIT STATE CALCULATIONS

Figure 17.12 shows the timing diagram of the 2764-45 EPROM with 450 ns access time; this is the time the memory takes to place the data byte on the data bus after receiving an address. The 8085 microprocessor, with a clock of 320 ns, has t_{AD} equal to 575 ns, meaning the 8085 will begin to read data 575 ns after the address is valid (see simplified timing sketch in Figure 17.13). This leaves 125 ns ($575 - 450 = 125$ ns) for delays in the decoding circuit. In Figure 17.11, the delay in the decoder will be 38 ns, and if the address bus has a driver such as 74LS244 (not shown in this figure), it will cause an 18 ns delay.

$$t_{AD} \text{ (8085)} \quad < t_{ACC} \text{ (memory)} + \text{Decoding delays} + \text{Bus driver delays}$$
$$575 \text{ ns} \qquad < 450 \text{ ns} + 38 \text{ ns} + 18 \text{ ns}$$
$$575 \text{ ns} \qquad < 506 \text{ ns}$$

In this particular circuit, the memory 2764 with 450 ns access time is adequate, and Wait states are unnecessary.

If the system has a faster version of the 8085 microprocessor, this memory may not be adequate. For example, the 8085A-2 can operate at 5 MHz and has $t_{AD} = 350$ ns. In such a case, we can use either the faster version of the 2764 (2764-25 with 250 ns access time) or the READY signal to extend t_{AD} as shown in Figure 17.14. These decisions are generally dependent on cost consideration: the cost of the faster version of memory vs. that of a flip-flop.

In Figure 17.14, the ALE is connected to the clock of the D flip-flop (F/F_1). When the ALE goes high, the output of F/F_1 goes high. At the next clock, the output ($\overline{Q}$) of F/F_2 goes low, which pulls the READY signal low and resets F/F_1. This makes $\overline{Q}$ high in the next clock period. Thus, the output ($\overline{Q}$) is low only for one clock period, adding one Wait state (clock = 200 ns) to every machine cycle, as shown in Figure 17.15. This technique allows enough time for memory to place data on the data bus.

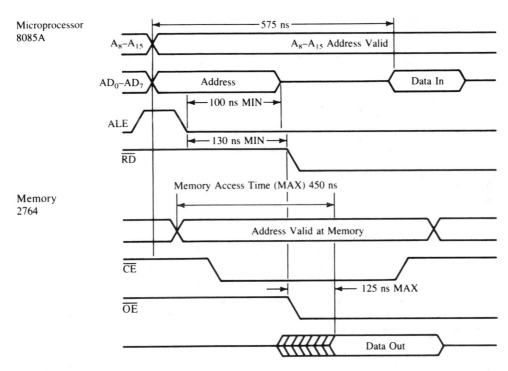

FIGURE 17.12
2764 Memory Chip and Its Read Timing

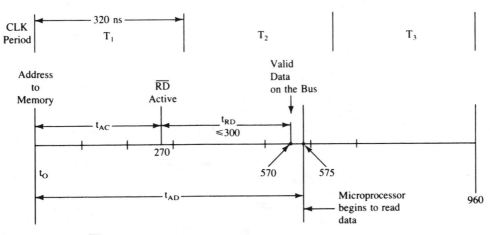

t_{AC} = Address to $\overline{RD}$
t_{RD} = $\overline{RD}$ to Valid Data
t_{AD} = Address to Valid Data In

FIGURE 17.13
Typical Timings for an 8085 System with 320 ns Clock Period

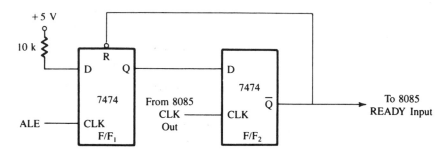

FIGURE 17.14
Circuit to Add One Wait State

Figure 17.15 shows two Memory Read machine cycles. The first machine cycle represents the Memory Read operation without the Wait state; the second includes one Wait state. During T_2 of the machine cycle, the microprocessor checks the READY line; if it is low, the microprocessor inserts one Wait state. Again it checks the READY line during the Wait state. If the line is high, the microprocessor goes to T_3, and if it is low, it inserts the next Wait state. Figure 17.15 shows how one Wait state extends the Read signal, thus extending t_{AD} by one clock period.

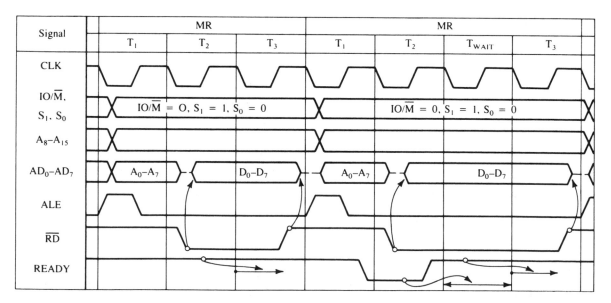

FIGURE 17.15
Memory Read Machine Cycles (with and without WAIT States)
SOURCE: Intel Corporation, *Family User's Manual* (Santa Clara, Calif.: 1979), p. 2–11.

MPU DESIGN 17.4

The MPU design is determined primarily by the system requirements, such as load on the system (driving capacity), process of data transfer (interrupt, DMA, etc.), and interfacing devices used in the system. For the time being, we assume that we will design the MPU for a general-purpose single-board microcomputer with appropriate driving capacity for the buses, and we will use general-purpose memory and I/O components.

17.41 8085 MPU

The MPU design can be divided into the following segments:

1. Address bus
2. Data bus
3. Control signals
4. Frequency and power requirements
5. Externally triggered signals (Reset, Interrupts, etc.)

ADDRESS BUS

The 8085 has a multiplexed bus; it must be demultiplexed so that general-purpose memory devices can be used. In addition, we must use bus drivers to provide sufficient driving capacity.

Figure 17.16 shows the 74LS244, an octal bus driver used with the high-order address bus to increase its driving capacity. Typically, the 8085 buses can source 400 μA (I_{OH} = −400 μA) and sink 2 mA (I_{OL} = 2 mA) of current; they can drive one TTL logic load. The 74LS244 driver is capable of sourcing 15 mA and sinking 24 mA of current.

The low-order address bus is demultiplexed by using the ALE signal (Address Latch Enable) and the latch 74LS373. At the beginning of each machine cycle, ALE goes high during T_1 (see Chapter 3, Figure 3.3), and this signal is connected to the Enable line G of the latch. As ALE goes low, the address on bus AD_7–AD_0 is latched, and the eight output lines of the 74LS373 serve as the low-order address bus (A_7–A_0). The address on the output of the 74LS373 remains latched until the next ALE signal. In addition to demultiplexing the address bus, the 74LS373 can serve as a bus driver.

DATA BUS

Figure 17.16 shows the 74LS245 as an 8-bit bidirectional bus driver to increase the driving capacity of the data bus. The 74LS245 can sink 24 mA and source 15 mA of current. The 74LS245 has eight bidirectional data lines; the direction of the data flow is determined by the direction control line (DIR). Figure 17.16 shows that the bus driver is enabled by grounding the enable signal ($\overline{G}$). The direction of the data flow is determined by connecting the $\overline{RD}$ signal from the MPU to the DIR signal. When the MPU is writing to peripherals, the $\overline{RD}$ is high and data flow from the MPU to peripherals. When it is reading from peripherals, the $\overline{RD}$ is low and data flow toward the MPU.

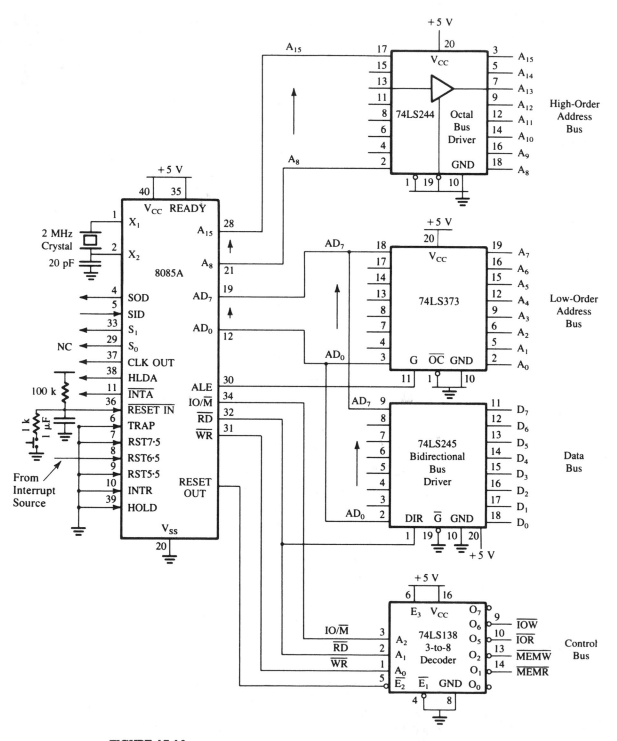

FIGURE 17.16
Schematic of the 8085 MPU with Demultiplex Address Bus and Control Signals

CONTROL SIGNALS

The 8085 generates three signals: IO/$\overline{\text{M}}$ (I/O or memory), $\overline{\text{RD}}$ (Read), and $\overline{\text{WR}}$ (Write). The IO/$\overline{\text{M}}$ signal differentiates between I/O and memory functions. When IO/$\overline{\text{M}}$ is high, it is an I/O-related function; when IO/$\overline{\text{M}}$ is low, it is a memory-related function. Therefore, by combining IO$\overline{\text{M}}$ with $\overline{\text{RD}}$ and $\overline{\text{WR}}$ signals, appropriate control signals can be generated.

Figure 17.16 shows that three signals—IO/$\overline{\text{M}}$, $\overline{\text{RD}}$, and $\overline{\text{WR}}$—are used as inputs to the 74LS138 3-to-8 decoder to generate four control signals—$\overline{\text{IOR}}$, $\overline{\text{IOW}}$, $\overline{\text{MEMR}}$, and $\overline{\text{MEMW}}$. These signals can be used for interfacing with any peripherals.

FREQUENCY AND POWER REQUIREMENTS

The 8085A can operate with a maximum clock frequency of 3 MHz. To obtain 3 MHz operating frequency, the clock logic should be driven by double the desired frequency (6 MHz). The 8085 has two clock inputs: X_1 and X_2 at pins 1 and 2. These inputs can be driven with a crystal, an LC tuned circuit, or an RC network.

Figure 17.17(a) shows a 2 MHz crystal with a 20 pF capacitor to drive the clock inputs. This input frequency is divided in half internally, and the system will run on 1 MHz clock frequency. The capacitor is required to assure oscillator start-up at the correct frequency. Figure 17.17(b) shows an alternative method of providing a clock input using an RC network.

The 8085 and other components used in this system require one power supply with +5 V. The current requirement of the power supply is determined primarily by the display load and the peripherals of the system; the MPU and memory components of the system require less than 400 mA.

EXTERNALLY TRIGGERED SIGNALS

As discussed in Chapter 2, the 8085 has provision for four external input signals: Reset, Interrupt, Ready, and Hold. Of these signals, RESET and one interrupt signal (RST 6.5) are used in this system, and the others are disabled.

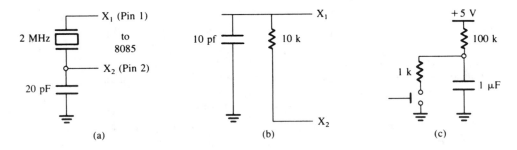

FIGURE 17.17
Clock and Reset Circuits: Clock Circuit with Crystal (a), RC Clock Circuit (b), and Reset Circuit (c)

Reset The $\overline{\text{RESETIN}}$ is an active low signal used to reset the system. When this pin goes low, the program counter is set to 0, the Interrupt Enable and HLDA flip-flops are reset, and all buses are placed in tri-state. The reset circuit shown in Figure 17.17(c) is an RC network with a sufficiently long time constant. When the Reset key is pushed, the $\overline{\text{RESETIN}}$ goes low and slowly rises to +5 V, providing sufficient time for the MPU to reset the system.

Interrupts The 8085 has five interrupt signals, all of them active high. In this system, RST 6.5 will be used; the others need to be grounded as shown in the circuit. Otherwise, the floating interrupt pins can cause the system to malfunction. To allow the use of interrupt signals for further expansion of the system, the pins can be grounded using switches.

HOLD This is an active high signal used in the DMA. This signal is also grounded in this system.

READY When this signal is high, it indicates that the memory or peripheral is ready to send or receive data. When READY goes low, the MPU enters the Wait state until READY goes high; then the MPU completes the Read or Write cycle. This signal is used primarily to synchronize slow peripherals with the MPU. To prevent the MPU from entering the Wait state, this pin is tied high.

    A complete single-board microcomputer can be built from this MPU, which will be discussed in the next section.

17.5 DESIGNING A SYSTEM: SINGLE-BOARD MICROCOMPUTER

In the last four sections, we discussed the designing and interfacing of various components of a system, such as a scanned display, a matrix keyboard, and memory. We also examined general requirements of the 8085 MPU. Now we will combine these components in a general-purpose single-board microcomputer system.

17.51 Project Statement

Design a single-board microcomputer to meet the following specifications:

☐ Input: Hex keyboard with minimum of 20 keys
☐ Output: Six seven-segment LEDs to display memory address and data
       : Two seven-segment LEDs to display results
☐ Memory: 8K of EPROM—2764 (8192×8) with 450 ns (or less) access time
      : 8K of R/W static memory—6264 (8192×8) or equivalent
☐ Microprocessor: 8085A
☐ System Frequency: 2 MHz
☐ Suggested Interfacing Devices: 8255A, bus drivers, 3-to-8 decoders, key encoder, segment and digit drivers, and Hex decoder/drivers.

The system should allow a user to enter and execute programs, and the buses should have enough driving capacity to interface with additional peripherals. While machine codes are being entered, the memory address and data should be displayed by seven-segment LEDs. A two-digit seven-segment LED port should be available as an output port to display the results when programs are executed.

17.52   Project Analysis

In analyzing the specifications of a microprocessor-based product, it is essential to consider hardware and software simultaneously. They are interrelated, and each will have an impact on the other.

The functions of the single-board microcomputer according to the specifications (given above) can be classified in three categories as follows:

1. Check the keyboard for data or functions.
2. Display memory address, data, and results.
3. Execute programs.

KEYBOARD

The keyboard in this design is an input port with keys arranged in the matrix format. When a key is pressed, the keyboard routine should provide a binary equivalent of the key. This can be accomplished in various ways: one is software approach, whereby a key closure is sensed, debounced, and identified and the key code is obtained by using the software. The other is the hardware approach, whereby all these key functions are performed through a programmable keyboard encoder.

The keys are divided into two groups: one group is for Hex digits from 0 to F, and the second is concerned with various functions. Basically, there are two approaches to recognizing Hex keys. One approach is to begin with Hex keys, identify the memory address, and then specify a function such as Examine Memory or Execute. In the second approach, a function is specified first and then Hex keys are accepted.

DISPLAY

This project has two types of display: the system display and the user display. The system display consists of four seven-segment LEDs for memory address and two seven-segment LEDs for data, and the user display consists of two seven-segment LEDs for results. We can explore both hardware and software approaches to design output ports for these displays.

EXECUTE

This is the simplest of the three functions and can be performed with one instruction: PCHL. When the user wants to execute a program, she/he provides the memory address where the program is stored and presses the Execute key. Assuming the memory address is stored in HL registers, the instruction PCHL simply loads the program counter with the specified memory address, and the program control is transferred from the monitor program to the user's program.

17.53   System Design

Figure 17.18 shows the block diagram of a single-board microcomputer. We can divide the project design into the following sections:

1. 8085 MPU design
2. Memory design
3. Display design
4. Keyboard interfacing
5. System software

We have designed these system components, except system software, in previous sections. Now our primary focus will be on system software and its implications for design of the components because of the interaction between the components in the system.

SYSTEM BUSES AND THEIR DRIVING CAPACITY

The 8085 has eight address lines $A_{15}-A_8$ and eight multiplexed lines AD_7-AD_0 with driving current capacity of I_{OH} 400 μa and sinking capacity of I_{OL} 2 mA. At this point, we do not know the total load the user may have on the buses, but by examining the block diagram, we can make some reasonable estimates of the load. Figure 17.18 shows that the address bus will drive two decode circuits (I/O and memory decoders) and two memory

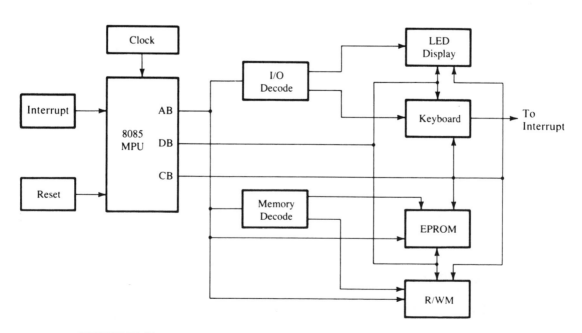

FIGURE 17.18
Block Diagram of a Single-Board Microcomputer

chips (CMOS 6264 and EPROM 2764). We can calculate the bus loading as follows (Figure 17.19):

	High-level input current I_{IH}	Low-level input current I_{IL}
Two decoders = 20 μA × 2 = 40 μA		400 μA × 2 = 800 μA
R/W memory	= 10 μA	= 10 μA
2764 EPROM	= 10 μA	= 10 μA
	60 μA	820 μA

By examining these load currents, we can conclude that the bus drivers are unnecessary for the address bus; we can even add a few decode circuits or gates. However, this single-board microcomputer is expected to be used for general-purpose interfacing; therefore, as a precaution, we will use the 74LS244 as a bus driver to increase the driving capacity. The 74LS244 is an octal buffer/driver, capable of sourcing 15 mA and sinking 24 mA of current. Thus, the 8085 address bus can drive additional devices (decoders, gates, etc.) with sufficient driving capacity.

The 8085 multiplexed bus has eight bidirectional lines with driving capacity similar to that of the address bus. Because the data bus is bidirectional, the loading on the bus varies considerably. When the 8085 is reading from memory, the memory chip that is enabled becomes the driving source and the microprocessor becomes the load, and when the 8085 is writing to an output port, the microprocessor is the source and the latches of the output port constitute the load. An octal latch, such as the 74LS373, requires 400 μA input current at the low-level logic; on the other hand, the 7475 requires 3–6 mA. Therefore, as a precaution, we will use a bidirectional buffer as a data bus driver. We designed the MPU with these considerations in Section 17.4.

KEYBOARD AND DISPLAYS

To interface displays and keyboards in a system, we have various approaches available to us; now we need to make some trade-offs between hardware and software and between

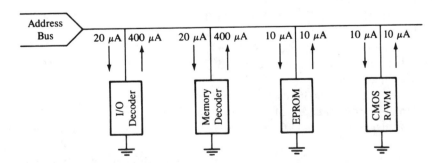

FIGURE 17.19
Loading on the Address Bus

production cost and development cost. The least expensive solution per production unit cost is to use software for refreshing the LED display and checking the keyboard. The other extreme is to use the 8279, the keyboard/display interface, to perform the refreshing of the LED display and checking of the keyboard.

In the software approach, we can minimize the interface devices by combining the matrix keyboard and the scanned display. The digit code driver of the scanned display can be used to connect the rows of the matrix keyboard; thus, we can reduce the number of I/O ports required from four to three. The program scans the display once, turns off the display temporarily by sending logic zero to the LED segments, and then grounds rows of the keyboard and checks the columns for a key closure. This approach requires one 8255A; for example, port A as the segment driver for the display, port B as an input port to read the columns of the keyboard, and port C as a digit code driver for the display as well as to ground the keys. However, we will need an additional port for the user display.

The second approach is to replace refreshing the display and checking the keyboard by a programmable device, such as the Intel 8279. This device removes the burden of scanning the display and checking the keyboard from the MPU. When a key is pressed, the MPU is informed by generating an interrupt. When the MPU reads the keyboard, it places the code in the encoder memory and informs the encoder how many LEDs to display. This simplifies the software necessary for the keyboard monitor.

For instructional purposes, we will take the middle road. We will replace the keyboard software by a keyboard encoder, such as the National MM74C923, and use software for refreshing the display. This approach reduces the software; the program has to continuously refresh the display, and, when a keyboard encoder generates an interrupt, the MPU can read the keyboard, process the key, and go back to refreshing the display. In this approach, we can use two ports of the 8255A for the scanned multiplexed display and one port for the user display. This approach can illustrate important concepts, such as interrupt, refreshing, and key encoding, without being excessively dependent on software.

Now we can summarize the specifications as follows:

1. Memory map: EPROM 0000H to 1FFFH (Figure 17.11)
                R/WM   2000H to 3FFFH (Figure 17.11)
2. System display: Scanned display with six LEDs using the 8255A
                Port A = 80H, port C = 82H
3. User display: Latched LED port using the port B (81H) of the 8255A and two Hex
                decoders 9370
4. Keyboard: 20-key matrix keyboard using the encoder MM74C923
                Port is interrupt-driven with the address 84H
                Four function keys and 16 data keys

On the basis of these specifications, we will illustrate an approach to software design in the next section.

SOFTWARE DESIGN 17.6

The most puzzling aspect of software design is where to begin and how to synthesize all functions in one program. The place to begin is the list of the functions to be performed. In the project analysis section, three functions are listed: check keyboard, display, and execute. The next place to look for clues is hardware. Examination of the hardware design reveals the following clues:

1. The program should begin at location 0000H.
2. Initial memory locations should be reserved for interrupt restarts.
3. Programmable peripherals need initialization instructions.
4. At the system turn-on, a message should be displayed.
5. Four keys are available to identify functions, and 16 keys are used as Hex digits from 0 to F.

By combining the functions to be performed and the clues obtained from hardware design, the task can be divided into the following steps:

1. Initialize programmable peripherals.
2. Display the sign-on message to indicate that the system is ready.
3. The keyboard is interrupt-driven; therefore, continue the display-scanning until an interrupt is generated.
4. When an interrupt is generated, turn off the sign-on message, process the key, and return to the display loop to display the new key or the error message, or blank the display.

The first three steps are fairly simple. The initialization is determined by peripheral devices and their decode logic (discussed in the previous sections). The display involves refreshing the sign-on message with the table look-up technique. The third step is to continue to refresh until an interrupt is generated.

The fourth step—determination of an appropriate key—is critical to the software design, and the appropriateness depends upon how the user is allowed to enter and execute a program. Basically, there are two approaches: one approach is to begin with a memory address and then specify the function to be performed; the second approach is to begin with a function, and then Hex keys are accepted. In addition to the Reset key, at least three keys are required: MEMEX (Memory Examine), NEXT (Next Memory Location), and EXEC (Execute). The MEMEX key allows the user to enter the memory address, examine the data stored in that memory, and enter new data. The NEXT key stores the new data byte and increments to the next memory location. The EXEC key allows the user to execute a program. With this minimum configuration, if an inappropriate sequence of keys is pressed, and the program displays the error message, it can be terminated only by the Reset key. Therefore, a key called MON (Monitor) is added to terminate a program.

In the first approach, in which the user begins with a function, we will use four keys: MON (Monitor), MEMEX (Memory Examine), NEXT (Next address), and EXEC (Execute). If the MON function is selected, the program goes to the beginning and displays a Ready message; if the Exec function is selected, the program transfers control to the user program. If the MEMEX key is pressed, the program displays the contents of the memory; if the NEXT key is pressed, the program increments to the next memory location. After the MEMEX key, the user is allowed to enter new data or go to any one of the functions except the EXEC function.

In the second approach, the user must begin with a memory address and then specify a function; otherwise, an error message is generated.

The proposed system includes a scanned multiplexed display, which needs refreshing at a regular interval, and a matrix keyboard that is interrupt-driven. Therefore, the main program revolves primarily around refreshing the display and waiting for an interrupt to occur, as shown in Figure 17.20. The primary task of the interrupt service routine is to read and process the key pressed and perform the designated function of the key. We can divide the software design into the following modules:

1. Initialize.
2. Display the sign-on message and wait for an interrupt.
3. When an interrupt occurs, read the key.
4. Decode the key and jump to the appropriate function.
5. Perform the function and return to refreshing the display.

17.61 Initialization

When a system is reset, the 8085 clears the program counter, and the program execution begins at location 0000H. If the system includes several sources of interrupts, the initial

FIGURE 17.20
Flowchart: Main Program

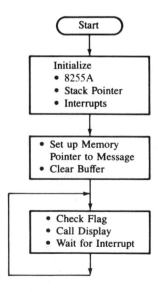

memory locations can be used for interrupts (RST 7 is 0038H). Therefore, the initialization program module can be written starting at 0040H, and the Jump instruction at 0000H can transfer the program to 0040H.

This module must initialize the programmable I/O device—the 8255A—and the stack pointer and enable the interrupts. In Section 17.1, we wrote the initialization instructions for the 8255A. The stack pointer is generally initialized at the top of the R/W memory; however, in this project, we will need six top locations as Display Buffer and one location to save input data. After reserving top locations (for example, from 3FF9H to 3FFFH) for the buffer (explained in the next section), the stack pointer can be initialized. In the 8085, RST 7.5, 6.5, or 5.5 interrupts are the simplest to implement because they do not require any external hardware; we will use RST 6.5 for this project. When an interrupt is generated by the keyboard encoder and accepted by the MPU, the program is automatically transferred to location 0034H. Thus, the keyboard service routine must begin at 0034H or a Jump instruction must be written at 0034H to locate the start of the interrupt service routine.

17.62 Display Module

We discussed the scanned multiplexed display routine in Section 17.1. That routine can be used for any fixed message, such as the sign-on message or the error message; the codes for these messages can be stored permanently in the EPROM. However, to display memory address and data that change with key strokes, we need to reserve memory locations: four for memory address and two for data in the R/W memory. These locations are called Display Buffer. For example, in this single-board microcomputer, the R/W memory ranges from 2000H to 3FFFH; we can reserve the last six locations 3FFAH to 3FFFH as the Display Buffer. The display routine must be modified to check the Display Buffer and get the segment code by using the table look-up technique, output the code, and turn on the corresponding digit (Figure 17.3). In addition, the routine must be informed of the number of digits to be refreshed. For example, when a memory address is entered, four digits are displayed; when the MEMEX key is pressed, six digits are displayed.

Now we need to find a way to inform the routine of the number of digits to be displayed and how to differentiate between data keys and address keys. This can be accomplished by using the flag concept. The routine that calls the codes to be displayed sets a flag when four digits are to be displayed. For example, bit D_7 in register B can be used as a flag. When D_7 is 0, the routine refreshes four memory digits; when D_7 is 1, the routine refreshes six digits. Another approach is to use the CY flag, instead of bit D_7 in register B, to perform the same function.

In block 3 of the main program (Figure 17.20), this flag concept is used. Before calling the Display routine, the program checks the flag to determine whether it should display four locations or six locations of the buffer.

17.63 Reading the Keyboard and Placing the Byte in the Buffer

When an interrupt request is generated by the keyboard encoder, the MPU should read the keyboard and save the reading in the input buffer (location 3FF9H). If it is a Hex key of a memory address or of a data byte, the MPU should place the binary value of the new key in the buffer. In addition, a register pair such as DE can be used as the Memory Register to save the memory address, and register C can be used as the Data Register. However, before the key code is placed in the buffer, the previous codes must be shifted by one location; in the process, the MSD (most significant digit) must be discarded (see Programming Assignment 21, in Chapter 10).

Figure 17.21 shows the partial flowchart of the interrupt service routine, which begins with reading a key and saving the data in the input buffer. Then the program checks

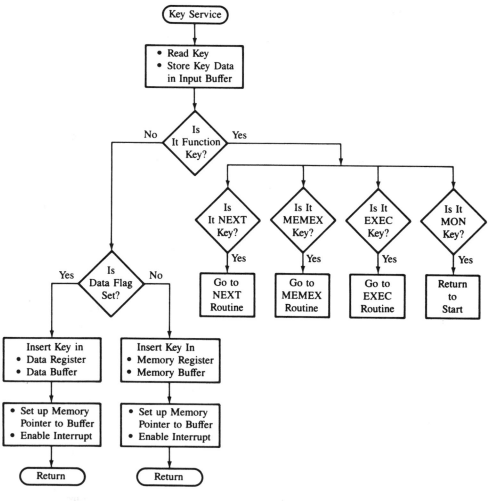

FIGURE 17.21
Flowchart: Key Service Routine

whether it is a function key or a Hex digit key; the keys with binary code 00 to 0FH are Hex digit keys; with binary code higher than 0FH, they are function keys. If the key is a function key, the program determines whether it is the MEMEX, NEXT, MON, or EXEC key and then jumps to appropriate locations. If the key is a Hex digit key, the program checks the flag (bit D_7 in register B) and identifies the key that is a part of a memory address or of a data byte. Then the program inserts the code of the new key as the least significant four bits in the Display Buffer and in the Memory Register or Data Register and returns from the interrupt service routine.

17.64 Performing Functions

The user must enter four Hex digits as a memory address. If more than four Hex keys are entered, the binary code of the last four is saved in register DE, and the Display buffer is updated accordingly. After entering an address, if a function key is pressed, the program checks which function key is pressed. If the MEMEX key is pressed, the memory address is already in the DE register. This address can be used as a memory pointer, and data from that memory location can be retrieved and placed in the Display Buffer (Figure 17.22). Similarly, when the NEXT key is pressed, the address in the DE register is incremented, the data byte in that location is obtained, and all six digits are placed in the display buffer.

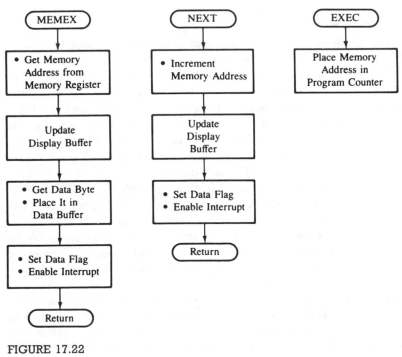

FIGURE 17.22
Function Routines

When a data byte is being entered, we can use register C to save data keys, as mentioned earlier. When a new key is pressed, the most significant nibble can be discarded, and the binary code of the new key can be entered as a least significant nibble (see Problem 3).

If the key is EXEC, the memory address where the execution should begin is already in the DE register. The program places the memory address in the program counter, and the control is transferred to the user program. If the key is MON, the program returns to the beginning and displays the System Ready message.

In writing this monitor program, the critical issue to remember is that the system uses the scanned display and needs continuous refreshing. Therefore, the main program consists primarily of calling the display routine. The next steps are to code this program in 8085 assembly language and to test it on prototype hardware, using such debugging tools as an in-circuit emulator and a logic analyzer (discussed later).

PROGRAM CODING

Assuming that program coding is to be performed by a team, it is necessary to break down the task into small, manageable, and independent modules. It is not always possible to break logic flow into independent subroutine modules. However, it is necessary to agree on symbols or labels that might be used by various members of the team; these are called global symbols.

17.65   Prototype Building and Testing

Microprocessor-based products are hardly ever built and tested as complete systems during the initial stages of design. If a system is completely built, it is difficult to troubleshoot. Traditional approaches, such as signal injection and isolating trouble spots, are ineffective for troubleshooting bus-oriented systems. Therefore a system is built and tested in stages. Each subsystem, such as keyboard, displays, and memory, should be built and tested separately as an independent module. Now, the question is: How to test a module without building a system? An answer can be found in such everyday incidents as testing a light bulb or starting a car with a dead battery. The light bulb can be tested by plugging it into a working socket, and the car can be started with a jumper cable. There are two principles involved in these examples: (1) borrowing resources from a working system and (2) substitution. These principles can be used in testing each separate subsystem of a microprocessor-based product. What is needed is a working system that can create an environment similar to the complete prototype system and that is generous enough to share its resources with hardware modules to be built. Such a working system is called an in-circuit emulator, and is described in Section 17.7.

Assuming that such an in-circuit emulator is available, subsystems of the single-board microcomputer can be built and tested one at a time. Similarly, as software modules are being written, they can be tested first on a software development system (discussed in Chapter 11). Finally, hardware and software can be integrated and tested using an in-circuit emulator.

DEVELOPMENT AND TROUBLESHOOTING TOOLS 17.7

In bus-oriented systems, there is constant flow of data, which continuously changes the logic states. This flow of data is controlled by software instructions. Therefore, to examine what is happening inside the system, special instruments, capable of capturing data in relation to instructions, are required. Two such instruments are discussed briefly in the next sections: In-circuit Emulator and Logic State Analyzer.

17.71 In-Circuit Emulator

The in-circuit emulation technique has become an essential part of the design process for microprocessor-based products. In-circuit emulation is the execution of a prototype software program in prototype hardware under the control of a software development system. To perform an in-circuit emulation, the microprocessor is removed from the prototype design board, and a 40-pin cable from an in-circuit emulator is plugged into the socket previously occupied by the microprocessor. The in-circuit emulator performs all the functions of the replaced microprocessor; in addition, it allows the prototype hardware to share all its resources, such as software, memory, and I/Os. It provides a window for looking into the dynamic, real-time operation of the prototype hardware. At present, a wide variety of in-circuit emulators are available, ranging from universal emulators with complete software development systems to stand-alone microprocessor units. Figure 17.23 shows a stand-alone in-circuit emulator (EM 188) designed by Applied Micro Systems.

EMULATION PROCESS

To test subsystems (such as I/O and memory) using an in-circuit emulator, the minimum prototype hardware required is a 40-pin microprocessor socket, without the microprocessor, and a power supply. All other resources can be borrowed from the in-circuit emulator. As more and more prototype hardware are built, fewer and fewer resources from the in-circuit emulator will be required. In the final stage, total software and hardware are integrated for testing. A hardware prototype can be viewed as a fetus growing in stages in the womb of an in-circuit emulator; until the fetus is fully developed and functioning independently, the in-circuit emulator provides the necessary environment and resources.

FEATURES OF IN-CIRCUIT EMULATOR

An in-circuit emulator is a software/hardware troubleshooting instrument. It can be a stand-alone unit or part of a software development system. A small program can be entered directly into the emulator, or a program can be transferred into the emulator from a host computer system through an RS-232 serial link. Once a program is loaded, a user can interact with the emulator through its keyboard or a terminal. The emulator has its own

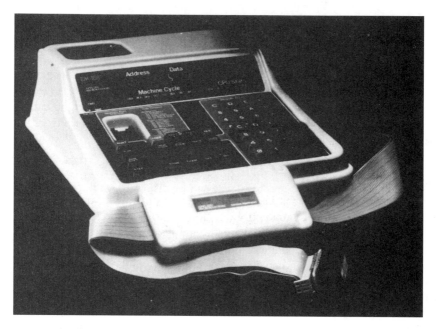

FIGURE 17.23
Stand-Alone In-Circuit Emulator—EM 188
SOURCE: Photograph courtesy of Applied Micro Systems.

software commands to perform various debugging functions. The main capabilities of an in-circuit emulator can be listed as follows:

☐ **Downloading:** Facilities are provided to transfer programs between a software development system or a host computer and the in-circuit emulator.
☐ **Resource Sharing:** The in-circuit emulator allows the system being tested to share its memory and I/O ports. The memory and I/O ports of the in-circuit emulator can be assigned any addresses, thus avoiding conflict with memory and I/Os of the prototype; this is called memory and I/O mapping.
☐ **Debugging Tools:**   Breakpoints
                        Mnemonic Display
                        Real-Time Trace
                        In-Line Assembly
                        Disassembly
                        Register Display/Modifications

DEBUGGING TOOLS

The debugging tools listed above are used in troubleshooting programs. Single-stepping and setting breakpoints have already been discussed in Chapter 7. The others are briefly discussed below.

Real-Time Trace   The in-circuit emulator has R/W memory used as a buffer to store the last several (such as 128) bus operations, and these can be displayed on the screen. The display is like a snapshot of all the bus operations in real time. The user can specify several requirements, such as a memory address and certain data conditions for recognizing an event, in order to trigger and display a trace. Similarly, a trace can be observed between two breakpoints or at a specified delay after a certain event. The real-time trace is a valuable tool in debugging microprocessor-based products.

In-Line Assembly   This allows the user to change data or instructions while the software is in the in-circuit emulator.

Disassembly   After instructions are changed in the in-circuit emulator, this facility can write mnemonics in software.

Register Display   This displays the register contents after the execution of instructions.

17.72   Logic State Analyzer

The logic state analyzer, also known as the logic analyzer, is a multitrace digital oscilloscope especially designed to use with microprocessor-related products. In a multitrace scope, the timing relationships of several signals can be observed with respect to some triggering event or events. For example, a four-trace scope can show the timing relationships of four signals. In a microprocessor-related product, the user is interested in observing digital signals on the address bus, the data bus, the control bus and, possibly, an external instrument relative to a specified triggering event or events. Furthermore, data display should be in a conveniently readable format, such as Hex or binary. The logic analyzer performs these functions.

A typical logic analyzer designed primarily to work with the microprocessor has a 40-pin probe plus an auxiliary probe to gather external information. It includes Read-Only memory (ROM) to store instructions related to the analyzer, R/W buffer memory to store data from a product under test, a microprocessor to monitor data gathering, and a keyboard to specify operations and enter data in Hex or octal format. The analyzer can be triggered to gather information at a specified event related to the microprocessor in the product under test or in relation to an external word. The analyzer in a trace mode takes a snapshot of real-time information at a specific trigger, stores it in its buffer memory, and displays it on its CRT.

The in-circuit emulator is a valuable tool during the initial stages of product development and, in later stages, the logic analyzer can perform some of the troubleshooting functions.

SUMMARY

In this chapter, various techniques of interfacing the scanned display and the matrix keyboard were illustrated, and the trade-offs between hardware and software were discussed. Then we used the scanned display and keyboard illustrations in designing a single-board microcomputer. In addition, debugging tools such as the in-circuit emulator and the logic analyzer were introduced.

The design of the single-board microcomputer integrates all the concepts of the microprocessor architecture, software, and interfacing discussed throughout this text. In this chapter, we discussed the necessary steps in designing hardware and software. The necessary software modules were illustrated with flowcharts; however, the coding of these modules is given as assignments.

QUESTIONS, PROBLEMS, AND PROGRAMMING ASSIGNMENTS

1. Draw a schematic to interface a 16-key matrix keyboard using port C of the 8255A. Write instructions to initialize the port.
2. Draw a schematic to interface a 30-key matrix board and a six-LED scanned display using the 8255A. Combine the matrix columns and the digit-driver lines, and explain why it is possible.
3. In a key monitor program, register E is used to save 4-bit codes of two data keys. Write a subroutine to insert a new 4-bit key code that is available in the accumulator; the new code must be inserted as a low-order nibble, and the most significant nibble in register E must be discarded.
4. Write instructions to unpack the data keys in Problem 3, and place the codes in two different memory locations of the output buffer.
5. In a monitor program, register BC is used to save a 16-bit memory address. Write instructions to insert a 4-bit code of a new key in the BC register as a least significant nibble.
6. In Problem 5, unpack all the codes, and store them in four memory locations in the output buffer.
7. In Section 17.2, modify the matrix keyboard routine to accommodate 30 keys (six rows and five columns).
8. Modify the program in Section 17.11 to display an error message as Err and blanks.
9. Write instructions for the EXEC module, assuming the memory address where execution should begin is in register DE.
10. Write a subroutine to transfer a 16-bit address from register DE and a data byte from register C into the display buffer (3FFAH to 3FFFH); the least significant nibble of the memory address should be placed in location 3FFAH, and the least significant nibble of the data byte in location 3FFEH.
11. Write a Display subroutine that takes the unpacked memory address and the byte from the buffer, looks up the seven-segment code, sends the code to the segment driver, and scans the digit code in a sequence to display the address and the byte.

Extending 8-Bit Microprocessor Concepts to Higher-Level Processors and Microcontrollers

The microprocessor has had an impact on industries as diversified as machine tools, chemical processes, medical instrumentation, and sophisticated guidance control. Some applications require simple timing and bit set/reset functions; others require high-speed data processing capability. Therefore, a number of different microprocessor families are being designed to meet these diversified requirements. Microprocessors range from single-chip microcontrollers (microcomputers) to general-purpose 32- and 64-bit microprocessors. In the present state of microprocessor technology, 4-bit microprocessors are used in high-volume products for simple functions. At the other extreme, 32- and 64-bit microprocessors are commercially available and have begun to invade the territories of traditional mainframe and minicomputers. This chapter will examine recent trends in this fast-changing technology and their implications for industry.

The microprocessor topics in this book have been discussed in the context of the widely used 8085 microprocessor. However, various other 8-bit microprocessors are available, with varying degrees of capability. In addition to general-purpose 8-bit microprocessors, microprocessor technology is evolving in three different directions. One direction is to design a complete microcomputer on a single

chip, called a microcontroller, geared toward dedicated applications. The second is to integrate peripheral devices, such as serial I/O and DMA controllers, with the microprocessor chip. And the third direction is to design 32-bit and 64-bit microprocessors with general-purpose capability similar to that of mini- and mainframe computers.

The high-end (32- and 64-bit) general-purpose microprocessors are divided into two distinct design philosophies known as CISC (complex instruction set computing) and RISC (reduced instruction set computing). High-performance workstations are generally designed with RISC processors. The focus of this chapter is to examine similarities and differences between 8-bit processors such as the 8085 and larger processors such as the 8088 to Pentium-type processors. We will focus on: (1) how concepts learned in the context of the 8085 are applicable to larger processors; (2) limitations of 8-bit processors; and (3) new concepts underlying larger processors.

OBJECTIVES

☐ List important characteristics of contemporary 8-bit microprocessors to the 8085, such as the Z80, the MC6800, and the Hitachi HD64180.

☐ Review the concepts of 8-bit microprocessors.

☐ Describe important features of Intel 8088/86 16-bit microprocessors and explain the concepts of memory segmentation, parallel processing, queuing, coprocessing, and memory and I/O interfacing.

☐ Compare the features of the Intel 8088/86 and the Motorola MC68000 16-bit microprocessors.

☐ Explain the design features of 32-bit microprocessors, and describe the characteristics of the Intel 80386, 80486, Pentium, and Pro-Pentium processors.

☐ List important features of RISC processors.

☐ List the elements of a single-chip microcontroller, and compare the characteristics of representative microcontrollers such as the Intel MCS®-51 and the Motorola MC68HC11.

18.1 CONTEMPORARY 8-BIT MICROPROCESSORS TO THE 8085

The Intel 8008, which was later superseded by the Intel 8080A, was the first 8-bit microprocessor. Just about the same time, Motorola brought out the MC6800 as an improvement over the first 8008, but with substantially different architecture. Within a few years, Zilog designed the Z80 and Intel came up with the 8085 as an improvement over the 8080A. Both are upward machine-language compatible with the 8080A. Later, National Semiconductor introduced an 8-bit microprocessor—the NSC800—combining the features of the 8085 and the Z80. In recent years, the trend seems to be toward integrated devices that reduce the chip count; these devices are known as integrated super 8-bit MPUs. One example of such devices is the Hitachi HD64180. The 8085 has been discussed throughout this text; other 8-bit microprocessors mentioned above will be discussed in the next few sections.

18.11 The Z80

The Z80 microprocessor is manufactured by Zilog, using N-channel MOS technology. It is upward software-compatible with the 8080A. Its instruction set has 158 basic instructions, which include the 8080A instruction set. However, the Zilog mnemonics are different from the Intel mnemonics, even though the machine codes are identical for the 8080A set. Furthermore, the Z80 instruction set does not include two serial I/O instructions (RIM and SIM) of the 8085. Figure 18.1 shows the Z80 signals and its internal registers. The Z80 is not pin-compatible with the 8080A or the 8085.

The Z80 microprocessor requires one +5 V power supply and the clock frequency of recent versions ranges from 4 MHz to 20 MHz. The Z80 chip has 16 address lines to address 64K memory and eight data lines. No lines are multiplexed.

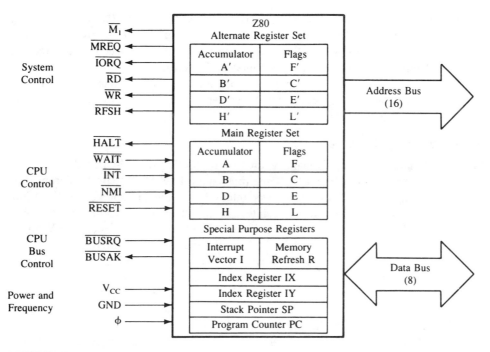

FIGURE 18.1

Z80 Microprocessor: Signals and Internal Registers

SOURCE: Adapted from Zilog, *Intelligent Peripheral Controllers* (Campbell, CA; Author, 1991).

The Z80 has two interrupt lines: one is compatible with the 8080A interrupt line, and the second is a nonmaskable interrupt (NMI). An additional significant feature of the Z80 is its on-board logic (RFSH) to refresh dynamic memories. This allows the user to use dynamic memories in a system without having to build an additional refresh circuit. Dynamic memory chips, in general, are much less expensive than static memory chips.

The internal architecture of the Z80 (Figure 18.1) includes all the 8085 registers: A, B, C, D, E, H, L, the flag register, the program counter, and the stack pointer. In addition, it has the entire set of 8-bit alternate registers, shown as A′ through L′. However, access to the alternate set of registers is only through an instruction called Exchange (EXX). This instruction exchanges the contents of registers A to L with the alternate registers A′ to L′. In essence, the alternate set of registers is used as temporary storage. Figure 18.1 shows two 16-bit index registers (IX and IY), one 8-bit Interrupt Vector Register (I), and one 7-bit Memory Refresh Register (R). The two index registers in the Z80 allow various types of memory addressing modes—a significant improvement over the 8085, in which memory addressing is restricted primarily to the HL register.

The instruction set of the Z80 is the most powerful set among the 8-bit micro-processors. It includes instructions to transfer data from one block of memory to another (LDIR = Load, Increment, and Repeat) and to search the entire memory for an 8-bit char-acter (CPIR = Compare, Increment, and Repeat). Some of its Jump instructions perform more than one function, such as Decrement B and Jump if Nonzero (DJNZ). The group of instructions called "Bit Manipulation" can test, set, or reset a bit in any register or a memory location. In addition, the Z80 has an extensive set of I/O instructions that include block input/output instructions and various modes of interrupts.

The Z80 microprocessor is supported by peripheral devices such as the parallel I/O (PIO), the clock timer circuit (CTC), the Direct Memory Access controller (DMA), and the serial I/O (SIO and DART).

18.12   The MC6800

The MC6800 microprocessor is manufactured by Motorola using N-channel MOS technology. It was developed at about the same time as the Intel 8080A. Both microprocessors were developed as improvements over Intel's first 8-bit microprocessor, the 8008. However, the MC6800 has a different architecture than its competitor, the Intel 8080A.

The MC6800 has 16 address lines and eight data lines. It requires one +5 V power supply—a significant improvement over the 8080A—and runs on a 1 MHz standard clock signal with two phases: Φ_1 and Φ_2. However, the chip does not include a clock logic. It has two interrupt lines: a regular interrupt and a nonmaskable interrupt.

The internal architecture of the MC6800 includes two 8-bit accumulators and one index register. The other three registers—the program counter, the stack pointer, and the status register—are similar to the registers in the 8085.

The MC6800 instruction set includes 72 basic instructions and makes extensive use of memory referencing. The set does not include typical direct I/O instructions (IN/OUT); it has only memory-mapped I/O. It has simple timing and control signals; the clock period is the same as the machine cycle. In general, the MC6800 is a much simpler microprocessor than the 8080A. It is currently being replaced by an 8-bit version (68008) of the MC6800 family of microprocessors discussed later.

18.13   Hitachi HD64180

This is a Z80 type upward-compatible 8-bit high-integration CMOS microprocessor in a 64-pin package, designed for applications with low power consumption, and it can operate with a 6 MHz clock. It includes a clock generator, an interrupt controller, and a memory management unit (MMU) as support devices for the microprocessor. It has 19 address lines that can address 512K bytes of physical memory, and the MMU translates internal 64K logical addressing into appropriate physical addressing. The interrupt controller is capable of handling four external and eight internal interrupting sources.

The HD64180 includes four I/O related devices: DMA controller (DMAC-two channels), asynchronous serial communication interface (ASCI-two channels), clocked serial I/O port (CSI/O), and programmable reload timer (PRT-two channels). The DMAC has two channels that support high-speed data transfer of 64K bytes per channel anywhere in the physical space of 512K bytes of memory. The ASCI has two separate channels for full duplex communication, and the CSI/O provides a half-duplex communication; it is used primarily for simple high-speed connection between microcomputers. Similarly, the timer has two channels with 16-bit counters, and one of the channels can be used for waveform generation.

The instruction set of the HD64180 is upward-compatible with the Z80 instruction set. The HD64180 has seven additional instructions, including 8-bit Multiply and Sleep. The Sleep instruction reduces the power consumption to 19 mW. One of the powerful features of this device is that the Opcode Fetch cycle of an instruction consists of three T-states versus four T-states in the Z80, resulting in faster program execution.

This device is in many ways similar to Zilog's Z180 and Z280 processors. Examination of this device indicates that the Z80-type microprocessors reasserted their presence in the form of integrated devices in industrial applications.

18.14 Review of 8-Bit Microprocessors

The architectures of the Z80 and the 8085 are register-oriented. The Z80 has a larger and more powerful instruction set than the 8085, and it is software-compatible with the 8085, except for serial I/O instructions. In contrast to the register-oriented Z80 and 8085, the 6800 is memory-reference oriented. It includes fewer registers in its architecture than the 8085.

Now the question is: Are 8-bit processors obsolete? To find an answer to this question, we must look at the worldwide sales volume of various processors. The number of 8-bit units (including microcontrollers discussed later) is much larger than that of 16- and 32-bit processors. The 8-bit processor is being used in a variety of applications such as appliances, automobiles, and industrial process and control applications. The Z80 type 8-bit microprocessors have reasserted their presence in the form of integrated devices. Similarly, the 8085 processor is being used in graphic calculators. For many industrial applications, the 8-bit microprocessor is too powerful in terms of its capability, and it is rarely used to its full capacity. On the other hand, 16-bit processors (discussed in Section 18.3) are squeezed out by 32- and 64-bit processors in PC systems; they are almost obsolete. Therefore, it appears that in an educational setting, 8-bit processors are well suited to teach basic microprocessor concepts.

REVIEW OF MICROPROCESSOR CONCEPTS 18.2

In Chapter 1, we defined the microprocessor as a clock-driven, register-based, programmable digital electronic device that reads binary instructions from memory, accepts data from input devices (or reads stored data from memory), processes the data according to the instructions, and displays the results at output devices or stores them in memory.

18.21 Microprocessor Architecture

To understand the architecture of a microprocessor, we studied the following concepts in the context of the 8085 microprocessor:

1. address bus, data bus, control and status signals
2. execution of an instruction with reference to a clock signal and contents of various buses at different times

3. interfacing of memory and I/O devices that includes address decoding and timing of various signals
4. processes of accepting external signals (requests) such as interrupt, hold, and ready and responding to those requests

The 8085 microprocessor signals are divided into six categories: (1) address bus, (2) data bus, (3) status and control signals, (4) external requests, (5) response to external requests, and (6) power and clock. These are typical signals any processor must have irrespective of its size. However, the 8085 is housed in a 40-pin package that limits the number of signals; therefore, in this processor, the data bus is multiplexed with the low-order address bus that must be demultiplexed. On the other hand, the Z80 and the MC6800 do not have any multiplexed signals. However, the concept of multiplexing signal lines is important; it is commonly used in 40-pin devices such as microcontrollers.

18.22 Programming Registers

The 8085 processor has six general-purpose registers, one accumulator, one flag register, and two memory address registers or pointers (SP and PC). The 8085 lacks memory pointers such as index registers; it relies heavily on the HL register as a memory pointer. The Z80 removed that deficiency by adding two indexed registers (IX and IY). The architectures of the Z80 and the 8085 are register oriented. In contrast to the register-oriented Z80 and 8085, the MC6800 is memory-reference oriented. It has two accumulators and no general-purpose registers; it uses memory registers as its general-purpose registers.

The 8085 instruction set is divided into five groups: (1) data copy (transfer), (2) arithmetic, (3) logic and bit manipulation, (4) program transfer (jump, call), and (5) machine (processor) control. The set does not include instructions such as multiply or divide and instructions that can handle multiple operations such as copying a block of data. However, these operations can be performed by writing a set of instructions.

The architecture of Intel 16- or 32-bit processors, discussed in the following section, is in many ways similar to the architecture of the 8085. They include larger registers and buses. The instruction sets of these processors include additional instructions that make programming tasks easier, such as multiply, divide, and string manipulation. These larger processors are designed to function in a multiuser and multiprocessor environment. Therefore, these processors need additional signals and instructions. However, the concepts of the 8085 microprocessor are applicable and can be easily extended to these larger processors.

18.3   16-BIT MICROPROCESSORS

The 16-bit microprocessor families are used primarily in microcomputers and are oriented toward high-level languages. Their applications sometimes overlap those of the 8-bit microprocessor. They have powerful instruction sets and are capable of addressing megabytes of memory. Typical examples of 16-bit microprocessors include the Intel

8086/8088 and 80186/286, Zilog Z8001/8002, Digital Equipment LSI-11, Motorola 68000, and National Semiconductor NS16000.

Apart from design concepts and instruction sets, one of the critical factors that decides the capability of the microprocessor is the number of pins available. In the 1970s, one trend was to stay within the 40-pin package size and take advantage of the existing production and testing facilities. The 40-pin package either limits the size of the memory that can be addressed or necessitates multiplexing of several functions. Intel (8086), Zilog (Z8002), and Digital Equipment (LSI-11) have stayed with the 40-pin package. Another option was to go beyond the 40-pin limit. National Semiconductor (NS16000) and Zilog (Z8001) chose the 48-pin package. Motorola selected the 64-pin package for the MC68000 and Intel chose the 68-pin package for its 80286 microprocessor. The primary objectives of these 16-bit microprocessors can be summarized as follows:

1. Increase memory addressing capacity.
2. Increase execution speed.
3. Provide a powerful instruction set.
4. Facilitate programming in high-level languages.
5. Function in a multiprocessor environment.

These objectives can be met by using various design concepts. To illustrate differences in design philosophies, the following microprocessors are described briefly: the Intel 8086/8088, the Intel 80186 and 80286, and the Motorola MC68000.

18.31 Intel 8086/8088

The Intel iAPX 8086/8088 is a 16-bit microprocessor housed in a 40-pin package and capable of addressing one megabyte of memory. Various versions of this chip can operate with clock frequencies from 5 MHz to 10 MHz.

The 8088 is functionally similar to the 8086, except that it has an 8-bit external data bus. Its internal architecture and instruction set are identical with those of the 8086. The only difference is that a 16-bit data word must be transferred in two segments in the 8088. The 8088 can be viewed as an 8-bit microprocessor with the execution power of a 16-bit microprocessor. The next few paragraphs describe the features of the 8086/88 architecture that meet the objectives described above.

8086/8088* ARCHITECTURE

Figure 18.2 shows logical signals of the 8088 and 8086 microprocessors. These processors have identical internal architecture; they differ only in external signals. The 8086 has a 16-bit external data bus, and the 8088 has an 8-bit external data bus. The signals shown in Figure 18.2 are classified in seven categories. Categories 1 through 6—(1) address bus, (2) data bus, (3) control and status signals, (4) external requests, (5) response to external requests, and (6) power and clock—are identical to that of the 8085 processor.

---

*The terms 8086 and 8088 are used synonymously; the discussion is applicable to both processors unless specifically mentioned otherwise.

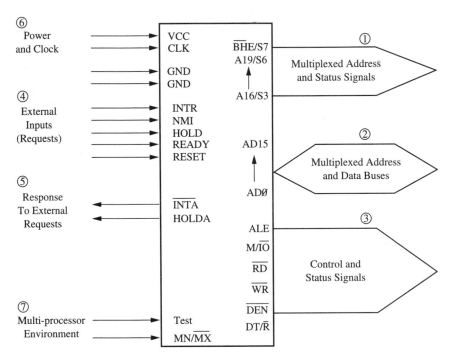

FIGURE 18.2
8086 Microprocessor—Logical Signals in minimum mode.

The seventh category, signals for multiprocessor environment, is new. It includes two signals: MN/$\overline{\text{MX}}$ (minimum or maximum mode) and Test. The data bus and status signals are multiplexed with the address bus. In the 8088 processor, eight data lines, AD7–AD0, and in the 8086 processor 16 data lines, AD15–AD0, are multiplexed. Signals that are different from the 8085 signals are explained as follows.

☐ **MN/$\overline{\text{MX}}$—Minimum/Maximum Mode:** This signal represents two operation modes of the processor: minimum and maximum. When the signal is high (connected to +5 V), the processor operates in the minimum mode, and when it is low (grounded), the processor operates in the maximum mode. The minimum mode is used for the single processor environment, and the maximum mode is used for the multiprocessor environment such as having a coprocessor in a system. In the maximum mode, eight pins are assigned different functions compared to that of the minimum mode, as shown in Table 18.1, and a bus controller (such as the 8288) is necessary to generate control signals.

TABLE 18.1
8086/8088 Minimum and Maximum Mode Control Signals

Pin	Minimum Mode Function	Maximum Mode Function
24	INTA: Interrupt Acknowledge	QS1: Queue Status Signal
25	ALE: Address Latch Enable	QS0: Queue Status Signal
26	DEN: Data Enable	S0: Input signals to 8288 to generate control signals
27	DT/R̄: Data Transmit/Recieve	S1
28	M/ĪŌ (8086): Memory or Input/Output IO/M̄ (8088): Input/Output or Memory	S2
29	WR: Write	LOCK: Lock—to prevent another processor from gaining control
30	HLDA: Hold Acknowledge	RQ/GT1: Request/Grant—enable another processor to gain control
31	Hold: Hold	RQ/GT0

☐ **TEST**—This signal is used to synchronize operations of multiple processors in a system. When the WAIT instruction is being executed, the processor checks this signal. If it is high, the processor interrupts the execution of the program, and if it is low, it continues the execution.

☐ **DEN—Data Enable:** This is an active low output signal that is generally connected to a bidirectional buffer (such as 74LS245) to isolate the MPU from the system bus.

☐ **DT/R—Data Transmit/Receive:** This is also connected to a bidirectional buffer to enable data flow.

☐ **M/IO and IO/M—Memory and I/O:** This signal indicates whether the processor cycle is an I/O operation or a memory operation. The active levels indicated by the symbols are exactly opposite of each other in the 8086 (M/ĪŌ) and 8088 (IO/M̄) processors.

☐ **BHE—Bus High Enable:** This is an active low signal used only in the 8086 processor to enable the high-order byte of 16-bit data. In the 8088, it is used as a status line (SSO) to decode the current cycle.

GENERATING MPU* SIGNALS

In the 8088/86 processor, the data bus and the status signals are multiplexed, like the data bus in the 8085. These signals must be demultiplexed to form a complete functional MPU that can be used in a system. Figure 18.3 shows a complete schematic of the demultiplexed bus of the 8086 processor in the minimum mode. In the 8088 processor, the address lines A15–A8 are not multiplexed. Therefore, the 8088 processor needs only two latches.

---

*The term *MPU* indicates that all signals are demultiplexed and appropriate control signals are available.

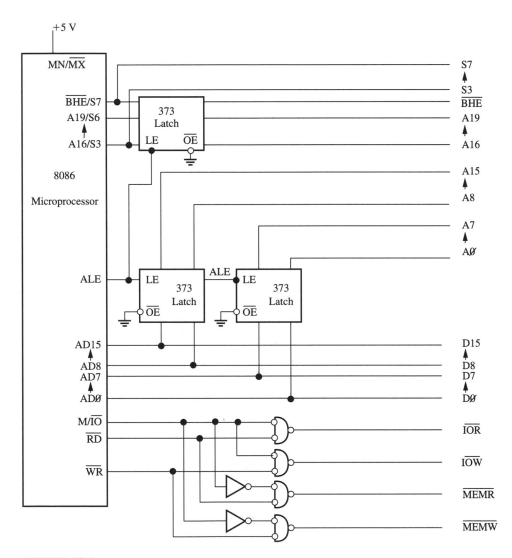

FIGURE 18.3
The 8086 MPU with demultiplexed address bus and control signals in the minimum mode.

The processor asserts the ALE signal high at the beginning of each machine cycle when it places the address on the address bus. This ALE signal is used to enable three latches ('373s) to separate the address lines from the multiplexed bus, as shown in Figure 18.3. The lower two latches in Figure 18.3 generate the address bus A15–A0, and the top latch separates status signals from the address signals and generates the address lines A19–A16 and the BHE signal.

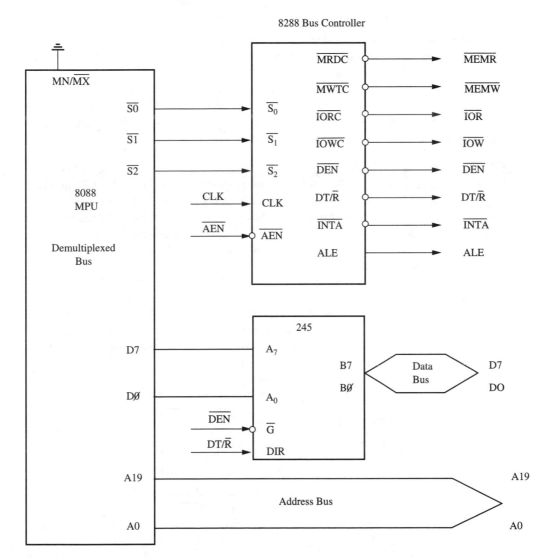

FIGURE 18.4
8088 MPU in the max mode and control signal generation.

Figure 18.4 shows the 8088 MPU (with demultiplexed address bus A19–A0) in the maximum mode. In this mode, eight signals are assigned the function of status signals that are used as inputs to the 8288 bus controller to replace the signals that are lost. The 8288 generates the necessary control signals as shown in Figure 18.4. In addition, Figure 18.4 also shows how DT/R̄ and D̄ĒN̄ are used to buffer the data bus to avoid excessive loading of the bus.

INTERNAL ARCHITECTURE

When the 8085 microprocessor fetches and executes an instruction, the buses are occupied during the fetch operation, but they are idle during the internal execution of an instruction. To speed up the execution, the 8086 processor includes two processing units called Execution Unit (EU) and Bus Interface Unit (BIU), shown in Figure 18.5. The concept of dividing work between two units and processing it simultaneously speeds up the execution. The BIU fetches the instructions and places them in a queue, and the EU continuously executes them until the queue is empty.

PROGRAMMING MODEL

The 8086 processor includes four 16-bit general purpose registers: AX, BX, CX, and DX. They are equivalent to four accumulators even though some registers are assigned additional special functions. Each register can also be used as two 8-bit registers, and their compatibility with the 8085 registers is shown in the gray shaded area in Figure 18.6. For example, the HL register pair in the 8085 is the same as the BX register in the 8086, and the BX register can be used as two 8-bit registers BH (B-High) and BL (B-Low).

The next four registers, SP (Stack Pointer), BP (Base Pointer), SI (Source Index), and DI (Destination Index), are 16-bit registers used as memory pointers. BP, SI, and DI are new additions compared to the 8085 registers, and these registers eliminate one of the drawbacks of not having index registers in the 8085 microprocessor. The next group of registers is called segment registers: CS (Code Segment), DS (Data Segment), SS (Stack

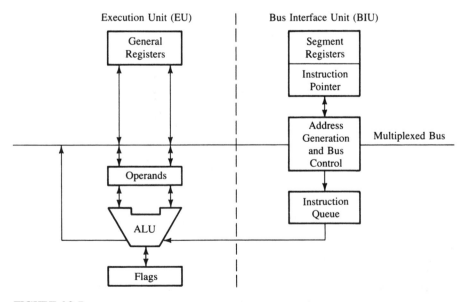

FIGURE 18.5

Execution and Bus Interface Units (EU and BIU) of the 8086

SOURCE: Intel Corporation, *iAPX 86, 88 User's Manual* (Santa Clara, Calif.: Author, 1981), p. 2–5.

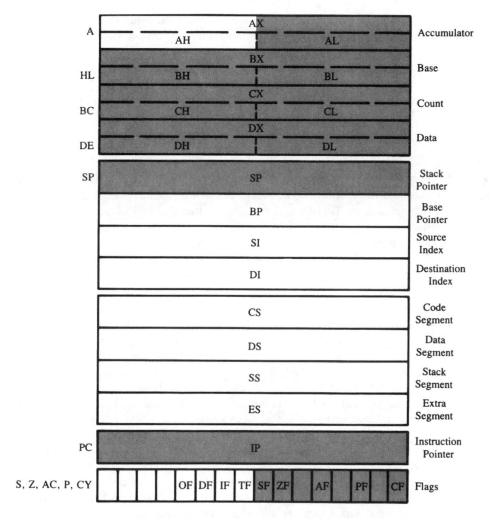

FIGURE 18.6

The 8086 Programming Registers (the Shaded Portions Show Areas Equivalent to the 8085)

SOURCE: Adapted from Intel Corporation, *Microprocessors Vol. II* (Santa Clara, Calif.: Author, 1993).

Segment), and ES (Extra Segment). The segment registers are combined with memory pointers to generate a memory address as explained in the next section. The last 16-bit register shown in Figure 18.6 is called IP (Instruction Pointer), which is the same as PC (Program Counter) in the 8085.

The flag register includes nine flags, four are new additions. These flags are divided into two groups: six data flags and three control flags. In the data flag category, the OF (Overflow) is new; the remaining are the same as in the 8085 processor. All control flags (DF, IF, and TF) are new additions. The new flags are defined as follows.

☐ **OF—Overflow:** This is used in signed numbers. When the result of a signed number operation is too large, causing the most significant bit to overflow into the sign bit, this flag is set.

☐ **DF—Direction Flag:** This flag is used to control the direction (increment/decrement) of string operations.

☐ **IF—Interrupt Flag:** This flag is used to enable or disable external maskable interrupt requests.

☐ **TF—Trap Flag:** This flag is used for single-stepping instructions.

MEMORY SEGMENTATION

The 8086 processor has a 20-bit address bus; however, its Instruction Pointer is 16-bit wide and can hold a 16-bit address. The processor must be able to place a 20-bit address to access 1M-byte memory. Therefore, the concept of memory segmentation is employed in the design of this processor as explained below.

The processor includes four segment registers that are used to assign memory base addresses: CS for instruction code, DS for data, SS for stack, and ES for additional data. The segment registers are combined with memory pointers to form a 20-bit memory address. The default combination is shown in Table 18.2; however, the instructions can override these default combinations and specify appropriate combinations. If the size of a program (including data and stack) is less than 64K, all segment registers can be defined at the same base address. If a program is large, all segments can be separate from each other or they can overlap. The memory addresses in the 8086 systems are specified in three formats: physical, logical, and offset. The physical address is a 20-bit address, the offset address is the address of an instruction (data) in reference to the base address in the segment register, and the logical address is the combination of a segment and an offset address. These definitions are clarified in the following example.

Example 18.1

Assume that all segment registers are initialized at 1200H. The instruction register IP holds the address 0000H, which is the beginning of the instruction code (program). The stack is initialized at FFFFH, and the SI register is initialized at 8000H to indicate where data bytes are stored. Calculate the beginning physical addresses of the instruction code, data, and stack. Explain what the physical address, logic address, and offset address are.

TABLE 18.2
Segment Registers and Associated Offset Registers (Memory Pointers)

Segment Registers	Offset Registers
Code Segments (CS)	Instruction Pointer (IP)
Stack Segment (SS)	Stack Pointer (SP) and Base Pointer (BP)
Data Segment (DS)	Source Index (SI), Destination Index (DI), and BX Register
Extra Segment (ES)	Source Index (SI), Destination Index (DI), and BX Register

The processor generates the beginning address of the instruction code by combining the addresses in CS and IP. However, it is not an addition. The processor shifts the address in the code segment by four bits (one Hex digit) to the left and adds the address in IP. Similarly, it combines DS and SI, and SS and SP.

Solution

Physical Addresses (PA)

Code Address	Stack Address	Data Address
(CS): 1 2 0 0	(SS): 1 2 0 0	(DS): 1 2 0 0
(IP) : + 0 0 0 0	+ F F F F	+ 8 0 0 0
(PA): 1 2 0 0 0	2 1 F F F	1 A 0 0 0

The physical address is a 20-bit address that the processor places on the address bus. In the 8086, it ranges from 00000 to FFFFFH. The logical addresses are: Code—1200:0000; Stack—1200:FFFF; and Data—1200:8000. The offset addresses are: Code—0000H; Stack—FFFFH; and Data—8000H.

INSTRUCTION SET

The 8086 has a large instruction set, consisting of 135 basic instructions, which can operate on individual bits, bytes, 16-bit words and 32-bit double words, signed numbers, ASCII characters, and BCD numbers. The instruction set can be divided into six categories: (1) data copy (MOV), (2) arithmetic, (3) logic, (4) program transfer (such as jump and call), (5) string manipulation, and (6) machine or processor control (such as halt and interrupt). We are already familiar with five categories out of these six in the 8085 microprocessor; the only new category is string manipulation. The set includes instructions such as multiply, divide, jumps based on multiple flags, and string manipulation. If one is familiar with the 8085 assembly language and basic programming concepts, the transition to the 8086 assembly language is not difficult and can be self-directed.

In addition to the powerful instruction set, the chip design is oriented toward modular programming, very desirable for high-level languages. The memory-segmentation concept facilitates programming of independent modules that can communicate with each other as well as share common data.

COPROCESSING

In addition to the 8086, Intel has designed a series of special-function devices such as the 8089 (1/0 Processor) and the 8087 (Numeric Processor). These processors are compatible with the 8086 in a master-slave relationship. They are designed with additional instructions and can be assigned dedicated functions to increase the overall execution speed of large systems.

MEMORY INTERFACING

The concepts of memory interfacing in the 8088 are identical with the 8085. It requires the address decoding of the high-order address bus. For example, to interface a 16K memory chip, we need 14 address lines for the chip, and the remaining six lines must be decoded by using a decoder. However, memory interfacing in the 8086 requires a different technique. The 8086 has a 16-bit data bus, but memory has 8 data lines. This raises a question: How do we transfer a 16-bit word over 8 data lines? The answer lies in connecting two memory chips in parallel, called high or odd and low or even bank, and accessing them by using the BHE (Bus High Enable) signal.

Figure 18.7 shows an illustration of interfacing two 32K-byte memory chips (total of 64K bytes) with the 8086. We need 16 address lines (A15–A0) from the processor to address 64K bytes of memory ($2^{16} = 64K$). The 32K memory chips have 15 address lines, A14–A0. Normally, we connect the processor address lines A14–A0 to the respective address lines of the memory chip. However, Figure 18.7 shows that the processor address lines A15–A1 are connected to lines A14–A0 of the memory chips. A0 of the processor is used to enable the 3-to-8 decoder of the low bank, and the BHE signal is used to enable the high bank. When A0 is at logic 0, the low bank is enabled, and the processor accesses the low-order data byte (D7–D0), and when BHE is low, the high bank is enabled, and the processor accesses the high-order data byte (D15–D8). When both are low, the processor reads (or writes) the entire data bus (D15–D0). The memory addresses in Figure 18.7 range from 40000H to 4FFFFH.

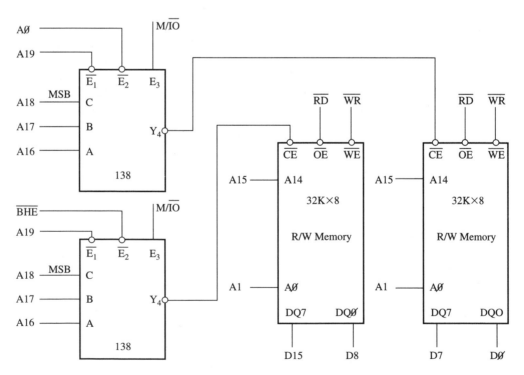

FIGURE 18.7
Interfacing byte-wide memory chips with the 16-bit 8086 processor.

I/O INTERFACING

The 8085 microprocessor has an 8-bit I/O space, separate from memory; thus the I/O addresses can range from 00 to FFH. The IN and OUT instructions are designed to use this I/O space. On the other hand, the 8088/86 processors can have 8-bit or 16-bit I/O space; thus the I/O addresses can range from 00 to FFH or from 0000 to FFFFH. The IN and OUT instructions are designed to use either 8-bit I/O space or 16-bit I/O space. Figure 18.8 shows the interfacing of the 8255 (Programmable Peripheral Interface) and the 8254 (Timer) in a personal computer (PC); it uses the 16-bit I/O space. However, the address lines A15–A10 and A4–A2 are not decoded, leaving them as don't care lines. Thus the port addresses of the timer range from 0040H to 0043H and of the 8255 from 0060H to 0063H (assuming all don't care lines at logic 0). The absolute decoding practice is generally not used for I/O interfacing in PCs to reduce hardware cost.

18.32 Intel 80186 and 80286

The Intel 80186 and 80286 are 16-bit microprocessors, extended versions of the 8086. One of the critical barriers of Intel's earlier microprocessors was the 40-pin package. Once that barrier was broken, it became easier to address large memory. These microprocessors are housed in 68-pin packages and use the concepts of prefetched pipeline structure, parallel processing, and memory management.

The 80186 is an improved version of the 8086, available in two speeds: an 8 MHz and a 10 MHz version. It is an integrated device designed to reduce the chip count rather than to increase the memory-addressing capacity. It has multiplexed address and data buses, and the additional lines of the bigger package are used to include devices such as a clock generator, interrupt controller, timers, DMA controller, and a chip select unit.

The 80286 is also a 16-bit microprocessor, an improved version of the 8086 but with a different architectural philosophy. It eliminates the multiplexing of the buses; it has

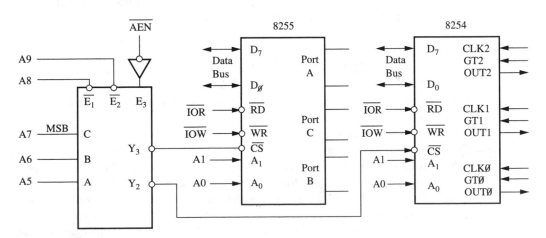

FIGURE 18.8
Interfacing I/O devices with the 8088 processor—schematic from PC.

a linear address bus with 24 address lines that can address 16M bytes of memory directly. It can also support a memory management unit, and through the memory management unit it can address 1G bytes of memory, also known as *virtual memory*. The processor includes various built-in mechanisms that can protect system software from the user programs, protect users' programs, and restrict access to some regions of memory. The 80286 is designed for multiuser systems; its architectural philosophy is closer to Intel's 80386 32-bit microprocessor, which is described in Section 18.4.

18.33   Motorola MC68000

The MC68000 is a family of microprocessors that includes 8-, 16-, and 32-bit versions with varying memory addressing capability. The 68000 (also 68010) is a 16-bit microprocessor with a 32-bit internal architecture housed in a 64-pin package. It is capable of addressing 16 megabytes of memory, and the clock frequency ranges from 4 MHz to 10 MHz for different versions of the chip. The internal architecture and registers are almost common to all versions, thus making them software compatible. The 60008 is an 8-bit microprocessor with two versions: one is capable of addressing 1 megabyte of memory and the other 4 megabytes of memory. The 68020, 68030, and 68040 have a 32-bit data bus and a 32-bit address bus with the memory addressing capability of 4 gigabytes.

Figure 18.9 shows the internal architecture of the device. It includes seventeen 32-bit, general-purpose registers, a 32-bit program counter, and a 16-bit status register. The general-purpose registers are divided into three groups: eight data registers, seven address registers, and two stack pointers. The contents of the data registers can be accessed as bytes, 16-bit words, or 32-bit words, and the contents of the address registers can be accessed as 16-bit or 32-bit addresses. The 68000 can operate in two different modes: the user mode and the supervisor mode. The supervisor mode is designed primarily for operating systems; in this mode, some privileged system control instructions can be used. Some of its other features can be described as follows.

NONSEGMENTED MEMORY

To increase the memory-addressing capacity, Motorola increased the number of pins in its package. The chip is designed with 23 separate lines to address eight megawords (16 megabytes). Similarly, its program counter is 32 bits long; only the low-order 24 bits are necessary to address the entire memory map.

INSTRUCTION SET

The 68000 has one of the most powerful yet simple instruction sets. It includes 56 basic instructions and can operate on five different types of data: bit, byte, BCD, 16-bit word, and 32-bit word. It has only memory-mapped I/O but includes 14 memory-addressing modes. To cite one example of its powerful set, its MOV instruction can transfer data from any source to any destination. It includes instructions such as Multiply and Divide and special instructions to deal with numbers longer than 32 bits. Its orientation toward high-level languages comes primarily from its instruction set.

FIGURE 18.9
Programming Registers of the
68000

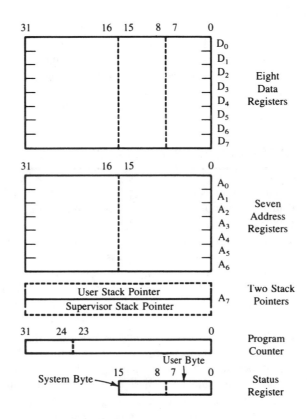

ASYNCHRONOUS AND SYNCHRONOUS CONTROL LINES

The 68000 has a special way of handling slow and fast peripherals. It has two sets of control signals, called asynchronous and synchronous signals. Communication with asynchronous peripherals is handled through the control lines called Upper Data Strobe (UDS), Lower Data Strobe (LDS), and Data Acknowledge (DACK). The DACK signal is similar to a handshake line; the bus cycle is not terminated until the signal, DACK, is received. The 68000 family offers some synchronous peripherals, and communication with these peripherals is handled through the control signals called Valid Peripherals Address (VPA), Valid Memory Address (VMA), and Enable (E).

18.34 Review of 16-Bit Microprocessors

The 8086 and the 68000 are designed with two distinct philosophies. The 8086 uses only 40 pins and multiplexes most of the functions. It has employed several new architectural concepts such as memory segmentation, parallel processing, queueing, and coprocessing. On the other hand, the 68000 has adopted a 64-pin package and simplified its architecture. However, both are oriented toward high-level languages. The 80186 is an integrated package, whereas the 80286 is essentially a faster version of the 8086 with the memory addressing capacity of 16M bytes without the multiplexed bus.

The 16-bit microprocessors are too powerful to perform the functions of general-purpose 8-bit microprocessors; therefore, they are less likely to replace 8-bit processors. In the early 1980s many microcomputers were designed based on 16-bit microprocessors. However, now they are being replaced by more powerful 32-bit and 64-bit processors.

18.4 32-BIT MICROPROCESSORS

At the high end of the microprocessor range, we have 32- and 64-bit microprocessors available; examples include such microprocessors as the Intel 80386 and 80486, Zilog Z80000, National Semiconductor NS32032, Motorola MC68020, and Digital Equipment Alpha 21064. We are interested not in discussing the details of these microprocessors, but in exploring trends in microprocessor technology. These microprocessors are not merely more of the same except bigger and faster; they offer some unique features not available in the earlier 16-bit microprocessors. The applications and the environments in which they operate are far different from those of the 8-bit microprocessor and the earlier 16-bit microprocessors. It appears that two trends are evolving: one is multiuser, multitasking, time-sharing environments, and the other is distributed processing, interconnected with networks. As soon as we move away from the single-user system, the demands on these microprocessors change drastically; the environment is more like that of minicomputers or mainframe computers. These microprocessors include many features that are normally available in maniframe computers.

In a single-user system, the user has unlimited access to all aspects of the system. The user need not be concerned with sharing the time or the resources of the system, but can schedule various tasks according to his or her convenience. The user has access to the operating system, can tamper with the system to include some personal conveniences, or in the process can lock up the system. However, the multiuser system cannot afford to provide the luxuries of unlimited access to all users. Some of the requirements of the multiuser system are as follows:

1. Higher speed of execution
2. Ability to handle different types of tasks efficiently
3. Large memory space that can be shared by multiple users
4. Appropriate memory allocations and the management of memory access
5. Data security and data access
6. Limited and selected access to part of the system
7. Resource (printer, hard disk, etc.) sharing and management

Some of these requirements must be managed by a multiuser operating system, and some should be facilitated by the architectural design of the microprocessor. The 32-bit microprocessors are designed to work in this type of environment. Some of the important features of the Intel 80386/486 are described in this section as a representative sample of 32-bit microprocessor technology.

18.41 The Intel 80386/80486

The 80386 is a 32-bit microprocessor with a nonmultiplexed 32-bit address bus and is housed in a 132-pin grid array package. It has versions that can operate from 20 MHz to 33 MHz. It is capable of addressing 4G bytes of physical memory and, through its memory management unit, can address 64 (2^{46}) terabytes of virtual memory. The processor can operate in two modes: Real Mode and Protected Mode. In Real Mode, physical address space is 1M bytes (20 address lines), which is extended to 4G bytes in Protected Mode (32 address lines). The primary difference between these modes is the availability of the memory space and the addressing scheme. The 80386 has 32-bit registers and is upward software-compatible with the 8086. The execution of instructions is highly pipelined, and the processor is designed to operate in a multiuser and multitasking environment. It has the protection mechanism necessary for this type of environment.

FUNCTIONAL SIGNAL GROUPS

Figure 18.10 shows the functional groups of the 80386 signals. The functional signal groups are in many ways similar to those of the 8085, except with larger buses: 32-bit address bus and 32-bit data bus. The 80386's arbitration and interrupt signals are similar to the external request signals, and its bus cycle definition and control signals are similar to the control signals and status signals in the 8085. However, the 80386 has a new group of signals that is used for interfacing a coprocessor. In addition, each group has some signals for specialized functions. For example, in the address bus, the signals BE0#–BE3#, called byte enable signals, are used to identify which groups of data lines are active in a transfer of 32-bit data. In the bus control signals, BS16# allows the processor to directly connect to 32-bit and 16-bit data buses.

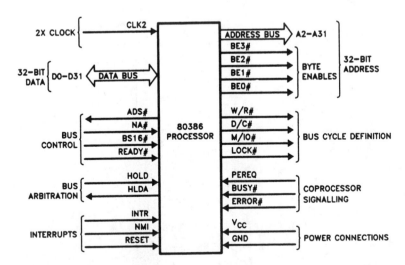

FIGURE 18.10

Functional Groups of the 80386 Signals

SOURCE: Intel Corporation, *Microprocessors Vol. II* (Santa Clara, Calif.: Author, 1993).

PROGRAMMING MODEL

Figure 18.11 shows a programming model of the 80386. It has eight general-purpose registers to hold 32-bit data or addresses; they can be used as 8-, 16-, or 32-bit registers. The 32-bit registers are named with the prefix E (EAX, etc.), and the least 16 bits (0–15) of these registers can be accessed with names such as AX or SI. Similarly, the lower eight bits (0–7) can be accessed with names such as AL and BL and the higher eight bits (8–15) with names such as AH and BH. Figure 18.11 shows six 16-bit registers called segment selector registers; these are used for determining memory addresses. The instruction pointer (EIP), known as a program counter in 8-bit microprocessors, is a 32-bit register to handle 32-bit memory addresses, and the lower 16-bit segment (IP) is used for 16-bit memory addresses.

The flag register is a 32-bit register; however, only 14 bits are being used at present for 13 different tasks; these flags are upward-compatible with those of the 8086 and 80286. The comparison of the available flags in 16-bit and 32-bit microprocessors may provide some clues related to capabilities of these processors. The 8086 has 9 flags, the 80286 has 11 flags, and the 80386 has 13 flags. All of these flag registers include six flags related to data conditions (Sign, Zero, Carry, Auxiliary Carry, Overflow, and Parity) and three flags related to machine operations (Interrupt, Single-Step, and Strings). The 80286 has two additional flags: I/O Privilege and Nested Task. The I/O Privilege uses two bits in Protected Mode to determine which I/O instructions can be used, and the Nested Task is used to show a link between two tasks. The 80386 added two more flags: VM (Virtual Mode) to run the 8086-compatible programs in this mode and RF (Resume Flag) to work with breakpoints.

GENERAL DATA AND ADDRESS REGISTERS   SEGMENT SELECTOR REGISTERS

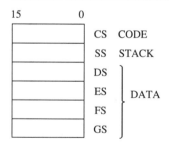

31	16	15	8	7	0	
		AH	AX	AL		EAX
		BH	BX	BL		EBX
		CH	CX	CL		ECX
		DH	DX	DL		EDX
			SI			ESI
			DI			EDI
			BP			EBP
			SP			ESP

15	0		
		CS	CODE
		SS	STACK
		DS	
		ES	DATA
		FS	
		GS	

31	16	15	0	
		IP		EIP INSTRUCTION POINTER
		FLAGS		EFLAGS FLAG REGISTER

FIGURE 18.11

The 80386 Programming Model

SOURCE: Intel Corporation, *Microprocessors Vol. II* (Santa Clara, Calif.: Author, 1993).

The processor also includes Control registers, System Address registers, and Debug and Test registers for system and debugging operations; these registers are shown in Figure 18.11.

INSTRUCTION SET AND ADDRESSING MODES

The instruction set is divided into nine categories of operations and has 11 addressing modes. In addition to commonly available instructions in an 8-bit microprocessor, the set includes operations, such as bit manipulation, string operations, high-level language support, and operating system support. An instruction may have 0 to 3 operands, and the operands can be 8, 16, or 32 bits long. The 80386 handles various types of data, such as a single bit, string of bits, signed and unsigned 8-, 16-, 32-, and 64-bit data, ASCII characters, and BCD numbers.

The high-level language support group includes instructions such as ENTER and LEAVE. The ENTER instruction is used to enter from a high-level language; it assigns memory locations on the stack for the routine being entered and manages the stack. On the other hand, the LEAVE generates a return procedure for a high-level language. The operating system support group includes several instructions such as APRL (Adjust Requested Privilege Level) and VERR/W (Verify Segment for Reading or Writing). The APRL is designed to prevent the operating system from gaining access to routines with a higher priority level, and the instructions VERR/W verify whether the specified memory address can be reached from the current privilege level.

MEMORY MANAGEMENT AND PROTECTION

The memory space available to the user is dependent on the mode of operation. In Real Mode, the maximum memory size is 1M byte; the processor uses 20 address lines. In Protected Mode, the processor has 32 lines available for addressing; thus, the physical memory size is extended to 4G bytes. This physical memory is translated into 64-terabyte virtual address space (logical addresses) by using the Memory Management Unit (MMU). In virtual memory systems, disk storage is used as memory by swapping its information between R/W memory and the disk; thus, the processor's R/W memory becomes a temporary holding area. In the 80386, the memory space can be divided into segments of various lengths or into pages of 4K bytes. The segmentation is suitable for various-sized logical addresses of programs, and the paging simplifies the swapping of information between the physical memory and the disk. In addition to managing the memory space, the MMU provides a four-level protection mechanism suitable for a multiuser system: the operating system can reside at the highest privilege level 0 and application programs at the least privilege level 3. The processor isolates tasks and prevents low-level software from having access to a higher level.

THE INTEL 80486

The 80486 is an upgraded faster version of the 80386. The DX type version is a 32-bit processor housed in a 168-pin grid array package and can operate with the clock frequencies from 25 MHz to 66 MHz. The design is based on 1.2 million transistors compared to 300 thousand transistors of the 386 processor. The important additional features of the 486 processor in comparison with the 386 processor are as follows. The 486 processor includes:

1. **Built-in math coprocessor.** In 386 systems, a math coprocessor is an external device. Therefore, the math instructions in 486 systems are executed three times faster than in 386 systems.
2. 8K-byte of code and data cache memory on the chip.
3. Highly pipelined execution unit. Therefore, the execution time for many instructions is one clock period.

In summary, the 486 is a high-speed, high-performance 32-bit microprocessor. It executes many of its instructions in one clock cycle by using highly pipelined execution units. It is designed to facilitate the execution of high-level languages and suited for multiprocessing and multiuser systems. The 486 is generally used in high-end microcomputers and network environments.

18.42   The Intel Pentium™ Processor

The Pentium processor has a 32-bit address bus and a 64-bit data bus and is designed to operate from 60 MHz to 233 MHz. It is upward software-compatible with the previous line of the Intel microprocessors from the 8088/86 to the 486, and it is also designed for easy upgradability. The chip has more than 3.1 million transistors and is housed in a 273-pin grid array package. It includes many advanced features normally available in mainframe computers. The processor is ideally suited for high-end desktop PCs, workstations, and network file servers where high-speed computation (such as 3D graphics) is needed. It can achieve performance levels equal to or better than some of today's workstations and can run an advanced operating system such as UNIX.

ADVANCED DESIGN FEATURES

The Pentium processor includes enhancement of some of the features available on the 486 processor. For example, it supports either the traditional memory page size 4K byte or the larger size of 4M byte. It has a 64-bit data bus, which increases the processing speed. Similarly, the architecture design makes the processor well suited to multiprocessing applications; it maintains data integrity in multiple processors. The processor also includes many advanced features; some of them are described as follows.

1. **Superscaler architecture.** RISC processors (described later) are generally designed with the superscaler architecture. The term *superscaler* refers to a microprocessor architecture that includes more than one execution unit. The Pentium has two execution units with dual-pipelined architecture; thus, it is able to execute two instructions simultaneously per clock cycle and achieve a high level of performance.
2. **On-chip cache memory for code and data.** The Pentium processor has two 8K byte of cache memory on the chip; one is used for code and the other is used for data. The cache memory is used for temporary storage of commonly used code and data copied from the system's (or the main) memory. For example, if the processor is executing a

loop, the codes within the loop are copied into the cache. This eliminates the need for the processor to go off the chip and access the main memory during the loop execution, thus improving the performance of the processor.

3. **Branch prediction.** This is a mainframe technique implemented in this processor to improve the performance. In this technique, the most likely set of instructions to be executed is predetermined, and the pipelines are kept full accordingly. For example, in a loop, a prediction is based on the assumption that the instructions in the loop are likely to be executed in the next interaction.

4. **High-performance floating-point unit.** This processor has an on-chip floating-point unit that incorporates highly sophisticated seven-stage pipelining and hardwired codes. This processor is capable of executing floating-point instructions five to ten times faster than is the 486 processor.

5. **Performance monitoring.** The processor design enables the user to monitor the performance of the processor and to optimize the performance by identifying potential bottlenecks in code execution. The user can observe and record time for internal events that affect the performance of data read and write, interrupts, cache hits and misses, and bus utilization. And based on these observations, the user can fine-tune programs.

18.43   The Intel Pentium Pro-Processor

The Pentium Pro-Processor has more than 4.2 million transistors and is housed in a 387-pin grid array package. It also has a 32-bit address bus and a 64-bit data bus, similar to the Pentium processor; however, there is provision for a 36-bit address bus, enabling the processor to access 64G bytes of physical memory in future. It has many versions based on its operating frequency; at present, its fastest version operates at 200 MHz.

One of the major differences of this processor compared to the Pentium processor is in the area of cache memory. The Pentium processor has two 8K memory cache on the chip known as level 1. The Pentium Pro-Processor has an additional 256K (or 512K) cache memory on the chip known as level 2 cache. This additional cache memory increases processing speed.

18.44   RISC Processors

The term *RISC* represents *reduced instruction set computing*. The Intel processors discussed in this text are considered CISC (complex instruction set computing) processors. As the processor technology began to evolve beginning with 4- and 8-bit processors, the trend was to add more and more instruction capability to the processor. As a consequence, the control unit and the instruction decoder became very complex. In some processors, the control unit occupies 50 to 60 percent of the chip area. This results in the reduction in space available to registers. Similarly, the instructions that access memory tend to slow down the code execution. Furthermore, many studies of application programs revealed that even if these processors include a large number of instructions, only a small number of them is used frequently. This fact gave rise to the RISC (reduced instruction set computing) design philosophy that focused on a small set of frequently used instructions, thus simplifying the hardware design and improving the processor performance. RISC processors do not have a set of prescribed design rules, but they are governed by a set of guidelines evolved over many years. Generally, RISC processors include the following features:

1. The number of instructions is minimized (less than 100).
2. The number of addressing modes is relatively few (less than 3).
3. The number of memory-reference instructions is minimized. Memory is accessed only by Load and Store instructions.
4. The processing is register intensive, meaning it includes many more registers and most of the computing is performed using registers rather than memory. The number of registers ranges from 32 registers to more than 100 registers.
5. The instruction format is simplified and each instruction is executed in one cycle.
6. Design considerations include support for high-level languages through appropriate selection of instructions and optimization for compilers.

At present, many processors designed with RISC philosophy are available. A list includes such processors as Alpha from Digital Equipment Corp., R4400SC from MIPS Technologies, PA7100 from Hewlett-Packard, PowerPC from Apple, IBM, and Motorola, and Super Sparc from Sun Microsystem. They are either 32- or 64-bit processors. The number of general-purpose registers ranges from 32 to 136 registers, and the clock frequency ranges from 50 MHz to 200 MHz; the Alpha processor (Digital Equipment) runs on 200 MHz frequency. They include multiple pipeline stages and support multiprocessing. Some of these processors execute more than one instruction per cycle. These processors are generally used in workstations and file servers. However, some of these processors (such as the PowerPC) are also gearing toward the high end of the personal computer (PC) market. To take advantage of the power and speed of these processors, an operating system that can handle 32- and 64-bit processors in multiuser and multiprocessing environments is needed. At present, operating systems such as UNIX (many variations), OS/2 (IBM), Windows 95 (Microsoft), Mac OS (Apple), and Solaris (Sun Microsystems) are also evolving to meet the demands of the changing microprocessor technology.

18.45   Trends in High-Performance Processors

The recent 32-bit and 64-bit processors address large memory space in gigabytes, execute instructions with high speed, and perform floating-point arithmetic operations with high precision. They are high-performance processors oriented toward high-level languages in multiuser and multiprocessing environments.

These processors are classified as CISC or RISC processors. The CISC processors are designed with a large set of instructions and many addressing modes. On the other hand, the RISC processors include a small set of instructions with relatively few addressing modes. These instructions are selected based on how frequently they are used in application programs and how they can optimize compiler designs in high-level languages. Most instructions are executed using registers and the need for memory access is minimized. The RISC processors include multiple execution units and pipeline stages that enable the execution of more than two or more instructions per cycle; this is known as superscaler operation. This operation sets these processors apart from the previous generation processors. At present, the trend is toward the RISC approach because of its relative simplicity and high performance.

SINGLE-CHIP MICROCONTROLLERS 18.5

Single-chip microcomputers, also known as microcontrollers, are used primarily to perform dedicated functions. They are used as independent controllers in machines or as slaves in distributed processing, as described in Chapter 15 (Section 15.3). Generally, they include all the essential elements of a computer on a single chip: MPU, R/W memory, ROM, and I/O lines. Typical examples of the single-chip microcomputers are the Intel MCS-48, 51, and 96 families, the Motorola MC68HC11 family, and the Zilog Z8.

Most of these microcontrollers have an 8-bit word size (except the MCS-96, with a 16-bit word size), at least 64 bytes of R/W memory, and 1K byte of ROM. The range of I/O lines varies considerably, from 16 to 40 lines. However, most of these devices cannot be easily programmed in college laboratories unless they include EPROM on the chip. A variety of single-chip microcontrollers is available in the market to meet diversified industry needs. To illustrate the trend, two different single-chip microcontrollers are described here.

18.51 Intel MCS®-51 Single-Chip Family

The Intel MCS®-51 is a widely used 8-bit single-chip microcontroller family. At the high end of the single-chip device spectrum in terms of its capability and versatility, it is designed for use in sophisticated real-time instrumentation and industrial control. It can operate with a 12 MHz clock and has a very powerful instruction set.

Figure 18.12 shows a block diagram of the chip. It includes the following features:

☐ 4K bytes of ROM or EPROM
☐ 128 bytes of data memory plus 21 special-function registers (SFR—not included in figure)
☐ four programmable I/O ports (32 I/O lines)
☐ two 16-bit timer/event counters
☐ a serial I/O port with a UART
☐ five interrupt lines: two for external signals and three for internal operations

The 8051 is known as a "bit and byte processor." The instruction set includes binary and BCD arithmetic operations, bit set/reset functions, and all logical functions. However, its real power comes from its ability to handle Boolean functions. On any addressable bit, the processor can perform functions such as Set, Clear, Complement, Jump If Set or Not Set, and Jump If Set Then Clear. It can also perform logical functions with two bits and place the result in the Carry flag.

The 8051 can use its 32 I/O lines as 32 individual bits or as four 8-bit parallel ports. It can service five interrupts: two external, two from the counters, and one from

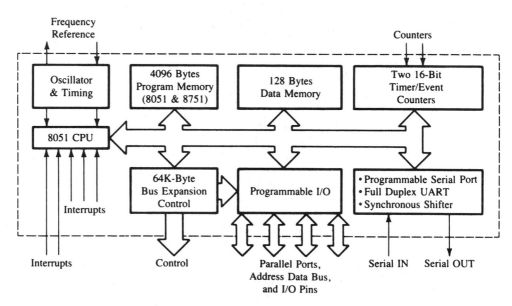

FIGURE 18.12
Block Diagram: The 8051

SOURCE: Adapted from Bob Koehler, "Microcontroller Doubles as Boolean Processors," *Electronic Designs,* vol. 28, no. 11;
copyright Hayden Publishing Company, Inc., 1980.

the serial I/O port. The chip includes two 16-bit counters that can operate in three different modes, and a serial I/O port that can operate in the full duplex mode.

18.52   Motorola MC68HC11 Microcontroller Family

The MC68HC11 is also a widely used single-chip microcontroller family in industrial applications. It is an advanced 8-bit microcontroller with highly sophisticated on-chip peripherals such as timers, serial I/O, and an eight-channel A/D converter and can operate in four different modes. It has 40 I/O lines, and they serve multiple functions depending on the operating mode. It includes 256 bytes of internal R/W memory that can be expanded to 64K system memory in the expanded mode. It is housed in a 48-pin package and can operate up to 2 MHz clock frequency. Its instruction set is upward-compatible with the older processors 6800 and 6801.

Figure 18.13 shows a block diagram of the MC68HC11A8 version; this version is used as a primary reference. Other versions differ primarily in terms of memory size, availability of on-chip A/D converter, and number of pins in a package. Some of the primary features of this microcontroller are as follows:

☐ Four operating modes
☐ 8K bytes of ROM and 512 bytes of EEPROM
☐ 256 bytes of RAM (R/W memory)
☐ 40 I/O pins with multiple functions

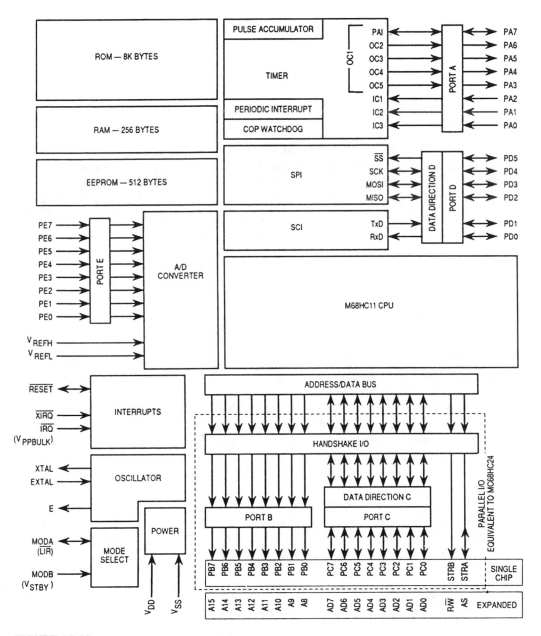

FIGURE 18.13
Block Diagram MC68HC11
SOURCE: Courtesy of Motorola, Inc.

☐ Eight-channel, 8-bit A/D converter
☐ 16-bit timer system
☐ 8-bit pulse accumulator circuit
☐ Serial communication and serial peripheral interface
☐ Computer operating properly (COP) watching system

This microcontroller has two fundamental modes of operation and their two variations: (1) single-chip and its variation special bootstrap, and (2) expanded and its variation special test. In the single-chip mode, it is a microcontroller with the features as shown in Figure 18.13 without any external memory. In the special bootstrap mode, programs can be downloaded through the on-chip serial communication interface into internal RAM (R/W memory) for execution. In the expanded mode, 18 lines (I/O ports B and C and two strobe lines) are used as multiplexed address and data buses and read/write control signals; thus memory can be expanded to 64K bytes. The special test mode is intended primarily for factory testing.

The 68HC11 includes an 8-channel A/D converter with 8-bit resolution. This is an important peripheral on the chip that makes the controller ideal for data acquisition applications. It also has a 16-bit programmable timer with four prescalers (that can set four different frequencies), three input capture signals, and five output compare signals. The input capture signals can provide information about the time reference of an event and the pulse width and/or period of a signal. The output compare signals are used to generate a pulse of a specific duration or a specified delay. Similarly, the pulse accumulator is essentially an 8-bit event counter or a timer.

The controller is designed to handle 17 hardware interrupts and one software interrupt. It is also capable of receiving and transmitting serial data through its serial communication interface (SCI) lines. Similarly, the serial peripheral interface (SPI) enables several SPI-type controllers and peripherals to be interconnected. The controller also includes self-monitoring circuitry to protect against system errors; it is known as the computer operating properly (COP) watchdog system.

The programming model is in many ways similar to that of the 6800 discussed earlier. It includes two 8-bit accumulators, A and B, and these can be combined as a 16-bit accumulator. It has an 8-bit condition code register (flag register) and four 16-bit registers: two index registers, X and Y, one stack pointer, and one program counter. On the other hand, the 6800 has only one index register and it cannot combine two accumulators. The instruction set includes 91 new opcodes in addition to the 72 instructions of the 6800 microprocessor.

18.53 Review of Single-Chip Microcontrollers

Examination of the examples discussed above shows that the single-chip microcontroller plays a vital role in control applications and is an important segment of microprocessor technology. These devices are designed for special-purpose applications, and the circuitry on the chip varies according to its objectives. Applications range from bit set/reset functions to processing high-speed analog signals.

Number Systems

Computers communicate and operate in binary digits 0 and 1; on the other hand, human beings generally use the decimal system with ten digits, from 0 to 9. Other number systems are also used, such as octal with eight digits, from 0 to 7, and hexadecimal (Hex) with digits from 0 to 15. In the hexadecimal system, digits 10 through 15 are designated as A through F, respectively, to avoid confusion with the decimal numbers 10 to 15.

A positional scheme is usually used to represent a number in any of the number systems. This means that each digit will have its value according to its position in a number. The number of digits in a position is also referred to as the base. For example, the binary system has base 2, the decimal system has base 10, and the hexadecimal system has base 16.

A.1 NUMBER CONVERSION

A number in any base system can be represented in a generalized format as follows:

$$N = A_n B^n + A_{n-1} B^{n-1} + \ldots + A_1 B^1 + A_0 B^0$$

N = number, B = base, A = any digit in that base

For example, number 154 can be represented in various number systems as follows:

Decimal: $154 = 1 \times 10^2 + 5 \times 10^1 + 4 \times 10^0 = 154$
Octal: $232 = 2 \times 8^2 + 3 \times 8^1 + 2 \times 8^0$
                $= 128 + 24 + 2 = 154$
Hexadecimal: $9A = 9 \times 16^1 + A \times 16^0$
                $= 144 + 10 = 154$
Binary: 10011010
$= 1 \times 2^7 + 0 \times 2^6 + 0 \times 2^5 + 1 \times 2^4 + 1 \times 2^3 + 0 \times 2^2 + 1 \times 2^1 + 0 \times 2^0$
$= 128 + 0 + 0 + 16 + 8 + 0 + 2 + 0 = 154$

The above example also shows how to convert a given number in any system into its decimal equivalent.

CONVERSION TABLE: DECIMAL, BINARY, OCTAL, AND HEXADECIMAL

Decimal	Hex	Binary	Octal
0	0	0000	00
1	1	0001	01
2	2	0010	02
3	3	0011	03
4	4	0100	04
5	5	0101	05
6	6	0110	06
7	7	0111	07
8	8	1000	10
9	9	1001	11
10	A	1010	12
11	B	1011	13
12	C	1100	14
13	D	1101	15
14	E	1110	16
15	F	1111	17

HOW TO CONVERT A NUMBER FROM BINARY INTO HEXADECIMAL AND OCTAL

Example A.1

Convert the binary number 1 0 0 1 1 0 1 0 into its Hex and octal equivalents.

Hexadecimal

Step 1: Starting from the right (LSB) arrange the binary digits in groups of four.

1 0 0 1   1 0 1 0

Step 2: Convert each group into its equivalent Hex number.

$$9 \quad A$$

Octal

Step 1: Starting from the right (LSB) arrange the binary digits in groups of three.

$$1\,0 \quad 0\,1\,1 \quad 0\,1\,0$$

Step 2: Convert each group into its equivalent octal number.

$$2 \quad 3 \quad 2$$

2'S COMPLEMENT AND ARITHMETIC OPERATIONS A.2

The 8085 microprocessor performs the subtraction of two binary numbers using the 2's complement method. In digital logic circuits, it is easier to design a circuit to add numbers than to design a circuit to subtract numbers. The 2's complement of a binary number is equivalent to its negative number; thus by adding the complement of the subtrahend (the number to be subtracted) to the minuend, a subtraction can be performed. The method of 2's complement is explained below with the examples from the decimal number system.

DECIMAL SUBTRACTION

Subtract the following two decimal numbers using the borrow method and the 10's complement method: $52 - 23$

Example A.2

Borrow Method

$$\text{Minuend:} \quad 52 = 5 \times 10 + 2$$
$$\text{Subtrahend:} \quad 23 = 2 \times 10 + 3$$

Step 1: To subtract 3 from 2, 10 must be borrowed from the second place of the minuend.

$$52 = 4 \times 10 + 12$$

Step 2: The subtraction of the digits in the first place and the second place is as follows.

$$\begin{array}{r} 52 = 4 \times 10 + 12 \\ -23 = \underline{2 \times 10 + 3} \\ 2 \times 10 + 9 = 29 \end{array}$$

**10's
Complement
Method**

Step 1: Find the 9's complement of the subtrahend (23), meaning subtract each digit of
the subtrahend from 9.

$$
\begin{array}{r}
\text{9's complement of 23:} \quad 9 \quad 9 \\
-2 \quad 3 \\
\hline
7 \quad 6
\end{array}
$$

Step 2: Add 1 to the 9's complement to find the 10's complement of the subtrahend.

$$
\begin{array}{r}
\text{10's complement of 23:} \quad 76 \\
+\ 1 \\
\hline
77
\end{array}
$$

The reason to find the 9's complement is to demonstrate a similar procedure to find the
2's complement of a binary number. However, in reality, the 10's complement of 23 is
equivalent to subtracting 23 from 100.

Step 3: Add the 10's complement of the subtrahend (77) to the minuend (52) to subtract 23 from 52.

$$
\begin{array}{r}
\text{10's complement of 23:} \quad 77 \\
\text{Minuend:} \quad +\ 52 \\
\hline
1\ 29 = 29 \quad \text{(by dropping the most significant digit)}
\end{array}
$$

The elimination of the most significant bit is equivalent to subtracting 100 from the sum.
This is necessary to compensate for the 100 that was added to find the 10's complement
of 23.

**Example
A.3**

Perform the subtraction of the following two numbers using the borrow method and the
10's complement method: 23 − 52.

**Borrow
Method**

$$
\begin{array}{r}
\text{Minuend:} \quad 2\ 3 \\
\text{Subtrahend:} \quad 5\ 2
\end{array}
$$

Step 1: The subtraction of the digits in the first place results in: 3 − 2 = 1.
Step 2: To subtract the digits in the second place, a borrow is required from the third
place. Assuming the borrow is available from the third place, the digit 5 can be
subtracted from 2 as follows:

$$
\begin{array}{r}
1\ 2 \\
-\ 5 \\
\hline
\overline{1}\ 7 \quad \text{(the nonexistent borrow is shown with the bar)}
\end{array}
$$

$$
\begin{array}{r}
\text{Result:} \quad 23 \\
-\ 52 \\
\hline
\overline{1}\ 71
\end{array}
$$

The same result is obtained with the 10's complement method, as shown below.

Step 1: Find the 9's complement of the subtrahend (52).

$$9\text{'s complement of } 52: \quad \begin{array}{cc} 9 & 9 \\ -5 & 2 \\ \hline 4 & 7 \end{array}$$

Step 2: Add 1 to the 9's complement to find 10's complement: $47 + 1 = 48$
Step 3: Add the 10's complement of the subtrahend to the minuend.

$$\begin{array}{ll} 10\text{'s complement of } 52: & 48 \\ \text{Minuend:} & \underline{23} \\ & 71 \text{ (this is negative 29, expressed in 10's complement)} \end{array}$$

By examining these two examples, the following conclusions can be drawn and these conclusions can be used for any number system.

1. The complement of a number is its equivalent negative number.
2. A number can be subtracted by using its complement.
3. The sum of a number and its complement results in 0 if the most significant digit of the sum is ignored.
4. When the subtrahend is larger than the minuend, the result of the 10's complement method is negative, and it is expressed in terms of 10's complement. The same result can be obtained by borrowing a digit from the most significant position.

PROCEDURE TO FIND THE 2'S COMPLEMENT OF A BINARY NUMBER

Step 1: Find the 1's complement. This amounts to replacing 0 by 1 and 1 by 0.
Step 2: To find the 2's complement, add 1 to the 1's complement. This is similar to the procedure of 10's complement.

Find the 2's complement of the binary number.

$$0\ 0\ 0\ 1 \quad 1\ 1\ 0\ 0 \quad (1CH \text{ or } 28_{10})$$

Step 1: Find the 1's complement, meaning replace 0 with 1 and 1 with 0.
$$1\text{'s complement} = \quad 1\ 1\ 1\ 0 \quad 0\ 0\ 1\ 1$$
Step 2: Add 1
$$\begin{array}{l} + \qquad\qquad\qquad 1 \\ \hline 2\text{'s complement} = \quad 1\ 1\ 1\ 0 \quad 0\ 1\ 0\ 0 \end{array}$$

By examining the result of the example, the following rule can be stated to find the 2's complement of a binary number, instead of the above procedure of the 1's complement.

Rule 1: Start at the LSB of a given number, and check all the bits to the left. Keep all
the bits as they are up to and including the least significant 1.

Rule 2: After the first 1, replace all 0s with 1s and 1s with 0s.

These rules can be applied to the given binary number (1CH), as illustrated below:

Binary Number: 0 0 0 1 1   1 0 0

                           ↓   ↓   Start Here
                      Replace 0 with 1   Keep as they are
                      and 1 with 0
2's complement   1 1 1 0 0   1 0 0

The 2's complement of the number can be verified by adding the complement to the original number as follows; the sum should be 0:

Binary Number: 0 0 0 1 1 1 0 0
2's Complement: 1 1 1 0 0 1 0 0
             1 0 0 0 0 0 0 0 0 (ignore the MSB)

BINARY SUBTRACTION USING 2'S COMPLEMENT

The binary subtraction can be performed by using the 2's complement method; if the result is negative, it is expressed in terms of 2's complement.

**Example
A.5**

Subtract 32H (0011 0010) from 45H (0100 0101).

Subtrahend: 32H =   0 0 1 1 0 0 1 0
2's complement of 32H =   1 1 0 0 1 1 1 0
                          +
Minuend: 45H =   0 1 0 0 0 1 0 1
                  CY   1   1 1
                          0 0 0 1 0 0 1 1 = 13H

**Example
A.6**

Subtract 45H (0100 0101) from 32H (0011 0010).

Subtrahend: 45H =   0 1 0 0 0 1 0 1
2's complement of 45H =   1 0 1 1 1 0 1 1
                          +
Minuend: 32H =   0 0 1 1 0 0 1 0
                  CY =   1 1   1
                          1 1 1 0 1 1 0 1 = EDH

The result is negative and it is expressed in 2's complement. This can be verified by taking the 2's complement of the result; the 2's complement of the result should be 13H as in Example A.5.

$$\text{Result EDH} = 1\ 1\ 1\ 0\quad 1\ 1\ 0\ 1$$
$$\text{Two's complement of EDH} = 0\ 0\ 0\ 1\quad 0\ 0\ 1\ 1 = \text{13H}$$

SIGNED NUMBERS

To perform the arithmetic operations with signed numbers (positive and negative), the sign must be indicated as well as the magnitude of the number. In 8-bit microprocessors, bit D_7 is used to indicate the sign of a number; 0 in D_7 indicates a positive number and 1 indicates a negative number. Bit D_7 can be used to indicate the sign of a number because

1. the 8085/8080A performs the subtraction of two numbers using 2's complement and, if the result is negative, it saves (shows) the result in the form of 2's complement.
2. the 2's complement of all the 7-bit numbers have 1 in D_7.

When a programmer uses bit D_7 to indicate the sign of a number, the magnitude of the number can be represented by seven bits (D_6–D_0). For example, number 74H is represented with a sign as follows:

$$
\begin{array}{llllllll}
D_7 & D_6 & D_5 & D_4 & D_3 & D_2 & D_1 & D_0
\end{array}
$$
$$+74H = 0 \quad 1 \quad 1 \quad 1 \quad 0 \quad 1 \quad 0 \quad 0$$
$$-74H = 1 \quad 0 \quad 0 \quad 0 \quad 1 \quad 1 \quad 0 \quad 0 \quad \text{(2's complement of 74H)}$$

sign   magnitude

However, the microprocessor cannot differentiate between a positive number and a negative number. For example, in the above illustration, −74H can be interpreted as the unsigned positive number 8CH or the bit pattern. It is the responsibility of the programmer to provide the necessary interpretation.

SUBTRACTION PROCESS IN THE 8085 MICROPROCESSOR

The 8085 performs the following operations when it subtracts (SUB or SUI) two binary numbers:

Step 1: Finds 1's complement of the subtrahend.
Step 2: Finds 2's complement of the subtrahend by adding 1 to the result of Step 1.
Step 3: Adds 2's complement of the subtrahend to the minuend.
Step 4: Complements the CY flag.

These steps are internal to the microprocessor and invisible to the user; only the result is available to the user.

Example A.7

Show the internal steps performed by the microprocessor to subtract the following unsigned numbers:

a. FAH − 62H

b. 62H − FAH

a. Minuend: FAH = 1 1 1 1   1 0 1 0
Subtrahend: 62H = 0 1 1 0   0 0 1 0

Step 1: 1's complement of 62H =	1 0 0 1 1 1 0 1
Step 2: Add 1	+ 1
2's complement of 62H =	1 0 0 1 1 1 1 0
Step 3: Add minuend (FAH)	+ 1 1 1 1 1 0 1 0
	1 1 0 0 1 1 0 0 0
Step 4: Complement CY	0 1 0 0 1 1 0 0 0
Result:	0 1 0 0 1 1 0 0 0 = 98H
Flags:	CY = 0, S = 1, Z = 0, P = 0

b. Minuend   62H = 0 1 1 0   0 0 1 0
Subtrahend: FAH = 1 1 1 1   1 0 1 0

Step 1: 1's complement of FAH =	0 0 0 0 0 1 0 1
Step 2: Add 1	+ 1
2's complement of FAH =	0 0 0 0 0 1 1 0
Step 3: Add minuend (62H)	+ 0 1 1 0 0 0 1 0
	0 0 1 1 0 1 0 0 0
Step 4: Complement CY	1 0 1 1 0 1 0 0 0
Result:	1 0 1 1 0 1 0 0 0 = 68H (CY = 1)
Flags:	CY = 1, S = 0, Z = 0, P = 0

This result is negative and is expressed in 2's complement of the magnitude.

Results

a. FAH − 62H = 98H (positive), CY = 0, S = 1

b. 62H − FAH = 68H (negative), CY = 1, S = 0

These results and associated flags appear to be confusing. In Example A.7a, the result is positive but the sign flag indicates that it is negative. On the other hand, in Example A.7b, the result is negative but the sign flag indicates that it is positive. This confusion can be explained as follows:

1. This subtraction is concerned with the unsigned numbers; therefore, the sign flag is irrelevant. In signed arithmetic, the number FAH is invalid because it is an 8-bit number.
2. The programmer can check whether the result indicates the true magnitude by checking the CY flag. If CY is reset, the result is positive, and if CY is set, the result is expressed in 2's complement.

In Example A.7a, assume that the numbers are signed numbers, and interpret the result.

Minuend: FAH   This is a negative number because $D_7 = 1$; therefore, this must be represented in 2's complement. The magnitude of the number can be found by taking the 2's complement of FAH:

$$FAH = 1\ 1\ 1\ 1\ \ 1\ 0\ 1\ 0$$
$$\text{2's complement of FAH} = 0\ 0\ 0\ 0\ \ 0\ 1\ 1\ 0$$
$$= 06H\ (\text{magnitude})$$

Subtrahend: 62H   This is a positive number because $D_7 = 0$. The problem given in 7a can be represented as follows:

$$FAH - 62H = (-06H) - (+62H)$$
$$= -68H$$

The final result is $-68H$, which will be in the form of its 2's complement:

$$-68H = -(0\ 1\ 1\ 0\ \ 1\ 0\ 0\ 0)$$
$$\text{2's complement of 68H} = \ \ \ 1\ 0\ 0\ 1\ \ 1\ 0\ 0\ 0$$
$$= 98H$$

The final answer is the same as before; however, it will be interpreted as a negative number with the magnitude of 68H. When signed numbers are used in arithmetic operations, the sign flag will indicate the proper sign of the result.

Add the following two positive numbers and interpret the sign flag: +41H, +54H.

$$41H = 0\ 1\ 0\ 0\ \ 0\ 0\ 0\ 1$$
$$+$$
$$\underline{54H = 0\ 1\ 0\ 1\ \ 0\ 1\ 0\ 0}$$
$$95H = 1\ 0\ 0\ 1\ \ 0\ 1\ 0\ 1\ \ \ S = 1, CY = 0, Z = 0$$

This is an addition of two positive numbers; therefore, the sign flag indicates that the sum is larger than seven bits. This is also known as overflow. If this had been the sum of two unsigned numbers, the sign flag would have had no significance.

Introduction to the EMAC Primer*

B.1 THE PRIMER TRAINER

B.11 System Description

The Primer is a low-cost single-board microcomputer from EMAC Inc., designed with the 8085 microprocessor. The system includes all the necessary features for training purposes in college laboratories. It will be referred to as the Primer or the trainer.

---

*Courtesy of EMAC Inc. This trainer is designed, manufactured, and marketed by EMAC Inc., 11 Emac Way, Carbondale, Illinois 62901 (Tel. 618-529-4525)/www.emacinc.com

Figure B.1 shows the functional block diagram of the basic system. It includes the 8085 microprocessor, EPROM (32K), the 8155 (R/W memory with I/O and timer), 20-key Hex keyboard, and six seven-segment LEDs for display. The keyboard and the display are interfaced with the 8085 through the keyboard/display controller, the 8279. In addition, the board includes an 8-position dip switch, 8 LEDs, D/A and A/D converters, and a speaker port.

The keyboard enables the user to enter and store the 8085 Hex machine code representing the 8085 assembly language programs in R/W memory. The seven-segment LEDs are used to display memory addresses and their contents while entering, modifying, or examining the programs. The 8-position dip switch can be used as an input port and 8 LEDs can be used as an output port. A program can be executed using the function keys on the keyboard, and results can be either displayed at 8 LEDs or stored in memory.

The Primer system can be expanded to include additional 32K EPROM, 32K R/W memory, the 8251 communication port, and I/O ports.

THE EPROM

This is a 32K memory with a memory address range from 0000H to 7FFFH. The system monitor program, called MOS (Monitor Operating System), is stored permanently in this memory. This program continuously monitors the keyboard and displays the Hex keys at

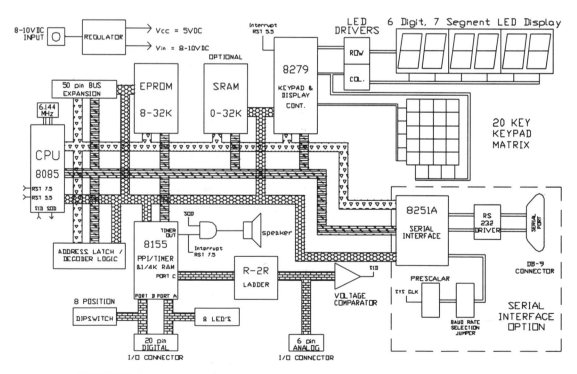

FIGURE B.1
Primer Functional Block Diagram

seven-segment LEDs. The primary function of the MOS is to enable the user to enter a program (instructions in Hex code) in the trainer's R/W memory, modify it if necessary, execute and debug the program.

THE 8155

This chip has 256 bytes of R/W memory, two I/O ports, and one timer. The R/W memory addresses of the 8155 range from FF00H to FFFFH; however, memory locations from FFD0H to FFFFH are reserved for the use of the monitor program. In this memory, the user can enter a program and execute it using the keyboard.

The two I/O ports of the 8155 can be used to interface additional devices, and the timer can be used to provide various timing functions such as time delays.

THE 8279

This is a programmable device used to handle keyboard and display. It displays memory addresses and data from its internal memory at the seven-segment LEDs using the multiplexed technique. When a key is pressed, the 8085 is interrupted, and a new key is stored in the internal memory of the 8279 and displayed.

B.12 Keyboard

The keyboard has 20 keys; 16 keys for the Hex digits 0 to F are used for entering data, and the remaining four keys are used to perform various functions. In addition, Hex keys are also assigned to perform other functions. The monitor program operates in two modes: data entry mode and function mode. The functions of these keys are described as follows.

1. **0 to F:** Enter Hex digits.
2. **Enter:** Stores the displayed data in memory, increments the address of the memory location, and displays the address and data of the new location.
3. **Dec:** Decrements the memory address and displays the new address and its data.
4. **Step:** Executes one instruction at a time, called single stepping.
5. **Func:** Selects the second function for Hex keys. The monitor reverts to the data entry mode when the key is pressed twice.

Additional functions of Hex keys:

6. **B.P. (Break Point):** Displays the current break point address.
7. **S.C. (Stack Contents):** Displays two bytes from the top of the stack.
8. **Run:** Executes the program from the displayed memory address.
9. **A.F., B.C., D.E., H.L., S.P., and P.C.:** Displays contents of the specified registers, high-order byte on the left and low-order byte on the right.
10. **Key 1:** Invokes the Primer diagnostics.
11. **Key 2:** Executes the MOS service selected by the value in register C.
12. **Key 3:** Enables serial I/O for downloading a program from the PC. (Requires upgrade option.)

TABLE B.1
Primer Memory Map

	Memory
0000	Monitor ROM
7FFF	Slot 0 (32K)
4000	Slot 1 (32K)
BFFF	
C000	Foldback or Mirror
EFFF	8155 Memory
FF00	User Memory
FFFF	8155 R/W Memory

13. Key 4: Invokes the EPROM programmer menu-driven interface. (Requires EPROM programmer option.)

B.13   Memory Map

The memory map of the system is shown in Table B.1. The user memory ranges from FF00H to FFFFH; however, locations FFD0H to FFFFH are reserved for user interrupts and the monitor stack.

B.14   Available Subroutines

The MOS program includes many subroutines called services that are available to the user. The user must supply a service number in register C and call location 1000H. See the Primer manual for details.

B.2   USING THE PRIMER

The Monitor Operating System (MOS) of the Primer can be used to

☐ enter programs in its R/W memory (F001H–FFCFH).
☐ examine and modify the contents of memory, registers, and stack.
☐ execute programs.
☐ debug programs using single step and breakpoint techniques.

The user can display results of a program at the LED port in binary and/or at the system's seven-segment LEDs in Hex or BCD by calling appropriate display routines from the monitor program.

B.21   How to Enter a Program

When the Primer is turned on, it shows the memory address F001H and random data at seven-segment LED display. The memory address is shown by the left four LEDs, called the address field, and the random data byte is shown on the rightmost two LEDs, called the data field. When the user presses various keys, displays will be as follows (each key press is indicated by a bracket around the key symbol).

1. To examine the contents of R/W memory:

	Displays		
Press	**Address Field**	**Data Field**	
[RESET]	F 0 0 1	XX	;Keys in data entry mode
[ENTER]	F 0 0 2	XX	;XX indicates random data
[ENTER]	F 0 0 3	XX	

2. To change the contents of memory and load a program:

Changing the contents of memory and loading a program are similar functions. When we load a program, we enter Hex codes in memory locations for given instructions. We will accomplish both functions in the following steps. If you are not familiar with the 8085 instruction set, the comments on the right will indicate the function of a given instruction.

The following three instructions load a byte (96H) in the accumulator and display the byte at the user LED port in binary.

[RESET]	F 0 0 1	XX		
[3][E]	F 0 0 1	3 E	MVI A, 96H	;Load 96H in A
[ENTER]	F 0 0 2	XX		
[9][6]	F 0 0 2	96		
[ENTER]	F 0 0 3	XX		
[D][3]	F 0 0 3	D 3	OUT 11	;Display 96H at port #11H
[ENTER]	F 0 0 4	XX		
[1][1]	F 0 0 4	11		
[ENTER]	F 0 0 5	XX		
[F][F]	F 0 0 5	F F	RST 7	;End of program
[ENTER]	F 0 0 6	XX		
[RESET]	F 0 0 1	3E		

The last key [REST] will take the display at the location F001H.

B.22   To Execute a Program

To execute a program, direct the processor to the first instruction of a program being executed by loading the address of the instruction in the program counter. In our example above, the program begins at location FF01H. The Primer is designed to begin at location FF01H. By pushing the RESET key, we direct the processor to the beginning of the program.

To execute a program, the following key sequence should be followed:

Press	Address Field	Data Field
[RESET]	F 0 0 1	3 E
[FUNC]	FUNC	
[STEP/RUN]		

To single step:

[RESET]	F 0 0 1	3 E
[STEP/RUN]		

B.23 How to Examine and Change Register Contents

To examine:

[RESET]	F 0 0 1	3 E
[FUNC]	FUNC	
[A.F.]	9 6 X X	A.F.

This shows 96H in the accumulator and a random byte in the flag register.
To change the contents, press four data keys as follows.

[3][5][0][0]		
[ENTER]	3 5 0 0	A.F.

These keys place 35H in A and clear the flag register. This is a right-entry mode; therefore, if you press two keys (such as 35), followed by the ENTER key, the contents of the flag register will be shifted into A and the flag register will show 35H.

B.24 How to Execute a Program Using the Program Counter (PC)

Another approach to execute a program is to enter the address of the first instruction in the program counter. This requires placing MOS in the function mode and entering the address in the PC as follows.

Let us assume that your program begins at memory location FF20H. Enter all the instructions of the program to be executed, and press the following key sequence.

[FUNC]	FUNC		
[P.C.]	X X X X	P.C.	;Address in PC will be where the last instruction is entered
[F][F][2][0]	F F 2 0	P.C.	
[STEP/RUN]			

B.25　How to Display Results at the System Seven-Segment LEDs

The Primer display is managed by the 8279 programmable keyboard/display interface device. The seven-segment LEDs are connected to use the multiplexed technique and cannot be accessed by simply writing the OUT instruction. However, a data byte can be displayed by using services (subroutines) provided by the MOS. The MOS includes 36 service routines that the user can call by loading the service number in register C (the Primer manual lists all the subroutines that are available to the user). An example is illustrated below.

To display 96H from the above illustration (B.21), we need to call service 1BH. This service displays the Hex byte stored in register E. The instructions are modified as follows:

Memory	Code	Mnemonics	
F001	3 E	MVI A, 96H	;Load the byte in A
F002	9 6		
F003	5 F	MOV E,A	;Service 1B displays what is in E
F004	0 E	MVI C, 1BH	;Load service # in C
F005	1 B		
F006	C D	CALL 1000H	;Call MOS
F007	0 0		
F008	1 0		
F009	F F	RST 7	;End of instructions

Enter these instructions starting at location F001H and execute the program. The program will display 96H in the data field of the seven-segment display. Similarly, the service 12H displays the contents of register DE in the address field.

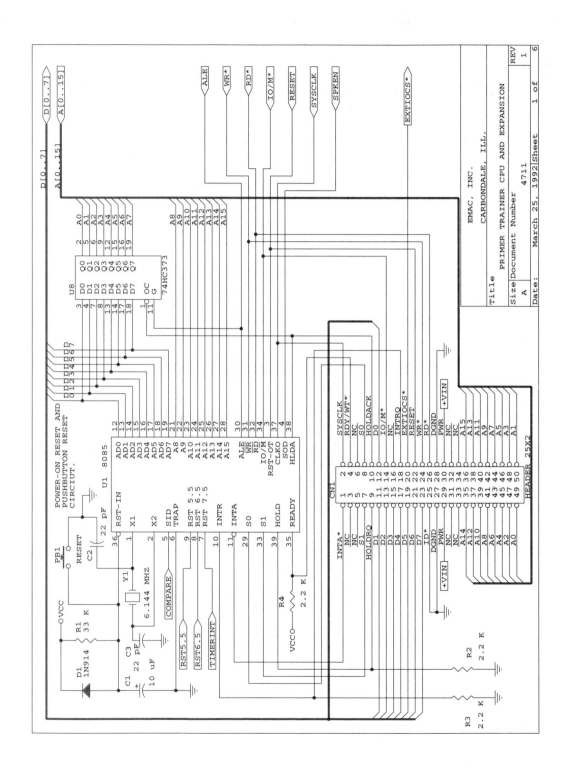

638

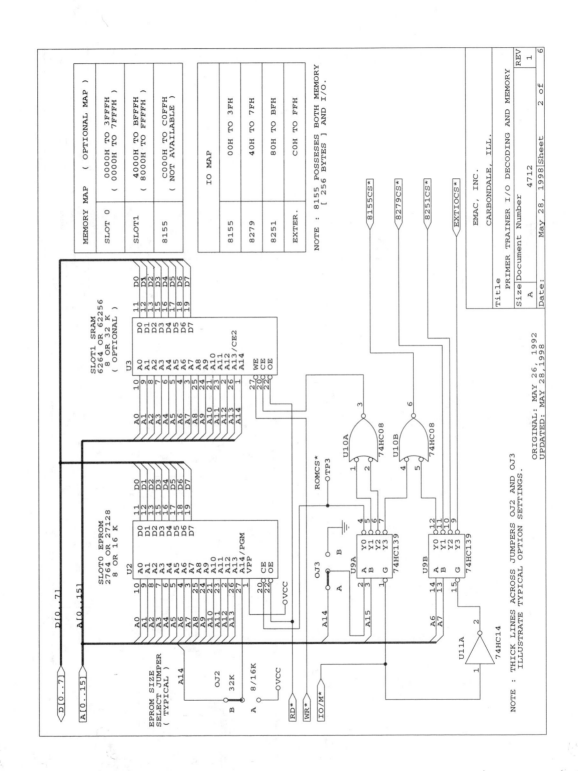

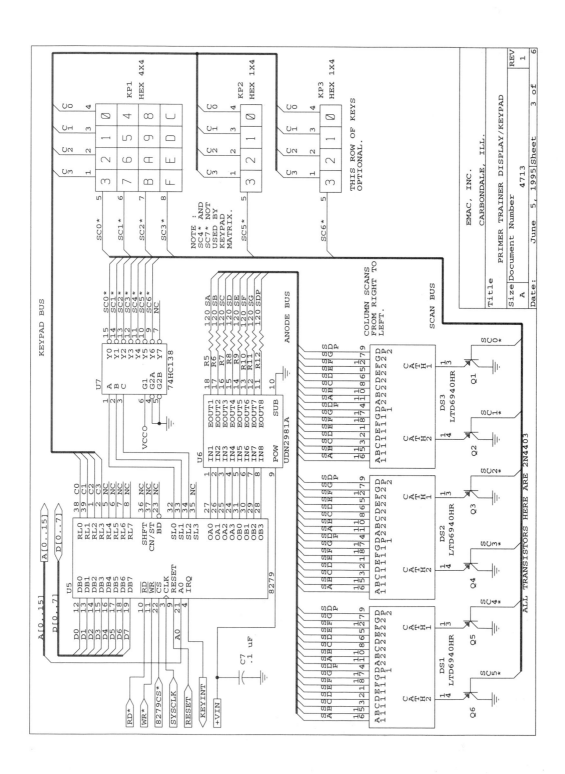

640

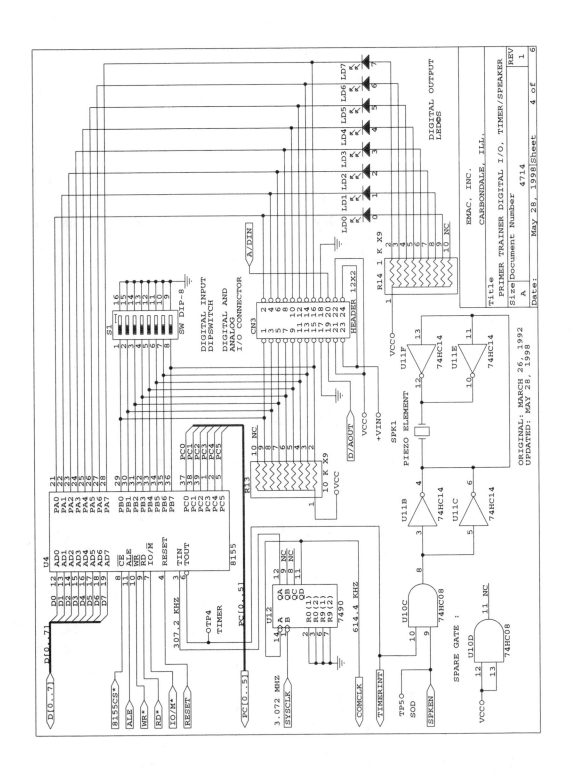

641

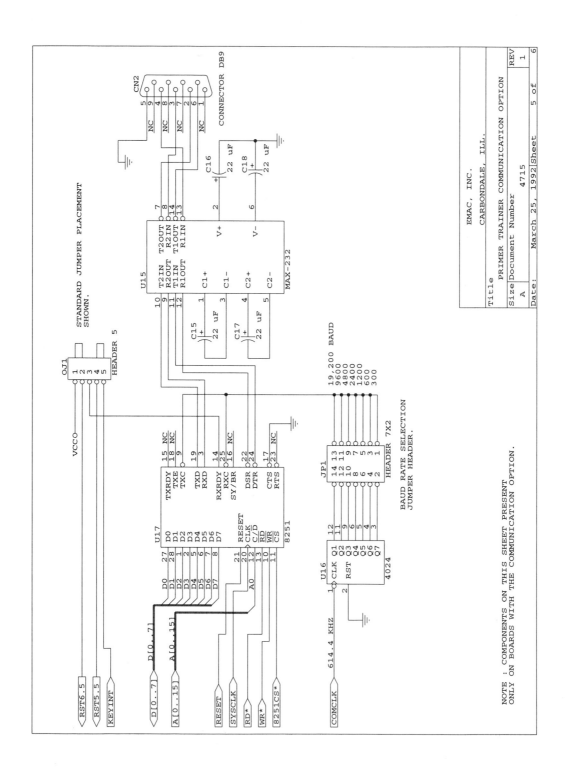

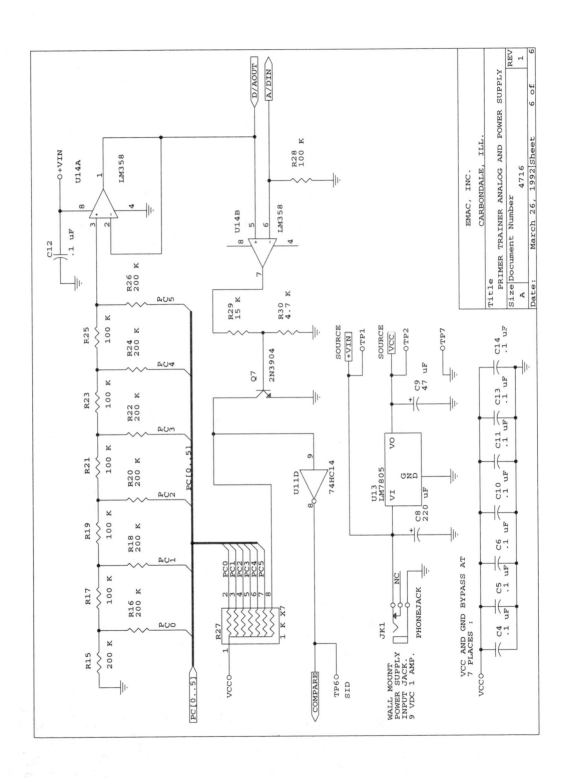

Pin Configuration
of Selected
Devices

C.1   SELECTED LOGIC AND DISPLAY DEVICES: PIN CONFIGURATION AND LOGIC SYMBOLS

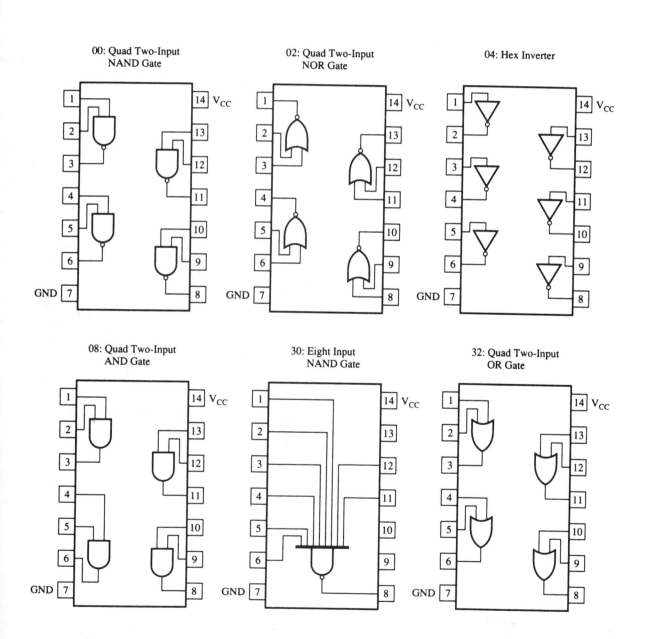

00: Quad Two-Input
NAND Gate

02: Quad Two-Input
NOR Gate

04: Hex Inverter

08: Quad Two-Input
AND Gate

30: Eight Input
NAND Gate

32: Quad Two-Input
OR Gate

74: DUAL D-TYPE POSITIVE EDGE-TRIGGERED FLIP-FLOP

Description This is a positive edge-triggered flip-flop with Set and Reset inputs and complementary outputs (Q and $\overline{Q}$).

Information at D input is transferred to Q output on the positive edge (Low-to-High) of the clock pulse (CP). Set (S) and Reset (R) are asynchronous active low inputs and operate independently of the clock; S input sets Q to logic 1 and R resets Q to logic 0.

Pin Configuration

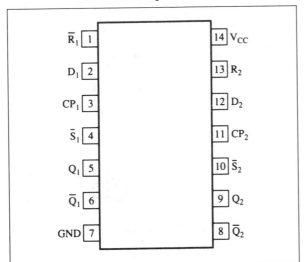

Logic Symbol

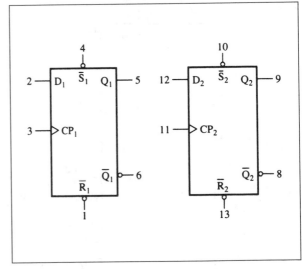

75: 4-BIT BISTABLE LATCH

Description '75 is a level sensitive 4-bit latch; each pair of bits is controlled by High Enable input E. It also has complementary outputs (Q and $\overline{Q}$).

Information at D input is transferred to Q output when E is High, and Q follows D input as long as E is High. When E goes Low, D input is latched at the output and remains latched until E goes High again.

Pin Configuration

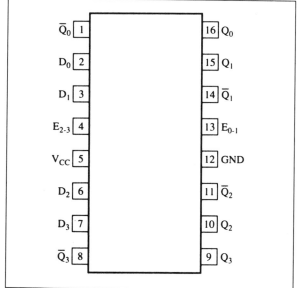

Logic Symbol

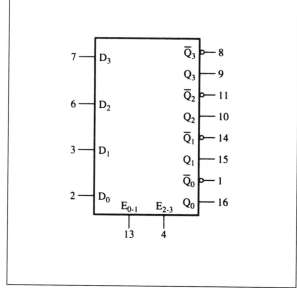

76: DUAL JK FLIP-FLOP

Description '76 is a dual JK flip-flop with J, K, Clock, Set, and Reset inputs for each flip-flop. JK information is loaded into the master when the Clock is high and transferred to the slave on the High-to-Low Clock transition.

'LS76 is a negative edge-triggered flip-flop.

Pin Configuration

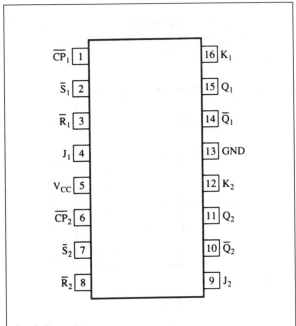

Logic Symbol

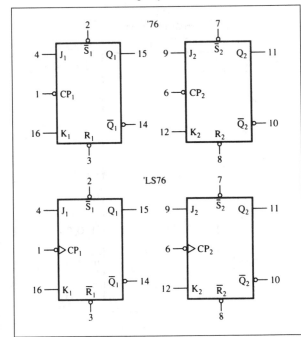

138: 3-TO-8 DECODER/DEMULTIPLEXER*

Description '138 has three binary weighted inputs (C, B, A or A_2, A_1, A_0) and eight mutually exclusive active Low outputs. It has three enable inputs, two active Low, and one active High; all three must be enabled to obtain an output.

When '138 is enabled, one of the output signals goes active Low corresponding to the decimal equivalent of the input. When it is not enabled, all output signals remain High.

Pin Configuration

A/A_0	1	16 V_{CC}
B/A_1	2	15 O_0/Y_0
C/A_2	3	14 O_1/Y_1
$\bar{E}_1$	4	13 O_2/Y_2
$\bar{E}_2$	5	12 O_3/Y_3
E_3	6	11 O_4/Y_4
Y_7/O_7	7	10 O_5/Y_5
GND	8	9 O_6/Y_5

Logic Symbol

*Note: Pin configuration shows notations commonly used by various manufacturers.

139: 2-TO-4 DECODER/DEMULTIPLEXER

Description '139 is a dual 2-to-4 decoder with two binary weighted inputs (B, A or A_1, A_0) and one active Low enable input. When it is enabled, one of four output signals, corresponding to the decimal equivalent of the input, goes active Low. When it is not enabled, all output signals remain High.

Pin Configuration

Logic Symbol

148: 8-INPUT PRIORITY ENCODER

Description   '148 has eight active Low inputs ($\bar{I}_7$ to $\bar{I}_0$), one enable input, and three active Low outputs. When it is enabled, the output provides the binary equivalent of the active input. If multiple inputs go active simultaneously, the input with the highest priority is encoded on the output; other inputs are ignored. $\bar{I}_7$ has the highest priority, and $\bar{I}_0$ has the lowest priority.

Pin Configuration   Logic Symbol

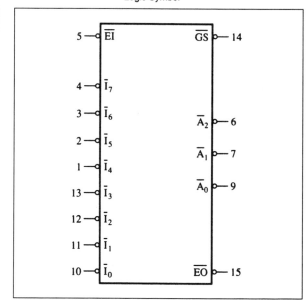

244: OCTAL BUFFER

Pin Configuration

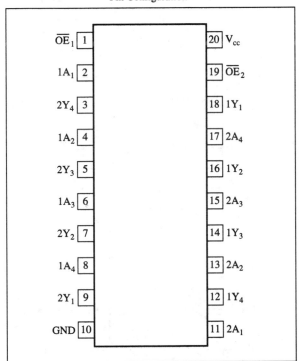

Logic Symbol

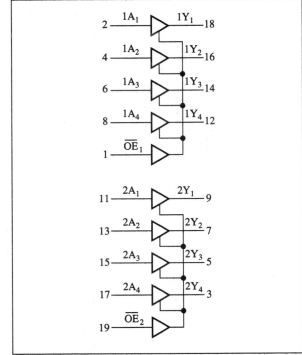

245: OCTAL TRANSCEIVER (BIDIRECTIONAL BUFFER)

Description '245 is a bidirectional buffer with $\overline{E}$ (Enable) and DIR as two control inputs. $\overline{E}$ is active Low, and it enables the buffer. DIR determines the direction of the data flow; if it is High, data flow from A to B and if it is low, data flow from B to A.

Pin Configuration

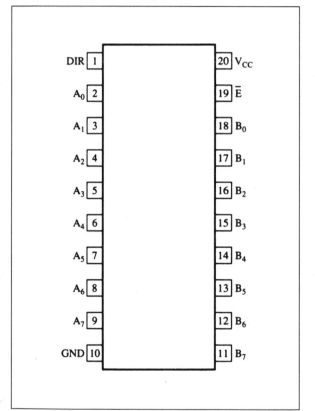

Logic Symbol

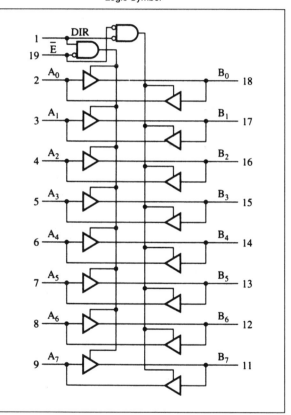

373: OCTAL TRANSPARENT LATCH

Description '373 is an octal transparent latch followed by tri-state output buffers. The data from D inputs are transferred to Q outputs when the latch enable (LE) signal is High and are latched when LE goes from High to Low. When $\overline{OE}$ goes Low, latched data appear on the outputs; otherwise, the outputs remain in high impedance.

Pin Configuration

Logic Symbol

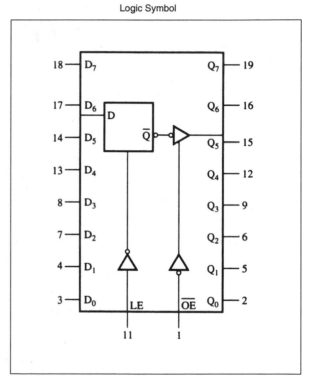

SEVEN-SEGMENT NUMERIC DISPLAY: (10-PIN PACKAGE)

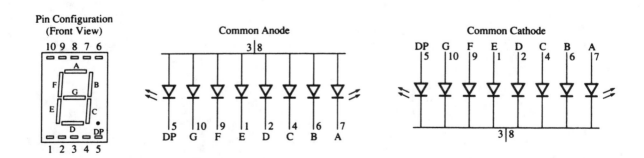

Specifications:
Data Converters*
and Peripheral
Devices

Appendix D specifications begin on the next page.

*SOURCES:

Pages 658–663: Courtesy of National Semiconductor Corporation, *Linear Databook 2* (Santa Clara, Calif.: Author, 1988).

Pages 664–706: Reprinted by permission of Intel Corporation, Copyright Intel Corporation, 1995.

Pages 707–716: Courtesy of Hitachi America, Ltd.

DAC0808, DAC0807, DAC0806 8-Bit D/A Converters

General Description

The DAC0808 series is an 8-bit monolithic digital-to-analog converter (DAC) featuring a full scale output current settling time of 150 ns while dissipating only 33 mW with ±5V supplies. No reference current (I_{REF}) trimming is required for most applications since the full scale output current is typically ±1 LSB of 255 I_{REF}/ 256. Relative accuracies of better than ±0.19% assure 8-bit monotonicity and linearity while zero level output current of less than 4 μA provides 8-bit zero accuracy for $I_{REF} \geq 2$ mA. The power supply currents of the DAC0808 series are independent of bit codes, and exhibits essentially constant device characteristics over the entire supply voltage range.

The DAC0808 will interface directly with popular TTL, DTL or CMOS logic levels, and is a direct replacement for the MC1508/MC1408. For higher speed applications, see DAC0800 data sheet.

Features

- Relative accuracy: ±0.19% error maximum (DAC0808)
- Full scale current match: ±1 LSB typ
- 7 and 6-bit accuracy available (DAC0807, DAC0806)
- Fast settling time: 150 ns typ
- Noninverting digital inputs are TTL and CMOS compatible
- High speed multiplying input slew rate: 8 mA/μs
- Power supply voltage range: ±4.5V to ±18V
- Low power consumption: 33 mW @ ±5V

Block and Connection Diagrams

Dual-In-Line Package

Order Number DAC0808, DAC0807, or DAC0806 See NS Package Number J16A, M16A or N16A

TL/H/5687-1

TOP VIEW

TL/H/5687-2

Small-Outline Package

Top View

TL/H/5687-13

Ordering Information

ACCURACY	OPERATING TEMPERATURE RANGE	ORDER NUMBERS				
		J PACKAGE (J16A)*		N PACKAGE (N16A)*		SO PACKAGE (M16A)
8-bit	−55°C ≤ T_A ≤ +125°C	DAC0808LJ	MC1508L8			
8-bit	0°C ≤ T_A ≤ +75°C	DAC0808LCJ	MC1408L8	DAC0808LCN	MC1408P8	DAC0808LCM
7-bit	0°C ≤ T_A ≤ +75°C	DAC0807LCJ	MC1408L7	DAC0807LCN	MC1408P7	DAC0807LCM
6-bit	0°C ≤ T_A ≤ +75°C	DAC0806LCJ	MC1408L6	DAC0806LCN	MC1408P6	DAC0806LCM

*Note. Devices may be ordered by using either order number.

DAC0808 is compatible with the 1408.

Absolute Maximum Ratings (Note 1)

If Military/Aerospace specified devices are required, contact the National Semiconductor Sales Office/Distributors for availability and specifications.

Power Supply Voltage	
V_{CC}	$+18\ V_{DC}$
V_{EE}	$-18\ V_{DC}$
Digital Input Voltage, V5–V12	$-10\ V_{DC}$ to $+18\ V_{DC}$
Applied Output Voltage, V_O	$-11\ V_{DC}$ to $+18\ V_{DC}$
Reference Current, I_{14}	5 mA
Reference Amplifier Inputs, V14, V15	V_{CC}, V_{EE}
Power Dissipation (Note 3)	1000 mW
ESD Susceptibility (Note 4)	TBD

Storage Temperature Range	$-65°C$ to $+150°C$
Lead Temp. (Soldering, 10 seconds)	
Dual-In-Line Package (Plastic)	260°C
Dual-In-Line Package (Ceramic)	300°C
Surface Mount Package	
Vapor Phase (60 seconds)	215°C
Infrared (15 seconds)	220°C

Operating Ratings

Temperature Range	$T_{MIN} \leq T_A \leq T_{MAX}$
DAC0808L	$-55°C \leq T_A \leq +125°C$
DAC0808LC Series	$0 \leq T_A \leq +75°C$

Electrical Characteristics

(V_{CC} = 5V, V_{EE} = $-15\ V_{DC}$, V_{REF}/R14 = 2 mA, DAC0808: T_A = $-55°C$ to $+125°C$, DAC0808C, DAC0807C, DAC0806C, T_A = 0°C to $+75°C$, and all digital inputs at high logic level unless otherwise noted.)

Symbol	Parameter	Conditions	Min	Typ	Max	Units
E_r	Relative Accuracy (Error Relative to Full Scale I_O)	(Figure 4)				%
	DAC0808L (LM1508-8), DAC0808LC (LM1408-8)				±0.19	%
	DAC0807LC (LM1408-7), (Note 5)				±0.39	%
	DAC0806LC (LM1408-6), (Note 5)				±0.78	%
	Settling Time to Within ½ LSB (Includes t_{PLH})	T_A = 25°C (Note 6), (Figure 5)		150		ns
t_{PLH}, t_{PHL}	Propagation Delay Time	T_A = 25°C, (Figure 5)		30	100	ns
TCI_O	Output Full Scale Current Drift			±20		ppm/°C
MSB V_{IH} V_{IL}	Digital Input Logic Levels High Level, Logic "1" Low Level, Logic "0"	(Figure 3)	2		0.8	V_{DC} V_{DC}
MSB	Digital Input Current High Level Low Level	(Figure 3) V_{IH} = 5V V_{IL} = 0.8V		0 −0.003	0.040 −0.8	mA mA
I_{15}	Reference Input Bias Current	(Figure 3)		−1	−3	μA
	Output Current Range	(Figure 3) V_{EE} = −5V V_{EE} = −15V, T_A = 25°C	0 0	2.0 2.0	2.1 4.2	mA mA
I_O	Output Current Output Current, All Bits Low	V_{REF} = 2.000V, R14 = 1000Ω, (Figure 3) (Figure 3)	1.9	1.99 0	2.1 4	mA μA
	Output Voltage Compliance (Note 2) V_{EE} = −5V, I_{REF} = 1 mA V_{EE} Below −10V	E_r ≤ 0.19%, T_A = 25°C			−0.55, +0.4 −5.0, +0.4	V_{DC} V_{DC}

DAC0808 is compatible with the 1408.

Electrical Characteristics (Continued)

(V_{CC} = 5V, V_{EE} = −15 V_{DC}, V_{REF}/R14 = 2 mA, DAC0808: T_A = −55°C to +125°C, DAC0808C, DAC0807C, DAC0806C, T_A = 0°C to +75°C, and all digital inputs at high logic level unless otherwise noted.)

Symbol	Parameter	Conditions	Min	Typ	Max	Units
SRI_{REF}	Reference Current Slew Rate	*(Figure 6)*	4	8		mA/μs
	Output Current Power Supply Sensitivity	−5V ≤ V_{EE} ≤ −16.5V		0.05	2.7	μA/V
I_{CC} I_{EE}	Power Supply Current (All Bits Low)	*(Figure 3)*		2.3 −4.3	22 −13	mA mA
V_{CC} V_{EE}	Power Supply Voltage Range	T_A = 25°C, *(Figure 3)*	4.5 −4.5	5.0 −15	5.5 −16.5	V_{DC} V_{DC}
	Power Dissipation All Bits Low	V_{CC} = 5V, V_{EE} = −5V		33	170	mW
		V_{CC} = 5V, V_{EE} = −15V		106	305	mW
	All Bits High	V_{CC} = 15V, V_{EE} = −5V		90		mW
		V_{CC} = 15V, V_{EE} = −15V		160		mW

Note 1: Absolute Maximum Ratings indicate limits beyond which damage to the device may occur. DC and AC electrical specifications do not apply when operating the device beyond its specified operating conditions.

Note 2: Range control is not required.

Note 3: The maximum power dissipation must be derated at elevated temperatures and is dictated by T_{JMAX}, θ_{JA}, and the ambient temperature, T_A. The maximum allowable power dissipation at any temperature is $P_D = (T_{JMAX} - T_A)/\theta_{JA}$ or the number given in the Absolute Maximum Ratings, whichever is lower. For this device, T_{JMAX} = 125°C, and the typical junction-to-ambient thermal resistance of the dual-in-line J package when the board mounted is 100°C/W. For the dual-in-line N package, this number increases to 175°C/W and for the small outline M package this number is 100°C/W.

Note 4: Human body model, 100 pF discharged through a 1.5 kΩ resistor.

Note 5: All current switches are tested to guarantee at least 50% of rated current.

Note 6: All bits switched.

Note 7: Pin-out numbers for the DAL080X represent the dual-in-line package. The small outline package pinout differs from the dual-in-line package.

Typical Application

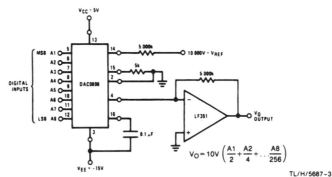

$$V_O = 10V \left(\frac{A1}{2} + \frac{A2}{4} + \ldots \frac{A8}{256} \right)$$

TL/H/5687–3

FIGURE 1. +10V Output Digital to Analog Converter (Note 7)

DAC0808 is compatible with the 1408.

**National
Semiconductor
Corporation**

ADC0801, ADC0802, ADC0803, ADC0804, ADC0805 8-Bit μP Compatible A/D Converters

General Description

The ADC0801, ADC0802, ADC0803, ADC0804 and ADC0805 are CMOS 8-bit successive approximation A/D converters that use a differential potentiometric ladder—similar to the 256R products. These converters are designed to allow operation with the NSC800 and INS8080A derivative control bus with TRI-STATE® output latches directly driving the data bus. These A/Ds appear like memory locations or I/O ports to the microprocessor and no interfacing logic is needed.

Differential analog voltage inputs allow increasing the common-mode rejection and offsetting the analog zero input voltage value. In addition, the voltage reference input can be adjusted to allow encoding any smaller analog voltage span to the full 8 bits of resolution.

Features

- Compatible with 8080 μP derivatives—no interfacing logic needed - access time - 135 ns
- Easy interface to all microprocessors, or operates "stand alone"

- Differential analog voltage inputs
- Logic inputs and outputs meet both MOS and TTL voltage level specifications
- Works with 2.5V (LM336) voltage reference
- On-chip clock generator
- 0V to 5V analog input voltage range with single 5V supply
- No zero adjust required
- 0.3" standard width 20-pin DIP package
- 20-pin molded chip carrier or small outline package
- Operates ratiometrically or with 5 V_{DC}, 2.5 V_{DC}, or analog span adjusted voltage reference

Key Specifications

- Resolution   8 bits
- Total error $\pm \frac{1}{4}$ LSB, $\pm \frac{1}{2}$ LSB and ± 1 LSB
- Conversion time   100 μs

Typical Applications

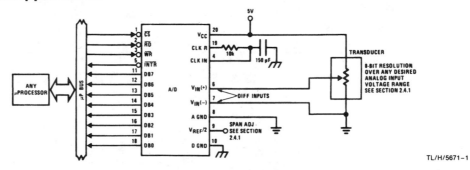

TL/H/5671-1

8080 Interface

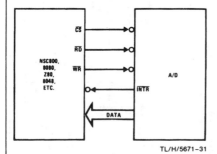

TL/H/5671-31

Error Specification (Includes Full-Scale, Zero Error, and Non-Linearity)			
Part Number	Full-Scale Adjusted	$V_{REF}/2 = 2.500\ V_{DC}$ (No Adjustments)	$V_{REF}/2 =$ No Connection (No Adjustments)
ADC0801	$\pm \frac{1}{4}$ LSB		
ADC0802		$\pm \frac{1}{2}$ LSB	
ADC0803	$\pm \frac{1}{2}$ LSB		
ADC0804		± 1 LSB	
ADC0805			± 1 LSB

Absolute Maximum Ratings (Notes 1 & 2)

If Military/Aerospace specified devices are required, contact the National Semiconductor Sales Office/Distributors for availability and specifications.

Supply Voltage (V_{CC}) (Note 3)	6.5V
Voltage	
Logic Control Inputs	−0.3V to +18V
At Other Input and Outputs	−0.3V to (V_{CC} + 0.3V)
Lead Temp. (Soldering, 10 seconds)	
Dual-In-Line Package (plastic)	260°C
Dual-In-Line Package (ceramic)	300°C
Surface Mount Package	
Vapor Phase (60 seconds)	215°C
Infrared (15 seconds)	220°C

Storage Temperature Range	−65°C to +150°C
Package Dissipation at T_A = 25°C	875 mW
ESD Susceptibility (Note 10)	800V

Operating Ratings (Notes 1 & 2)

Temperature Range	$T_{MIN} \leq T_A \leq T_{MAX}$
ADC0801/02LJ	−55°C ≤ T_A ≤ +125°C
ADC0801/02/03/04LCJ	−40°C ≤ T_A ≤ +85°C
ADC0801/02/03/05LCN	−40°C ≤ T_A ≤ +85°C
ADC0804LCN	0°C ≤ T_A ≤ +70°C
ADC0802/03/04LCV	0°C ≤ T_A ≤ +70°C
ADC0802/03/04LCWM	0°C ≤ T_A ≤ +70°C
Range of V_{CC}	4.5 V_{DC} to 6.3 V_{DC}

Electrical Characteristics

The following specifications apply for V_{CC} = 5 V_{DC}, $T_{MIN} \leq T_A \leq T_{MAX}$ and f_{CLK} = 640 kHz unless otherwise specified.

Parameter	Conditions	Min	Typ	Max	Units
ADC0801: Total Adjusted Error (Note 8)	With Full-Scale Adj. (See Section 2.5.2)			± 1/4	LSB
ADC0802: Total Unadjusted Error (Note 8)	V_{REF}/2 = 2.500 V_{DC}			± 1/2	LSB
ADC0803: Total Adjusted Error (Note 8)	With Full-Scale Adj. (See Section 2.5.2)			± 1/2	LSB
ADC0804: Total Unadjusted Error (Note 8)	V_{REF}/2 = 2.500 V_{DC}			± 1	LSB
ADC0805: Total Unadjusted Error (Note 8)	V_{REF}/2-No Connection			± 1	LSB
V_{REF}/2 Input Resistance (Pin 9)	ADC0801/02/03/05 ADC0804 (Note 9)	2.5 0.75	8.0 1.1		kΩ kΩ
Analog Input Voltage Range	(Note 4) V(+) or V(−)	Gnd−0.05		V_{CC} + 0.05	V_{DC}
DC Common-Mode Error	Over Analog Input Voltage Range		± 1/16	± 1/8	LSB
Power Supply Sensitivity	V_{CC} = 5 V_{DC} ± 10% Over Allowed V_{IN}(+) and V_{IN}(−) Voltage Range (Note 4)		± 1/16	± 1/8	LSB

AC Electrical Characteristics

The following specifications apply for V_{CC} = 5 V_{DC} and T_A = 25°C unless otherwise specified.

Symbol	Parameter	Conditions	Min	Typ	Max	Units
T_C	Conversion Time	f_{CLK} = 640 kHz (Note 6)	103		114	μs
T_C	Conversion Time	(Note 5, 6)	66		73	1/f_{CLK}
f_{CLK}	Clock Frequency Clock Duty Cycle	V_{CC} = 5V, (Note 5) (Note 5)	100 40	640	1460 60	kHz %
CR	Conversion Rate in Free-Running Mode	$\overline{INTR}$ tied to $\overline{WR}$ with $\overline{CS}$ = 0 V_{DC}, f_{CLK} = 640 kHz	8770		9708	conv/s
$t_{W(\overline{WR})L}$	Width of $\overline{WR}$ Input (Start Pulse Width)	$\overline{CS}$ = 0 V_{DC} (Note 7)	100			ns
t_{ACC}	Access Time (Delay from Falling Edge of $\overline{RD}$ to Output Data Valid)	C_L = 100 pF		135	200	ns
t_{1H}, t_{0H}	TRI-STATE Control (Delay from Rising Edge of $\overline{RD}$ to Hi-Z State)	C_L = 10 pF, R_L = 10k (See TRI-STATE Test Circuits)		125	200	ns
t_{WI}, t_{RI}	Delay from Falling Edge of $\overline{WR}$ or $\overline{RD}$ to Reset of $\overline{INTR}$			300	450	ns
C_{IN}	Input Capacitance of Logic Control Inputs			5	7.5	pF
C_{OUT}	TRI-STATE Output Capacitance (Data Buffers)			5	7.5	pF
CONTROL INPUTS [Note: CLK IN (Pin 4) is the input of a Schmitt trigger circuit and is therefore specified separately]						
V_{IN} (1)	Logical "1" Input Voltage (Except Pin 4 CLK IN)	V_{CC} = 5.25 V_{DC}	2.0		15	V_{DC}

AC Electrical Characteristics (Continued)

The following specifications apply for V_{CC} = 5 V_{DC} and $T_{MIN} \leq T_A \leq T_{MAX}$, unless otherwise specified.

Symbol	Parameter	Conditions	Min	Typ	Max	Units
CONTROL INPUTS [Note: CLK IN (Pin 4) is the input of a Schmitt trigger circuit and is therefore specified separately]						
V_{IN} (0)	Logical "0" Input Voltage (Except Pin 4 CLK IN)	V_{CC} = 4.75 V_{DC}			0.8	V_{DC}
I_{IN} (1)	Logical "1" Input Current (All Inputs)	V_{IN} = 5 V_{DC}		0.005	1	μA_{DC}
I_{IN} (0)	Logical "0" Input Current (All Inputs)	V_{IN} = 0 V_{DC}	−1	−0.005		μA_{DC}
CLOCK IN AND CLOCK R						
V_T+	CLK IN (Pin 4) Positive Going Threshold Voltage		2.7	3.1	3.5	V_{DC}
V_T-	CLK IN (Pin 4) Negative Going Threshold Voltage		1.5	1.8	2.1	V_{DC}
V_H	CLK IN (Pin 4) Hysteresis $(V_T+) - (V_T-)$		0.6	1.3	2.0	V_{DC}
V_{OUT} (0)	Logical "0" CLK R Output Voltage	I_O = 360 μA V_{CC} = 4.75 V_{DC}			0.4	V_{DC}
V_{OUT} (1)	Logical "1" CLK R Output Voltage	I_O = −360 μA V_{CC} = 4.75 V_{DC}	2.4			V_{DC}
DATA OUTPUTS AND $\overline{INTR}$						
V_{OUT} (0)	Logical "0" Output Voltage Data Outputs $\overline{INTR}$ Output	I_{OUT} = 1.6 mA, V_{CC} = 4.75 V_{DC} I_{OUT} = 1.0 mA, V_{CC} = 4.75 V_{DC}			0.4 0.4	V_{DC} V_{DC}
V_{OUT} (1)	Logical "1" Output Voltage	I_O = −360 μA, V_{CC} = 4.75 V_{DC}	2.4			V_{DC}
V_{OUT} (1)	Logical "1" Output Voltage	I_O = −10 μA, V_{CC} = 4.75 V_{DC}	4.5			V_{DC}
I_{OUT}	TRI-STATE Disabled Output Leakage (All Data Buffers)	V_{OUT} = 0 V_{DC} V_{OUT} = 5 V_{DC}	−3		3	μA_{DC} μA_{DC}
I_{SOURCE}		V_{OUT} Short to Gnd, T_A = 25°C	4.5	6		mA_{DC}
I_{SINK}		V_{OUT} Short to V_{CC}, T_A = 25°C	9.0	16		mA_{DC}
POWER SUPPLY						
I_{CC}	Supply Current (Includes Ladder Current)	f_{CLK} = 640 kHz, $V_{REF}/2$ = NC, T_A = 25°C and $\overline{CS}$ = 5V				
	ADC0801/02/03/04LCJ/05			1.1	1.8	mA
	ADC0804LCN/LCV/LCWM			1.9	2.5	mA

Note 1: Absolute Maximum Ratings indicate limits beyond which damage to the device may occur. DC and AC electrical specifications do not apply when operating the device beyond its specified operating conditions.

Note 2: All voltages are measured with respect to Gnd, unless otherwise specified. The separate A Gnd point should always be wired to the D Gnd.

Note 3: A zener diode exists, internally, from V_{CC} to Gnd and has a typical breakdown voltage of 7 V_{DC}.

Note 4: For $V_{IN}(-) \geq V_{IN}(+)$ the digital output code will be 0000 0000. Two on-chip diodes are tied to each analog input (see block diagram) which will forward conduct for analog input voltages one diode drop below ground or one diode drop greater than the V_{CC} supply. Be careful, during testing at low V_{CC} levels (4.5V), as high level analog inputs (5V) can cause this input diode to conduct–especially at elevated temperatures, and cause errors for analog inputs near full-scale. The spec allows 50 mV forward bias of either diode. This means that as long as the analog V_{IN} does not exceed the supply voltage by more than 50 mV, the output code will be correct. To achieve an absolute 0 V_{DC} to 5 V_{DC} input voltage range will therefore require a minimum supply voltage of 4.950 V_{DC} over temperature variations, initial tolerance and loading.

Note 5: Accuracy is guaranteed at f_{CLK} = 640 kHz. At higher clock frequencies accuracy can degrade. For lower clock frequencies, the duty cycle limits can be extended so long as the minimum clock high time interval or minimum clock low time interval is no less than 275 ns.

Note 6: With an asynchronous start pulse, up to 8 clock periods may be required before the internal clock phases are proper to start the conversion process. The start request is internally latched, see *Figure 2* and section 2.0.

Note 7: The $\overline{CS}$ input is assumed to bracket the $\overline{WR}$ strobe input and therefore timing is dependent on the $\overline{WR}$ pulse width. An arbitrarily wide pulse width will hold the converter in a reset mode and the start of conversion is initiated by the low to high transition of the $\overline{WR}$ pulse (see timing diagrams).

Note 8: None of these A/Ds requires a zero adjust (see section 2.5.1). To obtain zero code at other analog input voltages see section 2.5 and *Figure 5*.

Note 9: The $V_{REF}/2$ pin is the center point of a two resistor divider connected from V_{CC} to ground. Each resistor is 2.2k, except for the ADC0804LCJ where each resistor is 16k. Total ladder input resistance is the sum of the two equal resistors.

Note 10: Human body model, 100 pF discharged through a 1.5 kΩ resistor.

8237A/8237A-4/8237A-5
HIGH PERFORMANCE
PROGRAMMABLE DMA CONTROLLER

- **Enable/Disable Control of Individual DMA Requests**

- **Four Independent DMA Channels**

- **Independent Autoinitialization of all Channels**

- **Memory-to-Memory Transfers**

- **Memory Block Initialization**

- **Address Increment or Decrement**

- **High performance: Transfers up to 1.6M Bytes/Second with 5 MHz 8237A-5**

- **Directly Expandable to any Number of Channels**

- **End of Process Input for Terminating Transfers**

- **Software DMA Requests**

- **Independent Polarity Control for DREQ and DACK Signals**

- **Available in EXPRESS**
  – Standard Temperature Range

The 8237A Multimode Direct Memory Access (DMA) Controller is a peripheral interface circuit for microprocessor systems. It is designed to improve system performance by allowing external devices to directly transfer information from the system memory. Memory-to-memory transfer capability is also provided. The 8237A offers a wide variety of programmable control features to enhance data throughput and system optimization and to allow dynamic reconfiguration under program control.

The 8237A is designed to be used in conjunction with an external 8-bit address register such as the 8282. It contains four independent channels and may be expanded to any number of channels by cascading additional controller chips.

The three basic transfer modes allow programmability of the types of DMA service by the user. Each channel can be individually programmed to Autoinitialize to its original condition following an End of Process (EOP).

Each channel has a full 64K address and word count capability.

The 8237A-4 and 8237A-5 are 4 MHz and 5 MHz selected versions of the standard 3 MHz 8237A respectively.

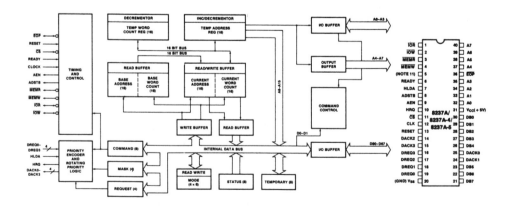

Figure 1. Block Diagram

Figure 2.
Pin Configuration

Table 1. Pin Description

Symbol	Type	Name and Function
V_{CC}		**Power:** +5 volt supply.
V_{SS}		**Ground:** Ground.
CLK	I	**Clock Input:** Clock Input controls the internal operations of the 8237A and its rate of data transfers. The input may be driven at up to 3 MHz for the standard 8237A and up to 5 MHz for the 8237A-5.
CS	I	**Chip Select:** Chip Select is an active low input used to select the 8237A as an I/O device during the Idle cycle. This allows CPU communication on the data bus.
RESET	I	**Reset:** Reset is an active high input which clears the Command, Status, Request and Temporary registers. It also clears the first/last flip/flop and sets the Mask register. Following a Reset the device is in the Idle cycle.
READY	I	**Ready:** Ready is an input used to extend the memory read and write pulses from the 8237A to accommodate slow memories or I/O peripheral devices. Ready must not make transitions during its specified setup/hold time.
HLDA	I	**Hold Acknowledge:** The active high Hold Acknowledge from the CPU indicates that it has relinquished control of the system busses.
DREQ0–DREQ3	I	**DMA Request:** The DMA Request lines are individual asynchronous channel request inputs used by peripheral circuits to obtain DMA service. In fixed Priority, DREQ0 has the highest priority and DREQ3 has the lowest priority. A request is generated by activating the DREQ line of a channel. DACK will acknowledge the recognition of DREQ signal. Polarity of DREQ is programmable. Reset intializes these lines to active high. DREQ must be maintained until the corresponding DACK goes active.
DB0–DB7	I/O	**Data Bus:** The Data Bus lines are bidirectional three-state signals connected to the system data bus. The outputs are enabled in the Program condition during the I/O Read to output the contents of an Address register, a Status register, the Temporary register or a Word Count register to the CPU. The outputs are disabled and the inputs are read during an I/O Write cycle when the CPU is programming the 8237A control registers. During DMA cycles the most significant 8 bits of the address are output onto the data bus to be strobed into an external latch by ADSTB. In mem-

Symbol	Type	Name and Function
		ory-to-memory operations, data from the memory comes into the 8237A on the data bus during the read-from-memory transfer. In the write-to-memory transfer, the data bus outputs place the data into the new memory location.
IOR	I/O	**I/O Read:** I/O Read is a bidirectional active low three-state line. In the Idle cycle, it is an input control signal used by the CPU to read the control registers. In the Active cycle, it is an output control signal used by the 8237A to access data from a peripheral during a DMA Write transfer.
IOW	I/O	**I/O Write:** I/O Write is a bidirectional active low three-state line. In the Idle cycle, it is an input control signal used by the CPU to load information into the 8237A. In the Active cycle, it is an output control signal used by the 8237A to load data to the peripheral during a DMA Read transfer.
EOP	I/O	**End of Process:** End of Process is an active low bidirectional signal. Information concerning the completion of DMA services is available at the bidirectional EOP pin. The 8237A allows an external signal to terminate an active DMA service. This is accomplished by pulling the EOP input low with an external EOP signal. The 8237A also generates a pulse when the terminal count (TC) for any channel is reached. This generates an EOP signal which is output through the EOP Line. The reception of EOP, either internal or external, will cause the 8237A to terminate the service, reset the request, and, if Autoinitialize is enabled, to write the base registers to the current registers of that channel. The mask bit and TC bit in the status word will be set for the currently active channel by EOP unless the channel is programmed for Autoinitialize. In that case, the mask bit remains clear. During memory-to-memory transfers, EOP will be output when the TC for channel 1 occurs. EOP should be tied high with a pull-up resistor if it is not used to prevent erroneous end of process inputs.
A0–A3	I/O	**Address:** The four least significant address lines are bidirectional three-state signals. In the Idle cycle they are inputs and are used by the 8237A to address the control register to be loaded or read. In the Active cycle they are outputs and provide the lower 4 bits of the output address.

Table 1. Pin Description (Continued)

Symbol	Type	Name and Function
A4-A7	O	**Address:** The four most significant address lines are three-state outputs and provide 4 bits of address. These lines are enabled only during the DMA service.
HRQ	O	**Hold Request:** This is the Hold Request to the CPU and is used to request control of the system bus. If the corresponding mask bit is clear, the presence of any valid DREQ causes 8237A to issue the HRQ. After HRQ goes active at least one clock cycle (TCY) must occur before HLDA goes active.
DACK0-DACK3	O	**DMA Acknowledge:** DMA Acknowledge is used to notify the individual peripherals when one has been granted a DMA cycle. The sense of these lines is programmable. Reset initializes them to active low.

Symbol	Type	Name and Function
AEN	O	**Address Enable:** Address Enable enables the 8-bit latch containing the upper 8 address bits onto the system address bus. AEN can also be used to disable other system bus drivers during DMA transfers. AEN is active HIGH.
ADSTB	O	**Address Strobe:** The active high, Address Strobe is used to strobe the upper address byte into an external latch.
MEMR	O	**Memory Read:** The Memory Read signal is an active low three-state output used to access data from the selected memory location during a DMA Read or a memory-to-memory transfer.
MEMW	O	**Memory Write:** The Memory Write is an active low three-state output used to write data to the selected memory location during a DMA Write or a memory-to-memory transfer.

FUNCTIONAL DESCRIPTION

The 8237A block diagram includes the major logic blocks and all of the internal registers. The data interconnection paths are also shown. Not shown are the various control signals between the blocks. The 8237A contains 344 bits of internal memory in the form of registers. Figure 3 lists these registers by name and shows the size of each. A detailed description of the registers and their functions can be found under Register Description.

Name	Size	Number
Base Address Registers	16 bits	4
Base Word Count Registers	16 bits	4
Current Address Registers	16 bits	4
Current Word Count Registers	16 bits	4
Temporary Address Register	16 bits	1
Temporary Word Count Register	16 bits	1
Status Register	8 bits	1
Command Register	8 bits	1
Temporary Register	8 bits	1
Mode Registers	6 bits	4
Mask Register	4 bits	1
Request Register	4 bits	1

Figure 3. 8237A Internal Registers

The 8237A contains three basic blocks of control logic. The Timing Control block generates internal timing and external control signals for the 8237A. The Program Command Control block decodes the various commands given to the 8237A by the microprocessor prior to servicing a DMA Request. It also decodes the Mode Control word used to select the type of DMA during the servicing. The Priority Encoder block resolves priority contention between DMA channels requesting service simultaneously.

The Timing Control block derives internal timing from the clock input. In 8237A systems this input will usually be the φ2 TTL clock from an 8224 or CLK from an 8085AH or 8284A. For 8085AH-2 systems above 3.9 MHz, the 8085 CLK(OUT) does not satisfy 8237A-5 clock LOW and HIGH time requirements. In this case, an external clock should be used to drive the 8237A-5.

DMA Operation

The 8237A is designed to operate in two major cycles. These are called Idle and Active cycles. Each device cycle is made up of a number of states. The 8237A can assume seven separate states, each composed of one full clock period. State I (SI) is the inactive state. It is entered when the 8237A has no valid DMA requests pending. While in SI, the DMA controller is inactive but may be in the Program Condition, being programmed by the processor. State S0 (S0) is the first state of a DMA service. The 8237A has requested a hold but the processor has not yet returned an acknowledge. The 8237A may still be programmed until it receives HLDA from the CPU. An acknowledge from the CPU will signal that DMA transfers may begin. S1, S2, S3 and S4 are the working states of the DMA service. If more time is needed to complete a transfer than is available with normal timing, wait states (SW) can be inserted between S2 or S3 and S4 by the use of the Ready line on the 8237A. Note that the data is transferred directly from the I/O device to memory (or vice versa) with IOR and MEMW (or MEMR and IOW) being active at the same time. The data is not read into or driven out of the 8237A in I/O-to-memory or memory-to-I/O DMA transfers.

Memory-to-memory transfers require a read-from and a write-to-memory to complete each transfer. The states, which resemble the normal working states, use two digit numbers for identification. Eight states are required for a single transfer. The first four states (S11, S12, S13, S14) are used for the read-from-memory half

666

and the last four states (S21, S22, S23, S24) for the write-to-memory half of the transfer.

IDLE CYCLE

When no channel is requesting service, the 8237A will enter the Idle cycle and perform "SI" states. In this cycle the 8237A will sample the DREQ lines every clock cycle to determine if any channel is requesting a DMA service. The device will also sample $\overline{CS}$, looking for an attempt by the microprocessor to write or read the internal registers of the 8237A. When $\overline{CS}$ is low and HLDA is low, the 8237A enters the Program Condition. The CPU can now establish, change or inspect the internal definition of the part by reading from or writing to the internal registers. Address lines A0–A3 are inputs to the device and select which registers will be read or written. The $\overline{IOR}$ and $\overline{IOW}$ lines are used to select and time reads or writes. Due to the number and size of the internal registers, an internal flip-flop is used to generate an additional bit of address. This bit is used to determine the upper or lower byte of the 16-bit Address and Word Count registers. The flip-flop is reset by Master Clear or Reset. A separate software command can also reset this flip-flop.

Special software commands can be executed by the 8237A in the Program Condition. These commands are decoded as sets of addresses with the $\overline{CS}$ and $\overline{IOW}$. The commands do not make use of the data bus. Instructions include Clear First/Last Flip-FLop and Master Clear.

ACTIVE CYCLE

When the 8237A is in the Idle cycle and a non-masked channel requests a DMA service, the device will output an HRQ to the microprocessor and enter the Active cycle. It is in this cycle that the DMA service will take place, in one of four modes:

Single Transfer Mode — In Single Transfer mode the device is programmed to make one transfer only. The word count will be decremented and the address decremented or incremented following each transfer. When the word count "rolls over" from zero to FFFFH, a Terminal Count (TC) will cause an Autoinitialize if the channel has been programmed to do so.

DREQ must be held active until DACK becomes active in order to be recognized. If DREQ is held active throughout the single transfer, HRQ will go inactive and release the bus to the system. It will again go active and, upon receipt of a new HLDA, another single transfer will be performed, in 8080A, 8085AH, 8088, or 8086 system this will ensure one full machine cycle execution between DMA transfers. Details of timing between the 8237A and other bus control protocols will depend upon the characteristics of the microprocessor involved.

Block Transfer Mode — In Block Transfer mode the device is activated by DREQ to continue making transfers during the service until a TC, caused by word count going to FFFFH, or an external End of Process ($\overline{EOP}$) is encountered. DREQ need only be held active until DACK

becomes active. Again, an Autoinitialization will occur at the end of the service if the channel has been programmed for it.

Demand Transfer Mode — In Demand Transfer mode the device is programmed to continue making transfers until a TC or external $\overline{EOP}$ is encountered or until DREQ goes inactive. Thus transfers may continue until the I/O device has exhausted its data capacity. After the I/O device has had a chance to catch up, the DMA service is re-established by means of a DREQ. During the time between services when the microprocessor is allowed to operate, the intermediate values of address and word count are stored in the 8237A Current Address and Current Word Count registers. Only an $\overline{EOP}$ can cause an Autoinitialize at the end of the service. $\overline{EOP}$ is generated either by TC or by an external signal.

Cascade Mode — This mode is used to cascade more than one 8237A together for simple system expansion. The HRQ and HLDA signals from the additional 8237A are connected to the DREQ and DACK signals of a channel of the initial 8237A. This allows the DMA requests of the additional device to propagate through the priority network circuitry of the preceding device. The priority chain is preserved and the new device must wait for its turn to acknowledge requests. Since the cascade channel of the initial 8237A is used only for prioritizing the additional device, it does not output any address or control signals of its own. These could conflict with the outputs of the active channel in the added device. The 8237A will respond to DREQ and DACK but all other outputs except HRQ will be disabled.

Figure 4 shows two additional devices cascaded into an initial device using two of the previous channels. This forms a two level DMA system. More 8237As could be added at the second level by using the remaining channels of the first level. Additional devices can also be added by cascading into the channels of the second level devices, forming a third level.

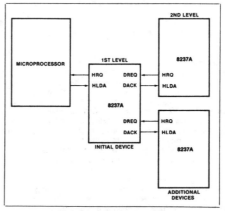

Figure 4. Cascaded 8237As

TRANSFER TYPES

Each of the three active transfer modes can perform three different types of transfers. These are Read, Write and Verify. Write transfers move data from an I/O device to the memory by activating MEMW and IOR. Read transfers move data from memory to an I/O device by activating MEMR and IOW. Verify transfers are pseudo transfers. The 8237A operates as in Read or Write transfers generating addresses, and responding to EOP, etc. However, the memory and I/O control lines all remain inactive. Verify mode is not permitted during memory to memory operation.

Memory-to-Memory — To perform block moves of data from one memory address space to another with a minimum of program effort and time, the 8237A includes a memory-to-memory transfer feature. Programming a bit in the Command register selects channels 0 and 1 to operate as memory-to-memory transfer channels. The transfer is initiated by setting the software DREQ for channel 0. The 8237A requests a DMA service in the normal manner. After HLDA is true, the device, using eight-state transfers in Block Transfer mode, reads data from the memory. The channel 0 Current Address register is the source for the address used and is decremented or incremented in the normal manner. The data byte read from the memory is stored in the 8237A internal Temporary register. Channel 1 then writes the data from the Temporary register to memory using the address in its Current Address register and incrementing or decrementing it in the normal manner. The channel 1 Current Word Count is decremented. When the word count of channel 1 goes to FFFFH, a TC is generated causing an EOP output terminating the service.

Channel 0 may be programmed to retain the same address for all transfers. This allows a single word to be written to a block of memory.

The 8237A will respond to external EOP signals during memory-to-memory transfers. Data comparators in block search schemes may use this input to terminate the service when a match is found. The timing of memory-to-memory transfers is found in Figure 12. Memory-to-memory operations can be detected as an active AEN with no DACK outputs.

Autoinitialize — By programming a bit in the Mode register, a channel may be set up as an Autoinitialize channel. During Autoinitialize initialization, the original values of the Current Address and Current Word Count registers are automatically restored from the Base Address and Base Word count registers of that channel following EOP. The base registers are loaded simultaneously with the current registers by the microprocessor and remain unchanged throughout the DMA service. The mask bit is not set when the channel is in Autoinitialize. Following Autoinitialize the channel is ready to perform another DMA service, without CPU intervention, as soon as a valid DREQ is detected.

Priority — The 8237A has two types of priority encoding available as software selectable options. The first is Fixed Priority which fixes the channels in priority order

based upon the descending value of their number. The channel with the lowest priority is 3 followed by 2, 1 and the highest priority channel, 0. After the recognition of any one channel for service, the other channels are prevented from interferring with that service until it is completed.

The second scheme is Rotating Priority. The last channel to get service becomes the lowest priority channel with the others rotating accordingly.

With Rotating Priority in a single chip DMA system, any device requesting service is guaranteed to be recognized after no more than three higher priority services have occurred. This prevents any one channel from monopolizing the system.

Compressed Timing — In order to achieve even greater throughput where system characteristics permit, the 8237A can compress the transfer time to two clock cycles. From Figure 11 it can be seen that state S3 is used to extend the access time of the read pulse. By removing state S3, the read pulse width is made equal to the write pulse width and a transfer consists only of state S2 to change the address and state S4 to perform the read/write. S1 states will still occur when A8–A15 need updating (see Address Generation). Timing for compressed transfers is found in Figure 14.

Address Generation — In order to reduce pin count, the 8237A multiplexes the eight higher order address bits on the data lines. State S1 is used to output the higher order address bits to an external latch from which they may be placed on the address bus. The falling edge of Address Strobe (ADSTB) is used to load these bits from the data lines to the latch. Address Enable (AEN) is used to enable the bits onto the address bus through a three-state enable. The lower order address bits are output by the 8237A directly. Lines A0–A7 should be connected to the address bus. Figure 11 shows the time relationships between CLK, AEN, ADSTB, DB0–DB7 and A0–A7.

During Block and Demand Transfer mode services, which include multiple transfers, the addresses generated will be sequential. For many transfers the data held in the external address latch will remain the same. This data need only change when a carry or borrow from A7 to A8 takes place in the normal sequence of addresses. To save time and speed transfers, the 8237A executes S1 states only when updating of A8–A15 in the latch is necessary. This means for long services, S1 states and Address Strobes may occur only once every 256 transfers, a savings of 255 clock cycles for each 256 transfers.

REGISTER DESCRIPTION

Current Address Register — Each channel has a 16-bit Current Address register. This register holds the value of the address used during DMA transfers. The address is automatically incremented or decremented after each transfer and the intermediate values of the address are stored in the Current Address register during the transfer. This register is written or read by the microprocessor in successive 8-bit bytes. It may also be reinitialized by an Autoinitialize back to its original value. Autoinitialize takes place only after an $\overline{EOP}$.

Current Word Register — Each channel has a 16-bit Current Word Count register. This register determines the number of transfers to be performed. The actual number of transfers will be one more than the number programmed in the Current Word Count register (i.e., programming a count of 100 will result in 101 transfers). The word count is decremented after each transfer. The intermediate value of the word count is stored in the register during the transfer. When the value in the register goes from zero to FFFFH, a TC will be generated. This register is loaded or read in successive 8-bit bytes by the microprocessor in the Program Condition. Following the end of a DMA service it may also be reinitialized by an Autoinitialization back to its original value. Autoinitialize can occur only when an $\overline{EOP}$ occurs. If it is not Autoinitialized, this register will have a count of FFFFH after TC.

Base Address and Base Word Count Registers — Each channel has a pair of Base Address and Base Word Count registers. These 16-bit registers store the original value of their associated current registers. During Autoinitialize these values are used to restore the current registers to their original values. The base registers are written simultaneously with their corresponding current register in 8-bit bytes in the Program Condition by the microprocessor. These registers cannot be read by the microprocessor.

Command Register — This 8-bit register controls the operation of the 8237A. It is programmed by the microprocessor in the Program Condition and is cleared by Reset or a Master Clear instruction. The following table lists the function of the command bits. See Figure 6 for address coding.

Mode Register — Each channel has a 6-bit Mode register associated with it. When the register is being written to by the microprocessor in the Program Condition, bits 0 and 1 determine which channel Mode register is to be written.

Request Register — The 8237A can respond to requests for DMA service which are initiated by software as well as by a DREQ. Each channel has a request bit associated with it in the 4-bit Request register. These are non-maskable and subject to prioritization by the Priority Encoder network. Each register bit is set or reset separately under software control or is cleared upon generation of a TC or external $\overline{EOP}$. The entire register is cleared by a Reset. To set or reset a bit, the software loads the proper form of the data word. See Figure 5 for register address coding. In order to make a software request, the channel must be in Block Mode.

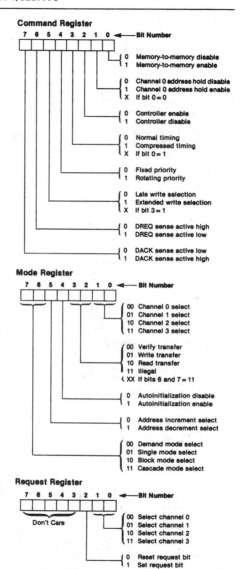

Command Register

Bit Number 7 6 5 4 3 2 1 0

0	Memory-to-memory disable
1	Memory-to-memory enable
0	Channel 0 address hold disable
1	Channel 0 address hold enable
X	If bit 0 = 0
0	Controller enable
1	Controller disable
0	Normal timing
1	Compressed timing
X	If bit 0 = 1
0	Fixed priority
1	Rotating priority
0	Late write selection
1	Extended write selection
X	If bit 3 = 1
0	DREQ sense active high
1	DREQ sense active low
0	DACK sense active low
1	DACK sense active high

Mode Register

Bit Number 7 6 5 4 3 2 1 0

00	Channel 0 select
01	Channel 1 select
10	Channel 2 select
11	Channel 3 select
00	Verify transfer
01	Write transfer
10	Read transfer
11	Illegal
XX	If bits 6 and 7 = 11
0	Autoinitialization disable
1	Autoinitialization enable
0	Address increment select
1	Address decrement select
00	Demand mode select
01	Single mode select
10	Block mode select
11	Cascade mode select

Request Register

Bit Number 7 6 5 4 3 2 1 0

Don't Care

00	Select channel 0
01	Select channel 1
10	Select channel 2
11	Select channel 3
0	Reset request bit
1	Set request bit

Mask Register — Each channel has associated with it a mask bit which can be set to disable the incoming DREQ. Each mask bit is set when its associated channel produces an EOP if the channel is not programmed for Autoinitialize. Each bit of the 4-bit Mask register may also be set or cleared separately under software control. The entire register is also set by a Reset. This disables all DMA requests until a clear Mask register instruction allows them to occur. The instruction to separately set or clear the mask bits is similar in form to that used with the Request register. See Figure 5 for instruction addressing.

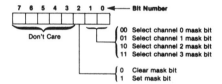

All four bits of the Mask register may also be written with a single command.

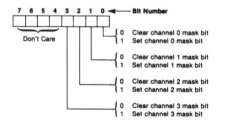

Register	Operation	Signals						
		CS	IOR	IOW	A3	A2	A1	A0
Command	Write	0	1	0	1	0	0	0
Mode	Write	0	1	0	1	0	1	1
Request	Write	0	1	0	1	0	0	1
Mask	Set/Reset	0	1	0	1	0	1	0
Mask	Write	0	1	0	1	1	1	1
Temporary	Read	0	0	1	1	1	0	1
Status	Read	0	0	1	1	0	0	0

Figure 5. Definition of Register Codes

Status Register — The Status register is available to be read out of the 8237A by the microprocessor. It contains information about the status of the devices at this point. This information includes which channels have reached a terminal count and which channels have pending DMA requests. Bits 0–3 are set every time a TC is reached by that channel or an external EOP is applied. These bits are cleared upon Reset and on each Status Read. Bits 4–7 are set whenever their corresponding channel is requesting service.

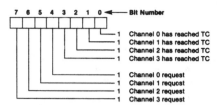

Temporary Register — The Temporary register is used to hold data during memory-to-memory transfers. Following the completion of the transfers, the last word moved can be read by the microprocessor in the Program Condition. The Temporary register always contains the last byte transferred in the previous memory-to-memory operation, unless cleared by a Reset.

Software Commands — These are additional special software commands which can be executed in the Program Condition. They do not depend on any specific bit pattern on the data bus. The two software commands are:

Clear First/Last Flip-Flop: This command is executed prior to writing or reading new address or word count information to the 8237A. This initializes the flip-flop to a known state so that subsequent accesses to register contents by the microprocessor will address upper and lower bytes in the correct sequence.

Master Clear: This software instruction has the same effect as the hardware Reset. The Command, Status, Request, Temporary, and Internal First/Last Flip-Flop registers are cleared and the Mask register is set. The 8237A will enter the Idle cycle.

Clear Mask Register: This command clears the mask bits of all four channels, enabling them to accept DMA requests.

Figure 6 lists the address codes for the software commands:

Signals						Operation
A3	A2	A1	A0	IOR	IOW	
1	0	0	0	0	1	Read Status Register
1	0	0	0	1	0	Write Command Register
1	0	0	1	0	1	Illegal
1	0	0	1	1	0	Write Request Register
1	0	1	0	0	1	Illegal
1	0	1	0	1	0	Write Single Mask Register Bit
1	0	1	1	0	1	Illegal
1	0	1	1	1	0	Write Mode Register
1	1	0	0	0	1	Illegal
1	1	0	0	1	0	Clear Byte Pointer Flip/Flop
1	1	0	1	0	1	Read Temporary Register
1	1	0	1	1	0	Master Clear
1	1	1	0	0	1	Illegal
1	1	1	0	1	0	Clear Mask Register
1	1	1	1	0	1	Illegal
1	1	1	1	1	0	Write All Mask Register Bits

Figure 6. Software Command Codes

Channel	Register	Operation	Signals							Internal Flip-Flop	Data Bus DB0–DB7
			CS̄	IOR̄	IOW̄	A3	A2	A1	A0		
0	Base and Current Address	Write	0	1	0	0	0	0	0	0	A0–A7
			0	1	0	0	0	0	0	1	A8–A15
	Current Address	Read	0	0	1	0	0	0	0	0	A0–A7
			0	0	1	0	0	0	0	1	A8–A15
	Base and Current Word Count	Write	0	1	0	0	0	0	1	0	W0–W7
			0	1	0	0	0	0	1	1	W8–W15
	Current Word Count	Read	0	0	1	0	0	0	1	0	W0–W7
			0	0	1	0	0	0	1	1	W8–W15
1	Base and Current Address	Write	0	1	0	0	0	1	0	0	A0–A7
			0	1	0	0	0	1	0	1	A8–A15
	Current Address	Read	0	0	1	0	0	1	0	0	A0–A7
			0	0	1	0	0	1	0	1	A8–A15
	Base and Current Word Count	Write	0	1	0	0	0	1	1	0	W0–W7
			0	1	0	0	0	1	1	1	W8–W15
	Current Word Count	Read	0	0	1	0	0	1	1	0	W0–W7
			0	0	1	0	0	1	1	1	W8–W15
2	Base and Current Address	Write	0	1	0	0	1	0	0	0	A0–A7
			0	1	0	0	1	0	0	1	A8–A15
	Current Address	Read	0	0	1	0	1	0	0	0	A0–A7
			0	0	1	0	1	0	0	1	A8–A15
	Base and Current Word Count	Write	0	1	0	0	1	0	1	0	W0–W7
			0	1	0	0	1	0	1	1	W8–W15
	Current Word Count	Read	0	0	1	0	1	0	1	0	W0–W7
			0	0	1	0	1	0	1	1	W8–W15
3	Base and Current Address	Write	0	1	0	0	1	1	0	0	A0–A7
			0	1	0	0	1	1	0	1	A8–A15
	Current Address	Read	0	0	1	0	1	1	0	0	A0–A7
			0	0	1	0	1	1	0	1	A8–A15
	Base and Current Word Count	Write	0	1	0	0	1	1	1	0	W0–W7
			0	1	0	0	1	1	1	1	W8–W15
	Current Word Count	Read	0	0	1	0	1	1	1	0	W0–W7
			0	0	1	0	1	1	1	1	W8–W15

Figure 7. Word Count and Address Register Command Codes

PROGRAMMING

The 8237A will accept programming from the host processor any time that HLDA is inactive; this is true even if HRQ is active. The responsibility of the host is to assure that programming and HLDA are mutually exclusive. Note that a problem can occur if a DMA request occurs, on an unmasked channel while the 8237A is being programmed. For instance, the CPU may be starting to reprogram the two byte Address register of channel 1 when channel 1 receives a DMA request. If the 8237A is enabled (bit 2 in the command register is 0) and channel 1 is unmasked, a DMA service will occur after only one byte of the Address register has been reprogrammed. This can be avoided by disabling the controller (setting bit 2 in the command register) or masking the channel before programming any other registers. Once the programming is complete, the controller can be enabled/unmasked.

After power-up it is suggested that all internal locations, especially the Mode registers, be loaded with some valid value. This should be done even if some channels are unused.

APPLICATION INFORMATION

Figure 8 shows a convenient method for configuring a DMA system with the 8237A controller and an 8080A/8085AH microprocessor system. The multimode DMA controller issues a HRQ to the processor whenever there is at least one valid DMA request from a peripheral device. When the processor replies with a HLDA signal, the 8237A takes control of the address bus, the data bus and the control bus. The address for the first transfer operation comes out in two bytes — the least significant 8 bits on the eight address outputs and the most significant 8 bits on the data bus. The contents of the data bus are then latched into the 8282 8-bit latch to complete the full 16 bits of the address bus. The 8282 is a high speed, 8-bit, three-state latch in a 20-pin package. After the initial transfer takes place, the latch is updated only after a carry or borrow is generated in the least significant address byte. Four DMA channels are provided when one 8237A is used.

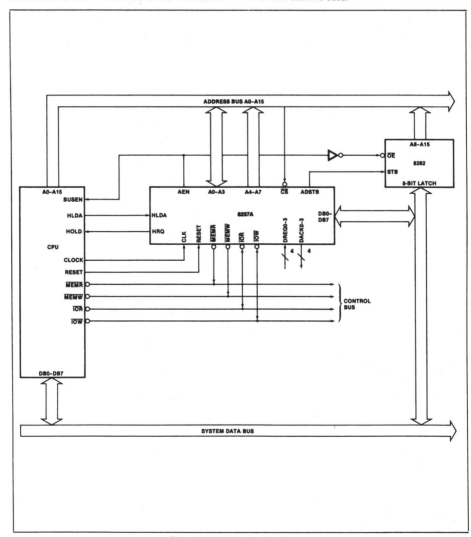

Figure 8. 8237A System Interface

ABSOLUTE MAXIMUM RATINGS*

Ambient Temperature under Bias ....0°C to 70°C
Storage Temperature .... − 65°C to + 150°C
Voltage on any Pin with
   Respect to Ground .... − 0.5 to 7V
Power Dissipation ....1.5 Watt

*NOTICE: Stresses above those listed under "Absolute Maximum Ratings" may cause permanent damage to the device. This is a stress rating only and functional operation of the device at these or any other conditions above those indicated in the operational sections of this specification is not implied. Exposure to absolute maximum rating conditions for extended periods may affect device reliability.

D.C. CHARACTERISTICS   (T_A = 0°C to 70°C, V_{CC} = 5.0V ±5%, GND = 0V)

Symbol	Parameter	Min.	Typ.(1)	Max.	Unit	Test Conditions
V_{OH}	Output HIGH Voltage	2.4			V	I_{OH} = − 200 µA
		3.3			V	I_{OH} = − 100 µA (HRQ Only)
V_{OL}	Output LOW Voltage			.45	V	I_{OL} = 2.0 mA (data bus) I_{OL} = 3.2 mA (other outputs)
V_{IH}	Input HIGH Voltage	2.0		V_{CC} + 0.5	V	
V_{IL}	Input LOW Voltage	− 0.5		0.8	V	
I_{LI}	Input Load Current			± 10	µA	0V ≤ V_{IN} ≤ V_{CC}
I_{LO}	Output Leakage Current			± 10	µA	0.45V ≤ V_{OUT} ≤ V_{CC}
I_{CC}	V_{CC} Supply Current		65	130	mA	T_A = +25°C
			75	150	mA	T_A = 0°C
C_O	Output Capacitance		4	8	pF	
C_I	Input Capacitance		8	15	pF	fc = 1.0 MHz, Inputs = 0V
C_{IO}	I/O Capacitance		10	18	pF	

NOTES:

1. Typical values are for T_A = 25°C, nominal supply voltage and nominal processing parameters.
2. Input timing parameters assume transition times of 20 ns or less. Waveform measurement points for both input and output signals are 2.0V for HIGH and 0.8V for LOW, unless otherwise noted.
3. Output loading is 1 TTL gate plus 50 pF capacitance, unless otherwise noted.
4. The net $\overline{IOW}$ or $\overline{MEMW}$ Pulse width for normal write will be TCY-100 ns and for extended write will be 2TCY-100 ns. The net $\overline{IOR}$ or $\overline{MEMR}$ pulse width for normal read will be 2TCY-50 ns and for compressed read will be TCY-50 ns.
5. TDQ is specified for two different output HIGH levels. TDQ1 is measured at 2.0V. TDQ2 is measured at 3.3V. The value for TDQ2 assumes an external 3.3 kΩ pull-up resistor connected from HRQ to V_{CC}.
6. DREQ should be held active until DACK is returned.
7. DREQ and DACK signals may be active high or active low. Timing diagrams assume the active high mode.
8. Output loading on the data bus is 1 TTL gate plus 100 pF capacitance.
9. Successive read and/or write operations by the external processor to program or examine the controller must be timed to allow at least 600 ns for the 8237A, at least 500 ns for the 8237A-4 and at least 400 ns for the 8237A-5, as recovery time between active read or write pulses.
10. Parameters are listed in alphabetical order.
11. Pin 5 is an input that should always be at a logic high level. An internal pull-up resistor will establish a logic high when the pin is left floating. Alternatively, pin 5 may be tied to V_{CC}.

A.C. TESTING INPUT, OUTPUT WAVEFORM

```
INPUT/OUTPUT

2.4 ──╲      ╱── 2.0        2.0 ──╲
       ╲    ╱                      
        ╳ TEST POINTS ╳            
       ╱    ╲                      
0.45 ──╱      ╲── 0.8        0.8 ──╱
```

A.C. TESTING: INPUTS ARE DRIVEN AT 2.4V FOR A LOGIC "1" AND 0.45V FOR A LOGIC "0". TIMING MEASUREMENTS ARE MADE AT 2.0V FOR A LOGIC "1" AND 0.8V FOR A LOGIC "0".

A.C. TESTING LOAD CIRCUIT

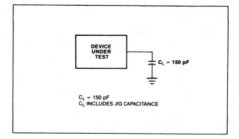

C_L = 150 pF
C_L INCLUDES JIG CAPACITANCE

A.C. CHARACTERISTICS—DMA (MASTER) MODE ($T_A = 0°C$ to $70°C$, $V_{CC} = +5V \pm 5\%$, $GND = 0V$)

Symbol	Parameter	8237A Min.	8237A Max.	8237A-4 Min.	8237A-4 Max.	8237A-5 Min.	8237A-5 Max.	Unit
TAEL	AEN HIGH from CLK LOW (S1) Delay Time		300		225		200	ns
TAET	AEN LOW from CLK HIGH (S1) Delay Time		200		150		130	ns
TAFAB	ADR Active to Float Delay from CLK HIGH		150		120		90	ns
TAFC	READ or WRITE Float from CLK HIGH		150		120		120	ns
TAFDB	DB Active to Float Delay from CLK HIGH		250		190		170	ns
TAHR	ADR from READ HIGH Hold Time	TCY–100		TCY–100		TCY–100		ns
TAHS	DB from ADSTB LOW Hold Time	50		40		30		ns
TAHW	ADR from WRITE HIGH Hold Time	TCY–50		TCY–50		TCY–50		ns
TAK	DACK Valid from CLK LOW Delay Time (Note 7)		250		220		170	ns
	EOP HIGH from CLK HIGH Delay Time		250		190		170	ns
	EOP LOW to CLK HIGH Delay Time		250		190		100	ns
TASM	ADR Stable from CLK HIGH		250		190		170	ns
TASS	DB to ADSTB LOW Setup Time	100		100		100		ns
TCH	Clock High Time (Transitions ≤ 10 ns)	120		100		80		ns
TCL	Clock LOW Time (Transitions ≤ 10 ns)	150		110		68		ns
TCY	CLK Cycle Time	320		250		200		ns
TDCL	CLK HIGH to READ or WRITE LOW Delay (Note 4)		270		200		190	ns
TDCTR	READ HIGH from CLK HIGH (S4) Delay Time (Note 4)		270		210		190	ns
TDCTW	WRITE HIGH from CLK HIGH (S4) Delay Time (Note 4)		200		150		130	ns
TDQ1	HRQ Valid from CLK HIGH Delay Time (Note 5)		160		120		120	ns
TDQ2			250		190		120	ns
TEPS	EOP LOW from CLK LOW Setup Time	60		45		40		ns
TEPW	EOP Pulse Width	300		225		220		ns
TFAAB	ADR Float to Active Delay from CLK HIGH		250		190		170	ns
TFAC	READ or WRITE Active from CLK HIGH		200		150		150	ns
TFADB	DB Float to Active Delay from CLK HIGH		300		225		200	ns
THS	HLDA Valid to CLK HIGH Setup Time	100		75		75		ns
TIDH	Input Data from MEMR HIGH Hold Time	0		0		0		ns
TIDS	Input Data to MEMR HIGH Setup Time	250		190		170		ns
TODH	Output Data from MEMW HIGH Hold Time	20		20		10		ns
TODV	Output Data Valid to MEMW HIGH	200		125		125		ns
TQS	DREQ to CLK LOW (SI, S4) Setup Time (Note 7)	0		0		0		ns
TRH	CLK to READY LOW Hold Time	20		20		20		ns
TRS	READY to CLK LOW Setup Time	100		60		60		ns
TSTL	ADSTB HIGH from CLK HIGH Delay Time		200		150		130	ns
TSTT	ADSTB LOW from CLK HIGH Delay Time		140		110		90	ns

A.C. CHARACTERISTICS—PERIPHERAL (SLAVE) MODE (T_A = 0°C to 70°C, V_{CC} = 5.0V ±5%, GND = 0V)

Symbol	Parameter	8237A Min.	8237A Max.	8237A-4 Min.	8237A-4 Max.	8237A-5 Min.	8237A-5 Max.	Unit
TAR	ADR Valid or $\overline{CS}$ LOW to $\overline{READ}$ LOW	50		50		50		ns
TAW	ADR Valid to $\overline{WRITE}$ HIGH Setup Time	200		150		150		ns
TCW	CS LOW to $\overline{WRITE}$ HIGH Setup Time	200		150		150		ns
TDW	Data Valid to $\overline{WRITE}$ HIGH Setup Time	200		150		150		ns
TRA	ADR or CS Hold from $\overline{READ}$ HIGH	0		0		0		ns
TRDE	Data Access from $\overline{READ}$ LOW (Note 8)		200		200		140	ns
TRDF	DB Float Delay from $\overline{READ}$ HIGH	20	100	20	100	0	70	ns
TRSTD	Power Supply HIGH to RESET LOW Setup Time	500		500		500		ns
TRSTS	RESET to First $\overline{IOWR}$	2TCY		2TCY		2TCY		ns
TRSTW	RESET Pulse Width	300		300		300		ns
TRW	$\overline{READ}$ Width	300		250		200		ns
TWA	ADR from $\overline{WRITE}$ HIGH Hold Time	20		20		20		ns
TWC	CS HIGH from $\overline{WRITE}$ HIGH Hold Time	20		20		20		ns
TWD	Data from $\overline{WRITE}$ HIGH Hold Time	30		30		30		ns
TWWS	Write Width	200		200		160		ns

WAVEFORMS

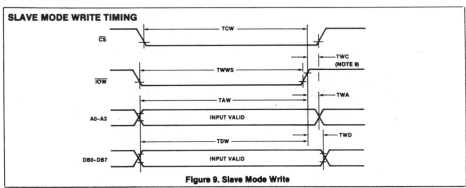

SLAVE MODE WRITE TIMING

Figure 9. Slave Mode Write

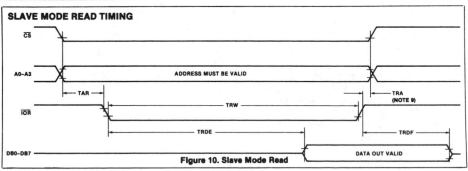

SLAVE MODE READ TIMING

Figure 10. Slave Mode Read

WAVEFORMS (Continued)

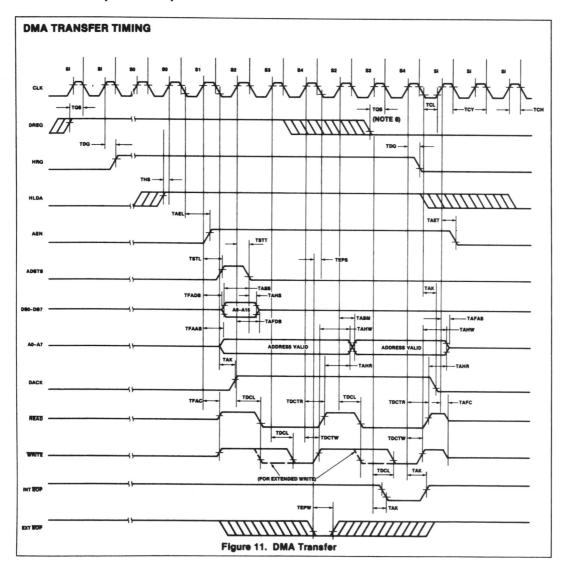

Figure 11. DMA Transfer

8259A/8259A-2/8259A-8
PROGRAMMABLE INTERRUPT CONTROLLER

- **IAPX 86, IAPX 88 Compatible**
- **MCS-80®, MCS-85® Compatible**
- **Eight-Level Priority Controller**
- **Expandable to 64 Levels**
- **Programmable Interrupt Modes**
- **Individual Request Mask Capability**
- **Single +5V Supply (No Clocks)**
- **28-Pin Dual-In-Line Package**
- **Available In EXPRESS**
  - **Standard Temperature Range**
  - **Extended Temperature Range**

The Intel® 8259A Programmable Interrupt Controller handles up to eight vectored priority interrupts for the CPU. It is cascadable for up to 64 vectored priority interrupts without additional circuitry. It is packaged in a 28-pin DIP, uses NMOS technology and requires a single +5V supply. Circuitry is static, requiring no clock input.

The 8259A is designed to minimize the software and real time overhead in handling multi-level priority interrupts. It has several modes, permitting optimization for a variety of system requirements.

The 8259A is fully upward compatible with the Intel® 8259. Software originally written for the 8259 will operate the 8259A in all 8259 equivalent modes (MCS-80/85, Non-Buffered, Edge Triggered).

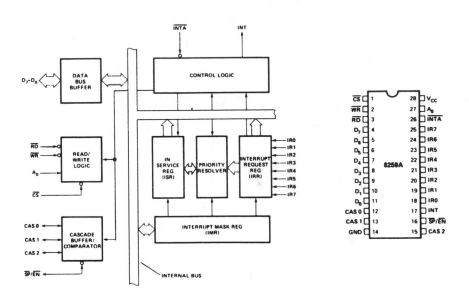

Figure 1. Block Diagram   Figure 2. Pin Configuration

677

Table 1. Pin Description

Symbol	Pin No.	Type	Name and Function
V_{CC}	28	I	**Supply:** +5V Supply.
GND	14	I	**Ground.**
$\overline{CS}$	1	I	**Chip Select:** A low on this pin enables $\overline{RD}$ and $\overline{WR}$ communication between the CPU and the 8259A. INTA functions are independent of CS.
$\overline{WR}$	2	O	**Write:** A low on this pin when CS is low enables the 8259A to accept command words from the CPU.
$\overline{RD}$	3	I	**Read:** A low on this pin when CS is low enables the 8259A to release status onto the data bus for the CPU.
D_7–D_0	4–11	I/O	**Bidirectional Data Bus:** Control, status and interrupt-vector information is transferred via this bus.
CAS_0–CAS_2	12, 13, 15	I/O	**Cascade Lines:** The CAS lines form a private 8259A bus to control a multiple 8259A structure. These pins are outputs for a master 8259A and inputs for a slave 8259A.
$\overline{SP}/\overline{EN}$	16	I/O	**Slave Program/Enable Buffer:** This is a dual function pin. When in the Buffered Mode it can be used as an output to control buffer transceivers (EN). When not in the buffered mode it is used as an input to designate a master (SP = 1) or slave (SP = 0).
INT	17	O	**Interrupt:** This pin goes high whenever a valid interrupt request is asserted. It is used to interrupt the CPU, thus it is connected to the CPU's interrupt pin.
IR_0–IR_7	18–25	I	**Interrupt Requests:** Asynchronous inputs. An interrupt request is executed by raising an IR input (low to high), and holding it high until it is acknowledged (Edge Triggered Mode), or just by a high level on an IR input (Level Triggered Mode).
$\overline{INTA}$	26	I	**Interrupt Acknowledge:** This pin is used to enable 8259A interrupt-vector data onto the data bus by a sequence of interrupt acknowledge pulses issued by the CPU.
A_0	27	I	**A0 Address Line:** This pin acts in conjunction with the $\overline{CS}$, $\overline{WR}$, and $\overline{RD}$ pins. It is used by the 8259A to decipher various Command Words the CPU writes and status the CPU wishes to read. It is typically connected to the CPU A0 address line (A1 for iAPX 86, 88).

FUNCTIONAL DESCRIPTION

Interrupts in Microcomputer Systems

Microcomputer system design requires that I/O devices such as keyboards, displays, sensors and other components receive servicing in an efficient manner so that large amounts of the total system tasks can be assumed by the microcomputer with little or no effect on throughput.

The most common method of servicing such devices is the *Polled* approach. This is where the processor must test each device in sequence and in effect "ask" each one if it needs servicing. It is easy to see that a large portion of the main program is looping through this continuous polling cycle and that such a method would have a serious, detrimental effect on system throughput, thus limiting the tasks that could be assumed by the microcomputer and reducing the cost effectiveness of using such devices.

A more desirable method would be one that would allow the microprocessor to be executing its main program and only stop to service peripheral devices when it is told to do so by the device itself. In effect, the method would provide an external asynchronous input that would inform the processor that it should complete whatever instruction that is currently being executed and fetch a new routine that will service the requesting device. Once this servicing is complete, however, the processor would resume exactly where it left off.

This method is called *Interrupt*. It is easy to see that system throughput would drastically increase, and thus more tasks could be assumed by the microcomputer to further enhance its cost effectiveness.

The Programmable Interrupt Controller (PIC) functions as an overall manager in an Interrupt-Driven system environment. It accepts requests from the peripheral equipment, determines which of the incoming requests is of the highest importance (priority), ascertains whether the incoming request has a higher priority value than the level currently being serviced, and issues an interrupt to the CPU based on this determination.

Each peripheral device or structure usually has a special program or "routine" that is associated with its specific functional or operational requirements; this is referred to as a "service routine". The PIC, after issuing an Interrupt to the CPU, must somehow input information into the CPU that can "point" the Program Counter to the service routine associated with the requesting device. This "pointer" is an address in a vectoring table and will often be referred to, in this document, as vectoring data.

The 8259A

The 8259A is a device specifically designed for use in real time, interrupt driven microcomputer systems. It manages eight levels or requests and has built-in features for expandability to other 8259A's (up to 64 levels). It is programmed by the system's software as an I/O peripheral. A selection of priority modes is available to the programmer so that the manner in which the requests are processed by the 8259A can be configured to match his system requirements. The priority modes can be changed or reconfigured dynamically at any time during the main program. This means that the complete interrupt structure can be defined as required, based on the total system environment.

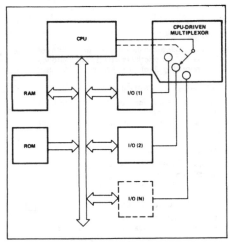

Figure 3a. Polled Method

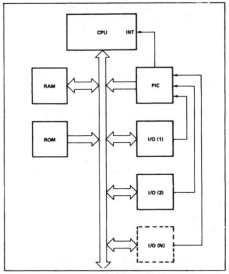

Figure 3b. Interrupt Method

679

INTERRUPT REQUEST REGISTER (IRR) AND IN-SERVICE REGISTER (ISR)

The interrupts at the IR input lines are handled by two registers in cascade, the Interrupt Request Register (IRR) and the In-Service Register (ISR). The IRR is used to store all the interrupt levels which are requesting service; and the ISR is used to store all the interrupt levels which are being serviced.

PRIORITY RESOLVER

This logic block determines the priorities of the bits set in the IRR. The highest priority is selected and strobed into the corresponding bit of the ISR during $\overline{INTA}$ pulse.

INTERRUPT MASK REGISTER (IMR)

The IMR stores the bits which mask the interrupt lines to be masked. The IMR operates on the IRR. Masking of a higher priority input will not affect the interrupt request lines of lower priority.

INT (INTERRUPT)

This output goes directly to the CPU interrupt input. The V_{OH} level on this line is designed to be fully compatible with the 8080A, 8085A and 8086 input levels.

$\overline{INTA}$ (INTERRUPT ACKNOWLEDGE)

$\overline{INTA}$ pulses will cause the 8259A to release vectoring information onto the data bus. The format of this data depends on the system mode (μPM) of the 8259A.

DATA BUS BUFFER

This 3-state, bidirectional 8-bit buffer is used to interface the 8259A to the system Data Bus. Control words and status information are transferred through the Data Bus Buffer.

READ/WRITE CONTROL LOGIC

The function of this block is to accept OUTput commands from the CPU. It contains the Initialization Command Word (ICW) registers and Operation Command Word (OCW) registers which store the various control formats for device operation. This function block also allows the status of the 8259A to be transferred onto the Data Bus.

$\overline{CS}$ (CHIP SELECT)

A LOW on this input enables the 8259A. No reading or writing of the chip will occur unless the device is selected.

$\overline{WR}$ (WRITE)

A LOW on this input enables the CPU to write control words (ICWs and OCWs) to the 8259A.

$\overline{RD}$ (READ)

A LOW on this input enables the 8259A to send the status of the Interrupt Request Register (IRR), In Service Register (ISR), the Interrupt Mask Register (IMR), or the interrupt level onto the Data Bus.

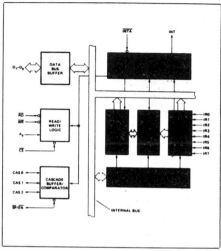

Figure 4a. 8259A Block Diagram

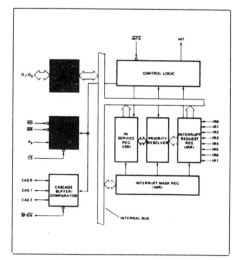

Figure 4b. 8259A Block Diagram

A_0

This input signal is used in conjunction with $\overline{WR}$ and $\overline{RD}$ signals to write commands into the various command registers, as well as reading the various status registers of the chip. This line can be tied directly to one of the address lines.

680

THE CASCADE BUFFER/COMPARATOR

This function block stores and compares the IDs of all 8259A's used in the system. The associated three I/O pins (CAS0–2) are outputs when the 8259A is used as a master and are inputs when the 8259A is used as a slave. As a master, the 8259A sends the ID of the interrupting slave device onto the CAS0–2 lines. The slave thus selected will send its preprogrammed subroutine address onto the Data Bus during the next one or two consecutive INTA pulses. (See section "Cascading the 8259A".)

INTERRUPT SEQUENCE

The powerful features of the 8259A in a microcomputer system are its programmability and the interrupt routine addressing capability. The latter allows direct or indirect jumping to the specific interrupt routine requested without any polling of the interrupting devices. The normal sequence of events during an interrupt depends on the type of CPU being used.

The events occur as follows in an MCS-80/85 system:

1. One or more of the INTERRUPT REQUEST lines (IR7–0) are raised high, setting the corresponding IRR bit(s).
2. The 8259A evaluates these requests, and sends an INT to the CPU, if appropriate.
3. The CPU acknowledges the INT and responds with an INTA pulse.
4. Upon receiving an INTA from the CPU group, the highest priority ISR bit is set, and the corresponding IRR bit is reset. The 8259A will also release a CALL instruction code (11001101) onto the 8-bit Data Bus through its D7–0 pins.
5. This CALL instruction will initiate two more INTA pulses to be sent to the 8259A from the CPU group.
6. These two INTA pulses allow the 8259A to release its preprogrammed subroutine address onto the Data Bus. The lower 8-bit address is released at the first INTA pulse and and the higher 8-bit address is released at the second INTA pulse.
7. This completes the 3-byte CALL instruction released by the 8259A. In the AEOI mode the ISR bit is reset at the end of the third INTA pulse. Otherwise, the ISR bit remains set until an appropriate EOI command is issued at the end of the interrupt sequence.

The events occurring in an iAPX 86 system are the same until step 4.

4. Upon receiving an INTA from the CPU group, the highest priority ISR bit is set and the corresponding IRR bit is reset. The 8259A does not drive the Data Bus during this cycle.
5. The iAPX 86/10 will initiate a second INTA pulse. During this pulse, the 8259A releases an 8-bit pointer onto the Data Bus where it is read by the CPU.
6. This completes the interrupt cycle. In the AEOI mode the ISR bit is reset at the end of the second INTA pulse. Otherwise, the ISR bit remains set until an appropriate EOI command is issued at the end of the interrupt subroutine.

If no interrupt request is present at step 4 of either sequence (i.e., the request was too short in duration) the 8259A will issue an interrupt level 7. Both the vectoring bytes and the CAS lines will look like an interrupt level 7 was requested.

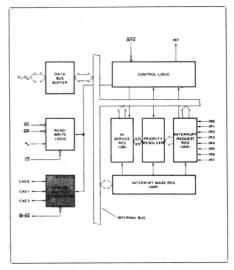

Figure 4c. 8259A Block Diagram

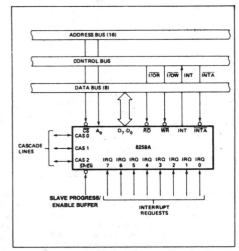

Figure 5. 8259A Interface to Standard System Bus

INTERRUPT SEQUENCE OUTPUTS
MCS-80®, MCS-85®

This sequence is timed by three $\overline{INTA}$ pulses. During the first $\overline{INTA}$ pulse the CALL opcode is enabled onto the data bus.

Content of First Interrupt Vector Byte

	D7	D6	D5	D4	D3	D2	D1	D0
CALL CODE	1	1	0	0	1	1	0	1

During the second $\overline{INTA}$ pulse the lower address of the appropriate service routine is enabled onto the data bus. When Interval = 4 bits A_5–A_7 are programmed, while A_0–A_4 are automatically inserted by the 8259A. When Interval = 8 only A_6 and A_7 are programmed, while A_0–A_5 are automatically inserted.

Content of Second Interrupt Vector Byte

IR	Interval = 4							
	D7	D6	D5	D4	D3	D2	D1	D0
7	A7	A6	A5	1	1	1	0	0
6	A7	A6	A5	1	1	0	0	0
5	A7	A6	A5	1	0	1	0	0
4	A7	A6	A5	1	0	0	0	0
3	A7	A6	A5	0	1	1	0	0
2	A7	A6	A5	0	1	0	0	0
1	A7	A6	A5	0	0	1	0	0
0	A7	A6	A5	0	0	0	0	0

IR	Interval = 8							
	D7	D6	D5	D4	D3	D2	D1	D0
7	A7	A6	1	1	1	0	0	0
6	A7	A6	1	1	0	0	0	0
5	A7	A6	1	0	1	0	0	0
4	A7	A6	1	0	0	0	0	0
3	A7	A6	0	1	1	0	0	0
2	A7	A6	0	1	0	0	0	0
1	A7	A6	0	0	1	0	0	0
0	A7	A6	0	0	0	0	0	0

During the third INTA pulse the higher address of the appropriate service routine, which was programmed as byte 2 of the initialization sequence (A_8–A_{15}), is enabled onto the bus.

Content of Third Interrupt Vector Byte

D7	D6	D5	D4	D3	D2	D1	D0
A15	A14	A13	A12	A11	A10	A9	A8

iAPX 86, iAPX 88

iAPX 86 mode is similar to MCS-80 mode except that only two Interrupt Acknowledge cycles are issued by the processor and no CALL opcode is sent to the processor. The first interrupt acknowledge cycle is similar to that of MCS-80, 85 systems in that the 8259A uses it to internally freeze the state of the interrupts for priority resolution and as a master it issues the interrupt code on the cascade lines at the end of the INTA pulse. On this first cycle it does

not issue any data to the processor and leaves its data bus buffers disabled. On the second interrupt acknowledge cycle in iAPX 86 mode the master (or slave if so programmed) will send a byte of data to the processor with the acknowledged interrupt code composed as follows (note the state of the ADI mode control is ignored and A_5–A_{11} are unused in iAPX 86 mode):

Content of Interrupt Vector Byte for iAPX 86 System Mode

	D7	D6	D5	D4	D3	D2	D1	D0
IR7	T7	T6	T5	T4	T3	1	1	1
IR6	T7	T6	T5	T4	T3	1	1	0
IR5	T7	T6	T5	T4	T3	1	0	1
IR4	T7	T6	T5	T4	T3	1	0	0
IR3	T7	T6	T5	T4	T3	0	1	1
IR2	T7	T6	T5	T4	T3	0	1	0
IR1	T7	T6	T5	T4	T3	0	0	1
IR0	T7	T6	T5	T4	T3	0	0	0

PROGRAMMING THE 8259A

The 8259A accepts two types of command words generated by the CPU:

1. *Initialization Command Words (ICWs):* Before normal operation can begin, each 8259A in the system must be brought to a starting point — by a sequence of 2 to 4 bytes timed by $\overline{WR}$ pulses.

2. *Operation Command Words (OCWs):* These are the command words which command the 8259A to operate in various interrupt modes. These modes are:
   a. Fully nested mode
   b. Rotating priority mode
   c. Special mask mode
   d. Polled mode

The OCWs can be written into the 8259A anytime after initialization.

INITIALIZATION COMMAND WORDS (ICWS)
GENERAL

Whenever a command is issued with A0 = 0 and D4 = 1, this is interpreted as Initialization Command Word 1 (ICW1). ICW1 starts the initialization sequence during which the following automatically occur.

a. The edge sense circuit is reset, which means that following initialization, an interrupt request (IR) input must make a low-to-high transition to generate an interrupt.

b. The Interrupt Mask Register is cleared.

c. IR7 input is assigned priority 7.

d. The slave mode address is set to 7.

e. Special Mask Mode is cleared and Status Read is set to IRR.

f. If IC4=0, then all functions selected in ICW4 are set to zero. (Non-Buffered mode*, no Auto-EOI, MCS-80, 85 system).

*Note: Master/Slave in ICW4 is only used in the buffered mode.

8259A/8259A-2/8259A-8

INITIALIZATION COMMAND WORDS 1 AND 2 (ICW1, ICW2)

A_5-A_{15}: *Page starting address of service routines.* In an MCS 80/85 system, the 8 request levels will generate CALLs to 8 locations equally spaced in memory. These can be programmed to be spaced at intervals of 4 or 8 memory locations, thus the 8 routines will occupy a page of 32 or 64 bytes, respectively.

The address format is 2 bytes long (A_0-A_{15}). When the routine interval is 4, A_0-A_4 are automatically inserted by the 8259A, while A_5-A_{15} are programmed externally. When the routine interval is 8, A_0-A_5 are automatically inserted by the 8259A, while A_6-A_{15} are programmed externally.

The 8-byte interval will maintain compatibility with current software, while the 4-byte interval is best for a compact jump table.

In an iAPX 86 system A_{15}-A_{11} are inserted in the five most significant bits of the vectoring byte and the 8259A sets the three least significant bits according to the interrupt level. A_{10}-A_5 are ignored and ADI (Address interval) has no effect.

LTIM: If LTIM = 1, then the 8259A will operate in the level interrupt mode. Edge detect logic on the interrupt inputs will be disabled.

ADI: CALL address interval, ADI = 1 then interval = 4; ADI = 0 then interval = 8.

SNGL: Single. Means that this is the only 8259A in the system. If SNGL = 1 no ICW3 will be issued.

IC4: If this bit is set — ICW4 has to be read. If ICW4 is not needed, set IC4 = 0.

INITIALIZATION COMMAND WORD 3 (ICW3)

This word is read only when there is more than one 8259A in the system and cascading is used, in which case SNGL = 0. It will load the 8-bit slave register. The functions of this register are:

a. In the master mode (either when SP = 1, or in buffered mode when M/S = 1 in ICW4) a "1" is set for each slave in the system. The master then will release byte 1 of the call sequence (for MCS-80/85 system) and will enable the corresponding slave to release bytes 2 and 3 (for iAPX 86 only byte 2) through the cascade lines.

b. In the slave mode (either when $\overline{SP}$ = 0, or if BUF = 1 and M/S = 0 in ICW4) bits 2–0 identify the slave. The slave compares its cascade input with these bits and, if they are equal, bytes 2 and 3 of the call sequence (or just byte 2 for iAPX 86 are released by it on the Data Bus.

INITIALIZATION COMMAND WORD 4 (ICW4)

SFNM: If SFNM = 1 the special fully nested mode is programmed.

BUF: If BUF = 1 the buffered mode is programmed. In buffered mode $\overline{SP}/\overline{EN}$ becomes an enable output and the master/slave determination is by M/S.

M/S: If buffered mode is selected: M/S = 1 means the 8259A is programmed to be a master, M/S = 0 means the 8259A is programmed to be a slave. If BUF = 0, M/S has no function.

AEOI: If AEOI = 1 the automatic end of interrupt mode is programmed.

μPM: Microprocessor mode: μPM = 0 sets the 8259A for MCS-80, 85 system operation, μPM = 1 sets the 8259A for iAPX 86 system operation.

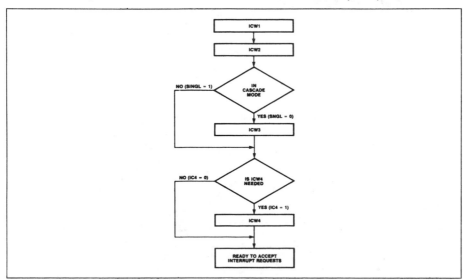

Figure 6. Initialization Sequence

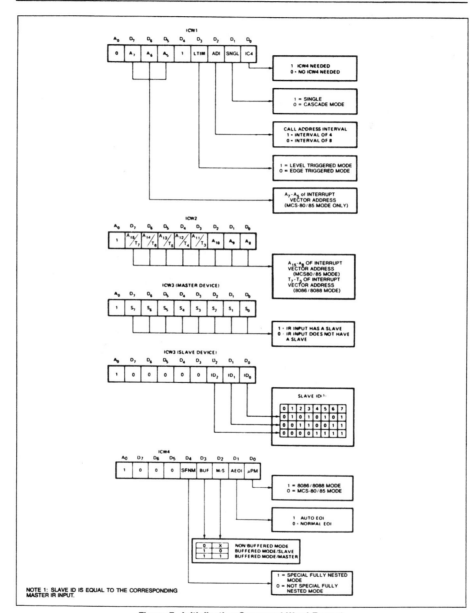

Figure 7. Initialization Command Word Format

OPERATION COMMAND WORDS (OCWs)

After the Initialization Command Words (ICWs) are programmed into the 8259A, the chip is ready to accept interrupt requests at its input lines. However, during the 8259A operation, a selection of algorithms can command the 8259A to operate in various modes through the Operation Command Words (OCWs).

OPERATION CONTROL WORDS (OCWs)

A0		D7	D6	D5	D4	D3	D2	D1	D0
					OCW1				
1		M7	M6	M5	M4	M3	M2	M1	M0

A0					OCW2				
0		R	SL	EOI	0	0	L2	L1	L0

A0					OCW3				
0		0	ESMM	SMM	0	1	P	RR	RIS

OPERATION CONTROL WORD 1 (OCW1)

OCW1 sets and clears the mask bits in the interrupt Mask Register (IMR). $M_7 - M_0$ represent the eight mask bits. $M = 1$ indicates the channel is masked (inhibited), $M = 0$ indicates the channel is enabled.

OPERATION CONTROL WORD 2 (OCW2)

R, SL, EOI — These three bits control the Rotate and End of Interrupt modes and combinations of the two. A chart of these combinations can be found on the Operation Command Word Format.

L_2, L_1, L_0—These bits determine the interrupt level acted upon when the SL bit is active.

OPERATION CONTROL WORD 3 (OCW3)

ESMM — Enable Special Mask Mode. When this bit is set to 1 it enables the SMM bit to set or reset the Special Mask Mode. When ESMM = 0 the SMM bit becomes a "don't care".

SMM — Special Mask Mode. If ESMM = 1 and SMM = 1 the 8259A will enter Special Mask Mode. If ESMM = 1 and SMM = 0 the 8259A will revert to normal mask mode. When ESMM = 0, SMM has no effect.

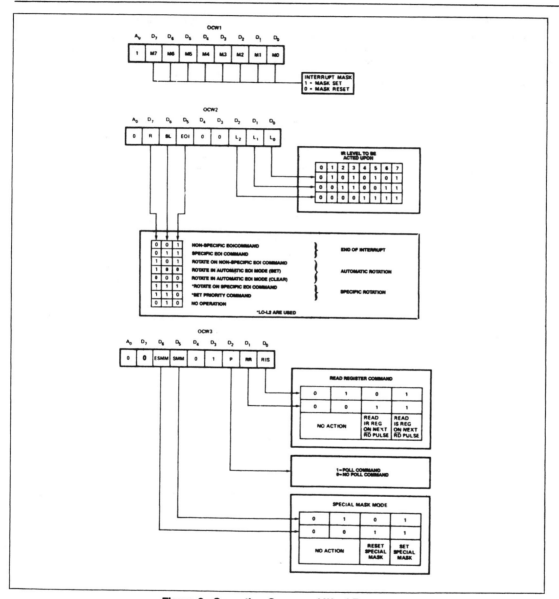

Figure 8. Operation Command Word Format

FULLY NESTED MODE

This mode is entered after initialization unless another mode is programmed. The interrupt requests are ordered in priority form 0 through 7 (0 highest). When an interrupt is acknowledged the highest priority request is determined and its vector placed on the bus. Additionally, a bit of the Interrupt Service register (ISO-7) is set. This bit remains set until the microprocessor issues an End of Interrupt (EOI) command immediately before returning from the service routine, or if AEOI (Automatic End of Interrupt) bit is set, until the trailing edge of the last INTA. While the IS bit is set, all further interrupts of the same or lower priority are inhibited, while higher levels will generate an interrupt (which will be acknowledged only if the microprocessor internal interrupt enable flip-flop has been re-enabled through software).

After the initialization sequence, IR0 has the highest priority and IR7 the lowest. Priorities can be changed, as will be explained, in the rotating priority mode.

END OF INTERRUPT (EOI)

The In Service (IS) bit can be reset either automatically following the trailing edge of the last in sequence INTA pulse (when AEOI bit in ICW1 is set) or by a command word that must be issued to the 8259A before returning from a service routine (EOI command). An EOI command must be issued twice if in the Cascade mode, once for the master and once for the corresponding slave.

There are two forms of EOI command: Specific and Non-Specific. When the 8259A is operated in modes which preserve the fully nested structure, it can determine which IS bit to reset on EOI. When a Non-Specific EOI command is issued the 8259A will automatically reset the highest IS bit of those that are set, since in the fully nested mode the highest IS level was necessarily the last level acknowledged and serviced. A non-specific EOI can be issued with OCW2 (EOI = 1, SL = 0, R = 0).

When a mode is used which may disturb the fully nested structure, the 8259A may no longer be able to determine the last level acknowledged. In this case a Specific End of Interrupt must be issued which includes as part of the command the IS level to be reset. A specific EOI can be issued with OCW2 (EOI = 1, SL = 1, R = 0, and LO-L2 is the binary level of the IS bit to be reset).

It should be noted that an IS bit that is masked by an IMR bit will not be cleared by a non-specific EOI if the 8259A is in the Special Mask Mode.

AUTOMATIC END OF INTERRUPT (AEOI) MODE

If AEOI = 1 in ICW4, then the 8259A will operate in AEOI mode continuously until reprogrammed by ICW4. In this mode the 8259A will automatically perform a non-specific EOI operation at the trailing edge of the last interrupt acknowledge pulse (third pulse in MCS-80/85, second in iAPX 86). Note that from a system standpoint, this mode should be used only when a nested multilevel interrupt structure is not required within a single 8259A.

The AEOI mode can only be used in a master 8259A and not a slave.

AUTOMATIC ROTATION
(Equal Priority Devices)

In some applications there are a number of interrupting devices of equal priority. In this mode a device, after being serviced, receives the lowest priority, so a device requesting an interrupt will have to wait, in the worst case until each of 7 other devices are serviced at most *once*. For example, if the priority and "in service" status is:

Before Rotate (IR4 the highest priority requiring service)

After Rotate (IR4 was serviced, all other priorities rotated correspondingly)

There are two ways to accomplish Automatic Rotation using OCW2, the Rotation on Non-Specific EOI Command (R = 1, SL = 0, EOI = 1) and the Rotate in Automatic EOI Mode which is set by (R = 1, SL = 0, EOI = 0) and cleared by (R = 0, SL = 0, EOI = 0).

SPECIFIC ROTATION
(Specific Priority)

The programmer can change priorities by programming the bottom priority and thus fixing all other priorities; i.e., if IR5 is programmed as the bottom priority device, then IR6 will have the highest priority.

The Set Priority command is issued in OCW2 where: R = 1, SL = 1; LO-L2 is the binary priority level code of the bottom priority device.

Observe that in this mode internal status is updated by software control during OCW2. However, it is independent of the End of Interrupt (EOI) command (also executed by OCW2). Priority changes can be executed during an EOI command by using the Rotate on Specific EOI command in OCW2 (R = 1, SL = 1, EOI = 1 and LO-L2 = IR level to receive bottom priority).

INTERRUPT MASKS

Each Interrupt Request input can be masked individually by the Interrupt Mask Register (IMR) programmed through OCW1. Each bit in the IMR masks one interrupt channel if it is set (1). Bit 0 masks IR0, Bit 1 masks IR1 and so forth. Masking an IR channel does not affect the other channels operation.

SPECIAL MASK MODE

Some applications may require an interrupt service routine to dynamically alter the system priority structure during its execution under software control. For example, the routine may wish to inhibit lower priority requests for a portion of its execution but enable some of them for another portion.

The difficulty here is that if an Interrupt Request is acknowledged and an End of Interrupt command did not reset its IS bit (i.e., while executing a service routine), the 8259A would have inhibited all lower priority requests with no easy way for the routine to enable them

That is where the Special Mask Mode comes in. In the special Mask Mode, when a mask bit is set in OCW1, it inhibits further interrupts at that level *and enables* interrupts from *all other* levels (lower as well as higher) that are not masked.

Thus, any interrupts may be selectively enabled by loading the mask register.

The special Mask Mode is set by OCW3 where: SSMM = 1, SMM = 1, and cleared where SSMM = 1, SMM = 0.

POLL COMMAND

In this mode the INT output is not used or the microprocessor internal Interrupt Enable flip-flop is reset, disabling its interrupt input. Service to devices is achieved by software using a Poll command.

The Poll command is issued by setting P = "1" in OCW3. The 8259A treats the next $\overline{RD}$ pulse to the 8259A (i.e., $\overline{RD} = 0$, $\overline{CS} = 0$) as an interrupt acknowledge, sets the appropriate IS bit if there is a request, and reads the priority level. Interrupt is frozen from $\overline{WR}$ to $\overline{RD}$.

The word enabled onto the data bus during $\overline{RD}$ is:

D7	D6	D5	D4	D3	D2	D1	D0
I	—	—	—	—	W2	W1	W0

W0-W2: Binary code of the highest priority level requesting service.

I: Equal to a "1" if there is an interrupt.

This mode is useful if there is a routine command common to several levels so that the $\overline{INTA}$ sequence is not needed (saves ROM space). Another application is to use the poll mode to expand the number of priority levels to more than 64.

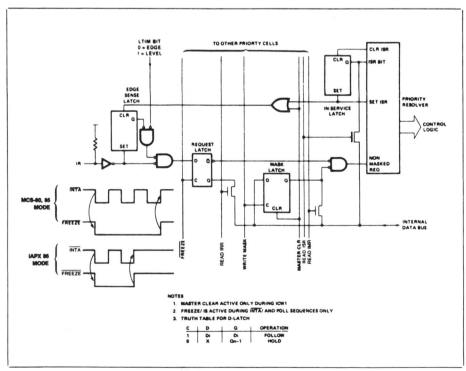

Figure 9. Priority Cell—Simplified Logic Diagram

READING THE 8259A STATUS

The input status of several internal registers can be read to update the user information on the system. The following registers can be read via OCW3 (IRR and ISR or OCW1 [IMR]).

Interrupt Request Register (IRR): 8-bit register which contains the levels requesting an interrupt to be acknowledged. The highest request level is reset from the IRR when an interrupt is acknowledged. (Not affected by IMR.)

In-Service Register (ISR): 8-bit register which contains the priority levels that are being serviced. The ISR is updated when an End of Interrupt Command is issued.

Interrupt Mask Register: 8-bit register which contains the interrupt request lines which are masked.

The IRR can be read when, prior to the RD pulse, a Read Register Command is issued with OCW3 (RR = 1, RIS = 0.)

The ISR can be read when, prior to the RD pulse, a Read Register Command is issued with OCW3 (RR = 1, RIS = 1).

There is no need to write an OCW3 before every status read operation, as long as the status read corresponds with the previous one; i.e., the 8259A "remembers" whether the IRR or ISR has been previously selected by the OCW3. This is not true when poll is used.

After initialization the 8259A is set to IRR.

For reading the IMR, no OCW3 is needed. The output data bus will contain the IMR whenever $\overline{RD}$ is active and AO = 1 (OCW1).

Polling overrides status read when P = 1, RR = 1 in OCW3.

EDGE AND LEVEL TRIGGERED MODES

This mode is programmed using bit 3 in ICW1.

If LTIM = '0', an interrupt request will be recognized by a low to high transition on an IR input. The IR input can remain high without generating another interrupt.

If LTIM = '1', an interrupt request will be recognized by a 'high' level on IR Input, and there is no need for an edge detection. The interrupt request must be removed before the EOI command is issued or the CPU interrupt is enabled to prevent a second interrupt from occurring.

The priority cell diagram shows a conceptual circuit of the level sensitive and edge sensitive input circuitry of the 8259A. Be sure to note that the request latch is a transparent D type latch.

In both the edge and level triggered modes the IR inputs must remain high until after the falling edge of the first INTA. If the IR input goes low before this time a DEFAULT IR7 will occur when the CPU acknowledges the interrupt. This can be a useful safeguard for detecting interrupts caused by spurious noise glitches on the IR inputs. To implement this feature the IR7 routine is used for "clean up" simply executing a return instruction, thus ignoring the interrupt. If IR7 is needed for other purposes a default IR7 can still be detected by reading the ISR. A normal IR7 interrupt will set the corresponding ISR bit, a default IR7 won't. If a default IR7 routine occurs during a normal IR7 routine, however, the ISR will remain set. In this case it is necessary to keep track of whether or not the IR7 routine was previously entered. If another IR7 occurs it is a default.

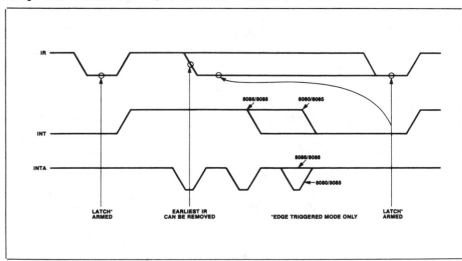

Figure 10. IR Triggering Timing Requirements

689

THE SPECIAL FULLY NESTED MODE

This mode will be used in the case of a big system where cascading is used, and the priority has to be conserved within each slave. In this case the fully nested mode will be programmed to the master (using ICW4). This mode is similar to the normal nested mode with the following exceptions:

a. When an interrupt request from a certain slave is in service this slave is not locked out from the master's priority logic and further interrupt requests from higher priority IR's within the slave will be recognized by the master and will initiate interrupts to the processor. (In the normal nested mode a slave is masked out when its request is in service and no higher requests from the same slave can be serviced.)

b. When exiting the Interrupt Service routine the software has to check whether the interrupt serviced was the only one from that slave. This is done by sending a non-specific End of Interrupt (EOI) command to the slave and then reading its In-Service register and checking for zero. If it is empty, a non-specific EOI can be sent to the master too. If not, no EOI should be sent.

BUFFERED MODE

When the 8259A is used in a large system where bus driving buffers are required on the data bus and the cascading mode is used, there exists the problem of enabling buffers.

The buffered mode will structure the 8259A to send an enable signal on $\overline{SP}/\overline{EN}$ to enable the buffers. In this mode, whenever the 8259A's data bus outputs are enabled, the $\overline{SP}/\overline{EN}$ output becomes active.

This modification forces the use of software programming to determine whether the 8259A is a master or a slave. Bit 3 in ICW4 programs the buffered mode, and bit 2 in ICW4 determines whether it is a master or a slave.

CASCADE MODE

The 8259A can be easily interconnected in a system of one master with up to eight slaves to handle up to 64 priority levels.

The master controls the slaves through the 3 line cascade bus. The cascade bus acts like chip selects to the slaves during the $\overline{INTA}$ sequence.

In a cascade configuration, the slave interrupt outputs are connected to the master interrupt request inputs. When a slave request line is activated and afterwards acknowledged, the master will enable the corresponding slave to release the device routine address during bytes 2 and 3 of INTA. (Byte 2 only for 8086/8088).

The cascade bus lines are normally low and will contain the slave address code from the trailing edge of the first INTA pulse to the trailing edge of the third pulse. Each 8259A in the system must follow a separate initialization sequence and can be programmed to work in a different mode. An EOI command must be issued twice: once for the master and once for the corresponding slave. An address decoder is required to activate the Chip Select (CS) input of each 8259A.

The cascade lines of the Master 8259A are activated only for slave inputs, non slave inputs leave the cascade line inactive (low).

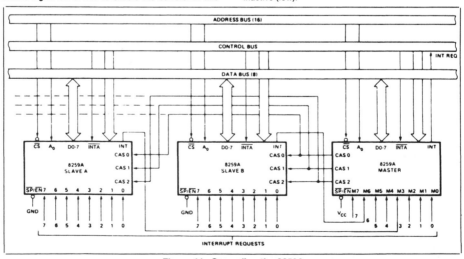

Figure 11. Cascading the 8259A

ABSOLUTE MAXIMUM RATINGS*

Ambient Temperature Under Bias .... 0°C to 70°C
Storage Temperature .... −65°C to +150°C
Voltage on Any Pin
  with Respect to Ground .... −0.5V to +7V
Power Dissipation .... 1 Watt

NOTICE: Stresses above those listed under "Absolute Maximum Ratings" may cause permanent damage to the device. This is a stress rating only and functional operation of the device at these or any other conditions above those indicated in the operational sections of this specification is not implied.

D.C. CHARACTERISTICS [T$_A$ = 0°C to 70°C, V$_{CC}$ = 5V ±5% (8259A-8), V$_{CC}$ = 5V ±10% (8259A, 8259A-2)]

Symbol	Parameter	Min.	Max.	Units	Test Conditions
V$_{IL}$	Input Low Voltage	−0.5	0.8	V	
V$_{IH}$	Input High Voltage	2.0*	V$_{CC}$ +0.5V	V	
V$_{OL}$	Output High Voltage		0.45	V	I$_{OL}$ = 2.2mA
V$_{OH}$	Output High Voltage	2.4		V	I$_{OH}$ = −400μA
V$_{OH(INT)}$	Interrupt Output High Voltage	3.5		V	I$_{OH}$ = −100μA
		2.4		V	I$_{OH}$ = −400μA
I$_{LI}$	Input Load Current	−10	+10	μA	0V ≤ V$_{IN}$ ≤ V$_{CC}$
I$_{LOL}$	Output Leakage Current	−10	+10	μA	0.45V ≤ V$_{OUT}$ ≤ V$_{CC}$
I$_{CC}$	V$_{CC}$ Supply Current		85	mA	
I$_{LIR}$	IR Input Load Current		−300	μA	V$_{IN}$ = 0
			10	μA	V$_{IN}$ = V$_{CC}$

*Note: For Extended Temperature EXPRESS V$_{IH}$ = 2.3V.

CAPACITANCE (T$_A$ = 25°C; V$_{CC}$ = GND = 0V)

Symbol	Parameter	Min.	Typ.	Max.	Unit	Test Conditions
C$_{IN}$	Input Capacitance			10	pF	fc = 1 MHZ
C$_{I/O}$	I/O Capacitance			20	pF	Unmeasured pins returned to V$_{SS}$

A.C. CHARACTERISTICS [T$_A$ = 0°C to 70°C, V$_{CC}$ = 5V ±5% (8259A-8), V$_{CC}$ = 5V ± 10% (8259A, 8259A-2)]

TIMING REQUIREMENTS

Symbol	Parameter	8259A-8 Min.	8259A-8 Max.	8259A Min.	8259A Max.	8259A-2 Min.	8259A-2 Max.	Units	Test Conditions
TAHRL	AO/$\overline{CS}$ Setup to $\overline{RD}$/$\overline{INTA}$↓	50		0		0		ns	
TRHAX	AO/$\overline{CS}$ Hold after $\overline{RD}$/$\overline{INTA}$↑	5		0		0		ns	
TRLRH	$\overline{RD}$ Pulse Width	420		235		160		ns	
TAHWL	AO/$\overline{CS}$ Setup to $\overline{WR}$↓	50		0		0		ns	
TWHAX	AO/$\overline{CS}$ Hold after $\overline{WR}$↑	20		0		0		ns	
TWLWH	$\overline{WR}$ Pulse Width	400		290		190		ns	
TDVWH	Data Setup to $\overline{WR}$↑	300		240		160		ns	
TWHDX	Data Hold after $\overline{WR}$↑	40		0		0		ns	
TJLJH	Interrupt Request Width (Low)	100		100		100		ns	See Note 1
TCVIAL	Cascade Setup to Second or Third $\overline{INTA}$↓ (Slave Only)	55		55		40		ns	
TRHRL	End of $\overline{RD}$ to next $\overline{RD}$ End of $\overline{INTA}$ to next $\overline{INTA}$ within an $\overline{INTA}$ sequence only	160		160		160		ns	
TWHWL	End of $\overline{WR}$ to next $\overline{WR}$	190		190		190		ns	

A.C. CHARACTERISTICS (Continued)

Symbol	Parameter	8259A-8		8259A		8259A-2		Units	Test Conditions
		Min.	Max.	Min.	Max.	Min.	Max.		
*TCHCL	End of Command to next Command (Not same command type)	500		500		500		ns	
	End of $\overline{\text{INTA}}$ sequence to next $\overline{\text{INTA}}$ sequence.								

*Worst case timing for TCHCL in an actual microprocessor system is typically much greater than 500 ns (i.e. 8085A = 1.6μs, 8085A-2 = 1μs, 8086 = 1μs, 8086-2 = 625 ns)
NOTE: This is the low time required to clear the input latch in the edge triggered mode.

TIMING RESPONSES

Symbol	Parameter	8259A-8		8259A		8259A-2		Units	Test Conditions
		Min.	Max.	Min.	Max.	Min.	Max.		
TRLDV	Data Valid from $\overline{\text{RD}}$/$\overline{\text{INTA}}$↓		300		200		120	ns	C of Data Bus = 100 pF
TRHDZ	Data Float after $\overline{\text{RD}}$/$\overline{\text{INTA}}$↑	10	200	10	100	10	85	ns	C of Data Bus Max text C = 100 pF Min. test C = 15 pF
TJHIH	Interrupt Output Delay		400		350		300	ns	
TIALCV	Cascade Valid from First $\overline{\text{INTA}}$↓ (Master Only)		565		565		360	ns	C_{INT} = 100 pF
TRLEL	Enable Active from $\overline{\text{RD}}$↓ or $\overline{\text{INTA}}$↓		160		125		100	ns	$C_{CASCADE}$ = 100 pF
TRHEH	Enable Inactive from $\overline{\text{RD}}$↑ or $\overline{\text{INTA}}$↑		325		150		150	ns	
TAHDV	Data Valid from Stable Address		350		200		200	ns	
TCVDV	Cascade Valid to Valid Data		300		300		200	ns	

A.C. TESTING INPUT, OUTPUT WAVEFORM

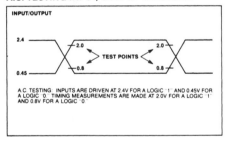

A.C. TESTING: INPUTS ARE DRIVEN AT 2.4V FOR A LOGIC "1" AND 0.45V FOR A LOGIC "0". TIMING MEASUREMENTS ARE MADE AT 2.0V FOR A LOGIC "1" AND 0.8V FOR A LOGIC "0".

A.C. TESTING LOAD CIRCUIT

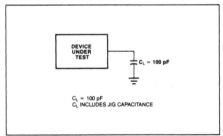

C_L = 100 pF
C_L INCLUDES JIG CAPACITANCE

WAVEFORMS

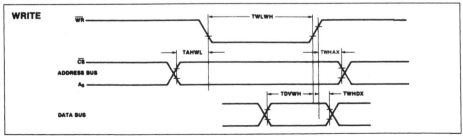

WAVEFORMS (Continued)

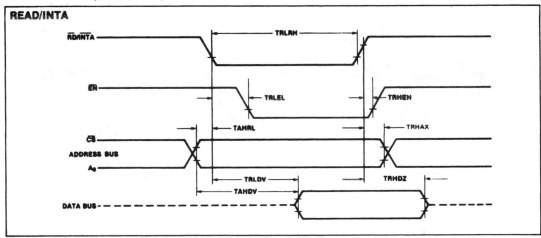

READ/INTA

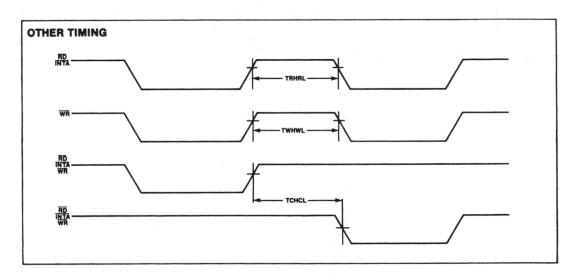

OTHER TIMING

WAVEFORMS (Continued)

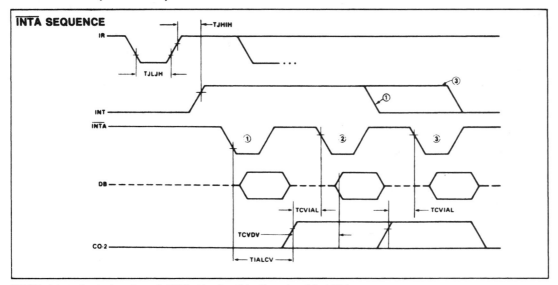

INTA SEQUENCE

NOTES: Interrupt output must remain HIGH at least until leading edge of first INTA.
1. Cycle 1 in iAPX 86, iAPX 88 systems, the Data Bus is not active.

intel®

8279/8279-5
PROGRAMMABLE KEYBOARD/DISPLAY INTERFACE

- **Simultaneous Keyboard Display Operations**
- **Scanned Keyboard Mode**
- **Scanned Sensor Mode**
- **Strobed Input Entry Mode**
- **8-Character Keyboard FIFO**
- **2-Key Lockout or N-Key Rollover with Contact Debounce**
- **Dual 8- or 16-Numerical Display**

- **Single 16-Character Display**
- **Right or Left Entry 16-Byte Display RAM**
- **Mode Programmable from CPU**
- **Programmable Scan Timing**
- **Interrupt Output on Key Entry**
- **Available in EXPRESS**
  —Standard Temperature Range
  —Extended Temperature Range

The Intel® 8279 is a general purpose programmable keyboard and display I/O interface device designed for use with Intel® microprocessors. The keyboard portion can provide a scanned interface to a 64-contact key matrix. The keyboard portion will also interface to an array of sensors or a strobed interface keyboard, such as the hall effect and ferrite variety. Key depressions can be 2-key lockout or N-key rollover. Keyboard entries are debounced and strobed in an 8-character FIFO. If more than 8 characters are entered, overrun status is set. Key entries set the interrupt output line to the CPU.

The display portion provides a scanned display interface for LED, incandescent, and other popular display technologies. Both numeric and alphanumeric segment displays may be used as well as simple indicators. The 8279 has 16X8 display RAM which can be organized into dual 16X4. The RAM can be loaded or interrogated by the CPU. Both right entry, calculator and left entry typewriter display formats are possible. Both read and write of the display RAM can be done with auto-increment of the display RAM address.

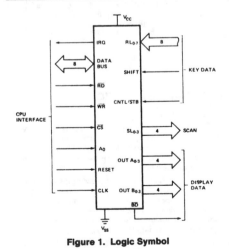

Figure 1. Logic Symbol

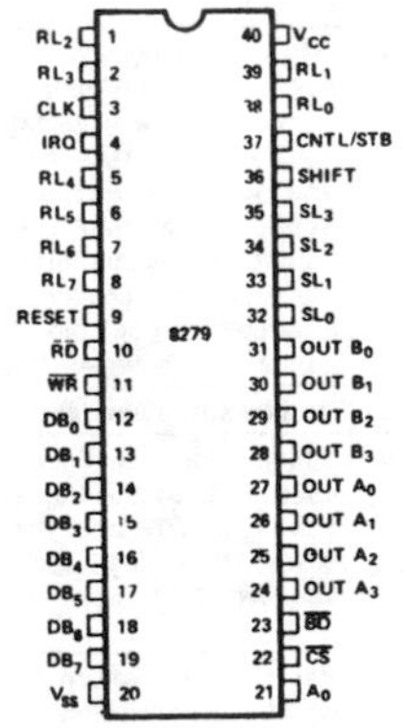

Figure 2. Pin Configuration

HARDWARE DESCRIPTION

The 8279 is packaged in a 40 pin DIP. The following is a functional description of each pin.

Table 1. Pin Descriptions

Symbol	Pin No.	Name and Function
DB_0-DB_7	8	**Bi-directional data bus:** All data and commands between the CPU and the 8279 are transmitted on these lines.
CLK	1	**Clock:** Clock from system used to generate internal timing.
RESET	1	**Reset:** A high signal on this pin resets the 8279. After being reset the 8279 is placed in the following mode: 1) 16 8-bit character display —left entry. 2) Encoded scan keyboard—2 key lockout. Along with this the program clock prescaler is set to 31.
CS	1	**Chip Select:** A low on this pin enables the interface functions to receive or transmit.
A_0	1	**Buffer Address:** A high on this line indicates the signals in or out are interpreted as a command or status. A low indicates that they are data.
RD, WR	2	**Input/Output Read and Write:** These signals enable the data buffers to either send data to the external bus or receive it from the external bus.
IRQ	1	**Interrupt Request:** In a keyboard mode, the interrupt line is high when there is data in the FIFO/Sensor RAM. The interrupt line goes low with each FIFO/Sensor RAM read and returns high if there is still information in the RAM. In a sensor mode, the interrupt line goes high whenever a change in a sensor is detected.
V_{SS}, V_{CC}	2	**Ground and power supply pins.**
SL_0-SL_3	4	**Scan Lines:** Scan lines which are used to scan the key switch or sensor matrix and the display digits. These lines can be either encoded (1 of 16) or decoded (1 of 4).
RL_0-RL_7	8	**Return Line:** Return line inputs which are connected to the scan lines through the keys or sensor switches. They have active internal pullups to keep them high until a switch closure pulls one low. They also serve as an 8-bit input in the Strobed Input mode.

Symbol	Pin No.	Name and Function
SHIFT	1	**Shift:** The shift input status is stored along with the key position on key closure in the Scanned Keyboard modes. It has an active internal pullup to keep it high until a switch closure pulls it low.
CNTL/STB	1	**Control/Strobed Input Mode:** For keyboard modes this line is used as a control input and stored like status on a key closure. The line is also the strobe line that enters the data into the FIFO in the Strobed Input mode. (Rising Edge). It has an active internal pullup to keep it high until a switch closure pulls it low.
OUT A_0–OUT A_3 OUT B_0–OUT B_3	4 4	**Outputs:** These two ports are the outputs for the 16 x 4 display refresh registers. The data from these outputs is synchronized to the scan lines (SL_0–SL_3) for multiplexed digit displays. The two 4 bit ports may be blanked independently. These two ports may also be considered as one 8-bit port.
BD	1	**Blank Display:** This output is used to blank the display during digit switching or by a display blanking command.

FUNCTIONAL DESCRIPTION

Since data input and display are an integral part of many microprocessor designs, the system designer needs an interface that can control these functions without placing a large load on the CPU. The 8279 provides this function for 8-bit microprocessors.

The 8279 has two sections: keyboard and display. The keyboard section can interface to regular typewriter style keyboards or random toggle or thumb switches. The display section drives alphanumeric displays or a bank of indicator lights. Thus the CPU is relieved from scanning the keyboard or refreshing the display.

The 8279 is designed to directly connect to the microprocessor bus. The CPU can program all operating modes for the 8279. These modes include:

AFN-00742B

Input Modes

- Scanned Keyboard — with encoded (8 x 8 key keyboard) or decoded (4 x 8 key keyboard) scan lines. A key depression generates a 6-bit encoding of key position. Position and shift and control status are stored in the FIFO. Keys are automatically debounced with 2-key lockout or N-key rollover.

- Scanned Sensor Matrix — with encoded (8 x 8 matrix switches) or decoded (4 x 8 matrix switches) scan lines. Key status (open or closed) stored in RAM addressable by CPU.

- Strobed Input — Data on return lines during control line strobe is transferred to FIFO.

Output Modes

- 8 or 16 character multiplexed displays that can be organized as dual 4-bit or single 8-bit ($B_0 = D_0$, $A_3 = D_7$).

- Right entry or left entry display formats.

Other features of the 8279 include.

- Mode programming from the CPU.
- Clock Prescaler
- Interrupt output to signal CPU when there is keyboard or sensor data available.
- An 8 byte FIFO to store keyboard information.
- 16 byte internal Display RAM for display refresh. This RAM can also be read by the CPU.

PRINCIPLES OF OPERATION

The following is a description of the major elements of the 8279 Programmable Keyboard/Display interface device. Refer to the block diagram in Figure 3.

I/O Control and Data Buffers

The I/O control section uses the $\overline{CS}$, A_0, $\overline{RD}$ and $\overline{WR}$ lines to control data flow to and from the various internal registers and buffers. All data flow to and from the 8279 is enabled by $\overline{CS}$. The character of the information, given or desired by the CPU, is identified by A_0. A logic one means the information is a command or status. A logic zero means the information is data. $\overline{RD}$ and $\overline{WR}$ determine the direction of data flow through the Data Buffers. The Data Buffers are bi-directional buffers that connect the internal bus to the external bus. When the chip is not selected ($\overline{CS}$ = 1), the devices are in a high impedance state. The drivers input during $\overline{WR} \cdot \overline{CS}$ and output during $\overline{RD} \cdot \overline{CS}$.

Control and Timing Registers and Timing Control

These registers store the keyboard and display modes and other operating conditions programmed by the CPU. The modes are programmed by presenting the proper command on the data lines with $A_0 = 1$ and then sending a WR. The command is latched on the rising edge of WR.

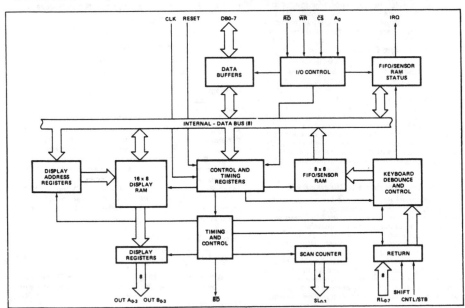

Figure 3. Internal Block Diagram

697

The command is then decoded and the appropriate function is set. The timing control contains the basic timing counter chain. The first counter is a ÷ N prescaler that can be programmed to yield an internal frequency of 100 kHz which gives a 5.1 ms keyboard scan time and a 10.3 ms debounce time. The other counters divide down the basic internal frequency to provide the proper key scan, row scan, keyboard matrix scan, and display scan times.

Scan Counter

The scan counter has two modes. In the encoded mode, the counter provides a binary count that must be externally decoded to provide the scan lines for the keyboard and display. In the decoded mode, the scan counter decodes the least significant 2 bits and provides a decoded 1 of 4 scan. Note than when the keyboard is in decoded scan, so is the display. This means that only the first 4 characters in the Display RAM are displayed.

In the encoded mode, the scan lines are active high outputs. In the decoded mode, the scan lines are active low outputs.

Return Buffers and Keyboard Debounce and Control

The 8 return lines are buffered and latched by the Return Buffers. In the keyboard mode, these lines are scanned, looking for key closures in that row. If the debounce circuit detects a closed switch, it waits about 10 msec to check if the switch remains closed. If it does, the address of the switch in the matrix plus the status of SHIFT and CONTROL are transferred to the FIFO. In the scanned Sensor Matrix modes, the contents of the return lines is directly transferred to the corresponding row of the Sensor RAM (FIFO) each key scan time. In Strobed Input mode, the contents of the return lines are transferred to the FIFO on the rising edge of the CNTL/STB line pulse.

FIFO/Sensor RAM and Status

This block is a dual function 8 x 8 RAM. In Keyboard or Strobed Input modes, it is a FIFO. Each new entry is written into successive RAM positions and each is then read in order of entry. FIFO status keeps track of the number of characters in the FIFO and whether it is full or empty. Too many reads or writes will be recognized as an error. The status can be read by an RD with CS low and A0 high. The status logic also provides an IRQ signal when the FIFO is not empty. In Scanned Sensor Matrix mode, the memory is a Sensor RAM. Each row of the Sensor RAM is loaded with the status of the corresponding row of sensor in the sensor matrix. In this mode, IRQ is high if a change in a sensor is detected.

Display Address Registers and Display RAM

The Display Address Registers hold the address of the word currently being written or read by the CPU and the two 4-bit nibbles being displayed. The read/write addresses are programmed by CPU command. They also can be set to auto increment after each read or write. The Display RAM can be directly read by the CPU after the correct mode and address is set. The addresses for the A and B nibbles are automatically updated by the 8279 to match data entry by the CPU. The A and B nibbles can be entered independently or as one word, according to the mode that is set by the CPU. Data entry to the display can be set to either left or right entry. See Interface Considerations for details.

SOFTWARE OPERATION

8279 commands

The following commands program the 8279 operating modes. The commands are sent on the Data Bus with CS low and A0 high and are loaded to the 8279 on the rising edge of WR.

Keyboard/Display Mode Set

Code:

	MSB						LSB
	0	0	0	D	D	K	K	K

Where DD is the Display Mode and KKK is the Keyboard Mode.

DD

0 0   8 8-bit character display — Left entry

0 1   16 8-bit character display — Left entry*

1 0   8 8-bit character display — Right entry

1 1   16 8-bit character display — Right entry

For description of right and left entry, see Interface Considerations. Note that when decoded scan is set in keyboard mode, the display is reduced to 4 characters independent of display mode set.

KKK

0 0 0   Encoded Scan Keyboard — 2 Key Lockout*

0 0 1   Decoded Scan Keyboard — 2-Key Lockout

0 1 0   Encoded Scan Keyboard — N-Key Rollover

0 1 1   Decoded Scan Keyboard — N-Key Rollover

1 0 0   Encoded Scan Sensor Matrix

1 0 1   Decoded Scan Sensor Matrix

1 1 0   Strobed Input, Encoded Display Scan

1 1 1   Strobed Input, Decoded Display Scan

Program Clock

Code:

	0	0	1	P	P	P	P	P

All timing and multiplexing signals for the 8279 are generated by an internal prescaler. This prescaler divides the external clock (pin 3) by a programmable integer. Bits PPPPP determine the value of this integer which ranges from 2 to 31. Choosing a divisor that yields 100 kHz will give the specified scan and debounce times. For instance, if Pin 3 of the 8279 is being clocked by a 2 MHz signal, PPPPP should be set to 10100 to divide the clock by 20 to yield the proper 100 kHz operating frequency.

Read FIFO/Sensor RAM

Code:

	0	1	0	AI	X	A	A	A

X = Don't Care

The CPU sets up the 8279 for a read of the FIFO/Sensor RAM by first writing this command. In the Scan Key-

*Default after reset.

698

board Mode, the Auto-Increment flag (AI) and the RAM address bits (AAA) are irrelevant. The 8279 will automatically drive the data bus for each subsequent read ($A_0 = 0$) in the same sequence in which the data first entered the FIFO. All subsequent reads will be from the FIFO until another command is issued.

In the Sensor Matrix Mode, the RAM address bits AAA select one of the 8 rows of the Sensor RAM. If the AI flag is set (AI = 1), each successive read will be from the subsequent row of the sensor RAM.

Read Display RAM

Code: | 0 | 1 | 1 | AI | A | A | A | A |

The CPU sets up the 8279 for a read of the Display RAM by first writing this command. The address bits AAAA select one of the 16 rows of the Display RAM. If the AI flag is set (AI = 1), this row address will be incremented after each following read *or write* to the Display RAM. Since the same counter is used for both reading and writing, this command sets the next read *or write* address and the sense of the Auto-Increment mode for both operations.

Write Display RAM

Code: | 1 | 0 | 0 | AI | A | A | A | A |

The CPU sets up the 8279 for a write to the Display RAM by first writing this command. After writing the command with $A_0 = 1$, all subsequent writes with $A_0 = 0$ will be to the Display RAM. The addressing and Auto-Increment functions are identical to those for the Read Display RAM. However, this command does not affect the source of subsequent Data Reads; the CPU will read from whichever RAM (Display or FIFO/Sensor) which was last specified. If, indeed, the Display RAM was last specified, the Write Display RAM will, nevertheless, change the next Read location.

Display Write Inhibit/Blanking

Code:

			A	**B**	**A**	**B**
1	0	1	X	IW	IW	BL	BL

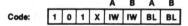

The IW Bits can be used to mask nibble A and nibble B in applications requiring separate 4-bit display ports. By setting the IW flag (IW = 1) for one of the ports, the port becomes marked so that entries to the Display RAM from the CPU do not affect that port. Thus, if each nibble is input to a BCD decoder, the CPU may write a digit to the Display RAM without affecting the other digit being displayed. It is important to note that bit B_0 corresponds to bit D_0 on the CPU bus, and that bit A_3 corresponds to bit D_7.

If the user wishes to blank the display, the BL flags are available for each nibble. The last Clear command issued determines the code to be used as a "blank." This code defaults to all zeros after a reset. Note that both BL flags must be set to blank a display formatted with a single 8-bit port.

Clear

Code: | 1 | 1 | 0 | C_D | C_D | C_D | C_F | C_A |

The C_D bits are available in this command to clear all rows of the Display RAM to a selectable blanking code as follows:

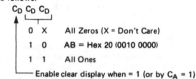

C_D C_D C_D

0	X	All Zeros (X = Don't Care)
1	0	AB = Hex 20 (0010 0000)
1	1	All Ones

Enable clear display when = 1 (or by $C_A = 1$)

During the time the Display RAM is being cleared ($\sim 160 \mu s$), it may not be written to. The most significant bit of the FIFO status word is set during this time. When the Display RAM becomes available again, it automatically resets.

If the C_F bit is asserted ($C_F = 1$), the FIFO status is cleared and the interrupt output line is reset. Also, the Sensor RAM pointer is set to row 0.

C_A, the Clear All bit, has the combined effect of C_D and C_F; it uses the C_D clearing code on the Display RAM and also clears FIFO status. Furthermore, it resynchronizes the internal timing chain.

End Interrupt/Error Mode Set

Code: | 1 | 1 | 1 | E | X | X | X | X | X = Don't care.

For the sensor matrix modes this command lowers the IRQ line and enables further writing into RAM. (The IRQ line would have been raised upon the detection of a change in a sensor value. This would have also inhibited further writing into the RAM until reset).

For the N-key rollover mode — if the E bit is programmed to "1" the chip will operate in the special Error mode. (For further details, see Interface Considerations Section.)

Status Word

The status word contains the FIFO status, error, and display unavailable signals. This word is read by the CPU when A_0 is high and $\overline{CS}$ and $\overline{RD}$ are low. See Interface Considerations for more detail on status word.

Data Read

Data is read when A_0, $\overline{CS}$ and $\overline{RD}$ are all low. The source of the data is specified by the Read FIFO or Read Display commands. The trailing edge of $\overline{RD}$ will cause the address of the RAM being read to be incremented if the Auto-Increment flag is set. FIFO reads always increment (if no error occurs) independent of AI.

Data Write

Data that is written with A_0, $\overline{CS}$ and $\overline{WR}$ low is always written to the Display RAM. The address is specified by the latest Read Display or Write Display command. Auto-Incrementing on the rising edge of $\overline{WR}$ occurs if AI set by the latest display command.

INTERFACE CONSIDERATIONS

Scanned Keyboard Mode, 2-Key Lockout

There are three possible combinations of conditions that can occur during debounce scanning. When a key is depressed, the debounce logic is set. Other depressed keys are looked for during the next two scans. If none are encountered, it is a single key depression and the key position is entered into the FIFO along with the status of CNTL and SHIFT lines. If the FIFO was empty, IRQ will be set to signal the CPU that there is an entry in the FIFO. If the FIFO was full, the key will not be entered and the error flag will be set. If another closed switch is encountered, no entry to the FIFO can occur. If all other keys are released before this one, then it will be entered to the FIFO. If this key is released before any other, it will be entirely ignored. A key is entered to the FIFO only once per depression, no matter how many keys were pressed along with it or in what order they were released. If two keys are depressed within the debounce cycle, it is a simultaneous depression. Neither key will be recognized until one key remains depressed alone. The last key will be treated as a single key depression.

Scanned Keyboard Mode, N-Key Rollover

With N-key Rollover each key depression is treated independently from all others. When a key is depressed, the debounce circuit waits 2 keyboard scans and then checks to see if the key is still down. If it is, the key is entered into the FIFO. Any number of keys can be depressed and another can be recognized and entered into the FIFO. If a simultaneous depression occurs, the keys are recognized and entered according to the order the keyboard scan found them.

Scanned Keyboard — Special Error Modes

For N-key rollover mode the user can program a special error mode. This is done by the "End Interrupt/Error Mode Set" command. The debounce cycle and key-validity check are as in normal N-key mode. If during a single debounce cycle, two keys are found depressed, this is considered a simultaneous multiple depression, and sets an error flag. This flag will prevent any further writing into the FIFO and will set interrupt (if not yet set). The error flag could be read in this mode by reading the FIFO STATUS word. (See "FIFO STATUS" for further details.) The error flag is reset by sending the normal CLEAR command with $C_F = 1$.

Sensor Matrix Mode

In Sensor Matrix mode, the debounce logic is inhibited. The status of the sensor switch is inputted directly to the Sensor RAM. In this way the Sensor RAM keeps an image of the state of the switches in the sensor matrix. Although debouncing is not provided, this mode has the advantage that the CPU knows how long the sensor was closed and when it was released. A keyboard mode can only indicate a validated closure. To make the software easier, the designer should functionally group the sensors by row since this is the format in which the CPU will read them.

The IRQ line goes high if any sensor value change is detected at the end of a sensor matrix scan. The IRQ line is cleared by the first data read operation if the Auto-Increment flag is set to zero, or by the End Interrupt command if the Auto-Increment flag is set to one.

Note: Multiple changes in the matrix Addressed by ($SL_{0-3} = 0$) may cause multiple interrupts. ($SL_0 = 0$ in the Decoded Mode). Reset may cause the 8279 to see multiple changes.

Data Format

In the Scanned Keyboard mode, the character entered into the FIFO corresponds to the position of the switch in the keyboard plus the status of the CNTL and SHIFT lines (non-inverted). CNTL is the MSB of the character and SHIFT is the next most significant bit. The next three bits are from the scan counter and indicate the row the key was found in. The last three bits are from the column counter and indicate to which return line the key was connected.

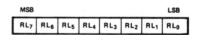

SCANNED KEYBOARD DATA FORMAT

In Sensor Matrix mode, the data on the return lines is entered directly in the row of the Sensor RAM that corresponds to the row in the matrix being scanned. Therefore, each switch postion maps directly to a Sensor RAM position. The SHIFT and CNTL inputs are ignored in this mode. Note that switches are not necessarily the only thing that can be connected to the return lines in this mode. Any logic that can be triggered by the scan lines can enter data to the return line inputs. Eight multiplexed input ports could be tied to the return lines and scanned by the 8279.

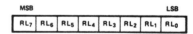

In Strobed Input mode, the data is also entered to the FIFO from the return lines. The data is entered by the rising edge of a CNTL/STB line pulse. Data can come from another encoded keyboard or simple switch matrix. The return lines can also be used as a general purpose strobed input.

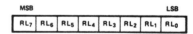

Display

Left Entry

Left Entry mode is the simplest display format in that each display position directly corresponds to a byte (or nibble) in the Display RAM. Address 0 in the RAM is the left-most display character and address 15 (or address 7 in 8 character display) is the right most display character. Entering characters from position zero causes the display to fill from the left. The 17th (9th) character is entered back in the left most position and filling again proceeds from there.

[Left column diagram: LEFT ENTRY MODE (AUTO INCREMENT)]

1st entry — 0 1 ... 14 15 ← Display RAM Address : [1]
2nd entry — 0 1 ... 14 15 : [1 2]
16th entry — 0 1 ... 14 15 : [1 2 ... 15 16]
17th entry — 0 1 ... 14 15 : [17 2 ... 15 16]
18th entry — 0 1 ... 14 15 : [17 18 ... 15 16]

LEFT ENTRY MODE (AUTO INCREMENT)

[Right column diagram: LEFT ENTRY MODE (AUTO INCREMENT)]

1st entry — 0 1 2 3 4 5 6 7 ← Display RAM Address : [1]
2nd entry — 0 1 2 3 4 5 6 7 : [1 2]
Command 10010101 — 0 1 2 3 4 5 6 7 : [1 2]
Enter next at Location 5 Auto Increment
3rd entry — 0 1 2 3 4 5 6 7 : [1 2 ... 3]
4th entry — 0 1 2 3 4 5 6 7 : [1 2 ... 3 4]

LEFT ENTRY MODE (AUTO INCREMENT)

Right Entry

Right entry is the method used by most electronic calculators. The first entry is placed in the right most display character. The next entry is also placed in the right most character after the display is shifted left one character. The left most character is shifted off the end and is lost.

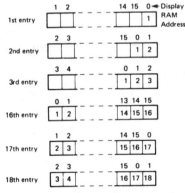

1st entry — 1 2 ... 14 15 0 ← Display RAM Address : [1]
2nd entry — 2 3 ... 15 0 1 : [1 2]
3rd entry — 3 4 ... 0 1 2 : [1 2 3]
16th entry — 0 1 ... 13 14 15 : [14 15 16]
17th entry — 1 2 ... 14 15 0 : [15 16 17]
18th entry — 2 3 ... 15 0 1 : [16 17 18]

RIGHT ENTRY MODE (AUTO INCREMENT)

Note that now the display position and register address do not correspond. Consequently, entering a character to an arbitrary position in the Auto Increment mode may have unexpected results. Entry starting at Display RAM address 0 with sequential entry is recommended.

Auto Increment

In the Left Entry mode, Auto Incrementing causes the address where the CPU will next write to be incremented by one and the character appears in the next location. With non-Auto Incrementing the entry is both to the same RAM address and display position. Entry to an arbitrary address in the Auto Increment mode has no undesirable side effects and the result is predictable:

In the Right Entry mode, Auto Incrementing and non Incrementing have the same effect as in the Left Entry except if the address sequence is interrupted:

1st entry — 1 2 3 4 5 6 7 0 ← Display RAM Address : [1]
2nd entry — 2 3 4 5 6 7 0 1 : [1 2]
Command 10010101 — 2 3 4 5 6 7 0 1 : [1 2]
Enter next at Location 5 Auto Increment
3rd entry — 3 4 5 6 7 0 1 2 : [3 ... 1 2]
4th entry — 4 5 6 7 0 1 2 3 : [3 4 ... 1 2]

RIGHT ENTRY MODE (AUTO INCREMENT)

Starting at an arbitrary location operates as shown below:

Command 10010101 — 0 1 2 3 4 5 6 7 ← Display RAM Address
Enter next at Location 5 Auto Increment

1st entry — 1 2 3 4 5 6 7 0 : [1]
2nd entry — 2 3 4 5 6 7 0 1 : [1 2]
8th entry — 4 5 6 7 8 1 2 3
9th entry — 5 6 7 8 9 2 3 4

RIGHT ENTRY MODE (AUTO INCREMENT)

Entry appears to be from the initial entry point.

8/16 Character Display Formats

If the display mode is set to an 8 character display, the on duty-cycle is double what it would be for a 16 character display (e.g., 5.1 ms scan time for 8 characters vs. 10.3 ms for 16 characters with 100 kHz internal frequency).

G. FIFO Status

FIFO status is used in the Keyboard and Strobed Input modes to indicate the number of characters in the FIFO and to indicate whether an error has occurred. There are two types of errors possible: overrun and underrun. Overrun occurs when the entry of another character into a full FIFO is attempted. Underrun occurs when the CPU tries to read an empty FIFO.

The FIFO status word also has a bit to indicate that the Display RAM was unavailable because a Clear Display or Clear All command had not completed its clearing operation.

In a Sensor Matrix mode, a bit is set in the FIFO status word to indicate that at least one sensor closure indication is contained in the Sensor RAM.

In Special Error Mode the S/E bit is showing the error flag and serves as an indication to whether a simultaneous multiple closure error has occurred.

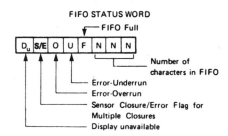

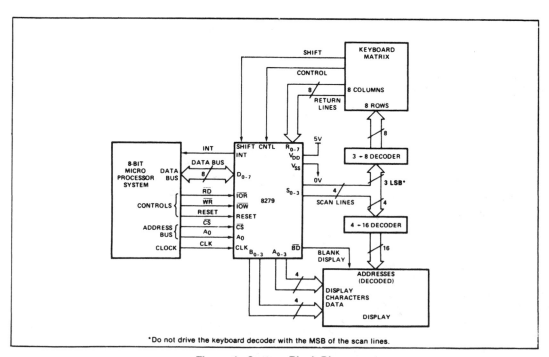

Figure 4. System Block Diagram

ABSOLUTE MAXIMUM RATINGS*

Ambient Temperature .... 0°C to 70°C
Storage Temperature .... -65°C to 125°C
Voltage on any Pin with
Respect to Ground .... -0.5V to +7V
Power Dissipation ....1 Watt

*NOTICE: Stresses above those listed under "Absolute Maximum Ratings" may cause permanent damage to the device. This is a stress rating only and functional operation of the device at these or any other conditions above those indicated in the operational sections of this specification is not implied. Exposure to absolute maximum rating conditions for extended periods may affect device reliability.

D.C. CHARACTERISTICS [T_A = 0°C to 70°C, V_{SS} = V_{CC} = +5V ± 5%, V_{CC} = +5V ±10% (8279-5)] *

Symbol	Parameter	Min.	Max.	Unit	Test Conditions
V_{IL1}	Input Low Voltage for Return Lines	-0.5	1.4	V	
V_{IL2}	Input Low Voltage for All Others	-0.5	0.8	V	
V_{IH1}	Input High Voltage for Return Lines	2.2		V	
V_{IH2}	Input High Voltage for All Others	2.0		V	
V_{OL}	Output Low Voltage		0.45	V	Note 1
V_{OH1}	Output High Voltage on Interrupt Line	3.5		V	Note 2
V_{OH2}	Other Outputs	2.4			I_{OH} = -400 μA
I_{IL1}	Input Current on Shift, Control and Return Lines		+10 -100	μA μA	V_{IN} = V_{CC} V_{IN} = 0V
I_{IL2}	Input Leakage Current on All Others		±10	μA	V_{IN} = V_{CC} to 0V
I_{OFL}	Output Float Leakage		±10	μA	V_{OUT} = V_{CC} to 0.45V
I_{CC}	Power Supply Current		120	mA	

CAPACITANCE

Symbol	Parameter	Typ.	Max.	Unit	Test Conditions
C_{IN}	Input Capacitance	5	10	pF	f_C = 1 MHz Unmeasured
C_{OUT}	Output Capacitance	10	20	pF	pins returned to V_{SS}

A.C. CHARACTERISTICS [T_A = 0°C to 70°C, V_{SS} = 0V, (Note 3)] *
Bus Parameters

READ CYCLE

Symbol	Parameter	8279		8279-5		Unit
		Min.	Max.	Min.	Max.	
t_{AR}	Address Stable Before READ	50		0		ns
t_{RA}	Address Hold Time for READ	5		0		ns
t_{RR}	READ Pulse Width	420		250		ns
t_{RD} [4]	Data Delay from READ		300		150	ns
t_{AD} [4]	Address to Data Valid		450		250	ns
t_{DF}	READ to Data Floating	10	100	10	100	ns
t_{RCY}	Read Cycle Time	1		1		μs

A.C. CHARACTERISTICS (Continued)

WRITE CYCLE

Symbol	Parameter	8279		8279-5		Unit
		Min.	Max.	Min.	Max.	
t$_{AW}$	Address Stable Before $\overline{WRITE}$	50		0		ns
t$_{WA}$	Address Hold Time for $\overline{WRITE}$	20		0		ns
t$_{WW}$	$\overline{WRITE}$ Pulse Width	400		250		ns
t$_{DW}$	Data Set Up Time for $\overline{WRITE}$	300		150		ns
t$_{WD}$	Data Hold Time for $\overline{WRITE}$	40		0		ns
t$_{WCY}$	Write Cycle Time	1		1		μs

OTHER TIMINGS

Symbol	Parameter	8279		8279-5		Unit
		Min.	Max.	Min.	Max.	
t$_{\phi W}$	Clock Pulse Width	230		120		nsec
t$_{CY}$	Clock Period	500		320		nsec

Keyboard Scan Time .... 5.1 msec
Keyboard Debounce Time .... 10.3 msec
Key Scan Time .... 80 μsec
Display Scan Time .... 10.3 msec

Digit-on Time .... 480 μsec
Blanking Time .... 160 μsec
Internal Clock Cycle[5] .... 10 μsec

NOTES:
1. 8279, I$_{OL}$ = 1.6mA; 8279-5, I$_{OL}$ = 2.2mA.
2. I$_{OH}$ = $-$100 μA
3. 8279, V$_{CC}$ = +5V ±5%; 8279-5, V$_{CC}$ = +5V ±10%.
4. 8279, C$_L$ = 100pF; 8279-5, C$_L$ = 150pF.
5. The Prescaler should be programmed to provide a 10 μs internal clock cycle.
* For Extended Temperature EXPRESS, use M8279A electrical parameters.

A.C. TESTING INPUT, OUTPUT WAVEFORM

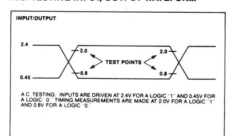

A.C. TESTING LOAD CIRCUIT

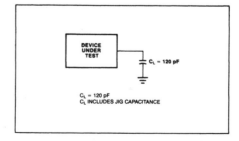

WAVEFORMS

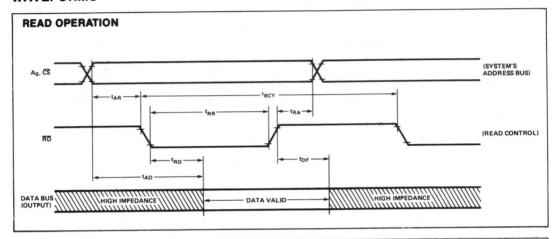

READ OPERATION

A₀, $\overline{CS}$ — (SYSTEM'S ADDRESS BUS)

t_{AR} t_{RCY} t_{RA}

t_{RR}

$\overline{RD}$ — (READ CONTROL)

t_{RD} t_{DF}

t_{AD}

DATA BUS (OUTPUT) — HIGH IMPEDANCE   DATA VALID   HIGH IMPEDANCE

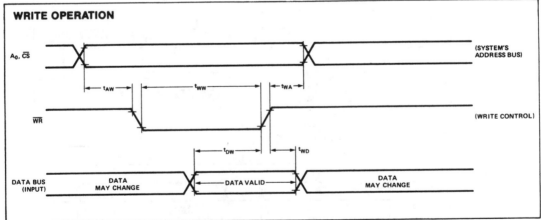

WRITE OPERATION

A₀, $\overline{CS}$ — (SYSTEM'S ADDRESS BUS)

t_{AW} t_{WW} t_{WA}

$\overline{WR}$ — (WRITE CONTROL)

t_{DW} t_{WD}

DATA BUS (INPUT) — DATA MAY CHANGE   DATA VALID   DATA MAY CHANGE

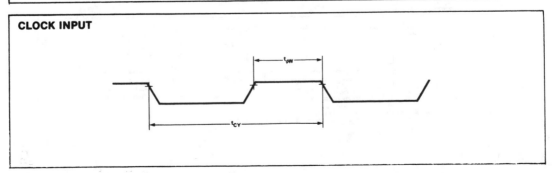

CLOCK INPUT

$t_{\phi W}$

t_{CY}

705

WAVEFORMS (Continued)

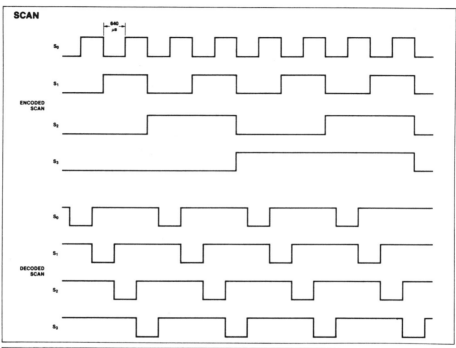

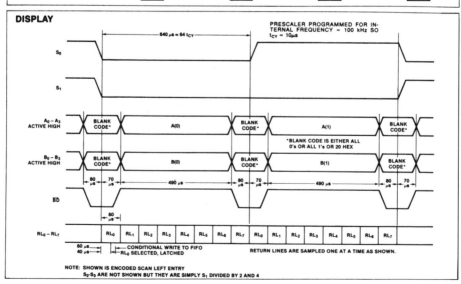

LM061L

SUMMARY
- 20 Character × 2 lines
- Built-in control LSI HD44780 type (see section 6).
- +5V single power supply

MECHANICAL DATA (Nominal dimensions)
Module size . . . . .115W × 39H × 13D (max.) mm
Effective display area . . . . 83W × 18.6H mm
Character size (5 × 7 dots) . . . . 3.2W × 4.85H mm
Pitch . . . . 3.7 mm
Dot size . . . . 0.6W × 0.65H mm
Weight . . . . about 50 g

ABSOLUTE MAXIMUM RATINGS   min.   max.
Power supply for logic ($V_{DD} - V_{SS}$) 0   6.5 V
Power supply for LCD drive
($V_{DD} - V_O$) 0   6.5 V
Input voltage (Vi) . . . . V_{SS} V_{DD} V
Operating temperature (Ta) 0   50°C
Storage temperature (Tstg) . . . . −20   70°C

ELECTRICAL CHARACTERISTICS
Ta = 25°C, V_{DD} = 5.0 V ± 0.25 V
Input "high" voltage (Vi_H) . . . . 2.2 V min.
Input "low" voltage (Vi_L) . . . . 0.6 V max.
Output high voltage (V_{OH}) ($-I_{OH}$ = 0.2 mA) . . 2.4 V min.
Output low voltage (V_{OL}) (I_{OL} = 1.2 mA) . . . . 0.4 V max.
Power supply current (I_{DD}) (V_{DD} = 5.0 V) . . . 1.0 mA typ.
3.0 mA max.
Power supply for LCD drive (Recommended) ($V_{DD} - V_O$)
Du=1/16
at Ta = 0°C . . . . .4.6 V typ.
at Ta = 25°C . . . . .4.4 V typ.
at Ta = 50°C . . . . .4.2 V typ.

OPTICAL DATA . . . . See page 15.

INTERNAL PIN CONNECTION

Pin No.	Symbol	Level	Function	
1	V_{SS}	–	0V	
2	V_{DD}	–	+5V	Power supply
3	V_O	–		
4	RS	H/L	L: Instruction code input H: Data input	
5	R/W	H/L	H: Data read (LCD module→MPU) L: Data write (LCD module←MPU)	
6	E	H, H→L	Enable signal	
7	DB0	H/L		
8	DB1	H/L		
9	DB2	H/L		
10	DB3	H/L	Data bus line Note (1), Note (2)	
11	DB4	H/L		
12	DB5	H/L		
13	DB6	H/L		
14	DB7	H/L		

Note:

In the HD44780, the data can be sent in either 4-bit 2-operation or 8-bit 1-operation so that it can interface to both 4 and 8 bit MPU's.

(1) When interface data is 4 bits long, data is transferred using only 4 buses of $DB_4 \sim DB_7$ and $DB_0 \sim DB_3$ are not used. Data transfer between the HD44780 and the MPU completes when 4-bit data is transferred twice. Data of the higher order 4 bits (contents of $DB_4 \sim DB_7$ when interface data is 8 bits long) is transferred first and then lower order 4 bits (contents of $DB_0 \sim DB_3$ when interface data is 8 bits long).

(2) When interface data is 8 bits long, data is transferred using 8 data buses of $DB_0 \sim DB_7$.

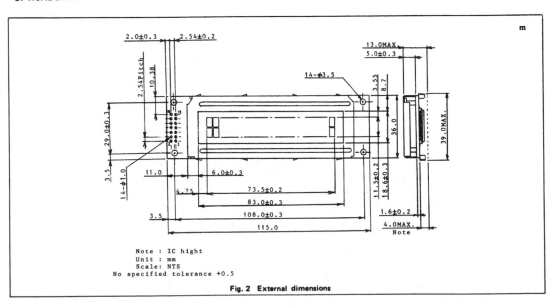

Note : IC hight
Unit : mm
Scale: NTS
No specified tolerance +0.5

Fig. 2 External dimensions

TIMING CHARACTERISTICS

Item	Symbol	Test condition	Min.	Typ.	Max.	Unit
Enable cycle time	t_{cyc}	Fig. 5, Fig. 6	1.0	—	—	μs
Enable pulse width	P_{wEH}	Fig. 5, Fig. 6	450	—	—	ns
Enable rise/fall time	t_{Er}, t_{Ef}	Fig. 5, Fig. 6	—	—	25	ns
RS, R/W set up time	t_{AS}	Fig. 5, Fig. 6	140	—	—	ns
Data delay time	t_{DDR}	Fig. 6	—	—	320	ns
Data set up time	t_{DSW}	Fig. 5	195	—	—	ns
Hold time	t_H	Fig. 5, Fig. 6	20	—	—	ns

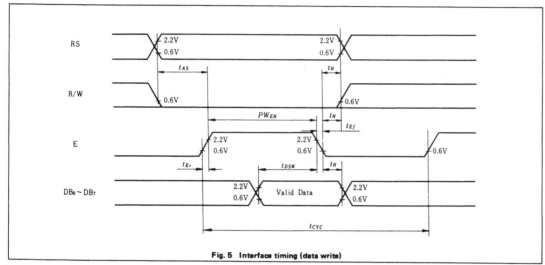

Fig. 5 Interface timing (data write)

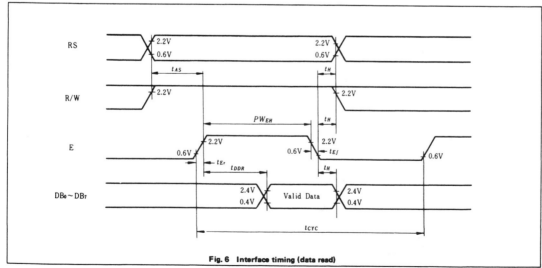

Fig. 6 Interface timing (data read)

HOW TO USE HITACHI'S BUILT-IN CONTROLLER DRIVER LCD-II (HD44780) DOT MATRIX LCD MODULE

■ INTRODUCTION

The LCD-II (HD44780) is a dot matrix liquid crystal display controller & driver LSI that displays alphanumerics, kana characters and symbols. It drives dot matrix liquid crystal display under 4-bit or 8-bit microcomputer or microprocessor control. All the functions required for dot matrix liquid crystal display drive are internally provided on one chip.

The user can complete dot matrix liquid crystal display systems with less number of chips by using the LCD-II (HD44780). If a driver LSI HD44100H is externally connected to the HD44780, up to 80 characters can be displayed.

The LCD-II is produced in the CMOS process. Therefore, the combination of the LCD-II with a CMOS microcomputer or microprocessor can accomplish a portable battery-drive device with lower power dissipation.

1. Applicable type

(1) 1 line series
LM054 • H2570 • LM015 • LM568AF • LM020L • LM070L • LM038 • LM027 • H2571 • LM058

(2) 2 line series
LM052L • LM016L • LM032L • LM060L • LM017L • LM018L • LM075L • LM074L • LM068L • LM061L

(3) 4 lines series
LM041L • LM044L

(4) Compact version
LM104L • LM105L • LM107L

2. Connecting MPU with LCM

2.1 Driver circuit block diagram

Figure 1 shows the driver circuit block diagram of LCM with built-in controller LSI. Controller LSI HD44780 (LCD-II) is built-in this LCM. Also extended LCD driver LSI is built in the LCM that displays more than 16 digits.

■ FEATURES

● Capable of interfacing to 4-bit or 8-bit MPU.
● Display data RAM .... 80 x 8 bits
    (80 characters, max.)
● Character generator ROM ....
    Character font 5 x 7 dots: 160 characters
    Character font 5 x 10 dots: 32 characters
● Both display data and character generator RAMs can be read from the MPU.
● Wide range of instruction functions
    Display clear, Cursor home, Display ON/OFF, Cursor ON/OFF, Display character blink, Cursor shift, Display shift
● Internal automatic reset circuit at power ON. (Internal reset circuit)

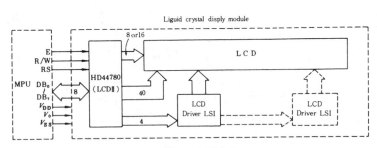

Liguid crystal disply module

Fig. 1 Driver circuit block diagram

2.2 Interfacing to MPU

In the HD44780, data can be sent in either 4-bit 2-operation or 8-bit 1-operation so it can interface to both 4 and 8 bit MPU's.

(1) When interface data is 4-bits long, data is transferred using only 4 buses: $DB_4 \sim DB_7$. $DB_0 \sim DB_3$ are not used. Data transfer between the HD44780 and the MPU completes when 4-bit data is transferred twice. Data of the higher order 4 bits (contents of $DB_4 \sim DB_7$ when interface data is 8 bits long) is transferred first, then the lower order 4 bits (content of $DB_0 \sim DB_3$ when interface data is 8 bits long) is transferred. Check the busy flag after 4-bit data has been transferred twice (one instruction). A 4-bit 2-operation will then transfer the busy flag and address counter data.

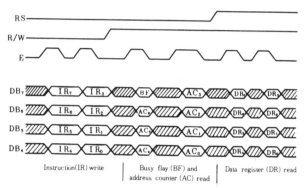

Fig. 2 4-bit data transfer example

(2) When interface data is 8 bit long, data is transferred using the 8 data buses of $DB_0 \sim DB_7$.

2.3 Interface to MPU

(1) Interface to 8-bit MPU

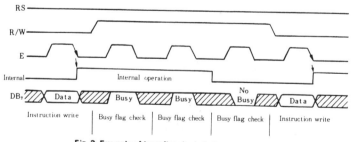

Fig. 3 Example of busy flag check timing sequence

Initializing by instruction

If the power supply conditions for correctly operating the internal reset circuit are not met, initialization by instruction is required.

Use the following procedure for initialization.

(1) When interface is 8 bits long;

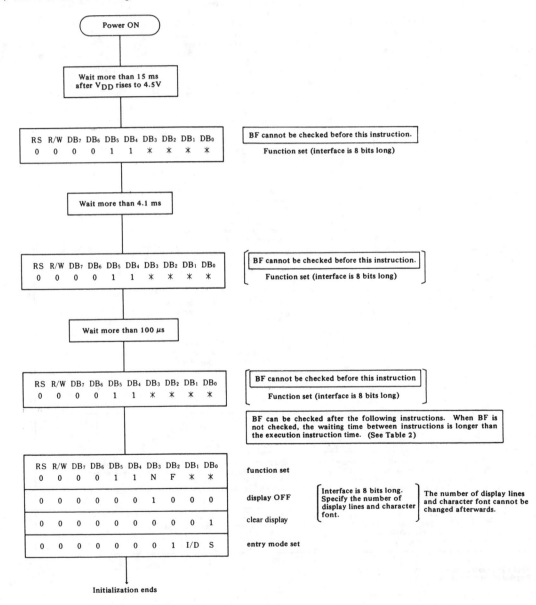

Power ON

Wait more than 15 ms
after V_{DD} rises to 4.5V

RS	R/W	DB7	DB6	DB5	DB4	DB3	DB2	DB1	DB0
0	0	0	0	1	1	✳	✳	✳	✳

BF cannot be checked before this instruction.

Function set (interface is 8 bits long)

Wait more than 4.1 ms

RS	R/W	DB7	DB6	DB5	DB4	DB3	DB2	DB1	DB0
0	0	0	0	1	1	✳	✳	✳	✳

BF cannot be checked before this instruction.

Function set (interface is 8 bits long)

Wait more than 100 μs

RS	R/W	DB7	DB6	DB5	DB4	DB3	DB2	DB1	DB0
0	0	0	0	1	1	✳	✳	✳	✳

BF cannot be checked before this instruction

Function set (interface is 8 bits long)

BF can be checked after the following instructions. When BF is not checked, the waiting time between instructions is longer than the execution instruction time. (See Table 2)

RS	R/W	DB7	DB6	DB5	DB4	DB3	DB2	DB1	DB0
0	0	0	0	1	1	N	F	✳	✳
0	0	0	0	0	0	1	0	0	0
0	0	0	0	0	0	0	0	0	1
0	0	0	0	0	0	0	1	I/D	S

function set

display OFF

clear display

entry mode set

Interface is 8 bits long. Specify the number of display lines and character font.

The number of display lines and character font cannot be changed afterwards.

Initialization ends

711

Table 2 Instructions

Instruction	Code										Description	Execution time (when fosc is 250 kHz) Note 1	Execution time (when fosc is 160 kHz) Note 2
	RS	R/W	DB7	DB6	DB5	DB4	DB3	DB2	DB1	DB0			
Clear display	0	0	0	0	0	0	0	0	0	1	Clears all display and returns the cursor to the home position (Address 0).	82 μs ~ 1.64 ms	120 μs ~ 4.9 ms
Return home	0	0	0	0	0	0	0	0	1	*	Returns the cursor to the home position (Address 0). Also returns the display being shifted to the original position. DD RAM contents remain unchanged.	40 μs ~ 1.6 ms	120 μs ~ 4.8 ms
Entry mode set	0	0	0	0	0	0	0	1	I/D	S	Sets the cursor move direction and specifies or not to shift the display. These operations are performed during data write and read.	40 μs	120 μs
Display ON/OFF control	0	0	0	0	0	0	1	D	C	B	Sets ON/OFF of all display (D), cursor ON/OFF (C), and blink of cursor position character (B).	40 μs	120 μs
Cursor and display shift	0	0	0	0	0	1	S/C	R/L	*	*	Moves the cursor and shifts the display without changing DD RAM contents	40 μs	120 μs
Function set	0	0	0	0	1	DL	N	F	*	*	Sets interface data length (DL) number of display lines (L) and character font (F).	40 μs	120 μs
Set CG RAM address.	0	0	0	1	A_{CG}						Sets the CG RAM address. CG RAM data is sent and received after this setting.	40 μs	120 μs
Set DD RAM address	0	0	1	A_{DD}							Sets the DD RAM address. DD RAM data is sent and received after this setting.	40 μs	120 μs
Read busy flag & address	0	1	BF	AC							Reads Busy flag (BF) indicating internal operation is being performed and reads address counter contents.	1 μs	1 μs
Write data to CG or DD RAM	1	0	Write Data								Writes data into DD RAM or CG RAM.	40 μs	120 μs
Read data to CG or DD RAM	1	1	Read Data								Reads data from DD RAM or CG RAM.	40 μs	120 μs

I/D = 1: Increment (+1) I/D = 0: Decrement (−1) S = 1: Accompanies display shift. S/C = 1: Display shift S/C = 0: Cursor move R/L = 1: Shift to the right. R/L = 0: Shift to the left. DL = 1: 8 bits DL = 0: 4 bits N = 1: 2 lines N = 0: 1 line F = 1: 5 x 10 dots F = 0: 5 x 7 dots BF = 1: Internally operating BF = 0: Can accept instruction	DD RAM: Display data RAM CG RAM: Character generator RAM A_{CG}: CG RAM address A_{DD}: DD RAM address Corresponds to cursor address. AC: Address counter used for both of DD and CG RAM address.	Execution time changes when frequency changes. (Example) When fosc is 270 kHz: $40\ \mu s \times \dfrac{250}{270} = 37\ \mu s$

*No effect
Notes 1. Applied to models driven by 1/8 duty or 1/11 duty.
      2. Applied to models driven by 1/16 duty.

Description of details

(1) Clear display

RS	R/W	DB₇							DB₀

Code: 0 0 0 0 0 0 0 0 0 1

Writes space code "20" (hexadecimal) (character pattern for character code "20" must be blank pattern) into all DD RAM addresses. Sets DD RAM address 0 in address counter. Returns display to its original status if it was shifted. In other words, the display disappears and the cursor or blink go to the left edge of the display (the first line if 2 lines are displayed). Set I/D = 1 (Increment Mode) of Entry Mode. S of Entry Mode doesn't change.

(2) Return home

RS	R/W	DB₇							DB₀

Code: 0 0 0 0 0 0 0 0 1 *

* No effect

Sets the DD RAM address 0 in address counter. Returns display to its original status if it was shifted. DD RAM contents do not change. The cursor or blink go to the left edge of the display (the first line if 2 lines are displayed).

(3) Entry mode set

RS	R/W	DB₇							DB₀

Code: 0 0 0 0 0 0 0 0 I/D S

I/D: Increments (I/D = 1) or decrements (I/D = 0) the DD RAM address by 1 when a character code is written into or read from the DD RAM. The cursor or blink moves to the right when incremented by 1 and to the left when decremented by 1. The same applies to writing and reading of CG RAM.

S: Shifts the entire display either to the right or to the left when S is 1; to the left when I/D = 1 and to the right when I/D = 0. Thus it looks as if the cursor stands still and the display moves. The display does not shift when reading from the DD RAM when writing into or reading out from the CG RAM does it shift when S = 0.

(4) Display ON/OFF control

RS	R/W	DB₇							DB₀

Code: 0 0 0 0 0 0 1 D C B

D: The display is ON when D = 1 and OFF when D = 0. When off due to D = 0, display data remains in the DD RAM. It can be displayed immediately by setting D = 1.

C: The cursor displays when C = 1 and does not display when C = 0. Even if the cursor disappears, the function of I/D, etc. does not change during display data write. The cursor is displayed using 5 dots in the 8th line when the 5 x 7 dot character font is selected and 5 dots in the 11th line when the 5 x 10 dot character font is selected.

B: The character indicated by the cursor blinks when B = 1. The blink is displayed by switching between all blank dots and display characters at 409.6 ms interval when f_{CP} or f_{osc} = 250 kHz. The cursor and the blink can be set to display simultaneously. (The blink frequency changes according to the reciprocal of f_{CP} or f_{osc}. $409.6 \times \dfrac{250}{270} = 379.2$ ms when f_{CP} = 270 kHz.)

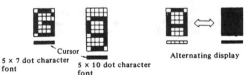

Cursor

5 x 7 dot character font

5 x 10 dot character font

Alternating display

(a) Cursor Display Example   (b) Blink Display Example

(5) Cursor or display shift

RS	R/W	DB₇							DB₀

Code: 0 0 0 0 0 1 S/C R/L * *

* No effect

Shifts cursor position or display to the right or left without writing or reading display data. This function is used to correct or search for the display. In a 2-line display, the cursor moves to the 2nd line when it passes the 40th digit of the 1st line. Notice that the 1st and 2nd line displays will shift at the same time. When the displayed data is shifted repeatedly each line only moves horizontally. The 2nd line display does not shift into the 1st line position.

S/C	R/L	
0	0	Shifts the cursor position to the left. (AC is decremented by one.)
0	1	Shifts the cursor position to the right. (AC is incremented by one.)
1	0	Shifts the entire display to the left. The cursor follows the display shift.
1	1	Shifts the entire display to the right. The cursor follows the display shift.

Address counter (AC) contents do not change if the only action performed is shift display.

(6) Function set

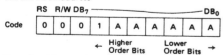

Code

* No effect

DL: Sets interface data length. Data is sent or received in 8 bit lengths ($DB_7 \sim DB_0$) when DL = 1 and in 4 bit lengths ($DB_7 \sim DB_4$) when DL = 0.

When the 4 bit length is selected, data must be sent or received twice.

N: Sets number of display lines.

F: Sets character font.

(Note) Perform the function at the head of the program before executing all instructions (except "Busy flag/address read"). From this point, the function set instruction cannot be executed unless the interface data length is changed.

N F	No. of display lines	Character font	Duty factor	Remarks
0 0	1	5 x 7 dots	1/8	
0 1	1	5 x 10 dots	1/11	
1 *	2	5 x 7 dots	1/16	Cannot display 2 lines with 5 x 10 dot character font.

* No effect

(7) Set CG RAM address

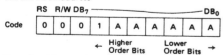

Code

Sets the CG RAM address into the address counter in binary AAAAAA. Data is then written or read from the MPU for the CG RAM.

(8) Set DD RAM address

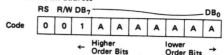

Code

Sets the DD RAM address into the address counter in binary AAAAAAA. Data is then written or read from the MPU for the DD RAM.

However, when N = 0 (1-line display), AAAAAAA is "00" ~ "4F" (hexadecimal),

when N = 1 (2-line display), AAAAAAA is "00" ~ "27" (hexadecimal) for the first line, and "40" ~ "67" (hexadecimal) for the second line.

(9) Read busy flag & address

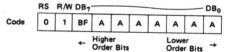

Code

Reads the busy flag (BF) that indicates the system is now internally operating by a previously received instruction. BF = 1 indicates that internal operation is in progress. The next instruction will not be accepted until BF is set to "0". Check the BF status before the next wire operation.

At the same time, the value of the address counter expressed in binary AAAAAAA is read out. The address counter is used by both CG and DD RAM addresses, and its value is determined by the previous instruction. Address contents are the same as in Items (7) and (8).

(10) Write data to CG or DD RAM

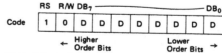

Code

Writes binary 8 bit data DDDDDDDD to the CG or the DD RAM. Whether the CG or DD RAM is to be written into is determined by the previous specification of CG RAM or DD RAM address setting. After write, the address is automatically incremented or decremented by 1 according to entry mode. The entry mode also determines display shift.

(11) Read data from CG or DD RAM

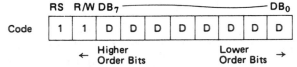

Reads binary 8 bit data DDDDDDDD from the CG or DD RAM. The previous designation determines whether the CG or DD RAM is to be read. Before entering the read instruction, you must execute either the CG RAM or DD RAM address set instruction. If you don't, the first read data will be invalidated. When serially executing the "read" instruction, the next address data is normally read from the second read. The "address set" instruction need not be executed just before the "read" instruction when shifting the cursor by cursor shift instruction (when reading out DD RAM). The cursor shift instruction operation is the same as that of the DD RAM's address set instruction.

After a read, the entry mode automatically increases or decreases the address by 1. However, display shift is not executed no matter what the entry mode is.

(Note) The address counter (AC) is automatically incremented or decremented by 1 after "write" instructions to either CG RAM or DD RAM. RAM data selected by the AC cannot than be read out even if "read" instructions are executed. The conditions for correct data read out are: execute either the address set instruction or cursor shift instruction (only with DD RAM), just before reading out execute the "read" instruction from the second time the "read" instruction is serial.

(3) 8-bit operation, 8-digit x 2-line display

For 2-line display, the cursor automatically moves from the first to the second line after the 40th digit of the 1st line has been written. Thus, if there are only 8 characters in the first line, the DD RAM address must again be set after the 8th character is completed. (See the following table) Note that the first and second lines of the display shift are performed. In the example, the display shift is performed when the cursor is on the second line. However, if shift operation is performed when the cursor is on the first line, both the first and second lines move together. When you repeat the shift, the display of the second display will only move within each line many times.

8 bit operation, 8-digit x 2-line display example (using internal reset)

No.	Instruction	Display	Operation
1	Power supply ON (HD44780 is initialized by the internal reset circuit)		Initialized. No display appears.
2	Function Set RS R/W DB$_7$ ————————— DB$_0$ 0 0 0 0 1 1 1 0 * *		Sets to 8-bit operation and selects 2-line display and 5 x 7 dot character font.
3	Display ON/OFF Control 0 0 0 0 0 0 1 1 1 0	—	Turns on display and cursor. All display is in space mode because of initialization.
4	Entry Mode Set 0 0 0 0 0 0 0 1 1 0	—	Sets mode to increment the address by one and to shift the cursor to the right, at the time of write, to the DD/CG RAM. Display is not shifted.
5	Write Data to CG RAM/DD RAM 1 0 0 1 0 0 1 0 0 0	H_	Write "H". The DD RAM has already been selected by initialization when the power is turned on. The cursor is incremented by one and shifted to the right.
6			
7	Write Data to CG RAM/DD RAM 1 0 0 1 0 0 1 0 0 1	H I T A C H I _	Writes "I".
8	Set DD RAM Address 0 0 1 1 0 0 0 0 0 0	H I T A C H I __	Sets RAM address so that the cursor is positioned at the head of the 2nd line.
9	Write Data to CG RAM/DD RAM 1 0 0 1 0 0 1 1 0 1	H I T A C H I M_	Writes "M".
10			
11	Write Data to CG RAM/DD RAM 1 0 0 1 0 0 1 1 1 1	H I T A C H I M I C R O C O _	Writes "O".
12	Entry Mode Set 0 0 0 0 0 0 0 1 1 1	H I T A C H I M I C R O C O _	Sets mode for display shift at the time of write.
13	Write Data to CG RAM/DD RAM 1 0 0 1 0 0 1 1 0 1	I T A C H I I C R O C O M _	Writes "M". Display is shifted to the right. The first and second lines' shift are operated at the same time.
14			
15	Return Home 0 0 0 0 0 0 0 0 1 0	H I T A C H I M I C R O C O M	Returns both display and cursor to the original position (Address 0).

American Standard Code for Information Interchange: ASCII Codes

	Graphic or Control	ASCII (Hexadecimal)
NUL	Null	00
SOH	Start of Heading	01
STX	Start of Text	02
ETX	End of Text	03
EOT	End of Transmission	04
ENQ	Enquiry	05
ACK	Acknowledge	06
BEL	Bell	07
BS	Backspace	08
HT	Horizontal Tabulation	09
LF	Line Feed	0A
VT	Vertical Tabulation	0B
FF	Form Feed	0C
CR	Carriage Return	0D
SO	Shift Out	0E
SI	Shift In	0F
DLE	Data Link Escape	10
DC1	Device Control 1	11
DC2	Device Control 2	12
DC3	Device Control 3	13
DC4	Device Control 4	14
NAK	Negative Acknowledge	15
SYN	Synchronous Idle	16
ETB	End of Transmission Block	17
CAN	Cancel	18
EM	End of Medium	19
SUB	Substitute	1A

	Graphic or Control	ASCII (Hexadecimal)
ESC	Escape	1B
FS	File Separator	1C
GS	Group Separator	1D
RS	Record Separator	1E
US	Unit Separator	1F
SP	Space	20
	!	21

Graphic or Control	ASCII (Hexadecimal)	Graphic or Control	ASCII (Hexadecimal)
"	22	Q	51
#	23	R	52
$	24	S	53
%	25	T	54
&	26	U	55
'	27	V	56
(	28	W	57
)	29	X	58
*	2A	Y	59
+	2B	Z	5A
'	2C	[	5B
−	2D	\	5C
.	2E	]	5D
/	2F	∧	5E
0	30	−	5F
1	31	`	60
2	32	a	61
3	33	b	62
4	34	c	63
5	35	d	64
6	36	e	65
7	37	f	66
8	38	g	67
9	39	h	68
:	3A	i	69
;	3B	j	6A
<	3C	k	6B
=	3D	l	6C
>	3E	m	6D
?	3F	n	6E
@	40	o	6F
A	41	p	70
B	42	q	71
C	43	r	72
D	44	s	73
E	45	t	74
F	46	u	75
G	47	v	76
H	48	w	77
I	49	x	78
J	4A	y	79
K	4B	z	7A
L	4C	{	7B
M	4D	|	7C
N	4E	}	7D
O	4F	~	7E
P	50	DEL Delete	7F

8085
Instruction Set

Appendix F describes each instruction fully in terms of its operation and the operand, in-cluding details such as number of bytes, machine cycles, T-states, Hex code, and affected flags. The instructions appear in alphabetical order and are illustrated with examples.

The following abbreviations are used in the description of the instruction set.

Flags

Reg. = 8080A/8085 Register   S = Sign
Mem. = Memory Location   Z = Zero
R   = Register   AC = Auxiliary Carry
Rs   = Register Source   P = Parity
Rd   = Register Destination   CY = Carry
M   = Memory
( ) = Contents of
XX   = Random Information

ACI: Add Immediate to Accumulator with Carry

Opcode	Operand	Bytes	M-Cycles	T-States	Hex Code
ACI	8-bit data	2	2	7	CE

Description   The 8-bit data (operand) and the Carry flag are added to the contents of the accumulator, and the result is stored in the accumulator.

Flags   All flags are modified to reflect the result of the addition.

Example   Assuming the accumulator contains 26H and the previous operation has set the Carry flag, add byte 57H to the accumulator.

Instruction: ACI 57H   Hex Code: CE 57

Addition:

$$
\begin{array}{rl}
\text{(A): } 26H = & 0\ 0\ 1\ 0\ \ 0\ 1\ 1\ 0 \\
\text{(Data): } 57H = & 0\ 1\ 0\ 1\ \ 0\ 1\ 1\ 1 \\
\text{CY 1} = & \underline{\hspace{4.5cm}1} \\
7EH = & 0\ 1\ 1\ 1\ \ 1\ 1\ 1\ 0 \\
\end{array}
$$

Flags: S = 0   Z = 0   AC = 0
                P = 1   CY = 0

Comments:

1. After addition the previous Carry flag is cleared.
2. This instruction is commonly used in 16-bit addition. This instruction should not be used to account for a carry generated by 8-bit numbers.

ADC: Add Register to Accumulator with Carry

Opcode	Operand	Bytes	M-Cycles	T-States	Reg.	Hex
ADC	Reg.	1	1	4	B	88
	Mem.	1	2	7	C	89
					D	8A
					E	8B
					H	8C
					L	8D
					M	8E
					A	8F

Description The contents of the operand (register or memory) and the Carry flag are added to the contents of the accumulator and the result is placed in the accumulator. The contents of the operand are not altered; however, the previous Carry flag is reset.

Flags All flags are modified to reflect the result of the addition.

Example Assume register pair BC contains 2498H and register pair DE contains 54A1H. Add these 16-bit numbers and save the result in BC registers.

The steps in adding 16-bit numbers are as follows:

1. Add the contents of registers C and E by placing the contents of one register in the accumulator. This addition generates a Carry. Use instruction ADD (explained on the next page) and save the low-order 8-bits in register C.

```
      98H =       1 0 0 1  1 0 0 0
      A1H =       1 0 1 0  0 0 0 1
   1  39H =  1    0 0 1 1  1 0 0 1    Store in register C
  CY         CY
```

2. Add the contents of registers B and D by placing the contents of one register in the accumulator. Use instruction ADC.

The result will be as follows.

```
  24H = 0 0 1 0  0 1 0 0
  54H = 0 1 0 1  0 1 0 0
   1  =                1   (Carry from the previous addition)
  79H = 0 1 1 1  1 0 0 1   Store in register B
```

Comments: This instruction is generally used in 16-bit addition. For example, to add the contents of BC registers to the contents of DE registers this instruction is used to account for the carry generated by low-order bytes.

ADD: Add Register to Accumulator

Opcode	Operand	Bytes	M-Cycles	T-States	Reg.	Hex
ADD	Reg.	1	1	4	B	80
	Mem.	1	2	7	C	81
					D	82
					E	83
					H	84
					L	85
					M	86
					A	87

Description   The contents of the operand (register or memory) are added to the contents of the accumulator and the result is stored in the accumulator. If the operand is a memory location, that is indicated by the 16-bit address in the HL register.

Flags   All flags are modified to reflect the result of the addition.

Example   Register B has 51H and the accumulator has 47H. Add the contents of register B to the contents of the accumulator.

Instruction: ADD B   Hex Code: 80

Register contents
before instruction   Addition

Register contents
after instruction

A	47	X	F
B	51	X	C

$$47H = 0\ 1\ 0\ 0\quad 0\ 1\ 1\ 1$$
$$51H = 0\ 1\ 0\ 1\quad 0\ 0\ 0\ 1$$
$$\overline{98H = 1\ 0\ 0\ 1\quad 1\ 0\ 0\ 0}$$

SZ AC P CY

A	98	1 0 0 0 0	F
B	51	X	C

Flags: S = 1, Z = 0, AC = 0
          P = 0, CY = 0

Example   Memory location 2050H has data byte A2H and the accumulator has 76H. Add the contents of the memory location to the contents of the accumulator.

Instruction: ADD M   Hex Code: 86
Before this instruction is executed, registers HL should be loaded with data 2050H.

Register contents
before instruction

```
    A | 76 | X | F
    B |  X | X | C          2050 | A2 |
    D |  X | X | E
    H | 20 | 50 | L
```

Addition:

Register contents
after instruction

$$
\begin{array}{ll}
(A) & 76H = 0\ 1\ 1\ 1\quad 0\ 1\ 1\ 0 \\
(2050H)_{Mem} & A2H = 1\ 0\ 1\ 0\quad 0\ 0\ 1\ 0 \\
\hline
& 1/18H = 1/0\ 0\ 0\ 1\quad 1\ 0\ 0\ 0 \\
& CY \qquad CY
\end{array}
$$

```
                        S  Z  AC  P  CY
    A |  18  | 0,0, 0, 1, 1 | F
    B |  18  |      X       | C
    D |  18  |      X       | E
    H |  18  |     50       | L
```

Flags: S = 0, Z = 0, AC = 0,
P = 1, CY = 1

ADI: Add Immediate to Accumulator

Opcode	Operand	Bytes	M-Cycles	T-States	Hex Codes
ADI	8-bit data	2	2	7	C6

Description The 8-bit data (operand) are added to the contents of the accumulator, and the result is placed in the accumulator.

Flags All flags are modified to reflect the result of the addition.

Example The accumulator contains 4AH. Add the data byte 59H to the contents of the accumulator.

Instruction: ADI 59H   Hex Code: C6 59

Addition:

$$
\begin{array}{ll}
(A)\ :\ & 4AH = 0\ 1\ 0\ 0\quad 1\ 0\ 1\ 0 \\
& + \\
(Data)\ :\ & 59H = 0\ 1\ 0\ 1\quad 1\ 0\ 0\ 1 \\
\hline
& A3H = 1\ 0\ 1\ 0\quad 0\ 0\ 1\ 1
\end{array}
$$

Flags: S = 1, Z = 0, AC = 1
P = 1, CY = 0

ANA: Logical AND with Accumulator

Opcode	Operand	Bytes	M-Cycles	T-States	Hex Codes	
ANA	Reg.	1	1	4	**Reg.**	**Hex**
	Mem.	1	2	7	B	A0
					C	A1
					D	A2
					E	A3
					H	A4
					L	A5
					M	A6
					A	A7

Description   The contents of the accumulator are logically ANDed with the contents of the operand (register or memory), and the result is placed in the accumulator. If the operand is a memory location, its address is specified by the contents of HL registers.

Flags   S, Z, P are modified to reflect the result of the operation. CY is reset. In 8085, AC is set, and in 8080A AC is the result of ORing bits D_3 of the operands.

Example   The contents of the accumulator and the register D are 54H and 82H, respectively. Logically AND the contents of register D with the contents of the accumulator. Show the flags and the contents of each register after ANDing.

Instruction: ANA D   Hex Code: A2

Register contents before instruction	Logical AND	Register contents after instruction

SZ AC P CY

A 54 X F   54H = 0 1 0 1   0 1 0 0   A [00   ] [0 1 1 1 0] F
                        AND
B 82 X E   82H = 1 0 0 0   0 0 1 0   D [82   ] [   ] E
                        ‾‾‾‾‾‾‾‾‾‾‾‾‾‾‾‾‾
                        0 0 0 0   0 0 0 0

Flags: S = 0, Z = 1, P = 1
        AC = 1, CY = 0
        (for 8080A, AC = 0)

ANI: AND Immediate with Accumulator

Opcode	Operand	Bytes	M-Cycles	T-States	Hex Code
ANI	8-bit data	2	2	7	E6

Description   The contents of the accumulator are logically ANDed with the 8-bit data (operand) and the results are placed in the accumulator.

Flags   S, Z, P are modified to reflect the results of the operation. CY is reset. In 8085, AC is set.

Example   AND data byte 97H with the contents of the accumulator, which contains A3H.

Instruction: ANI 97H   Hex Code: E6 97

Logical AND:

```
  (A)   : A3H = 1 0 1 0  0 0 1 1
                AND
(Data)  : 97H = 1 0 0 1  0 1 1 1
                1 0 0 0  0 0 1 1
```

A [83]

S Z AC P CY

| 1, 0, | 1, | 0, | 0 | F

CALL: Unconditional Subroutine Call

Opcode	Operand	Bytes	M-Cycles	T-States	Hex Code
CALL	16-bit address	3	5	18	CD

Description   The program sequence is transferred to the address specified by the operand. Before the transfer, the address of the next instruction to CALL (the contents of the program counter) is pushed on the stack. The sequence of events is described in the example below.

Flags   No flags are affected.

Example   Write CALL instruction at memory location 2010H to call a subroutine located at 2050H. Explain the sequence of events when the stack pointer is at location 2099H.

Memory Address	Hex Code	Mnemonics
2010	CD	CALL 2050H
2011	50	
2012	20	

Note: See the difference between writing a 16-bit address as mnemonics and code. In the code, the low-order byte (50) is entered first, then the high-order byte (20) is entered. However, in mnemonics the address is shown in the proper sequence. If an assembler is used to obtain the codes, it will automatically reverse the sequence of the mnemonics.

Execution of CALL: The address in the program counter (2013H) is placed on the stack as follows.

Stack pointer is decremented to 2098H

MSB is stored

Stack pointer is again decremented   SP→

LSB is stored

Call address (2050H) is temporarily stored in internal WZ registers and placed on the bus for the fetch cycle

2097	13
2098	20
2099	

Comments: The CALL instruction should be accompanied by one of the return (RET or conditional return) instructions in the subroutine.

Conditional Call to Subroutine   **Operand—16-Bit Address**

Op Code	Description	Flag Status	Hex Code	M-Cycles T-States
CC	Call on Carry	CY = 1	DC	2/9 (if condition is not true)
CNC	Call with No Carry	CY = 0	D4	5/18 (if condition is true)
CP	Call on positive	S = 0	F4	*Note:* If condition is not true it continues
CM	Call on minus	S = 1	FC	the sequence, and thus requires
CPE	Call on Parity Even	P = 1	EC	fewer T-states.
CPO	Call on Parity Odd	P = 0	E4	If condition is true it calls the
CZ	Call on Zero	Z = 1	CC	subroutine, thus requires more
CNZ	Call on No Zero	Z = 0	C4	T-states.

Flags   No flags are affected.

CMA: Complement Accumulator

Opcode	Operand	Bytes	M-Cycles	T-States	Hex Code
CMA	None	1	1	4	2F

Description   The contents of the accumulator are complemented.

Flags   No flags are affected.

Example   Complement the accumulator, which has data byte 89H.

Instruction: CMA   Hex Code: 2F

Before instruction   After instruction

A $\boxed{1 \quad 0 \quad 0 \quad 0 \quad 1 \quad 0 \quad 0 \quad 1}$ = 89H   A $\boxed{0 \quad 1 \quad 1 \quad 1 \quad 0 \quad 1 \quad 1 \quad 0}$ = 76H

CMC: Complement Carry

Opcode	Operand	Bytes	M-Cycles	T-States	Hex Code
CMC	None	1	1	4	3F

Description   The Carry flag is complemented.

Flags   The Carry flag is modified, no other flags are affected.

CMP: Compare with Accumulator

Opcode	Operand	Bytes	M-Cycles	T-States	Hex Codes	
					Reg.	**Hex**
CMP	Reg.	1	1	4	B	B8
	Mem.	1	2	7	C	B9
					D	BA
					E	BB
					H	BC
					L	BD
					M	BE
					A	BF

Description   The contents of the operand (register or memory) are compared with the contents of the accumulator. Both contents are preserved and the comparison is shown by setting the flags as follows:

☐ If (A) < (Reg/Mem): Carry flag is set and Zero flag is reset.
☐ If (A) = (Reg/Mem): Zero flag is set and Carry flag is reset.
☐ If (A) > (Reg/Mem): Carry and Zero flags are reset.

The comparison of two bytes is performed by subtracting the contents of the operand from the contents of the accumulator; however, neither contents are modified.

Flags   S, P, AC are also modified in addition to Z and CY to reflect the results of the operation.

Example   Register B contains data byte 62H and the accumulator contains data byte 57H. Compare the contents of register B with those of the accumulator.

Instruction: CMP B   Hex Code: B8

Before instruction

A | 57 | XX | F
B | 62 | XX | C

After instruction

A | 57 | |1| F
B | 62 | XX | C

Flags: S = 1, Z = 0, AC = 1
P = 1, CY = 1

Results after executing the instruction:

☐ No contents are changed.
☐ Carry flag is set because (A) < (B).
☐ S, Z, P, AC flags will also be modified as listed above.

CPI: Compare Immediate with Accumulator

Opcode	Operand	Bytes	M-Cycles	T-States	Hex Code
CPI	8-bit	2	2	7	FE

Description The second byte (8-bit data) is compared with the contents of the accumulator. The values being compared remain unchanged and the results of the comparison are indicated by setting the flags as follows.

☐ If (A) < Data: Carry flag is set and Zero flag is reset.
☐ If (A) = Data: Zero flag is set and Carry flag is reset.
☐ If (A) > Data: Carry and Zero flags are reset.

The comparison of two bytes is performed by subtracting the data byte from the contents of the accumulator; however, neither contents are modified.

Flags S, P, AC are also modified in addition to Z and CY to reflect the result of the operation.

Example Assume the accumulator contains data byte C2H. Compare 98H with the accumulator contents.

Instruction: CPI 98H   Hex Code: FE 98

Results after executing the instruction:

☐ The accumulator contents remain unchanged.
☐ Z and CY flags are reset because (A) > Data.
☐ Other flags: S = 0, AC = 0, P = 0.

Example Compare data byte C2H with the contents of the accumulator in the above example.

Instruction: CPI C2H   Hex Code: FE C2

Results after executing the instruction:

☐ The accumulator contents remain unchanged.
☐ Zero flag is set because (A) = Data.
☐ Other flags: S = 0, AC = 1, P = 1, CY = 0.

DAA: Decimal-Adjust Accumulator

Opcode	Operand	Bytes	M-Cycles	T-States	Hex Code
DAA	None	1	1	4	27

Description The contents of the accumulator are changed from a binary value to two 4-bit binary-coded decimal (BCD) digits. This is the only instruction that uses the auxiliary flag (internally) to perform the binary-to-BCD conversion; the conversion procedure is described below.

Flags S, Z, AC, P, CY flags are altered to reflect the results of the operation. Instruction DAA converts the binary contents of the accumulator as follows:

1. If the value of the low-order four bits (D_3–D_0) in the accumulator is greater than 9 or if AC flag is set, the instruction adds 6 (06) to the low-order four bits.
2. If the value of the high-order four bits (D_7–D_4) in the accumulator is greater than 9 or if the Carry flag is set, the instruction adds 6 (60) to the high-order four bits.

Example Add decimal 12_{BCD} to the accumulator, which contains 39_{BCD}.

$$
\begin{aligned}
(A) = \quad 39_{BCD} &= 0\ 0\ 1\ 1 \quad 1\ 0\ 0\ 1 \\
+12_{BCD} &= 0\ 0\ 0\ 1 \quad 0\ 0\ 1\ 0 \\
\hline
51_{BCD} &= 0\ 1\ 0\ 0 \quad 1\ 0\ 1\ 1 \\
&\quad\ \ 4 \qquad\quad\ B
\end{aligned}
$$

The binary sum is 4BH. The value of the low-order four bits is larger than 9. Add 06 to the low-order four bits.

$$
\begin{aligned}
4B &= 0\ 1\ 0\ 0 \quad 1\ 0\ 1\ 1 \\
+\ 06 &= 0\ 0\ 0\ 0 \quad 0\ 1\ 1\ 0 \\
&\qquad\qquad\ \ 1 \quad 1\ 1 \\
\hline
51 &= 0\ 1\ 0\ 1 \quad 0\ 0\ 0\ 1
\end{aligned}
$$

Example Add decimal 68_{BCD} to the accumulator, which contains 85_{BCD}.

$$
\begin{aligned}
(A) = \quad\ \ 85_{BCD} &= 1\ 0\ 0\ 0 \quad 0\ 1\ 0\ 1 \\
+\ \ 68_{BCD} &= 0\ 1\ 1\ 0 \quad 1\ 0\ 0\ 0 \\
\hline
153_{BCD} &= 1\ 1\ 1\ 0 \quad 1\ 1\ 0\ 1
\end{aligned}
$$

The binary sum is EDH. The values of both, low-order and high-order, four bits are higher than 9. Add 6 to both.

$$
\begin{aligned}
=\quad ED &= \quad 1\ 1\ 1\ 0 \quad 1\ 1\ 0\ 1 \\
+\ 66 &= \quad 0\ 1\ 1\ 0 \quad 0\ 1\ 1\ 0 \\
&\qquad\ 1\ 1 \quad\ 1\ 1 \\
\hline
\boxed{1}\ 53 &= \boxed{1}\ 0\ 1\ 0\ 1 \quad 0\ 0\ 1\ 1 \\
\text{CY} &\qquad \text{CY}
\end{aligned}
$$

The accumulator contains 53 and the Carry flag is set to indicate that the sum is larger than eight bits (153). The program should keep track of the Carry; otherwise it may be altered by the subsequent instructions.

DAD: Add Register Pair to H and L Registers

Opcode	Operand	Bytes	M-Cycles	T-States	Hex Codes	
DAD	Reg. pair	1	3	10	**Reg. Pair**	**Hex**
					B	09
					D	19
					H	29
					SP	39

Description   The 16-bit contents of the specified register pair are added to the contents of the HL register and the sum is saved in the HL register. The contents of the source register pair are not altered.

Flags   If the result is larger than 16 bits the CY flag is set. No other flags are affected.

Example   Assume register pair HL contains 0242H. Multiply the contents by 2.

Instruction: DAD H   Hex Code: 29

Before instruction	DAD operation	After instruction
	0242	
H 02 42 L	+0242	
	0484	H 04 84 L

Example   Assume register pair HL is cleared. Transfer the stack pointer (register) that points to memory location 2099H to the HL register pair.

Instruction: DAD SP   Hex Code: 39

Before instruction	DAD operation	After instruction
H 00 00 L	0000	H 20 99 L
SP 2099	+2099	SP 2099
	2099	

Note: After the execution of the instruction, the contents of the stack pointer register are not altered.

DCR: Decrement Source by 1

Opcode	Operand	Bytes	M-Cycles	T-States	Hex Codes	
DCR	Reg.	1	1	4	**Reg.**	**Hex**
					B	05
	Mem.	1	3	10	C	0D
					D	15
					E	1D
					H	25
					L	2D
					M	35
					A	3D

Description The contents of the designated register/memory is decremented by 1 and the results are stored in the same place. If the operand is a memory location, it is specified by the contents of the HL register pair.

Flags S, Z, P, AC are modified to reflect the result of the operation. CY is not modified.

Example Decrement register B, which is cleared, and specify its contents after the decrement.

Instruction: DCR B   Hex Code: 05

Before instruction		Decrement operation
A		XX	F	$(B) = 0\ 0\ 0\ 0\quad 0\ 0\ 0\ 0$
B	00	XX	C	$-01 = 0\ 0\ 0\ 0\quad 0\ 0\ 0\ 1$

Subtraction is performed in 2's complement:

$$(B) = \quad 0\ 0\ 0\ 0\quad 0\ 0\ 0\ 0$$
$$+$$
$$\text{2's complement of } 1 = \quad \underline{1\ 1\ 1\ 1\quad 1\ 1\ 1\ 1}$$
$$(B) = \quad 1\ 1\ 1\ 1\quad 1\ 1\ 1\ 1$$

After the execution of the DCR instruction register B will contain FFH; however, this instruction does not modify the CY flag.

Example Decrement the contents of memory location 2085, which presently holds A0H.

Assume the HL register contains 2085H.

Instruction: DCR M   Hex Code: 35

Before instruction   Memory

H | 20 | 85 | L   2084 |___|
                     2085 | A0 |
                     2086 |___|

After instruction

H | 20 | 85 | L   2084 |___|
                     2085 | 9F |
                     2086 |___|

DCX: Decrement Register Pair by 1

Opcode	Operand	Bytes	M-Cycles	T-States	Hex Codes	
DCX	Reg. pair	1	1	6	**Reg.**	
					Pair	**Hex**
					B	0B
					D	1B
					H	2B
					SP	3B

Description   The contents of the specified register pair are decremented by 1. This instruction views the contents of the two registers as a 16-bit number.

Flags   No flags are affected.

Example   Register pair DE contains 2000H. Specify the contents of the entire register if it is decremented by 1.

Instruction: DCX D   Hex Code: 1B

After subtracting 1 from the DE register pair the answer is

D | 1F | FF | E

Example   Write instructions to set the Zero flag when a register pair (such as BC) is used as a down-counter.

   To decrement the register pair, instruction DCX is necessary; instruction DCR is used for one register. However, instruction DCX does not set the Zero flag when the register pair goes to 0 and it continues counting indefinitely. The Zero flag can be set by using the following instructions.

For BC pair:

```
 ┌→ DCX B        ;Decrement register pair BC
 ├  MOV A,C      ;Load accumulator with the contents of register C
 │  ORA B        ;Set Zero flag if B and C are both 0
 └→ JNZ          ;If Zero flag is not set, go back and decrement the contents of BC
                 ;pair
```

DI: Disable Interrupts

Opcode	Operand	Bytes	M-Cycles	T-States	Hex Code
DI	None	1	1	4	F3

Description The Interrupt Enable flip-flop is reset and all the interrupts except the TRAP (8085) are disabled.

Flags No flags are affected.

Comments: This instruction is commonly used when the execution of a code sequence cannot be interrupted. For example, in critical time delays, this instruction is used at the beginning of the code and the interrupts are enabled at the end of the code. The 8085 TRAP cannot be disabled.

EI: Enable Interrupts

Opcode	Operand	Bytes	M-Cycles	T-States	Hex Code
EI	None	1	1	4	FB

Description The Interrupt Enable flip-flop is set and all interrupts are enabled.

Flags No flags are affected.

Comments: After a system reset or the acknowledgment of an interrupt, the Interrupt Enable flip-flop is reset, thus disabling the interrupts. This instruction is necessary to reenable the interrupts (except TRAP).

HLT: Halt and Enter Wait State

Opcode	Operand	Bytes	M-Cycle	T-States	Hex Code
HLT	None	1	2 or more	5 or more	76

Description The MPU finishes executing the current instruction and halts any further execution. The MPU enters the Halt Acknowledge machine cycle and Wait states are inserted in every clock period. The address and the data bus are placed in the high impedance state. The contents of the registers are unaffected during the HLT state. An interrupt or reset is necessary to exit from the Halt state.

Flags No flags are affected.

IN: Input Data to Accumulator from a Port with 8-bit Address

Opcode	Operand	Bytes	M-Cycles	T-States	Hex Code
IN	8-bit port address	2	3	10	DB

Description The contents of the input port designated in the operand are read and loaded into the accumulator.

Flags No flags are affected.

Comments: The operand is an 8-bit address; therefore, port addresses can range from 00H to FFH. While executing the instruction, a port address is duplicated on low-order (A_7–A_0) and high-order (A_{15}–A_8) address buses. Any one of the sets of address lines can be decoded to enable the input port.

INR: Increment Contents of Register/Memory by 1

Opcode	Operand	Bytes	M-Cycles	T-States	Reg.	Hex
INR	Reg.	1	1	4		
	Mem.	1	3	10	B	04
					C	0C
					D	14
					E	1C
					H	24
					L	2C
					M	34
					A	3C

Description   The contents of the designated register/memory are incremented by 1 and the results are stored in the same place. If the operand is a memory location, it is specified by the contents of HL register pair.

Flags   S, Z, P, AC are modified to reflect the result of the operation. CY is not modified.

Example   Register D contains FF. Specify the contents of the register after the increment.

Instruction: INR D   Hex Code: 14

$$
\begin{array}{rl}
(D) = & 1\ 1\ 1\ 1\quad 1\ 1\ 1\ 1 \\
+ 1 = & 0\ 0\ 0\ 0\quad 0\ 0\ 0\ 1 \\
\hline
& 1\quad 1\ 1\ 1\ 1\quad 1\ 1\ 1 \quad\text{Carry} \\
00 = & \boxed{0}\ \ 0\ 0\ 0\ 0\quad 0\ 0\ 0\ 0 \\
& \text{CY}
\end{array}
$$

After the execution of the INR instruction, register D will contain 00H; however, no Carry flag is set.

Example   Increment the contents of memory location 2075H, which presently holds 7FH. Assume the HL register contains 2075H.

Instruction: INR M   Hex Code: 34

Before instruction	Memory	
H 20 75 L	2074	
	2075	7F
	2076	

After instruction	Memory	
H 20 75 L	2074	
	2075	80
	2076	

INX: Increment Register Pair by 1

Opcode	Operand	Bytes	M-Cycles	T-States	Hex Codes	
INX	Reg. pair	1	1	6	**Reg.**	
					Pair	**Hex**
					B	03
					D	13
					H	23
					SP	33

Description   The contents of the specified register pair are incremented by 1. The instruction views the contents of the two registers as a 16-bit number.

Flags   No flags are affected.

Example   Register pair HL contains 9FFFH. Specify the contents of the entire register if it is incremented by 1.

Instruction: INX H   Hex Code: 23

After adding 1 to the contents of the HL pair the answer is

H | A0   00 | L

JMP: Jump Unconditionally

Opcode	Operand	Bytes	M-Cycles	T-States	Hex Code
JMP	16-bit	3	3	10	C3

Description   The program sequence is transferred to the memory location specified by the 16-bit address. This is a 3-byte instruction; the second byte specifies the low-order byte and the third byte specifies the high-order byte.

Example   Write the instruction at location 2000H to transfer the program sequence to memory location 2050H.

Instruction:

Memory Address	Code	Mnemonics
2000	C3	JMP 2050H
2001	50	
2002	20	

Comments: The 16-bit address of the operand is entered in memory in reverse order, the low-order byte first, followed by the high-order byte.

Jump Conditionally

Operand: 16-bit address

Op Code	Description	Flag Status	Hex Code	M-Cycles/T-States
JC	Jump on Carry	CY = 1	DA	2M/7T (if condition
JNC	Jump on No Carry	CY = 0	D2	is not true)
JP	Jump on positive	S = 0	F2	3M/10T (if condition
JM	Jump on minus	S = 1	FA	is true)
JPE	Jump on Parity Even	P = 1	EA	
JPO	Jump on Parity Odd	P = 0	E2	
JZ	Jump on Zero	Z = 1	CA	
JNZ	Jump on No Zero	Z = 0	C2	

Flags   No flags are affected.

Comments: The 8085 requires only seven T-states when condition is not true. For example, instruction JZ 2050H will transfer the program sequence to location 2050H when the Zero flag is set (Z = 1) and the execution requires ten T-states. When the Zero flag is reset (Z = 0), the execution sequence will not be changed and this requires seven T-states.

LDA: Load Accumulator Direct

Opcode	Operand	Bytes	M-Cycles	T-States	Hex Code
LDA	16-bit address	3	4	13	3A

Description The contents of a memory location, specified by a 16-bit address in the operand, are copied to the accumulator. The contents of the source are not altered. This is a 3-byte instruction; the second byte specifies the low-order address and the third byte specifies the high-order address.

Flags No flags are affected.

Example Assume memory location 2050H contains byte F8H. Load the accumulator with the contents of location 2050H.

Instruction: LDA 2050H   Hex Code: 3A 50 20   (note the reverse order)

A $\boxed{\text{F8} \mid \text{X}}$ F   2050 $\boxed{\text{F8}}$

LDAX: Load Accumulator Indirect

Opcode	Operand	Bytes	M-Cycles	T-States	Hex Code	
LDAX	B/D reg. pair	1	2	7	**Reg.**	**Hex**
					BC	0A
					DE	1A

Description The contents of the designated register pair point to a memory location. This instruction copies the contents of that memory location into the accumulator. The contents of either the register pair or the memory location are not altered.

Flags No flags are affected.

Example Assume the contents of register B = 20H, C = 50H, and memory location 2050H = 9FH. Transfer the contents of the memory location 2050H to the accumulator.

Instruction: LDAX B   Hex Code: 0A

Register contents before instruction				Memory contents		Register contents after instruction		
A	XX	XX	F			A	9F	XX	F
B	20	50	C ⟶ 2050	9F		B	20	50	C

LHLD: Load H and L Registers Direct

Opcode	Operand	Bytes	M-Cycles	T-States	Hex Code
LHLD	16-bit address	3	5	16	2A

Description   The instruction copies the contents of the memory location pointed out by the 16-bit address in register L and copies the contents of the next memory location in register H. The contents of source memory locations are not altered.

Flags   No flags are affected.

Example   Assume memory location 2050H contains 90H and 2051H contains 01H. Transfer memory contents to registers HL.

Instruction: LHLD 2050H   Hex Code: 2A 50 20

Memory contents before instruction		Register contents after instruction
2050	90	
2051	01	
		H 01 90 L

LXI: Load Register Pair Immediate

Opcode	Operand	Bytes	M-Cycles	T-States	Hex Code	
LXI	Reg. pair, 16-bit data	3	3	10	**Reg. Pair**	**Hex**
					B	01
					D	11
					H	21
					SP	31

Description   The instruction loads 16-bit data in the register pair designated in the operand. This is a 3-byte instruction; the second byte specifies the low-order byte and the third byte specifies the high-order byte.

Flags   No flags are affected.

Example Load the 16-bit data 2050H in register pair BC.

Instruction: LXI B,2050H   Hex Code: 01 50 20
This instruction loads 50H in register C and 20H in register B.

Comments: Note the reverse order in entering the code of 16-bit data. This is the only instruction that can directly load a 16-bit address in the stack pointer register.

MOV: Move—Copy from Source to Destination

Opcode	Operand	Bytes	M-Cycles	T-States	Hex Code
MOV	Rd,Rs	1	1	4	See table below
MOV	M,Rs	2		7	
MOV	Rd,M				

Description   This instruction copies the contents of the source register into the destination register; the contents of the source register are not altered. If one of the operands is a memory location, it is specified by the contents of HL registers.

Flags   No flags are affected.

Hex Code

		Source Location							
		B	C	D	E	H	L	M	A
	B	40	41	42	43	44	45	46	47
	C	48	49	4A	4B	4C	4D	4E	4F
	D	50	51	52	53	54	55	56	57
Destination	E	58	59	5A	5B	5C	5D	5E	5F
Location	H	60	61	62	63	64	65	66	67
	L	68	69	6A	6B	6C	6D	6E	6F
	M	70	71	72	73	74	75		77
	A	78	79	7A	7B	7C	7D	7E	7F

Example   Assume register B contains 72H and register C contains 9FH. Transfer the contents of register C to register B.

Instruction: MOV B,C   Hex Code: 41
Note the first operand B specifies the destination and the second operand C specifies the source.

	Register contents before instruction					Register contents after instruction		
B	72	9F	C		B	9F	9F	C

Example   Assume the contents of registers HL are 20H and 50H, respectively. Memory location 2050H contains 9FH. Transfer the contents of the memory location to register B.

Instruction: MOV B,M   Hex Code: 46

	Register contents before instruction			Memory contents		Register contents after instruction		
B	XX	XX	C		B	9F	XX	C
D	XX	XX	E	2050 9F	D	XX	XX	E
H	20	50	L		H	20	50	L

MVI: Move Immediate 8-Bit

Opcode	Operand	Bytes	M-Cycles	T-States	Hex Code	
MVI	Reg., Data	2	2	7	**Reg.**	**Hex**
	Mem., Data	2	3	10	B	06
					C	0E
					D	16
					E	1E
					H	26
					L	2E
					M	36
					A	3E

Description   The 8-bit data is stored in the destination register or memory. If the operand is a memory location, it is specified by the contents of HL registers.

Flags   No flags are affected.

Example   Load 92H in register B.

Instruction: MVI B,92H   Hex Code: 06 92
This instruction loads 92H in register B.

Example   Assume registers H and L contain 20H and 50H, respectively. Load 3AH in memory location 2050H.

Instruction: MVI M,3AH   Hex Code: 36

	Contents before instruction				Contents after instruction				
H	20	50	L	→ 2050 3A	H	20	50	L	

NOP: No Operation

Opcode	Operand	Bytes	M-Cycles	T-States	Hex Code
NOP	None	1	1	4	00

Description   No operation is performed. The instruction is fetched and decoded; however, no operation is executed.

Flags   No flags are affected.

Comments: The instruction is used to fill in time delays or to delete and insert instructions while troubleshooting.

ORA: Logically OR with Accumulator

Opcode	Operand	Bytes	M-Cycles	T-States	Hex Code	
ORA	Reg.	1	1	4	**Reg.**	**Hex**
	Mem.	1	2	7	B	B0
					C	B1
					D	B2
					E	B3
					H	B4
					L	B5
					M	B6
					A	B7

Description   The contents of the accumulator are logically ORed with the contents of the operand (register or memory), and the results are placed in the accumulator. If the operand is a memory location, its address is specified by the contents of HL registers.

Flags   Z, S, P are modified to reflect the results of the operation. AC and CY are reset.

Example   Assume the accumulator has data byte 03H and register C holds byte 81H. Combine the bits of register C with the accumulator bits.

Instruction: ORA C   Hex Code: B1

Register contents
before instruction   Logical OR

Register contents
after instruction
S Z AC P CY

A $\boxed{03}$ $\boxed{XX}$ F   03H = 0 0 0 0 0 0 1 1   A $\boxed{83}$ $\boxed{1\ 0\ \ 0\ \ 0\ \ 0}$ F
B $\boxed{XX}$ $\boxed{81}$ C   81H = 1 0 0 0 0 0 0 1   B $\boxed{XX}$ $\boxed{81}$ C
                                  83H = 1 0 0 0 0 0 1 1
                                  S = 1, Z = 0, P = 0
                                  Flags: CY = 0, AC = 0

Comments: The instruction is commonly used to

☐ reset the CY flag by ORing the contents of the accumulator with itself.
☐ set the Zero flag when 0 is loaded into the accumulator by ORing the contents of the accumulator with itself.
☐ combine bits from different registers.

ORI: Logically OR Immediate

Opcode	Operand	Bytes	M-Cycles	T-States	Hex Code
ORI	8-bit data	2	2	7	F6

Description   The contents of the accumulator are logically ORed with the 8-bit data in the operand and the results are placed in the accumulator.

Flags   S, Z, P are modified to reflect the results of the operation. CY and AC are reset.

OUT: Output Data from Accumulator to a Port with 8-Bit Address

Opcode	Operand	Bytes	M-Cycles	T-States	Hex Code
OUT	8-bit port address	2	3	10	D3

Description   The contents of the accumulator are copied into the output port specified by the operand.

Flags   No flags are affected.

Comments: The operand is an 8-bit address; therefore, port addresses can range from 00H to FFH. While executing the instruction, a port address is placed on the low-order address bus (A_7–A_0) as well as the high-order address bus (A_{15}–A_8). Any of the sets of address lines can be decoded to enable the output port.

PCHL: Load Program Counter with HL Contents

Opcode	Operand	Bytes	M-Cycles	T-States	Hex Code
PCHL	None	1	1	6	E9

Description   The contents of registers H and L are copied into the program counter. The contents of H are placed as a high-order byte and of L as a low-order byte.

Flags   No flags are affected.

Comments: This instruction is equivalent to a 1-byte unconditional Jump instruction. A program sequence can be changed to any location by simply loading the H and L registers with the appropriate address and by using this instruction.

POP: Pop off Stack to Register Pair

Opcode	Operand	Bytes	M-Cycles	T-States	Hex Code	
					Reg.	**Hex**
POP	Reg. pair	1	3	10	B	C1
					D	D1
					H	E1
					PSW	F1

Description The contents of the memory location pointed out by the stack pointer register are copied to the low-order register (such as C, E, L, and flags) of the operand. The stack pointer is incremented by 1 and the contents of that memory location are copied to the high-order register (B, D, H, A) of the operand. The stack pointer register is again incremented by 1.

Flags No flags are modified.

Example Assume the stack pointer register contains 2090H, data byte F5 is stored in memory location 2090H, and data byte 01H is stored in location 2091H. Transfer the contents of the stack to register pair H and L.

Instruction: POP H   Hex Code: E1

Register contents before instruction	Stack contents	Register contents after instruction
H	XX	XX	L	2090	F5	H	01	F5	L
SP	2090			2091	01	SP	2092		
				2092					

Comments: Operand PSW (Program Status Word) represents the contents of the accumulator and the flag register; the accumulator is the high-order register and the flags are the low-order register.

  Note that the contents of the source, stack locations, are not altered after the POP instruction.

PUSH: Push Register Pair onto Stack

Opcode	Operand	Bytes	M-Cycles	T-States	Hex Code	
					Reg.	**Hex**
PUSH	Reg. pair	1	3	12	B	C5
					D	D5
					H	E5
					PSW	F5

Description The contents of the register pair designated in the operand are copied into the stack in the following sequence. The stack pointer register is decremented and the contents of the high-order register (B, D, H, A) are copied into that location. The stack pointer register is decremented again and the contents of the low-order register (C, E, L, flags) are copied to that location.

Flags No flags are modified.

Example Assume the stack pointer register contains 2099H, register B contains 32H and register C contains 57H. Save the contents of the BC register pair on the stack.

Instruction: PUSH B   Hex Code: C5

Register contents
before instruction
B [32 | 57] C

SP [2099]

Stack contents
after instruction
2097 [57]
2098 [32]
2099 [XX]

Register contents
after instruction
B [32 | 57] C

SP [2097]

Comments: Operand PSW (Program Status Word) represents the contents of the accumulator and the flag register; the accumulator is the high-order register and the flags are the low-order register.

Note that the contents of the source registers are not altered after the PUSH instruction.

RAL: Rotate Accumulator Left through Carry

Opcode	Operand	Bytes	M-Cycles	T-States	Hex Code
RAL	None	1	1	4	17

Description Each binary bit of the accumulator is rotated left by one position through the Carry flag. Bit D_7 is placed in the bit in the Carry flag and the Carry flag is placed in the least significant position D_0.

Flags CY is modified according to bit D_7. S, Z, AC, P are not affected.

Example   Rotate the contents of the accumulator through Carry, assuming the accumulator has A7H and the Carry flag is reset.

Instruction: RAL   Hex Code: 17

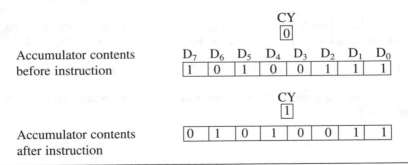

CY

$\boxed{0}$

Accumulator content
before instruction

D_7	D_6	D_5	D_4	D_3	D_2	D_1	D_0
1	0	1	0	0	1	1	1

CY

$\boxed{1}$

Accumulator contents
after instruction

0	1	0	0	1	1	1	0

Comment: This instruction effectively provides a 9-bit accumulator. The original contents of the accumulator can be restored by using instruction RAR (Rotate Accumulator Right through Carry). However; the contents will be modified if the instruction RRC (Rotate Accumulator Right) is used to restore the contents.

RAR: Rotate Accumulator Right through Carry

Opcode	Operand	Bytes	M-Cycles	T-States	Hex Code
RAR	None	1	1	4	1F

Description   Each binary bit of the accumulator is rotated right by one position through the Carry flag. Bit D_0 is placed in the Carry flag and the bit in the Carry flag is placed in the most significant position, D_7.

Flags   CY is modified according to bit D_0. S, Z, P, AC are not affected.

Example   Rotate the contents of the accumulator assuming it contains A7H and the Carry flag is reset to 0.

Instruction: RAR   Hex Code: 1F

CY

$\boxed{0}$

Accumulator contents
before instruction

D_7	D_6	D_5	D_4	D_3	D_2	D_1	D_0
1	0	1	0	0	1	1	1

CY

$\boxed{1}$

Accumulator contents
after instruction

0	1	0	1	0	0	1	1

RLC: Rotate Accumulator Left

Opcode	Operand	Bytes	M-Cycles	T-States	Hex Code
RLC	None	1	1	4	07

Description   Each binary bit of the accumulator is rotated left by one position. Bit D_7 is placed in the position of D_0 as well as in the Carry flag.

Flags   CY is modified according to bit D_7. S, Z, P, AC are not affected.

Example   Rotate the contents of the accumulator left, assuming it contains A7H and the Carry flag is reset to 0.

Instruction: RLC   Hex Code: 07

CY
0

Accumulator contents D_7 D_6 D_5 D_4 D_3 D_2 D_1 D_0
before instruction   1   0   1   0   0   1   1   1

CY
1

Accumulator contents   0   1   0   0   1   1   1   1
after instruction

Comments: The contents of bit D_7 are placed in bit D_0, and the Carry flag is modified accordingly. However, the contents of the Carry are not placed in bit D_0 as in instruction RAL.

RRC: Rotate Accumulator Right

Opcode	Operand	Bytes	M-Cycles	T-States	Hex Code
RRC	None	1	1	4	0F

Description   Each binary bit of the accumulator is rotated right by one position. Bit D_0 is placed in the position of D_7 as well as in the Carry flag.

Flags   CY is modified according to bit D_0. S, Z, P, AC are not affected.

Example   Rotate the contents of the accumulator right, if it contains A7H and the Carry flag is reset to 0.

Instruction: RRC   Hex Code: 0F

CY

$\boxed{0}$

Accumulator contents D_7 D_6 D_5 D_4 D_3 D_2 D_1 D_0
before instruction | 1 | 0 | 1 | 0 | 0 | 1 | 1 | 1 |

CY

$\boxed{1}$

Accumulator contents | 1 | 1 | 0 | 1 | 0 | 0 | 1 | 1 |
after instruction

Comments: The contents of bit D_0 are placed in bit D_7, and the Carry flag is modified accordingly. However, the contents of the Carry are not placed in bit D_7, as in the instruction RAR.

RET: Return from Subroutine Unconditionally

Opcode	Operand	Bytes	M-Cycles	T-States	Hex Code
RET	None	1	3	10	C9

Description   The program sequence is transferred from the subroutine to the calling program. The two bytes from the top of the stack are copied into the program counter and the program execution begins at the new address. The instruction is equivalent to POP Program Counter.

Flags   No flags are affected.

Example   Assume the stack pointer is pointing to location 2095H. Explain the effect of the RET instruction if the contents of the stack locations are as follows:

2095 $\boxed{50}$
2096 $\boxed{20}$

After instruction RET, the program execution is transferred to location 2050H and the stack pointer is shifted to location 2097H.

Comments: This instruction is used in conjunction with CALL or conditional call instructions.

Return Conditionally

Op Code	Description	Flag Status	Hex Code	M-Cycles/T-States
RC	Return on Carry	CY = 1	D8	
RNC	Return with No Carry	CY = 0	D0	1/6 (if condition is not true)
RP	Return on positive	S = 0	F0	3/12 (if condition is true)
RM	Return on minus	S = 1	F8	*Note:* If condition is not true, it continues
RPE	Return on Parity Even	P = 1	E8	the sequence and thus requires
RPO	Return on Parity Odd	P = 0	E0	fewer T-states.
RZ	Return on Zero	Z = 1	C8	If condition is true, it returns to the
RNZ	Return on No Zero	Z = 0	C0	calling program and thus requires
				more T-states.

Flags   No flags are affected.

RIM: Read Interrupt Mask

Opcode	Operand	Bytes	M-Cycles	T-States	Hex Code
RIM	None	1	1	4	20

Description   This is a multipurpose instruction used to read the status of interrupts 7.5, 6.5, 5.5 and to read serial data input bit. The instruction loads eight bits in the accumulator with the following interpretations:

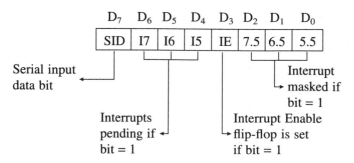

Flags   No flags are affected.

Example   After the execution of instruction RIM, the accumulator contained 49H. Explain the accumulator contents.

(A):   49H = 0 1 0 0 1 0 0 1

RST 7.5 is pending ⟵⎦

Interrupt Enable
flip-flop is set ⟵⎦

⎣⟶ RST 5.5 masked

RST 7.5 and 6.5
⟶ are enabled

RST: Restart

Bytes	M-Cycles	T-States
1	3	12

Opcode/Operand	Binary Code	Hex Code	Restart Address (H)
RST 0	1 1 0 0 0 1 1 1	C7	0000
RST 1	1 1 0 0 1 1 1 1	CF	0008
RST 2	1 1 0 1 0 1 1 1	D7	0010
RST 3	1 1 0 1 1 1 1 1	DF	0018
RST 4	1 1 1 0 0 1 1 1	E7	0020
RST 5	1 1 1 0 1 1 1 1	EF	0028
RST 6	1 1 1 1 0 1 1 1	F7	0030
RST 7	1 1 1 1 1 1 1 1	FF	0038

Description   The RST instructions are equivalent to 1-byte call instructions to one of the eight memory locations on page 0. The instructions are generally used in conjunction with interrupts and inserted using external hardware. However, these can be used as software instructions in a program to transfer program execution to one of the eight locations.

Flags   No flags are affected.

Additional 8085 Interrupts   The 8085 has four additional interrupts and these interrupts generate RST instructions internally and thus do not require any external hardware. These instructions and their Restart addresses are as follows:

Interrupts	Restart Address
TRAP	24H
RST 5.5	2CH
RST 6.5	34H
RST 7.5	3CH

SBB: Subtract Source and Borrow from Accumulator

Opcode	Operand	Bytes	M-Cycles	T-States	Hex Code	
SBB	Reg.	1	1	4	**Reg.**	**Hex**
	Mem.	1	2	7	B	98
					C	99
					D	9A
					E	9B
					H	9C
					L	9D
					M	9E
					A	9F

Description   The contents of the operand (register or memory) and the Borrow flag are subtracted from the contents of the accumulator and the results are placed in the accumulator. The contents of the operand are not altered; however, the previous Borrow flag is reset.

Flags   All flags are altered to reflect the result of the subtraction.

Example   Assume the accumulator contains 37H, register B contains 3FH, and the Borrow flag is already set by the previous operation. Subtract the contents of B with the borrow from the accumulator.

Instruction: SBB B   Hex Code: 98
The subtraction is performed in 2's complement; however, the borrow needs to be added first to the subtrahend:

$$
\begin{array}{rl}
\text{(B):} & \text{3F} \\
\text{Borrow:} & +\ \ 1 \\
\hline
\text{Subtrahend:} & 40\text{H} = 0\ 1\ 0\ 0\quad 0\ 0\ 0\ 0 \\
\text{2's complement of} & 40\text{H} = 1\ 1\ 0\ 0\quad 0\ 0\ 0\ 0 \\
\text{(A)} & = 0\ 0\ 1\ 1\quad 0\ 1\ 1\ 1 \\
\hline
& 0/1\ 1\ 1\ 1\quad 0\ 1\ 1\ 1 = \text{F7H} \\
\text{Complement Carry:} & 1/1\ 1\ 1\ 1\quad 0\ 1\ 1\ 1
\end{array}
$$

The Borrow flag is set to indicate the result is in 2's complement. The previous Borrow flag is reset during the subtraction.

SBI: Subtract Immediate with Borrow

Opcode	Operand	Bytes	M-Cycles	T-States	Hex Code
SBI	8-bit data	2	2	7	DE

Description The 8-bit data (operand) and the borrow are subtracted from the contents of the accumulator, and the results are placed in the accumulator.

Flags All flags are altered to reflect the result of the operation.

Example Assume the accumulator contains 37H and the Borrow flag is set. Subtract 25H with borrow from the accumulator.

Instruction: SBI 25H   Hex Code: DE 25

$$
\begin{array}{rl}
\text{(Data):} & 25H \\
+ \text{(Borrow):} & \underline{1H} \\
\text{Subtrahend:} & 26H = 0\ 0\ 1\ 0\ \ 0\ 1\ 1\ 0 \\
\text{2's complement of } 26H = & 1\ 1\ 0\ 1\ \ 1\ 0\ 1\ 0 \\
\text{(A) } 37H = & \underline{0\ 0\ 1\ 1\ \ 0\ 1\ 1\ 1} \\
& 1/0\ 0\ 0\ 1\ \ 0\ 0\ 0\ 1 = 11H \\
\text{Complement Carry: } 0/0\ 0\ 0\ 1 & \ \ 0\ 0\ 0\ 1 = 11H \\
\text{Flags: } & S = 0, Z = 0, AC = 1 \\
& P = 1, CY = 0
\end{array}
$$

SHLD: Store H and L Registers Direct

Opcode	Operand	Bytes	M-Cycles	T-States	Hex Code
SHLD	16-bit address	3	5	16	22

Description The contents of register L are stored in the memory location specified by the 16-bit address in the operand, and the contents of H register are stored in the next memory location by incrementing the operand. The contents of registers HL are not altered. This is a 3-byte instruction; the second byte specifies the low-order address and the third byte specifies the high-order address.

Flags No flags are affected.

Example Assume the H and L registers contain 01H and FFH, respectively. Store the contents at memory locations 2050H and 2051H.

Instruction: SHLD 2050H   Hex Code: 22 50 20

Register contents before instruction		Memory and register contents after instruction
H │ 01 , FF │ I	2050 │ FF │ 2051 │ 01 │	H │ 01 , FF │ L

SIM: Set Interrupt Mask

Opcode	Operand	Bytes	M-Cycles	T-States	Hex Code
SIM	None	1	1	4	30

Description   This is a multipurpose instruction and used to implement the 8085 interrupts (RST 7.5, 6.5, and 5.5) and serial data output.

The instruction interprets the accumulator contents as follows:

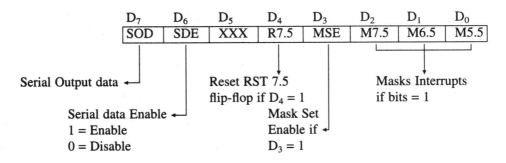

- ☐ SOD—Serial Output Data: Bit D_7 of the accumulator is latched into the SOD output line and made available to a serial peripheral if bit $D_6 = 1$.
- ☐ SDE—Serial Data Enable: If this bit = 1, it enables the serial output. To implement serial output, this bit needs to be enabled.
- ☐ XXX—Don't Care
- ☐ R7.5—Reset RST 7.5: If this bit = 1, RST 7.5 flip-flop is reset. This is an additional control to reset RST 7.5.
- ☐ MSE—Mask Set Enable: If this bit is high, it enables the functions of bits D_2, D_1, D_0. This is a master control over all the interrupt masking bits. If this bit is low, bits D_2, D_1, and D_0 do not have any effect on the masks.
- ☐ M7.5—$D_2 = 0$, RST 7.5 is enabled
         $= 1$, RST 7.5 is masked or disabled
- ☐ M6.5—$D_1 = 0$, RST 6.5 is enabled
         $= 1$, RST 6.5 is masked or disabled
- ☐ M5.5—$D_0 = 0$, RST 5.5 is enabled
         $= 1$, RST 5.5 is masked or disabled

Example   Write instructions to enable interrupt RST 5.5 and mask other interrupts.

Instruction: MVI A, 0EH   ;Bits $D_3 = 1$ and $D_0 = 0$
               SIM   ;Enable RST 5.5

Example   A TTY receiver line is connected to the SOD pin of the 8085. Disable all interrupts and send START bit (logic 0) to TTY without affecting interrupt masks.

Instructions: MVI A,40H   ;D$_7$ = 0, START bit at logic 0
                              ;D$_6$ = 1, Enables serial output bit D$_7$
                              ;D$_3$ = 0, Does not affect masks
                SIM   ;Send START bit

Comments: This instruction does not affect TRAP interrupt.

SPHL: Copy H and L Registers to the Stack Pointer

Opcode	Operand	Bytes	M-Cycles	T-States	Hex Code
SPHL	None	1	1	6 (8085) 5 (8080)	F9

Description   The instruction loads the contents of the H and L registers into the stack pointer register; the contents of the H register provide the high-order address, and the contents of the L register provide the low-order address. The contents of the H and L registers are not altered.

Flags   No flags are affected.

STA: Store Accumulator Direct

Opcode	Operand	Bytes	M-Cycles	T-States	Hex Code
STA	16-bit	3	4	13	32

Description   The contents of the accumulator are copied to a memory location specified by the operand. This is a 3-byte instruction; the second byte specifies the low-order address and the third byte specifies the high-order address.

Flags   No flags are affected.

Example   Assume the accumulator contains 9FH. Load the accumulator contents into memory location 2050H.

Instruction: STA 2050H   Hex Code: 32 50 20

Register contents
before instruction

A | 9F | XX | F

Memory contents
after instruction

2050 | 9F |

STAX: Store Accumulator Indirect

Opcode	Operand	Bytes	M-Cycles	T-States	Hex Code	
STAX	B/D reg. pair	1	2	7	**Reg.** B D	**Hex** 02 12

Description   The contents of the accumulator are copied into the memory location specified by the contents of the operand (register pair). The contents of the accumulator are not altered.

Flags   No flags are affected.

Example   Assume the contents of the accumulator are F9H and the contents of registers B and C are 20H and 50H, respectively. Store the accumulator contents in memory location 2050H.

Instruction: STAX B   Hex Code: 02

Register contents before instruction

Register and memory contents after instruction

A [F9 | XX] F
B [20 | 50] C

2050 [F9]

A [F9 | XX] F
B [20 | 50] C

Comments: This instruction performs the same function as MOV A,M except this instruction uses the contents of BC or DE as memory pointers.

STC: Set Carry

Opcode	Operand	Bytes	M-Cycles	T-States	Hex Code
STC	None	1	1	4	37

Description   The Carry flag is set to 1.

Flags   No other flags are affected.

SUB: Subtract Register or Memory from Accumulator

Opcode	Operand	Bytes	M-Cycles	T-States	Hex Code	
SUB	Reg.	1	1	4	**Reg.**	**Hex**
	Mem.	1	2	7	B	90
					C	91
					D	92
					E	93
					H	94
					L	95
					M	96
					A	97

Description The contents of the register or the memory location specified by the operand are subtracted from the contents of the accumulator, and the results are placed in the accumulator. The contents of the source are not altered.

Flags All flags are affected to reflect the result of the subtraction.

Example Assume the contents of the accumulator are 37H and the contents of register C are 40H. Subtract the contents of register C from the accumulator.

Instruction: SUB C   Hex Code: 91

$$
\begin{array}{rl}
\text{(C):}\quad 40H = & 0\,1\,0\,0 \quad 0\,0\,0\,0 \\
\text{2's complement (C):}\quad = & 1\,1\,0\,0 \quad 0\,0\,0\,0 \\
\text{(A):}\quad 37H = & \underline{0\,0\,1\,1 \quad 0\,1\,1\,1} \\
& 0/\ 1\,1\,1\,1 \quad 0\,1\,1\,1 = F7H \\
\text{Complement Carry:}\quad & 1/\ 1\,1\,1\,1 \quad 0\,1\,1\,1
\end{array}
$$

Flags: S = 1, Z = 0, AC = 0
P = 0, CY = 1

The result, as a negative number, will be in 2's complement and thus the Carry (Borrow) flag is set.

SUI: Subtract Immediate from Accumulator

Opcode	Operand	Bytes	M-Cycles	T-States	Hex Code
SUI	8-bit data	2	2	7	D6

Description The 8-bit data (the operand) are subtracted from the contents of the accumulator, and the results are placed in the accumulator.

Flags   All flags are modified to reflect the results of the subtraction.

Example   Assume the accumulator contains 40H. Subtract 37H from the accumulator.

Instruction: SUI 37H   Hex Code: D6 37

$$
\begin{array}{rl}
\text{Subtrahend: } 37\text{H} = & 0\ 0\ 1\ 1\ \ 0\ 1\ 1\ 1 \\
\text{2's complement of } 37\text{H} = & 1\ 1\ 0\ 0\ \ 1\ 0\ 0\ 1 \\
+ & \\
(A)\text{: } 40\text{H} = & 0\ 1\ 0\ 0\ \ 0\ 0\ 0\ 0 \\
1/ & \overline{0\ 0\ 0\ 0\ \ 1\ 0\ 0\ 1} \\
\text{Complement Carry: } 0/ & 0\ 0\ 0\ 0\ \ 1\ 0\ 0\ 1 \ = \ 09\text{H}
\end{array}
$$

Flags: S = 0, Z = 0, AC = 0
P = 1, CY = 0

XCHG: Exchange H and L with D and E

Opcode	Operand	Bytes	M-Cycles	T-States	Hex Code
XCHG	None	1	1	4	EB

Description   The contents of register H are exchanged with the contents of register D, and the contents of register L are exchanged with the contents of register E.

Flags   No flags are affected.

XRA: Exclusive OR with Accumulator

Opcode	Operand	Bytes	M-Cycles	T-States	Reg.	Hex
XRA	Reg.	1	1	4	B	A8
	Mem.	1	2	7	C	A9
					D	AA
					E	AB
					H	AC
					L	AD
					M	AE
					A	AF

Description   The contents of the operand (register or memory) are Exclusive ORed with the contents of the accumulator, and the results are placed in the accumulator. The contents of the operand are not altered.

Flags Z, S, P are altered to reflect the results of the operation. CY and AC are reset.

Example Assume the contents of the accumulator are 77H and of register D are 56H. Exclusive OR the contents of the register D with the accumulator.

Instruction: XRA D   Hex Code: AA

$$(A): 77H = 0\ 1\ 1\ 1\ \ 0\ 1\ 1\ 1$$
$$(D): 56H = \underline{0\ 1\ 0\ 1\ \ 0\ 1\ 1\ 0}$$
$$\text{Exclusive OR:}\qquad 0\ 0\ 1\ 0\ \ 0\ 0\ 0\ 1$$

Flags: S = 0, Z = 0, P = 1,
CY = 0, AC = 0

XRI: Exclusive OR Immediate with Accumulator

Opcode	Operand	Bytes	M-Cycles	T-States	Hex Code
XRI	8-bit data	2	2	7	EE

Description The 8-bit data (operand) are Exclusive ORed with the contents of the accumulator, and the results are placed in the accumulator.

Flags Z, S, P are altered to reflect the results of the operation. CY and AC are reset.

Example Assume the contents of the accumulator are 8FH. Exclusive OR the contents of the accumulator with A2H.

Instruction: XRI A2H   Hex Code: EE A2

$$(A): 8FH = 1\ 0\ 0\ 0\ \ 1\ 1\ 1\ 1$$
$$(Data): A2H = \underline{1\ 0\ 1\ 0\ \ 0\ 0\ 1\ 0}$$
$$\text{Exclusive OR:}\qquad 0\ 0\ 1\ 0\ \ 1\ 1\ 0\ 1$$

Flags: S = 0, Z = 0, P = 1
CY = 0, AC = 0

XTHL: Exchange H and L with Top of Stack

Opcode	Operand	Bytes	M-Cycles	T-States	Hex Code
XTHL	None	1	5	16	E3

Description The contents of the L register are exchanged with the stack location pointed out by the contents of the stack pointer register. The contents of the H register are exchanged with the next stack location (SP + 1); however, the contents of the stack pointer register are not altered.

Flags No flags are affected.

Example The contents of various registers and stack locations are as shown:

					Stacks
H	A2	57	L	2095	38
SP	2095			2096	67

Illustrate the contents of these registers after instruction XTHL.

Register contents
after XTHL

					Stacks
H	67	38	L	2095	57
SP	2095			2096	A2

8085
Instruction Summary: Hexademical Order

Hex	Mnemonic	Hex	Mnemonic	Hex	Mnemonic	Hex	Mnemonic
00	NOP	11	LXI D	21	LXI H	31	LXI SP
01	LXI B	12	STAX D	22	SHLD	32	STA
02	STAX B	13	INX D	23	INX H	33	INX SP
03	INX B	14	INR D	24	INR H	34	INR M
04	INR B	15	DCR D	25	DCR H	35	DCR M
05	DCR B	16	MVI D	26	MVI H	36	MVI M
06	MVI B	17	RAL	27	DAA	37	STC
07	RLC	19	DAD D	29	DAD H	39	DAD SP
09	DAD B	1A	LDAX D	2A	LHLD	3A	LDA
0A	LDAX B	1B	DCX D	2B	DCX H	3B	DCX SP
0B	DCX B	1C	INR E	2C	INR L	3C	INR A
0C	INR C	1D	DCR E	2D	DCR L	3D	DCR A
0D	DCR C	1E	MVI E	2E	MVI L	3E	MVI A
0E	MVI C	1F	RAR	2F	CMA	3F	CMC
0F	RRC	20	RIM	30	SIM	40	MOV B,B

DATA TRANSFER GROUP

ARITHMETIC AND LOGICAL GROUP

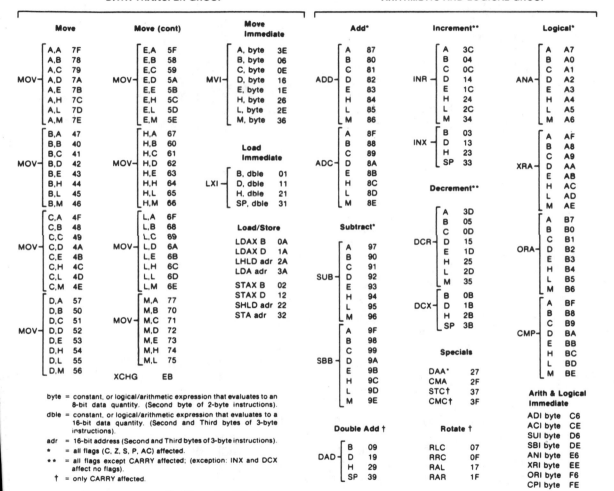

Move

MOV
A,A	7F
A,B	78
A,C	79
A,D	7A
A,E	7B
A,H	7C
A,L	7D
A,M	7E

MOV
B,A	47
B,B	40
B,C	41
B,D	42
B,E	43
B,H	44
B,L	45
B,M	46

MOV
C,A	4F
C,B	48
C,C	49
C,D	4A
C,E	4B
C,H	4C
C,L	4D
C,M	4E

MOV
D,A	57
D,B	50
D,C	51
D,D	52
D,E	53
D,H	54
D,L	55
D,M	56

Move (cont)

MOV
E,A	5F
E,B	58
E,C	59
E,D	5A
E,E	5B
E,H	5C
E,L	5D
E,M	5E

MOV
H,A	67
H,B	60
H,C	61
H,D	62
H,E	63
H,H	64
H,L	65
H,M	66

MOV
L,A	6F
L,B	68
L,C	69
L,D	6A
L,E	6B
L,H	6C
L,L	6D
L,M	6E

MOV
M,A	77
M,B	70
M,C	71
M,D	72
M,E	73
M,H	74
M,L	75

XCHG EB

Move Immediate

MVI
A, byte	3E
B, byte	06
C, byte	0E
D, byte	16
E, byte	1E
H, byte	26
L, byte	2E
M, byte	36

Load Immediate

LXI
B, dble	01
D, dble	11
H, dble	21
SP, dble	31

Load/Store

LDAX B	0A
LDAX D	1A
LHLD adr	2A
LDA adr	3A
STAX B	02
STAX D	12
SHLD adr	22
STA adr	32

Add*

ADD
A	87
B	80
C	81
D	82
E	83
H	84
L	85
M	86

ADC
A	8F
B	88
C	89
D	8A
E	8B
H	8C
L	8D
M	8E

Subtract*

SUB
A	97
B	90
C	91
D	92
E	93
H	94
L	95
M	96

SBB
A	9F
B	98
C	99
D	9A
E	9B
H	9C
L	9D
M	9E

Double Add †

DAD
B	09
D	19
H	29
SP	39

Increment**

INR
A	3C
B	04
C	0C
D	14
E	1C
H	24
L	2C
M	34

INX
B	03
D	13
H	23
SP	33

Decrement**

DCR
A	3D
B	05
C	0D
D	15
E	1D
H	25
L	2D
M	35

DCX
B	0B
D	1B
H	2B
SP	3B

Specials

DAA*	27
CMA	2F
STC†	37
CMC†	3F

Rotate †

RLC	07
RRC	0F
RAL	17
RAR	1F

Logical*

ANA
A	A7
B	A0
C	A1
D	A2
E	A3
H	A4
L	A5
M	A6

XRA
A	AF
B	A8
C	A9
D	AA
E	AB
H	AC
L	AD
M	AE

ORA
A	B7
B	B0
C	B1
D	B2
E	B3
H	B4
L	B5
M	B6

CMP
A	BF
B	B8
C	B9
D	BA
E	BB
H	BC
L	BD
M	BE

Arith & Logical Immediate

ADI byte	C6
ACI byte	CE
SUI byte	D6
SBI byte	DE
ANI byte	E6
XRI byte	EE
ORI byte	F6
CPI byte	FE

byte = constant, or logical/arithmetic expression that evaluates to an 8-bit data quantity. (Second byte of 2-byte instructions).

dble = constant, or logical/arithmetic expression that evaluates to a 16-bit data quantity. (Second and Third bytes of 3-byte instructions).

adr = 16-bit address (Second and Third bytes of 3-byte instructions).

* = all flags (C, Z, S, P, AC) affected.

** = all flags except CARRY affected; (exception: INX and DCX affect no flags).

† = only CARRY affected.

All mnemonics copyright ©Intel Corporation 1976.

Hex	Mnemonic	Hex	Mnemonic	Hex	Mnemonic	Hex	
41	MOV B,C	70	MOV M,B	9F	SBB A	CF	RST 1
42	MOV B,D	71	MOV M,C	A0	ANA B	D0	RNC
43	MOV B,E	72	MOV M,D	A1	ANA C	D1	POP D
44	MOV B,H	73	MOV M,E	A2	ANA D	D2	JNC
45	MOV B,L	74	MOV M,H	A3	ANA E	D3	OUT
46	MOV B,M	75	MOV M,L	A4	ANA H	D4	CNC
47	MOV B,A	76	HLT	A5	ANA L	D5	PUSH D
48	MOV C,B	77	MOV M,A	A6	ANA M	D6	SUI
49	MOV C,C	78	MOV A,B	A7	ANA A	D7	RST 2
4A	MOV C,D	79	MOV A,C	A8	XRA B	D8	RC
4B	MOV C,E	7A	MOV A,D	A9	XRA C	DA	JC
4C	MOV C,H	7B	MOV A,E	AA	XRA D	DB	IN
4D	MOV C,L	7C	MOV A,H	AB	XRA E	DC	CC
4E	MOV C,M	7D	MOV A,L	AC	XRA H	DE	SBI
4F	MOV C,A	7E	MOV A,M	AD	XRA L	DF	RST 3
50	MOV D,B	7F	MOV A,A	AE	XRA M	E0	RPO
51	MOV D,C	80	ADD B	AF	XRA A	E1	POP H
52	MOV D,D	81	ADD C	B0	ORA B	E2	JPO
53	MOV D,E	82	ADD D	B1	ORA C	E3	XTHL
54	MOV D,H	83	ADD E	B2	ORA D	E4	CPO
55	MOV D,L	84	ADD H	B3	ORA E	E5	PUSH H
56	MOV D,M	85	ADD L	B4	ORA H	E6	ANI
57	MOV D,A	86	ADD M	B5	ORA L	E7	RST 4
58	MOV E,B	87	ADD A	B6	ORA M	E8	RPE
59	MOV E,C	88	ADC B	B7	ORA A	E9	PCHL
5A	MOV E,D	89	ADC C	B8	CMP B	EA	JPE
5B	MOV E,E	8A	ADC D	B9	CMP C	EB	XCHG
5C	MOV E,H	8B	ADC E	BA	CMP D	EC	CPE
5D	MOV E,L	8C	ADC H	BB	CMP E	EE	XRI
5E	MOV EM	8D	ADC L	BC	CMP H	EF	RST 5
5F	MOV EA	8E	ADC M	BD	CMP L	F0	RP
60	MOV H,B	8F	ADC A	BE	CMP M	F1	POP PSW
61	MOV H,C	90	SUB B	BF	CMP A	F2	JP
62	MOV H,D	91	SUB C	C0	RNZ	F3	DI
63	MOV H,E	92	SUB D	C1	POP B	F4	CP
64	MOV H,H	93	SUB E	C2	JNZ	F5	PUSH PSW
65	MOV H,L	94	SUB H	C3	JMP	F6	ORI
66	MOV H,M	95	SUB L	C4	CNZ	F7	RST 6
67	MOV H,A	96	SUB M	C5	PUSH B	F8	RM
68	MOV L,B	97	SUB A	C6	ADI	F9	SPHL
69	MOV L,C	98	SBB B	C7	RST 0	FA	JM
6A	MOV L,D	99	SBB C	C8	RZ	FB	EI
6B	MOV L,E	9A	SBB D	C9	RET	FC	CM
6C	MOV L,H	9B	SBB E	CA	JZ	FE	CPI
6D	MOV L,L	9C	SBB H	CC	CZ	FF	RST 7
6E	MOV L,M	9D	SBB L	CD	CALL		
6F	MOV L,A	9E	SBB M	CE	ACI		

BRANCH CONTROL GROUP

Jump

JMP adr	C3
JNZ adr	C2
JZ adr	CA
JNC adr	D2
JC adr	DA
JPO adr	E2
JPE adr	EA
JP adr	F2
JM adr	FA
PCHL	E9

Call

CALL adr	CD
CNZ adr	C4
CZ adr	CC
CNC adr	D4
CC adr	DC
CPO adr	E4
CPE adr	EC
CP adr	F4
CM adr	FC

Return

RET	C9
RNZ	C0
RZ	C8
RNC	D0
RC	D8
RPO	E0
RPE	E8
RP	F0
RM	F8

Restart

RST	0	C7
	1	CF
	2	D7
	3	DF
	4	E7
	5	EF
	6	F7
	7	FF

I/O AND MACHINE CONTROL

Stack Ops

PUSH	B	C5
	D	D5
	H	E5
	PSW	F5

POP	B	C1
	D	D1
	H	E1
	PSW*	F1

XTHL	E3
SPHL	F9

Input/Output

OUT byte	D3
IN byte	DB

Control

DI	F3
EI	FB
NOP	00
HLT	76

New Instructions (8085 Only)

RIM	20
SIM	30

ASSEMBLER REFERENCE (Cont.)

Pseudo Instruction

General:

ORG
END
EQU
SET
DS
DB
DW

Macros:

MACRO
ENDM
LOCAL
REPT
IRP
IRPC
EXITM

Relocation:

ASEG	NAME
DSEG	STKLN
CSEG	STACK
PUBLIC	MEMORY
EXTRN	

Conditional Assembly:

IF
ELSE
ENDIF

RESTART TABLE

Name	Code	Restart Address
RST 0	C7	0000_{16}
RST 1	CF	0008_{16}
RST 2	D7	0010_{16}
RST 3	DF	0018_{16}
RST 4	E7	0020_{16}
TRAP	Hardware* Function	0024_{16}
RST 5	EF	0028_{16}
RST 5.5	Hardware* Function	$002C_{16}$
RST 6	F7	0030_{16}
RST 6.5	Hardware* Function	0034_{16}
RST 7	FF	0038_{16}
RST 7.5	Hardware* Function	$003C_{16}$

*NOTE: The hardware functions refer to the on-chip Interrupt feature of the 8085 only.

8085

Instruction Set with Machine Cycles and Flag Status (see notes at end of table)

Mnemonic	Operand	Instruction	Code[1]	B/M/T[2]	Machine[3] Cycles	S D7	Z D6	AC D4	P D2	CY D0
ACI	DATA	: Add 8-bit and CY to A	CE data	2/2/7	F R	✓	✓	✓	✓	✓
ADC	REG	: Add Reg. and CY to A	1000 1SSS	1/1/4	F	✓	✓	✓	✓	✓
ADC	M	: Add Mem. and CY to A	8E	1/2/7	F R	✓	✓	✓	✓	✓
ADD	REG	: Add Reg. to A	1000 0SSS	1/1/4	F	✓	✓	✓	✓	✓
ADD	M	: Add Mem. to A	86	1/2/7	F R	✓	✓	✓	✓	✓
ADI	DATA	: ADD 8-BIT TO A	C6 DATA	2/2/7	F R	✓	✓	✓	✓	✓
ANA	REG	: AND Reg. with A	1010 0SSS	1/1/4	F	✓	✓	1	✓	0
ANA	M	: AND Mem. with A	A6	1/2/7	F R	✓	✓	1	✓	0
ANI	DATA	: AND 8-bit with A	E6 data	2/2/7	F R	✓	✓	1	✓	0
CALL	ADDR	: Call Unconditional	CD addr	3/5/18	S R R W W					
CC	ADDR	: Call On CY	DC addr	3/5/9–18	S R R W W					
CM	ADDR	: Call On Minus	FC addr	3/5/9–18	S R R W W					
CMA		: Complement A	2F	1/1/4	F					
CMC		: Complement CY	3F	1/1/4	F					✓
CMP	REG	: Compare Reg. with A	1011 1SSS	1/1/4	F	✓	✓	✓	✓	✓
CMP	M	: Compare Mem. with A	BE	1/2/7	F R	✓	✓	✓	✓	✓
CNC	ADDR	: Call On No CY	D4 addr	3/5/9–18	S R R W W					
CNZ	ADDR	: Call On No Zero	C4 addr	3/5/9–18	S R R W W					
CP	ADDR	: Call On Positive	F4 addr	3/5/9–18	S R R W W					
CPE	ADDR	: Call On Parity Even	EC addr	3/5/9–18	S R R W W					
CPI	DATA	: Compare 8-bit with A	FE data	2/2/7	F R	✓	✓	✓	✓	✓
CPO	ADDR	: Call On Parity Odd	E4 addr	3/5/9–18	S R R W W					
CZ	ADDR	: Call On Zero	CC addr	3/5/9–18	S R R W W					

Flags[4]

Mnemonic		Description	Code	B/M/T	Machine Cycles[3]	S	Z	AC	P	CY
DAA		Decimal-Adjust A	27	1/1/4	F	✓	✓	✓	✓	✓
DAD	Rp	Add Reg. Pair to HL	00Rp 1001	1/3/10	F B B					✓
DCR	REG	Decrement Reg.	00SS S101	1/1/4	F	✓	✓	✓	✓	
DCR	M	Decrement Mem. Contents	35	1/3/10	F R W	✓	✓	✓	✓	
DCX	Rp	Decrement Reg. Pair	00Rp 1011	1/1/6	S					
DI		Disable Interrupt	F3	1/1/4	F					
EI		Enable Interrupt	FB	1/1/4	F					
HLT		Halt	76	1/2/5	F B					
IN	PORT	Input from 8-bit Port	DB data	2/3/10	F R I					
INR	REG	Increment Reg.	00SS S100	1/1/4	F	✓	✓	✓	✓	
INR	M	Increment Mem. Contents	34	1/3/10	F R W	✓	✓	✓	✓	
INX	Rp	Increment Reg. Pair	00Rp 0011	1/1/6	S					
JC	ADDR	Jump On Carry	DA addr	3/3/7-10	F R R					
JM	ADDR	Jump On Minus	FA addr	3/3/7-10	F R R					
JMP	ADDR	Unconditional Jump	C3 addr	3/3/10	F R R					
JNC	ADDR	Jump On No Carry	D2 addr	3/3/7-10	F R R					
JNZ	ADDR	Jump On No Zero	C2 addr	3/3/7-10	F R R					
JP	ADDR	Jump On Positive	F2 addr	3/3/7-10	F R R					
JPE	ADDR	Jump On Parity Even	EA addr	3/3/7-10	F R R					
JPO	ADDR	Jump On Parity Odd	E2 addr	3/3/7-10	F R R					
JZ	ADDR	Jump On Zero	CA addr	3/3/7-10	F R R					
LDA	ADDR	Load A Direct	3A addr	3/4/13	F R R R					
LDAX	Rp	Load A from M; memory address is in BC/DE	000X 1010	1/2/7	F R					

Codes[1]

DDD = Binary digits identifying a destination register
SSS = Binary digits identifying a source register
B = 000, C = 001, D = 010, Memory = 110
E = 001, H = 100, L = 101, A = 111
Rp = Register Pair   BC = 00, HL = 10
   DE = 01, SP = 11

B/M/T[2]

B = Bytes
M = Machine cycles
T = T-states

Machine Cycles[3]

F = Fetch with 4 T-states
S = Fetch with 6 T-states
R = Memory Read
I = I/O Read
W = Memory Write
O = I/O Write
B = Bus Idle

Flags[4]

✓ = Flag is modified according to result
0 = Flag is cleared
1 = Flag is set
Blank = No change in flag, remains in previous state

S = Sign
Z = Zero
AC = Auxiliary Carry
P = Parity
CY = Carry

8085

Instruction Set with Machine Cycles and Flag Status (see notes at end of table)

Instruction		Code[1]	B/M/T[2]	Machine[3] Cycles	Flags[4] S D7	Z D6	AC D4	P D2	CY D0
LHLD ADDR	: Load HL Direct	2A addr	3/5/16	F R R R					
LXI Rp, 16-bit	: Load 16-bit in Reg. Pair	00Rp 0001 16-bit	3/3/10	F R R					
MOV Rd,Rs	: Move from Reg. R_s to Reg. R_d	01DD DSSS	1/1/4	F					
MOV M,R	: Move from Reg. to Mem.	0111 0SSS	1/2/7	F W					
MOV R,M	: Move from Mem. to Reg.	01DD D110	1/2/7	F R					
MVI R,DATA	: Load 8-bit in Reg.	00DD D110 data	2/2/7	F R					
MVI M,DATA	: Load 8-bit in Mem.	36 data	2/3/10	F R W					
NOP	: No Operation	00	1/1/4	F					
ORA R	: OR Reg. with A	1011 0SSS	1/1/4	F	✓	✓	0	✓	0
ORA M	: OR Mem. Contents with A	B6	1/2/7	F R	✓	✓	0	✓	0
ORI DATA	: OR 8-bit with A	F6 data	2/2/7	F R	✓	✓	0	✓	0
OUT PORT	: Output to 8-bit Port	D3 data	2/3/10	F R O					
PCHL	: Move HL to Program Counter	E9	1/1/6	S					
POP Rp	: Pop Reg. Pair	11Rp 0001	1/3/10	F R R					
PUSH Rp	: Push Reg. Pair	11Rp 0101	1/3/12	S W W					
RAL	: Rotate A Left through CY	17	1/1/4	F					✓
RAR	: Rotate A Right through CY	1F	1/1/4	F					✓
RC	: Return On Carry	D8	1/3/6–12	S R R					
RET	: Return	C9	1/3/10	F R R					
RIM	: Read Interrupt Mask	20	1/1/4	F					
RLC	: Rotate A Left	07	1/1/4	F					✓
RM	: Return On Minus	F8	1/3/6–12	S R R					
RNC	: Return On No Carry	D0	1/3/6–12	S R R					
RNZ	: Return On No Zero	C0	1/3/6–12	S R R					
RP	: Return On Positive	F0	1/3/6–12	S R R					

Mnemonic	Operand	Description	Code	B/M/T[2]	Machine Cycles[3]	S	Z	AC	P	CY
RPE		: Return On Parity Even	E8	1/3/6–12	S R R					
RPO		: Return On Parity Odd	E0	1/3/6–12	S R R					
RRC		: Rotate A to Right	0F	1/1/4	F					✓
RST	N	: Restart	11XX X111	1/3/12	S W W					
RZ		: Return On Zero	C8	1/3/6–12	S R R					
SBB	R	: Subtract Reg. from A with Borrow	1001 1SSS	1/1/4	F	✓	✓	✓	✓	✓
SBB	M	: Subtract Mem. Contents from A with Borrow	9E	1/2/7	F R	✓	✓	✓	✓	✓
SBI	DATA	: Subtract 8-bit from A	DE data	2/2/7	F R	✓	✓	✓	✓	✓
SHLD	ADDR	: Store HL Direct	22 addr	3/5/16	F R R W W					
SIM		: Set Interrupt Mask	30	1/1/4	F					
SPHL		: Move HL to Stack Pointer	F9	1/1/6	S					
STA	ADDR	: Store A Direct	32 addr	3/4/13	F R R W					
STAX	Rp	: Store A in M, memory address is in BC/DE	000X 0010	1/2/7	F W					
STC		: Set Carry	37	1/1/4	F					1
SUB	R	: Subtract Reg. from A	1001 0SSS	1/1/4	F	✓	✓	✓	✓	✓
SUB	M	: Subtract Mem. from A	96	1/2/7	F R	✓	✓	✓	✓	✓
SUI	DATA	: Subtract 8-bit from A	D6 data	2/2/7	F R	✓	✓	✓	✓	✓
XCHG		: Exchange DE with HL	EB	1/1/4	F					
XRA	R	: Exclusive OR Reg. with A	1010 1SSS	1/1/4	F	✓	✓	0	✓	0
XRA	M	: Exclusive OR Mem. with A	AE	1/2/7	F R	✓	✓	0	✓	0
XRI	DATA	: Exclusive OR 8-bit with A	EE data	2/2/7	F R	✓	✓	0	✓	0
XTHL		: Exchange Stack with HL	E3	1/4/16	F R R W W					

Codes[1]
DDD = Binary digits identifying a destination register
SSS = Binary digits identifying a source register
B = 000, C = 001, D = 010, Memory = 110
E = 001, H = 100, L = 101, A = 111
Rp = Register Pair BC = 00, HL = 10
DE = 01, SP = 11

B/M/T[2]
B = Bytes
M = Machine cycles
T = T-states

Machine Cycles[3]
F = Fetch with 4 T-states
S = Fetch with 6 T-states
R = Memory Read
I = I/O Read
W = Memory Write
O = I/O Write
B = Bus Idle

S = Sign.
Z = Zero
AC = Auxiliary Carry
P = Parity
CY = Carry

Flags[4]
✓ = Flag is modified according to result
0 = Flag is cleared
1 = Flag is set
Blank = No change in flag, remains in previous state

Solutions to Selected Questions, Problems, and Programming Assignments

CHAPTER 1

1. Components of a computer: ALU and Control Unit (CPU), Memory, Input and Output.

3. A microprocessor functions as the CPU of a microcomputer, and includes the ALU, register arrays, and the control unit on one chip; it is manufactured using the LSI technology. On the other hand, the CPU is designed with various discrete boards. Functionally, both are similar; however, technology and processes used for designing are different.

4. A microprocessor is one component of a microcomputer, and the microcomputer is a complete computer consisting of a microprocessor, memory, input and output.

7. Four bytes.

8. The machine language of the 8085 is the commands to the microprocessor given in binary. These are the binary instructions the processor can understand and execute. The assembly language is comprised of mnemonics (group of letters to represent commands) assigned by the manufacturer for the convenience of the users.

12. The assembly language mnemonics represent instructions to the microprocessor; therefore, when they are translated into machine language, there is one-to-one correspondence between the mnemonics and the machine code. The assembly language programs are compact, require less memory space, and are efficient. The high-level languages are written in English-like statements, and when these statements are translated in machine language, the object code tends to be large, and requires large memory. The execution of the programs written in high-level languages is less efficient than that of assembly language programs.

CHAPTER 2

1. Memory Read, Memory Write, I/O Read, and I/O Write.

4. A microprocessor with 14 address lines is capable of addressing 16K (2^{14}) memory locations.

5. 21 address lines.

7. $\overline{\text{IOR}}$ (I/O Read), $\overline{\text{IOW}}$ (I/O Write), $\overline{\text{MEMR}}$ (Memory Read), and $\overline{\text{MEMW}}$ (Memory Write).

8. In memory write operation, the control signal required is $\overline{\text{MEMW}}$, and the direction of the data flow is from the MPU to memory.

10. A flag is the output of a given flip-flop to indicate certain data conditions.

11. The program counter and the stack pointer are used as memory pointers. In this microprocessor the size of the memory address is 16 bits; thus these registers are required to store 16-bit addresses.

12. The program counter always points to the next memory location; therefore, the content of the program counter will be 2058H.

13. 128 registers and $128 \times 4 = 512$ memory cells.

15. 8-bit word size.

18. 32 chips.

20. 11 address lines.

23. The starting address is: E000H.

26. The memory map ranges from 2000H to 23FFH.

28. 8 address lines are required for a peripheral I/O port, and 16 address lines are required for a memory-mapped I/O port.

31. From B to A.

32. None. The decoder is not enabled; all output lines will be high.

33. The line 6 (O_6).

35. A transparent latch is a flip-flop; its output changes according to input when the clock signal is high, and it latches the input on the trailing edge of the clock (high to low). The latch is necessary for output devices to retain the result; otherwise, the result will disappear.

38. The memory occupies the memory space from F000H to FFFFH. The don't care line A_{11} generates additional address range from F800 to FFFFH when it is assumed to be at logic 1. This is a 2K memory chip that occupies 4K of memory space in the map, thus wasting 2K of memory space.

CHAPTER 3

1. The ALE signal goes high at the beginning of each machine cycle indicating the availability of an address on the address bus, and the signal is used to latch the low-order address bus. The IO/$\overline{\text{M}}$ signal is a status signal indicating whether the machine cycle is I/O or memory operation. The IO/$\overline{\text{M}}$ signal is combined with the $\overline{\text{RD}}$ and $\overline{\text{WR}}$ control signals to generate $\overline{\text{IOR}}$, $\overline{\text{IOW}}$, $\overline{\text{MEMR}}$, and $\overline{\text{MEMW}}$ control signals.

3. In Fig. 3.22, the input signals $\overline{\text{RD}}$ and $\overline{\text{WR}}$ cannot be low at the same time. Therefore, the valid combinations of the input signals are:

IO/$\overline{M}$	$\overline{RD}$	$\overline{WR}$	Decoder Output
1	0	1	$O_5 - \overline{IOR}$
1	1	0	$O_6 - \overline{IOW}$
0	0	1	$O_1 - \overline{MEMR}$
0	1	0	$O_2 - \overline{MEMW}$
0	0	0	$-$ Invalid
1	0	0	$-$ Invalid

The remaining two output signals O_3 and O_7 do not represent any operation.

5. In Fig. 3.23, the 74LS139 is enabled when IO/$\overline{M}$ is low. Therefore, the following memory control signals can be generated.

$\overline{RD}$	$\overline{WR}$	Decoder Output
0	0	$O_0 -$ Invalid
0	1	$O_1 - \overline{MEMR}$
1	0	$O_2 - \overline{MEMW}$
1	1	$O_3 -$ No operation

6. The output of the latch will be 05H; however, it will not be latched until the ALE goes low.

10. The sum of 87H + 79H = 100H. Therefore, the accumulator will have 00H, and the flags will be S = 0, CY = 1, Z = 1.

12. 18T × .2 micro-sec = 3.6 micro-sec.

13. $(A_{15}-A_8) = 20H$, $(AD_7-AD_0) = 47H$.

15. The second machine cycle is Memory Read; the processor reads the contents of memory in register B, and the control signal is $\overline{RD}$.

17. $(A_{15}-A_0) = 2050H$.
(AD_7-AD_0) as data bus = Contents of location 2050H.

19. Memory map: 6000H to 6FFFH.

23. Memory map: 28000H to 2FFFH.

24. Total range = 16K. Map = 8000H to BFFFH.

26. Memory map: 0800H–08FFH, and the foldback memory ranges from 0900 to 0FFFH.

29. ROM1: 0000H – 1FFFH, ROM2: E000H – FFFFH, R/WM1: 8000H – 83FFH.

32. Address range: 4000H to 7FFFH.

36. When $A_{14} = 1$, Y_1 is active. Address range: 4000H to 7FFFH. When $A_{14} = 0$, Y_2 is active. Address range: 8000H to BFFFH.

38. The last MEMR is the third byte of the STA Instruction. It reads FFH.

CHAPTER 4

1. The number of output ports in the peripheral I/O is restricted to 256 ports because the operand of the OUT instruction is 8 bits; it can have only 256 combinations.

3. The 8085 differentiates between the input and the output ports of the same address by the control signal. The input port requires the $\overline{RD}$ and the output port requires the $\overline{WR}$ control signals.

6. Trailing edge.

9. A latch is necessary to hold the output data for display; however, the input data byte is obtained by enabling a tri-state buffer and placed in the accumulator.

10. $\overline{RD}$, $\overline{WR}$, and IO/$\overline{M}$ (low).

12. 78H.

15. 8000H.

17. If $A_7 = 0$, port address = 75H, and if $A_7 = 1$, address = F5H.

18. If IO/$\overline{M}$ is connected to $\overline{E1}$ (active low), it will be a memory-mapped I/O. The port address = 00F5H.

    Replace OUT F5H by STA 00F5H.

19. Output code either 40H or C0H.

21. If $A_7 = 0$, the addresses are: 04H, 0CH, 14H, and 1CH.

    If $A_7 = 1$, the addresses are: 84H, 8CH, 94H, and 9CH (as shown in Section 4.34).

23. Port A = Memory-mapped Output Port (0085H).

    Port B = Memory-mapped Input Port (0085H).

25. In Figure 4.10, the output O_5 is enabled by the address, which is active for three T-states. On the other hand, the $\overline{IOW}$ signal requires $\overline{WR}$ signal, which is active for approximately one and one-half T-states.

27. **a.** LDA FFF9H $\rightarrow$ OF, MR, MR, and MR

    STA FFF8H $\rightarrow$ OF, MR, MR, and MW

    MOV B,A $\rightarrow$ OF

    JMP START $\rightarrow$ OF, MR, and MR

    b. FFF9H.

    c. $\overline{RD}$ = 11 times and $\overline{WR}$ = 1 time.

    d. 40T × 0.5 micro-sec = 20 micro-sec.

28. Assuming $A_4 = 0$: Input Port = 2FH and Output Port = 8FH.

    Assuming $A_4 = 1$: Input Port = 3FH and Output Port = 9FH.

29. START: IN 2FH   ;Read input port

    ANI 00000011B   ;Mask all bits except D1 and D0

    JNZ START   ;If a switch is open, read again

    MVI A, 00   ;Unnecessary instruction, used for clarity

    OUT 8FH   ;Turn on all LEDs

    HLT

32. MVI A, 98H   ;Code for '9' to upper LED

    OUT F5H

    MVI A, F8H   ;Code for '7' to lower LED

    OUT F4H

    HLT

CHAPTER 5

2. Opcode: MOV and Operand: H, L.

5. (a) Hex Code = 32 50 20, Opcode = STA Operands = 2050H.

   (b) Hex Code = C2 70 20, Opcode = JNZ Operands = 2070H.

6.

Mnemonics		Bytes	Hex Code
MVI	B,4FH	(2)	064F
MVI	C,78H	(2)	0E78
MOV	A,C	(1)	79
ADD	B	(1)	80
OUT	07H	(2)	D307
HLT		(1)	76

11. MVI B, A2H ;Load bytes
MVI C, 18H
MOV A,B
ADD C   ;The result is in A
HLT

13. The processor assumes the code of the next instruction (HLT:76H) as the port address, outputs the contents of A to the address 76H, and continues the execution; the result is unpredictable.

CHAPTER 6

3. MVI   C,65H
MVI   A,92H
OUT PORT1 ; DISPLAY 92H
MOV A,C   ; Copy C into A for display
OUT PORT0 ; Display 65H
HLT

5. 82H.

7. Both will be 80H.

9. The instruction ADD A will add the content of the accumulator to itself; this is equivalent to multiplying by 2.

10. The instruction SUB A will clear the accumulator. The flag status will be:
$CY = 0$, $Z = 1$.

14. MVI   A, 00H   (A = 0 0 0 0 0 0 0 0
DCR   A   — 0 0 0 0 0 0 0 1
OUT PORT#   —————————————
HLT   1 1 1 1 1 1 1 1 = FFH
The instruction DCR does not set the CY flag.

15. $A = 8FH$, $S = 1$ and $CY = 0$.
The S flag has no meaning when subtracting unsigned numbers.

18. SUB   A   ; Clear accumulator
ADI   47H
SUI   92H ; A = B5H, CY = 1 (Borrow Flag)
OUT PORT0 ; Display B5H
ADI   64H   ; A = 19H
OUT PORT1
HLT

20. The instruction XRA A will clear the accumulator, and the flag status will be: CY = 0, Z = 1.

23. The instruction ORA A will set the flag without affecting the content of the accumulator.

26.
```
MVI   C, A8H
MOV   A,C
ANI   0FH      ; Masking byte to mask D₇–D₄
OUT   PORT0
HLT
```

28.
```
MVI   B, 91H
MVI   C, 87H
MOV   A,B
ANI   01H      ; Mask all bits of 91H except D₀
MOV   B,A      ; Save D₀ from first byte
MOV   A,C
ANI   01H      ; Mask all bits of 87H except D₀
ANA   B        ; AND bits D₀ of 91H and 87H
OUT   PORT1    ; Turn on/off light connected to D₀
HLT
```

29.
```
IN    07H
CMA            ; Complement data from port 07H
ORA A          ; Set Z flag if all switches are open
 |             ; Continue
```

32. In this problem, the range of bytes that will be displayed at PORT2 is 50H to 7FH.

34. 00.

35. This routine displays the absolute value (magnitude) of BYTE1.

38.
```
         XRA   A        ; Clear CY
         MVI   B, FFH
         INR   B
         MOV   A, B
         JNC   DSPLAY
         MVI   A, 01H
DSPLAY:  OUT   PORT#     ; Output = 00H because INR does not
         HLT             ; set CY flag.
```
To clear the CY flag, the instructions such as ANA A, SUB A, ORA A can be used instead of the instruction XRA A.

40.
```
         MVI   B, BYTE1
         MVI   C, BYTE2
         MOV   A, B
         SUB   C
         JNC   DSPLAY    ; Jump if result is positive
         CMA             ; Take one's complement
         ADI   01H       ; Find two's complement
DSPLAY:  OUT PORT1
         HLT
```

CHAPTER 7

The following programs assume the system's R/W memory begins at location 2000H.

The symbols XX in the assignments are assumed as memory page 20H.

5. Location 2075H will contain F7H.

7. A = FFH and (2070H) = FFH.

9. A = 00H, D = 00H, HL = 209FH.

11.

	LXI	B,2090H
	SUB	A
	MVI	D, 0FH
LOOP:	STAX	B
	INX	B
	DCR	D
	JNZ	LOOP
	HLT	

13. 7 times.

15.

START:	LXI	H, 2055H	; Index for data source
	LXI	D, 2085H	; Index for data destination, starting
			; at last location
	MVI	B, 06H	; Byte counter
NEXT:	MOV	A, M	; Get data byte
	STAX	D	; Store data byte
	INX	H	; Next location
	DCX	D	
	DCR	B	; Next count
	JNZ	NEXT	; If counter is not 0, go back
			; to transfer next byte
	HLT		

18.

START:	MVI	B, 6	; Byte count
	LXI	H, 2050H	; Source
	LXI	D, 2050H	; Destination
LOOP:	MOV	A, M	; Get byte
	ORA	A	; Set flag in zero
	JNZ	SKIP	
	STAX	D	; Not zero, so store it
	INX	D	
SKIP:	INX	H	; Go on to next
	DCR	B	
	JNZ	LOOP	
	HLT		

22. Locations 2070H to 2074H contain 01, 02, 03, 04, and 05 respectively.

25. S = 0, Z = 1, CY is unchanged.

34.

(a)	A	CY		(b)	A	CY
MVI A, C5H	C5	NA		MVI A, A7H	A7	NA
ORA A	C5	0		ORA A	A7	0

RAL	8A	1	RAR	4E	1
RRC	45	0	RAL	A7	0

35. These instructions will move the MSD of a BCD number (7 in this case) to the unit's position. A = 07H.

37. Multiply by 10.

41.

	A	S	Z	CY
MVI A, 7FH	7F	NA	NA	NA
ORA A	7F	0	0	0
CPI A2H	7F	1	0	1

43. 00, 00, 7A, 87, 00, 00.

49.
```
START:    LXI    H, 2050H    ; Set index to point to data location
          MVI    C, 08H      ; Set up counter
          MVI    B, 00H      ; Clear (B) to save the highest reading
NXTBYTE:  MOV    A, M        ; Get data byte
          CMP    B           ; Is (B) > (A)?
          JNC    NEXT        ; If yes, replace (B) with (A)
          MOV    B, A        ; Save the larger number
NEXT:     INX    H           ; Point to next data byte
          DCR    C           ; Next count
          JNZ    NXTBYTE     ; Jump to get next byte
          MOV    A, B        ; Load the largest byte
          OUT    PORT1       ; Display the largest byte in the string
          HLT
```

53.
```
START:    LXI    H, 2070H    ; Source pointer
          LXI    D, 2090H    ; Save pointer
LOOP:     MOV    A, M
          CPI    0DH         ; Check for end of string
          JZ     ENDS        ; If = 0DH, then end of string
          CPI    30H
          JC     REJECT      ; Reject if < 30H
          CPI    3AH         ; Note: if subtract 39H, would reject 39H
          JNC    REJECT      ; Reject if < 39H
          STAX   D           ; OK, so save it
          INX    D
REJECT:   INX    H           ; Loop for next
          JMP    LOOP
ENDS:     HLT
```

55.
```
START:    LXI H,XXXX   ; Set up HL as a pointer to the data set
          SUB A        ; Clear A
          MOV  C, A    ; Clear C to set up as a counter
AGAIN:    CMP M        ; Is this end of data (00)?
          JZ DSPLAY    ; If yes, this is end of counting
          MVI A, A3H   ; Load television code
          CMP M        ; Is this television code?
```

```
                 JNZ NEXT      ; If not, go to the next data byte
                 INR C         ; Update the count
        NEXT:    INX H         ; Next data
                 SUB A         ; Clear A
                 JMP AGAIN     ; Go back to check next byte
        DSPLAY:  MOV   A, C    ; Get count
                 OUT PORT1
                 HLT
```

57. (a) 256 bytes (b) 2090H–2091H (c) A must be cleared before addition and INX H is missing in the loop.

CHAPTER 8

2. 234.67 mSec.

3. 468.58 mSec.

5. 234.13 mSec.

7. (a) infinite   (b) infinite   (c) 1

9. If the system frequency is 3.072 MHz, the clock period will be 325 ns. This will reduce the delay to .325 s.

11. BC = 34965_{10}. Insignificant difference when the delay is calculated with JNZ = 7 T-states in the last iteration.

16.
```
START:   MVI    D, 00H       ; Load bit pattern
ROTATE:  MOV    A, D                                          4
         CMA                 ; Complement bit pattern          4
         MOV    D, A                                          4
         ANI    01H          ; Mask D₇–D₁                      7
         OUT PORT1                                           10
         MVI    B, COUNT                                      7
LOOP:    DCR    B                                             4
         JNZ    LOOP                                        10/7
         JMP ROTATE                                         10
```

Total Delay T_D = $T_O + T_L$ (System Frequency = 3.072 MHz)

200s   = 46 × 325.5 ns + 4 × 325.5 ns × COUNT

COUNT   = 42.5 ≈ 42

18.
```
START:   MVI    L, 10101010B ; Alternating light pattern
LIGHTS:  MOV    A, L
         RRC
         OUT PORT
         MOV    L, A
         MVI    B, 50        ; 20 × 50 mSec DELAY = 1Sec
OUTER:   LXI    D, 2559      ; 20 mSec DELAY
INNER:   DCX    D
         MOV    A, D
         ORA    E
         JNZ    INNER
```

```
DCR    B
JNZ    OUTER
JMP    LIGHTS
```

CHAPTER 9

1. (a) 3   (b) 1   (c) 4   (d) 3   (e) 2   (f) 2

3. (a) The address in the program counter will be stored on the stack: (20CCH) = 20H, (20CBH) = 0BH.

The program counter will have the same address as the stack.

The contents of the stack pointer will be incremented to 20CBH.

(b) (20CAH) = 00H (H), (20C9H) = 08H (L)

(20C8H) = 0FH (B), (20C7H) = XX (C)

(c) (stack pointer) = 20C7H.

(d) (stack pointer) = 20CDH.

5. (a) All flags are cleared, and A = 00H.

(b) 20FEH = 0EH, 20FFH = 20H.

(c) SP = 20FEH, PC = 2064H.

(d) 200BH.

(e) Endless loop.

10. This is a 20 ms delay subroutine. It clears Z flag before returning to the calling program. However, the contents of the accumulator are destroyed.

```
DELAY:  PUSH   B           ; Save register contents from
        PUSH   H           ; main or calling program
        LXI    B, COUNT    ; Load delay count for 20 ms
LOOP:   DCX    B
        MOV    A, C
        ORA    B           ; Set Z flag if (B) and (C) = 0
        JNZ    LOOP
        PUSH   PSW         ; Save flag status on stack
        POP    H           ; Copy flag status in register L
        MOV    A, L
        ANI    BFH         ; Set bit D6 (Z flag) = 0
        MOV    L, A
        PUSH   H           ; Save flag status with Z flag reset
        POP    PSW         ; Clear Z flag
        POP    H
        POP    B
        RET                ; Return to calling program
```

Where ANI BFH comment reads: Set bit D_6 (Z flag) = 0

CHAPTER 10

2. This program converts a set of BCD numbers into binary equivalents and stores them in the Output Buffer.

```
; Main program—Uses subroutine BCDBIN (9-1)
;
START:    LXI    SP, STACK     ; Initialize stack pointer
          LXI    H, INBUF      ; HL pair points to input buffer
          LXI    B, OUTBUF     ; BC pair points to output buffer
          MVI    D, COUNT      ; How many numbers
LOOP:     CALL   BCDBIN
          INX    H             ; Bump pointers
          INX    B
          DCR    D             ; Decrement count
          JNZ    LOOP          ; Loop until done
          HLT
```

4. BCD to binary conversion.

```
; Input: 2 Digit BCD in accumulator
; Output: Binary in accumulator
; No registers are changed
;
BCDBIN:
    PUSH   D       ; Save contents of D & E registers
    MOV    E, A    ; Save accumulator
    ANI    0F0H    ; Mask MSD
    RRC
    MOV    D, A    ; D = 8 * MSD
    RRC
    RRC
    ADD    D
    MOV    D, A    ; D = 10 * MSD
    MOV    A, E    ; Restore accumulator
    ANI    0FH     ; Mask LSD
    ADD    D       ; Add in 10 * MSD
    POP    D       ; Restore DE
    RET
```

6. This program supplies the powers of ten in registers B and C.

```
STACK     EQU    0FFFFH
BINBYT    EQU    1000H
OUTBUF    EQU    2000H
PWRTEN    EQU    640AH
          ORG    100H
START:    LXI    SP, STACK
          LXI    H, BINBYT
          MOV    A, M
          LXI    H, OUTBUF
          LXI    B, PWRTEN
          CALL   BINBCD
```

```
                    MOV    B,C
                    CALL   BINBCD
                    MOV    M, A
                    HLT
```
BINBCD: No changes in this routine.

12. This is a modified version of the LEDCOD subroutine (SECTION 10.31) that can be used irrespective of code locations.

LEDCOD:
```
                    PUSH   H
                    LXI    H, CODE
                    ADD    L
                    JNC    NOCY
                    INR    H
     NOCY:          MOV    A, L
                    MOV    A, M
                    STAX   B
                    POP    H
                    RET
```

16. This program converts the unpacked BCD digits (from the Illustrative Program—Section 10.51) into ASCII characters.

Main Program: Same as in SECTION 10.51

Subroutine BCDADD: Same as in SECTION 10.51

```
UNPAK:     MOV        D, A
           CALL       UNPAKI
           MOV        A, D
           RRC
           RRC
           RRC
           RRC
UNPAK1:    ANI 0FH
           ORI '0'         ;This is ASCII 0 = 30H
           MOV        M, A
           DCX        H
           RET
```

18. This program subtracts two-digit BCD numbers stored in two consecutive memory locations. It uses a programming trick to perform these BCD subtractions rather than following the hints given in the assignment.

```
START:     LXI    H, NUMBERS
           MOV    A, M
           INX    H
           SUB    M
           MOV    B, A
           MVI    A, 0      ; Do not change flags of subtraction
           DAA             ; If CY was = 1, A = 60 and AC = 1, A = 06
           MOV    C, A      ; C = 06 or 60
```

```
                MOV     A, B
                SUB     C
                OUT     PORT
                HLT
```

19. This program adds 16-bit numbers; each number is stored in two consecutive memory locations. The total numbers to be added are specified by the contents of register B, and the maximum sum can be 24 bits.

```
START:  LXI     SP, 2050H
        MVI     B, NUMCNT
        XRA     A
        LXI     H, 0            ; Clear HL
LOOP:   XCHG                    ; Save (HL) in DE
        POP     H
        DAD     D
        ACI     0               ; if there is a 17th bit, and 1 to A.
        DCR     B
        JNZ     LOOP
        SHLD    OUTBUF
        STA     OUTBUF + 2      ; Store the MSB
        HLT
        END
```

CHAPTER 12

2. (a) RST 6   (b) 0030H   (c) JMP 2075H

3. (a) 03FF = 01 Memory address of the instruction.
       03FE = 23 following the CALL instruction.
   (b) 03FD = (B) Contents of registers B and C.
       03FC = (C).
   (c) 03FB = 01 Memory address of the instruction.
       03FC = 54 following the LXI B instruction.

5. (a) 11 110 111 = F7H (RST 6).
   (b) The priority encoder accepts the request from I_5 and rejects all other low priority requests. The instruction that will be placed on the data bus is 11 101 111 = EFH (RST 5).

CHAPTER 13

1. $I_O = 2$ mA $(1/2 + 1/128) = 1.0156$ mA.

3. $V_O = (2$ mA $(1/4 + 1/64 + 1/256) - 1$ mA$) \times 5k = -2.305$ V.

5.
```
            LXI SP, STACK
START:  MVI     A, 00H
        OUT     FFH     ; Output 0 V
        CALL    DELAY   ; Wait 0.5 ms
        MVI     A, 80H  ; Load A for 5 V output
        OUT     FFH     ; Output 5 V
```

```
CALL    DELAY    ; Wait 0.5 ms
JMP     START
```

7. If input = 01H, Output = 10 V/4096 = 2.44 mV

If input = 82H, Output = 10/(1/32 + 1/2048) = 317.38 mV

9. START: OUT 80H

```
TEST:      IN 80H      ; Read Data Ready status
           RAL         ; Rotate D7 into CY
           JC TEST     ; If D0 = 1, wait
           IN 81H
           RET
```

CHAPTER 14

3. The handshake signals are used to verify the readiness (status) of peripherals.

5. See Section 14.25.

8. Control/Status Port = 38H, Port A = 39H, Port B = 3AH

Port C = 3BH, Timer (LSB) = 3CH, Timer (MSB) = 3DH

11. The control word = FFF8H and the instructions are as follows:

```
MVI    A, F8H
OUT    LO        ; Load the LSB timer register
MVI    A, FFH
OUT    HI        ; Load the MSB timer in mode 3
```

13. To obtain the square wave of 500 μs with 3.072 MHz clock frequency, the count should be 1536 = 0600H, and the timer should be set up in Mode 1. The control word = 4600H.

CHAPTER 15

3. BSR mode D7 = 0, and I/O mode D7 = 1.

5. Yes. If the port A is set up as input port and INTR (bit D3) of the status word is high, IBF signal (bit D5) should be 1.

If port A is set up as output port and INTR (bit D3) of the status world is high, INTE (bit D6) should be set to 1.

9. To start the timer, appropriate control word and count should be loaded, and gate should be high.

11. The most significant two bits (D_{15} and D_{14}) of the count register (Fig. 15.34) are used to specify DMA function and the remaining fourteen bits are used to specify the number of bytes to be transferred. The maximum number of bytes that can be specified with a 14-bit count is 16K.

13. Assuming the don't care lines are at logic 0, the port addresses are as follows:

Port A = 8000H, Port B = 8100H, Port C = 8200H, and Control Register = 8300H

CHAPTER 16

1. (a) F   (b) F   (c) F   (d) T   (e) T   (f) F   (g) T

(h) F   (i) T   (j) F   (k) F   (l) T   (m) F

2. BITTIME delay = 1/1200 = .833 mS and Half BITTIME = .416 mS.

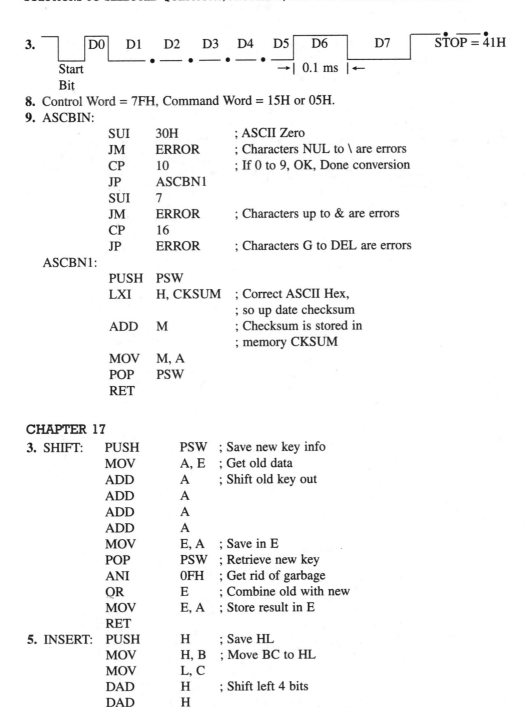

3.

Start Bit

D0 D1 D2 D3 D4 D5 D6 D7   STOP = 41H

$\rightarrow$| 0.1 ms |$\leftarrow$

8. Control Word = 7FH, Command Word = 15H or 05H.

9. ASCBIN:

	SUI	30H	; ASCII Zero
	JM	ERROR	; Characters NUL to \ are errors
	CP	10	; If 0 to 9, OK, Done conversion
	JP	ASCBN1	
	SUI	7	
	JM	ERROR	; Characters up to & are errors
	CP	16	
	JP	ERROR	; Characters G to DEL are errors

ASCBN1:

	PUSH	PSW	
	LXI	H, CKSUM	; Correct ASCII Hex, ; so up date checksum
	ADD	M	; Checksum is stored in ; memory CKSUM
	MOV	M, A	
	POP	PSW	
	RET		

CHAPTER 17

3. SHIFT:

PUSH	PSW	; Save new key info
MOV	A, E	; Get old data
ADD	A	; Shift old key out
ADD	A	
ADD	A	
ADD	A	
MOV	E, A	; Save in E
POP	PSW	; Retrieve new key
ANI	0FH	; Get rid of garbage
OR	E	; Combine old with new
MOV	E, A	; Store result in E
RET		

5. INSERT:

PUSH	H	; Save HL
MOV	H, B	; Move BC to HL
MOV	L, C	
DAD	H	; Shift left 4 bits
DAD	H	
DAD	H	
DAD	H	

```
            ANI        0FH    ; Remove high order garbage
            OR         L      ; Combine
            MOV        B, H   ; Store back in BC
            MOV        C, A
            POP        H      ; Restore HL
            RET
7. Change   ANI        00001111B  to  ANI   00011111B  (3 places)
            CPI        0FH        to  CPI   1FH        (2 places)
            MVI        B, 05H     to  MVI   B, 06H
            MVI        C, 04H     to  MVI   C, 05H
9. EXEC:    PUSH D
            RET
```

Index